T0189089

Lecture Notes in Computer Science 12821

More information about this subseries at http://www.springer.com/series/7412

Josep Lladós · Daniel Lopresti ·
Seiichi Uchida (Eds.)

Document Analysis and Recognition – ICDAR 2021

16th International Conference
Lausanne, Switzerland, September 5–10, 2021
Proceedings, Part I

 Springer

Editors
Josep Lladós ⓘ
Universitat Autònoma de Barcelona
Barcelona, Spain

Daniel Lopresti ⓘ
Lehigh University
Bethlehem, PA, USA

Seiichi Uchida ⓘ
Kyushu University
Fukuoka-shi, Japan

ISSN 0302-9743   ISSN 1611-3349   (electronic)
Lecture Notes in Computer Science
ISBN 978-3-030-86548-1   ISBN 978-3-030-86549-8   (eBook)
https://doi.org/10.1007/978-3-030-86549-8

LNCS Sublibrary: SL6 – Image Processing, Computer Vision, Pattern Recognition, and Graphics

This Springer imprint is published by the registered company Springer Nature Switzerland AG
The registered company address is: Gewerbestrasse 11, 6330 Cham, Switzerland

Foreword

Our warmest welcome to the proceedings of ICDAR 2021, the 16th IAPR International Conference on Document Analysis and Recognition, which was held in Switzerland for the first time. Organizing an international conference of significant size during the COVID-19 pandemic, with the goal of welcoming at least some of the participants physically, is similar to navigating a rowboat across the ocean during a storm. Fortunately, we were able to work together with partners who have shown a tremendous amount of flexibility and patience including, in particular, our local partners, namely the Beaulieu convention center in Lausanne, EPFL, and Lausanne Tourisme, and also the international ICDAR advisory board and IAPR-TC 10/11 leadership teams who have supported us not only with excellent advice but also financially, encouraging us to setup a hybrid format for the conference.

We were not a hundred percent sure if we would see each other in Lausanne but we remained confident, together with almost half of the attendees who registered for on-site participation. We relied on the hybridization support of a motivated team from the Lule University of Technology during the pre-conference, and professional support from Imavox during the main conference, to ensure a smooth connection between the physical and the virtual world. Indeed, our welcome is extended especially to all our colleagues who were not able to travel to Switzerland this year. We hope you had an exciting virtual conference week, and look forward to seeing you in person again at another event of the active DAR community.

With ICDAR 2021, we stepped into the shoes of a longstanding conference series, which is the premier international event for scientists and practitioners involved in document analysis and recognition, a field of growing importance in the current age of digital transitions. The conference is endorsed by IAPR-TC 10/11 and celebrates its 30th anniversary this year with the 16th edition. The very first ICDAR conference was held in St. Malo, France in 1991, followed by Tsukuba, Japan (1993), Montreal, Canada (1995), Ulm, Germany (1997), Bangalore, India (1999), Seattle, USA (2001), Edinburgh, UK (2003), Seoul, South Korea (2005), Curitiba, Brazil (2007), Barcelona, Spain (2009), Beijing, China (2011), Washington DC, USA (2013), Nancy, France (2015), Kyoto, Japan (2017), and Syndey, Australia (2019).

The attentive reader may have remarked that this list of cities includes several venues for the Olympic Games. This year the conference was be hosted in Lausanne, which is the headquarters of the International Olympic Committee. Not unlike the athletes who were recently competing in Tokyo, Japan, the researchers profited from a healthy spirit of competition, aimed at advancing our knowledge on how a machine can understand written communication. Indeed, following the tradition from previous years, 13 scientific competitions were held in conjunction with ICDAR 2021 including, for the first time, three so-called "long-term" competitions addressing wider challenges that may continue over the next few years.

Other highlights of the conference included the keynote talks given by Masaki Nakagawa, recipient of the IAPR/ICDAR Outstanding Achievements Award, and Mickaël Coustaty, recipient of the IAPR/ICDAR Young Investigator Award, as well as our distinguished keynote speakers Prem Natarajan, vice president at Amazon, who gave a talk on "OCR: A Journey through Advances in the Science, Engineering, and Productization of AI/ML", and Beta Megyesi, professor of computational linguistics at Uppsala University, who elaborated on "Cracking Ciphers with 'AI-in-the-loop': Transcription and Decryption in a Cross-Disciplinary Field".

A total of 340 publications were submitted to the main conference, which was held at the Beaulieu convention center during September 8–10, 2021. Based on the reviews, our Program Committee chairs accepted 40 papers for oral presentation and 142 papers for poster presentation. In addition, nine articles accepted for the ICDAR-IJDAR journal track were presented orally at the conference and a workshop was integrated in a poster session. Furthermore, 12 workshops, 2 tutorials, and the doctoral consortium were held during the pre-conference at EPFL during September 5–7, 2021, focusing on specific aspects of document analysis and recognition, such as graphics recognition, camera-based document analysis, and historical documents.

The conference would not have been possible without hundreds of hours of work done by volunteers in the organizing committee. First of all we would like to express our deepest gratitude to our Program Committee chairs, Joseph Lladós, Dan Lopresti, and Seiichi Uchida, who oversaw a comprehensive reviewing process and designed the intriguing technical program of the main conference. We are also very grateful for all the hours invested by the members of the Program Committee to deliver high-quality peer reviews. Furthermore, we would like to highlight the excellent contribution by our publication chairs, Liangrui Peng, Fouad Slimane, and Oussama Zayene, who negotiated a great online visibility of the conference proceedings with Springer and ensured flawless camera-ready versions of all publications. Many thanks also to our chairs and organizers of the workshops, competitions, tutorials, and the doctoral consortium for setting up such an inspiring environment around the main conference. Finally, we are thankful for the support we have received from the sponsorship chairs, from our valued sponsors, and from our local organization chairs, which enabled us to put in the extra effort required for a hybrid conference setup.

Our main motivation for organizing ICDAR 2021 was to give practitioners in the DAR community a chance to showcase their research, both at this conference and its satellite events. Thank you to all the authors for submitting and presenting your outstanding work. We sincerely hope that you enjoyed the conference and the exchange with your colleagues, be it on-site or online.

September 2021

Andreas Fischer
Rolf Ingold
Marcus Liwicki

Preface

It gives us great pleasure to welcome you to the proceedings of the 16th International Conference on Document Analysis and Recognition (ICDAR 2021). ICDAR brings together practitioners and theoreticians, industry researchers and academics, representing a range of disciplines with interests in the latest developments in the field of document analysis and recognition. The last ICDAR conference was held in Sydney, Australia, in September 2019. A few months later the COVID-19 pandemic locked down the world, and the Document Analysis and Recognition (DAR) events under the umbrella of IAPR had to be held in virtual format (DAS 2020 in Wuhan, China, and ICFHR 2020 in Dortmund, Germany). ICDAR 2021 was held in Lausanne, Switzerland, in a hybrid mode. Thus, it offered the opportunity to resume normality, and show that the scientific community in DAR has kept active during this long period.

Despite the difficulties of COVID-19, ICDAR 2021 managed to achieve an impressive number of submissions. The conference received 340 paper submissions, of which 182 were accepted for publication (54%) and, of those, 40 were selected as oral presentations (12%) and 142 as posters (42%). Among the accepted papers, 112 had a student as the first author (62%), and 41 were identified as coming from industry (23%). In addition, a special track was organized in connection with a Special Issue of the International Journal on Document Analysis and Recognition (IJDAR). The Special Issue received 32 submissions that underwent the full journal review and revision process. The nine accepted papers were published in IJDAR and the authors were invited to present their work in the special track at ICDAR.

The review model was double blind, i.e. the authors did not know the name of the reviewers and vice versa. A plagiarism filter was applied to each paper as an added measure of scientific integrity. Each paper received at least three reviews, totaling more than 1,000 reviews. We recruited 30 Senior Program Committee (SPC) members and 200 reviewers. The SPC members were selected based on their expertise in the area, considering that they had served in similar roles in past DAR events. We also included some younger researchers who are rising leaders in the field.

In the final program, authors from 47 different countries were represented, with China, India, France, the USA, Japan, Germany, and Spain at the top of the list. The most popular topics for accepted papers, in order, included text and symbol recognition, document image processing, document analysis systems, handwriting recognition, historical document analysis, extracting document semantics, and scene text detection and recognition. With the aim of establishing ties with other communities within the concept of reading systems at large, we broadened the scope, accepting papers on topics like natural language processing, multimedia documents, and sketch understanding.

The final program consisted of ten oral sessions, two poster sessions, three keynotes, one of them given by the recipient of the ICDAR Outstanding Achievements Award, and two panel sessions. We offer our deepest thanks to all who contributed their time and effort to make ICDAR 2021 a first-rate event for the community. This year's ICDAR had a large number of interesting satellite events as well: workshops, tutorials, competitions, and the doctoral consortium. We would also like to express our sincere thanks to the keynote speakers, Prem Natarajan and Beta Megyesi.

Finally, we would like to thank all the people who spent time and effort to make this impressive program: the authors of the papers, the SPC members, the reviewers, and the ICDAR organizing committee as well as the local arrangements team.

September 2021

Josep Lladós
Daniel Lopresti
Seiichi Uchida

Organization

Organizing Committee

General Chairs

Andreas Fischer	University of Applied Sciences and Arts Western Switzerland, Switzerland
Rolf Ingold	University of Fribourg, Switzerland
Marcus Liwicki	Luleå University of Technology, Sweden

Program Committee Chairs

Josep Lladós	Computer Vision Center, Spain
Daniel Lopresti	Lehigh University, USA
Seiichi Uchida	Kyushu University, Japan

Workshop Chairs

Elisa H. Barney Smith	Boise State University, USA
Umapada Pal	Indian Statistical Institute, India

Competition Chairs

Harold Mouchère	University of Nantes, France
Foteini Simistira	Luleå University of Technology, Sweden

Tutorial Chairs

Véronique Eglin	Institut National des Sciences Appliquées, France
Alicia Fornés	Computer Vision Center, Spain

Doctoral Consortium Chairs

Jean-Christophe Burie	La Rochelle University, France
Nibal Nayef	MyScript, France

Publication Chairs

Liangrui Peng	Tsinghua University, China
Fouad Slimane	University of Fribourg, Switzerland
Oussama Zayene	University of Applied Sciences and Arts Western Switzerland, Switzerland

Sponsorship Chairs

David Doermann	University at Buffalo, USA
Koichi Kise	Osaka Prefecture University, Japan
Jean-Marc Ogier	University of La Rochelle, France

Local Organization Chairs

Jean Hennebert	University of Applied Sciences and Arts Western Switzerland, Switzerland
Anna Scius-Bertrand	University of Applied Sciences and Arts Western Switzerland, Switzerland
Sabine Süsstrunk	École Polytechnique Fédérale de Lausanne, Switzerland

Industrial Liaison

Aurélie Lemaitre	University of Rennes, France

Social Media Manager

Linda Studer	University of Fribourg, Switzerland

Program Committee

Senior Program Committee Members

Apostolos Antonacopoulos	University of Salford, UK
Xiang Bai	Huazhong University of Science and Technology, China
Michael Blumenstein	University of Technology Sydney, Australia
Jean-Christophe Burie	University of La Rochelle, France
Mickaël Coustaty	University of La Rochelle, France
Bertrand Coüasnon	University of Rennes, France
Andreas Dengel	DFKI, Germany
Gernot Fink	TU Dortmund University, Germany
Basilis Gatos	Demokritos, Greece
Nicholas Howe	Smith College, USA
Masakazu Iwamura	Osaka Prefecture University, Japan
C. V. Javahar	IIIT Hyderabad, India
Lianwen Jin	South China University of Technology, China
Dimosthenis Karatzas	Computer Vision Center, Spain
Laurence Likforman-Sulem	Télécom ParisTech, France
Cheng-Lin Liu	Chinese Academy of Sciences, China
Angelo Marcelli	University of Salerno, Italy
Simone Marinai	University of Florence, Italy
Wataru Ohyama	Saitama Institute of Technology, Japan
Luiz Oliveira	Federal University of Parana, Brazil
Liangrui Peng	Tsinghua University, China
Ashok Popat	Google Research, USA
Partha Pratim Roy	Indian Institute of Technology Roorkee, India
Marçal Rusiñol	Computer Vision Center, Spain
Robert Sablatnig	Vienna University of Technology, Austria
Marc-Peter Schambach	Siemens, Germany

Srirangaraj Setlur	University at Buffalo, USA
Faisal Shafait	National University of Sciences and Technology, India
Nicole Vincent	Paris Descartes University, France
Jerod Weinman	Grinnell College, USA
Richard Zanibbi	Rochester Institute of Technology, USA

Program Committee Members

Sébastien Adam
Irfan Ahmad
Sheraz Ahmed
Younes Akbari
Musab Al-Ghadi
Alireza Alaei
Eric Anquetil
Srikar Appalaraju
Elisa H. Barney Smith
Abdel Belaid
Mohammed Faouzi Benzeghiba
Anurag Bhardwaj
Ujjwal Bhattacharya
Alceu Britto
Jorge Calvo-Zaragoza
Chee Kheng Ch'Ng
Sukalpa Chanda
Bidyut B. Chaudhuri
Jin Chen
Youssouf Chherawala
Hojin Cho
Nam Ik Cho
Vincent Christlein
Christian Clausner
Florence Cloppet
Donatello Conte
Kenny Davila
Claudio De Stefano
Sounak Dey
Moises Diaz
David Doermann
Antoine Doucet
Fadoua Drira
Jun Du
Véronique Eglin
Jihad El-Sana
Jonathan Fabrizio

Nadir Farah
Rafael Ferreira Mello
Miguel Ferrer
Julian Fierrez
Francesco Fontanella
Alicia Fornés
Volkmar Frinken
Yasuhisa Fujii
Akio Fujiyoshi
Liangcai Gao
Utpal Garain
C. Lee Giles
Romain Giot
Lluis Gomez
Petra Gomez-Krämer
Emilio Granell
Mehdi Hamdani
Gaurav Harit
Ehtesham Hassan
Anders Hast
Sheng He
Jean Hennebert
Pierre Héroux
Laurent Heutte
Nina S. T. Hirata
Tin Kam Ho
Kaizhu Huang
Qiang Huo
Donato Impedovo
Reeve Ingle
Brian Kenji Iwana
Motoi Iwata
Antonio Jimeno
Slim Kanoun
Vassilis Katsouros
Ergina Kavallieratou
Klara Kedem

Kazem Taghva
Ryohei Tanaka
Christopher Tensmeyer
Kengo Terasawa
Ruben Tolosana
Alejandro Toselli
Cao De Tran
Szilard Vajda
Ernest Valveny
Marie Vans
Eduardo Vellasques
Ruben Vera-Rodriguez
Christian Viard-Gaudin
Mauricio Villegas
Qiu-Feng Wang

Da-Han Wang
Curtis Wigington
Liang Wu
Mingkun Yang
Xu-Cheng Yin
Fei Yin
Guangwei Zhang
Heng Zhang
Xu-Yao Zhang
Yuchen Zheng
Guoqiang Zhong
Yu Zhou
Anna Zhu
Majid Ziaratban

Contents – Part I

Handwriting Recognition

Scene Text Detection and Recognition

NLP for Document Understanding

Graphics, Diagram, and Math Recognition

Historical Document Analsyis 1

BoundaryNet: An Attentive Deep Network with Fast Marching Distance Maps for Semi-automatic Layout Annotation

Abhishek Trivedi🆔 and Ravi Kiran Sarvadevabhatla(✉)🆔

Centre for Visual Information Technology (CVIT), International Institute of Information Technology, Hyderabad 500032, India
`abhishek.trivedi@research.iiit.ac.in, ravi.kiran@iiit.ac.in`
`http://ihdia.iiit.ac.in/BoundaryNet/`

Abstract. Precise boundary annotations of image regions can be crucial for downstream applications which rely on region-class semantics. Some document collections contain densely laid out, highly irregular and overlapping multi-class region instances with large range in aspect ratio. Fully automatic boundary estimation approaches tend to be data intensive, cannot handle variable-sized images and produce sub-optimal results for aforementioned images. To address these issues, we propose BoundaryNet, a novel resizing-free approach for high-precision semi-automatic layout annotation. The variable-sized user selected region of interest is first processed by an attention-guided skip network. The network optimization is guided via Fast Marching distance maps to obtain a good quality initial boundary estimate and an associated feature representation. These outputs are processed by a Residual Graph Convolution Network optimized using Hausdorff loss to obtain the final region boundary. Results on a challenging image manuscript dataset demonstrate that BoundaryNet outperforms strong baselines and produces high-quality semantic region boundaries. Qualitatively, our approach generalizes across multiple document image datasets containing different script systems and layouts, all without additional fine-tuning. We integrate BoundaryNet into a document annotation system and show that it provides high annotation throughput compared to manual and fully automatic alternatives.

Keywords: Document layout analysis · Interactive · Deep learning

1 Introduction

Document images exhibit incredible diversity in terms of language [6,36,40], content modality (printed [12,34], handwritten [4,16,32,35]), writing surfaces (paper, parchment [29], palm-leaf [17,30]), semantic elements such as text, tables,

© Springer Nature Switzerland AG 2021
J. Lladós et al. (Eds.): ICDAR 2021, LNCS 12821, pp. 3–18, 2021.
https://doi.org/10.1007/978-3-030-86549-8_1

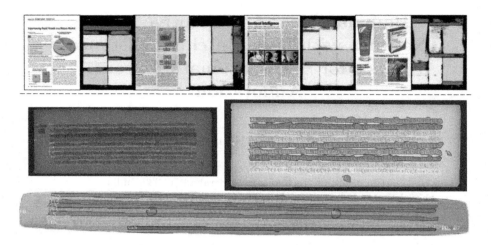

Fig. 1. Compare the contours of semantic region instances for printed documents (top) [41] and historical document images (bottom). The latter are very diverse, often found damaged, contain densely laid out overlapping region instances (lines, holes) with large range in aspect ratios and high local curvature. These factors pose a challenge for region annotation.

photos, graphics [6,34,43] and other such attributes. Within this variety, hand-written and historical documents pose the toughest challenges for tasks such as Optical Character Recognition (OCR) and document layout parsing.

In this work, we focus on historical documents. These documents form an important part of world's literary and cultural heritage. The mechanised process of machine printing imparts structure to modern-era paper documents. In contrast, historical document images are typically handwritten, unstructured and often contain information in dense, non-standard layouts (Fig. 1). Given the large diversity in language, script and non-textual elements in these documents, accurate spatial layout parsing can assist performance for other document-based tasks such as word-spotting [21], optical character recognition (OCR), style or content-based retrieval [39,44]. Despite the challenges posed by such images, a number of deep-learning based approaches have been proposed for fully automatic layout parsing [3,26,30,35]. However, a fundamental trade off exists between global processing and localized, compact nature of semantic document regions. For this reason, fully automatic approaches for documents with densely spaced, highly warped regions often exhibit false negatives or imprecise region boundaries. In practice, correction of predicted boundaries can be more burdensome than manual annotation itself.

Therefore, we propose an efficient semi-automatic approach for parsing images with dense, highly irregular layouts. The user selected bounding-box enclosing the region of interest serves as a weak supervisory input. Our proposed deep neural architecture, BoundaryNet, processes this input to generate precise region contours which require minimal to no manual post-processing.

Numerous approaches exist for weakly supervised bounding-box based semantic parsing of scene objects [2,37]. However, the spatial dimensions and aspect ratios of semantic regions in these datasets are less extreme compared to ones found in handwritten documents (Fig. 1). More recently, a number of approaches model the annotation task as an active contour problem by regressing boundary points on the region's contour [1,7,10,24,27]. However, the degree of curvature for document region contours tends to be larger compared to regular object datasets. The image content and associated boundaries are also distorted by the standard practice of resizing the image to a common height and width. For these reasons, existing approaches empirically tend to produce imprecise contours, especially for regions with high warp, extreme aspect ratio and multiple curvature points (as we shall see).

To address these shortcomings, we propose a two-stage approach (Sect. 3). In the first stage, the variable-sized input image is processed by an attention-based fully convolutional network to obtain a region mask (Sect. 3.1). The region mask is morphologically processed to obtain an initial set of boundary points (Sect. 3.2). In the second stage, these boundary points are iteratively refined using a Residual Graph Convolutional Network to generate the final semantic region contour (Sect. 3.3). As we shall show, our design choices result in a high-performing system for accurate document region annotation.

Qualitatively and quantitatively, BoundaryNet outperforms a number of strong baselines for the task of accurate boundary generation (Sect. 4). BoundaryNet handles variable-sized images without resizing, in real-time, and generalizes across document image datasets with diverse languages, script systems and dense, overlapping region layouts (Sect. 4). Via end-to-end timing analysis, we showcase BoundaryNet's superior annotation throughput compared to manual and fully-automatic approaches (Sect. 5.2).

Source code, pre-trained models and associated documentation are available at http://ihdia.iiit.ac.in/BoundaryNet/.

2   Related Work

Annotating spatial regions is typically conducted in three major modes – manual, fully automatic and semi-automatic. The manual mode is obviously labor-intensive and motivates the existence of the other two modes. Fully automatic approaches fall under the task categories of semantic segmentation [8] and instance segmentation [13]. These approaches work reasonably well for printed [6,43] and structured handwritten documents [3], but have been relatively less successful for historical manuscripts and other highly unstructured documents containing distorted, high-curvature regions [16,30].

Given the challenges with fully automatic approaches, semi-automatic variants operate on the so-called 'weak supervision' provided by human annotators. The weak supervision is typically provided as class label [11,38], scribbles [5,9,15] or bounding box [2] for the region of interest with the objective of predicting the underlying region's spatial support. This process is repeated for all image regions relevant to the annotation task. In our case, we assume box-based weak supervision. Among box-based weakly supervised approaches, spatial support is typically predicted as a 2-D mask [2] or a boundary contour [1, 7, 10, 24, 27].

Contour-based approaches generally outperform mask-based counterparts and provide the flexibility of semi-automatic contour editing [1, 7, 10, 24, 27]. We employ a contour-based approach. However, unlike existing approaches, (i) BoundaryNet efficiently processes variable-sized images without need for resizing (ii) Boundary points are adaptively initialized from an inferred estimate of region mask instead of a fixed geometrical shape (iii) BoundaryNet utilizes skip connection based attentional guidance and boundary-aware distance maps to semantically guide region mask production (iv) BoundaryNet also produces region class label reducing post-processing annotation efforts. Broadly, our choices help deal with extreme aspect ratios and highly distorted region boundaries typically encountered in irregularly structured images.

3   BoundaryNet

Overview: Given the input bounding box, our objective is to obtain a precise contour of the enclosed semantic region (e.g. text line, picture, binding hole). BoundaryNet's processing pipeline consists of three stages – see Fig. 2. In the first stage, the bounding box image is processed by a Mask-CNN (MCNN) to obtain a good quality estimate of the underlying region's spatial mask (Sect. 3.1). Morphological and computational geometric procedures are used to sample contour points along the mask boundary (Sect. 3.2). A graph is constructed with contour points as nodes and edge connectivity defined by local neighborhoods of each contour point. The intermediate skip attention features from MCNN and contour point location are used to construct feature representations for each graph node. Finally, the feature-augmented contour graph is processed by a Graph Convolutional Network (Anchor GCN - Sect. 3.3) iteratively to obtain final set of contour points which define the predicted region boundary.

Semantic regions in documents are often characterized by extreme aspect ratio variations across region classes and uneven spatial distortion. In this context, it is important to note that BoundaryNet processes the input as-is without any resizing to arbitrarily fixed dimensions. This helps preserve crucial appearance detail. Next, we describe the components of BoundaryNet.

3.1   Mask-CNN (MCNN)

As the first step, the input image is processed by a backbone network ('Skip Attention Backbone' in Fig. 2). The backbone has U-Net style long-range skip connections with the important distinction that no spatial downsampling or upsampling is involved. This is done to preserve crucial boundary information. In the first part of the backbone, a series of residual blocks are used to obtain progressively refined feature representations (orange blocks). The second part of the backbone contains another series of blocks we refer to as Skip Attentional

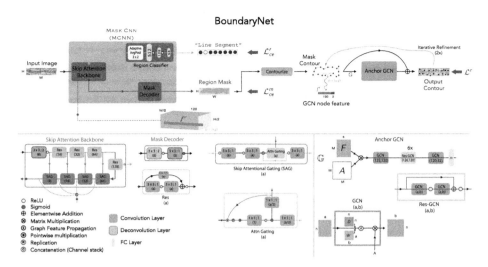

Fig. 2. The architecture of BoundaryNet (top) and various sub-components (bottom). The variable-sized $H \times W$ input image is processed by Mask-CNN (MCNN) which predicts a region mask estimate and an associated region class (Sect. 3.1). The mask's boundary is determined using a contourization procedure (light brown) applied on the estimate from MCNN. M boundary points are sampled on the boundary (Sect. 3.2). A graph is constructed with the points as nodes and edge connectivity defined by $\leqslant k$-hop neighborhoods of each point. The spatial coordinates of a boundary point location $p = (x, y)$ and corresponding backbone skip attention features from MCNN f^r are used as node features for the boundary point. The feature-augmented contour graph $\mathbb{G} = (F, A)$ is iteratively processed by Anchor GCN (Sect. 3.3) to obtain the final output contour points defining the region boundary. Note that all filters in MCNN have a 3×3 spatial extent. The orange lock symbol on region classifier branch indicates that it is trained standalone, i.e. using pre-trained MCNN features. (Color figure online)

Guidance (SAG). Each SAG block produces increasingly compressed (channel-wise) feature representations of its input. To accomplish this feat without losing crucial low-level feature information, the output from immediate earlier SAG block is fused with skip features originating from a lower-level residual block layer (refer to 'Skip Attention Backbone' and its internal module diagrams in Fig. 2). This fusion is modulated via an attention mechanism (gray 'Attn Gating' block) [28].

The final set of features generated by skip-connection based attentional guidance (magenta) are provided to the 'Mask Decoder' network which outputs a region mask binary map. In addition, features from the last residual block (Res-128) are fed to 'Region Classifier' sub-network which predicts the associated region class. Since input regions have varying spatial dimensions, we use adaptive average pooling [14] to ensure a fixed-dimensional fully connected layer output (see 'Region Classifier' in Fig. 2).

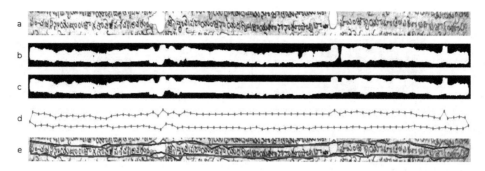

Fig. 3. Contourization (Sect. 3.2): a - input image, b - thresholded initial estimate from MCNN, c - after area-based thresholding and joining centroids of largest connected components by an adaptive $m = \frac{H}{7}$ pixel-thick line where H is the height of the input image, d - after morphological closing, contour extraction, b-spline fitting and uniform point sampling, e - estimated contour (red) and ground-truth (blue) overlaid on input image. (Color figure online)

The input image is processed by an initial convolutional block with stride 2 filters before the resulting features are relayed to the backbone residual blocks. The spatial dimensions are restored via a transpose convolution upsampling within 'Mask Decoder' sub-network. These choices help keep the feature representations compact while minimizing the effect of downsampling.

3.2   Contourization

The pixel predictions in the output from 'Mask Decoder' branch are thresholded to obtain an initial estimate of the region mask. The result is morphologically processed, followed by the extraction of mask contour. A b-spline representation of mask contour is further computed to obtain a smoother representation. M locations are uniformly sampled along the b-spline contour curve to obtain the initial set of region boundary points. Figure 3 illustrates the various steps.

An advantage of the above procedure is that the set of mask-based boundary points serves as a reasonably accurate estimate of the target boundary. Therefore, it lowers the workload for the subsequent GCN stage which can focus on refining the boundary estimate.

3.3   Anchor GCN

The positional and appearance-based features of boundary points from the contourization stage (Sect. 3.2) are used to progressively refine the region's boundary estimate. For this, the boundary points are first assembled into a contour graph. The graph's connectivity is defined by $\leqslant k$-hop neighbors for each contour point node. The node's s-dimensional feature representation is comprised of (i) the contour point 2-D coordinates $p = (x, y)$ (ii) corresponding skip attention features from MCNN f^r - refer to 'GCN node feature' in Fig. 2 for a visual illustration.

The contour graph is represented in terms of two matrices - feature matrix F and adjacency matrix A [18, 42]. F is a $M \times s$ matrix where each row corresponds to the s-dimensional boundary point feature representation described previously. The $M \times M$ binary matrix A encodes the $\leqslant k$-hop connectivity for each boundary point. Thus, we obtain the contour graph representation $G = (F, A)$ (denoted 'Residual Graph' at bottom-right of Fig. 2). We briefly summarize GCNs next.

Graph Convolutional Network (GCN): A GCN takes a graph G as input and computes hierarchical feature representations at each node in the graph while retaining the original connectivity structure. The feature representation at the $(i + 1)$-th layer of the GCN is defined as $H_{i+1} = f(H_i, A)$ where H_i is a $p \times F_i$ matrix whose j-th row contains the i-th layer's feature representation for node indexed by j ($1 \leqslant j \leqslant N$). f (the so-called propagation rule) determines the manner in which node features of previous layer are aggregated to obtain current layer's feature representation. We use the following propagation rule [19]:

$$f(H_i, A) = \sigma(D^{\frac{-1}{2}} \widetilde{A} D^{\frac{-1}{2}} H_i W_i) \tag{1}$$

where $\widetilde{A} = A + I$ represents the adjacency matrix modified to include self-loops, D is a diagonal node-degree matrix (i.e. $D_{jj} = \sum_m \widetilde{A}_{jm}$) and W_i are the trainable weights for i-th layer. σ represents a non-linear activation function (ReLU in our case). Also, $H_0 = F$ (input feature matrix).

Res-GCN: The residual variant of GCN operates via an appropriate 'residual' modification to the GCN layer's feature representation and is defined as $H_{i+1} = f(H_i, A) + H_i$.

The input contour graph features are processed by a series of Res-GCN blocks [22] sandwiched between two GCN blocks. The Anchor GCN module culminates in a 2-dimensional fully connected layer whose output constitutes per-point displacements of the input boundary locations. To obtain the final boundary, we perform iterative refinement of predicted contour until the net displacements are negligibly small by re-using GCN's prediction for the starting estimate at each iteration [24].

3.4   Training and Inference

We train BoundaryNet in three phases.

First Phase: In this phase, we aim to obtain a good quality estimate of the boundary contour from MCNN. For this, the binary prediction from MCNN is optimized using per-pixel class-weighted binary focal loss [23]:

$$l_{BFL} = \alpha_c y (1 - p)^\gamma \cdot \log p + (1 - y) p^\gamma \cdot \log(1 - p) \tag{2}$$

where $y \in \{0, 1\}$ is ground-truth label, p is the corresponding pixel-level prediction, $\alpha_c = N_b/N_f$ is the ratio of background to foreground (region mask)

Table 1. Train, validation and test split for different region types.

Split	Total	Hole	Line segment	Degradation	Character	Picture	Decorator	Library marker	Boundary line
Train	6491	422	3535	1502	507	81	48	116	280
Validation	793	37	467	142	75	5	12	18	37
Test	1633	106	836	465	113	9	5	31	68

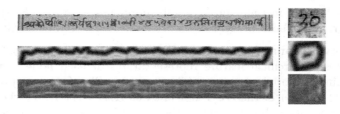

Fig. 4. Some examples of region image (top) and color-coded versions of fast marching distance map (middle) and the attention map computed by final SAG block of BoundaryNet (bottom). The relatively larger values at the crucial boundary portions can be clearly seen in the attention map.

pixel counts and γ is the so-called focusing hyperparameter in focal loss. The class-weighting ensures balanced optimization for background and foreground, indirectly aiding contour estimation. The focal loss encourages the optimization to focus on the harder-to-classify pixels.

To boost the precision of estimate in a more boundary-aware manner, we first construct a distance map using a variant of the Fast Marching method [33]. The procedure assigns a distance of 0 to boundary pixels and progressively higher values to pixels based on contours generated by iterative erosion and dilation of ground-truth region mask (see Fig. 4. The distance map is inverted by subtracting each entry from the global maximum within the map. Thus, the highest weights are assigned to boundary pixels, with the next highest set of values for pixels immediately adjacent to the boundary. The inverted distance map is then normalized ($[0, 1]$) to obtain the final map Ψ. The class-weighted binary focal loss spatial map $\mathcal{L}_{BFL}$ is constituted from per-pixel losses l_{BFL} (Eq. 2) and further weighted by Ψ as follows:

$$\mathcal{L}_{FM} = (1 + \Psi) \odot \mathcal{L}_{BFL} \tag{3}$$

where $\odot$ stands for the Hadamard product. The above formulation is preferred to solely weighting $\mathcal{L}_{BFL}$ with Ψ to mitigate the vanishing gradient issue.

Second Phase: In this phase of training, MCNN's weights are frozen and the estimate of region mask is obtained as described previously (Sect. 3.1). The contour graph constructed from points on region mask boundary (Sect. 3.2) is fed to Anchor GCN. The output nodes from Anchor GCN are interpolated 10× through grid sampling. This ensures maximum optimal shifts towards ground-truth contour for original graph nodes and avoids graph distortion.

Let $\mathcal{G}$ be the set of points (x-y locations) in ground-truth contour and $\mathcal{B}$, the point set predicted by Anchor GCN. Let E_1 be the list of minimum Euclidean distances calculated per ground-truth point $g_i \in \mathcal{G}$ to a point in $\mathcal{B}$, i.e. $e_i = \min_j \| g_i - b_j \|, e_i \in E_1, b_j \in \mathcal{B}$. Let E_2 be a similar list obtained by flipping the roles of ground-truth and predicted point sets. The Hausdorff Distance loss [31] for optimizing Anchor GCN is defined as:

$$L_C(E_1, E_2) = 0.5(\sum_i e_i + \sum_j e_j) \tag{4}$$

where $e_j \in E_2$.

Third Phase: In this phase, we jointly fine-tune the parameters of both MCNN and Anchor GCN in an end-to-end manner. The final optimization is performed by minimizing L_{FT} loss defined as: $L_{FT} = L_C + \lambda \, \mathcal{L}_{FM}$. As we shall see, the end-to-end optimization is crucial for improved performance (Table 3).

The region classification sub-branch is optimized using categorical cross-entropy loss ($\mathcal{L}_{CE}^r$) after all the phases mentioned above. During this process, the backbone is considered as a pre-trained feature extractor, i.e. backpropagation is not performed on MCNN backbone's weights.

3.5 Implementation Details

MCNN: The implementation details of MCNN can be found in Fig. 2. The input $H \times W \times 3$ RGB image is processed by MCNN to generate a corresponding $H \times W$ region mask representation (magenta branch in Fig. 2) and a region class prediction (orange branch) determined from the final 8-way softmax layer of the branch. In addition, the outputs from the SAG blocks are concatenated and the resulting $\frac{H}{2} \times \frac{W}{2} \times 120$ output (shown at the end of dotted green line in Fig. 2) is used to determine the node feature representations f^r used in the downstream Anchor GCN module.

For MCNN training, the focal loss (Eq. 2) is disabled at the beginning, i.e. $\gamma = 0$. The batch size is set to 1 with an initial learning rate of $3e^{-5}$. A customized variant of Stochastic Gradient Descent with Restart [25] is conducted. Two fresh restarts are performed by increasing learning rate 5× for 3 epochs and dropping it back to counter potential loss saturation. The focal loss is invoked with $\gamma = 2$ when $\mathcal{L}_{FM}$ (Eq. 3) starts to plateau. At this stage, the learning rate is set to decay by 0.5 every 7 epochs.

Contourization: The region formed by pixels labelled as region interior in MCNN's output is morphologically closed using a 3×3 disk structuring element. Major regions are extracted using area-based thresholding. The final region interior mask is obtained by connecting all the major sub-regions through their centroids. A b-spline curve is fit to the boundary of the resulting region and $M = 200$ boundary points are uniformly sampled along the curve - this process is depicted in Fig. 3.

Table 2. Region-wise average and overall Hausdorff Distance (HD) for different baselines and BoundaryNet on Indiscapes dataset.

	HD ↓	Hole	Line segment	Degradation	Character	Picture	Decorator	Library marker	Boundary line
BoundaryNet	**17.33**	6.95	20.37	10.15	7.58	51.58	20.17	16.42	5.45
Polygon-RNN++ [1]	30.06	5.59	66.03	7.74	5.11	105.99	25.11	9.97	15.01
Curve-GCN [24]	39.87	8.62	142.46	14.55	10.25	68.64	32.11	19.51	22.85
DACN [7]	41.21	8.48	105.61	14.10	11.42	91.18	26.55	22.24	50.16
DSAC [27]	54.06	14.34	237.46	10.40	8.27	65.81	39.36	23.34	33.53

Anchor GCN: Each boundary point's 122-dimensional node feature is obtained by concatenating the 120-dimensional feature column (f^r in Fig. 2) and the point's 2-D coordinates $p = (x, y)$ (normalized to a $[0, 1] \times [0, 1]$ grid). Each contour point is connected to its 20 nearest sequential neighbors in the contour graph, ten on each side along the contour (see 'Mask Contour' in Fig. 2), i.e. maximum hop factor $k = 10$. The graph representation is processed by two GCN and six residual GCN layers (see 'Residual GCN' in Fig. 2 for architectural details). The resulting features are processed by a fully connected layer to produce 2-D residuals for each of the boundary points. The iterative refinement of boundary points is performed two times. During training, the batch size is set to 1 with a learning rate of $1e^{-3}$.

End-to-End Framework: For joint optimization, the batch size set to 1 with a relatively lower learning rate of $1e^{-5}$. Weighting coefficient λ (in Eq. 4) is set to 200.

Throughout the training phases and for loss computation, the predicted points and ground-truth are scaled to a unit normalized ($[0, 1] \times [0, 1]$) canvas. Also, to ensure uniform coverage of all region classes, we perform class-frequency based mini-batch resampling and utilize the resultant sequences for all phases of training.

4   Experimental Setup

Performance Measure: As performance measure, we use the Hausdorff Distance (HD) [20] between the predicted contour and its ground-truth counterpart (Sect. 3.5). Note that smaller the HD, the better is the boundary prediction. The per-region HD is obtained as the average over the HD of associated region instances.

For all the models, we use performance on the validation set to determine the optimal hyperparameters and determine architectural choices. Subsequently, we optimize the models on the combined training and validation splits and conduct a one-time evaluation on the test split.

Baselines: To perform a comparative evaluation of BoundaryNet, we include multiple state-of-the-art semi-automatic annotation approaches - DSAC [27], Polygon-RNN++ [1], Curve-GCN [24] and DACN [7]. These approaches exhibit

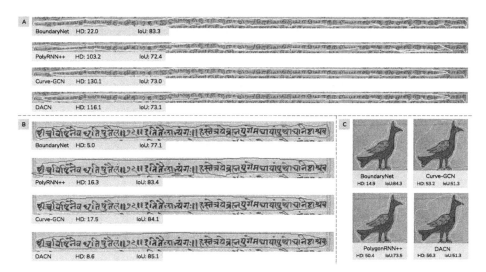

Fig. 5. Qualitative comparison of BoundaryNet with baselines on sample test images from Indiscapes dataset. For each region, the ground-truth contour is outlined in white. The IoU score is also mentioned for reference (see Sect. 5).

impressive performance for annotating semantic regions in street-view dataset and for overhead satellite imagery. However, directly fine-tuning the baselines resulted in bad performance due to the relatively fewer annotation nodes regression and domain gap between document images and imagery (street-view, satellite) for which the baselines were designed. Therefore, we use the original approaches as a guideline and train modified versions of the baseline deep networks.

Evaluation Dataset: For training and primary evaluation, we use Indiscapes [30], a challenging historical document dataset of handwritten manuscript images. It contains 526 diverse document images containing 9507 regions spanning the following categories: Holes, Line Segments, Physical Degradation, Character Component, Picture, Decorator, Library Marker, Boundary Line, Page Boundary (omitted for our evaluation). Details of the training, validation and test splits can be viewed in Table 1.

5 Results

5.1 Indiscapes

Quantitative Baseline Comparison: As Table 2 shows, BoundaryNet outperforms other baselines by a significant margin in terms of overall Hausdorff Distance (HD). Considering that images in the test set have widths as large as 6800 pixels, the results indicate a high degree of precision for obtained contours.

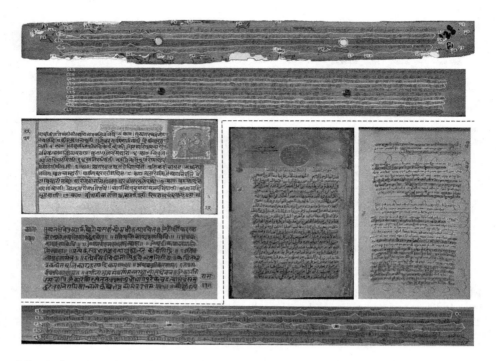

Fig. 6. Semantic region boundaries predicted by BoundaryNet. The colors distinguish instances – they are *not* region labels (written in shorthand alongside the regions). The dotted line separates Indiscapes dataset images (top) and those from other document collections (bottom). Note: BoundaryNet has been trained only on Indiscapes.

The performance of BoundaryNet is slightly lower than the best on regions such as 'Holes', 'Library Marker' and 'Degradation' due to the filtering induced by the GCN. However, notice that the performance for region present most frequently - 'Line Segment' - is markedly better than other baselines.

Qualitative Baseline Comparison: The performance of BoundaryNet and top three baseline performers for sample test images can be viewed in Fig. 5. In addition to HD, we also mention the IoU score. As the results demonstrate, HD is more suited than IoU for standalone and comparative performance assessment of boundary precision. The reason is that IoU is an area-centric measure, suited for annotating objects in terms of their rigid edges (e.g. objects found in real-world scenery). As example B in Fig. 5 shows, a boundary estimate which fails to enclose the semantic content of the region properly can still have a high IoU. In contrast, semantic regions found in documents, especially character lines, typically elicit annotations which aim to minimally enclose the region's semantic content in a less rigid manner. Therefore, a contour-centric measure which penalizes boundary deviations is more suitable.

Table 3. Performance for ablative variants of BoundaryNet. The + refers to MCNN's output being fed to mentioned ablative variants of AGCN.

Component	Ablation type	Default configuration in BoundaryNet	HD ↓
MCNN	Max residual channels = 64	Max residual channels = 128	21.77
MCNN	No focal loss	Focal loss	22.98
MCNN	No fast marching weights penalization	Fast marching weights penalization	23.96
MCNN	Normal skip connection, no attention gating	Skip connection with attention gating	28.27
MCNN	No AGCN	AGCN	19.17
+AGCN	⩽5-hop neighborhood	⩽10-hop neighborhood	19.26
+AGCN	⩽15-hop neighborhood	⩽10-hop neighborhood	20.82
+AGCN	1× spline interpolation	10× interpolation	20.48
+AGCN	1 iteration	2 iterations	19.31
+AGCN	100 graph nodes	200 graph nodes	20.37
+AGCN	300 graph nodes	200 graph nodes	19.98
+AGCN	Node features: backbone only $f^r(x, y)$	$f^r(x, y)$, (x, y)	20.50
+Fine-tuning	No end-to-end finetuning	End to end finetuning	18.79
BoundaryNet	–	original	**17.33**

Qualitative Results (Image-level): Examples of document images with BoundaryNet predictions overlaid can be seen in Fig. 6. The images above the dotted line are from the Indiscapes dataset. The documents are characterized by dense layouts, degraded quality (first image), ultra wide character lines (second image). Despite this, BoundaryNet provides accurate annotation boundaries. Note that BoundaryNet also outputs region labels. This results in amortized time and labor savings for the annotator since region label need not be provided separately. Region Classifier performance can be seen in Fig. 7 (left).

Performance on Other Document Collections: To determine its general utility, we used BoundaryNet for semi-automatic annotation of documents from other historical manuscript datasets (South-East Asian palm leaf manuscripts, Arabic and Hebrew documents). The results can be viewed in the images below the dotted line in Fig. 6. Despite not being trained on images from these collections, it can be seen that BoundaryNet provides accurate region annotations.

Ablations: To determine the contribution of various architectural components, we examined lesioned variants of BoundaryNet for ablation analysis. The results can be viewed in Table 3. As can be seen, the choices related to the MCNN's loss function, presence of error penalizing distance maps, number of points sampled on mask boundary estimate, spline interpolation, all impact performance in a significant manner.

5.2 Timing Analysis

To determine BoundaryNet utility in a practical setting, we obtained document-level annotations for test set images from Indiscapes dataset. The annotations

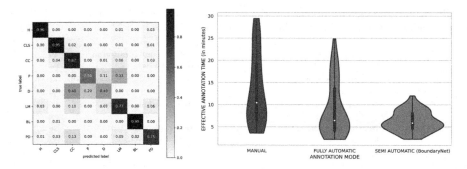

Fig. 7. (left) Confusion Matrix from Region classifier branch, (right) Document-level end-to-end annotation duration distribution for various approaches depicted as a violin plot (the white dot represents mean duration - see Sect. 5.2).

for each image were sourced using an in-house document annotation system in three distinct modes: Manual Mode (hand-drawn contour generation and region labelling), Fully Automatic Mode (using an existing instance segmentation approach [30] with post-correction using the annotation system) and Semi-Automatic Mode (manual input of region bounding boxes which are subsequently sent to BoundaryNet, followed by post-correction). For each mode, we recorded the end-to-end annotation time at per-document level, including manual correction time. The distribution of annotation times for the three modes can be seen in Fig. 7 (right). As can be seen, the annotation durations for the BoundaryNet-based approach are much smaller compared to the other approaches, despite BoundaryNet being a semi-automatic approach. This is due to the superior quality contours generated by BoundaryNet which minimize post-inference manual correction burden.

6   Conclusion

In this paper, we propose BoundaryNet, a novel architecture for semi-automatic layout annotation. The advantages of our method include (i) the ability to process variable dimension input images (ii) accommodating large variation in aspect ratio without affecting performance (iii) adaptive boundary estimate refinement. We demonstrate the efficacy of BoundaryNet on a diverse and challenging document image dataset where it outperforms competitive baselines. Finally, we show that BoundaryNet readily generalizes to a variety of other historical document datasets containing dense and uneven layouts. Going ahead, we plan to explore the possibility of incorporating BoundaryNet into popular instance segmentation frameworks in an end-to-end manner.

References

1. Acuna, D., et al.: Efficient interactive annotation of segmentation datasets with Polygon-RNN++. In: CVPR, pp. 859–868 (2018)

2. Bonechi, S., Andreini, P., et al.: COCO_TS dataset: pixel-level annotations based on weak supervision for scene text segmentation. In: ICANN, pp. 238–250 (2019)
3. Breuel, T.M.: Robust, simple page segmentation using hybrid convolutional MDL-STM networks. ICDAR **01**, 733–740 (2017)
4. Buss, J.F., Rosenberg, A.L., Knott, J.D.: Vertex types in book-embeddings. Tech. rep., Amherst, MA, USA (1987)
5. Can, Y.B., Chaitanya, K., Mustafa, B., Koch, L.M., Konukoglu, E., Baumgartner, C.F.: Learning to segment medical images with scribble-supervision alone. In: Stoyanov, D., et al. (eds.) DLMIA/ML-CDS -2018. LNCS, vol. 11045, pp. 236–244. Springer, Cham (2018). https://doi.org/10.1007/978-3-030-00889-5_27
6. Clausner, C., Antonacopoulos, A., Derrick, T., Pletschacher, S.: ICDAR 2019 competition on recognition of early Indian printed documents-REID2019. In: ICDAR, pp. 1527–1532 (2019)
7. Dong, Z., Zhang, R., Shao, X.: Automatic annotation and segmentation of object instances with deep active curve network. IEEE Access **7**, 147501–147512 (2019)
8. Fu, J., Liu, J., Wang, Y., Zhou, J., Wang, C., Lu, H.: Stacked deconvolutional network for semantic segmentation. In: IEEE TIP (2019)
9. Garz, A., Seuret, M., Simistira, F., Fischer, A., Ingold, R.: Creating ground truth for historical manuscripts with document graphs and scribbling interaction. In: DAS, pp. 126–131 (2016)
10. Gur, S., Shaharabany, T., Wolf, L.: End to end trainable active contours via differentiable rendering. In: ICLR (2020)
11. Gurjar, N., Sudholt, S., Fink, G.A.: Learning deep representations for word spotting under weak supervision. In: DAS, pp. 7–12 (2018)
12. Harley, A.W., Ufkes, A., Derpanis, K.G.: Evaluation of deep convolutional nets for document image classification and retrieval. In: ICDAR (2015)
13. He, K., Gkioxari, G., Dollár, P., Girshick, R.: Mask R-CNN. In: ICCV (2017)
14. He, K., Zhang, X., Ren, S., Sun, J.: Spatial pyramid pooling in deep convolutional networks for visual recognition. IEEE TPAMI **37**(9), 1904–1916 (2015)
15. Kassis, M., El-Sana, J.: Scribble based interactive page layout segmentation using Gabor filter. In: ICFHR, pp. 13–18 (2016)
16. Kassis, M., Abdalhaleem, A., Droby, A., Alaasam, R., El-Sana, J.: VML-HD: the historical Arabic documents dataset for recognition systems. In: 1st International Workshop on Arabic Script Analysis and Recognition, pp. 11–14. IEEE (2017)
17. Kesiman, M.W.A., et al.: Benchmarking of document image analysis tasks for palm leaf manuscripts from Southeast Asia. J. Imaging **4**(2), 43 (2018)
18. Kipf, T.N., Welling, M.: Variational graph auto-encoders. arXiv preprint arXiv:1611.07308 (2016)
19. Kipf, T.N., Welling, M.: Semi-supervised classification with graph convolutional networks. In: ICLR (2017)
20. Klette, R., Rosenfeld, A. (eds.): Digital Geometry. The Morgan Kaufmann Series in Computer Graphics. Morgan Kaufmann, San Francisco (2004)
21. Lais Wiggers, K., de Souza Britto Junior, A., Lameiras Koerich, A., Heutte, L., Soares de Oliveira, L.E.: Deep learning approaches for image retrieval and pattern spotting in ancient documents. arXiv e-prints (2019)
22. Li, G., Muller, M., Thabet, A., Ghanem, B.: DeepGCNs: can GCNs go as deep as CNNs? In: ICCV, pp. 9267–9276 (2019)
23. Lin, T.Y., Goyal, P., Girshick, R., He, K., Dollár, P.: Focal loss for dense object detection. In: ICCV, pp. 2980–2988 (2017)
24. Ling, H., Gao, J., Kar, A., Chen, W., Fidler, S.: Fast interactive object annotation with curve-GCN. In: CVPR, pp. 5257–5266 (2019)

25. Loshchilov, I., Hutter, F.: SGDR: stochastic gradient descent with warm restarts. arXiv preprint arXiv:1608.03983 (2016)
26. Ma, L., Long, C., Duan, L., Zhang, X., Li, Y., Zhao, Q.: Segmentation and recognition for historical Tibetan document images. IEEE Access **8**, 52641–52651 (2020)
27. Marcos, D., Tuia, D., et al.: Learning deep structured active contours end-to-end. In: CVPR, pp. 8877–8885 (2018)
28. Oktay, O., et al.: Attention U-Net: learning where to look for the pancreas. In: Medical Imaging with Deep Learning (2018)
29. Pal, K., Terras, M., Weyrich, T.: 3D reconstruction for damaged documents: imaging of the great parchment book. In: Frinken, V., Barrett, B., Manmatha, R., Märgner, V. (eds.) HIP@ICDAR 2013, pp. 14–21. ACM (2013)
30. Prusty, A., Aitha, S., Trivedi, A., Sarvadevabhatla, R.K.: Indiscapes: instance segmentation networks for layout parsing of historical Indic manuscripts. In: ICDAR, pp. 999–1006 (2019)
31. Ribera, J., Güera, D., Chen, Y., Delp, E.J.: Locating objects without bounding boxes. In: CVPR, June 2019
32. Saini, R., Dobson, D., et al.: ICDAR 2019 historical document reading challenge on large structured Chinese family records. In: ICDAR, pp. 1499–1504 (2019)
33. Sethian, J.A.: A fast marching level set method for monotonically advancing fronts. PNAS **93**(4), 1591–1595 (1996)
34. Shahab, A., Shafait, F., et al.: An open approach towards benchmarking of table structure recognition systems. In: DAS, pp. 113–120 (2010)
35. Simistira, F., Seuret, M., Eichenberger, N., Garz, A., Liwicki, M., Ingold, R.: DIVA-HisDB: a precisely annotated large dataset of challenging medieval manuscripts. In: ICFHR, pp. 471–476 (2016)
36. Slimane, F., Ingold, R., Kanoun, S., Alimi, A.M., Hennebert, J.: Database and evaluation protocols for Arabic printed text recognition. DIUF-University of Fribourg-Switzerland (2009)
37. Song, C., Huang, Y., Ouyang, W., Wang, L.: Box-driven class-wise region masking and filling rate guided loss for weakly supervised semantic segmentation. In: CVPR, pp. 3136–3145 (2019)
38. Tang, M., Perazzi, F., Djelouah, A., Ayed, I.B., Schroers, C., Boykov, Y.: On regularized losses for weakly-supervised CNN segmentation. In: Ferrari, V., Hebert, M., Sminchisescu, C., Weiss, Y. (eds.) ECCV 2018. LNCS, vol. 11220, pp. 524–540. Springer, Cham (2018). https://doi.org/10.1007/978-3-030-01270-0_31
39. Wiggers, K.L., Junior, A.S.B., Koerich, A.L., Heutte, L., de Oliveira, L.E.S.: Deep learning approaches for image retrieval and pattern spotting in ancient documents. ArXiv abs/1907.09404 (2019)
40. Yalniz, I.Z., Manmatha, R.: A fast alignment scheme for automatic OCR evaluation of books. In: ICDAR, pp. 754–758 (2011)
41. Yang, X., et al.: Learning to extract semantic structure from documents using multimodal fully convolutional neural networks. In: CVPR, pp. 5315–5324 (2017)
42. Zhang, L., Song, H., Lu, H.: Graph node-feature convolution for representation learning. arXiv preprint arXiv:1812.00086 (2018)
43. Zhong, X., Tang, J., Jimeno Yepes, A.: PubLayNet: largest dataset ever for document layout analysis. In: ICDAR, pp. 1015–1022 (2019)
44. Zhou, J., Guo, B., Zheng, Y.: Document image retrieval based on convolutional neural network. In: Advances in Intelligent Information Hiding and Multimedia Signal Processing, pp. 221–229 (2020)

Pho(SC)Net: An Approach Towards Zero-Shot Word Image Recognition in Historical Documents

Anuj Rai[1(✉)], Narayanan C. Krishnan[1], and Sukalpa Chanda[2]

[1] Department of Computer Science and Engineering,
Indian Institute of Technology, Ropar, India
{2019aim1003,ckn}@iitrpr.ac.in
[2] Department of Computer Science and Communication,
Østfold University College, Halden, Norway
sukalpa@ieee.org

Abstract. Annotating words in a historical document image archive for word image recognition purpose demands time and skilled human resource (like historians, paleographers). In a real-life scenario, obtaining sample images for all possible words is also not feasible. However, Zero-shot learning methods could aptly be used to recognize unseen/out-of-lexicon words in such historical document images. Based on previous state-of-the-art methods for word spotting and recognition, we propose a hybrid representation that considers the character's shape appearance to differentiate between two different words and has shown to be more effective in recognizing unseen words. This representation has been termed as Pyramidal Histogram of Shapes (PHOS), derived from PHOC, which embeds information about the occurrence and position of characters in the word. Later, the two representations are combined and experiments were conducted to examine the effectiveness of an embedding that has properties of both PHOS and PHOC. Encouraging results were obtained on two publicly available historical document datasets and one synthetic handwritten dataset, which justifies the efficacy of "Phos" and the combined "Pho(SC)" representation.

Keywords: PHOS · Pho(SC) · Zero-shot word recognition · Historical documents · Zero-shot learning · Word recognition

1 Introduction

Historical documents could provide information and conditions about human societies in the past. Due to easy availability and usability of image acquisition devices such documents are being digitized and archived nowadays. Searching for important and relevant information from the large pool of images in the digital archive is a challenging task. Earlier, end-to-end transcription of the text using OCR was a popular way to achieve this goal. However, the performance of OCR

© Springer Nature Switzerland AG 2021
J. Lladós et al. (Eds.): ICDAR 2021, LNCS 12821, pp. 19–33, 2021.
https://doi.org/10.1007/978-3-030-86549-8_2

often depends on character-segmentation accuracy, which is itself error prone, specially in the context of cursive handwritten text. Moreover, end-users of such digital archives (historians, paleographers etc.) are often not interested in an end-to-end transcription of the text, rather they are interested in specific document pages where a query incident, place name, person name has been mentioned. To cater to this requirement, word-spotting and recognition techniques play an important role as they help directly in document indexing and retrieval.

Deep learning models have been quite successful in many document analysis problems. Thus, it is a natural fit for word recognition in historical documents as well. However, training deep networks for this problem is a challenging task due to many reasons. The lack of a large corpus of word images, partly due to the changes in the appearance of characters and spellings of words over centuries, makes it difficult to train a deep model. The complexity is further compounded by the requirement of learning large number of word labels, using only a small set of samples. In addition, the historical documents exhibit many undesirable characteristics such as torn and blurred segments, unwanted noise and faded regions, handwritten annotations by historians and artefacts; all of them contributing to the difficulty of the task.

Word spotting and recognition sound similar to each other but are two different tasks. This is illustrated in Fig. 1. Given an image containing a word as input, a word recognition system identifies the word from the lexicon that is present in the image. On the other hand, word spotting refers to detecting other image segments in the document exhibiting patterns similar to the query image. This work is focused on the former problem and extends it to the zero-shot learning (ZSL) setting. Classical word recognition involves training a machine learning model to recognise the words given the images containing them. It is assumed that the test query images also contain only the words that were presented during the training. However, in the ZSL setting, the test query images can contain words that the model did not see during training. This is a more challenging task requiring a visual representation (akin to the semantic embedding in ZSL literature) that can bridge the set of seen and unseen words.

In this work, we propose a visual characterization of words to learn a mapping between word-images and their corresponding word labels such that it can also be used to recognise out-of-lexicon words. We call this characterization the Pyramidal Histogram of Shapes (PHOS). The PHOS representation encodes the primary shapes of character strokes of different word segments in a hierarchical manner. We use a deep convolutional network to learn the PHOS representation from images of words present in the training lexicon.

Overall we make the following contributions

- We present a novel representation of words (the PHOS representation) that encodes the visual features of the characters.
- The PHOS representation is used to perform zero-shot word recognition.
- Experiments on the PHOS and the popular PHOC representations suggest that PHOS encodes visual shape features of the characters that is missed by PHOC and therefore is more suitable to recognize unseen words.

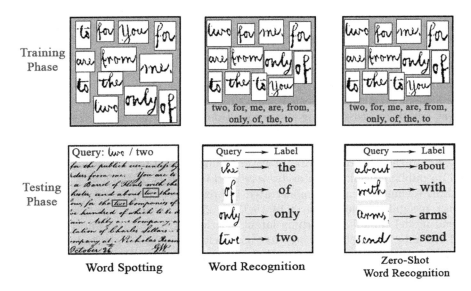

Fig. 1. Difference between word spotting and recognition: Given a word image as input, word recognition is finding the word in the lexicon that is the most likely to be the word present in the image. Word spotting involves finding image segments similar to the query image. Document image belongs to the GW dataset.

– Combining both PHOC and PHOS representations achieves the highest zero-shot word recognition accuracy on a synthetic and two real-world (George Washington and IAM Handwriting) datasets. However, for the generalized ZSL setting, the combined model is only able to outperform PHOS for the synthetic and IAM Handwriting datasets.

2   Related Work

Word spotting and recognition has been well explored over the last 25 years, with a spurt in deep learning based solutions in the recent past [4–7,9–12,20–22]. The seminal work [20] on word spotting involved training of a CNN for predicting the PHOC representation [2]. This work considered both contemporary as well as historical document images in their experiments. The proposed system can be used in both "Query By Example" (QBE) and "Query By String" (QBS) settings. In [11], the authors proposed an End2End embedding scheme to learn a common representation for word images and its labels, whereas in [7], a CNN-RNN hybrid model was proposed for handwritten word recognition. The method proposed in [4] uses a deep recurrent neural network (RNN) and a statistical character language model to attain high accuracy for word spotting and indexing. A recent work on robust learning of embeddings presents a generic deep CNN learning framework that includes pre-training with a large synthetic corpus and augmentation methods to generate real document images, achieving state-of-the-art performance for word spotting on handwritten and historical document images [12]. In [13], the authors introduce a novel semantic representation for word images that uses linguistic and contextual information of words to perform semantic word spotting on Latin and Indic scripts. From the brief discussion it is evident that though deep learning based methods have been used in the recent past for word spotting and recognition tasks, zero-shot word recognition - recognizing a word image without having seen examples of the word during training; has not been studied. Only [6] explore Latin script word recognition problem in the ZSL framework; however the number of test classes were limited and no publicly available datasets were used in their experiments.

ZSL techniques have demonstrated impressive performances in the field of object recognition/detection [3,14,15,17,24,25]. Li et al. [15] propose an end-to-end model capable of learning latent discriminative features jointly in visual and semantic space. They note that user-defined signature attributes loose its discriminativeness in classification because they are not exhaustive even though they are semantically descriptive. Zhang et al. [25], use a Graph Convolutional Network (GCN) along with semantic embeddings and the categorical relationships to train the classifiers. This approach takes as input semantic embeddings for each node (representing the visual characteristic). It predicts the visual classifier for each category after undergoing a series of graph convolutions. During training, the visual classifiers for a few categories are used for learning the GCN parameters. During the test phase, these filters are used to predict the visual classifiers of unseen categories [25]. In [3], the objective functions were customized to preserve the relations between the labels in the embedding space. In [1], attribute label embedding methods for zero-shot and few-shot learning systems were investigated. Later, a benchmark and systematical evaluation of zero-shot learning w.r.t. three aspects, i.e. methods, datasets and evaluation protocol was done in [23]. In [18], the authors propose a conditional generative model to learn latent representations of unseen classes using the generator trained for the seen classes. The synthetically generated features are used to train a classifier for predicting the labels of images from the unseen object category.

In summary, the current ZSL methods are focused towards object detection. There is no work on ZSL for word recognition. One must note that semantic attribute space in ZSL-based object detection is rich as attributes like colour and texture pattern play a crucial role. But in case of ZSL-based word recognition the semantic attribute space is rather constrained due to the absence of such rich visual features. Further the major bottleneck for ZSL based word recognition is the absence of the semantic embedding or attribute signature that establishes the relationship between the various word labels. This is further challenged by large number of word classes with relatively few examples. In this work, we propose a novel attribute signature and validate its effectiveness for ZSL-based word image recognition using standard benchmark datasets.

3   Methodology

We begin by defining our problem of interest. Let $\mathcal{S} = \{(x_i, y_i, c(y_i))\}_{i=1}^N$, where $x_i \in \mathcal{X}, y_i \in \mathcal{Y}^s, c(y_i) \in \mathcal{C}$, $\mathcal{S}$ stands for the training examples of seen word labels, x_i is the image of the word, y_i is the corresponding word label in $\mathcal{Y}^s = \{s_1, s_2, \ldots, s_K\}$ consisting of K discrete seen word labels, and $c(y_i) \in \mathcal{R}^Q$ is the unique word label embedding or attribute signature that models the visual relationship between the word labels. In addition, we have a disjoint word label set $\mathcal{Y}^U = \{u_1, \ldots, u_L\}$ of unseen labels, whose attribute signature set $U = \{u_l, c(u_l)\}_{l=1}^L, c(u_l) \in \mathcal{C}$ is available, but the corresponding images are missing. Given $\mathcal{S}$ and $\mathcal{U}$, the task of zero-shot word recognition is to learn a classifier $f_{zsl} : \mathcal{X} \rightarrow \mathcal{Y}^u$ and in the generalized zero-shot word recognition, the objective is to learn the classifier $f_{gzsl} : \mathcal{X} \rightarrow \mathcal{Y}^u \cup \mathcal{Y}^s$. In the absence of training images from the unseen word labels, it is difficult to directly learn f_{zsl} and f_{gzsl}. Instead, we learn a mapping (ϕ) from the input image space $\mathcal{X}$ to the attribute signature space $\mathcal{C}$ that is shared between $\mathcal{Y}^s$ and $\mathcal{Y}^u$. The word label for the test image x is obtained by performing a nearest neighbor search in the attribute signature space using $\phi(x)$. Thus, the critical features of zero-shot word recognition are the attribute signature space $\mathcal{C}$ that acts as a bridge between the seen and unseen word labels and the mapping ϕ. In this work, we propose a novel attribute signature representation that can effectively model the visual similarity between seen and unseen word labels. The mapping ϕ is modeled as a deep neural network.

A very popular word label representation that can serve as the attribute signature for our problem is the pyramidal histogram of characters (PHOC). A PHOC is a pyramidal binary vector that contains information about the occurrence of characters in a segment of the word. It encodes the presence of a character in a certain split of the string representation of the word. The splits of different lengths result in the pyramidal representation. The PHOC allows to transfer information about the attributes of words present in the training set to the test set as long as all attributes in the test set are also present in the training set. However, this constraint may be violated in the context of zero-shot word recognition. Further, the PHOC also misses the visual shape features of the characters as they appear in a word image. We also observe these limitations of PHOC from our experiments on unseen word recognition. Literature also suggests parity between various representations that only encode the occurrence and position of characters within a word [19]. Thus we are motivated to present a novel attribute signature representation that complements the existing word label characterizations.

3.1   The Pyramidal Histogram of Shapes

We propose the pyramidal histogram of shapes (PHOS) as a robust bridge between seen and unseen words. Central to the PHOS representation is the assumption that every character can be described in terms of a set of primitive shape attributes [6]. We consider the following set of primitive shapes:- ascender, descender, left small semi-circle, right small semi-circle, left large semi-circle, right large semi-circle, circle, vertical line, diagonal line, diagonal line at a slope of 135°, and horizontal line. These shapes are illustrated in Fig. 2. Only the counts of these shapes is insufficient to adequately characterize each word uniquely. Inspired by the pyramidal capture of occurrence and position of characters in a word, we propose the pyramidal histogram of shapes that helps in characterizing each word uniquely.

Fig. 2. 11 primary shape attributes: ascender, descender, left small semi-circle, right small semi-circle, left large semi-circle, right large semi-circle, circle, vertical line, diagonal line, diagonal line at a slope of 135°, and horizontal line

The process of capturing the PHOS representation for a word is illustrated in Fig. 3. At the highest level of the pyramid, there exists only a single segment, which is the entire word. At every level h of the pyramid, we divide the word into h equal segments. Further, at every level h, we count the occurrence of the 11 primary shapes in every h segments of the word. The concatenation of the count vectors for every segment in a level and across all the levels of the pyramid results in the PHOS representation of the word. In this work, we have used levels 1 through 5, resulting in a PHOS vector of length $(1 + 2 + 3 + 4 + 5) * 11 = 165$. Thus the PHOS vector encodes the occurrence and relative position of the shapes in the word string.

For example, let us consider a pair of anagrams "listen" and "silent" for 3 levels of segmentation. The segments at three levels for "listen" are: $L_1 = \{listen\}, L_2 = \{lis, ten\}, L_3 = \{li, st, en\}$. Similarly for "silent": $L_1 = \{silent\}$, $L_2 = \{sil, ent\}$, $L_3 = \{si, le, nt\}$. The corresponding shape counts and their PHOS vector at each level for both words has been illustrated in Fig. 3.

For the zero-shot word recognition problem, it is important to encode the occurrence and relative position of characters within a word, as well as that of visual shapes. Therefore, we propose to use the concatenated PHOC and PHOS vector of a word as its attribute signature representation $\mathcal{C}$. Thus, the attribute signature representation for the word label y_i is $[c_c(y_i), c_s(y_i)]$, where $c_c(y_i)$ and $c_s(y_i)$ are the PHOC and PHOS representations respectively.

3.2   Pho(SC)Net Architecture

Having defined the augmented attribute signature space $\mathcal{C}$, our next objective is to learn the mapping ϕ to transform an input word image into its corresponding attribute signature representation - PHOC+PHOS vector. We use the architecture of SPP-PhocNet [20] as the backbone for the Pho(SC)Net (ϕ) that is used

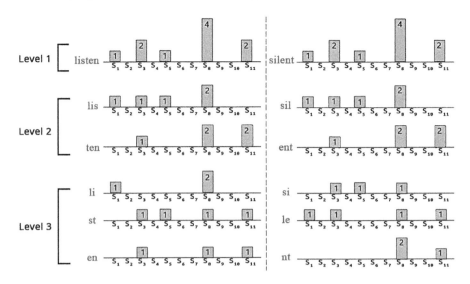

Fig. 3. Pyramidal structure of PHOS representation

to predict the combined representation. The Pho(SC)Net is a multi-task network with shared feature extraction layers between the two tasks (PHOC and PHOS). The shared feature extraction network is a series of convolution layers, followed by a spatial pyramid pooling (SPP) layer. The SPP layer facilitates the extraction of features across multiple image resolutions. The Pho(SC)Net separates out into two branches after the SPP layer to output the two representations. The two branches contain two independent fully connected layers. As the PHOC representation is a binary vector, the PHOC branch ends with a sigmoid activation layer. On the other hand the PHOS representation being a non-negative vector, the PHOS branch ends with a ReLU activation layer. The multi-task Pho(SC)Net architecture is illustrated in Fig. 4.

The output of the Pho(SC)Net for an input word image is the vector $\phi(x) = [\phi_C(x), \phi_S(x)]$, where $\phi_C(x)$ and $\phi_S(x)$ are the predicted PHOC and PHOS representations respectively. Given a mini batch of B instances from the training set consisting of seen word images and their labels, we minimize the following loss function during training.

$$L = \sum_{i=1}^{B} \lambda_c L_c(\phi_c(x_i), c_c(y_i)) + \lambda_s L_s(\phi_s(x_i), c_s(y_i)) \qquad (1)$$

where $L_c(\phi_c(x_i), c_c(y_i))$ is the cross entropy loss between the predicted and actual PHOC representations, $L_s(\phi_s(x_i), c_s(y_i))$ is the squared loss between the predicted and actual PHOS representations, and λ_c, λ_s are hyper-parameters used to balance the contribution of the two loss functions.

Given a test image x, the Pho(SC)Net is used to predict the PHOC and PHOS representations to obtain the predicted attribute signature representation

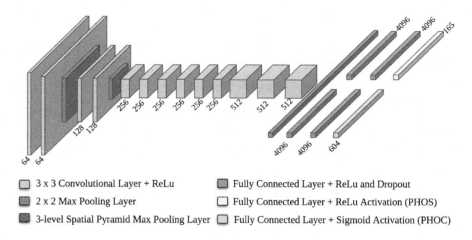

3 x 3 Convolutional Layer + ReLu   Fully Connected Layer + ReLu and Dropout

2 x 2 Max Pooling Layer   Fully Connected Layer + ReLu Activation (PHOS)

3-level Spatial Pyramid Max Pooling Layer Fully Connected Layer + Sigmoid Activation (PHOC)

Fig. 4. Architecture of the multi-task Pho(SC)Net

$[\phi_c(x), \phi_s(x)]$. The word whose attribute signature representation has the highest similarity (measured as cosine similarity) with $[\phi_c(x), \phi_s(x)]$ is the predicted word label for the test image, in the conventional ZSL setting, as defined below

$$\hat{y} = \underset{k \in \mathcal{Y}^U}{\operatorname{argmax}} \cos([\phi_c(x), \phi_s(x)]^T [c_c(k), c_s(k)]) \qquad (2)$$

4   Experiments

4.1   Datasets

We validate the effectiveness of the Pho(SC) representation for the zero-shot word recognition problem on the following three datasets.

Most Frequently Used Words (MFU) Dataset. A synthetic dataset was created from the most frequently used English words list. $\mathcal{Y}^s$ was chosen to be the first 2000 words and the subsequent 1000 words were made part of $\mathcal{Y}^u$. Eight handwriting fonts were used to generate a total of 16000-word images (split into 12000 for training and 4000 for testing) for $\mathcal{Y}^s$ and 8000 for $\mathcal{Y}^u$. We ensure that the word labels of the 4000 test images of $\mathcal{Y}^s$ are present in the corresponding training set. These details are summarized in Table 1.

George Washington (GW) Dataset. The George Washington dataset [8] exhibits homogeneous writing style and contains 4894 images of 1471 word labels. We used the lower-case word images from the standard four-fold cross-validation splits accompanying the dataset to evaluate the Pho(SC)Net. We modified the validation and test sets to suit the zero-shot word recognition problem. Specifically, the test sets in each split was further divided into two parts: seen and unseen word label images, and the validation set contained only seen word label images. The details for each set across all the splits are presented in Table 1.

IAM Handwriting (IAM) Dataset. IAM handwriting dataset [16] is a multi-writer dataset that consists of 115320 word-images, from 657 different writers. We used the lowercase word-class images from this dataset to create train, validation, and test sets. Specifically we created two different splits. In the first split (overlapping writers ZSL split), we ensured that the writers in the test set are also part of the training set. Further, the test set contained both unseen and seen word labels, while the train and validation sets contained only seen word images. In the second split (standard ZSL split) derived from the standard split accompanying the dataset we removed the unseen word images from the validation set, and divided the test split further into seen word and unseen word images. The standard ZSL split has an additional challenge as the writers in the training, validation, and test sets are non-overlapping. The number of images and the (seen and unseen) word labels for each split is presented in Table 1.

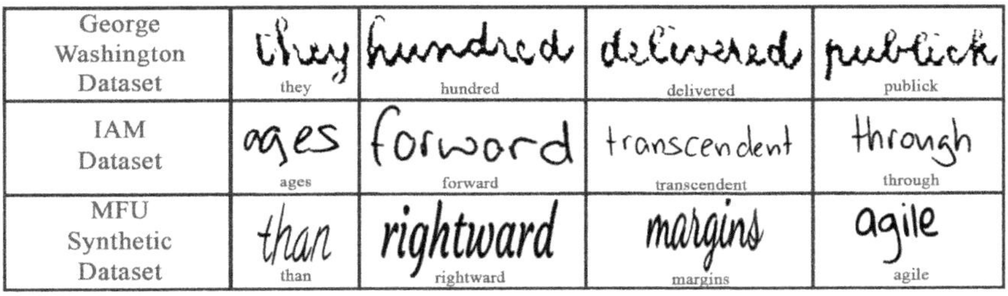

George Washington Dataset	they	hundred	delivered	publick
IAM Dataset	ages	forward	transcendent	through
MFU Synthetic Dataset	than	rightward	margins	agile

Fig. 5. Examples of word images and their labels from the three datasets

The training set of all the three datasets were further augmented through shearing, and addition of Gaussian noise. The size of a word image (in the training, validation, and test sets) depends on the length of the word and the handwriting style (font), but we needed images of uniform size for training. Hence, the binarized images were resized to the best fitting sizes (without changing the aspect ratio) and then padded with white pixels to get images of size 250 * 50. Figure 5 presents a few examples of word images from the three datasets.

4.2 Training and Baselines

The Pho(SC)Net was trained using an Adam optimizer with learning rate $1e-4$, weight decay set as $5e-5$, momentum at 0.9. The batch size is kept as 16. The hyper-parameters λ_c and λ_s were fine tuned using the validation set. The final values chosen for these parameters are 1 and 4.5 respectively. We also conducted ablation studies with the individual PHOCNet (SPP-PHOCNet) and the PHOSNet counter part. These two networks were also trained in a similar fashion for all the three datasets. This helps to investigate the effect of adding visual shape representations to the default attribute signature vector (PHOC vector). Early stopping was applied to the training process using reducedLR

Table 1. Details of dataset used for experiments *Numbers inside the parentheses represent the number of word classes in the set.*

Split	Train Set	Validation Set	Test (Seen Classes)	Test (Unseen Classes)
MFU Dataset				
MFU 2000	36000(2000)	3600	4000	8000(1000)
George Washington (GW Dataset)				
Split 1	1585(374)	662(147)	628(155)	121(110)
Split 2	1657(442)	637(165)	699(155)	114(100)
Split 3	1634(453)	709(164)	667(148)	105(89)
Split 4	1562(396)	668(149)	697(163)	188(152)
IAM Handwriting Dataset				
ZSL Split	30414(7898)	2500(1326)	1108(748)	538(509)
Standard Split	34549(5073)	9066(1499)	8318(1355)	1341(1071)

on plateau on the validation set. As this is the first work to perform zero-shot word recognition on publicly available benchmark datasets, we compare Pho(SC)Net with the classical SPP-PHOCNet model (trained using the settings in [20]). An implementation of the proposed method can be found in github.com/anuj-rai-23/PHOSC-Zero-Shot-Word-Recognition.

4.3   Performance Metrics

The top-1 accuracy of the model's prediction is used as the performance metric for all the experiments. The top-1 accuracy measures the proportion of test instances whose predicted attribute signature vector is closest to the true attribute signature vector. At test time, in the ZSL setting, the aim is to assign an unseen class label, i.e. $\mathcal{Y}^u$ to the test word image and in the generalized ZSL setting (GZSL), the search space includes both seen and unseen word labels i.e. $\mathcal{Y}^u \cup \mathcal{Y}^s$. Therefore, in the ZSL setting, we estimate the top-1 accuracy over $\mathcal{Y}^u$. In the GZSL setting, we determine the top-1 accuracy for both $\mathcal{Y}^u$ and $\mathcal{Y}^s$ independently and then compute their harmonic mean.

5   Results and Discussion

The accuracies for the unseen word labels under the conventional ZSL setting is presented in Table 2. Overall, it is observed that the PHOC representation is not well-suited for predicting unseen word labels. However, the PHOS representation is more accurate in predicting the unseen word labels. Further, the combination of both the vectors (Pho(SC)) results in a significant improvement (on an average >5%) in the unseen word prediction accuracy. The MFU dataset, on the account of being synthetically generated and noise-free, has the highest unseen word recognition accuracy. We obtained the least accuracy on the GW dataset split 4. We attribute this low accuracy to a rather large number of unseen word classes in the test set for this split.

Table 2. ZSL accuracy on all the splits

Split	PHOC	PHOS	Pho(SC)
MFU Dataset			
MFU 2000	.94	.96	**.98**
GW Dataset			
GW Split 1	.46	.61	**.68**
GW Split 2	.64	.72	**.79**
GW Split 3	.65	.71	**.80**
GW Split 4	.35	**.62**	.60
IAM Handwriting Dataset			
ZSL Split	.78	.79	**.86**
Standard Split	.89	.88	**.93**

It is also observed that the accuracy of the model on MFU and IAM datasets are significantly higher than that of any of the GW splits. This is explained by observing the number of seen classes the model is presented during training. Both MFU and IAM datasets have more than 2000 seen word labels, allowing the model to learn the rich relationships between the word labels as encoded by the attribute signatures. Learning this relationship is essential for the model to perform well on unseen word images.

We also observe that using only PHOS as the attribute signature results in better performance than using PHOC on datasets that have homogeneous writing style. This is inferred by noticing the significant increase in the unseen word accuracy of over 14% on the GW dataset splits by PHOS over PHOC.

Table 3. Generalized ZSL accuracy on various splits. A_u = Accuracy with unseen word classes, A_s = Accuracy with seen word classes, Generalized ZSL accuracy, h = Harmonic mean of A_u and A_s.

Split	PHOC			PHOS			Pho(SC)		
	A_u	A_s	h	A_u	A_s	h	A_u	A_s	h
MFU Dataset									
MFU 2000	.74	.99	.85	.92	.93	.92	.92	.99	**.96**
GW Dataset									
GW Split 1	.01	.96	.03	.24	.95	**.39**	.15	.97	.27
GW Split 2	.10	.98	.19	.30	.98	.46	.30	.98	**.46**
GW Split 3	.09	.97	.17	.40	.94	**.56**	.35	.96	.51
GW Split 4	.04	.94	.08	.31	.92	**.47**	.25	.95	.39
IAM Dataset									
ZSL Split	.58	.88	.70	.71	.82	.76	.77	.93	**.84**
Standard Split	.46	.87	.61	.64	.82	.72	.70	.90	**.79**

Fig. 6. Examples of correct and incorrect predictions in Generalized ZSL setting from the standard IAM Split

Table 3 presents the test seen and unseen word accuracies, along with the harmonic mean for the GZSL setting. The high accuracies on the seen word labels and low accuracies on unseen word labels for the model using only PHOC seems to suggest that the PHOC representation is more suitable for scenarios where the word labels across the train and test are similar. In contrast, the PHOS model has marginally lower accuracies on the test seen word labels (in comparison to PHOC), but significantly higher accuracies on the unseen word labels (again in comparison to PHOC). This indicates that looking at visual shapes is more reliable when the train and test sets contain different word labels. Further, we also observe that the combined model yields higher performance for both MFU and IAM datasets that have a large number of seen classes. However, on the GW dataset, that has a significantly smaller number of seen word classes, the PHOS model is able to even outperform the combined model. The poor performance of the PHOC model that is pulling down the performance of the combined model on the unseen classes for the GW dataset can also be attributed to the small number of seen word classes that the model is exposed to during training. However, note that the PHOC model achieves significantly higher accuracies on the test images of the seen word classes, indicating that the model has not overfit, but is biased towards these classes.

Error Analysis. Figure 6 illustrates a few predictions by the Pho(SC) model on the IAM dataset under the GZSL setting. It can observed that even when the model incorrectly predicts the word, there is a good overlap between the characters of the predicted and true word label.

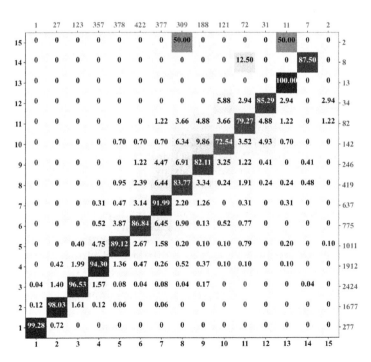

Fig. 7. Lengthwise confusion matrix (normalised) for predicted word length (left axis) and true word length (bottom) for Standard IAM Split. Labels on top (red) indicate the number of word classes in lexicon of that length while on right (green) represent the number of samples in test set for the corresponding word length. (Color figure online)

Figure 7 presents the confusion matrix for predictions on the IAM dataset standard split in the GZSL setting. The confusion matrix has been computed between words of different lengths to uncover any biases of the model (if any). It is difficult to visualize the class specific confusion matrix as there are over 1000 word labels with very few (often only 1) images per word label. The length of the predicted word labels is mostly within a range of the length of the true word label. In general, the model is not biased towards words of any specific length. The high values along the diagonal indicates that the model is often predicting the word of the correct length (except when the word length is 15, for which there are only 2 samples and 2 classes).

6   Conclusion and Future Work

In this paper, we present the first exhaustive study on zero-shot word recognition for historical document images. We propose a novel attribute signature representation (PHOS) that characterizes the occurrence and position of elementary visual shapes in a word. Our experiments demonstrate the effectiveness of PHOS for predicting unseen word labels in the ZSL setting, while the classical PHOC representation is more suitable for seen word labels. Further, we combine both the representations to train a multi-task model- Pho(SC)Net that achieves superior performance over the individual representations. We validate the performance of the models on standard benchmark datasets as well as on a synthetically generated handwritten words dataset. Future directions to this work include extending the PHOS representations to include other non-Latin scripts like Arabic, Chinese, and Indic scripts.

References

1. Akata, Z., Perronnin, F., Harchaoui, Z., Schmid, C.: Label-embedding for image classification. IEEE Trans. Pattern Anal. Mach. Intell. **38**(7), 1425–1438 (2016)
2. Almazán, J., Gordo, A., Fornés, A., Valveny, E.: Word spotting and recognition with embedded attributes. IEEE Trans. Pattern Anal. Mach. Intell. **36**(12), 2552–2566 (2014)
3. Annadani, Y., Biswas, S.: Preserving semantic relations for zero-shot learning. In: The IEEE Conference on Computer Vision and Pattern Recognition (2018)
4. Bluche, T., et al.: Preparatory KWS experiments for large-scale indexing of a vast medieval manuscript collection in the HIMANIS project. In: International Conference on Document Analysis and Recognition, pp. 311–316 (2017)
5. Carbonell, M., Fornés, A., Villegas, M., Lladós, J.: A neural model for text localization, transcription and named entity recognition in full pages. Pattern Recogn. Lett. **136**, 219–227 (2020)
6. Chanda, S., Baas, J., Haitink, D., Hamel, S., Stutzmann, D., Schomaker, L.: Zero-shot learning based approach for medieval word recognition using deep-learned features. In: International Conference on Frontiers of Handwriting Recognition, pp. 345–350 (2018)
7. Dutta, K., Krishnan, P., Mathew, M., Jawahar, C.V.: Improving CNN-RNN hybrid networks for handwriting recognition. In: International Conference on Frontiers of Handwriting Recognition, pp. 80–85 (2018)

8. Fischer, A., Keller, A., Frinken, V., Bunke, H.: Lexicon-free handwritten word spotting using character HMMS. Pattern Recogn. Lett. **33**(7), 934–942 (2012)
9. Graves, A., Schmidhuber, J.: Offline handwriting recognition with multidimensional recurrent neural networks. In: Advances in Neural Information Processing Systems, pp. 545–552 (2009)
10. Kang, L., Toledo, J.I., Riba, P., Villegas, M., Fornés, A., Rusiñol, M.: Convolve, attend and spell: an attention-based sequence-to-sequence model for handwritten word recognition. In: German Conference on Pattern Recognition, pp. 459–472 (2018)
11. Krishnan, P., Dutta, K., Jawahar, C.V.: Word spotting and recognition using deep embedding. In: Document Analysis Systems, pp. 1–6 (2018)
12. Krishnan, P., Jawahar, C.: HWNet v2: an efficient word image representation for handwritten documents. Int. J. Doc. Anal. Recogn. **22**(4), 387–405 (2019)
13. Krishnan, P., Jawahar, C.: Bringing semantics into word image representation. Pattern Recogn. **108**, 107542 (2020)
14. Li, K., Min, M.R., Fu, Y.: Rethinking zero-shot learning: a conditional visual classification perspective. In: IEEE International Conference on Computer Vision, pp. 3583–3592, October 2019
15. Li, Y., Zhang, J., Zhang, J., Huang, K.: Discriminative learning of latent features for zero-shot recognition. In: The IEEE Conference on Computer Vision and Pattern Recognition, pp. 7463–7471 (2018)
16. Marti, U.V., Bunke, H.: The IAM-database: an English sentence database for offline handwriting recognition. Int. J. Doc. Anal. Recogn. **5**(1), 39–46 (2002)
17. Niu, L., Veeraraghavan, A., Sabharwal, A.: Webly supervised learning meets zero-shot learning: a hybrid approach for fine-grained classification. In: IEEE Conference on Computer Vision and Pattern Recognition (2018)
18. Paul, A., Krishnan, N.C., Munjal, P.: Semantically aligned bias reducing zero shot learning. In: IEEE/CVF Conference on Computer Vision and Pattern Recognition, pp. 7056–7065 (2019)
19. Sudholt, S., Fink, G.A.: Evaluating word string embeddings and loss functions for CNN-based word spotting. In: International Conference on Document Analysis and Recognition, pp. 493–498 (2017)
20. Sudholt, S., Fink, G.A.: PHOCNet: a deep convolutional neural network for word spotting in handwritten documents. In: 15th International Conference on Frontiers in Handwriting Recognition, 2016, pp. 277–282 (2016)
21. Wilkinson, T., Lindström, J., Brun, A.: Neural ctrl-f: segmentation-free query-by-string word spotting in handwritten manuscript collections. In: International Conference on Computer Vision, pp. 4443–4452 (2017)
22. Wolf, F., Fink, G.A.: Annotation-free learning of deep representations for word spotting using synthetic data and self labeling. In: Bai, X., Karatzas, D., Lopresti, D. (eds.) DAS 2020. LNCS, vol. 12116, pp. 293–308. Springer, Cham (2020). https://doi.org/10.1007/978-3-030-57058-3_21
23. Xian, Y., Schiele, B., Akata, Z.: Zero-shot learning - the good, the bad and the ugly. In: 2017 IEEE Conference on Computer Vision and Pattern Recognition, pp. 3077–3086 (2017)
24. Xie, G.S., et al.: Attentive region embedding network for zero-shot learning. In: IEEE Conference on Computer Vision and Pattern Recognition (2019)
25. Zhang, H., Koniusz, P.: Zero-shot kernel learning. In: IEEE Conference on Computer Vision and Pattern Recognition, pp. 7670–7679 (2018)

Versailles-FP Dataset: Wall Detection in Ancient Floor Plans

Wassim Swaileh[1（✉)], Dimitrios Kotzinos[1], Suman Ghosh[3], Michel Jordan[1], Ngoc-Son Vu[1], and Yaguan Qian[2]

[1] ETIS Laboratory UMR 8051, CY Cergy Paris University, ENSEA, CNRS, Cergy-Pontoise, France
{wassim.swaileh,dimitrios.kotzinos,michel.jordan,ngoc-son.vu}@ensea.fr
[2] Zhejiang University of Sciences and Technologies, Hangzhou, China
qianyaguan@zust.edu.cn
[3] RACE, UKAEA, Oxford, UK
suman.ghosh@ukaea.uk

Abstract. Access to the floor plans of historical monuments over a time period is necessary in order to understand the architectural evolution and history. Such knowledge also helps to review (rebuild) the history by establishing connections between different events, persons and facts which were once part of the buildings. Since the two-dimensional plans do not capture the entire space, 3D modeling sheds new light on these unique archives and thus opens up great perspectives for understanding the ancient states of the monument. The first step towards generating the 3D model of the buildings and/or monuments is the wall detection inside the floor plan. Henceforth, the current work introduces a novel Versailles-FP dataset consisting Versailles Palace floor plan images and groundtruth in the form of wall masks regarding architectural developments during 17th and 18th century. The wall masks of the dataset are generated using an automated multi-directional steerable filters approach. The generated wall masks are then validated and corrected manually. We validate our approach of wall-mask generation in state-of-the-art modern datasets. Finally we propose a U-net based convolutional framework for wall detection. We have empirically shown that our U-net based method architecture achieves state-of-the-art results surpassing fully connected network based approach.

Keywords: Ancient floor plan dataset · Wall segmentation · U-net neural network model · Sequential training · Steerable filters

1 Introduction

Time and tide wait for none, however we can try to rebuild the time from the evidences which have underlying features to depict the stories from the past. Historical documents about ancient architecture provide important evidences regarding the time, planning, design and understanding of scientific methods practiced.

© Springer Nature Switzerland AG 2021
J. Lladós et al. (Eds.): ICDAR 2021, LNCS 12821, pp. 34–49, 2021.
https://doi.org/10.1007/978-3-030-86549-8_3

These documents can be used for different purposes like understanding and construction of historical events. These constructions facilitate academic research, restoration, preservation and tourism. For example, the unique 10D viewing experience inside Casa Batlló[1] in Barcelona have been constructed through specific feature extractions from the architectural plan of the building which bore unique semantic information about the novelty of the design. Architectural plans are scaled drawings of apartments or buildings. They contain structural and semantic information, for example, room types and sizes, and the location of doors, windows, and fixtures. Modern real-estate industry uses floor plans to generate 3D viewing for prospective clients. Analogously, ancient architecture plans can also be used to create 3D visualisations for interested audiences (academics, tourists etc.). These documents not only depict the stories of a certain time but also depict the altercations and modifications throughout history, thus, enabling the audience with a timeline, where one can navigate through and get a glimpse of the past. Recently this type of research activities and projects have been undertaken for different cities throughout Europe[2], however they rely mainly on historical information (text documents) rather than architectural plans.

One major and crucial component of floor plan analysis is wall detection as walls are the building blocks of rooms and define the structure of a building. Thus wall (or structure of walls) depicts the evolution of the architecture of the building and the way the different rooms started emerging and playing a specific role in the life of the structure. Thus wall detection is one of the most crucial step in analysis of floor plans and often features as sub-step (implicit or explicit) of many other research areas involving floor plan analysis.

However, wall detection is a difficult task to automate due to the fact that walls can be easily confused with different lines existing on the ancient floor plans and thus leading to many false identifications.

One of the main bottlenecks in this regard is the unavailability of floor plan documents regarding ancient architectures. In this paper, we propose a new dataset Versailles-FP to mitigate that. In particular we propose a dataset consisting of floor plans of the Versailles Palace drawn up in the period between 17th and 18th century. The dataset consists of 500 floor plan (FP) images associated with their corresponding wall mask ground truth. We expect the dataset to be used for investigations in different domains and further extension is both planned and encouraged. The current work is focused on wall detection. The dataset is made available in the url https://www.etis-lab.fr/versailles-fp/. The Fig. 1, shows some samples of the Versailles-FP dataset.

There has been attempts to use document analysis and computer vision techniques to study floor plans, except [20], others dealt with modern architectures.

In summary our contribution can be stated as:

- A novel floor plan dataset for ancient architecture (Versailles-FP)
- A method to automatic generation of wall masks and creation of groundtruth

---

[1] https://www.casabatllo.es/en/experience/.
[2] https://www.timemachine.eu/.

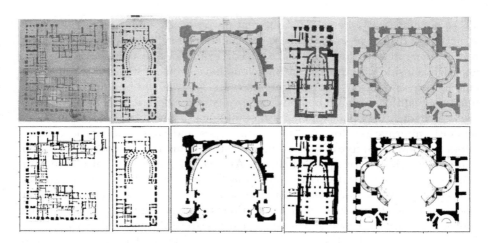

Fig. 1. Samples of Versailles palace floor plans dataset (Versailles-FP) annotated automatically and corrected manually. Original image and their corresponding masks are in the top and down rows of images respectively

- A CNN based wall detection pipeline and baseline results in proposed Versailles-FP dataset.
- Analysis and comparison of proposed wall detection and mask generation approaches with state-of-the-art techniques
- Our code is publicly available for researchers[3,4]

In rest of the paper, Sect. 2 elaborates a review of most related articles. The Sect. 3 details the proposed dataset along with semi-automatic groundtruth generation procedure. The proposed CNN framework for wall detection is described in Sect. 4. Finally in Sect. 5 and 6 we provide experimental validation and conclusion respectively.

2   Related Works

The research on architectural layout documents mainly focused on modern architectures due to lack of ancient architectural documents. In the context of modern architectural images different research problems has been proposed which involves wall detection. In the following two subsections, we briefly describe prior works involving wall detection, namely 3D reconstruction and wall detection.

2.1   Automatic 3D Model Generation from Floor Plans

Tombre's group was one of the pioneer in this field. In [1], they tackle the problem of floor plan 3D reconstruction. They continued their research in this direction

---

[3] https://github.com/swaileh/Bw-Mask-Gen.git.
[4] https://github.com/swaileh/U-net_wall_Segmentation_FP_Keras_model.git.

in [4] and [5]. In these works, they first use a prepossessing step on paper printed drawings to separate text and graphics information. Subsequently a graphical layer separates thick and thin lines and vectorizes the floor plan. The thick lines are considered source of walls whereas the rest of the symbols (e.g. windows and doors) originate from thin lines. Different geometric features are used to detect walls, doors and windows. For example doors are sought by detecting arcs, windows by finding small loops, and rooms are composed by even bigger loops. At the end, 3D reconstruction of a single level [1] is performed. This is extended in [4] and [5] by learning correspondence of several floors of the same building by finding special symbols as staircases, pipes, and bearing walls. They also observe the need of human feedback when dealing with complex plans in [4] and [5]. Moreover, the symbol detection strategies are oriented to one specific notation, thus clearly insufficient for ancient architecture, where plans are consistently changing for 100's of years.

Similarly, Or *et al.* in [12] focus on 3D model generation from a 2D plan. However, they use QGAR tools [15] to separate graphics from text. One major disadvantage of their work is the need of manual interactions to delete the graphical symbols as cupboards, sinks, etc. and other lines which can disturb the detection of the plan structure. Once only lines belonging to walls, doors, and windows remain, a set of polygons is generated using each polyline of the vectorized image. At the end, each polygon represents a particular block. Similar to [4] Or et al. also used geometric features to detect doors, windows etc. This system is able to generate a 3D model of one-stage buildings for plans of a predefined notation. Again this poses a serious problem in case of ancient architectural plans where the notation styles are varied in nature.

More recently Taiba *et al.* [20] proposed a method for 3D reconstruction from 2D drawings for ancient architectures. There approach also follows similar pipelines, more specifically they used specially designed morphological operations for wall detection tasks and they joined different walls to form a graph which then utilized to form the 3D visualizations. However their method is not fully automatic and need to be parameterize for every types of floor plan.

2.2   Wall Detection

Similar to other tasks, wall detection also relies traditionally on a sequence of low level image processing techniques including preprocessing to de-noise the floor plan image, text/graphics separation, symbol recognition and vectorization.

Ryall *et al.* [17] applied a semi-automatic method for room segmentation. Mace *et al.* uses Hough transform [10] to detect walls and doors. Subsequently Wall polygons are created by the hough lines and are partitioned iteratively into rooms, assuming convex room shapes. Ahmed *et al.* [2] process high-resolution images by segmenting lines according to their thickness, followed by geometrical reasoning to segment rooms. Doors are detected using SURF descriptors. More recently researchers adopted convolutional neural networks e.g. Dodge *et al.* in [3] used a simple neural network for the removal of noise followed by a fully convolutional neural network (FCN) for wall segmentation. Liu *et al.* [8] used a combination of deep neural networks and integer programming, where they first identify junction points in a given floor plan image, and then join the junctions to locate the walls in the floor plan. The method can only handle walls that align with the two principal axes in the floor plan image. Hence, it can recognize layouts with only rectangular rooms and walls of uniform thickness, this is a critical limitation in the context of historical layouts where rooms shapes are not always rectangular and often round shaped rooms are used. In [21], also trained a FCN to label the pixels in a floor plan with several classes. The classified pixels formed a graph model and were used to retrieve houses of similar structures.

As stated earlier all these works use different methods ranging from heuristics to geometric features, from low level image processing to deep learning and obtained noteworthy success. But, the efficacy and success of these techniques is a subject of investigation for ancient architectural documents. This concern is further emphasized due to the presence of noise, low differences between the FP image background and foreground colors, image size can vary between very big and very small image size, etc. In Sect. 3.1, we provide a detailed discussion on the challenges posed by this.

3   The Versailles-FP Dataset

In 2013 the research project VERSPERA[5] was started with the aim of digitizing a large amount of graphical documents related to the construction of the Versailles palace during the 17th and 18th centuries. There is a large corpus of floor plans including elevations and sketches present in the collection of French National Archives. The total number of documents in this archive is around 6500 among which about 1500 are floor plans. An ambitious project to digitize this varied corpus is taken in 2017, extraordinary technical capability is needed to achieve this task due to fragile and varied nature of the paper documents (for example some document can be as big as 3 m × 4 m). This project helped to provide free digital access to these stored knowledge, which were bind in paper form. This digital access open enormous possibilities for further research in various domains from history, to digital humanity, to architecture and of course to computer vision and document analysis. The idea of VERSPERA project can also be extended to other monuments of historical importance in the same way.

The digitized plans of Versailles Palace consists of graphics that illustrate the building architecture such as walls, stairs, halls, king rooms, royal apartments, et cetera in addition to texts and decorations. Since Versailles Palace digitized floor plans cover 120 years (1670–1790) of architectural design, different drawing styles clearly appear in the corpus. The obtained document images open new challenging fields of research for the scientific community in document analysis. Possible field of research can be driven for printed and handwritten text localization and recognition. Another interesting research direction can be thought of regarding automatic understanding of small parts of the building plans in a complete view of Versailles palace floors.

---

[5] https://verspera.hypotheses.org/.

Automatic plan change detection which can utilize floor plan over a time period among other sources to establish different changes in building which took place and which didn't (but planned) is another interesting research direction. Similarly floor plan captioning or question answering tasks can be thought of which will involve visual reasoning and understanding of different knowledge sources. Thus, different researchers or research groups can utilize this dataset for different academic purposes, but among them wall detection is one of the primary task which needs to be solved (either implicitly or explicitly) in order to tackle any of these tasks. Thus, in this paper we design our dataset primarily focused on wall detection for ancient architectural floor plans. For constructing the Versailles-FP dataset, we selected 500 images from the available digital archive of Versailles palace floor plans.

In the following sections we first describe different unique challenges posed by ancient floor plans (Sect. 3.1) Then briefly define the wall detection task (Sect. 3.2) and describe how we generate groundtruth using a semi-automatic procedure for Versallies-FP dataset (Sect. 3.3)

3.1   Challenges in Wall Detection Vis-a-Vis Ancient Floor Plans

The wall detection task poses several challenges due to similar form, shape and nature of the walls with other drawings in floor plans. Furthermore, background color of the enclosed spaces sometimes become quite similar to the walls (for example in R-FP-500 dataset), thus become difficult to separate using simple image processing techniques. Compared to modern floor plans, the ancient ones poses another set of problems. Firstly they are hand made/written and does not follow any specific uniform style or format. In addition, the drawing paper sheets varies wildly in size leading to a great deal of difficulty in case of training (specialy in convolutional neural networks)

Moreover, scanning process of such old documents add some additional noise. Sometimes modification to a plan is introduced by overwriting the original plan complicating the automatic processing even further. Figure 2 shows some typical samples of Versailles palace floor plan images and difficulties with regard to wall detection; (a) first floor FP image where the black blocks represent existing walls and red ones represent planed ones; (b) illustrate the capturing image luminance problem with black background border; (c & d) sample of overlapped floor plans.

3.2   Wall Detection Task

In this section, we define the wall detection tasks and discuss specifics of wall to be detected throughout this paper. Later we also discuss some challenges of wall detection both in case of modern and specially for ancient floor plans.

In the wall detection task, the aim is to identify the set of pixels corresponding to the walls. Depending on the drawing style of architects and types of walls, the representation of walls can vary. In [7] authors has identified four different types of walls in modern floor plans, however by analysing ancient floor plans we have seen that two types of walls are sufficient to correctly identify most

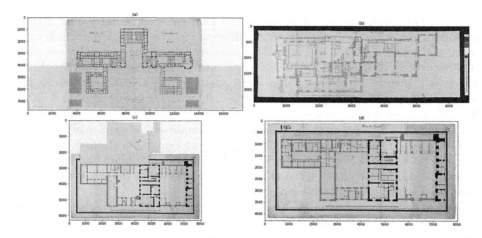

Fig. 2. Challenges posed by Versailles-FP (a) Black blocks represent existing walls and red ones represents planed ones; (b) Spatially varying background luminance/intensity (c & d) sample of overlapped floor plans

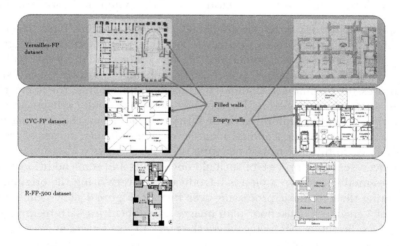

Fig. 3. Samples of filled and empty wall drawings (top) Versailles palace dataset (ancient), (middle) CVC-FP dataset [7] and (down) R-FP-500 modern datasets [3].

walls. Thus following two types of wall drawings are taken into consideration 1) Filled wall drawing arising from thick and dark strokes with straight and rectangular form in general. 2) Empty (Hollow) wall drawing arising from twins of parallel, thin and dark strokes and sometimes connected with diagonal lines. We use this consideration for both ancient and modern drawings as our main goal is to analyse ancient floor plans. Figure 3 illustrates the two types of filled and empty wall drawings in the modern and ancient floor plans.

3.3   Automatic Wall Masks Generation

Pixel level groundtruth is necessary for supervised learning schemes for wall detection tasks. Towards this goal we propose a semi-automatic procedure. First an automatic segmentation method is employed to produce binary mask corresponding to walls. Finally binary masks are checked and corrected manually to produce desired groundtruth. We propose a smearing segmentation [18,19] method based on steerable filters. This method has been shown to be less sensitive to variations in writing style in case of text line segmentation. In case of wall mask generation this is also desirable as ancient floor plan diagrams has considerable variation in drawing styles, thus a method in order to be successful needs to be less sensitive to drawing styles.

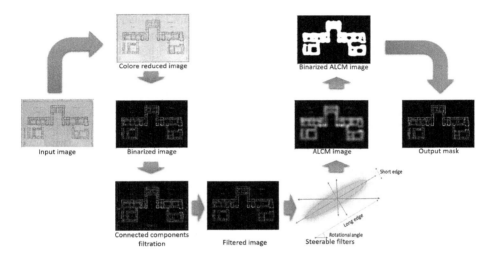

Fig. 4. Mask generation processing chain for Versailles palace plans' images

The proposed approach is explained in Fig. 4 and consists of following three steps:

I. Preprocessing: color floor plan image, the algorithm first corrects its contrast and luminance using exponential and gamma auto correction algorithms [14]. Color clustering is then applied to get 3 tone color images from standard RGB images, we use standard K-means clustering algorithm to achieve this. Next, the obtained grayscale image is binarized using Otsu [13] thresholding algorithm. This binarized image contains walls but also contains texts, stamps, and marks from pins, etc. which should be considered as noise for wall detection tasks. In order to remove those noises, we use heuristics based on connected component analysis. As the noise typically have less area then the wall masks, we need to find a threshold which can be used to filter out connected components. We propose to use the following as threshold.

Let MD and AV denote the median and the average area of the connected components respectively. We have empirically observed that a typical noise is always less than the summation of MD and AV, thus we define another parameter EC, which denotes estimated threshold of wall's CC area and calculated as $EC = MD + AV$. Every CC that have area less than the ECA is removed from the binary image. This process of noise filtration is explained in Fig. 4 where red particles represent the removed CCs.

II. Adaptive Local Connectivity Map Based on Steerable Filters: We apply a set of steerable filters to obtain the adaptive local connectivity map (ALCM) of the filtered FP image. Steerable filters [18] in different orientation are applied to increase the intensity of the foreground at every pixel of the filtered binary image. A steerable filter has an elliptical shape (see Fig. 4), where the short edge represents the filter's height (FH), the long edge represents the filter's width (FW) and the rotation angle (θ) represent the filter's orientation. The most effective parameter on the global performance of the mask generation method is FH as established in [19]. High value of FH leads to increase the intensity of all CCs including walls and unfiltered noise components, and consequently we obtain less clean walls masks. On the other hand, low value of FH can lead to many disjoint small wall masks. Thus there is a trade-off between high and low value and performance of wall mask generation process. An empirical investigation is needed to find the optimum value of FH and FW parameters. We tried different values between $0.7 \times EC$ and $1.2 \times EC$ in a reduced set of FP images from CVC-FP and R-FP-500 datasets. By carefully observing the results, we empirically set FH as equal to EC and $FW = 2 \times FH$. The filtered image is passed through steerable filters of different orientations. In total we use 11 directions of steerable filters with $\theta \in [-25°, 25°]$ in $5°$ intervals. The ALCM image is obtained then by applying the steerable filters with the previous configuration of FH, FW and θ (See right side of Fig. 4).

III. Thresholding and Superimposition: Finally the obtained ALCM image is binarized using the Otsu thresholding [13] algorithm. The final mask image is obtained by superimposing the filtered binary image and the binarized ALCM. Every pixel in the filtered binary image is turned off while its corresponding pixel in the thesholded ALCM image is off. This operation can be easily done by using logical and operation.

This approach for generating wall masks automatically showed good performance on filled walls FP images of CVC-FP and R-FP-500 datasets (see Sect. 5) However, the approach in its current form can not deal with binarization problems related to the quality or nature of the original plan image (stamps, graphical symbols, ink drop traces, folding marks, etc.). For this reason, we proceed to correct manually such error for obtaining valid wall ground truth.

4 U-net Based Deep Neural Network Model for Wall Detection

In this section, we describe another major contribution of the paper. We propose to use a U-net [16] based convolutional neural network for wall detection task. We provide experimental validation of our approach in both proposed ancient architecture dataset and state-of-the-art modern datasets. To train the proposed CNN model, we use sequential training strategy that relies on data augmentation (described in Sect. 4.2) which efficiently use the available annotated samples. In Fig. 5, the architecture of our model is illustrated. Our CNN model is derived from U-net neural network [16]. Our choice is motivated from the perceived advantages of U-net in case of medical image processing [6,16] vis-a-vis speed and scarcity of training data.

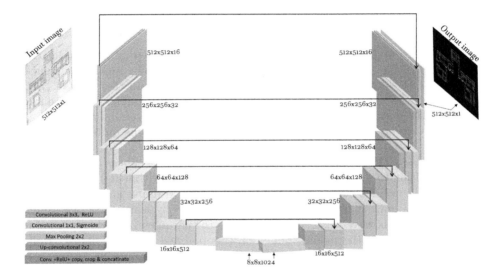

Fig. 5. Illustration of the used U-net architecture.

Gray scale floor plans images are scaled to 512×512 pixels of size at the input of the U-net model. The U-net model consists of two concatenated and symmetric paths; the contracting path and the expanding path. The model uses the contracting path to capture the context and semantic information and the expanding path to achieve precise localization, by copying, cropping and concatenating image information observed in the contracting path toward the expanding path.

We have used the dice loss function as introduced by Fausto, *et al.* [11]. In this study, we used the dice coefficient as a segmentation performance evaluation metric in one hand and in the other hand we used it to compute the loss function.

The dice coefficient D_{coef} is also called *F-score* and calculated as shown in Eq. 1.

$$D_{coef} = \frac{2 \times |Y \cap \hat{Y}|}{|Y| + |\hat{Y}|} \tag{1}$$

where Y is the neural network prediction about the segmentation and $\hat{Y}$ is the ground truth value. Equation 2 shows that if the prediction value is close to the ground truth value while the dice coefficient is close to 1. The dice loss function is defined with regards to the dice coefficient by the following equation:

$$\ell_D = 1 - \frac{2 \times |Y \cap \hat{Y}| + 1}{|Y| + |\hat{Y}| + 1} = 1 - D_{coef} \tag{2}$$

In [16] authors also advocated the use of data augmentation (see Sect. 4.2) in case of U-net based architecture In this paper we also used a similar strategy and performed data augmentation technique to train our model.

4.1 Training Strategy

We use a two step sequential learning process: First step consists of learning the model on the modern CVC-FP dataset samples [7]. As the Versailles palace dataset samples have quite different background than that of CVC-FP, we replaced the white background in the CVC original samples by Versailles colored plans background. This step provides a good pre-training for our final model. In fact we will see (Sect. 5) this initial model obtains quite impressive result

The choice of the learning rate, batch size and number of epochs hyper parameters in addition to the most convenient optimisation algorithm and loss function is empirically decided during this step of the model training and used for the rest of the training. Next, we obtain our final model by learning on Versailles dataset samples, however instead of initializing the parameters randomly, we use the parameters from the model trained on the modified CVC-FP dataset from the first step.

4.2 Training and Implementation Details

Best performance of the initial model on CVC-FP dataset is obtained with learning rate of $10e^{-4}$, batch size of 16 with 100 training epochs with active early stopping after 50 epochs. The model learned with *RMSprop* optimizer with the dice loss function ℓ_D outperforms the models learned with SGD, Adam or Ftrl optimizers and binary cross-entropy, weight cross-entropy or focal loss functions. In the first step of the sequential training strategy we followed, an initial U-net model is trained on the modified CVC-FP dataset with Versailles-FP background. We step up the training for 100 epochs with early stopping after 50 epochs. The training data are augmented five times its size divided by the batch size where the batch size is equal to 16. We observed that the most effective data augmentation settings are zoom = 0.3, right/left shift = 0.2 with activated horizontal and vertical flipping.

5   Experiments

In this section, we present the experiments carried out to validate our proposals regarding automatic mask generation and U-net based wall detection. In the following we first describe in Sect. 5.1 different experimental protocols that are followed for different datasets and then provide experimental validation for both mask generation in Sect. 5.2 and wall detection in Sect. 5.3.

5.1   Dataset and Evaluation Protocol

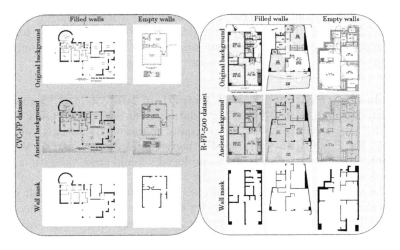

Fig. 6. Samples of CVC-FP and R-FP-500 data set; first image row show the original floor plan, second row show the floor plans with Versailles background and last row show the wall masks.

I. Datasets: Only a few modern floor plan datasets are available for research due to the requirement of challenging collaborative work. In our experiments, we considered the CVC-FP [7] and R-FP-500 [3] modern and real floor plan datasets for evaluating the performance of the proposed wall mask generation approach and the U-net wall segmentation model. CVC-FP dataset is a collection of 122 real floor plan scanned documents compiled and groundtruthed using SGT toolkit[6]. The dataset consists in four sets of plans that have different drawing styles, image quality and resolution incorporating graphics, text descriptions and tables. R-FP-500 dataset [3] is collected from a public real estate website. It consists of 500 floor plan images that varies in size and contains different sort of information, decorative elements, shading schemes and colored backgrounds.

---

[6] http://dag.cvc.uab.es/resources/floorplans/.

To the best of our knowledge, there is no available annotated dataset for ancient floor plans. We introduce in this study, first ancient floor plan dataset, named Versailles-FP dataset which is based on the french VERSPERA research project. The dataset is collected from scanned floor plans that belongs to Versailles palace constructions dated of 17th and 18th century. We used the proposed wall mask generation approach (Sect. 5.2) to annotate 500 images of Versailles palace floor plans. Figure 1 shows samples of this dataset.

II. Evaluation Protocol - Mask Generation: We evaluated the performance of the proposed wall mask generation approach on modified CVC-FP and R-FP-500 datasets with Versailles-FP background. Mean dice coefficient and mean IoU metrics are used as evaluation criteria. In order to critically analyze we further split every dataset into two evaluation sets based on wall types 1) filled wall evaluation set that contains floor plan images with filled walls. 2) empty wall evaluation set where the walls in the floor plan images are empty.

III. Evaluation Protocol - Wall Detection: We conducted experiments and reported results on the following 5 datasets; 1) original CVC-FP dataset, 2) CVC-FP with Versailles-FP background dataset, 3) original R-FP-500 dataset, 4) R-FP-500 with Versailles-FP background dataset and 5) Versailles-FP dataset. For these five datasets, we employed a 5-fold cross validation scheme, where each datasets is split into 3 parts: taking 3/5 samples for training, 1/5 samples for validating and rest 1/5 samples for testing. The final result is calculated by taking mean over all test folds for each datasets. As evaluation metrics, we use the Dice coefficient defined in Eq. 2 and mean Intersection-over-Union (IoU) as in [3] and [9].

5.2 Wall Mask Generation Evaluation

To validate our semi-automatic ground truth generation approach, we introduce this pseudo-task to analyze it's efficacy and correctness.

Table 1. Evaluation results of filled and empty Wall's masks generation

	CVC dataset (122 FP-image)		R-FP-500 dataset (500 FP-image)	
	Filled walls (70% of the dataset)	Empty walls (30% of the dataset)	Filled walls (73% of the dataset)	Empty walls (27% of the dataset)
Dice score %	90.75%	34.45%	85.09%	26.20%
IoU score %	83.05%	21.78%	76.36%	19%

Table 1 introduce the evaluation results on the filled and empty walls evaluation sets of CVC-FP and R-FP-500 dataset. We can observe from Fig. 6 that wall masks are always filled masks even for hollow (empty) walls in the floor plan images, we consider this as an annotation error. However, the proposed approach produce wall mask identical to the walls in the input image. Due to this, we obtained very low dice and IoU scores when evaluating on hollow (empty) walls evaluation sets.

On the other hand, the proposed mask generation approach performs well on 70% of CVC-FP and 73% of R-FP-500 plan images that belongs to the filled walls evaluation sets. From Table 1, we observe that approach achieves 90.75% and 83.05% of dice coefficient and IoU scores on the filled walls evaluation set of CVC-FP dataset. At same time, we observe a performance degradation on the filled walls of the R-FP-500 dataset measured by 5.66% and 6.69% of dice and IoU scores respectively. This degradation effect can be explained by the multi-color backgrounds of the R-FP-500 dataset images. By analysing the masks that are generated by the approach, we observed that the approach successes in eliminating most of texts, description tables and decoration elements but failed in case of orientation sign graphics and trees.

Another important factor to consider in this case is the time taken for finding groundtruth masks for a given FP image. Our approach takes only around 3.5 min (not considering the time taken for manual correction) for one image of size 17000×8000 in comparison to 30–35 min on average by human annotator. Thus this provides a faster annotation process, achieving 10X speed-up in annotation process.

5.3   Analysis of Wall Detection Experiment

In wall detection we have obtained the model in two steps, first we pre-train the proposed U-net model by training only on modified CVC-FP dataset and this alone produce impressive results for modified CVC-FP dataset. Our model achieve 99.57% accuracy and can separate walls with considerable difficulties. Next, we further train our detection model on 5 different datasets, where parameters are initialized using the pretrained model in previous step. The result of this model is provided in Table 2.

Table 2. Wall segmentation evaluation and comparison with the state-of-the-art

Dataset	Model	Mean accuracy %	Mean IoU %	Mean dice %
Original CVC-FP	FCN-2s [3]	97.3	94.4	NA
Original CVC-FP	U-net	**99.73**	**95.01**	**95.65**
Modified CVC-FP	U-net	99.71	93.45	95.01
Original R-FP-500	FCN-2s [3]	94.0	89.7	NA
Original R-FP-500	U-net	98.98	90.94	91.29
Modified R-FP-500	U-net	98.92	90.21	91.00
Versailles-FP	U-net	97.15	88.14	93.32

It can be observed from Table 2 that our U-net model outperform state-of-the-art FCN-2s model [3] on the modern floor plan datasets of CVC-FP and R-FP-500 with original (white) background. Thus providing a better state-of-the-art result in wall detection task. It should also be noted that the proposed convolutional model is more efficient, than a fully connected model as it uses fewer weights, consequently the proposed model should be faster in respect to compuational speed.

6   Conclusion

In this paper we introduced the Versailles-FP dataset that consists of 500 ancient floor plan images and wall masks of The Versailles Palace dated of 17th and 18th century. At first, the wall masks are generated automatically using multi-orientation steerable filters based approach. Then, the generated wall masks are validated and corrected manually. We showed that the mask generator performs well enough on the filled walls of CVC-FP and R-FP-500 datasets. A U-net architecture is then used to achieve the wall detection task on CVC-FP, R-FP-500 and Versailles-FP datasets. The cross-validation results confirmed that the convoulutional architecture of the U-net model performs better than the fully connected network architecture.

Acknowledgements. We thank our colleagues of the VERSPERA research project in the Research Center of Château de Versailles, French national Archives and French national Library, and the Fondation des sciences du patrimoine which supports VERSPERA.

References

1. Ah-Soon, C., Tombre, K.: Variations on the analysis of architectural drawings. In: Proceedings of the Fourth ICDAR, vol. 1, pp. 347–351 (1997)
2. Ahmed, S., Liwicki, M., Weber, M., Dengel, A.: Automatic room detection and room labeling from architectural floor plans. In: 2012 10th IAPR DAS, pp. 339–343 (2012)
3. Dodge, S., Xu, J., Stenger, B.: Parsing floor plan images. In: MVA, May 2017
4. Dosch, P., Masini, G.: Reconstruction of the 3D structure of a building from the 2D drawings of its floors. In: Proceedings of the Fifth ICDAR, pp. 487–490 (1999)
5. Dosch, P., Tombre, K., Ah-Soon, C., Masini, G.: A complete system for the analysis of architectural drawings. ICDAR **3**(2), 102–116 (2000)
6. Gao, H., Yuan, H., Wang, Z., Ji, S.: Pixel deconvolutional networks. arXiv preprint arXiv:1705.06820 (2017)
7. de las Heras, L.P., Terrades, O.R., Robles, S., Sánchez, G.: CVC-FP and SGT: a new database for structural floor plan analysis and its groundtruthing tool. IJDAR **18**(1), 15–30 (2015)
8. Liu, C., Wu, J., Kohli, P., Furukawa, Y.: Raster-to-vector: revisiting floorplan transformation. In: Proceedings of the IEEE ICCV, pp. 2195–2203 (2017)
9. Long, J., Shelhamer, E., Darrell, T.: Fully convolutional models for semantic segmentation. In: CVPR, vol. 3, p. 4 (2015)

10. Macé, S., Locteau, H., Valveny, E., Tabbone, S.: A system to detect rooms in architectural floor plan images. In: Proceedings of the 9th IAPR DAS, pp. 167–174 (2010)

11. Milletari, F., Navab, N., Ahmadi, S.: V-Net: fully convolutional neural networks for volumetric medical image segmentation. In: Proceedings of the 2016 Fourth International Conference on 3D Vision (3DV), pp. 565–571 (2016)

12. Or, S.H., Wong, K.H., Yu, Y.K., Chang, M.M.Y., Kong, H.: Highly automatic approach to architectural floorplan image understanding & model generation. Pattern Recogn. **V**, 25–32 (2005)

13. Otsu, N.: A threshold selection method from gray-level histograms. IEEE Trans. Syst. Man Cybern. **9**(1), 62–66 (1979). https://doi.org/10.1109/TSMC.1979.4310076

14. Reinhard, E., Heidrich, W., Debevec, P., Pattanaik, S., Ward, G., Myszkowski, K.: High Dynamic Range Imaging: Acquisition, Display, and Image-Based Lighting. Morgan Kaufmann (2010)

15. Rendek, J., Masini, G., Dosch, P., Tombre, K.: The search for genericity in graphics recognition applications: Design issues of the Qgar software system. In: DAS, pp. 366–377 (2004)

16. Ronneberger, O., Fischer, P., Brox, T.: U-Net: convolutional networks for biomedical image segmentation. In: Navab, N., Hornegger, J., Wells, W.M., Frangi, A.F. (eds.) MICCAI 2015. LNCS, vol. 9351, pp. 234–241. Springer, Cham (2015). https://doi.org/10.1007/978-3-319-24574-4_28

17. Ryall, K., Shieber, S., Marks, J., Mazer, M.: Semi-automatic delineation of regions in floor plans. In: Proceedings of 3rd ICDAR, vol. 2, pp. 964–969 (1995)

18. Shi, Z., Setlur, S., Govindaraju, V.: A steerable directional local profile technique for extraction of handwritten Arabic text lines. In: 2009 10th ICDAR, pp. 176–180 (2009)

19. Swaileh, W., Mohand, K.A., Paquet, T.: Multi-script iterative steerable directional filtering for handwritten text line extraction. In: 2015 13th ICDAR, pp. 1241–1245 (2015)

20. Tabia, H., Riedinger, C., Jordan, M.: Automatic reconstruction of heritage monuments from old architecture documents. J. Electron. Imaging **26**(1), 011006 (2016)

21. Yang, S.T., Wang, F.E., Peng, C.H., Wonka, P., Sun, M., Chu, H.K.: DuLa-Net: a dual-projection network for estimating room layouts from a single RGB panorama. In: Proceedings of the IEEE/CVF Conference on CVPR, pp. 3363–3372 (2019)

Graph Convolutional Neural Networks for Learning Attribute Representations for Word Spotting

Fabian Wolf[1]([✉])[iD], Andreas Fischer[2,3][iD], and Gernot A. Fink[1][iD]

[1] Department of Computer Science,
TU Dortmund University, 44227 Dortmund, Germany
{fabian.wolf,gernot.fink}@cs.tu-dortmund.de
[2] Department of Informatics, DIVA Group,
University of Fribourg, Fribourg, Switzerland
andreas.fischer@unifr.ch
[3] Institute of Complex Systems, University of Applied Sciences and Arts
Western Switzerland, Fribourg, Switzerland

Abstract. Graphs are an intuitive and natural way of representing handwriting. Due to their high representational power, they have shown high performances in different learning-free document analysis tasks. While machine learning is rather unexplored for graph representations, geometric deep learning offers a novel framework that allows for convolutional neural networks similar to the image domain. In this work, we show that the concept of attribute prediction can be adapted to the graph domain. We propose a graph neural network to map handwritten word graphs to a symbolic attribute space. This mapping allows to perform *query-by-example* word spotting as it was also tackled by other learning-free approaches in the graph domain. Furthermore, our model is capable of *query-by-string*, which is out of scope for other graph-based methods in the literature. We investigate two variants of graph convolutional layers and show that learning improves performances considerably on two popular graph-based word spotting benchmarks.

Keywords: Graph neural networks · Geometric deep learning · Word spotting

1 Introduction

The field of pattern recognition distinguishes the two principles of statistical and structural approaches [4]. For any application both approaches need to solve the problem of how to measure the similarity of different objects. Statistical approaches usually rely on numerically representing an object in the form of a high-dimensional feature vector. Measuring similarity is then feasible by common vector distances. Statistical approaches offer a mature framework of algorithms for clustering, retrieval or classification and have strongly benefited from the

© Springer Nature Switzerland AG 2021
J. Lladós et al. (Eds.): ICDAR 2021, LNCS 12821, pp. 50–64, 2021.
https://doi.org/10.1007/978-3-030-86549-8_4

uprise of deep neural networks. However, the representational power of a vector is limited, which motivates the structural approach. In this case, data is represented based on symbolic structures such as graphs. While graphs offer a more powerful data representation, they often lack the mathematical simplicity of Euclidean data. Already basic operations such as computing a distance between two graphs constitute a complex problem with high computational demand.

If data can be represented in a Euclidean manner, statistical approaches often dominate the field as in the case of computer vision [9,22]. Structural approaches are more common in areas, where relational data is essential and a non-Euclidean data structure is the obvious choice. Popular areas in this regard are the analysis of chemical molecules and social or citation networks [3,11,34]. Looking at application areas of statistical and structural pattern recognition, handwriting analysis holds a special position. Document analysis methods are highly focused on the image domain as image acquisition is easy and a huge amount of well researched statistical approaches exist. Nonetheless, a structural representation is inherent to any image of handwriting. The underlying structure of a handwritten word can naturally be captured in the form of a graph. This makes it an open question whether handwriting analysis can benefit from a structural approach based on graph representations.

Word spotting is a task that attracted a lot of attention in the document analysis community and also represents an area where structural and statistical approaches coexist [9]. In general, the problem of word spotting is a well researched field in document image analysis and many mature methods exit. However, word spotting has also been a topic of interest with respect to graph representations [2,18,26]. As word spotting essentially requires to measure the similarity of a word to a query, many graph-based methods explored how to efficiently compute a distance between graphs [2,25]. In terms of performance, a significant gap between image and graph-based approaches exists. This gap can be explained by the fact that most methods in the image domain heavily rely on learning and on training powerful models on labelled data. Additionally, handwriting graphs are usually extracted from images relying on binarization and skeletonization methods [24,26]. This step might limit performances compared to models of the image domain working with unprocessed images.

Learning in the graph domain recently gained a lot of attraction with the generalization of the convolution operation to graph structures. Geometric deep learning [3] provides a framework similar to deep convolutional neural networks that led to significant performance gains for different benchmarks [11]. In [18], the authors propose a learning-based model for word spotting in the graph domain to estimate a graph distance with graph neural networks.

In this work, we propose a graph convolutional neural network to predict an attribute representation from handwritten word graphs. This approach has been proven to show high performances on word spotting benchmarks in the image domain [28]. Compared to other methods considering the graph domain, our learning-based approach exploits character level instead of only word class information. Furthermore, mapping graphs to an attribute space allows to query by string, which is not the case for other methods in the literature.

The remainder of the paper is organized as follows. Sect. 2 presents related works on the topic of word spotting in the image and graph domain. The proposed graph convolutional neural network is discussed in Sect. 3. In our experimental evaluation in Sect. 4, we investigate the influence of key components of our model on four different benchmarks. Finally, we compare the proposed model to other graph and image-based word spotting methods known from the literature.

2   Related Work

Word Spotting describes the task of retrieving regions from a document collection that are similar to a query [9]. In contrast to handwriting recognition, the result of a word spotting system is not an explicit transcription result, but a ranked list of possible word occurrences. This retrieval approach allows for interpretation by the user, making word spotting an attractive alternative especially for information retrieval from historic document collections. For an extensive overview on word spotting methodology and taxonomy, see [9].

2.1   Document Image Analysis

Traditionally, word spotting methods have been highly focused on the image domain. Different methods either work on entire document images [14,20,32] or segmented regions such as word images [13,16,28]. Several works on word spotting exploit the sequential structure of handwriting. Models such as recurrent neural networks [14] and Hidden Markov Models (HMM) [5,20] were applied successfully and are still popular.

Traditional feature based approaches also attracted attention and were usually combined with models such as spatial pyramids [21] or HMMs to encode spatial information [16,20]. As these methods measure visual similarity based on a designed representation, they usually are not capable to generalize well across high variations in writing styles. To overcome this drawback, Almazan et al. proposed to predict certain image properties so called attributes from word images in [1]. In this influential work, the authors proposed the *Pyramidal Histogram of Characters* (PHOC) that encodes the occurrence and spatial position of characters in a pyramidal fashion. By mapping word images to an attribute vector space, *query-by-example* word spotting boils down to the computation of simple vector distances. Since it is straightforward to derive a PHOC vector from a string, *query-by-string* is easily possible.

In [27], the attribute-based approach of [1] was adopted using a convolutional neural network that replaced the formerly used combination of Fisher Vectors and SVMs. Training a neural network to predict an attribute representations from word images, resulted in high performances on almost all popular benchmarks [15,28]. Recently, image-based word spotting got increasingly more focused on methods that either do not rely on annotated training data [16,29,33] or are capable to jointly solve the segmentation problem [14,32].

2.2   Graph Representations

While word spotting in the image domain is a highly researched topic that resulted in well performing methods, significantly less works considered the task from a structural perspective. Here, we focus on methods that first extract a graph representation, in order to tackle the word spotting problem in the graph domain. In [30], the authors propose a graph representation that extracts vertices and edges from skeletonised word images to represent the structural properties of a handwritten word. This representation is enriched by using the Shape Context Descriptor as an additional node feature vector. In order to measure similarity, an approach based on dynamic time warping (DTW) and an approximated Graph Edit Distance (GED) is proposed. Most graph-based methods follow a similar approach. First, a graph representation is extracted from a word image, for example by computing keypoints [6] or projection profiles [24, 25]. Then, the similarity of a query graph to all word graphs is estimated by a graph distance. As common graph distances such as the GED are highly computational demanding, most graph-based word spotting methods rely on an approximation. Popular examples in this regard are bipartite matching (BP) [19], also known as assignment edit distance (AED) [24], or Hausdorff edit distance (HED) [7]. As deep learning and neural networks have drastically increased word spotting performances in the image domain, this approach was just recently investigated for graph representations. The Geometric Deep Learning [3] framework allows to build neural networks similar to CNNs that operate on graphs. In [17], the authors propose a graph neural network that learns an enriched graph representation with a siamese approach. Based on the enriched representation, a graph distance similar to the HED is defined, resulting in a fast and efficient similarity measure. This method is extended in [18] to a triplet approach achieving competitive results on multiple graph-based word spotting benchmarks.

2.3   Geometric Deep Learning

Neural networks for graph representations were first proposed in [23]. Motivated by the success in the image domain, generalizing the convolutional operation has been of significant interest [34]. As a general taxonomy, *spectral* and *non-spectral* methods are distinguished. In contrast to *spectral* approaches, which are motivated by the formulation of a graph signal, *non-spectral* approaches are defined for the entire graph representation and usually work on spatially close neighbourhoods [34]. Most graph neural networks share a common structure that can be summarized in the general framework of a *Message Passing Neural Network* [8]. Each layer is defined by a message and an update function. The message function aggregates information from neighbouring nodes, while the update function computes a node embedding based on the aggregated representations. Finally, a readout function is defined, which computes a feature vector for the entire graph. If all three functions are differentiable, the resulting model may be trained in a supervised manner. For an extensive review of graph neural networks, see [3, 34].

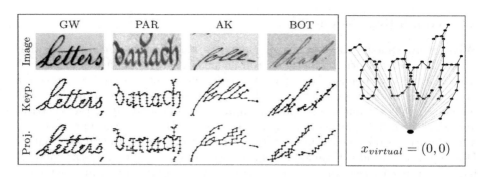

Fig. 1. (Left) Examples of keypoint (Keyp.) and projection (Proj.) graphs for George Washington (GW), Parzival (PAR), Alvermann Konzillsprotokolle (AK) and Botany (Bot) [26]. (Right) Representation enriched with virtual node.

3 Method

In the following section, we discuss the proposed graph convolutional network for graph-based word spotting. Following the approach of [28], we aim at predicting an attribute representation from a handwritten word graph. Predicting attributes from a graph is similar to the problem of graph property prediction, which is a popular task tackled with graph convolutional networks [11].

3.1 Graph Representations

In this work, a handwritten word is represented as a set of nodes V and undirected edges that are expressed by a binary adjacency matrix $\boldsymbol{A}$. Each node has a feature vector x_v, which represents its spatial position. Multiple extraction methods exist that extract such graph representations from word images [26]. Here, we focus on graphs extracted by means of identifying *keypoints* or by a segmentation resulting from *projection* profiles. See Fig. 1, for examples of keypoint and projection graphs for four different datasets from [26].

Several works in the literature introduce virtual nodes [8,11] to allow each node to receive context information from the entire graph. We investigate the use of this enriched graph representation, by introducing an additional node to each word graph. The virtual node is introduced with a zero vector as a feature vector and is connected to every node of the graph. Figure 1 visualizes this enriched graph representation.

3.2 Convolutional Layers

In order to map the node features to a hidden state h_v, we use convolutional layers that can be formulated in the message passing neural network framework [8] as follows. A message passing network performs a message passing phase for T

time steps that is defined by its message passing function M_t aggregating information from the node neighbourhood $\mathcal{N}(v)$. The update function U_t computes a hidden state of the node based on the received message m_v^{t+1}:

$$m_v^{t+1} = \sum_{w \in \mathcal{N}(v)} M_t(h_v^t, h_w^t) \qquad h_v^{t+1} = U_t(h_v^t, m_v^{t+1}) \tag{1}$$

Graph Convolutional Networks (GCN)

In [12], Kipf and Welling propose a convolutional layer for graphs that is based on an approximation of spectral graph convolutions. Despite its spectral nature, the GCN layer can be interpreted as a spatial method. Due to the simplifications introduced in [12], the resulting convolutional layer aggregates information of a node neighbourhood that is transformed using a layer-specific weight matrix $\boldsymbol{W}$. The resulting message and update functions can be expressed as

$$m_v^{t+1} = \sum_{w \in \mathcal{N}(v)} \frac{1}{\sqrt{deg(v) \cdot deg(w)}} \cdot A_{vw} \cdot h_w^t \tag{2}$$

$$h_v^{t+1} = \boldsymbol{W}^t m_v^{t+1}, \tag{3}$$

with $deg(\cdot)$ denoting the degree of a note. As in [12], we limit the message passing time steps to $T = 1$ and only consider a binary adjacency matrix. Therefore, a multi layer graph convolutional network can be mathematically formulated by the computation of a hidden state h_v^k at layer k:

$$h_v^{(k+1)} = \sum_{w \in \mathcal{N}(v) \cup \{v\}} \frac{1}{\sqrt{deg(v) \cdot deg(w)}} \cdot (\boldsymbol{W}^{(k)} \cdot h_v^{(k)}) \tag{4}$$

Sample and Aggregate (SAGE)

The spatial approach to generalize the convolution operation to graphs proposed in [10] relies on a sampling and aggregation strategy. First, the neighbourhood of a node is sampled followed by the generation of a neighbourhood embedding by means of an aggregation function. Hereby, the model learns a function on how to aggregate neighbourhood information. In this work, we consider the direct neighbourhood at each layer, resulting in $T = 1$ with respect to the message passing framework. As a simple aggregation function, we use the mean over the neighbourhood node embeddings. This results in a formulation of a convolutional layer that is similar to the GCN approach with the following message and update functions:

$$m_v^{k+1} = \text{mean}_{w \in \mathcal{N}(v)}(h_w^{(k)}) \tag{5}$$

$$h_v^{(k+1)} = \boldsymbol{W_1}^{(k)} h_v^{(k)} + \boldsymbol{W_2}^{(k)} \cdot m_v^{k+1} \tag{6}$$

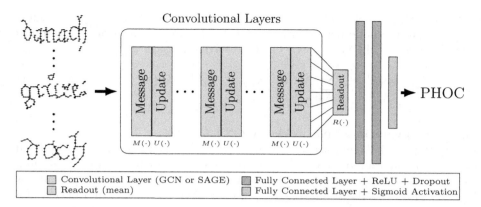

Fig. 2. Overview of the proposed graph neural network. Each graph is propagated through K convolutional layers, followed by a readout function to generate a graph embedding. A multilayer perceptron predicts respective PHOC vectors.

3.3   Architecture

Figure 2 presents an overview of the overall architecture. The extracted word graphs are propagated through the network to predict a PHOC representation similar to [28]. The backbone of the architecture is a number of K convolutional layers, as described in section Sect. 3.2. Following the message passing neural network framework, a readout function R generates an embedding $\hat{y}$ for the entire graph. We use the mean over all node embeddings as a readout function:

$$\hat{y} = R(\{h_v^K | v \in V\}) = \frac{1}{|V|} \sum_{v \in V} h_v^K \tag{7}$$

In analogy to the attribute CNN approach in the image domain [28], we use a multilayer perceptron with sigmoid activations to predict a PHOC representation from the learned graph embedding. The model is then optimized fully supervised in an end-to-end manner.

4   Experiments

In our experiments, we investigate the proposed model on datasets for graph-based word spotting (Sect. 4.1). We focus on the influence of increasing number of layers for GCN and SAGE convolutions and the introduction of a virtual node. Finally, we compare our model to other graph-based methods, as well as methods for *segmentation-based* word spotting in the image domain (Sect. 4.3).

In all experiments, we use a PHOC vector with splits [2, 3, 4, 5] and the cosine similarity to measure similarity between the representations in the attribute space. The size of the node embedding is set to 256 and both fully connected layers consist of 1024 neurons. All models are trained with ADAM optimization,

Table 1. Number of samples for different dataset splits.

Split	GW	PAR	AK	BOT
Train	2447	11468	1849	1684
Validation	1224	4621	3734	3380
Test	1224	6869	–	–
Keywords	105	1217	200	150

binary cross entropy loss, a learning rate of 0.001 and a batch size of 64. We train our model for 500 epochs on the designated training splits of the datasets. Performance is measured with mean average precision (mAP) [9].

Due to the large number of different works that have been published on the topic of word spotting, several evaluation protocols exist, despite most works focus on the same datasets. We follow the nomenclature proposed in [18], distinguishing the two ways of an *individual* and *combined* query representation. In this regard, individual means that a query is represented by a single graph. This representation is used in most image-based protocols where queries are usually represented by a single exemplar image. As most graph-based methods measure the structural similarity between graphs, retrieval performance is quite sensitive with respect to writing style variations. In the *combined* query protocol, this problem is mitigated by combining multiple graphs of the same keyword to represent a single query. The resulting similarity measure is then based on the most similar keyword graph. For our attribute model, this corresponds to the minimum over all distances between the estimated PHOC vector of a word graph and all query graphs corresponding to a single keyword. In case of *query-by-string*, each keyword is used as query once.

4.1 Datasets

In our experimental evaluation, we rely on the Histograph database [24,26]. The database provides graph representations for four popular manuscripts known from *segmentation-based* word spotting in the image domain. The authors provide multiple datasets that result from different graph extraction strategies, with keypoint and projection graphs being the most popular. See Table 1 for an overview of the datasets and corresponding numbers of samples.

The *George Washington* (GW) dataset has traditionally been a key benchmark dataset in the word spotting community. The historic dataset shows almost no degradation and variations in handwriting style are limited.

The *Parzival* (PAR) dataset is a historic dataset written in German. 45 pages are available and show degradations. Despite the fact that is was written by three authors, variations in writing style are comparable low.

The *Alvermann Konzilsprotokolle* (AK) dataset emerged from the keyword spotting competition at the International Conference of Frontiers in Handwriting Recognition 2016 [15]. The corresponding competition had a focus on evaluating

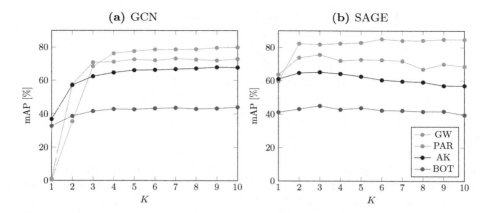

Fig. 3. *Query-by-example* performances for different numbers of convolutional layers K. Results reported as mAP [%].

the influence of different amounts of training data. The histograph database provides graph representation for period I of the competition which contains only the smallest number of training samples.

Botany (BOT) stems from the same competition as the AK dataset [15]. The manuscript shows significant marks of degradation such as fading. Furthermore, the rather artistic style of the botanical records lead to significant variations in writing style and also in scale.

4.2   Results

In order to investigate the depth of the model and the introduction of a virtual node, we evaluate performances in terms of mean average precision in the *query-by-example* scenario on the keypoint graph representations. We follow the combined query protocol in all experiments. As training data is highly limited in all cases except PAR, we include also validation data for GW during training. This is similar to the evaluation protocol in the image domain, where usually no designated validation split is used for GW [1]. Note that all evaluated models are also capable to perform *query-by-string* as the model estimates PHOC representations. Quantitative results for *query-by-string* are provided in Sect. 4.3.

Depth

In the application of convolutional neural networks, we observe a trend towards increasingly deep architectures. While it seems that depth often leads to superior performances in the image domain, most graph neural networks are quite shallow [12,18,34]. In this set of experiments, we vary the number of layers from one to ten for GCN and SAGE convolutions, as introduced in Sect. 3.

Table 2. Evaluation of Virtual Node (VN). Results reported as mAP [%].

Convolution	Layer	VN	GW QbE	PAR QbE	AK QbE	BOT QbE
SAGE	3	No	75.60	81.81	65.17	**45.01**
SAGE	3	Yes	**75.84**	86.59	63.62	44.69
SAGE	6	No	72.49	85.05	58.24	42.28
SAGE	6	Yes	73.63	**88.93**	60.43	41.78
GCN	10	No	73.00	79.87	**67.82**	44.07
GCN	10	Yes	71.18	81.55	63.52	41.79

Figure 3a shows performances for GCN layer. We are able to observe performance improvements for increasing depth for all datasets. This is especially interesting as often no performance gains are reported in the literature for models with more than three layers [12,18]. While we observe minor performance gains for increasing numbers of GCN layers, this is not the case for SAGE layers, see Fig. 3b. Only for PAR, using more than three convolutional layers improves performances and we do not observe any further gains beyond six layers. For all other datasets where considerably less training data is available, performances degrade after three layers. These results indicate that the availability of training data determines in how far the model can benefit from the increased complexity. In general, the models based on SAGE convolutions yield higher performances for GW and PAR despite using fewer layers. For AK and BOT the deep GCN models with ten layers, show slightly higher performances.

Virtual Node

Introducing a virtual node to a graph, allows message flow between all nodes. Motivated by the previously discussed analysis on increasing depth, we investigate the influence on a GCN model with ten layers and two SAGE models with three and six convolutional layers. Table 2 presents *query-by-example* performances on all datasets. For GW, we observe only a limited influence on the SAGE models of different depth, while performances decrease in case of the GCN model. In contrast to GW, we observe some clear performance gains for all models in case of PAR. For AK and Botany the results are not conclusive. It seems that the GCN models do not benefit from the introduction of a virtual node. However, when it is feasible to train an accurate SAGE model, a virtual node fosters performances. As in the case of GW and PAR, the highest performances are reported for SAGE models including virtual nodes.

4.3   Comparison to the Literature

In this section, we compare our model to other graph and image-based word spotting methods. To allow for a fair comparison, we only train our model on

Table 3. *Query-by-example* performances based on keypoint (Keyp.) and projection (Proj.) graphs of the Histograph DB. All method follow the individual query protocol. Results reported as mAP [%].

	Method	GW QbE	PAR QbE	AK QbE	BOT
Keyp.	Ours	**77.99**	**93.98**	55.37	34.21
	Riba et al. [18]	76.92	73.14	**62.90**	**41.52**
Proj.	Ours	**77.82**	**95.48**	59.92	33.31
	Riba et al. [18]	70.25	75.19	**65.04**	**42.83**

the designated training splits, if not further noted. In this work, we focus on *segmentation-based* methods, as the graph extraction methods that underly the Histograph database requires an independent segmentation step.

Graph Domain

A first comparison can be drawn between the proposed model and [18]. Table 3 presents *query-by-example* performances for keypoint and projection graphs following the individual query protocol. Our proposed attribute-based approach compares well on GW and PAR and improves performances in all cases. Performance gains are especially striking in case of PAR where a large training set is available. In [18], the authors propose a graph convolutional network trained with triplets in order to learn a graph distance. This approach only considers word class information during training, as opposed to attribute learning that relies on transcriptions. Exploiting the richer annotation offers an explanation for the observed performance gains in cases where a sufficiently large training set is available. For the smaller datasets of AK and BOT, the metric learning approach presented in [18] seems to be beneficial.

Table 4 compares our model to other graph-based methods from the literature under the combined query protocol. Except for [18], all other method are *learning-free*. Additionally, we report numbers for an extended training set, where we included the validation data during training for GW and PAR. The proposed attribute approach outperforms all *learning-free* methods, given enough training data as in case of PAR or the extended GW dataset. This emphasizes that the proposed method scales fairly well with the availability of labeled data. Furthermore, we see that end-to-end learning is feasible in the graph domain and considerable performance gains can be achieved. When data is highly limited as in the case of AK or BOT similarity-based approaches seem to be advantageous.

Another interesting observation can be made, comparing the results of the individual (Table 3) and combined (Table 4) protocol. While [18] reports higher performances under the combined protocol, this is not the case for our model on GW and PAR. As the PHOC vector is independent from the structural characteristics of the query graphs, our model does not benefit from combining multiple query graphs, if an accurate PHOC estimation is possible. In order to show this

Table 4. *Query-by-example* and *string* performances based on keypoint and projection graphs of the Histograph DB. All methods from the literature follow the combined query protocol. Results reported as mAP [%].

	Method	GW		PAR		AK		BOT	
		QbE	QbS	QbE	QbS	QbE	QbS	QbE	QbS
Keypoint	Ours	66.73	66.57	89.03	88.80	67.82	38.68	45.01	8.48
	Ours[†]	66.71	66.57	88.93	88.80	59.14	**38.68**	36.61	**8.48**
	Ours*	75.84	**75.74**	**90.61**	**90.66**	–	–	–	–
	Riba et al. [18]	**78.48**	–	79.29	–	78.64	–	**51.90**	–
	AED [2]	68.42	–	55.03	–	77.24	–	50.94	–
	HED [2]	69.28	–	69.23	–	**79.72**	–	51.74	–
Projection	Ours	68.28	68.20	90.55	90.51	71.98	39.98	44.43	7.77
	Ours[†]	68.17	68.20	90.59	90.51	63.25	**39.98**	37.15	**7.77**
	Ours*	**73.61**	**73.21**	**92.77**	**92.80**	–	–	–	–
	Riba et al. [18]	73.03	–	79.95	–	79.55	–	**52.83**	–
	AED [2]	60.83	–	63.35	–	76.02	–	50.49	–
	HED [2]	66.71	–	72.82	–	**81.06**	–	51.69	–
	Ensemble [26]	70.56	–	79.38	–	84.77	–	68.88	–

(∗) extended training data   (†) no query combination

characteristics, we report results under a changed combined protocol in row two of Table 4. Instead of taking the minimal distance to all query instances, we do not combine queries, but average over the average precisions for each keyword. This accounts for the different query counts per keyword under the individual protocol. It can be concluded that the performance loss of our model is a result of the change of query distributions. On the other hand, our model does not benefit from combing query instances in case of GW and PAR. This result underlines the potential power of the proposed model, as performances are expected to strongly decrease for the *learning-free* methods without query combination. As gathering multiple instances of a keyword is a high demand, we advocate to report results under an individual protocol.

While *query-by-string* is out of scope for all other graph-based methods, it is easily possible with the proposed attribute approach. Our model is capable of mapping graphs to an attribute space. This representation is more powerful than a simple numeric vector embedding, as it encodes symbolic information. In case of AK and BOT *query-by-string* performance is comparable poor, illustrating that the model is not capable to learn the desired character models.

Image Domain

Table 5 compares our method to image-based word spotting systems from the literature and illustrates the existing performance gap between structural and statistical approaches. An interesting observation can be made with respect to the TPP-PHOCNet, which follows an attribute learning approach. While the PHOCNet clearly outperforms the other models that reported results in the

Table 5. Image and graph-based *query-by-example* performances. All method follow an individual query protocol. Results reported as mAP [%].

	Method	GW	PAR	AK			BOT		
				I	II	III	I	II	III
Graph	Ours	77.99	**95.48**	59.92	–	–	34.21	–	–
	Riba et al. [18]	76.92	79.95	64.42	–	–	41.52	–	–
Image	TPP-PHOCNet [28]	**97.98**	–	86.01	**97.05**	**98.11**	47.75	**83.51**	**96.05**
	CNN & HMM [31]	85.06	94.57	–	–	–	–	–	–
	CVCDAG [15]	–	–	77.91	–	–	**75.77**	–	–
	TAU [15]	–	–	71.11	–	–	50.64	–	–
	QTOB [15]	–	–	82.15	–	–	54.95	–	–

competition based on the highest number of training samples, the performance gain is smaller in case of the smallest training set. Especially in case of BOT, the performance of the attribute-based approach degrades strongly, indicating the high complexity of the attribute prediction task and its sensitivity to training data. A similar observation can be made with respect to the proposed graph convolutional neural network, which performs comparable poor in these cases.

Overall, the attribute-based approach improves performances given enough training data in the graph domain and contributes to closing the performance gap between the graph and image domain. In case of PAR, we achieve comparable performances to the image domain motivating the further exploration of *learning-based* approaches for structural pattern recognition.

5 Conclusions

In this work, we propose a graph convolutional neural network for predicting attribute representations from handwritten word graphs. By mapping a graph to an attribute vector space, the word spotting problem can be solved with the help of a simple vector distance. We are able to show that a fully supervised learning approach is feasible in the graph domain and achieves considerable performance gains when sufficient training data is available. As performance depends on the availability of labeled samples, the exploration of techniques such as synthetic data, semi-supervised or transfer learning is a future line of research. In these limited data cases, methods potentially increasingly benefit from the high representational power of graphs. Our work constitutes a step towards bridging the performance gap between structural and statistical pattern recognition approaches for word spotting. A *learning-based* unification of both paradigms offers the potential to combine the representational power of graphs with the benefits from statistical approaches.

Acknowledgement. This work has been supported by the Swiss Hasler Foundation (project 20008).

References

1. Almazán, J., Gordo, A., Fornés, A., Valveny, E.: Word spotting and recognition with embedded attributes. TPAMI **36**(12), 2552–2566 (2014). https://doi.org/10.1109/TPAMI.2014.2339814
2. Ameri, M.R., Stauffer, M., Riesen, K., Bui, T.D., Fischer, A.: Graph-based keyword spotting in historical manuscripts using Hausdorff edit distance. Pattern Recogn. Lett. **121**, 61–67 (2019). https://doi.org/10.1016/j.patrec.2018.05.003
3. Bronstein, M.M., Bruna, J., LeCun, Y., Szlam, A., Vandergheynst, P.: Geometric deep learning: going beyond Euclidean data. IEEE Signal Process. Mag. **34**(4), 18–42 (2017). https://doi.org/10.1109/MSP.2017.2693418
4. Bunke, H., Riesen, K.: Towards the unification of structural and statistical pattern recognition. Pattern Recogn. Lett. **33**(7), 811–825 (2012). https://doi.org/10.1016/j.patrec.2011.04.017
5. Fischer, A., Keller, A., Frinken, V., Bunke, H.: Lexicon-free handwritten word spotting using character HMMs. Pattern Recogn. Lett. **33**(7), 934–942 (2012). https://doi.org/10.1016/j.patrec.2011.09.009
6. Fischer, A., Riesen, K., Bunke, H.: Graph similarity features for HMM-based handwriting recognition in historical documents. In: ICFHR, Kolkata, India, pp. 253–258 (2010). https://doi.org/10.1109/ICFHR.2010.47
7. Fischer, A., Suen, C.Y., Frinken, V., Riesen, K., Bunke, H.: Approximation of graph edit distance based on Hausdorff matching. Pattern Recogn. **48**(2), 331–343 (2015). https://doi.org/10.1016/j.patcog.2014.07.015
8. Gilmer, J., Schoenholz, S.S., Vinyals, O., Dahl, G.E.: Neural message passing for quantum chemistry. In: ICML, Sydney, Australia, vol. 70, pp. 1263–1272 (2017)
9. Giotis, A.P., Sfikas, G., Gatos, B., Nikou, C.: A survey of document image word spotting techniques. Pattern Recogn. **68**, 310–332 (2017). https://doi.org/10.1016/j.patcog.2017.02.023
10. Hamilton, W.L., Ying, Z., Leskovec, J.: Inductive representation learning on large graphs. In: NIPS, Long Beach, CA, USA, pp. 1024–1034 (2017)
11. Hu, W., et al.: Open graph benchmark: datasets for machine learning on graphs. In: NIPS (2020)
12. Kipf, T.N., Welling, M.: Semi-supervised classification with graph convolutional networks. In: ICLR, Toulon, France (2017)
13. Krishnan, P., Jawahar, C.V.: HWNet v2: an efficient word image representation for handwritten documents. IJDAR **22**(4), 387–405 (2019). https://doi.org/10.1007/s10032-019-00336-x
14. Lang, E., Puigcerver, J., Toselli, A.H., Vidal, E.: Probabilistic indexing and search for information extraction on handwritten German parish records. In: ICFHR, Niagara Falls, NY, USA, pp. 44–49 (2018). https://doi.org/10.1109/ICFHR-2018.2018.00017
15. Pratikakis, I., Zagoris, K., Gatos, B., Puigcerver, J., Toselli, A.H., Vidal, E.: ICFHR2016 handwritten keyword spotting competition (H-KWS 2016). In: ICFHR, Shenzhen, China, pp. 613–618 (2016). https://doi.org/10.1109/ICFHR.2016.0117
16. Retsinas, G., Louloudis, G., Stamatopoulos, N., Gatos, B.: Efficient learning-free keyword spotting. TPAMI **41**(7), 1587–1600 (2019). https://doi.org/10.1109/TPAMI.2018.2845880
17. Riba, P., Fischer, A., Lladós, J., Fornés, A.: Learning graph distances with message passing neural networks. In: ICPR, Beijing, China, pp. 2239–2244 (2018). https://doi.org/10.1109/ICPR.2018.8545310

18. Riba, P., Fischer, A., Lladós, J., Fornés, A.: Learning graph edit distance by graph neural networks. arXiv preprint CoRR abs/2008.07641 (2020)
19. Riesen, K., Bunke, H.: Approximate graph edit distance computation by means of bipartite graph matching. Image Vision Comput. **27**(7), 950–959 (2009). https://doi.org/10.1016/j.imavis.2008.04.004
20. Rothacker, L., Wolf, F., Fink, G.A.: Annotation-free word spotting with bag-of-features HMMs. In: IJPRAI, p. 2153001 (2020)
21. Rusiñol, M., Aldavert, D., Toledo, R., Lladós, J.: Efficient segmentation-free keyword spotting in historical document collections. Pattern Recogn. **48**(2), 545–555 (2015). https://doi.org/10.1016/j.patcog.2014.08.021
22. Russakovsky, O., et al.: ImageNet large scale visual recognition challenge. Int. J. Comput. Vision **115**(3), 211–252 (2015). https://doi.org/10.1007/s11263-015-0816-y
23. Scarselli, F., Gori, M., Tsoi, A.C., Hagenbuchner, M., Monfardini, G.: The graph neural network model. IEEE Trans. Neural Netw. **20**(1), 61–80 (2009). https://doi.org/10.1109/TNN.2008.2005605
24. Stauffer, M., Fischer, A., Riesen, K.: A novel graph database for handwritten word images. In: S+SSPR, Mérida, Mexico, pp. 553–563 (2016). https://doi.org/10.1007/978-3-319-49055-7_49
25. Stauffer, M., Fischer, A., Riesen, K.: Graph-based keyword spotting in historical documents using context-aware Hausdorff edit distance. In: DAS, Vienna, Austria, pp. 49–54 (2018). https://doi.org/10.1109/DAS.2018.31
26. Stauffer, M., Fischer, A., Riesen, K.: Keyword spotting in historical handwritten documents based on graph matching. Pattern Recogn. **81**, 240–253 (2018). https://doi.org/10.1016/j.patcog.2018.04.001
27. Sudholt, S., Fink, G.A.: PHOCNet: a deep convolutional neural network for word spotting in handwritten documents. In: ICFHR, Shenzhen, China, pp. 277–282 (2016). https://doi.org/10.1109/ICFHR.2016.0060
28. Sudholt, S., Fink, G.A.: Attribute CNNs for word spotting in handwritten documents. IJDAR **21**(3), 199–218 (2018). https://doi.org/10.1007/s10032-018-0295-0
29. Vats, E., Hast, A., Fornés, A.: Training-free and segmentation-free word spotting using feature matching and query expansion. In: ICDAR, Sydney, NSW, Australia, pp. 1294–1299 (2019). https://doi.org/10.1109/ICDAR.2019.00209
30. Wang, P., Eglin, V., Garcia, C., Largeron, C., Lladós, J., Fornés, A.: A novel learning-free word spotting approach based on graph representation. In: DAS, Tours, France, pp. 207–211 (2014). https://doi.org/10.1109/DAS.2014.46
31. Wicht, B., Fischer, A., Hennebert, J.: Deep learning features for handwritten keyword spotting. In: ICPR, Cancún, Mexico, pp. 3434–3439 (2016). https://doi.org/10.1109/ICPR.2016.7900165
32. Wilkinson, T., Lindström, J., Brun, A.: Neural Ctrl-F: segmentation-free query-by-string word spotting in handwritten manuscript collections. In: ICCV, Venice, Italy, pp. 4443–4452 (2017). https://doi.org/10.1109/ICCV.2017.475
33. Wolf, F., Fink, G.A.: Annotation-free learning of deep representations for word spotting using synthetic data and self labeling. In: DAS, Wuhan, China, pp. 293–308 (2020). https://doi.org/10.1007/978-3-030-57058-3_21
34. Wu, Z., Pan, S., Chen, F., Long, G., Zhang, C., Yu, P.S.: A comprehensive survey on graph neural networks. IEEE Trans. Neural Netw. Learn. Syst. **32**(1), 4–24 (2021). https://doi.org/10.1109/TNNLS.2020.2978386

Context Aware Generation
of Cuneiform Signs

Kai Brandenbusch$^{(\boxtimes)}$, Eugen Rusakov , and Gernot A. Fink

Department of Computer Science, TU Dortmund University,
44227 Dortmund, Germany
{kai.brandenbusch,eugen.rusakov,gernot.fink}@tu-dortmund.de

Abstract. With the advent of deep learning in many research areas, the need for very large datasets is emerging. Especially in more specific domains like cuneiform script, annotated data is scarce and costly to obtain by experts only. Therefore, approaches for automatically generating labeled training data are of high interest. In this paper, we present an approach for generating cuneiform signs in larger images of cuneiform tablets. The proposed method allows to control the class of the generated sample as well as the visual appearance by considering context information from surrounding pixels. We evaluate our method on different numbers of cuneiform tablets for training and examine methods for determining the number of training iterations. Besides generating images of promising visual quality, we are able to improve classification performance by augmenting original data with generated samples. Additionally, we demonstrate that our approach is applicable to other domains as well, like digit generation in house number signs.

Keywords: Document synthesis · Cuneiform script · Handwriting recognition · Text and symbol recognition

1 Introduction

Cuneiform is a writing system used in the ancient near east from the 4th millennium BC until the 1st century. The script is written by pressing a stylus into moist clay tablets. A large number of fragments of cuneiform tablets has been preserved over time [23], many of which are available in digitized form. These tablets are an invaluable source of information and their analysis could provide insight into the respective time in history. With the huge amount of data to be analyzed, the need for automatic and computer aided methods arises. Many state-of-the-art approaches for image and document analysis are based on deep neural networks. While these models achieve outstanding performance, training them requires a vast amount of annotated data.

Manually annotating cuneiform tablets is an effortful task and needs to be done by domain experts. Thus, annotated data is scarce and the automatic generation of annotated training material is of high interest. In other domains of

© Springer Nature Switzerland AG 2021
J. Lladós et al. (Eds.): ICDAR 2021, LNCS 12821, pp. 65–79, 2021.
https://doi.org/10.1007/978-3-030-86549-8_5

document analysis, like handwritten documents, a common way to approach this problem is the utilization of synthetic training data. Handwriting like text can be rendered from computer fonts [14]. However, this is not applicable to cuneiform script for different reasons. Cuneiform script consists of wedges impressed in moist clay and therefore exhibits a 3-dimensional surface structure. Additionally, complex variations in cuneiform script, like shape or depth of the impressions, degradation and lighting conditions when creating the photographs of the tablets, are hard to mimic by rendering computer fonts.

We propose a method for generating annotated data based on a conditional generative adversarial network (cGAN) with an auxiliary classifier. The variations in appearance are considered implicitly by providing context when solving an image completion task instead of generating isolated signs. Therefore, the method does not rely on a definition of style by domain experts. The auxiliary classifier and the conditioning of the generator on class labels allow for controlling content of the generated image. We demonstrate an application for data generation with our proposed framework by training a classifier on generated cuneiform signs, leading to an improvement of its performance. We show that the approach can be transferred to other domains exposing similar properties, like house number signs in the SVHN dataset [18].

2   Related Work

In 2014, Goodfellow et al. [3] introduced generative adversarial networks (GANs), consisting of two models with adversarial objectives. The generator network $G : G(\mathbf{z}) \mapsto \mathbb{R}^{|\mathbf{x}|}$ takes as an input a vector $\mathbf{z} \in \mathbb{R}^{|\mathbf{z}|}$ drawn from a latent space and tries to generate a realistic looking image wrt. the original data distribution. The discriminator $D : D(\mathbf{x}) \mapsto (0, 1)$ tries to distinguish between images from the original distribution and samples generated by the generator. Both networks are trained to optimize adversarial objectives:

$$\min_{G} \max_{D} \mathcal{L}(G, D) = \mathbb{E}_{\mathbf{x} \sim p_{data}(\mathbf{x})}[\log(D(\mathbf{x}))] + \mathbb{E}_{\mathbf{z} \sim p_{\mathbf{z}}(\mathbf{z})}[\log(1 - D(G(\mathbf{z})))]. \quad (1)$$

In many cases, it is not desirable to generate arbitrary images from the whole training data distribution but from a conditional distribution. Especially when targeting at the generation of training data for classification models, the generated images must contain an instance of the desired class. In order to achieve this, the generator as well as the discriminator are conditioned on additional information $\mathbf{y}$ (e.g. one-hot encoded class label) [17]:

$$\min_{G} \max_{D} \mathcal{L}(G, D) = \mathbb{E}_{\mathbf{x} \sim p_{data}(\mathbf{x})}[\log(D(\mathbf{x}|\mathbf{y}))] + \mathbb{E}_{\mathbf{z} \sim p_{\mathbf{z}}(\mathbf{z})}[\log(1 - D(G(\mathbf{z}|\mathbf{y})))] \quad (2)$$

In addition to the latent vector $\mathbf{z}$ for the generator and the (generated) image for the discriminator, respectively, the encoded class label is passed to both models. Instead of using the label as an input for the discriminator, augmenting the discriminator for an additional classification task improves performance and helps to stabilize the training [19]. Li et al. [15] extended the number of predicted classes for the auxiliary classifier by a *fake class* for generated images, which leads to further improved and accelerated training of the conditional GAN (cGAN).

Alonso et al. [1] used a cGAN to generate images containing handwriting. The input to the generator is a vector of noise and the string to be generated, encoded by a BiLSTM. Besides a discriminator for ensuring the generation of realistic looking images, a recognizer was utilized to control the textual content. Augmenting the training data with generated samples results in a slight increase in recognition performance. A similar setup was used by Kang et al. [11,12] for generating defined words with a calligraphic style of a given handwritten text image. The generator is conditioned to the textual content either by an encoded string [12] or an embedding extracted from another input image [11]. In both cases, the style is controlled by an embedding extracted from an image of the desired calligraphic style. The generated images are passed to a discriminator, a recognition model and a writer classifier to control both content and style.

Another way of conditioning GANs is by context information from surrounding pixels. Pathak et al. [20] trained an encoder-decoder network for reconstructing missing parts of images. Besides a reconstruction (L2) loss, they utilized an adversarial loss to enhance generation of realistic looking images. They showed that the learned embedding can be used to initialize models for different tasks on the respective datasets. A similar image completion approach was followed by Denton et al. [2]. However, instead of using the embedding learned by the generator, they train the discriminator for an additional classification task. Iizuka et al. [8] used a GAN for image completion, using a local discriminator in addition to the global discriminator in order to further increase the quality of the generated images.

Besides generating training data with cGANs, the idea of transferring images from one to another (closely) related domain can be utilized for training data generation. Instead of a latent vector and conditioning information, an image from one domain is used as an input for the generator and transferred to the other domain. A second generator is trained for the reverse transfer. This allows for the definition of a cycle consistency loss and to guide training without requiring pairs of images from both domains. Therefore, annotated images from the source domain can be transferred and used for training in the target domain, where only unlabeled samples are available. Liu et al. [16] applied the idea of domain transfer combined with partial weight sharing for the discriminators to train a classifier on MNIST using only labeled samples from SVHN or USPS and vice-versa. In [7] the cycle-consistent GANs were augmented by a task model, used to preserve content during domain transfer, and a feature discriminator to map features for the task model in a common space for both domains. In addition to a classification task, they also reported results for semantic segmentation. Rusakov et al. [21] implemented a cycle-consistent GAN on cuneiform data, transferring between hand drawn autographs with available labels and 2D-projections of 3D-scanned cuneiform signs. Using transferred data for training a retrieval model, performance was improved compared to a training on the source domain or on only small portions of data from the target domain.

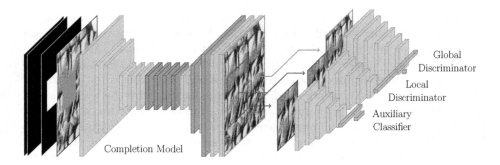

Fig. 1. Architecture of the conditional image completion system. Convolutional layers are depicted in green and orange with a stride of 1 and 2. Dilated convolutions are shown in purple, transpose convolutions in red and fully connected layers in blue. (Color figure online)

3   Method

The proposed method for semantic and context aware image completion is based on a system of multiple deep convolutional neural networks, each of which serves a specific task. The goal is to obtain a completion model which is capable of generating signs of any specified class (seen during training) adapting to the appearance of a larger image (tablet). The overall system architecture is shown in Fig. 1. An encoder-decoder network is used for the image completion task. Two discriminator networks are utilized to ensure local and global consistency of the generated image. To guide the generation semantically, a fourth network classifies the generated cuneiform sign.

3.1   Completion Model

For the image completion network, we use the encoder-decoder architecture proposed in [8]. The image is downsampled twice by strided convolutions. In order to increase the context for the feature representation, a block of dilated convolutions (with dilation factor $\eta = 2, 4, 8, 16$) is appended to the encoder. The feature maps are upsampled twice by transpose convolutions to generate the final image. We apply batch normalization [9] and a ReLU activation function after every convolution. After the final convolutional layer, we apply a sigmoid function to map the output to $[0, 1]$.

The model receives a masked input image $\tilde{\mathbf{x}}$ where the completion region is filled with the mean value of the training images (channel-wise mean for RGB images). In [8], a binary mask $\mathbf{m}$ of the same size as the input image is passed to the model as an additional channel. The mask is filled with 1 in the region to be completed and 0 elsewhere, not containing any information determining the class to be generated. In order to condition the completion model on a class label, we use a stack of such binary masks instead and encode the class label in a one hot fashion in the channel dimension. All masks only contain 0 except for the mask $\mathbf{m}_c$, where c is the class to be generated.

3.2 Discriminators and Auxiliary Classifier

Following [8], we use a local and a global discriminator to decide whether the image contains patches generated by the completion model. While the global discriminator computes features based on the full 256×256 pixel image, the local discriminator only takes a 128×128 pixel crop as input, centered around the completed patch. The features are concatenated and fed to a single neuron with sigmoid activation to compute the output of the discriminator.

The result of the discriminators is computed without considering any class information. To guide the generation process semantically, we follow the idea of using an auxiliary classifier [19]. We consider feeding the complete image as an input to the classifier as inappropriate, since in the completion task the image is expected to contain multiple signs of possibly different classes. For this reason, we decided to use the same 128×128 pixel crop for the classifier as for the local discriminator.

The classifier can be built by adding another linear layer with a softmax activation function on the features of the local discriminator [19]. However, the auxiliary classifier implemented this way never achieved a satisfactory classification performance on the training images. Therefore, we did not expect the auxiliary classifier to guide the generation semantically well. Instead of sharing the weights for the feature extraction with the local discriminator, we employed a separated model as an auxiliary classifier, keeping its architecture identical to the local discriminator except for the output layer.

3.3 Classification Models

In order to demonstrate a possible application for our conditional image completion method, we assume a classification task on domains with a limited amount of training data. Additional training data for the classifier is generated by completing patches in larger images with signs of selected classes.

We select two different architectures for the task models. We use a CNN with the same architecture as the auxiliary classifier (referred to as Simple-Classifier in the following). Additionally, we employ a modified ResNet-34 [5] as a classification model (referred to as ResNet-Classifier). To adapt the architecture to the comparably small cuneiform images, we remove all pooling and downsampling layers except for the first max pooling layer and the downsampling after three residual blocks [22]. The size of the output layer after the global max pooling is set to the number of classes for the respective splitting of the dataset (cf. Table 1).

3.4 Training

The completion model, the discriminators and the auxiliary classifier are trained in three stages. As training is performed with adversarial objectives, it was suggested in [8] to pretrain the models separately in order to improve the stability of the training process. At first, the completion model is pretrained for t_c iterations for reconstruction of patches. We calculate the difference between the original and the completed patch using the L1 norm as a loss function as it leads to less blurry results than using the L2 norm [10]:

$$\mathcal{L}_{\text{rec}}(\mathbf{x}, \tilde{\mathbf{x}}, \mathbf{m}_c, C) = \|\mathbf{x} \odot \mathbf{m}_c - C(\tilde{\mathbf{x}}, \mathbf{m}_c) \odot \mathbf{m}_c\|_1. \tag{3}$$

Note, that the reconstruction loss only considers pixels of the completed patch by a pixelwise multiplication $\odot$ with the binary mask $\mathbf{m}_c$ for class c. Since the model is trained for reconstruction, c is always identical to the original sign class in this stage.

In the second training stage, the dicriminators and the auxiliary classifier are pretrained for t_d iterations. The discriminator is trained using images from the original training data and images generated by the pretrained completion model using an adversarial loss:

$$\mathcal{L}_{\text{dis}}(\mathbf{x}, \tilde{\mathbf{x}}, \mathbf{m}_c, G, D) = -\log D(\mathbf{x}, \mathbf{m}_c) - \log(1 - D(C(\tilde{\mathbf{x}}, \mathbf{m}_c), \mathbf{m}_c)). \tag{4}$$

The images fed to the discriminator and the auxiliary classifier are compiled from the completed patch and the original image:

$$\hat{\mathbf{x}} = \mathbf{x} \odot (1 - \mathbf{m}_c) + C(\tilde{\mathbf{x}}, \mathbf{m}_c) \odot \mathbf{m}_c. \tag{5}$$

In contrast to [8], we do not perform any additional post-processing. For the training of the auxiliary classifier, 128×128 pixel crops from original images are used, centered around a cuneiform sign. The loss is computed using the cross entropy loss, scaled by a factor λ:

$$\mathcal{L}_{\text{cls}}(\mathbf{x}, \mathbf{m}_c, A) = -\lambda \cdot \log(A(\mathbf{x}, \mathbf{m}_c)_c), \tag{6}$$

where $A(\mathbf{x}, \mathbf{m}_c)_c$ is the softmax output of the auxiliary classifier for class c.

In the third stage, all models are trained jointly. The completion model is updated to optimize multiple loss functions. The reconstruction loss (Eq. 3) is used for improved training stability [8]. However, using only the reconstruction loss (L1 or L2) leads to blurry results in image generation [10,20]. In order to capture high frequency information, an adversarial loss is used in addition to the reconstruction loss:

$$\mathcal{L}_{\text{adv}}(\tilde{\mathbf{x}}, \mathbf{m}_c, C, D) = -\log(D(C(\tilde{\mathbf{x}}, \mathbf{m}_c), \mathbf{m}_c). \tag{7}$$

To guide the completion model in generating a sign of the class specified by the class mask, the generated input is passed to the auxiliary classifier:

$$\mathcal{L}_{\text{sem}}(\tilde{\mathbf{x}}, \mathbf{m}_c, C, A) = -\log(A(C(\tilde{\mathbf{x}}, \mathbf{m}_c)\mathbf{m}_c)_c). \tag{8}$$

The completion model is then trained to optimize:

$$
\begin{aligned}
\mathcal{L}_{\mathrm{com}}(\mathbf{x}, \tilde{\mathbf{x}}, \mathbf{m}_c, C, D, A) = \mathcal{L}_{\mathrm{rec}}(\mathbf{x}, \tilde{\mathbf{x}}, \mathbf{m}_c, C) \\
+ \alpha \cdot \mathcal{L}_{\mathrm{adv}}(\tilde{\mathbf{x}}, \mathbf{m}_c, C, D) \\
+ \lambda \cdot \mathcal{L}_{\mathrm{sem}}(\tilde{\mathbf{x}}, \mathbf{m}_c, C, A),
\end{aligned}
\tag{9}
$$

where α and λ are weighting the contribution of the different objectives with respect to the overall loss. Since we remove the complete sign from the image, it should be hard to infer the class for completing the patch only from the context. However, we alter target classes for generation to encourage the model to learn the generation of arbitrary classes in any patch. The original class labels are only retained for the first half of the batch. The target class labels for completion of the second half are sampled uniformly from all class labels. The reconstruction loss (Eq. 3) is only computed for the first half of the batch and omitted for generated images of randomly sampled classes. For updating the weights of the completion model, the weights of the discriminator and the auxiliary classifier are fixed. The discriminator and the auxiliary classifier are updated like in the second training stage, optimizing Eq. 4 and Eq. 6 respectively.

4   Evaluation

We evaluated our proposed method on the Cuneiform database, used in [22], and the Street View House Numbers (SVHN) dataset [18]. The cuneiform database contains photographs of tablets, providing images of signs in the context of the surrounding tablet. The SVHN dataset exhibits similar properties. Images of digits are not isolated but embedded in a house number sign of a specific color and appearance. While the generation of the correct class can only be verified by experts for the cuneiform dataset, the correctness of the generated class can be assessed easily in the case of digits in house numbers. Additionally, the varying styles of the house numbers in the RGB images of the SVHN dataset are well suited for a qualitative evaluation of the style of generated digits.

4.1   Cuneiform

Our experiments on cuneiform script are based on the database used in [22]. The database is a collection of 131 photographs of cuneiform tablets. $44,910$ occurrences of signs are annotated with bounding boxes and categorized into one out of 291 classes. Additionally, the Gottstein-Representation (GR) [4] is given for every sign. This representation determines the number of wedges of different direction and type in the sign. In [22], this dataset was split into four partitions of equal size, each containing a similar distribution of sign classes. As we assumed a scenario with a very small amount of annotated training data, this approach for splitting the dataset was not appropriate for our evaluation. Instead we split the data into a training and test set by a defined number of tablets. For our evaluation, we assumed having $10, 25, 50$ and 75 tablets with annotations for training. The remaining tablets were used for testing.

Table 1. Number of classes, training and test samples for a different number of training tablets.

#Train tablets	10	25	50	75
#Classes	40	52	55	57
#Train samples	1001	2997	6530	11333
#Test samples	17733	16152	12682	7890

As cuneiform signs can become very complex when composed of a large number of wedges, we expected the generation of those signs to become very hard. Therefore, we chose a subset of the sign classes by limiting the number of wedges according to the GR. We set the maximal number of wedges of type a and b to 2, c to 1 and d to 0. To ensure that the set of occurring cuneiform classes in the training and test set is identical, we discarded classes occurring in only one of the partitions. This resulted in a varying number classes depending on the number of tablets used for training (cf. Table 1). For the training of the GAN, we cropped patches of 256×256 pixels from the photographs of the cuneiform tablets placed randomly around some sign. We erased the respective sign in the crop, filled the region with the mean grey value and appended the class masks as additional channels (cf. Sect. 3.1).

Training Setup. We pretrained the completion model for $t_c = 5,000$ iterations for reconstruction. With the weights of the completion model kept frozen, we then pretrained the discriminators and the auxiliary classifier for $t_d = 3,000$ iterations. Afterwards, we trained all models jointly for another $32,000$ iterations, summing up to a total of $40,000$ training iterations. We set the mini-batch size to 32 and used ADAM [13] with a learning rate of 0.0002 and momentums set to 0.5 and 0.999. For weighting the losses, we set $\alpha = 0.05$ and $\lambda = 0.1$.

For the quantitative evaluation of the models, we used the completion model to generate images of cuneiform signs. We then trained a classifier (cf. Sect. 3.3) with both original and generated samples. Mini-batches were composed of 32 real and 32 generated samples. We trained the Simple-Classifier for $20,000$ and the ResNet-Classifier for $80,000$ iterations using ADAM [13].

Baseline. In order to compare the performance of classifiers trained with generated data, we trained the models using original training samples only and report the classification accuracy. Table 2 shows the results for both architectures being trained with a different number of tablets. Reducing the number of tablets used for training, considerably reduces the classification performance. The ResNet-Classifier achieves a significantly higher accuracy than the Simple-Classifier for all training sets. Since the 128×128 pixel crops contain distractors, we also trained models using images precisely cut out. The results using the exact crops are worse, leading to the assumption that either a small context around the sign is important for classification or that the cropping serves as a kind of augmentation. In order to compare the augmentation of training data by generated

Table 2. Accuracy [%] for the Simple-Classifier and the ResNet-Classifier trained on original images from a different number of tablets. ResNet-Cls. (aug.) uses homography augmentation during training and ResNet-Cls. (exact) used crops precisely placed on the signs instead of 128×128 pixel crops.

#Train tablets	10	25	50	75
Simple-Cls.	37.4	63.3	76.7	79.3
ResNet-Cls.	49.4	71.9	84.4	86.8
ResNet-Cls. (aug.)	46.5	75.2	84.6	87.6
ResNet-Cls. (exact)	29.6	61.0	74.4	79.2

examples with conventional augmentation, we report results for the ResNet-Classifier being trained using a homography augmentation (commonly used for handwritten word images, e.g. in [24]). Except for training with images from 10 cuneiform tablets, using augmentation helps to increase performance, though only for a little margin for 50 and 75 training tablets.

Quantitative Evaluation. Our quantitative evaluation was performed by training a classification model on a dataset augmented by generated samples. A major problem in this scenario is the selection of a suitable checkpoint during the training of the completion model. The quality of generated samples of GANs is often assessed by reporting the Frechet Inception Distance (FID) [6] between real and generated data. As the FID is class agnostic, it is only capable of measuring visual quality. When generating training data for a classifier, it is also important that the images show instances of the correct class. Therefore we did not use the FID to evaluate our model or to select a training checkpoint. Instead, we used the auxiliary classifier (trained jointly with the completion model) and measured its accuracy on generated images. Thereby, we hoped to choose a model that generates instances of the correct classes. This evaluation can be carried out on the training data. Building a validation set is problematic when only very limited data is available. To examine an upper bound for this kind of selecting a training checkpoint, we additionally deployed the evaluation with the auxiliary classifier on generated samples from the test set. Additionally, we evaluated the completion model every $10{,}000$ iterations in order to assess whether the previously described selection approach is suitable.

The results for the different checkpoint selection approaches are shown in Table 3 for the Simple-Classifier and Table 4 for the ResNet-Classifier. The quality of generated images varies during training of the completion model resulting in volatile classification performance. When choosing a completion model after a fixed number of training iterations (e.g. 40k) and training the Simple-Classifier, the accuracy could be improved in most cases compared to training on original data only. However, when selecting the model rated best by the auxiliary classifier on the test data, we could choose the best performing completion model for all three training sets. Using the auxiliary classifier for rating generation

Table 3. Accuracy [%] for the Simple-Classifier trained on real data augmented by generated data. The data was generated at different training iterations. For the selection by the auxiliary classifier, the selected iteration is reported in brackets. The last column shows the baseline results, when training on real data only.

#Train tablets	GAN iteration				Aux. Class. selection		Real data
	10k	20k	30k	40k	Train	Test	Only
10	34.9	40.5	40.1	39.6	34.7 *(12k)*	**42.7** *(24k)*	37.4
25	61.1	65.5	66.9	**67.9**	64.9 *(35k)*	**67.9** *(40k)*	63.3
50	75.8	75.8	76.9	76.3	**78.0** *(39k)*	**78.0** *(39k)*	76.7
75	–	–	–	–	–	–	79.3

Table 4. Accuracy [%] for the ResNet-Classifier trained on cuneiform data augmented by generated data. The last columns show the baseline results, when training on real data only.

#Train tablets	GAN iteration				Aux. Class. selection		Real data	
	10k	20k	30k	40k	Train	Test	Only	Aug
10	44.2	52.8	53.8	**55.0**	50.4	51.0	49.4	46.5
25	74.0	74.4	73.9	**76.3**	76.1	**76.3**	71.9	75.2
50	84.1	85.2	85.2	**86.7**	85.5	85.5	84.4	84.6
75	–	–	–	–	–	–	86.8	87.6

quality on the training data, the selected model does not achieve satisfactory results except for training with 50 tablets. As test data is unknown in practice, this selection strategy could only be applied, when a validation set is available, though data is scarce. Since the generation of samples does only require annotated boxes for the areas to be completed, labeling images for this evaluation only is still easier than annotating signs with class labels for classifier training.

For the ResNet-Classifier, we achieved best results when training on generated data from the completion model after 40,000 iterations. Since the generation quality wrt. the use as training data starts oscillating before 40,0000 iterations, we decided to stop training at that point. However, it was a design decision and not derived by any optimization. The results emphasize that finding a suitable method for selecting the number of training iterations is important but still an open problem.

Choosing a well performing completion model for training data generation could improve classification results for all three training sets. Even compared to using homography augmentation during classifier training, adding generated samples to the training data results in a better classification performance. Especially for the smallest set, containing only 10 tablets, the largest increase in performance can be observed. For the training set with 50 tablets the accuracy could still be enhanced, though only marginally. Therefore, we did not train a completion model on 75 tablets, using this set only for comparison of classifier training with original data.

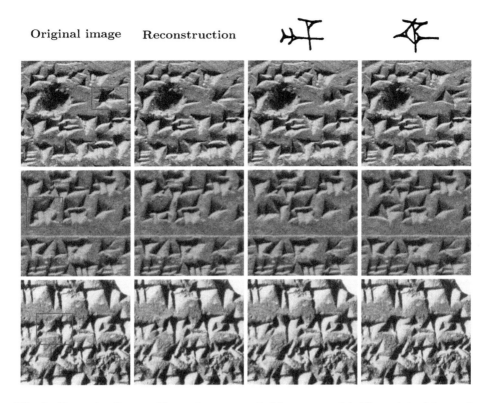

Fig. 2. Examples for cuneiform signs generated by our model. The original image is shown in the first column. The erased region is indicated with a red box. In the second column a generated sign of the same class is displayed. Columns 3 and 4 show generated signs for the classes *AN* and *IGI*. (Color figure online)

Qualitative Evaluation. Figure 2 shows examples for images containing generated signs. The first column shows the images before erasing the patch to be completed. To give a comparison of generated signs of the same class in the same context, the images with the original sign reconstructed by the completion model are displayed in the second column. Note that the process for generating the reconstructed sign is identical to the generation of a sign of any other arbitrary class. The remaining two columns present examples for signs of the classes *AN* and *IGI*, generated in the respective position. To allow for an assessment, whether an instance of the correct class was generated, we show autographs on top of the columns. The autographs indicate the impressed wedges of the respective classes. We applied the completion model trained on 25 tablets for 40,000 iterations for the all examples. The patches generated by the model exhibit a high visual quality and embed smoothly in the surrounding context. Lighting conditions and surface structure of the tablet are perfectly mimicked. Although we did not use any post-processing, there are no visual discontinuities at the borders of the completed patches.

4.2 SVHN

The Street View House Numbers dataset (SVHN) [18] is a collection of RGB images of house number signs from Google Street View. For our scenario, we only used a portion of the training data and omitted the extra training data provided. In order to further reduce the amount of training data, we only used the first 5,000 and the first 10,000 annotated digits for training. In contrast to the cuneiform dataset, all 10 digit classes occur in both training and test partition (no classes had to be discarded). For the SVHN dataset, training images are crops of 128×128 pixels randomly positioned on the respective digit (or resized to 128 × 128 px. if necessary). As the images are smaller than those of the cuneiform database, we removed one downsampling step from each discriminator and the classifier. The models were trained with the same setup as described for the cuneiform database. Note that we used SVHN in order to visualize the results in an easier comprehensible domain and to show applicability of the proposed approach in another domain. Therefore we did not compare to state-of-the-art classification results for SVHN classification.

Quantitative Evaluation. The quantitative results for the SVHN dataset are shown in Table 5. For SVHN, we also examined the selection of the number of training iterations for the completion model. In both cases (5,000 and 10,000 training samples) choosing the model after 40,000 iterations leads to the best results when training a classifier with the generated samples. The performance increases by more than 10 percentage points, using our completion model after 40,000 training iterations. Adding generated samples to the original data in these cases, we were able to compensate for half of difference in performance between training with very limited data and using the full set. However the results with 10,000 training samples show that during the GAN training the quality of generated samples varies. At around 20.000 iterations, the generated images are better suited as training data than those after 30,000 iterations. Furthermore, for SVHN none of the training checkpoints selected by the auxiliary classifier achieves competitive results (in all cases the model after 9,000 training iterations was chosen). Therefore it is questionable whether the use of an additional validation set (instead of using the test set) would have helped in this case.

Table 5. Accuracy [%] for the Simple-Classifier trained on SVHN data augmented by generated data. For the selection by the auxiliary classifier, the selected iteration is reported in brackets. The last column shows the baseline results, when training on real data only.

	GAN iteration				Aux. Class. selection		Real data
#Train samples	10k	20k	30k	40k	Train	Test	Only
5k	53.1	59.7	60.5	**66.6**	45.5 *(9k)*	45.5 *(9k)*	55.9
10k	61.1	66.3	64.9	**67.3**	56.7 *(9k)*	56.7 *(9k)*	56.6
All	–	–	–	–	–	–	76.7

Fig. 3. Examples for digits generated by our model in house number signs from [18]. (Color figure online)

Qualitative Evaluation. For the qualitative evaluation, examples for generated images of the SVHN dataset are presented in Fig. 3. The first column shows the original image. In the following columns one digit per row was chosen and replaced with generated digits from 0 to 9. In the examples of the first 3 rows, the model is capable of mimicking the style of the house number signs as well as generating readable instances of the respective digits in most cases. Row 4 shows an example where the color and the line width are imitated well. However, the overall quality of generation is poor for almost all digits. A failure case is depicted in the last row. The model is neither able to extract the correct style nor to generate well readable digits.

5   Conclusion

In this work, we presented a method for context aware document image completion. Our method is based on a conditional generative adversarial network with an auxiliary classifier. A completion model is generating instances of a defined class within larger images and is conditioned on the style by the surrounding pixels. A local and a global discriminator force the model to mimic the appearance while an auxiliary classifier guides the training semantically, checking generated samples for the correct class. We evaluated our proposed method on a collection of cuneiform tablets. Training our model with very limited data, we were able to generate realistic looking cuneiform signs of specified classes. Using the generated images in addition to samples from the original dataset for training a classifier led to a significant improvement in classification performance. By applying our approach to generation of digits in house number signs of the SVHN dataset [18], we demonstrated that the method is not limited to the domain of cuneiform script.

References

1. Alonso, E., Moysset, B., Messina, R.: Adversarial generation of handwritten text images conditioned on sequences. In: ICDAR, Sydney, NSW, Australia, pp. 481–486 (2019). https://doi.org/10.1109/ICDAR.2019.00083
2. Denton, E., Gross, S., Fergus, R.: Semi-supervised learning with context-conditional generative adversarial networks. arXiv preprint arXiv:1611.06430 (2016)
3. Goodfellow, I., et al.: Generative adversarial nets. In: Ghahramani, Z., Welling, M., Cortes, C., Lawrence, N.D., Weinberger, K.Q. (eds.) NIPS, Montreal, Canada, pp. 2672–2680 (2014)
4. Gottstein, N.: Ein stringentes Identifikations- und Suchsystem für Keilschriftzeichen. Online (2012). http://www.materiale-textkulturen.de/mtcblog/2012_005_Gottstein.pdf
5. He, K., Zhang, X., Ren, S., Sun, J.: Deep residual learning for image recognition. In: CVPR, Las Vegas, NV, USA, pp. 770–778 (2016). https://doi.org/10.1109/CVPR.2016.90
6. Heusel, M., Ramsauer, H., Unterthiner, T., Nessler, B., Hochreiter, S.: GANs trained by a two time-scale update rule converge to a local nash equilibrium. In: Guyon, I., et al.: (eds.) NIPS, Long Beach, CA, USA, pp. 6626–6637 (2017)
7. Hoffman, J., et al.: CyCADA: cycle-consistent adversarial domain adaptation. In: ICML, pp. 1989–1998. Stockholmsmässan, Stockholm (2018)
8. Iizuka, S., Simo-Serra, E., Ishikawa, H.: Globally and locally consistent image completion. ACM Trans. Graph. **36**(4) (2017). https://doi.org/10.1145/3072959.3073659
9. Ioffe, S., Szegedy, C.: Batch normalization: accelerating deep network training by reducing internal covariate shift. In: ICML, pp. 448–456 (2015)
10. Isola, P., Zhu, J.Y., Zhou, T., Efros, A.A.: Image-to-image translation with conditional adversarial networks. In: CVPR, Honolulu, HI, USA, pp. 1125–1134 (2017). https://doi.org/10.1109/CVPR.2017.632
11. Kang, L., Riba, P., Rusiñol, M., Fornés, A., Villegas, M.: Distilling content from style for handwrittenword recognition. In: ICFHR, Dortmund, Germany, pp. 139–144 (2020). https://doi.org/10.1109/ICFHR2020.2020.00035
12. Kang, L., Riba, P., Wang, Y., Rusiñol, M., Fornés, A., Villegas, M.: GANwriting: content-conditioned generation of styled handwritten word images. In: Vedaldi, A., Bischof, H., Brox, T., Frahm, J.-M. (eds.) ECCV 2020. LNCS, vol. 12368, pp. 273–289. Springer, Cham (2020). https://doi.org/10.1007/978-3-030-58592-1_17
13. Kingma, D.P., Ba, J.: Adam: a method for stochastic optimization. In: ICLR (2015)
14. Krishnan, P., Jawahar, C.V.: HWNet v2: an efficient word image representation for handwritten documents. IJDAR **22**(4), 387–405 (2019). https://doi.org/10.1007/s10032-019-00336-x
15. Li, C., Wang, Z., Qi, H.: Fast-converging conditional generative adversarial networks for image synthesis. In: ICIP, Athens, Greece, pp. 2132–2136 (2018). https://doi.org/10.1109/icip.2018.8451161
16. Liu, M.Y., Breuel, T., Kautz, J.: Unsupervised image-to-image translation networks. In: Guyon, I., et al. (eds.) NIPS, Long Beach, CA, USA, pp. 700–708 (2017)
17. Mirza, M., Osindero, S.: Conditional generative adversarial nets. arXiv preprint arXiv:1411.1784 (2014)
18. Netzer, Y., Wang, T., Coates, A., Bissacco, A., Wu, B., Ng, A.Y.: Reading digits in natural images with unsupervised feature learning. In: NIPS Workshop on Deep Learning and Unsupervised Feature Learning, Granada, Spain (2011)

19. Odena, A., Olah, C., Shlens, J.: Conditional image synthesis with auxiliary classifier GANs. In: ICML, Sydney, NSW, Australia, pp. 2642–2651 (2017)
20. Pathak, D., Krahenbuhl, P., Donahue, J., Darrell, T., Efros, A.A.: Context encoders: feature learning by inpainting. In: CVPR, Las Vegas, NV, USA, pp. 2536–2544 (2016). https://doi.org/10.1109/CVPR.2016.278
21. Rusakov, E., et al.: Generating cuneiform signs with cycle-consistent adversarial networks. In: Proceedings of International Workshop on Historical Document Imaging and Processing (HIP), Sydney, NSW, Australia, pp. 19–24 (2019). https://doi.org/10.1145/3352631.3352632
22. Rusakov, E., Somel, T., Fink, G.A., Müller, G.G.W.: Towards query-by-eXpression retrieval of cuneiform signs. In: ICFHR, Dortmund, NRW, Germany, pp. 43–48 (2020). https://doi.org/10.1109/icfhr2020.2020.00019
23. Streck, M.P.: Großes Fach Altorientalistik: Der Umfang des keilschriftlichen Textkorpus. Mitteilungen der Deutschen Orient-Gesellschaft zu Berlin **142**, 35–58 (2010)
24. Sudholt, S., Fink, G.A.: Attribute CNNs for word spotting in handwritten documents. IJDAR **21**(3), 199–218 (2018). https://doi.org/10.1007/s10032-018-0295-0

Adaptive Scaling for Archival Table Structure Recognition

Xiao-Hui Li[1,2], Fei Yin[1], Xu-Yao Zhang[1,2], and Cheng-Lin Liu[1,2,3](✉)

[1] National Laboratory of Pattern Recognition, Institute of Automation of Chinese Academy of Sciences, 95 Zhongguancun East Road, Beijing 100190, People's Republic of China
{xiaohui.li,fyin,xyz,liucl}@nlpr.ia.ac.cn
[2] School of Artificial Intelligence, University of Chinese Academy of Sciences, Beijing 100049, People's Republic of China
[3] CAS Center for Excellence of Brain Science and Intelligence Technology, Beijing, People's Republic of China

Abstract. Table detection and structure recognition from archival document images remain challenging due to diverse table structures, complex document layouts, degraded image qualities and inconsistent table scales. In this paper, we propose an instance segmentation based approach for archival table structure recognition which utilizes both foreground cell content and background ruling line information. To overcome the influence from inconsistent table scales, we design an adaptive image scaling method based on average cell size and density of ruling lines inside each document image. Different from previous multi-scale training and testing approaches which usually slow down the speed of the whole system, our adaptive scaling resizes each image to a single optimal size which can not only improve overall model performance but also reduce memory and computing overhead on average. Extensive experiments on cTDaR 2019 Archival dataset show that our method can outperform the baselines and achieve new state-of-the-art performance, which demonstrates the effectiveness and superiority of the proposed method.

Keywords: Archival document · Table detection · Table structure recognition · Adaptive scaling

1 Introduction

Automatic analysis of table images is an important task in field of document image analysis and understanding. Many works have been proposed in recent years to solve the tasks of table detection and structure recognition using deep learning, but most of them focus on tables from modern or born-digital document images which usually have good image qualities and clean backgrounds.

Compared with modern tables, automatic analysis of archival tables is a much more challenging problem due to diversified table structures and document layouts, degraded image qualities and inconsistent table scales. As we can see

J. Lladós et al. (Eds.): ICDAR 2021, LNCS 12821, pp. 80–95, 2021.
https://doi.org/10.1007/978-3-030-86549-8_6

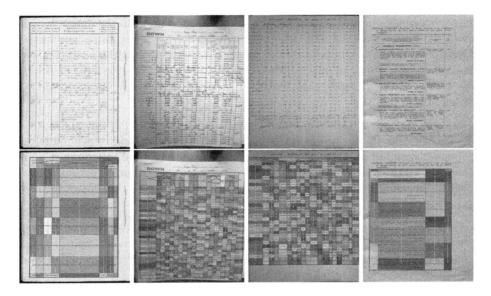

Fig. 1. Archival table examples. The first row: original images; the second row: annotation visualization. Cells are depicted with random colors to give better illustrations. Purple and cyan lines denote horizontal and vertical cell relations, respectively. (Color figure online)

from Fig. 1, archival tables may have printed or handwritten contents, arbitrary cell shapes, dense or sparse or none ruling lines, spanning row and column cells, complete or empty cells, single text line or multiple text line cells, tabular or less tabular structures.

In this paper, we propose an instance segmentation based method for archival table detection and structure recognition which utilizes both foreground cell content and background ruling line information. After multi-scale feature extraction using feature pyramid network (FPN) [24], cell masks and ruling line masks are predicted using label pyramid network (LPN) [23]. Pixels in the predicted cell masks and ruling line masks have higher scores when they are near to cell centers or line centers, otherwise, low scores. Then we perform watershed transform on cell masks and line masks to obtain cell boundaries. We show that cell masks and line masks contain complementary information and can help each other to obtain better cell segmentation results.

Although deep neural networks can learn a large receptive field for original image, however, the receptive field is identical for different images. This actually has bad influence on the performance, because the tables in different images usually have large variances in their sizes. A widely-used solution is using multiple scales in both training and test, and then fusing the results from multi-scale images. However, this approach is actually heuristic, sub-optimal, and also time-consuming. In this paper, we propose to use adaptive scaling to resize each image to a single optimal scale according to the table content in the image, which would significantly and consistently improve both the efficiency and effectiveness of deep neural network based table recognition. To be specific, the optimal scale for each image is estimated based on the mean cell size and density of ruling line in that image. Actually, the motivation of this approach is driven from two perspectives:

- Images with small cells should be enlarged to keep more details and images with big cells should be shrunk to enlarge the receptive field of the model.
- Images with full or dense ruling lines could be enlarged to obtain more precise segmentation results and images with sparse or none ruling lines should be shrunk to enlarge the receptive field of the model.

During training, the optimal scale for each image is computed using ground truth annotations; while during testing, the optimal scale for each image is predicted by the scale prediction network. With this adaptive scaling strategy, the overall performance of cell extraction is improved and the average memory and computing overhead for each image is reduced.

Since in this work cell regions are represented using polygons surrounded by ruling lines or virtual ruling lines, they are actually a dense partition of the entire table regions, see Fig. 1. Thus relations between adjacent cells can be directly obtained using some simple heuristic rules based on their common edges.

The rest of this paper is organized as follows. Section 2 briefly reviews related works. Section 3 and 4 gives details of the proposed adaptive scaling and table structure recognition network. Section 5 presents the experimental results, and Sect. 6 draws concluding remarks.

2   Related Work

Great attention has been paid to table analysis and understanding over the last few decades and plenty of methods have been proposed. We first briefly review these methods, then introduce some other works related to our work.

Table Detection. A straightforward idea for table detection is adapting off-the-shelf object detectors designed for natural scene images to document images. Guided by this idea, [1,11,18,20,36,43] use Faster R-CNN [29], Mask R-CNN [14] or YOLO [28] to detect tables and other types of regions in document images. Some other methods such as [10,21,22,30] conduct table detection through sub-region classification and clustering using conditional random fields or graph neural networks. Compared with whole tables, sub-regions with more constant appearance are easier to extract and cluster, thus improving the overall table detection performance.

Table Structure Recognition. As to table structure recognition, there exist mainly three strategies: top-down ones, bottom-up ones and end-to-end ones. Top-down methods [19,33,34,37] try to split entire table images into rows and columns using detection or segmentation models, then cells can be obtained

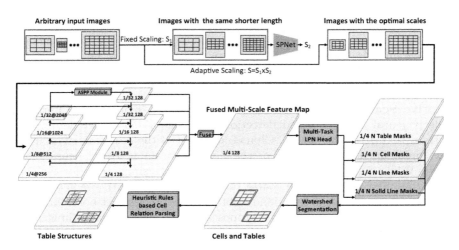

Fig. 2. Overview of the proposed method. Given input images with arbitrary sizes and resolutions, we first resize them using adaptive scaling strategy according to their contents. After multi-scale feature extraction using FPN, table, cell and line score maps are predicted using multi-task LPN. Then watershed transform is applied on these score maps to obtain table and cell boundaries. After cell segmentation, cell relations are analyzed based on some simple rules.

through row-column intersection. This kind of methods can only handle tables with clear tabular structures and become less effective when large amount of empty or row-span or column-span cells exits in tables. Bottom-up methods [5,26,27,40] first detect cells or text segments using detection models or OCR engines, then analyze the relations between neighbouring cells using GNN or LSTM. This kind of methods are more robust than top-down ones but rely heavily on the on cell detection and relation classification performance. End-to-end ones [7,20,42] adopt the image-to-text framework for table transcription based on encoder-decoder and attention model. These methods can not give cell bounding boxes and are less interpretable.

Instance Segmentation. Archival document images may suffer from severe degradation and deformation and cells in archival tables may have arbitrary shapes. For these reasons, instance segmentation becomes a more suitable way for cell extraction in archival document images. Our method is inspired by the previous works of [3,13,23,39] for instance segmentation in natural scene or document images. They use truncated distance transform or subsequently polygon shrinking to implicitly encode object boundary information and apply watershed transform or progressive scale expansion on the predicted score maps to obtain instance boundaries.

Ruling Line Extraction. Ruling lines and junctions (line intersections) are extremely important clues in table detection and structure recognition. Previous works [2,9,32,38,41] utilize them for table analysis and form understanding.

However, the ruling lines and junctions were usually extracted based on hand crafted features and heuristic rules, which could be unreliable for documents with bad image qualities or tables with sparse or none ruling lines. On the contrary, our deep learning based method is data driven and is more robust against image degeneration and missing of ruling lines.

Adaptive Scaling. Finding the optimal scale for an image is very important for deep neural network based detection tasks like video object detection [6] and scene text detection [16,31]. Chin et al. [6] select the best scale of an images from a pre-defined scale set based on the minimum training loss of some detected objects. While in [16] and [31], instead of predicting a single scale for the entire image, they first extract some coarse text regions along with predicted scales, then apply a second stage detection to obtain more precise text bounding boxes. Inspired by their work, our paper proposes a new method for estimating the optimal scale for each table image. Specifically, the optimal scale of each image is estimated directly based on the average cell size and density of ruling lines inside each document image. To the best of our knowledge, we are the first to consider optimal scale estimation in table image analysis, which would achieve significant advantages compared with other state-of-the-art approaches.

3   Adaptive Scaling

Tables in different document images usually have large variance in their sizes. To improve our model's performance and reduce the memory and computation cost in the meantime, we propose to use adaptive scaling to resize each image to a single optimal scale according to the table contents in the image, which would significantly and consistently improve both the efficiency and effectiveness of deep neural network based table recognition.

3.1   Optimal Scale Estimation

To estimate the optimal scale for each document image, we design three strategies based on the average cell size and density of solid ruling lines, introduced as below.

Strategy 1. The first one is to resize each image to a proper size so that cells inside each image roughly have the same average sizes. Given an image with arbitrary size, the desired scaling factor is defined as below:

$$S = S_1 \cdot S_{2C} = \frac{1024}{min(I_h, I_w)} \cdot \frac{64}{\frac{1}{N}\sum_{i=1}^{N} min(C_{ih}, C_{iw})}, \tag{1}$$

where S_1 is used to resize the original image to a fixed shorter length (*e.g.*, 1024), and S_{2C} is used to further resize the image so that the mean size of cells after resizing will have a roughly fixed value (*e.g.*, 64), see Fig. 2. We limit S_{2C} to be in the range of $[1.0, 3.0]$ so that images after resizing will have shorter sizes between $[1024, 3072]$ pixels. I_h and I_w are height and width of the original image, C_{ih} and C_{iw} are the height and width of C_i after first time resizing, N is the cell number.

Strategy 2. The second one is to resize images with full or dense ruling lines to larger sizes and images with sparse or none ruling lines to smaller sizes. The desired scaling factor is defined as below:

$$S = S_1 \cdot S_{2L} = \frac{1024}{min(I_h, I_w)} \cdot (2 \times \frac{L_{SL}}{L_L} + 1), \tag{2}$$

where S_1 is the same as that in Eq. 1, and S_{2L} is used to further resize the image so that images with denser solid ruling lines will be resized to larger sizes. S_{2L} is also in the range of $[1.0, 3.0]$. L_{SL} is the total length of solid ruling lines and L_L is the total length of all ruling lines including virtual lines.

Strategy 3. The third one is the combination of the first two strategies. The desired scaling factor is defined as below:

$$S = S_1 \cdot S_2 = \frac{1024}{min(I_h, I_w)} \cdot (\lambda S_{2C} + (1 - \lambda)S_{2L}), \tag{3}$$

where S_2 is a linear combination of S_{2C} and S_{2L}, and λ ($=0.5$ in this work) is a combination coefficient.

3.2 Scale Prediction Network

During training, for each image in the training set the optimal scale S_{2C}, S_{2L} and S_2 can be computed based on annotation using Eqs. 1, 2 and 3. While during testing, we need to predict them using a scale prediction network (SPNet) trained from the training set using smooth L1 loss [12]. The SPNet used in this work is ResNet-50 with two way outputs as the predicted S_{2C} and S_{2L}, S_2 can be computed from Eq. 3. We resize all images to the same shorter length of 1024 pixels and randomly cut sub-images of 512×512 pixels as the inputs of SPNet. For each test image, we conduct several times of prediction (*e.g.*, 16) and adopt the average of all outputs as the final predicted scale.

4 Table Structure Recognition Network

Overview of the proposed method can be seen in Fig. 2. After adaptive scaling for each input image, we extract feature maps using FPN and predict several score maps using multi-task LPN with table, cell and ruling line masks. Then watershed transform is applied on these score maps to obtain cell boundaries. Either cell masks or line masks can be used to segment cells, or they can be used together to obtain better results. Since cells in archival tables are closely located next to each other, we can used some simple rules to obtain the relations between neighbouring cells. Detailed descriptions are as below.

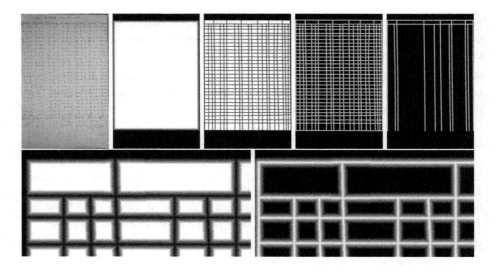

Fig. 3. Segmentation masks. The first row, from left to right: original image; table mask; cell mask; ruling line mask; solid ruling line mask. The second row: cropped and enlarged cell mask and ruling line mask for better illustration.

4.1   Feature Extraction

We use FPN [24] and ASPP [4] module for multi-scale feature extraction. The backbone of FPN can be any appropriate networks such as VGGNets [35], ResNets [15] and DenseNets [17]. After up-stream convolutions and down-stream deconvolutions, we reduce feature map channels of each scale to a lower dimension (*e.g.*, 128) using 1×1 convolutions. Then we interpolate feature maps of each scale to the same resolution (*e.g.*, quarter of the original size) and fuse them to a single feature map with fixed dimension (*e.g.*, 128) using 1×1 convolutions. The obtained feature map can capture information from multiple scales, thus benefiting the following tasks.

4.2   Score Map Prediction

To extract tables and cells from archival document images, we first use multi-task LPN [23] to predict a series of pyramid score maps including table mask, cell mask and ruling line mask and solid ruling line mask, see Fig. 3. By saying "ruling line mask", we mean all ruling lines no matter they are solid or virtual; while "solid ruling line mask" only consider solid ruling lines. Please see Fig. 3 for the difference.

The LPN is actually a multi-task fully convolutional networks (FCN) [25], and the output score map of each task contains gradually shrunken regions or gradually thicken lines. Ground-truth pyramid masks for archival tables are generated in the similar way with [23] using distance transformation, truncation and normalization. Value of each pixel on the distance map stands for the distance from this pixel to the nearest region boundaries or ruling lines. For table mask and line mask, we use global truncation thresholds $T_T = 64$ and $T_L = 16$; while for cell mask, we compute a separate truncation threshold T_{Ci} for cell C_i as:

$$T_{Ci} = min(T_C, D_{i\_max}/4), \tag{4}$$

where $T_C = 16$ is the global threshold and D_{i_max} is the max distance from a pixel inside cell C_i to its nearest boundary. During training, each pyramid mask is split into several binary masks through multi-level thresholding; while during testing, we averaged the score maps predicted by the LPN pixel-wisely into a single probability map on which watershed transform is carried out to segment table or cell boundaries.

The loss function of segmentation head during training is calculated as:

$$L_{seg}(\{Y\}, \{X\}) = \frac{1}{M} \sum_{i=1}^{M} L_{ce}(Y, X), \tag{5}$$

where M is task number and L_{ce} is cross entropy loss:

$$L_{ce}(Y, X) = \frac{1}{N} \sum_{i=1}^{N} - \sum_{c=1}^{C} w_c \cdot y_{ic} log(x_{ic}), \tag{6}$$

where Y and X are labels and predicted scores, N and C are pixel number and label number, w_c is weight coefficient for class c computed based on inverse class frequency.

4.3 Cell Segmentation

Although either cell or line mask can be used alone to extract cells, we found combining them together can further improve the performance because they can provide complementary information. To be concrete, we first reverse the values of line mask and use watershed transform to extract center line pixels, then we set the corresponding pixels in cell mask to be zero and apply watershed transform on it to extract more precise cell boundaries. After that, cells can be extracted using image binarization and connected component analysis (CCA). Figure 4 shows an example of cell region segmentation. As we can see, our method can extract cells of arbitrary shapes with precise boundaries.

4.4 Cell Relation Parsing

Once cells are extracted, cell relations can be obtained through some simple heuristic rules. Since cells in archival documents are closely located next to each other, and neighboring cells usually share common boundaries which can be utilized to determine the relation between them. If the direction of this common boundary is roughly vertical, then the relation of this two cells are *left-right*, otherwise, *top-down*. Table boundaries extracted from pyramid table masks are also utilized to assist cell relation parsing. After cell segmentation and relation parsing, whole table regions can be obtained by merging connected cells.

Fig. 4. Cell segmentation examples. From left to right: line mask with watershed boundaries (red pixels); cell mask with watershed boundaries; segmented cell regions. (Color figure online)

5   Experiments

5.1   Datasets

We conduct experiments on the cTDaR Archival dataset [8] to evaluate the performance of our proposed method. This dataset has been used for ICDAR 2019 competition on table detection and recognition. It contains 600 images for training and 499 images for testing. The whole test set is split into three subsets: TrackA (199 images), TrackB1 (150 images) and TrackB2 (150 images). TrackA is used for table region detection and only table coordinate annotations are given, TrackB1 and TrackB2 are used for table structure recognition and both table and cell coordinate annotations are given. We make no difference between TrackB1 and TrackB2, and don't use any prior information (*e.g.*, table coordinates) when recognizing table structures.

To evaluate the performance of our model for tables of different characteristics, we split TrackB into three subsets: Big (100 images), Mid (100 images) and Small (100 images), according to average cell sizes inside each image; and three other subsets: Full ($R_{SL} = 1.0$, 169 images), Dense ($0.5 \leq R_{SL} < 1.0$, 43 images) and Sparse ($R_{SL} < 0.5$, 88 images), according to the ratio of solid ruling lines which is computed as: $R_{SL} = L_{SL}/L_L$, where L_{SL} is total length of solid ruling lines, and L_L is total length of all ruling lines including virtual ones. Since the original cTDaR dataset doesn't provide line type annotation, we manually annotated them for experiments.

5.2   Evaluation Metrics

For table detection and structure recognition, we use evaluation metrics proposed in [8]. Specifically, precision, recall and F1 value of table detection, cell detection and cell pair relation prediction are calculated at IoU thresholds of 0.6, 0.7, 0.8 and 0.9, and the weighted average F1 (WAvg. F1) value of the whole dataset for each track is defined as:

$$W Avg.F1 = \frac{\sum_{i=1}^{4} IoU_i \cdot F1@IoU_i}{\sum_{i=1}^{4} IoU_i}. \tag{7}$$

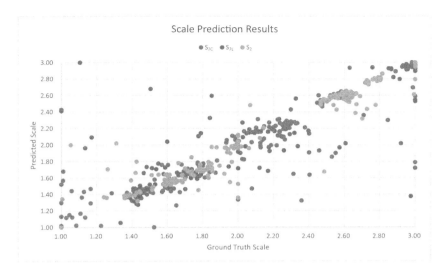

Fig. 5. Ground truth scales and predicted scales of all images in test set TrackB.

5.3 Implementation Details

We choose ResNet-50 [15] as the backbone of FPN with its parameters pre-trained on ImageNet. The first layer with kernel size of 7×7 is replaced with three layers with kernel size of 3×3 for better maintaining image details. Dilation rates of each block of ResNet are set as (1,2), (1,2), (2,4) and (4,8), and dilation rates of ASPP [4] module are set as (4,4), (8,8) and (12,12). Using larger dilation rates at horizontal direction can better explore horizontal contexts, thus benefiting long cell detection and relation parsing. Our model is implemented on the PyTorch[1] platform.

A training instance is a random 512×512 crop of the resized image followed by random rotation in the range of $-15°$ to $15°$. No other data augmentation techniques are used. The learning rate is updated using the *poly* strategy which can be formulated as:

$$lr = base\_lr \cdot (1 - iter/max\_iter)^{power}, \tag{8}$$

where $base\_lr = 0.01$, $max\_iter = 20k$ (for ablation studies) or $80k$ (for comparison with the state-of-the-art) and $power = 0.9$ in this work. We use the momentum of 0.9 and a weight decay of 0.0005. We employ one TITAN RTX GPU for training and use batch size of 4.

---

[1] https://pytorch.org/get-started/locally/.

Table 1. Cell segmentation results on TrackB using foreground and background clues. 2048: shorter length of fixed scaling; C: cell mask; L: line mask; L_S: solid line mask.

Scale	Method			F1@IoU				WAvg. F1
	C	L	L_S	0.6	0.7	0.8	0.9	
2048	✓	×	×	0.8873	0.8308	0.7010	0.3783	0.6717
	×	✓	×	0.8621	0.8080	0.6867	0.3791	0.6578
	✓	✓	×	0.8880	0.8334	0.7109	0.3963	0.6805
	✓	×	✓	0.8898	0.8363	0.7144	0.3977	0.6829

Table 2. Cell segmentation results on TrackB using different scaling strategies. 1204, 2048 and 3072: shorter length of fixed scaling; *AdScale_Cell*: adaptive scaling using strategy 1; *AdScale_Line*: adaptive scaling using strategy 2; *AdScale*: adaptive scaling using strategy 3; *AdScale**: model trained for more iterations.

Scale	Mean Size	WAvg.F1						
		Small	Mid	Big	Full	Dense	Sparse	All
1024	(1259,1182)	0.6463	0.6132	0.6421	0.6799	0.5266	0.4270	0.6365
2048	(2518,2364)	0.7267	0.6046	0.6009	0.7435	0.4551	0.3891	0.6829
3072	(3777,3546)	0.7617	0.5928	0.5409	0.7690	0.4199	0.3464	0.6973
AdScale_Cell	(2455,2338)	0.7609	0.6105	0.6203	0.7738	0.4914	0.3940	0.7073
AdScale_Line	(3176,2990)	0.7466	0.6347	0.6153	0.7632	0.5009	0.4228	0.7038
AdScale	(2816,2664)	0.7657	0.6159	0.6302	0.7738	0.5011	0.4269	0.7116
AdScale*	(2816,2664)	0.7741	0.6203	0.6286	0.7816	0.5047	0.4252	0.7182

5.4   Experimental Results

Ablation Study. We did a series of experiments to verify the rationalisation of our model design and the effect of difference cell segmentation methods and adaptive scaling. Experimental results are shown in Table 1 and Table 2.

Cell Segmentation Methods. To evaluate the effects of different cell segmentation methods, we first resize all images to the same shorter length of 2048 pixels, then compare the cell segmentation performance using foreground (cell mask) or background (ruling line mask) information. As we can see from Table 1, using both cell mask and ruling line mask for cell segmentation can obtain better cell segmentation results than using only one of them. That is because they contain complementary information and can help each other to get better results. To be specific, cell masks focus on locating cells and line masks focus on finding more precise boundaries.

What's more, using only solid ruling lines is better than using both solid and virtual ruling lines. Compared with solid ruling lines, virtual ruling lines are harder to extract and the segmentation results are noisy and unreliable. This is more significant for tables having neither ruling lines nor tabular structures. In

Table 3. Table structure recognition results on cTDaR Archival datasets. CD: cell detection; SR: structure recognition.

Method	Task	Subset	F1@IoU				WAvg. F1
			0.6	0.7	0.8	0.9	
NLPR-PAL [8]	SR	TrackB1	0.7439	0.6865	0.5076	0.1342	0.4846
		TrackB2	0.7458	0.6780	0.4844	0.1329	0.4764
TabStruct-Net [27]	CD	TrackB2	0.8380	0.6870	0.4410	0.1740	0.4977
	SR	TrackB2	0.8040	0.6330	0.3890	0.1540	0.4584
Proposed	CD	TrackB1	0.9067	0.8574	0.7643	0.4555	0.7219
		TrackB2	0.9235	0.8715	0.7538	0.4191	0.7148
	SR	TrackB1	0.8820	0.8290	0.7253	0.3846	0.6786
		TrackB2	0.9000	0.8403	0.7017	0.3443	0.6665

Table 4. Table detection results on cTDaR Archival datasets.

Method	Subset	F1@IoU				WAvg. F1
		0.6	0.7	0.8	0.9	
TableRadar [8]	TrackA	0.9716	0.9641	0.9527	0.9112	0.9467
NLPR-PAL [8]	TrackA	0.9624	0.9549	0.9351	0.8947	0.9333
Lenovo Ocean [8]	TrackA	0.9407	0.9369	0.9140	0.8604	0.9086
Proposed	TrackA	0.9715	0.9639	0.9564	0.9146	0.9486
	TrackB1	0.9215	0.9063	0.8861	0.8304	0.8812
	TrackB2	0.9362	0.9255	0.9255	0.8989	0.9197

the following, we choose to use the combination of cell mask and solid ruling line mask for cell region segmentation.

Adaptive Scaling. Scale prediction results of each image in test set TrackB are shown in Fig. 5. For most images in TrackB, our SPNet can provide precise scale prediction results. It's worth noting that S_{2L} is more difficult to predict compared with S_{2C}, thus the data points of S_{2L} are more scattered in Fig. 5.

Experimental results on different subsets of TrackB under different scaling strategies can be found in Table 2. We only report WAvg.F1 for saving space. As we can see from the first three rows of Table 2, enlarging image sizes can obtain higher cell segmentation performance for tables with small cells or full ruling lines, but will degrade the performance for tables with big cells or sparse ruling lines. On the contrary, adopting adaptive scaling for each image can improve the overall performance for the entire test set, which can be seen from row 3~5 of Table 2. To be concrete, adaptive scaling using strategy 1 can improve the cell segmentation performance for tables with small cells without degrading too much the performance for tables with big cells; and adaptive scaling using strategy 2 can improve the cell segmentation performance for tables with full

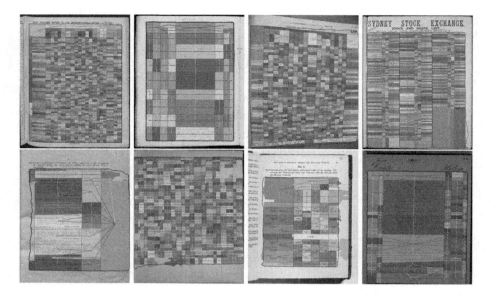

Fig. 6. Table detection and structure recognition results. The first row: correct ones; the second row: results with errors.

ruling lines without degrading too much the performance for tables with sparse ruling lines. Adaptive scaling using strategy 3 can combine the strengths of both thus improve the overall performance.

What's more, compared with resizing all images to the same shorter length, our adaptive scaling strategies can not only improve the overall performance, but also reduce the memory and computation overhead on average. This can be proved by the average image sizes after scaling shown in the second column of Table 2. Adaptive scaling only requires almost the same memeory and computation cost as resizing all images to shorter length of 2048 pixels, but can obtain higher performance than resizing all images to shorter length of 3072 pixels.

Comparison with the State-of-the-Art. In the end, we also compare our results with the state-of-the-art methods, shown in Table. 3 and 4. For table structure recognition, NLPR-PAL [8] extract cell regions based only on ruling line segmentation results under the fixed scaling strategy. Our method utilizes both cell mask and line mask for cell extraction and take advantage of adaptive scaling to improve the overall performance, thus outperforming their results by large margin. TabStruct-Net [27] extract cell regions in object detection manner and they can only predict rectangular cell bounding boxes. However, cells in archival documents can actually have arbitrary shapes due to the deformation of document images, which greatly limits the performance of their method. On the contrary, our instance segmentation based method can segment cell regions of arbitrary shapes.

We also report table detection results in Table 4. TableRadar, NLPR-PAL and Lenovo Ocean are three top ranking methods from the ICDAR 2019 cTDaR competition [8]. Our method can achieve higher or comparable results on cTDaR Archival datasets compared with these state-of-the-art methods. It is worth noting that the TableRadar method can only detect up-right rectangular tables because it is based on Faster R-CNN [29], while our method has no such limitation. Although our approach focus more on table structure recognition which is a much more challenging task, we find table detection results produced by our method are also fairly good.

In Fig. 6 we present some representative table analysis results. For most kinds of tables with various structures and appearances, our model can generate correct table detection and structure recognition results. However, there still exist some errors especially for tables with neither ruling lines nor tabular structures or tables with too many empty cells or ambiguous cells, showing that structure analysis for complex tables remains a very challenging task worthy further study.

A promising direction for improvement may be utilizing text line information for cell extraction. Original annotations of cTDaR Archival dataset are based on cell boundaries, but lots of pixels inside cells actually belong to background. This inconsistence brings difficulty for model training. On the contrary, text line regions are more compact and easier to extract and cluster, which may facilitate the procedure of cell extraction.

6 Conclusion

In this paper, we propose an instance segmentation based approach for archival table structure recognition utilizing both foreground cell content and background ruling line clues. Moreover, we design an adaptive image scaling method based on average cell size and density of ruling lines inside each document image to overcome the influence from inconsistent image scales. With adaptive scaling, each image is resized to an optimal size which can not only improve overall model performance but also reduce memory and computing overhead on average. Experimental results show that the proposed method can obtain correct table structure recognition results for most kinds of tables and outperform previous state-of-the-art methods. In the future, we plan to utilize more information such as text lines to improve the performance for complex tables and extend our adaptive scaling strategy to more document analysis tasks.

Acknowledgments. This work has been supported by the National Key Research and Development Program Grant 2020AAA0109702, the National Natural Science Foundation of China (NSFC) grants 61733007, 61721004.

References

1. Agarwal, M., Mondal, A., Jawahar, C.: Cdec-net: composite deformable cascade network for table detection in document images. arXiv:2008.10831 (2020)

2. Arias, J.F., Kasturi, R.: Efficient extraction of primitives from line drawings composed of horizontal and vertical lines. Mach. Vis. Appl. **10**(4), 214–221 (1997)
3. Bai, M., Urtasun, R.: Deep watershed transform for instance segmentation. In: CVPR, pp. 5221–5229 (2017)
4. Chen, L.C., Papandreou, G., Schroff, F., Adam, H.: Rethinking atrous convolution for semantic image segmentation. arXiv:1706.05587 (2017)
5. Chi, Z., Huang, H., Xu, H.D., Yu, H., Yin, W., Mao, X.L.: Complicated table structure recognition. arXiv:1908.04729 (2019)
6. Chin, T.W., Ding, R., Marculescu, D.: Adascale: towards real-time video object detection using adaptive scaling. arXiv:1902.02910 (2019)
7. Deng, Y., Rosenberg, D., Mann, G.: Challenges in end-to-end neural scientific table recognition. In: ICDAR, pp. 894–901. IEEE (2019)
8. Gao, L., et al.: Icdar 2019 competition on table detection and recognition (ctdar). In: ICDAR, pp. 1510–1515. IEEE (2019)
9. Gatos, B., Danatsas, D., Pratikakis, I., Perantonis, S.J.: Automatic table detection in document images. In: Singh, S., Singh, M., Apte, C., Perner, P. (eds.) ICAPR 2005. LNCS, vol. 3686, pp. 609–618. Springer, Heidelberg (2005). https://doi.org/10.1007/11551188_67
10. Ghanmi, N., Belaid, A.: Table detection in handwritten chemistry documents using conditional random fields. In: ICFHR, pp. 146–151. IEEE (2014)
11. Gilani, A., Qasim, S.R., Malik, I., Shafait, F.: Table detection using deep learning. In: ICDAR, vol. 1, pp. 771–776. IEEE (2017)
12. Girshick, R.: Fast r-cnn. In: ICCV (2015)
13. Hayder, Z., He, X., Salzmann, M.: Boundary-aware instance segmentation. In: CVPR, pp. 5696–5704 (2017)
14. He, K., Gkioxari, G., Dollár, P., Girshick, R.: Mask r-cnn. In: ICCV, pp. 2961–2969 (2017)
15. He, K., Zhang, X., Ren, S., Sun, J.: Deep residual learning for image recognition. In: CVPR, pp. 770–778 (2016)
16. He, W., Zhang, X.Y., Yin, F., Luo, Z., Ogier, J.M., Liu, C.L.: Realtime multi-scale scene text detection with scale-based region proposal network. Pattern Recogn. **98**, 107026 (2020)
17. Huang, G., Liu, Z., Van Der Maaten, L., Weinberger, K.Q.: Densely connected convolutional networks. In: CVPR, pp. 4700–4708 (2017)
18. Huang, Y., et al.: A yolo-based table detection method. In: ICDAR, pp. 813–818. IEEE (2019)
19. Khan, S.A., Khalid, S.M.D., Shahzad, M.A., Shafait, F.: Table structure extraction with bi-directional gated recurrent unit networks. In: ICDAR, pp. 1366–1371. IEEE (2019)
20. Li, M., Cui, L., Huang, S., Wei, F., Zhou, M., Li, Z.: Tablebank: Table benchmark for image-based table detection and recognition. arXiv:1903.01949 (2019)
21. Li, X.H., Yin, F., Liu, C.L.: Page object detection from pdf document images by deep structured prediction and supervised clustering. In: ICPR, pp. 3627–3632. IEEE (2018)
22. Li, X.-H., Yin, F., Liu, C.-L.: Page segmentation using convolutional neural network and graphical model. In: Bai, X., Karatzas, D., Lopresti, D. (eds.) DAS 2020. LNCS, vol. 12116, pp. 231–245. Springer, Cham (2020). https://doi.org/10.1007/978-3-030-57058-3_17
23. Li, X.H., Yin, F., Xue, T., Liu, L., Ogier, J.M., Liu, C.L.: Instance aware document image segmentation using label pyramid networks and deep watershed transformation. In: ICDAR, pp. 514–519. IEEE (2019)

24. Lin, T.Y., Dollár, P., Girshick, R., He, K., Hariharan, B., Belongie, S.: Feature pyramid networks for object detection. In: CVPR, pp. 2117–2125 (2017)
25. Long, J., Shelhamer, E., Darrell, T.: Fully convolutional networks for semantic segmentation. In: CVPR, pp. 3431–3440 (2015)
26. Qasim, S.R., Mahmood, H., Shafait, F.: Rethinking table recognition using graph neural networks. In: ICDAR, pp. 142–147. IEEE (2019)
27. Raja, S., Mondal, A., Jawahar, C.: Table structure recognition using top-down and bottom-up cues. arXiv:2010.04565 (2020)
28. Redmon, J., Farhadi, A.: Yolov3: An incremental improvement. arXiv:1804.02767 (2018)
29. Ren, S., He, K., Girshick, R., Sun, J.: Faster r-cnn: Towards real-time object detection with region proposal networks. In: NIPS, pp. 91–99 (2015)
30. Riba, P., Dutta, A., Goldmann, L., Fornés, A., Ramos, O., Lladós, J.: Table detection in invoice documents by graph neural networks. In: ICDAR, pp. 122–127. IEEE (2019)
31. Richardson, E., et al.: It's all about the scale-efficient text detection using adaptive scaling. In: WACV, pp. 1844–1853 (2020)
32. Seo, W., Koo, H.I., Cho, N.I.: Junction-based table detection in camera-captured document images. IJDAR **18**(1), 47–57 (2015)
33. Siddiqui, S.A., Fateh, I.A., Rizvi, S.T.R., Dengel, A., Ahmed, S.: Deeptabstr: deep learning based table structure recognition. In: ICDAR, pp. 1403–1409. IEEE (2019)
34. Siddiqui, S.A., Khan, P.I., Dengel, A., Ahmed, S.: Rethinking semantic segmentation for table structure recognition in documents. In: ICDAR, pp. 1397–1402. IEEE (2019)
35. Simonyan, K., Zisserman, A.: Very deep convolutional networks for large-scale image recognition. arXiv:1409.1556 (2014)
36. Sun, N., Zhu, Y., Hu, X.: Faster r-cnn based table detection combining corner locating. In: ICDAR, pp. 1314–1319. IEEE (2019)
37. Tensmeyer, C., Morariu, V.I., Price, B., Cohen, S., Martinez, T.: Deep splitting and merging for table structure decomposition. In: ICDAR, pp. 114–121. IEEE (2019)
38. Tseng, L.Y., Chen, R.C.: Recognition and data extraction of form documents based on three types of line segments. Pattern Recogn. **31**(10), 1525–1540 (1998)
39. Wang, W., et al.: Shape robust text detection with progressive scale expansion network. In: CVPR, pp. 9336–9345 (2019)
40. Xue, W., Li, Q., Tao, D.: Res2tim: reconstruct syntactic structures from table images. In: ICDAR, pp. 749–755. IEEE (2019)
41. Zheng, Y., Liu, C., Ding, X., Pan, S.: Form frame line detection with directional single-connected chain. In: ICDAR, pp. 699–703. IEEE (2001)
42. Zhong, X., ShafieiBavani, E., Yepes, A.J.: Image-based table recognition: data, model, and evaluation. arXiv:1911.10683 (2019)
43. Zhong, X., Tang, J., Yepes, A.J.: Publaynet: largest dataset ever for document layout analysis. In: ICDAR, pp. 1015–1022. IEEE (2019)

Document Analysis Systems

LGPMA: Complicated Table Structure Recognition with Local and Global Pyramid Mask Alignment

Liang Qiao[1], Zaisheng Li[1], Zhanzhan Cheng[1,2(✉)], Peng Zhang[1], Shiliang Pu[1], Yi Niu[1], Wenqi Ren[1], Wenming Tan[1], and Fei Wu[2]

[1] Hikvision Research Institute, Hangzhou, China
{qiaoliang6,lizaisheng,chengzhanzhan,zhangpeng23,pushiliang.hri,niuyi, renwenqi,tanwenming}@hikvision.com
[2] Zhejiang University, Hangzhou, China
wufei@cs.zju.edu.cn

Abstract. Table structure recognition is a challenging task due to the various structures and complicated cell spanning relations. Previous methods handled the problem starting from elements in different granularities (rows/columns, text regions), which somehow fell into the issues like lossy heuristic rules or neglect of empty cell division. Based on table structure characteristics, we find that obtaining the aligned bounding boxes of text region can effectively maintain the entire relevant range of different cells. However, the aligned bounding boxes are hard to be accurately predicted due to the visual ambiguities. In this paper, we aim to obtain more reliable aligned bounding boxes by fully utilizing the visual information from both text regions in proposed local features and cell relations in global features. Specifically, we propose the framework of *Local and Global Pyramid Mask Alignment*, which adopts the soft pyramid mask learning mechanism in both the local and global feature maps. It allows the predicted boundaries of bounding boxes to break through the limitation of original proposals. A pyramid mask re-scoring module is then integrated to compromise the local and global information and refine the predicted boundaries. Finally, we propose a robust table structure recovery pipeline to obtain the final structure, in which we also effectively solve the problems of empty cells locating and division. Experimental results show that the proposed method achieves competitive and even new state-of-the-art performance on several public benchmarks.

Keywords: Table structure recognition · Aligned bounding box · Empty cell

1 Introduction

Table is one of the rich-information data formats in many real documents like financial statements, scientific literature, purchasing lists, etc. Besides the text

L. Qiao and Z. Li—Contributed equally. Supported by National Key R&D Program of China (Grant No. 2018YFC0831601).

J. Lladós et al. (Eds.): ICDAR 2021, LNCS 12821, pp. 99–114, 2021.
https://doi.org/10.1007/978-3-030-86549-8_7

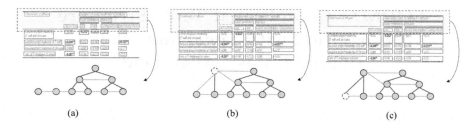

Fig. 1. (a) The visualized results without considering empty cells. (b) The ground-truth of aligned bounding boxes and nodes relations. (c) A false example due to the ambiguity between the empty cell and cross-column cell. The cells and their relations are represented as nodes and connected lines (red: vertical, green: horizontal). Empty cells are displayed in dashed circles. (Color figure online)

content, the table structure is vital for people to do the key information extraction. Thus, table structure recognition [4,5,10,21,30,34,39] becomes one of the important techniques in current document understanding systems.

From the global perspective, early table structure recognition processes usually depend on the detection of the grid's boundaries [18,19]. However, these methods can not handle tables without grid boundaries, such as three-line tables. Though recent works [22,30–32] attempt to predict row/column regions or even invisible grid lines [33], they are limited to handle tables that cross span multiple rows/columns. The row/column splitting operation might also cut cells that contain text in multiple lines.

Another group of methods solves the above problems in a bottom-up way to firstly detect the text blocks' positions and then recover the bounding-boxes' relations by heuristic rules [38] or GNN (Graph Neural Networks) [2,14,24,26,29]. However, rules designed based on bounding boxes of text regions are vulnerable to handling complicated matching situations. GNN-based methods not only bring extra network cost but also depend on more expensive training cost such as the data volume. Another issue is that these methods are difficult to obtain the empty cells because they usually fall into the visual ambiguity problem with the cross-row/column cells. The prediction of empty cells directly affects the correctness of table structure, as illustrated in Fig. 1(a). Moreover, how to split or merge these empty regions is still a challenging problem that cannot be neglected, because different division results will generate different editable areas when the image is transferred into digital format.

Notice that the structure of the table itself is a human-made rule-based data form. Under the situation that tables are without visual rotation or perspective transformation, if we could obtain all of the perfect aligned cell regions rather than the text regions [26], the structure inference will be easy and almost lossless, as illustrated in Fig. 1(b). Nevertheless, acquiring such information is not easy. On the one hand, the annotations of text regions [2,5,39] are much easier to get than cell regions. On the other hand, the aligned boxes are difficult to be accurately learned since there is usually no visible texture of boundaries at the region's periphery. Multi-row/column cells are easy to be confused with the empty cell regions. For example, in Fig. 1(c), the network usually falls into the situation that the predicted aligned boxes are not large enough and results in the wrong cell matching. Although [26] designs an alignment loss to assist the bounding boxes learning, it only considers the relative relations between boxes and fails to capture the cell's absolute coverage area.

In this paper, we aim to train the network to obtain more reliable aligned cell regions and solve the problems of empty cell generation and partition in one model. Observing that people perceive visual information from both local text region and global layout when they read, we propose a uniform table structure recognition framework to compromise the benefits from both local and global information, called LGPMA (Local and Global Pyramid Mask Alignment) Network. Specifically, the model simultaneously learns a local Mask-RCNN-based [6] aligned bounding boxes detection task and a global segmentation task. In both tasks, we adopt the pyramid soft mask supervision [17] to help obtain more accurate aligned bounding boxes. In LGPMA, the local branch (LPMA) acquires more reliable text region information through visible texture perceptron, while the global branch (GPMA) can learn more legible spatial information of cells' range or division. The two branches help the network learn better-fused features via jointly learning and effectively refine the detected aligned bounding boxes through a proposed mask re-scoring strategy. Based on the refined results, we design a robust and straightforward table structure recovery pipeline, which can effectively locate empty cells and precisely merge them according to the guidance of global segmentation.

The major contributions of this paper are as follows: (1) We propose a novel framework called LGPMA Network that compromises the visual features from both local and global perspectives. The model makes full use of the information from the local and global features through a proposed mask re-scoring strategy, which can obtain more reliable aligned cell regions. (2) We introduce a uniform table structure recovering pipeline, including cell matching, empty cell searching, and empty cell merging. Both non-empty cells and empty cells can be located and split efficaciously. (3) Extensive experiments show that our method achieves competitive and even state-of-the-art results on several popular benchmarks.

2   Related Works

Traditional table recognition researches mainly worked with hand-crafted features and heuristic rules [3,8,10,18,19,34]. These methods are mostly applied to simple table structures or specific data formats, such as PDFs. The early techniques about table detection and recognition can be found in the comprehensive survey [37]. With the great success of deep neural network in computer vision field, works began to focus on the image-based table with more general structures [9,13,14,21,23,24,30,33,36]. According to the basic components granularities, we roughly divide previous methods into two types: global-object-based methods and local-object-based methods.

Global-object-based methods mainly focus on the characteristics of global table components and mostly started from row/column or grid boundaries detection. Works of [30–32] firstly obtain the rows and columns regions using the detection or segmentation models and then intersect these two regions to obtain the grids of cells. [22] handles the table detection and table recognition tasks in an end-to-end manner by the table region mask learning and table's row/column mask learning. [33] detects the rows and columns by learning the interval areas' segmentation between rows/columns and then predicting the indicator to merge the separated cells.

There also exist some methods [13,39] that directly perceive the whole image information and output table structures as text sequence in an encoder-decoder framework. Although these methods look graceful and entirely avoid human being involved, the models are usually challenging to be trained and rely on a large amount of training data. Global-object-based methods usually have difficulties in handling various complicated table structures, such as cells spanning multiple rows/columns or containing text in multi-lines.

Local-object-based methods begin from the smallest fundamental element, cells. Given the cell-level text region annotation, the text detection task is relatively easy to finish by the general detection methods like Yolo [27], Faster R-CNN [28], etc. After that, a group of methods [23,36,38] tries to recover the cell relations based on some heuristic rules and algorithms. Another type of methods [2,11,14,24,26] treat the detected boxes as nodes in a graph and attempt to predict the relations based on techniques of Graph Neural Networks [29]. [14] predicts the relations between nodes in three classes (the horizontal connection, the vertical connection, no connection) using several features such as visual features, text positions, word embedding, etc. [2] adopts graph attention mechanism to enhance the predicting accuracy. [24] alleviates the problem of large graph nodes numbers by the pair sampling strategy. The above three works [2,14,24] also published new table datasets for this research area. Since there is no empty cell detected, local-object based-methods usually fall into empty cell ambiguity.

In this paper, we try to compromise the advantages of both global and local features. Based on the local detection results, we integrate the global information to refine the detected bounding boxes and provide a straightforward guide for empty cell division.

3   Methodology

3.1   Overview

We propose the model LGPMA, whose overall workflow is shown in Fig. 2.

The model is built based on the existing Mask-RCNN [6]. The bounding box branch directly learns the detection task of aligned bounding boxes for non-empty cells. The network simultaneously learns a Local Pyramid Mask Alignment (LPMA) task based on the local feature extracted by the RoI-Align operation and a Global Pyramid Mask Alignment (GPMA) task based on the global feature map.

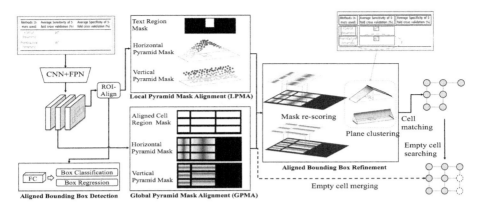

Fig. 2. The workflow of LGPMA. The network simultaneously learns a local aligned bounding boxes detection task (LPMA) and a global segmentation task (GPMA). We adopt the pyramid mask learning mechanisms in both branches and use a mask re-scoring strategy to refine the predicted bounding boxes. Finally, the table structure can be uniformly recovered by a pipeline, including cell matching, empty cell searching, and empty cell merging.

In LPMA, in addition to the binary segmentation task that learns the text region mask, the network is also trained with the pyramid soft mask supervision in both horizontal and vertical directions.

In GPMA, the network learns a global pyramid mask for all aligned bounding boxes of non-empty cells. To obtain more information about empty cell splitting, the network also learns the global binary segmentation task that considers both non-empty and empty cells.

A pyramid mask re-scoring module is then adopted to refine the predicted pyramid labels. The accurate aligned bounding boxes can be obtained by the process of plane clustering. Finally, a uniform structure recovering pipeline containing cell matching, empty cell searching, empty cell merging is integrated to obtain the final table structure.

3.2 Aligned Bounding Box Detection

The difficulty of accurate text region matching mainly comes from the covered range gap between text regions and the real cell regions. Real cell regions may contain empty spaces for row/column alignment, especially for those cells crossing span multiple rows/columns. Inspired by [26, 36], with the annotations of text regions and row/column indices, we can easily generate the aligned bounding box annotations according to the maximum box height/width in each row/column. The regions of aligned bounding boxes approximately equal to that of real cells. For the table images in print format and without visual rotation or perspective transformation, if we could obtain the aligned cell regions and assume there is

Fig. 3. (a) shows the original aligned bounding box (blue) and text region box (red). (b) shows the pyramid mask labels in horizontal and vertical direction, respectively. (Color figure online)

no empty cell, it is easy to infer the cell relations according to the coordinate overlapping information in horizontal and vertical directions.

We adopt Mask-RCNN [6] as the base model. In the bounding box branch, the network is trained based on the aligned bounding box supervision. However, the aligned bounding box learning is not easy because cells are easy to be confused with empty regions. Motivated by the advanced pyramid mask text detector [17], we find that using the soft-label segmentation may break through the proposed bounding box's limitation and provide more accurate aligned bounding boxes. To fully utilize the visual features from both local texture and global layout, we propose to learn the pyramid mask alignment information in these two folds simultaneously.

3.3 Local Pyramid Mask Alignment

In the mask branch, the model is trained to learn both a binary segmentation task and a pyramid mask regression task, which we call Local Pyramid Mask Alignment (LPMA).

The binary segmentation task is the same as the original model, in which only the text region is labeled as 1 and others are labeled as 0. The detected mask regions can be used in the following text recognition task.

For the pyramid mask regression, we assign the pixels in the proposal bounding box regions with the soft-label in both horizontal and vertical directions, as shown in Fig. 3. The middle point of text will have the largest regressed target 1. Specifically, we assume the proposed aligned bounding box has the shape of $H \times W$. The top-left point and bottom right point of the text region are denoted as $\{(x_1, y_1), (x_2, y_2)\}$, respectively, where $0 \leq x_1 < x_2 \leq W$ and $0 \leq y_1 < y_2 \leq H$. Therefore, the target of the pyramid mask is in shape $\mathbb{R}^{2 \times H \times W} \in [0, 1]$, in which the two channels represent the target map of the horizontal mask and vertical mask, respectively. For every pixel (h, w), these two targets can be formed as:

$$t_h^{(w,h)} = \begin{cases} w/x_1 & w \leq x_{mid} \\ \frac{W-w}{W-x_2} & w > x_{mid} \end{cases}, \quad t_v^{(w,h)} = \begin{cases} h/y_1 & h \leq y_{mid} \\ \frac{H-h}{H-y_2} & h > y_{mid} \end{cases}, \quad (1)$$

where $0 \leq w < W$, $0 \leq h < H$, and $x_{mid} = \frac{x_1+x_2}{2}$, $y_{mid} = \frac{y_1+y_2}{2}$. In this way, every pixel in the proposal region takes part in predicting the boundaries.

3.4   Global Pyramid Mask Alignment

Although LPMA allows the predicted mask to break through the proposal bounding boxes, the local region's receptive fields are limited. To determine the accurate coverage area of a cell, the global feature might also provide some visual clues. Inspired by [25,40], learning the offsets of each pixel from a global view could help locate more accurate boundaries. However, bounding boxes in cell-level might be varied in width-height ratios, which leads to the unbalance problem in regression learning. Therefore, we use the pyramid labels as the regressing targets for each pixel, named Global Pyramid Mask Alignment (GPMA).

Like LPMA, the GPMA learns two tasks simultaneously: a global segmentation task and a global pyramid mask regression task. In the global segmentation task, we directly segment all aligned cells, including non-empty and empty cells. The ground-truth of empty cells are generated according to the maximum height/width of the non-empty cells in the same row/column. Notice that only this task learns empty cell division information since empty cells don't have visible text texture that might influence the region proposal networks to some extent. We want the model to capture the most reasonable cell division pattern during the global boundary segmentation according to the human's reading habit, which is reflected by the manually labeled annotations. For the global pyramid mask regression, since only the text region could provide the information of distinct 'mountain top,' all non-empty cells will be assigned with the soft labels similar to LPMA. All of the ground-truths of aligned bounding boxes in GPMA will be shrunk by 5% to prevent boxes from overlapping.

3.5   Optimization

The proposed network is trained end-to-end with multiple optimization tasks. The global optimization can be written as:

$$\mathcal{L} = \mathcal{L}_{rpn} + \lambda_1(\mathcal{L}_{cls} + \mathcal{L}_{box}) + \lambda_2(\mathcal{L}_{mask} + \mathcal{L}_{LPMA}) + \lambda_3(\mathcal{L}_{seg} + \mathcal{L}_{GPMA}), \quad (2)$$

where $\mathcal{L}_{rpn}, \mathcal{L}_{cls}, \mathcal{L}_{box}, \mathcal{L}_{mask}$ are the same losses with that of Mask-RCNN, which represent the region proposal network loss, the bounding box classification loss, the bounding boxes regression loss and the segmentation loss of mask in proposals, respectively. $\mathcal{L}_{seg}$ is the global binary segmentation loss that is implemented in Dice coefficient loss [20], $\mathcal{L}_{LPMA}$ and $\mathcal{L}_{GPMA}$ are the pyramid label regression losses which are optimized by pixel-wise L1 loss. $\lambda_1, \lambda_2, \lambda_3$ are weighted parameters.

3.6   Inference

The inference process can be described in two stages. We first obtain the refined aligned bounding boxes according to the pyramid mask prediction and then generate the final table structure by the proposed structure recovery pipeline.

Aligned Bounding Box Refine. In addition to the benefits generated via joint training, the local and global features also exhibit various advantages in object perceiving [35]. In our setting, we find that local features predict more reliable text region masks, while global prediction can provide more credible long-distance visual information. To compromise both levels' merits, we propose a pyramid mask re-scoring strategy to compromise predictions from LPMA and GPMA. For any proposal region with local pyramid mask prediction, we add the information that comes from the global pyramid mask to adjust these scores. We use some dynamic weights to balance the impacts from LPMA and GPMA.

Specifically, for a predicted aligned bounding box $B = \{(x_1, y_1), (x_2, y_2)\}$, we firstly obtain the bounding box of the text region mask, denoted as $B_t = \{(x_1', y_1'), (x_2', y_2')\}$. Then, we can find a matched connected region $P = \{p_1, p_2, ..., p_n\}$ in the global segmentation map, where $p = (x, y)$ represents a pixel. We use $P_o = \{p | x_1 \leq p.x \leq x_2, y_1 \leq p.y \leq y_2, \forall p \in P\}$ to represent the overlap region. Then the predicted pyramid label of point $(x, y) \in P_o$ can be re-scored as follows.

$$
F(x) = \begin{cases} \frac{x-x_1}{x_{mid}-x_1} F_{hor}^{(L)}(x, y) + \frac{x_{mid}-x}{x_{mid}-x_1} F_{hor}^{(G)}(x, y) & x_1 \leq x \leq x_{mid} \\ \frac{x-x_2}{x_{mid}-x_2} F_{hor}^{(L)}(x, y) + \frac{x_{mid}-x}{x_{mid}-x_2} F_{hor}^{(G)}(x, y) & x_{mid} < x \leq x_2 \end{cases}, \quad (3)
$$

$$
F(y) = \begin{cases} \frac{y-y_1}{y_{mid}-y_1} F_{ver}^{(L)}(x, y) + \frac{y_{mid}-y}{y_{mid}-y_1} F_{ver}^{(G)}(x, y) & y_1 \leq y \leq y_{mid} \\ \frac{y-y_2}{y_{mid}-y_2} F_{ver}^{(L)}(x, y) + \frac{y_{mid}-y}{y_{mid}-y_2} F_{ver}^{(G)}(x, y) & y_{mid} < y \leq y_2 \end{cases}, \quad (4)
$$

where $x_{mid} = \frac{x_1'+x_2'}{2}, y_{mid} = \frac{y_1'+y_2'}{2}, F_{hor}^{(L)}(x, y), F_{hor}^{(G)}(x, y), F_{ver}^{(L)}(x, y), F_{ver}^{(G)}(x, y)$ are the local horizontal, global horizontal, local vertical and global vertical pyramid label prediction, respectively.

Next, for any proposal region, the horizontal and vertical pyramid mask labels (corresponding to the z-coordinate) can be used to fit two planes in the 3-dimensional space, respectively. All the four planes' intersection lines with the zero plane are the refined boundaries. For example, to refine the right boundary of the aligned box, we select all pixels that $P_r = \{p | x_{mid} \leq p.x \leq x_2, p \in P_o\}$ with the refined pyramid mask prediction $F(x, y)$ to fit the plane. If we formed the plane as $ax + by + c - z = 0$, using the least square method, the problem is equal to minimize the equation of:

$$
\min \sum_{y_i=y_1}^{y_2} \sum_{x_i=x_{mid}}^{x_2} (ax_i + by_i + c - F(x_i, y_i))^2, \quad \forall p = (x_i, y_i) \in P_r. \quad (5)
$$

The parameters of a, b, c can be calculated by the matrix as follows:

$$
\begin{pmatrix} a \\ b \\ c \end{pmatrix} = \begin{bmatrix} \sum x_i^2 & \sum x_i y_i & \sum x_i \\ \sum x_i y_i & \sum y_i^2 & \sum y_i \\ \sum x_i & \sum y_i & ||P_o|| \end{bmatrix}^{-1} \begin{pmatrix} x_i F(x_i, y_i) \\ y_i F(x_i, y_i) \\ F(x_i, y_i) \end{pmatrix}, \quad (6)
$$

where $||.||$ is the set size. Then we calculate the intersection line between the fitting plane with the plane of $z = 0$. Given that the bounding boxes are axis-aligned, we calculate the refined x-coordinate as the average value:

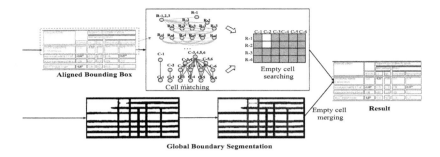

Fig. 4. The illustration of table structure recovery pipeline.

$$x_{refine} = -\frac{1}{y_2 - y_1 + 1} \sum_{y_i=y_1}^{y_2} \frac{by_i + c}{a} \tag{7}$$

Similarly, we can obtain the other three refined boundaries. Notice that the refining process can optionally be conducted iteratively refer to [17].

Table Structure Recovery. Based on the refined aligned bounding boxes, the table structure recovery pipeline aims to obtain the final table structure, including three steps: *cell matching, empty cell searching and empty cell merging*, as illustrated in Fig. 4.

Cell Matching. In the situation that all of the aligned bounding boxes are axis-aligned, the cells matching process is pretty simple but robust. Following the same naming convention with [2,14,24], the connecting relations can be divided into horizontal and vertical types. The main idea is that if two aligned bounding boxes has enough overlap in x/y-coordinate, we will match them in vertical/horizontal direction. Mathematically, for every two aligned bounding boxes, $\{(x_1, y_1), (x_2, y_2)\}$ and $\{(x_1', y_1'), (x_2', y_2')\}$, they will be horizontally connected if $y_1' \leq \frac{y_1+y_2}{2} \leq y_2'$ or $y_1 \leq \frac{y_1'+y_2'}{2} \leq y_2$. Similarly, they will be vertically connected if $x_1' \leq \frac{x_1+x_2}{2} \leq x_2'$ or $x_1 \leq \frac{x_1'+x_2'}{2} \leq x_2$.

Empty Cell Searching. After obtaining the relations between the detected aligned bounding boxes, we treat them as nodes in a graph, and the connected relations are edges. All of the nodes in the same row/column make up a complete subgraph. Inspired by [24], we adopt the algorithm of Maximum Clique Search [1] to find all maximum cliques in the graph. Take the row searching process as an example, every node that belongs to the same row will be in the same clique. For the cell that crosses span multiple rows, the corresponding node will appear multiple times in different cliques. After sorting these cliques by the average y-coordinate, we can easily label each node with its row index. Nodes that appear in multiple cliques will be labeled with multiple row indices. We can easily find those vacant positions, which are corresponding to the empty cells.

Empty Cell Merging. By now, we have obtained the empty cells at the smallest level (occupies 1 row and 1 column). To merge these cells more feasibly, we first assign the single empty cells with the aligned bounding box shape as the cell's maximum height/width in the same row/column. Thanks to the visual clues learned by the global segmentation task, we can design the simple merging strategy following the segmentation result. We compute the ratio of pixels that are predicted as 1 in the interval region for every two neighbor empty cells, as the red region illustrated in Fig. 4. If the ratio is larger than the preset threshold, we will merge these two cells. As we can see, the empty regions' visual ambiguity always exists, and the segmentation task can hardly be learned perfectly. That is why many segmentation-based methods [22–24] struggle with complicated post-processing, such as fracture completion and threshold setting. The proposed method straightforwardly adopts the original visual clue provided by global segmentation and uses pixel voting to obtain a more reliable result.

4   Experiments

4.1   Datasets

We evaluate our proposed framework on following popular benchmarks that contain the annotations of both text content bounding boxes and cell relations.

ICDAR 2013 [5]. This dataset contains 98 training samples and 156 testing samples that cropped from the PDF of government reports.

SciTSR [2]. This dataset contains 12,000 training images and 3,000 testing images cropped from PDF of scientific literature. Authors also select a subset of complicated samples that contains 2,885 training images and 716 testing images, called SciTSR-COMP.

PubTabNet [39]. It is a large-scale complicated table collection that contains 500,777 training images, 9,115 validating images and 9,138 testing images. This dataset contains a large amount of three-lines tables with multi-row/column cells, empty cells, etc.

4.2   Implementation Details

All experiments are implemented in Pytorch with 8×32 GB-Tesla-V100 GPUs. The deep features are extracted and aggregated through the backbone of ResNet-50 [7] with Feature Pyramid Network (FPN) [15]. The weights of the backbone are initialized from the pre-trained model of MS-COCO [16]. In LPMA, the model generates anchors in six different ratios $[1/20, 1/10, 1/5, 1/2, 1, 2]$ for capturing the different shapes of bounding boxes. The Non-Maximal suppression (NMS) IoU threshold of RCNN is 0.1 in the testing phase.

For all benchmarks, the model is trained by the SGD optimizer with batch-size = 4, momentum = 0.9, and weight-decay = 1×10^{-4}. The initial learning ratio of 1×10^{-2} is divided by 10 every 5 epochs. The model's training on SciTSR

Table 1. Results on ICDAR 2013, SciTSR, SciTSR-COMP datasets. P, R, F1 represent Precision, Recall, F1-Score, respectively. Symbol of † means pre-trained data are used.

Methods	Training dataset	ICDAR 2013			SciTSR			SciTSR-COMP		
		P	R	F1	P	R	F1	P	R	F1
DeepDeSRT [30]	–	0.959	0.874	0.914	0.906	0.887	0.890	0.863	0.831	0.846
Split [33]	Private	0.869	0.866	0.868	–	–	–	–	–	–
DeepTabStR [31]	ICDAR 2013	0.931	0.930	0.930	–	–	–	–	–	–
Siddiqui et al. [32]	Synthetic 500k	0.934	0.934	0.934	–	–	–	–	–	–
ReS2TIM [36]	ICDAR 2013†	0.734	0.747	0.740	–	–	–	–	–	–
GTE [38]	ICDAR 2013†	0.944	0.927	0.935	–	–	–	–	–	–
GraphTSR [2]	SciTSR	0.885	0.860	0.872	0.959	0.948	0.953	0.964	0.945	0.955
TabStruct-Net [26]	SciTSR	0.915	0.897	0.906	0.927	0.913	0.920	0.909	0.882	0.895
LGPMA	SciTSR	0.930	0.977	0.953	**0.982**	**0.993**	**0.988**	**0.973**	**0.987**	**0.980**
LGPMA	ICDAR 2013†	**0.967**	**0.991**	**0.979**	–	–	–	–	–	–

and PubTabNet lasts for 12 epochs, and the fine-tuning process on ICDAR 2013 lasts for 25 epochs. We also randomly scale the longer side of the input images to the lengths in the range [480, 1080] for all training processes. In the testing phase, we set the longer side of the input image as 768. We empirically set all weight parameters as $\lambda_1 = \lambda_2 = \lambda_3 = 1$.

4.3   Results on Table Structure Recognition Benchmarks

We first conduct experiments on the datasets of ICDAR 2013 and SciTSR, and the evaluation metric follows [5] (counting the micro-averaged correctness of neighboring relations). Notice that since ICDAR 2013 has very few samples, many previous works used different training or pre-training data. To be comparable to [2] and [26], our model is only trained by the training set of SciTSR. We also report the result of the model that is then fine-tuned on the training set of ICDAR2013, labeled by †. The results of DeepDeSRT on SciTSR come from [2]. The results are demonstrated in Table 1, from which we can see that the proposed LGP-TabNet vastly surpasses previous advances on these three benchmarks by 4.4%, 3.5%, 2.5%, respectively. Beside, LGPMA shows no performance decline on the complicated testing dataset SciTSR-COMP, which demonstrates its powerful ability to perceive spatial relations.

We also test our model in a more challenging benchmark of PubTabNet, whose results are demonstrated in Table 2. Since the corresponding evaluation metric of TEDS [39] considers both table structure and text content, we simply adopt an attention-based model [12] to recognize the text recognition. In the results, our method surpasses the previous SOTA by 1.6 in TEDS. We also report the results that only considers the table structure, denoted as TEDS-Struc. The performance gap between TEDS-Struc and TEDS mainly comes from the recognition error and annotation ambiguities.

Table 2. Results on PubTabNet. TESDS-Struc only considers the table structures.

Methods	Training dataset	Tesing dataset	TEDS (All)	TEDS-struc. (All)
EDD [39]	PTN-train	PTN-val	88.3	–
TabStruct-Net [26]	SciTSR	PTN-val	90.1	–
GTE [38]	PTN-train	PTN-val	93.0	–
LGPMA (ours)	PTN-train	PTN-val	**94.6**	**96.7**

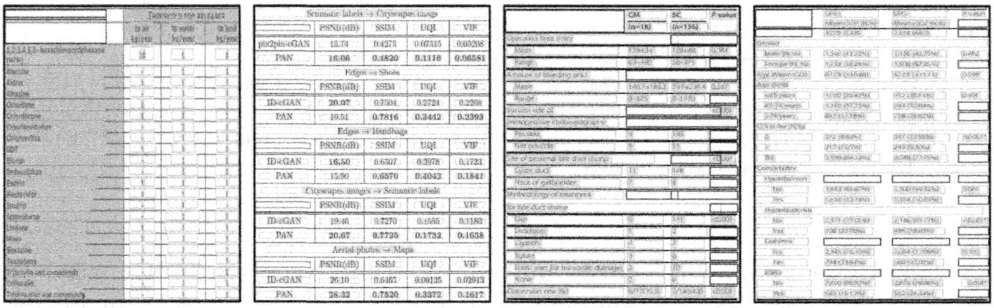

Fig. 5. Visualization results on ICDAR2013, SciTSR, PubTabNet. Green boxes are detected aligned bounding boxes, and blue boxes are empty cells generated by the proposed table structure recovery pipeline. (Color figure online)

Visualization Results. We demonstrate some of the visualization results in Fig. 5, in which the green boxes denote the predicted non-empty boxes, and blue boxes are the recovered empty boxes. We can see that our model can predict cells' accurate boundaries, even for those cross span multiple rows/columns.

Figure 6 demonstrates an example that is successfully predicted due to the correct refinement. We only show the LPMA and GPMA maps in the horizontal direction, where the LPMA map is generated by overlapping all proposals' local maps. In the initially proposed bounding boxes, some boxes do not have enough breadth, which would lead to wrong matching results. After refining by LPMA and GPMA, these boundaries can reach the more feasible positions. The empty cells can also be merged feasibly according to the predicted segmentation map during the table structure recovery.

4.4 Ablation Studies

We randomly select 60,000 training images and 1,000 validation images from PubTabNet to conduct the subsequent ablation studies.

Effectiveness of Aligned Bounding Box Detection. To verify the effectiveness of the designed modules, we conduct a group of ablation experiments, as shown in Table 3. Besides the TEDS-Struc metric, we also report the text region's detection results and aligned bounding boxes, where the detection IoU threshold is 0.7, and empty cells are ignored. From the results, we can easily

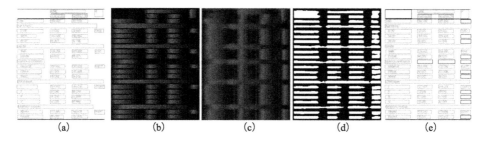

(a)   (b)   (c)   (d)   (e)

Fig. 6. Visualization of an example that is successfully refined. (a) The aligned bounding boxes before refinement. (b) LPMA (in horizontal). (c) GPMA (in horizontal). (d) Global binary segmentation. (e) Final result after refinement and empty cell merging.

Table 3. Ablation experiments on that how different modules effect the aligned bounding box detection

Models	Modules			Det of text regions			Det of non-empty aligned bounding boxes			TEDS- Struc.
	LPMA	GPMA	AL[26]	Precision	Recall	Hmean	Precision	Recall	Hmean	
Faster R-CNN				–	–	–	81.32	81.31	81.31	94.63
Mask R-CNN				91.71	91.53	91.62	81.83	81.82	81.83	94.65
	✓			91.92	91.66	91.79	84.29	84.10	84.20	95.22
		✓		91.98	91.50	91.74	83.48	83.18	83.33	95.04
	✓	✓		**92.27**	**91.86**	**92.06**	**85.14**	**84.77**	**84.95**	**95.53**
Mask R-CNN			✓	92.11	91.85	91.98	81.91	81.79	81.85	94.94
	✓	✓	✓	92.05	91.65	91.85	84.87	84.50	84.68	95.31

find that both LPMA and GPMA can vastly enhance the aligned bounding box detection performance, which is also proportional to the performance of TEDS-Struc. Although these modules are designed for aligned bounding boxes, they also slightly enhance the performance of text region detection results. The performance gap between text region detection and aligned bounding box detection again demonstrates the latter task is much more difficult and challenging.

We also evaluate the effectiveness of Alignment Loss (abbr. AL) proposed by [26]. Although solely adopting AL achieves better performance than the original Mask-RCNN, the performance is even lower than the best results of LGPMA when compromising all three modules. It means AL might bring adverse impact on LGPMA. Compared to AL, our proposed LGPMA can obtain more performance gain by 3.1% in aligned bounding box detection and 0.59 in TEDS-Struc.

Effectiveness of Table Structure Recovery. To verify the proposed table recovery pipeline's effectiveness, we conduct experiments to compare different empty cell merging strategies, as illustrated in Table 4. The evaluations contain the detection results that only consider empty cells, the detection results of all aligned bounding boxes, and TEDS-Struc. The strategy of using minimum cells means the direct results after *Empty Cell Searching*, and using maximum cells

Table 4. Ablation experiments on different empty cells merging strategy. The bottom part shows the results given non-empty aligned bounding boxes ground-truth.

Structure recovery strategies	with non-empty box GT?	Det of empty aligned bounding boxes			Det of all aligned bounding boxes			TEDS- Struc.
		Precision	Recall	Hmean	Precision	Recall	Hmean	
Minimum empty cells	✗	56.17	63.04	59.40	82.36	82.74	82.55	95.50
Maximum empty cells	✗	23.19	10.76	14.70	82.30	77.39	79.77	92.57
Proposed LGPMA	✗	**68.12**	**72.43**	**70.21**	**82.69**	**83.17**	**82.93**	**95.53**
Minimum cells	✓	96.14	97.00	96.56	99.60	99.69	99.64	99.52
Maximum cells	✓	41.13	15.22	22.22	97.58	91.19	94.28	95.60
Proposed LGPMA	✓	**97.26**	**97.68**	**97.47**	**99.85**	**99.88**	**99.87**	**99.77**

means merging all neighboring empty cells with the same height/width. From the results, we can see that our strategy using the visual information from GPMA can correctly merge many empty cells and obtain the highest performances in both detection and TED-Struc metrics. Compared with Strategy of *Minimum empty cells*, the promotion of *Empty Cell Merging by LGPMA* on TEDS-Struc is relatively small. This is because the number of empty cells minority, and most of them in this dataset are labeled in the smallest shapes. Nevertheless, our proposed empty cell merging strategy is more robust to adapt to any possibility.

Suppose the aligned bounding boxes of non-empty cells are detected perfectly, which equals the situation given ground-truths, as shown in the bottom part of Table 4. In this case, we can easily find that using the strategy of whether *Minimum cells* or the proposed *LGPMA* can almost achieve 100% accuracy in table structure recovery, and many errors come from the noisy labels. It demonstrates the robustness of our table structure recovery pipeline, and the performance mainly depends on the correctness of aligned bounding box detection.

5   Conclusion

In this paper, we present a novel framework for table structure recognition named LGPMA. We adopt the local and global pyramid mask learning to compromises advantages from both local texture and global layout information. In the inference stage, fusing the two levels' predictions via a mask re-scoring strategy, the network generates more reliable aligned bounding boxes. Finally, we propose a uniform table structure recovery pipeline to get the final results, which can also predict the feasible empty cell partition. Experimental results demonstrate our method has achieved the new state-of-the-art in three public benchmarks.

References

1. Bron, C., Kerbosch, J.: Finding all cliques of an undirected graph (algorithm 457). Commun. ACM **16**(9), 575–576 (1973)
2. Chi, Z., Huang, H., Xu, H., Yu, H., Yin, W., Mao, X.: Complicated table structure recognition. CoRR abs/1908.04729 (2019)
3. Doush, I.A., Pontelli, E.: Detecting and recognizing tables in spreadsheets. In: IAPR, pp. 471–478 (2010)
4. Gao, L., et al.: ICDAR 2019 competition on table detection and recognition (ctdar). In: ICDAR, pp. 1510–1515 (2019)
5. Göbel, M.C., Hassan, T., Oro, E., Orsi, G.: ICDAR 2013 table competition. In: ICDAR, pp. 1449–1453 (2013)
6. He, K., Gkioxari, G., Dollár, P., Girshick, R.B.: Mask R-CNN. In: ICCV pp. 2980–2988 (2017)
7. He, K., Zhang, X., Ren, S., Sun, J.: Deep residual learning for image recognition. In: CVPR, pp. 770–778 (2016)
8. Itonori, K.: Table structure recognition based on textblock arrangement and ruled line position. In: ICDAR, pp. 765–768 (1993)
9. Khan, S.A., Khalid, S.M.D., Shahzad, M.A., Shafait, F.: Table structure extraction with bi-directional gated recurrent unit networks. In: ICDAR, pp. 1366–1371 (2019)
10. Kieninger, T.: Table structure recognition based on robust block segmentation. Document Recogn. **3305**, 22–32 (1998)
11. Koci, E., Thiele, M., Lehner, W., Romero, O.: Table recognition in spreadsheets via a graph representation. In: IAPR, pp. 139–144 (2018)
12. Lee, C., Osindero, S.: Recursive recurrent nets with attention modeling for OCR in the wild. In: CVPR, pp. 2231–2239 (2016)
13. Li, M., Cui, L., Huang, S., Wei, F., Zhou, M., Li, Z.: Tablebank: table benchmark for image-based table detection and recognition. In: LREC, pp. 1918–1925 (2020)
14. Li, Y., Huang, Z., Yan, J., Zhou, Y., Ye, F., Liu, X.: GFTE: graph-based financial table extraction. In: ICPR Workshops, vol. 12662, pp. 644–658 (2020)
15. Lin, T., Dollár, P., Girshick, R.B., He, K., Hariharan, B., Belongie, S.J.: Feature pyramid networks for object detection. In: CVPR, pp. 936–944 (2017)
16. Lin, T.-Y., et al.: Microsoft COCO: common objects in context. In: Fleet, D., Pajdla, T., Schiele, B., Tuytelaars, T. (eds.) ECCV 2014. LNCS, vol. 8693, pp. 740–755. Springer, Cham (2014). https://doi.org/10.1007/978-3-319-10602-1_48
17. Liu, J., Liu, X., Sheng, J., Liang, D., Li, X., Liu, Q.: Pyramid mask text detector. CoRR abs/1903.11800 (2019)
18. Liu, Y., Bai, K., Mitra, P., Giles, C.L.: Improving the table boundary detection in pdfs by fixing the sequence error of the sparse lines. In: ICDAR, pp. 1006–1010 (2009)
19. Liu, Y., Mitra, P., Giles, C.L.: Identifying table boundaries in digital documents via sparse line detection. In: CIKM, pp. 1311–1320 (2008)
20. Milletari, F., Navab, N., Ahmadi, S.: V-net: fully convolutional neural networks for volumetric medical image segmentation. In: 3DV, pp. 565–571 (2016)
21. Nishida, K., Sadamitsu, K., Higashinaka, R., Matsuo, Y.: Understanding the semantic structures of tables with a hybrid deep neural network architecture. In: AAAI, pp. 168–174 (2017)
22. Paliwal, S.S., D, V., Rahul, R., Sharma, M., Vig, L.: Tablenet: deep learning model for end-to-end table detection and tabular data extraction from scanned document images. In: ICDAR, pp. 128–133 (2019)

23. Prasad, D., Gadpal, A., Kapadni, K., Visave, M., Sultanpure, K.: Cascadetabnet: an approach for end to end table detection and structure recognition from image-based documents. In: CVPR Workshops, pp. 2439–2447 (2020)
24. Qasim, S.R., Mahmood, H., Shafait, F.: Rethinking table recognition using graph neural networks. In: ICDAR, pp. 142–147 (2019)
25. Qiao, L., et al.: Text perceptron: towards end-to-end arbitrary-shaped text spotting. In: AAAI, pp. 11899–11907 (2020)
26. Raja, S., Mondal, A., Jawahar, C.V.: Table structure recognition using top-down and bottom-up cues. In: Vedaldi, A., Bischof, H., Brox, T., Frahm, J.-M. (eds.) ECCV 2020. LNCS, vol. 12373, pp. 70–86. Springer, Cham (2020). https://doi.org/10.1007/978-3-030-58604-1_5
27. Redmon, J., Divvala, S.K., Girshick, R.B., Farhadi, A.: You only look once: unified, real-time object detection. In: CVPR, pp. 779–788 (2016)
28. Ren, S., He, K., Girshick, R.B., Sun, J.: Faster R-CNN: towards real-time object detection with region proposal networks. In: NeurIPS, pp. 91–99 (2015)
29. Scarselli, F., Gori, M., Tsoi, A.C., Hagenbuchner, M., Monfardini, G.: The graph neural network model. IEEE Trans. Neural Networks **20**(1), 61–80 (2009)
30. Schreiber, S., Agne, S., Wolf, I., Dengel, A., Ahmed, S.: Deepdesrt: deep learning for detection and structure recognition of tables in document images. In: ICDAR, pp. 1162–1167 (2017)
31. Siddiqui, S.A., Fateh, I.A., Rizvi, S.T.R., Dengel, A., Ahmed, S.: Deeptabstr: deep learning based table structure recognition. In: ICDAR, pp. 1403–1409 (2019)
32. Siddiqui, S.A., Khan, P.I., Dengel, A., Ahmed, S.: Rethinking semantic segmentation for table structure recognition in documents. In: ICDAR, pp. 1397–1402 (2019)
33. Tensmeyer, C., Morariu, V.I., Price, B.L., Cohen, S., Martinez, T.R.: Deep splitting and merging for table structure decomposition. In: ICDAR, pp. 114–121 (2019)
34. Wang, Y., Phillips, I.T., Haralick, R.M.: Table structure understanding and its performance evaluation. Pattern Recognit. **37**(7), 1479–1497 (2004)
35. Xie, E., Zang, Y., Shao, S., Yu, G., Yao, C., Li, G.: Scene text detection with supervised pyramid context network. In: AAAI, pp. 9038–9045 (2019)
36. Xue, W., Li, Q., Tao, D.: Res2tim: reconstruct syntactic structures from table images. In: ICDAR, pp. 749–755 (2019)
37. Zanibbi, R., Blostein, D., Cordy, J.R.: A survey of table recognition. Int. J. Document Anal. Recognit. **7**(1), 1–16 (2004)
38. Zheng, X., Burdick, D., Popa, L., Wang, N.X.R.: Global table extractor (GTE): A framework for joint table identification and cell structure recognition using visual context. CoRR abs/2005.00589 (2020)
39. Zhong, X., ShafieiBavani, E., Jimeno Yepes, A.: Image-based table recognition: data, model, and evaluation. In: Vedaldi, A., Bischof, H., Brox, T., Frahm, J.-M. (eds.) ECCV 2020. LNCS, vol. 12366, pp. 564–580. Springer, Cham (2020). https://doi.org/10.1007/978-3-030-58589-1_34
40. Zhou, X., et al.: EAST: an efficient and accurate scene text detector. In: CVPR, pp. 2642–2651 (2017)

VSR: A Unified Framework for Document Layout Analysis Combining Vision, Semantics and Relations

Peng Zhang[1], Can Li[1], Liang Qiao[1], Zhanzhan Cheng[1,2(✉)], Shiliang Pu[1], Yi Niu[1], and Fei Wu[2]

[1] Hikvision Research Institute, Hangzhou, China
{zhangpeng23,lican9,qiaoliang6,chengzhanzhan,
pushiliang.hri,niuyi}@hikvision.com
[2] Zhejiang University, Hangzhou, China
wufei@cs.zju.edu.cn

Abstract. Document layout analysis is crucial for understanding document structures. On this task, *vision* and *semantics* of documents, and *relations* between layout components contribute to the understanding process. Though many works have been proposed to exploit the above information, they show unsatisfactory results. NLP-based methods model layout analysis as a sequence labeling task and show insufficient capabilities in layout modeling. CV-based methods model layout analysis as a detection or segmentation task, but bear limitations of inefficient modality fusion and lack of relation modeling between layout components. To address the above limitations, we propose a unified framework VSR for document layout analysis, combining *vision, semantics* and *relations*. VSR supports both NLP-based and CV-based methods. Specifically, we first introduce *vision* through document image and *semantics* through text embedding maps. Then, modality-specific visual and semantic features are extracted using a *two-stream* network, which are *adaptively* fused to make full use of complementary information. Finally, given component candidates, a *relation module* based on graph neural network is incorporated to model relations between components and output final results. On three popular benchmarks, VSR outperforms previous models by large margins. Code will be released soon.

Keywords: Vision · Semantics · Relations · Document layout analysis

1 Introduction

Document layout analysis is a crucial step in automatic document understanding and enables many important applications, such as document retrieval [4], digitization [7] and editing. Its goal is to identify the regions of interest in unstructured document and recognize the role of each region. This task is challenging due to the diversity and complexity of document layouts.

© Springer Nature Switzerland AG 2021
J. Lladós et al. (Eds.): ICDAR 2021, LNCS 12821, pp. 115–130, 2021.
https://doi.org/10.1007/978-3-030-86549-8_8

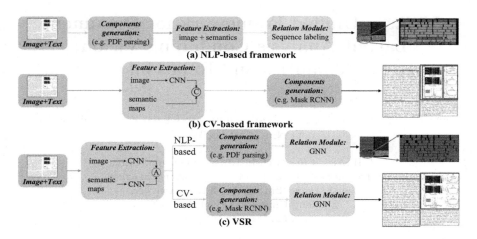

Fig. 1. Comparison of multimodal document layout analysis frameworks. VSR supports both NLP-based and CV-based frameworks. ⓒ and Ⓐ denote *concatenation* and *adaptive aggregation* fusion strategies. Different colored regions in the prediction results indicate different semantic labels (Paragraph, **Figure**, Figure Caption, Table Caption). (Color figure online)

Many deep learning models have been proposed on this task in both computer vision (CV) and natural language processing (NLP) communities. Most of them consider either only visual features [5,10,12,19,21,34,36,41] or only semantic features [6,17,27]. However, information from both modalities could help recognize the document layout better. Some regions (*e.g.,*Figure, Table) can be easily identified by visual features, while semantic features are important for separating visually similar regions (*e.g.,*Abstract and Paragraph). Therefore, some recent efforts try to combine both modalities [1,3,20,39]. Here we summarize them into two categories.

NLP-based methods (Fig. 1(a)) model layout analysis as a sequence labeling task and apply a bottom-up strategy. They first serialize texts into 1D token sequence[1]. Then, using both semantic and visual features (such as coordinates and image embedding) of each token, they determine token labels sequentially through a sequence labeling model. However, NLP-based methods show insufficient capabilities in layout modeling. For example in Fig. 1(a), all texts in a paragraph should have consistent semantic labels (Paragraph), but some of them are recognized as Figure Caption, which are the labels of adjacent texts.

CV-based methods (Fig. 1(b)) model layout analysis as object detection or segmentation task, and apply a top-down strategy. They first extract visual features by convolutional neural network and introduce semantic features through text embedding maps (at sentence-level [39] *or* character-level [3]), which are

---

[1] In the rest of this paper, we assume text is available. There are tools available to extract text from PDF documents (*e.g.,* PDFMiner [28]) and document images (*e.g.,* OCR engine [30]).

directly concatenated as the representation of document. Then, detection or segmentation models (*e.g.,* Mask RCNN [13]) are used to generate layout component candidates (coordinates and semantic labels). While capturing spatial information better compared to NLP-based methods, CV-based methods still have 3 limitations: (1) *limited semantics.* Semantic information are embedded in text at different granularities, including characters (or words) and sentences, which could help identify different document elements. For example, character-level features are better for recognizing components which need less context (*e.g.,* Author) while sentence-level features are better for contextual components (*e.g.,* Table caption). Exploiting semantics at one granularity could not achieve optimal performances. (2) *simple and heuristic modality fusion strategy.* Features from different modalities contribute differently to component recognition. Visual features contribute more to recognizing visually rich components (such as Figure and Table), while semantic features are better at distinguishing text-based components (Abstract and Paragraph). Simple and heuristic modality fusion by concatenation can not fully make use of complementary information between two modalities. (3) *lack of relation modeling between components.* Strong relations exist in documents. For example, "Figure" and "Figure Caption" often appear together, and "Paragraph"'s have aligned bounding box coordinates. Such relations could be utilized to boost layout analysis performances.

In this paper, we propose a unified framework VSR for document layout analysis, combining *V ision, S emantics* and *R elation modeling*, as shown in Fig. 1(c). This framework can be applied to both NLP-based and CV-based methods. First, documents are fed into VSR in the form of images (vision) and text embedding maps (semantics at both character-level and sentence-level). Then, modality-specific visual and semantic features are extracted through a two-stream network, which are effectively combined later in a multi-scale adaptive aggregation module. Finally, a GNN(Graph Neural Network)-based relation module is incorporated to model relations between component candidates, and generate final results. Specifically, for NLP-based methods, text tokens serve as the component candidates and relation module predicts their semantic labels. While for CV-based methods, component candidates are proposed by detection or segmentation model (*e.g.,* Faster RCNN/ Mask RCNN) and relation module generates their refined coordinates and semantic labels.

Our work makes four key contributions:

- We propose a unified framework VSR for document layout analysis, combining *vision, semantics* and *relations* in documents.
- To exploit vision and semantics effectively, we propose a *two-stream* network to extract modality-specific visual and semantic features, and fuse them *adaptively* through an adaptive aggregation module. Besides, we also explore document semantics at different granularities.
- A GNN-based relation module is incorporated to model relations between document components, and it supports relation modeling in both NLP-based and CV-based methods.

– We perform extensive evaluations of VSR, and on three public benchmarks, VSR shows significant improvements compared to previous models.

2   Related Works

Document Layout Analysis. In this paper, we try to review layout analysis works from the perspective of modality used, namely, *unimodal layout analysis* and *multimodal layout analysis*.

Unimodal layout analysis exploits either only visual features [19,21] (document image) or only semantic features (document texts) to understand document structures. Using visual features, several works [5,36] have been proposed to apply CNN to segment various objects, *e.g.,* text blocks [10], text lines [18,34], words [41], figures or tables [12,29]. At the same time, there are also methods [6,17,27] which try to address the layout analysis problem using semantic features. However, all the above methods are strictly restricted to visual *or* semantic features, and thus are not able to exploit complementary information from other modalities.

Multimodal layout analysis tries to combine information from both visual and semantic modalities. Related methods can be further divided into two categories, NLP-based and CV-based methods. NLP-based methods work on low-level elements (*e.g.,* tokens) and model layout analysis as a sequence labeling task. MMPAN [1] is presented to recognize form structures. DocBank [20] is proposed as a large scale dataset of multimodal layout analysis and several NLP baselines have been released. However, the above methods show insufficient capabilities in layout modeling. CV-based methods introduce document semantics through text embedding maps, and model layout analysis as object detection or segmentation task. MFCN [39] introduces sentence granularity semantics and inserts the text embedding maps at the decision-level (end of network), while *dhSegment*^{T2} [3] introduces character granularity semantics and inserts text embedding maps at the input-level. Though showing great success, the above methods also bear the following limitations: limited semantics used, simple modality fusion strategy and lack of relation modeling between components.

To remedy the above limitations, we propose a unified framework VSR to exploit vision, semantics and relations in documents.

Two-stream Networks. Two-stream networks are widely used to combine features in different modalities or representations [2] effectively. In action recognition, two-stream networks are used to capture the complementary *spatial* and *temporal* information [9]. In RGB-D saliency detection, the complete representations are fused from the deep features of the *RGB* stream and *depth* stream [11]. Also, two-stream networks are used to fuse different features of same input sample in sound event classification and image recognition [23]. Motivated by their successes, we apply two-stream networks to capture complementary *vision* and *semantics* information in documents.

---

^{2} *dhSegment*^{T} means *dhSegment* with inputs of image and text embedding maps.

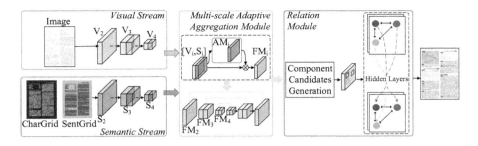

Fig. 2. The architecture overview of VSR. Best viewed in color. (Color figure online)

Relation Modeling. Relation modeling is a broad topic and has been studing for decades. In natural language processing, dependencies between sequential texts are captured through *RNN* [15] or *Transformer* [32] architectures. In computer vision, non-local networks [35] and relation networks [16] are presented to model long-range dependencies between pixels and objects. Besides, in document image processing, relations between text and layout [38] or relations between document entities [24,40,42] are explored. As to multimodal layout analysis, NLP-based methods model it as a sequence labeling task and use *RNN* to capture component relations, while CV-based methods model it as object detection task but lack relation modeling between layout components. In this paper, we propose a GNN-based relation module, supporting relation modeling in both NLP-based or CV-based methods.

3   Methodology

3.1   Architecture Overview

Our proposed framework has three parts: two-stream ConvNets, a multi-scale adaptive aggregation module and a relation module (as shown in Fig. 2). First, a two-stream convolutional network extracts modality-specific visual and semantic features, where visual stream and semantic stream take images and text embedding maps as input, respectively (Sect. 3.2). Next, instead of simply concatenating the visual and semantic features, we aggregate them via a multi-scale adaptive aggregation module (Sect. 3.3). Then, a set of component candidates are generated. Finally, a relation module is incorporated to model relations among those candidates and generate final results (Sect. 3.4).

Notice that multimodal layout analysis can be modeled as sequence labeling (NLP-based methods) or object detection tasks (CV-based methods). Our framework supports both modeling types. The only difference is what the component candidates are and how to generate them. Component candidates are low-level elements (*e.g.,* text tokens) in NLP-based methods and can be generated by parsing PDFs, while candidates are high-level elements (regions) generated by detection or segmentation model (*e.g.,* Mask RCNN) in CV-based methods. In the rest of this paper, we will illustrate how VSR is applied to CV-based methods, and show it can be easily adapted to NLP-based methods in experiments on DocBank benchmark (Sect. 4.3).

3.2 Two-Stream ConvNets

CNN is known to be good at learning deep features. However, previous multimodal layout analysis works [3,39] only apply it to extract visual features. Text embedding maps are directly used as semantic features. This single-stream network design could not make full use of document semantics. Motivated by great success of two-stream network in various multimodal applications [9,23], we apply it to extract deep visual and semantic features.

Visual Stream ConvNet. This stream directly takes document images as input and extracts multi-scale deep features using CNN backbones like ResNet [14]. Specifically, for an input image $x \in \mathbb{R}^{H \times W \times 3}$, multi-scale features maps (denoted by $\{V_2, V_3, V_4, V_5\}$) are extracted, where each $V_i \in \mathbb{R}^{\frac{H}{2^i} \times \frac{W}{2^i} \times C_i^V}$. H and W are the height and width of input image x, C_i^V is the channel dimension of feature map V_i, and $V_0 = x$.

Semantic Stream ConvNet. Similar to [3,39], we introduce document semantics through text embedding maps $S_0 \in \mathbb{R}^{H \times W \times C_0^S}$, which are the input of semantic stream ConvNet. S_0 have same spatial sizes with document image x (V_0) and C_0^S denotes the initial channel dimension. This type of representation not only encodes text content, but also preserves the 2D layout of a document. Previously, only semantics at one granularity is used (character-level [3] *or* sentence-level[3] [39]). However, semantics at different granularities contribute to identification of different components. Thus, S_0 consists of both character and sentence level semantics. Next, we show how we build text embedding maps S_0.

The characters and sentences of a document page are denoted as $\mathbb{D}_c = \{(c_k, b_k^c) \,|\, k = 0, \cdots, n\}$ and $\mathbb{D}_s = \{(s_k, b_k^s) \,|\, k = 0, \cdots, m\}$, where n and m are the total number of characters and sentences. c_k and $b_k^c = (x_0, y_0, x_1, y_1)$ are the k-th character and its associated box, where (x_0, y_0) and (x_1, y_1) are top-left and bottom-right pixel coordinates. Similarly, s_k and b_k^s are the k-th sentence and its box location. Next, character embedding maps $CharGrid \in \mathbb{R}^{H \times W \times C_0^S}$ and sentence embedding maps $SentGrid \in \mathbb{R}^{H \times W \times C_0^S}$ can be constructed as follows.

$$CharGrid_{ij} = \begin{cases} E^c(c_k) & \text{if } (i,j) \in b_k^c \\ 0 & \text{otherwise} \end{cases} \tag{1}$$

$$SentGrid_{ij} = \begin{cases} E^s(s_k) & \text{if } (i,j) \in b_k^s \\ 0 & \text{otherwise} \end{cases} \tag{2}$$

All pixels in each b_k^c (b_k^s) share the same character (sentence) embedding vector. E^c and E^s are the mapping functions of $c_k \to \mathbb{R}^{C_0^S}$ and $s_k \to \mathbb{R}^{C_0^S}$.

---

[3] Sentence is a group of words or phrases, which usually ends with a period, question mark or exclamation point. For simplicity, we approximate it with text lines.

In our implementation, E^c is a typical word embedding layer and we adopt pretrained language model BERT [8] as E^s. Finally, the text embedding maps S_0 can be constructed by applying LayerNorm normalization to the summation of $CharGrid$ and $SentGrid$, as shown in Eq. (3).

$$S_0 = LayerNorm\,(CharGrid + SentGrid) \tag{3}$$

Similar to the visual stream, semantic stream ConvNet then takes text embedding maps S_0 as input and extracts multi-scale features $\{S_2, S_3, S_4, S_5\}$, which have the same spatial sizes and channel dimension with $\{V_2, V_3, V_4, V_5\}$.

3.3   Multi-scale Adaptive Aggregation

Features from different modalities are important for identifying different objects. Modality fusion strategy should adaptively aggregate visual and semantic features. Thus, we design a multi-scale adaptive aggregation module that learns an attention map to combine visual features $\{V_2, V_3, V_4, V_5\}$ and semantic features $\{S_2, S_3, S_4, S_5\}$ adaptively. At scale i, this module first concatenates V_i and S_i, and then feed it into a convolutional layer to learn an attention map AM_i. Finally, aggregated multi-modal features FM_i is obtained. All operations in this module are formulated by:

$$AM_i = h\,(g\,([V_i, S_i])) \tag{4}$$

$$FM_i = AM_i \odot V_i + (1 - AM_i) \odot S_i \tag{5}$$

where $[\cdot]$ denotes the concatenation operation, $g\,(\cdot)$ is a convolutional layer with kernel size $1 \times 1 \times \left(C_i^V + C_i^S\right) \times C_i^S$ and $h\,(\cdot)$ is a non-linear activation function. $\odot$ denotes the element-wise multiplication. Through this module, a set of fused multi-modal features $FM = \{FM_2, FM_3, FM_4, FM_5\}$ are generated, which serve as the multimodal multi-scale features of a document. Then, FPN [22] (feature pyramid network) is applied on FM and provides enhanced representations.

3.4   Relation Module

Given aggregated features $FM = \{FM_2, FM_3, FM_4, FM_5\}$, a standard object detection or segmentation model (*e.g.*, Mask RCNN [26]) can be used to generate component candidates in a document. Previous works directly take those predictions as final results. However, strong relations exist between layout components. For example, bounding boxes of Paragraphs in the same column should be aligned; Table and Table Caption often appear together; there is no overlap between components. We find that such relations can be utilized to further refine predictions, as shown in Fig. 3, *i.e.*, adjusting regression coordinates for aligned bounding boxes, correcting wrong prediction labels based on co-occurrence of components and removing false predictions based on non-overlapping property.

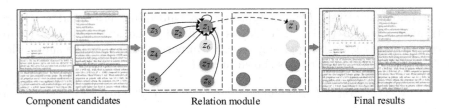

| Component candidates | Relation module | Final results |

Fig. 3. Illustration of relation module. It captures relations between component candidates, and thus improves detection results (remove false Figure prediction, correct Table Caption label and adjust Paragraph coordinates). The colors of semantic labels are: Figure, Paragraph, Figure Caption, Table, Table Caption. Best viewed in color. (Color figure online)

Next, we show how we use GNN (graph neural network) to model component relations and how to use it to refine prediction results.

We represent a document as a graph $G = (O, E)$, where $O = \{o_1, \cdots, o_N\}$ is the node set and E is the edge set. Each node o_j represents a component candidate generated by the object detection model previously, and each edge represents the relation between two component candidates. Since remote regions in a document may also bear close dependencies (*e.g.*, a paragraph spans two columns), all regions constitute a neighbor relationship. Thus, the document graph is a fully-connected graph and $E \subseteq O \times O$. The key idea of our relation module is to update the hidden representations of each node by attending over its neighbors ($z_1, z_2, \cdots, z_8 \rightarrow z_1'$, as shown in Fig. 3). With updated node features, we could predict its refined label and position coordinates.

Initially, each node, denoted by $o_j = (b_j, f_j)$, includes two pieces of information: position coordinates b_j and deep features $f_j = RoIAlign(FM, b_j)$. In order to incorporate both of them into node representation, we construct new node feature z_j as follows,

$$z_j = LayerNorm(f_j + e_j^{pos}(b_j)) \tag{6}$$

where $e_j^{pos}(b_j)$ is the position embedding vectors of j-th node.

Then, instead of explicitly specifying the relations between nodes, inspired by [33], we apply *self-attention* mechanism to automatically learn the relations, which has already shown great success in NLP and document processing [24, 38, 40, 42]. Specifically, we adopt the popular scaled dot-product attention [32] to obtain sufficient expressive power. Scaled dot-product attention consists of queries Q and keys K of dimension d_k, and values V of dimension d_v. The output $\widehat{O}$ is obtained by weighted sum over all values in V, where the attention weights are obtained using Q and K, as shown in Eq. (7). Please refer to [32] for details.

$$\widehat{O} = Attention(Q, K, V) = softmax(\frac{QK^\mathsf{T}}{\sqrt{d_k}})V \tag{7}$$

In our context, node feature set $Z = \{z_1, \cdots, z_N\}$ serves as K, Q and V and updated node feature set $Z' = \{z'_1, \cdots, z'_N\}$ is the output $\widehat{O}$. We apply multi-head attention to further improve representation capacity of node features.

Finally, given updated node features Z', refined detection results of j-th node (j-th layout component candidate) $\widetilde{o}_j = \left(\widetilde{p}^c_j, \widetilde{b}_j\right)$ is computed as,

$$\widetilde{p}^c_j = Softmax(Linear_{cls}\left(z'_j\right)) \tag{8}$$

$$\widetilde{b}_j = Linear_{reg}\left(z'_j\right) \tag{9}$$

where $\widetilde{p}^c_j$ is the probability of belonging to c-th class, $\widetilde{b}_j$ is its refined regression coordinates. $Linear_{cls}$ and $Linear_{reg}$ are projection layers.

Relation module can be easily applied to NLP-based methods. In this case, node feature z_j in Eq.(6) is the representation of j-th low-level elements (e.g., tokens). Then, GNN models pairwise relations between tokens and predicts their semantic labels ($\widetilde{p}^c_j$).

3.5 Optimization

Since multimodal layout analysis can be modeled as sequence labeling or object detection tasks, their optimization losses are different.

Layout Analysis as Sequence Labeling. The loss function is formulated as,

$$\mathcal{L} = -\frac{1}{T} \sum_{j=1}^{T} log\ \widetilde{p}_j\left(y_j\right) \tag{10}$$

where, T is the number of low-level elements and y_j is the groundtruth semantic label of j-th element.

Layout Analysis as Object Detection. The loss function is generated from two parts,

$$\mathcal{L} = \mathcal{L}_{DET} + \lambda\mathcal{L}_{RM} \tag{11}$$

where $\mathcal{L}_{DET}$ and $\mathcal{L}_{RM}$ are the losses used in candidate generation process and relation module. Both $\mathcal{L}_{DET}$ and $\mathcal{L}_{RM}$ consist of a cross entropy loss (classification) and a smooth L_1 loss (coordinate regression), as defined in [26]. Hyperparameters λ controls the trade-off between two losses.

4 Experiments

4.1 Datasets

All three benchmarks provide document images and their original PDFs. Therefore, text could be directly obtained by parsing PDFs, allowing the explorations of multi-modal techniques. To compare with existing solutions on each benchmark, we use the same evaluation metrics as used by each benchmark.

Article Regions [31] consists of 822 document samples and 9 region classes are annotated (Title, Authors, Abstract, Body, Figure, Figure Caption, Table, Table Caption and References). The annotation is in object detection format and the evaluation metric is mean average precision (mAP).

PubLayNet [43] is a large-scale document dataset recently released by IBM. It consists of 360K document samples and 5 region classes are annotated (Text, Title, List, Figure, and Table). The annotation is also in object detection format. They use the same evaluation metric as used in the COCO competition, *i.e.,* the mean average precision (AP) @ intersection over union (IOU) [0.50:0.95].

DocBank [20] is proposed by Microsoft. It contains 500K document samples with 12 region classes (Abstract, Author, Caption, Equation, Figure, Footer, List, Paragraph, Reference, Section, Table and Title). It provides token-level annotations, and use F1 score as official evaluation metric. Also, it provides object detection annotations, supporting object detection method.

4.2 Implementation Details

Document image is directly used as input for visual stream. For semantic stream, we extract embedding maps (*SentGrid* and *CharGrid*) from text as input, where *SentGrid* is generated by pretrained BERT model [8] and *CharGrid* is obtained from a word embedding layer. They all have the same channel dimension size ($C_0^S = 64$). ResNeXt-101 [37] is used as backbone to extract both visual and semantic features (unless otherwise specified), which are later fused by a multi-scale adaptive aggregation and feature pyramid network.

For CV-based multimodal layout analysis methods, fused features are fed into RPN, followed by RCNN, to generate component candidates. In RPN, 7 anchor ratios (0.02, 0.05, 0.1, 0.2, 0.5, 1.0, 2.0) are adopted to handle document elements that vary in sizes and scales. In relation module, dimension of each candidate is set to 1024 and 2 layers of multi-head attention with 16 heads are used to model relations. We set λ in Eq. (11) to be 1 in all our experiments. For NLP-based multimodal layout analysis methods, low-level elements parsed from PDFs (*e.g.,* tokens) serve as component candidates, and relation module predicts their semantic labels.

Our model is implemented under the PyTorch framework. It is trained by the SGD optimizer with batchsize=2, momentum=0.9 and weight-decay=10^{-4}. The initial learning rate is set to 10^{-3}, which is divided by 10 every 10 epochs on Article Regions dataset and 3 epochs on the other two benchmarks. The training of model on Article Regions lasts for 30 epochs while on the other two benchmarks lasts for 6. All the experiments are carried out on Tesla-V100 GPUs. Source code will be released in the near future.

4.3 Results

Article Regions. We compare the performance of VSR on this dataset with two models: Faster RCNN and Faster RCNN *with context* [31]. Faster RCNN

Table 1. Performance comparisons on Article Regions dataset

Method	Title	Author	Abstract	Body	Figure	Figure caption	Table	Table caption	Reference	mAP
Faster RCNN [31]	–	1.22	–	87.49	–	–	–	–	–	46.38
Faster RCNN w/ context [31]	–	10.34	–	93.58	–	–	–	30.8	–	70.3
Faster RCNN reimplement	100.0	51.1	94.8	98.9	94.2	91.8	97.3	67.1	90.8	87.3
Faster RCNN w/ context reimplement [31]	100.0	60.5	90.8	98.5	**96.2**	91.5	**97.5**	64.2	91.2	87.8
VSR	**100.0**	**94**	**95**	**99.1**	95.3	**94.5**	96.1	**84.6**	**92.3**	**94.5**

Note: missing entries are because those results are not reported in their original papers.

Table 2. Performance comparisons on PubLayNet dataset.

Method	Dataset	Text	Title	List	Table	Figure	AP
Faster RCNN [43]	Val	91	82.6	88.3	95.4	93.7	90.2
Mask RCNN [43]		91.6	84	88.6	96	94.9	91
VSR		**96.7**	**93.1**	**94.7**	**97.4**	**96.4**	**95.7**
Faster RCNN [43]	Test	91.3	81.2	88.5	94.3	94.5	90
Mask RCNN [43]		91.7	82.8	88.7	94.7	95.5	90.7
DocInsightAI		94.51	88.31	94.84	95.77	97.52	94.19
SCUT		94.3	89.72	94.25	96.62	97.68	94.51
SRK		94.65	89.98	**95.14**	**97.16**	**97.95**	94.98
SiliconMinds		96.2	89.75	94.6	96.98	97.6	95.03
VSR		**96.69**	**92.27**	94.55	97.03	97.90	**95.69**

with context adds limited context (page numbers, region-of-interest position and size) as input in addition to document images.

In Table 1, we first show mAP as reported in their original papers [31]. For fair comparison, we reimplement those two models using the same backbone (ResNet-101) and neck configuration as used in VSR. We also report their performance after reimplementation. We can see that our reimplemented models have much higher mAP than their original models. We believe this is mainly because we use multiple anchor ratios in RPN, thus achieve better detection results on document elements with various sizes. VSR makes full use of vision, semantics and relations between components, showing highest mAP on most classes. On Figure and Table categories, VSR achieves comparable results and the slight performance drop will be further discussed in Sect. 4.4.

PubLayNet. In Table 2, we compare the performance of VSR on this dataset with two pure image-based methods, Faster RCNN [26] and Mask RCNN [13]. While those two models present promising results (AP>90%) on validation dataset, VSR improves the performance on all classes and increases the final AP by 4.7%. VSR shows large performance improvements on text-related classes

Table 3. Performance comparisons on DocBank dataset in F1 Score.

Method	Abstract	Author	Caption	Equation	Figure	Footer	List	Paragraph	Reference	Section	Table	Title	Macro average
BERT$_{base}$	92.94	84.84	86.29	81.52	100.0	78.05	71.33	96.19	93.10	90.81	82.96	94.42	87.70
RoBERTa$_{base}$	92.88	86.18	89.44	82.48	100.0	80.14	73.53	96.46	93.41	93.37	83.89	95.11	88.91
LayoutLM$_{base}$	98.16	85.95	95.97	89.47	100.0	89.57	89.48	97.88	93.38	95.98	86.33	95.79	93.16
BERT$_{large}$	92.86	85.77	86.50	81.77	100.0	78.14	69.60	96.19	92.84	90.65	83.20	94.30	87.65
RoBERTa$_{large}$	94.79	87.24	90.81	83.70	100.0	83.92	74.51	96.65	93.34	94.07	84.94	94.61	89.88
LayoutLM$_{large}$	97.84	87.83	95.56	89.74	100.0	91.46	90.04	97.90	93.32	95.96	86.79	95.52	93.50
X101	97.17	82.27	94.35	89.38	88.12	90.29	90.51	96.82	87.98	94.12	83.53	91.58	90.51
X101+LayoutLM$_{base}$	98.15	89.07	96.69	94.30	99.90	92.92	93.00	98.43	94.37	96.64	88.18	95.75	94.78
X101+LayoutLM$_{large}$	98.02	89.64	96.66	94.40	99.94	93.52	92.93	98.44	94.30	96.70	88.75	95.31	94.88
VSR	98.29	91.19	96.32	95.84	99.96	95.11	94.66	98.66	95.05	97.11	89.24	95.63	95.59

Table 4. Performance comparisons on DocBank dataset in mAP.

Models	Abstract	Author	Caption	Equation	Figure	Footer	List	Paragraph	Reference	Section	Table	Title	mAP
Faster RCNN	96.2	88.9	93.9	78.1	85.4	93.4	86.1	67.8	89.9	76.7	77.2	95.3	86.3
VSR	96.3	89.2	94.6	77.3	97.8	93.2	86.2	69.0	90.3	79.2	77.5	94.9	87.6

(Text, Title and List) since it also utilizes document semantics in addition to document image. On test dataset (also known as *leaderboard of ICDAR2021 layout analysis recognition competition*[4]), VSR surpasses all participating teams and ranks first, with 4.99% increase on AP compared with Mask RCNN baseline.

DocBank. This dataset offers both token and detection annotations. Therefore, we could treat layout analysis task either as sequence labeling task or as object detection task, then compare VSR with existing solutions in both cases.

Layout Analysis as Sequence Labeling. Using token-level annotations, we compare VSR with BERT [8], RoBERTa [25], LayoutLM [38], Faster RCNN with ResNeXt-101 [37] and ensemble models (ResNeXt-101+LayoutLM) in Table 3. Even though highest F1 score of Caption and Figure are achieved by ensemble model (ResNeXt-101+LayoutLM) and LayoutLM respectively, VSR achieves comparable results with small gaps ($\leq 0.37\%$). More importantly, VSR gets the highest scores on all other classes. This indicates that VSR is significantly better than BERT, RoBERTa and LayoutLM architectures on document layout analysis task.

Layout Analysis as Object Detection. Since both VSR and Faster RCNN with ResNeXt-101 can provide object detection results, we further compare them in object detection format using mAP as evaluation metric. Results in Table 4 show that VSR outperforms Faster RCNN on most classes, except Equation, Footer and Title. Overall, VSR shows 1.3% gains in final mAP.

4.4   Ablation Studies

VSR introduces multi-granularity semantics, two-stream network with adaptive aggregation, and relation module. Now we explore how each of them contributes to VSR's performance improvement on Article Regions dataset.

---

[4] https://icdar2021.org/competitions/competition-on-scientific-literature-parsing/.

Table 5. Effects of semantic features at different granularities.

Vision	Semantics		Title	Author	Abstract	Body	Figure	Figure Caption	Table	Table Caption	Reference	mAP
	Char	Sentence										
✓			100.0	51.1	94.8	98.9	94.2	91.8	97.3	67.1	90.8	87.3
✓	✓		100.0	71.4	**96.5**	98.9	95.6	**93.6**	96.9	68.6	89.9	90.2
✓		✓	100.0	60.2	95.5	**99.0**	**97.8**	93.2	98.9	**73.0**	91.2	89.8
✓	✓	✓	**100.0**	**84.3**	96.1	98.7	95.7	92.5	**99.4**	71.4	**92.4**	**92.3**

Table 6. Effects of two-stream network with adaptive aggregation.

Method		Title	Author	Abstract	Body	Figure	Figure Caption	Table	Table Caption	Reference	mAP	FPS
Single-stream at input level	R101	94.7	58.7	82.7	98.1	97.9	**96.3**	91.8	63.7	91.5	86.2	19.07
	R152	100.0	50.5	85.3	97.9	**98.0**	94.4	93.3	62.6	90.5	85.8	18.15
Single-stream at decision level	R101	99.5	67.6	95.1	98.8	95.0	93.2	96.6	70.7	91.3	89.8	**19.79**
	R152	100.0	80.2	91.0	**99.4**	96.0	92.4	98.3	**73.8**	91.7	91.4	16.43
VSR	R101	**100.0**	**84.3**	**96.1**	98.7	95.7	92.5	**99.4**	71.4	**92.4**	**92.3**	13.94

Effects of Multi-granularity Semantic Features. To understand whether multi-granularity semantic features indeed improve VSR's performance, we compare 4 versions of VSR (*vision-only, vision+character, vision+sentence, vision+character+sentence*) in Table 5. Here *character* and *sentence* refer to semantic features at two different granularities. We can see that, introducing document semantics at each granularity alone can boost analysis performance while combining both of them leads to highest mAP. This is consistent with how humans comprehend documents. Humans can better recognize regions which require little context from characters/words (*e.g.,*Author) and those which need context from sentences (*e.g.,*Table caption).

Effects of Two-stream Network with Adaptive Aggregation. We propose a two-stream network with adaptive aggregation module to combine vision and semantics of document. To verify its effectiveness, we compare our VSR with its multimodal single-stream counterparts in Table 6. Instead of using extra stream to extract semantic features, single-stream networks directly use text embedding maps and concatenate them with visual features at input-level [3] or decision-level [39]. [3] performs concatenation fusion in the input level and shows worse performances, while [39] fuses multimodal features in the decision level and achieves impressive performances (89.8 mAP). VSR first extracts visual and semantic features separately using two-stream network, and then fuses them adaptively. This leads to highest mAP (92.3). At the same time, VSR can run at real-time (13.94 frames per second). We also experiment on larger backbone (ResNet-152) and reach consistent conclusions as shown in Table 6.

Effects of Relation Module. To verify the effectiveness of relation module (*RM*), we compare two versions of Faster RCNN and VSR in Table 7, *i.e.,* with *RM* and without *RM*. Since both labels and position coordinates can be refined in *RM*, both unimodal Faster RCNN and VSR show consistent improvements

Table 7. Effects of relation module.

Method		Title	Author	Abstract	Body	Figure	Figure caption	Table	Table caption	Reference	mAP
Faster RCNN	w/o RM	1	51.1	94.8	98.9	**94.2**	91.8	97.3	67.1	90.8	87.3
	w/ RM	1	**88.4**	**99.1**	**99.1**	85.4	**92.6**	**98.0**	**79.2**	**91.6**	**92.6**
VSR	w/o RM	1	84.3	**96.1**	98.7	**95.7**	92.5	**99.4**	71.4	**92.4**	92.3
	w/ RM	1	**94**	95	**99.1**	95.3	**94.5**	96.1	**84.6**	92.3	**94.5**

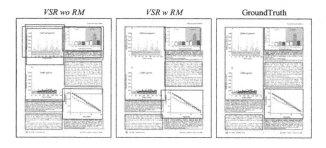

Fig. 4. Qualitative comparison between *VSR w/wo RM*. Introducing *RM* effectively removes duplicate predictions and provides more accurate detection results (both labels and coordinates). The colors of semantic labels are: Figure, Body, Figure Caption. (Color figure online)

after incorporating relation module, with 5.3% and 2.2% increase respectively. Visual examples are given in Fig. 4. However, for Figure component, performance may slightly drop after introducing *RM*. The reason is that, while removing duplicate predictions, our relation module may also risk removing correct predictions. But still, we see improvements on overall performances, showing the benefits of introducing relations.

Limitations. As mentioned above, in addition to document images, VSR also requires the positions and contents of texts in the document. Therefor, the generalization of VSR may be not good enough compared with its unimodal counterparts, which we'll address in the future.

5   Conclusion

In this paper, we present a unified framework VSR for multimodal layout analysis combining vision, semantics and relations. We first introduce semantics of document at character and sentence granularities. Then, a two-stream convolutional network is used to extract modality-specific visual and semantic features, which are further fused in the adaptive aggregation module. Finally, given component candidates, a relation module is adopted to model relations between them and output final results. On three benchmarks, VSR outperforms its unimodal and multimodal single-stream counterparts significantly. In the future, we will investigate pre-training models with VSR and extend it to other tasks, such as information extraction.

References

1. Aggarwal, M., Sarkar, M., Gupta, H., Krishnamurthy, B.: Multi-modal association based grouping for form structure extraction. In: WACV, pp. 2064–2073 (2020)
2. Baltrusaitis, T., Ahuja, C., Morency, L.: Multimodal machine learning: a survey and taxonomy. IEEE Trans. Pattern Anal. Mach. Intell. **41**(2), 423–443 (2019)
3. Barman, R., Ehrmann, M., Clematide, S., Oliveira, S.A., Kaplan, F.: Combining visual and textual features for semantic segmentation of historical newspapers. CoRR https://arxiv.org/abs/2002.06144 (2020)
4. BinMakhashen, G.M., Mahmoud, S.A.: Document layout analysis: a comprehensive survey. ACM Comput. Surv. **52**(6), 109:1–109:36 (2020)
5. Chen, K., Seuret, M., Liwicki, M., Hennebert, J., Ingold, R.: Page segmentation of historical document images with convolutional autoencoders. In: ICDAR, pp. 1011–1015 (2015)
6. Conway, A.: Page grammars and page parsing. A syntactic approach to document layout recognition. In: ICDAR, pp. 761–764 (1993)
7. Corbelli, A., Baraldi, L., Grana, C., Cucchiara, R.: Historical document digitization through layout analysis and deep content classification. In: ICPR, pp. 4077–4082 (2016)
8. Devlin, J., Chang, M., Lee, K., Toutanova, K.: BERT: pre-training of deep bidirectional transformers for language understanding. In: NAACL-HLT, pp. 4171–4186 (2019)
9. Feichtenhofer, C., Pinz, A., Zisserman, A.: Convolutional two-stream network fusion for video action recognition. In: CVPR, pp. 1933–1941 (2016)
10. Gatos, B., Louloudis, G., Stamatopoulos, N.: Segmentation of historical handwritten documents into text zones and text lines. In: ICFHR, pp. 464–469 (2014)
11. Han, J., Chen, H., Liu, N., Yan, C., Li, X.: CNNs-based RGB-D saliency detection via cross-view transfer and multiview fusion. IEEE Trans. Cybern. **48**(11), 3171–3183 (2018)
12. He, D., Cohen, S., Price, B.L., Kifer, D., Giles, C.L.: Multi-scale multi-task FCN for semantic page segmentation and table detection. In: ICDAR, pp. 254–261 (2017)
13. He, K., Gkioxari, G., Dollár, P., Girshick, R.B.: Mask R-CNN. In: ICCV, pp. 2980–2988 (2017)
14. He, K., Zhang, X., Ren, S., Sun, J.: Deep residual learning for image recognition. In: CVPR, pp. 770–778 (2016)
15. Hochreiter, S., Schmidhuber, J.: Long short-term memory. Neural Comput. **9**(8), 1735–1780 (1997)
16. Hu, H., Gu, J., Zhang, Z., Dai, J., Wei, Y.: Relation networks for object detection. In: CVPR, pp. 3588–3597 (2018)
17. Krishnamoorthy, M.S., Nagy, G., Seth, S.C., Viswanathan, M.: Syntactic segmentation and labeling of digitized pages from technical journals. IEEE Trans. Pattern Anal. Mach. Intell. **15**(7), 737–747 (1993)
18. Lee, J., Hayashi, H., Ohyama, W., Uchida, S.: Page segmentation using a convolutional neural network with trainable co-occurrence features. In: ICDAR, pp. 1023–1028 (2019)
19. Li, K., et al.: Cross-domain document object detection: benchmark suite and method. In: CVPR, pp. 12912–12921 (2020)
20. Li, M., et al.: Docbank: a benchmark dataset for document layout analysis. In: COLING, pp. 949–960 (2020)

21. Li, X., Yin, F., Xue, T., Liu, L., Ogier, J., Liu, C.: Instance aware document image segmentation using label pyramid networks and deep watershed transformation. In: ICDAR, pp. 514–519 (2019)
22. Lin, T., et al.: Feature pyramid networks for object detection. In: CVPR, pp. 936–944 (2017)
23. Lin, T., RoyChowdhury, A., Maji, S.: Bilinear CNN models for fine-grained visual recognition. In: ICCV, pp. 1449–1457 (2015)
24. Liu, X., Gao, F., Zhang, Q., Zhao, H.: Graph convolution for multimodal information extraction from visually rich documents. In: NAACL-HLT, pp. 32–39 (2019)
25. Liu, Y., et al.: Roberta: a robustly optimized BERT pretraining approach. CoRR https://arxiv.org/abs/1907.11692 (2019)
26. Ren, S., He, K., Girshick, R.B., Sun, J.: Faster R-CNN: towards real-time object detection with region proposal networks. In: NeurIPS, pp. 91–99 (2015)
27. Shilman, M., Liang, P., Viola, P.A.: Learning non-generative grammatical models for document analysis. In: ICCV, pp. 962–969 (2005)
28. Shinyama, Y.: Pdfminer: python pdf parser and analyzer. Retrieved on 11 (2015)
29. Siegel, N., Lourie, N., Power, R., Ammar, W.: Extracting scientific figures with distantly supervised neural networks. In: JCDL, pp. 223–232 (2018)
30. Smith, R.: An overview of the tesseract OCR engine. In: ICDAR, pp. 629–633 (2007)
31. Soto, C., Yoo, S.: Visual detection with context for document layout analysis. In: EMNLP-IJCNLP, pp. 3462–3468 (2019)
32. Vaswani, A., et al.: Attention is all you need. In: NeurIPS, pp. 5998–6008 (2017)
33. Velickovic, P., Cucurull, G., Casanova, A., Romero, A., Liò, P., Bengio, Y.: Graph attention networks. In: ICLR (2018)
34. Vo, Q.N., Lee, G.: Dense prediction for text line segmentation in handwritten document images. In: ICIP, pp. 3264–3268 (2016)
35. Wang, X., Girshick, R.B., Gupta, A., He, K.: Non-local neural networks. In: CVPR, pp. 7794–7803 (2018)
36. Wick, C., Puppe, F.: Fully convolutional neural networks for page segmentation of historical document images. In: DAS, pp. 287–292 (2018)
37. Xie, S., Girshick, R.B., Dollár P., He, K.: Aggregated residual transformations for deep neural networks. In: CVPR, pp. 5987–5995 (2017)
38. Xu, Y., Li, M., Cui, L., Huang, S., Wei, F., Zhou, M.: Layoutlm: pre-training of text and layout for document image understanding. In: KDD, pp. 1192–1200 (2020)
39. Yang, X., Yumer, E., Asente, P., Kraley, M., Kifer, D., Giles, C.L.: Learning to extract semantic structure from documents using multimodal fully convolutional neural networks. In: CVPR, pp. 4342–4351 (2017)
40. Yu, W., Lu, N., Qi, X., Gong, P., Xiao, R.: PICK: processing key information extraction from documents using improved graph learning-convolutional networks. In: ICPR, pp. 4363–4370 (2020)
41. Zagoris, K., Pratikakis, I., Gatos, B.: Segmentation-based historical handwritten word spotting using document-specific local features. In: ICFHR, pp. 9–14 (2014)
42. Zhang, P., et al.: TRIE: end-to-end text reading and information extraction for document understanding. In: MM, pp. 1413–1422 (2020)
43. Zhong, X., Tang, J., Jimeno-Yepes, A.: Publaynet: largest dataset ever for document layout analysis. In: ICDAR, pp. 1015–1022 (2019)

LayoutParser: A Unified Toolkit for Deep Learning Based Document Image Analysis

Zejiang Shen[1(✉)], Ruochen Zhang[2], Melissa Dell[3],
Benjamin Charles Germain Lee[4], Jacob Carlson[3], and Weining Li[5]

[1] Allen Institute for AI, Seattle, USA
shannons@allenai.org
[2] Brown University, Providence, USA
ruochen_zhang@brown.edu
[3] Harvard University, Cambridge, USA
{melissadell,jacob_carlson}@fas.harvard.edu
[4] University of Washington, Seattle, USA
bcgl@cs.washington.edu
[5] University of Waterloo, Waterloo, Canada
w422li@uwaterloo.ca

Abstract. Recent advances in document image analysis (DIA) have been primarily driven by the application of neural networks. Ideally, research outcomes could be easily deployed in production and extended for further investigation. However, various factors like loosely organized codebases and sophisticated model configurations complicate the easy reuse of important innovations by a wide audience. Though there have been on-going efforts to improve reusability and simplify deep learning (DL) model development in disciplines like natural language processing and computer vision, none of them are optimized for challenges in the domain of DIA. This represents a major gap in the existing toolkit, as DIA is central to academic research across a wide range of disciplines in the social sciences and humanities. This paper introduces `LayoutParser`, an open-source library for streamlining the usage of DL in DIA research and applications. The core `LayoutParser` library comes with a set of simple and intuitive interfaces for applying and customizing DL models for layout detection, character recognition, and many other document processing tasks. To promote extensibility, `LayoutParser` also incorporates a community platform for sharing both pre-trained models and full document digitization pipelines. We demonstrate that `LayoutParser` is helpful for both lightweight and large-scale digitization pipelines in real-word use cases. The library is publicly available at https://layout-parser. github.io.

Keywords: Document image analysis · Deep learning · Layout analysis · Character recognition · Open source library · Toolkit

© Springer Nature Switzerland AG 2021
J. Lladós et al. (Eds.): ICDAR 2021, LNCS 12821, pp. 131–146, 2021.
https://doi.org/10.1007/978-3-030-86549-8_9

1   Introduction

Deep Learning (DL)-based approaches are the state-of-the-art for a wide range of document image analysis (DIA) tasks including document image classification [11,37], layout detection [22,38], table detection [26], and scene text detection [4]. A generalized learning-based framework dramatically reduces the need for the manual specification of complicated rules, which is the status quo with traditional methods. DL has the potential to transform DIA pipelines and benefit a broad spectrum of large-scale document digitization projects.

However, there are several practical difficulties for taking advantages of recent advances in DL-based methods: 1) DL models are notoriously convoluted for reuse and extension. Existing models are developed using distinct frameworks like TensorFlow [1] or PyTorch [24], and the high-level parameters can be obfuscated by implementation details [8]. It can be a time-consuming and frustrating experience to debug, reproduce, and adapt existing models for DIA, and *many researchers who would benefit the most from using these methods lack the technical background to implement them from scratch.* 2) Document images contain diverse and disparate patterns across domains, and customized training is often required to achieve a desirable detection accuracy. Currently *there is no full-fledged infrastructure for easily curating the target document image datasets and fine-tuning or re-training the models.* 3) DIA usually requires a sequence of models and other processing to obtain the final outputs. Often research teams use DL models and then perform further document analyses in separate processes, and these pipelines are not documented in any central location (and often not documented at all). This makes it *difficult for research teams to learn about how full pipelines are implemented* and *leads them to invest significant resources in reinventing the DIA wheel.*

LayoutParser provides a unified toolkit to support DL-based document image analysis and processing. To address the aforementioned challenges, LayoutParser is built with the following components:

1. An off-the-shelf toolkit for applying DL models for layout detection, character recognition, and other DIA tasks (Sect. 3)
2. A rich repository of pre-trained neural network models (Model Zoo) that underlies the off-the-shelf usage
3. Comprehensive tools for efficient document image data annotation and model tuning to support different levels of customization
4. A DL model hub and community platform for the easy sharing, distribution, and discussion of DIA models and pipelines, to promote reusability, reproducibility, and extensibility (Sect. 4)

The library implements simple and intuitive Python APIs without sacrificing generalizability and versatility, and can be easily installed via pip. Its convenient functions for handling document image data can be seamlessly integrated with existing DIA pipelines. With detailed documentations and carefully curated tutorials, we hope this tool will benefit a variety of end-users, and will lead to advances in applications in both industry and academic research.

LayoutParser is well aligned with recent efforts for improving DL model reusability in other disciplines like natural language processing [8, 34] and computer vision [35], but with a focus on unique challenges in DIA. We show LayoutParser can be applied in sophisticated and large-scale digitization projects that require precision, efficiency, and robustness, as well as simple and light-weight document processing tasks focusing on efficacy and flexibility (Sect. 5). LayoutParser is being actively maintained, and support for more deep learning models and novel methods in text-based layout analysis methods [34, 37] is planned.

The rest of the paper is organized as follows. Section 2 provides an overview of related work. The core LayoutParser library, DL Model Zoo, and customized model training are described in Sect. 3, and the DL model hub and community platform are detailed in Sect. 4. Section 5 shows two examples of how LayoutParser can be used in practical DIA projects, and Sect. 6 concludes.

2   Related Work

Recently, various DL models and datasets have been developed for layout analysis tasks. The dhSegment [22] utilizes fully convolutional networks [20] for segmentation tasks on historical documents. Object detection-based methods like Faster R-CNN [28] and Mask R-CNN [12] are used for identifying document elements [38] and detecting tables [26, 30]. Most recently, Graph Neural Networks [29] have also been used in table detection [27]. However, these models are usually implemented individually and there is no unified framework to load and use such models.

There has been a surge of interest in creating open-source tools for document image processing: a search of document image analysis in Github leads to 5M relevant code pieces[1]; yet most of them rely on traditional rule-based methods or provide limited functionalities. The closest prior research to our work is the OCR-D project[2], which also tries to build a complete toolkit for DIA. However, similar to the platform developed by Neudecker et al. [21], it is designed for analyzing historical documents, and provides no supports for recent DL models. The DocumentLayoutAnalysis project[3] focuses on processing born-digital PDF documents via analyzing the stored PDF data. Repositories like DeepLayout[4] and Detectron2-PubLayNet[5] are individual deep learning models trained on layout analysis datasets without support for the full DIA pipeline. The Document Analysis and Exploitation (DAE) platform [15] and the DeepDIVA project [2] aim to improve the reproducibility of DIA methods (or DL models), yet they are not actively maintained. OCR engines like Tesseract [14], easyOCR[6] and paddleOCR[7] usually do not come with comprehensive functionalities for other DIA tasks like layout analysis.

---

[1] The number shown is obtained by specifying the search type as 'code'.
[2] https://ocr-d.de/en/about.
[3] https://github.com/BobLd/DocumentLayoutAnalysis.
[4] https://github.com/leonlulu/DeepLayout.
[5] https://github.com/hpanwar08/detectron2.
[6] https://github.com/JaidedAI/EasyOCR.
[7] https://github.com/PaddlePaddle/PaddleOCR.

134   Z. Shen et al.

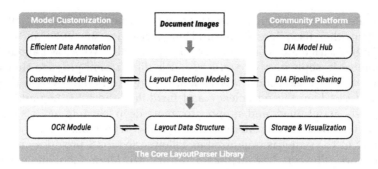

Fig. 1. The overall architecture of `LayoutParser`. For an input document image, the core `LayoutParser` library provides a set of off-the-shelf tools for layout detection, OCR, visualization, and storage, backed by a carefully designed layout data structure. `LayoutParser` also supports high level customization via efficient layout annotation and model training functions. These improve model accuracy on the target samples. The community platform enables the easy sharing of DIA models and whole digitization pipelines to promote reusability and reproducibility. A collection of detailed documentation, tutorials and exemplar projects make `LayoutParser` easy to learn and use.

Recent years have also seen numerous efforts to create libraries for promoting reproducibility and reusability in the field of DL. Libraries like Dectectron2 [35], AllenNLP [8] and transformers [34] have provided the community with complete DL-based support for developing and deploying models for general computer vision and natural language processing problems. `LayoutParser`, on the other hand, specializes specifically in DIA tasks. `LayoutParser` is also equipped with a community platform inspired by established model hubs such as `Torch Hub` [23] and `TensorFlow Hub` [1]. It enables the sharing of pretrained models as well as full document processing pipelines that are unique to DIA tasks.

There have been a variety of document data collections to facilitate the development of DL models. Some examples include PRImA [3](magazine layouts), PubLayNet [38](academic paper layouts), Table Bank [18](tables in academic papers), Newspaper Navigator Dataset [16,17](newspaper figure layouts) and `HJDataset` [31](historical Japanese document layouts). A spectrum of models trained on these datasets are currently available in the `LayoutParser` model zoo to support different use cases.

3   The Core `LayoutParser` Library

Figure 1 illustrates the key components in the `LayoutParser` library. At the core is an off the shelf toolkit that streamlines DL-based document image analysis. Five components support a simple interface with comprehensive functionalities: 1) The *layout detection models* enable using pre-trained or self-trained DL models for layout detection with just four lines of code. 2) The detected layout

Table 1. Current layout detection models in the **LayoutParser** model zoo

Dataset	Base model[1]	Large model	Notes
PubLayNet [38]	F / M	M	Layouts of modern scientific documents
PRImA [3]	M	–	Layouts of scanned modern magazines and scientific reports
Newspaper [17]	F	–	Layouts of scanned US newspapers from the 20th century
TableBank [18]	F	F	Table region on modern scientific and business document
HJDataset [31]	F / M	–	Layouts of history Japanese documents

[1]For each dataset, we train several models of different sizes for different needs (the trade-off between accuracy vs. computational cost). For "base model" and "large model", we refer to using the ResNet 50 or ResNet 101 backbones [13], respectively. One can train models of different architectures, like Faster R-CNN [28] (F) and Mask R-CNN [12] (M). For example, an F in the Large Model column indicates it has a Faster R-CNN model trained using the ResNet 101 backbone. The platform is maintained and a number of additions will be made to the model zoo in coming months.

information is stored in carefully engineered *layout data structures*, which are optimized for efficiency and versatility. 3) When necessary, users can employ existing or customized OCR models via the unified API provided in the *OCR module*. 4) **LayoutParser** comes with a set of utility functions for the *visualization and storage* of the layout data. 5) **LayoutParser** is also highly customizable, via its integration with functions for *layout data annotation and model training*. We now provide detailed descriptions for each component.

3.1 Layout Detection Models

In **LayoutParser**, a layout model takes a document image as an input and generates a list of rectangular boxes for the target content regions. Different from traditional methods, it relies on deep convolutional neural networks rather than manually curated rules to identify content regions. It is formulated as an object detection problem and state-of-the-art models like Faster R-CNN [28] and Mask R-CNN [12] are used. This yields prediction results of high accuracy and makes it possible to build a concise, generalized interface for layout detection. **LayoutParser**, built upon Detectron2 [35], provides a minimal API that can perform layout detection with only four lines of code in Python:

```
1  import layoutparser as lp
2  image = cv2.imread("image_file") # load images
3  model = lp.Detectron2LayoutModel(
4      "lp://PubLayNet/faster_rcnn_R_50_FPN_3x/config")
5  layout = model.detect(image)
```

LayoutParser provides a wealth of pre-trained model weights using various datasets covering different languages, time periods, and document types. Due to domain shift [7], the prediction performance can notably drop when

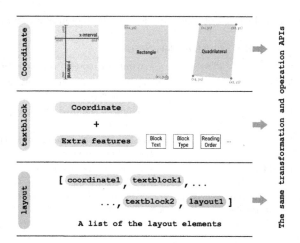

Fig. 2. The relationship between the three types of layout data structures. `Coordinate` supports three kinds of variation; `TextBlock` consists of the coordinate information and extra features like block text, types, and reading orders; a `Layout` object is a list of all possible layout elements, including other `Layout` objects. They all support the same set of transformation and operation APIs for maximum flexibility.

models are applied to target samples that are significantly different from the training dataset. As document structures and layouts vary greatly in different domains, it is important to select models trained on a dataset similar to the test samples. A semantic syntax is used for initializing the model weights in `LayoutParser`, using both the dataset name and model name `lp://<dataset-name>/<model-architecture-name>`. Shown in Table 1, `LayoutParser` currently hosts 9 pre-trained models trained on 5 different datasets. Description of the training dataset is provided alongside with the trained models such that users can quickly identify the most suitable models for their tasks. Additionally, when such a model is not readily available, `LayoutParser` also supports training customized layout models and community sharing of the models (detailed in Sect. 3.5).

3.2   Layout Data Structures

A critical feature of `LayoutParser` is the implementation of a series of data structures and operations that can be used to efficiently process and manipulate the layout elements. In document image analysis pipelines, various post-processing on the layout analysis model outputs is usually required to obtain the final outputs. Traditionally, this requires exporting DL model outputs and then loading the results into other pipelines. All model outputs from `LayoutParser` will be stored in carefully engineered data types optimized for further processing, which makes it possible to build an end-to-end document digitization pipeline within `LayoutParser`. There are three key components in the data structure, namely

Table 2. All operations supported by the layout elements. The same APIs are supported across different layout element classes including `Coordinate` types, `TextBlock` and `Layout`.

Operation name	Description
`block.pad(top, bottom, right, left)`	Enlarge the current block according to the input
`block.scale(fx, fy)`	Scale the current block given the ratio in x and y direction
`block.shift(dx, dy)`	Move the current block with the shift distances in x and y direction
`block1.is_in(block2)`	Whether block1 is inside of block2
`block1.intersect(block2)`	Return the intersection region of block1 and block2. Coordinate type to be determined based on the inputs
`block1.union(block2)`	Return the union region of block1 and block2. Coordinate type to be determined based on the inputs
`block1.relative_to(block2)`	Convert the absolute coordinates of block1 to relative coordinates to block2
`block1.condition_on(block2)`	Calculate the absolute coordinates of block1 given the canvas block2's absolute coordinates
`block.crop_image(image)`	Obtain the image segments in the block region

the `Coordinate` system, the `TextBlock`, and the `Layout`. They provide different levels of abstraction for the layout data, and a set of APIs are supported for transformations or operations on these classes.

Coordinates are the cornerstones for storing layout information. Currently, three types of `Coordinate` data structures are provided in `LayoutParser`, shown in Fig. 2. `Interval` and `Rectangle` are the most common data types and support specifying 1D or 2D regions within a document. They are parameterized with 2 and 4 parameters. A `Quadrilateral` class is also implemented to support a more generalized representation of rectangular regions when the document is skewed or distorted, where the 4 corner points can be specified and a total of 8 degrees of freedom are supported. A wide collection of transformations like `shift`, `pad`, and `scale`, and operations like `intersect`, `union`, and `is_in`, are supported for these classes. Notably, it is common to separate a segment of the image and analyze it individually. `LayoutParser` provides full support for this scenario via image cropping operations `crop_image` and coordinate transformations like `relative_to` and `condition_on` that transform coordinates to and from their relative representations. We refer readers to Table 2 for a more detailed description of these operations[8].

---

[8] This is also available in the `LayoutParser` documentation pages.

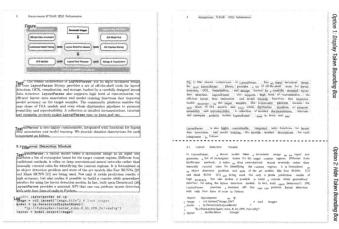

Mode I: Showing Layout on the Original Image   Mode II: Drawing OCR'd Text at the Correspoding Position

Fig. 3. Layout detection and OCR results visualization generated by the `LayoutParser` APIs. Mode I directly overlays the layout region bounding boxes and categories over the original image. Mode II recreates the original document via drawing the OCR'd texts at their corresponding positions on the image canvas. In this figure, tokens in textual regions are filtered using the API and then displayed.

Based on `Coordinate`s, we implement the `TextBlock` class that stores both the positional and extra features of individual layout elements. It also supports specifying the reading orders via setting the `parent` field to the index of the parent object. A `Layout` class is built that takes in a list of `TextBlock`s and supports processing the elements in batch. `Layout` can also be nested to support hierarchical layout structures. They support the same operations and transformations as the `Coordinate` classes, minimizing both learning and deployment effort.

3.3   OCR

`LayoutParser` provides a unified interface for existing OCR tools. Though there are many OCR tools available, they are usually configured differently with distinct APIs or protocols for using them. It can be inefficient to add new OCR tools into an existing pipeline, and difficult to make direct comparisons among the available tools to find the best option for a particular project. To this end, `LayoutParser` builds a series of wrappers among existing OCR engines, and provides nearly the same syntax for using them. It supports a plug-and-play style of using OCR engines, making it effortless to switch, evaluate, and compare different OCR modules:

```
ocr_agent = lp.TesseractAgent()
# Can be easily switched to other OCR software
tokens = ocr_agent.detect(image)
```

The OCR outputs will also be stored in the aforementioned layout data structures and can be seamlessly incorporated into the digitization pipeline. Currently LayoutParser supports the Tesseract and Google Cloud Vision OCR engines.

LayoutParser also comes with a DL-based CNN-RNN OCR model [6] trained with the Connectionist Temporal Classification (CTC) loss [10]. It can be used like the other OCR modules, and can be easily trained on customized datasets.

3.4   Storage and Visualization

The end goal of DIA is to transform the image-based document data into a structured database. LayoutParser supports exporting layout data into different formats like JSON, csv, and will add the support for the METS/ALTO XML format[9]. It can also load datasets from layout analysis-specific formats like COCO [38] and the Page Format [25] for training layout models (Sect. 3.5).

Visualization of the layout detection results is critical for both presentation and debugging. LayoutParser is built with an integrated API for displaying the layout information along with the original document image. Shown in Fig. 3, it enables presenting layout data with rich meta information and features in different modes. More detailed information can be found in the online LayoutParser documentation page.

3.5   Customized Model Training

Besides the off-the-shelf library, LayoutParser is also highly customizable with supports for highly unique and challenging document analysis tasks. Target document images can be vastly different from the existing datasets for training layout models, which leads to low layout detection accuracy. Training data can also be highly sensitive and not sharable publicly. To overcome these challenges, LayoutParser is built with rich features for efficient data annotation and customized model training.

LayoutParser incorporates a toolkit optimized for annotating document layouts using object-level active learning [32]. With the help from a layout detection model trained along with labeling, only the most important layout objects within each image, rather than the whole image, are required for labeling. The rest of the regions are automatically annotated with high confidence predictions from the layout detection model. This allows a layout dataset to be created more efficiently with only around 60% of the labeling budget.

After the training dataset is curated, LayoutParser supports different modes for training the layout models. *Fine-tuning* can be used for training models on a *small* newly-labeled dataset by initializing the model with existing pre-trained weights. *Training from scratch* can be helpful when the source dataset and target are significantly different and a large training set is available. However, as

---

[9] https://altoxml.github.io.

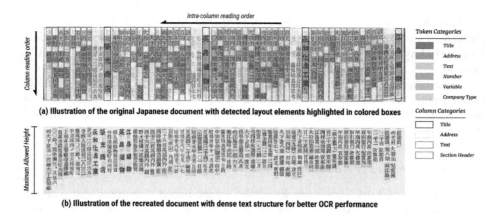

(a) Illustration of the original Japanese document with detected layout elements highlighted in colored boxes

(b) Illustration of the recreated document with dense text structure for better OCR performance

Fig. 4. Illustration of (a) the original historical Japanese document with layout detection results and (b) a recreated version of the document image that achieves much better character recognition recall. The reorganization algorithm rearranges the tokens based on the their detected bounding boxes given a maximum allowed height.

suggested in Studer et al.'s work [33], loading pre-trained weights on large-scale datasets like ImageNet [5], even from totally different domains, can still boost model performance. Through the integrated API provided by `LayoutParser`, users can easily compare model performances on the benchmark datasets.

4   LayoutParser Community Platform

Another focus of `LayoutParser` is promoting the reusability of layout detection models and full digitization pipelines. Similar to many existing deep learning libraries, `LayoutParser` comes with a community model hub for distributing layout models. End-users can upload their self-trained models to the model hub, and these models can be loaded into a similar interface as the currently available `LayoutParser` pre-trained models. For example, the model trained on the News Navigator dataset [17] has been incorporated in the model hub.

Beyond DL models, `LayoutParser` also promotes the sharing of entire document digitization pipelines. For example, sometimes the pipeline requires the combination of multiple DL models to achieve better accuracy. Currently, pipelines are mainly described in academic papers and implementations are often not publicly available. To this end, the `LayoutParser` community platform also enables the sharing of layout pipelines to promote the discussion and reuse of techniques. For each shared pipeline, it has a dedicated project page, with links to the source code, documentation, and an outline of the approaches. A discussion panel is provided for exchanging ideas. Combined with the core `LayoutParser` library, users can easily build reusable components based on the shared pipelines and apply them to solve their unique problems.

5   Use Cases

The core objective of LayoutParser is to make it easier to create both large-scale and light-weight document digitization pipelines. Large-scale document processing focuses on precision, efficiency, and robustness. The target documents may have complicated structures, and may require training multiple layout detection models to achieve the optimal accuracy. Light-weight pipelines are built for relatively simple documents, with an emphasis on development ease, speed and flexibility. Ideally one only needs to use existing resources, and model training should be avoided. Through two exemplar projects, we show how practitioners in both academia and industry can easily build such pipelines using LayoutParser and extract high-quality structured document data for their downstream tasks. The source code for these projects will be publicly available in the LayoutParser community hub.

5.1   A Comprehensive Historical Document Digitization Pipeline

The digitization of historical documents can unlock valuable data that can shed light on many important social, economic, and historical questions. Yet due to scan noises, page wearing, and the prevalence of complicated layout structures, obtaining a structured representation of historical document scans is often extremely complicated.

In this example, LayoutParser was used to develop a comprehensive pipeline, shown in Fig. 5, to generate high-quality structured data from historical Japanese firm financial tables with complicated layouts. The pipeline applies two layout models to identify different levels of document structures and two customized OCR engines for optimized character recognition accuracy.

As shown in Fig. 4(a), the document contains columns of text written vertically[10], a common style in Japanese. Due to scanning noise and archaic printing technology, the columns can be skewed or have variable widths, and hence cannot be easily identified via rule-based methods.

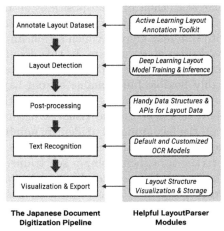

The Japanese Document Digitization Pipeline   **Helpful LayoutParser Modules**

Fig. 5. Illustration of how LayoutParser helps with the historical document digitization pipeline.

Within each column, words are separated by white spaces of variable size, and the vertical positions of objects can be an indicator of their layout type.

---

[10] A document page consists of eight rows like this. For simplicity we skip the row segmentation discussion and refer readers to the source code when available.

To decipher the complicated layout structure, two object detection models have been trained to recognize individual columns and tokens, respectively. A small training set (400 images with approximately 100 annotations each) is curated via the active learning based annotation tool [32] in `LayoutParser`. The models learn to identify both the categories and regions for each token or column via their distinct visual features. The layout data structure enables easy grouping of the tokens within each column, and rearranging columns to achieve the correct reading orders based on the horizontal position. Errors are identified and rectified via checking the consistency of the model predictions. Therefore, though trained on a small dataset, the pipeline achieves a high level of layout detection accuracy: it achieves a 96.97 AP [19] score across 5 categories for the column detection model, and a 89.23 AP across 4 categories for the token detection model.

A combination of character recognition methods is developed to tackle the unique challenges in this document. In our experiments, we found that irregular spacing between the tokens led to a low character recognition recall rate, whereas existing OCR models tend to perform better on densely-arranged texts. To overcome this challenge, we create a document reorganization algorithm that rearranges the text based on the token bounding boxes detected in the layout analysis step. Figure 4(b) illustrates the generated image of dense text, which is sent to the OCR APIs as a whole to reduce the transaction costs. The flexible coordinate system in `LayoutParser` is used to transform the OCR results relative to their original positions on the page.

Additionally, it is common for historical documents to use unique fonts with different glyphs, which significantly degrades the accuracy of OCR models trained on modern texts. In this document, a special flat font is used for printing numbers and could not be detected by off-the-shelf OCR engines. Using the highly flexible functionalities from `LayoutParser`, a pipeline approach is constructed that achieves a high recognition accuracy with minimal effort. As the characters have unique visual structures and are usually clustered together, we train the layout model to identify number regions with a dedicated category. Subsequently, `LayoutParser` crops images within these regions, and identifies characters within them using a self-trained OCR model based on a CNN-RNN [6]. The model detects a total of 15 possible categories, and achieves a 0.98 Jaccard score[11] and a 0.17 average Levinstein distances[12] for token prediction on the test set.

Overall, it is possible to create an intricate and highly accurate digitization pipeline for large-scale digitization using `LayoutParser`. The pipeline avoids specifying the complicated rules used in traditional methods, is straightforward to develop, and is robust to outliers. The DL models also generate fine-grained results that enable creative approaches like page reorganization for OCR.

---

[11] This measures the overlap between the detected and ground-truth characters, and the maximum is 1.

[12] This measures the number of edits from the ground-truth text to the predicted text, and lower is better.

5.2   A Light-Weight Visual Table Extractor

Detecting tables and parsing their structures (table extraction) are of central importance for many document digitization tasks. Many previous works [26, 27, 30] and tools[13] have been developed to identify and parse table structures. Yet they might require training complicated models from scratch, or are only applicable for born-digital PDF documents. In this section, we show how `LayoutParser` can help build a light-weight accurate visual table extractor for legal docket tables using the existing resources with minimal effort.

(a) Partial table at the bottom   (b) Full page table   (c) Partial table at the top   (d) Mis-detected text line

Fig. 6. This lightweight table detector can identify tables (outlined in red) and cells (shaded in blue) in different locations on a page. In very few cases (d), it might generate minor error predictions, e.g., failing to capture the top text line of a table. (Color figure online)

The extractor uses a pre-trained layout detection model for identifying the table regions and some simple rules for pairing the rows and the columns in the PDF image. Mask R-CNN [12] trained on the PubLayNet dataset [38] from the `LayoutParser` Model Zoo can be used for detecting table regions. By filtering out model predictions of low confidence and removing overlapping predictions, `LayoutParser` can identify the tabular regions on each page, which significantly simplifies the subsequent steps. By applying the line detection functions within the tabular segments, provided in the utility module from `LayoutParser`, the pipeline can identify the three distinct columns in the tables. A row clustering method is then applied via analyzing the y coordinates of token bounding boxes in the left-most column, which are obtained from the OCR engines. A non-maximal suppression algorithm is used to remove duplicated rows with extremely small gaps. Shown in Fig. 6, the built pipeline can detect tables at different positions on a page accurately. Continued tables from different pages are concatenated, and a structured table representation has been easily created.

---

[13] https://github.com/atlanhq/camelot. https://github.com/tabulapdf/tabula.

6   Conclusion

LayoutParser provides a comprehensive toolkit for deep learning-based document image analysis. The off-the-shelf library is easy to install, and can be used to build flexible and accurate pipelines for processing documents with complicated structures. It also supports high-level customization and enables easy labeling and training of DL models on unique document image datasets. The LayoutParser community platform facilitates sharing DL models and DIA pipelines, inviting discussion and promoting code reproducibility and reusability. The LayoutParser team is committed to keeping the library updated continuously and bringing the most recent advances in DL-based DIA, such as multi-modal document modeling [9,36,37] (an upcoming priority), to a diverse audience of end-users.

Acknowledgements. We thank the anonymous reviewers for their comments and suggestions. This project is supported in part by NSF Grant OIA-2033558 and funding from the Harvard Data Science Initiative and Harvard Catalyst. Zejiang Shen thanks Doug Downey for suggestions.

References

1. Abadi, M., et al.: TensorFlow: large-scale machine learning on heterogeneous systems (2015). https://www.tensorflow.org/. software available from tensorflow.org
2. Alberti, M., Pondenkandath, V., Würsch, M., Ingold, R., Liwicki, M.: Deepdiva: a highly-functional python framework for reproducible experiments. In: 2018 16th International Conference on Frontiers in Handwriting Recognition (ICFHR), pp. 423–428. IEEE (2018)
3. Antonacopoulos, A., Bridson, D., Papadopoulos, C., Pletschacher, S.: A realistic dataset for performance evaluation of document layout analysis. In: 2009 10th International Conference on Document Analysis and Recognition, pp. 296–300. IEEE (2009)
4. Baek, Y., Lee, B., Han, D., Yun, S., Lee, H.: Character region awareness for text detection. In: Proceedings of the IEEE/CVF Conference on Computer Vision and Pattern Recognition, pp. 9365–9374 (2019)
5. Deng, J., Dong, W., Socher, R., Li, L.J., Li, K., Fei-Fei, L.: ImageNet: a large-scale hierarchical image database. In: CVPR09 (2009)
6. Deng, Y., Kanervisto, A., Ling, J., Rush, A.M.: Image-to-markup generation with coarse-to-fine attention. In: International Conference on Machine Learning, pp. 980–989. PMLR (2017)
7. Ganin, Y., Lempitsky, V.: Unsupervised domain adaptation by backpropagation. In: International Conference on Machine Learning, pp. 1180–1189. PMLR (2015)
8. Gardner, M., et al.: Allennlp: a deep semantic natural language processing platform. arXiv preprint arXiv:1803.07640 (2018)
9. Garncarek, L., Powalski, R., Stanisław ek, T., Topolski, B., Halama, P., Graliński, F.: Lambert: layout-aware (language) modeling using bert for information extraction (2020)

10. Graves, A., Fernández, S., Gomez, F., Schmidhuber, J.: Connectionist temporal classification: labelling unsegmented sequence data with recurrent neural networks. In: Proceedings of the 23rd International Conference on Machine Learning, pp. 369–376 (2006)
11. Harley, A.W., Ufkes, A., Derpanis, K.G.: Evaluation of deep convolutional nets for document image classification and retrieval. In: 2015 13th International Conference on Document Analysis and Recognition (ICDAR), pp. 991–995. IEEE (2015)
12. He, K., Gkioxari, G., Dollár, P., Girshick, R.: Mask R-CNN. In: Proceedings of the IEEE International Conference on Computer Vision, pp. 2961–2969 (2017)
13. He, K., Zhang, X., Ren, S., Sun, J.: Deep residual learning for image recognition. In: Proceedings of the IEEE Conference on Computer Vision and Pattern Recognition, pp. 770–778 (2016)
14. Kay, A.: Tesseract: an open-source optical character recognition engine. Linux J. **2007**(159), 2 (2007)
15. Lamiroy, B., Lopresti, D.: An open architecture for end-to-end document analysis benchmarking. In: 2011 International Conference on Document Analysis and Recognition, pp. 42–47. IEEE (2011)
16. Lee, B.C., Weld, D.S.: Newspaper navigator: open faceted search for 1.5 million images. In: Adjunct Publication of the 33rd Annual ACM Symposium on User Interface Software and Technology, UIST 2020 Adjunct, pp. 120–122. Association for Computing Machinery, New York (2020). https://doi.org/10.1145/3379350.3416143
17. Lee, B.C.G., et al.: The newspaper navigator dataset: extracting headlines and visual content from 16 million historic newspaper pages in chronicling America, pp. 3055–3062. Association for Computing Machinery, New York (2020). https://doi.org/10.1145/3340531.3412767
18. Li, M., Cui, L., Huang, S., Wei, F., Zhou, M., Li, Z.: Tablebank: table benchmark for image-based table detection and recognition. arXiv preprint arXiv:1903.01949 (2019)
19. Lin, T.-Y., et al.: Microsoft COCO: common objects in context. In: Fleet, David, Pajdla, Tomas, Schiele, Bernt, Tuytelaars, Tinne (eds.) ECCV 2014. LNCS, vol. 8693, pp. 740–755. Springer, Cham (2014). https://doi.org/10.1007/978-3-319-10602-1_48
20. Long, J., Shelhamer, E., Darrell, T.: Fully convolutional networks for semantic segmentation. In: Proceedings of the IEEE Conference on Computer Vision and Pattern Recognition, pp. 3431–3440 (2015)
21. Neudecker, C., et al.: An experimental workflow development platform for historical document digitisation and analysis. In: Proceedings of the 2011 Workshop on Historical Document Imaging and Processing, pp. 161–168 (2011)
22. Oliveira, S.A., Seguin, B., Kaplan, F.: dhSegment: a generic deep-learning approach for document segmentation. In: 2018 16th International Conference on Frontiers in Handwriting Recognition (ICFHR), pp. 7–12. IEEE (2018)
23. Paszke, A., et al.: Automatic differentiation in pytorch (2017)
24. Paszke, A., et al.: Pytorch: an imperative style, high-performance deep learning library. arXiv preprint arXiv:1912.01703 (2019)
25. Pletschacher, S., Antonacopoulos, A.: The page (page analysis and ground-truth elements) format framework. In: 2010 20th International Conference on Pattern Recognition, pp. 257–260. IEEE (2010)

26. Prasad, D., Gadpal, A., Kapadni, K., Visave, M., Sultanpure, K.: Cascadetabnet: an approach for end to end table detection and structure recognition from image-based documents. In: Proceedings of the IEEE/CVF Conference on Computer Vision and Pattern Recognition Workshops, pp. 572–573 (2020)

27. Qasim, S.R., Mahmood, H., Shafait, F.: Rethinking table recognition using graph neural networks. In: 2019 International Conference on Document Analysis and Recognition (ICDAR), pp. 142–147. IEEE (2019)

28. Ren, S., He, K., Girshick, R., Sun, J.: Faster R-CNN: towards real-time object detection with region proposal networks. In: Advances in Neural Information Processing Systems, pp. 91–99 (2015)

29. Scarselli, F., Gori, M., Tsoi, A.C., Hagenbuchner, M., Monfardini, G.: The graph neural network model. IEEE Trans. Neural Netw. **20**(1), 61–80 (2008)

30. Schreiber, S., Agne, S., Wolf, I., Dengel, A., Ahmed, S.: Deepdesrt: deep learning for detection and structure recognition of tables in document images. In: 2017 14th IAPR International Conference on Document Analysis and Recognition (ICDAR), vol. 1, pp. 1162–1167. IEEE (2017)

31. Shen, Z., Zhang, K., Dell, M.: A large dataset of historical Japanese documents with complex layouts. In: Proceedings of the IEEE/CVF Conference on Computer Vision and Pattern Recognition Workshops, pp. 548–549 (2020)

32. Shen, Z., Zhao, J., Dell, M., Yu, Y., Li, W.: Olala: object-level active learning based layout annotation. arXiv preprint arXiv:2010.01762 (2020)

33. Studer, L., et al.: A comprehensive study of imagenet pre-training for historical document image analysis. In: 2019 International Conference on Document Analysis and Recognition (ICDAR), pp. 720–725. IEEE (2019)

34. Wolf, T., et al.: Huggingface's transformers: state-of-the-art natural language processing. arXiv preprint arXiv:1910.03771 (2019)

35. Wu, Y., Kirillov, A., Massa, F., Lo, W.Y., Girshick, R.: Detectron2. https://github.com/facebookresearch/detectron2 (2019)

36. Xu, Y., et al.: Layoutlmv2: multi-modal pre-training for visually-rich document understanding. arXiv preprint arXiv:2012.14740 (2020)

37. Xu, Y., Li, M., Cui, L., Huang, S., Wei, F., Zhou, M.: Layoutlm: pre-training of text and layout for document image understanding (2019)

38. Zhong, X., Tang, J., Yepes, A.J.: Publaynet: largest dataset ever for document layout analysis. In: 2019 International Conference on Document Analysis and Recognition (ICDAR), pp. 1015–1022. IEEE (2019). https://doi.org/10.1109/ICDAR.2019.00166

Understanding and Mitigating the Impact of Model Compression for Document Image Classification

Shoaib Ahmed Siddiqui[1,2](✉) , Andreas Dengel[1,2] , and Sheraz Ahmed[1,3]

[1] German Research Center for Artificial Intelligence (DFKI),
67663 Kaiserslautern, Germany
{shoaibahmed.siddiqui,andreas.dengel,sheraz.ahmed}@dfki.de
[2] TU Kaiserslautern, 67663 Kaiserslautern, Germany
[3] DeepReader GmbH, 67663 Kaiserlautern, Germany

Abstract. Compression of neural networks has become a common norm in industrial settings to reduce the cost of inference and deployment. As document classification is a common process in business workflows, there is a dire need of analyzing the potential of compressed models for the task of document image classification. Surprisingly, no such analysis has been done in the past. Furthermore, once a compressed model is obtained using a particular compression algorithm (which achieves similar accuracy to the uncompressed model), the model is directly deployed without much consideration. However, such compression process results in its own set of challenges which have large implications, especially when the uncompressed model is ensured to be fair or unbiased such as a fair model making critical hiring decisions based on the applicant's resume. In this paper, we show that current state-of-the-art compression algorithms can be successfully applied for the task of document image classification. We further analyze the impact of model compression on network outputs and highlight the discrepancy that arises during the compression process. Building on recent findings in this direction, we employ a principled approach based on logit-pairing for minimizing this deviation from the functional outputs of the uncompressed model. Our results on Tobacco-3482 dataset show that we can reduce the number of mismatches between the compressed and the uncompressed model drastically in both structured and unstructured cases (by more than 2x in some cases) by employing label-preservation-aware loss functions. These findings shed light on some important considerations before the compressed models are ready for deployment in the real world.

Keywords: Document image classification · Network compression · Deep learning · Convolutional Neural Networks · Model pruning · Compression Identified Examples (CIEs)

This work was supported by the BMBF project DeFuseNN (Grant 01IW17002) and the NVIDIA AI Lab (NVAIL) program.

© Springer Nature Switzerland AG 2021
J. Lladós et al. (Eds.): ICDAR 2021, LNCS 12821, pp. 147–159, 2021.
https://doi.org/10.1007/978-3-030-86549-8_10

1   Introduction

In this era of automation, more and more emphasis is being laid on automating document processing pipelines in business workflows [1–3,5,7,13,31,32]. One such simple process is of document image classification [1–3,5,7,13]. Seemingly a benign task, such systems have also been a great source of unfairness and bias towards individuals belonging to a particular ethnic group or a protected group. Numerous sources have analyzed the unfairness and biasness of these automated decision-making systems [10,23]. One such critical system is the one that makes hiring decisions based on resumes. It has been shown that such systems show biases towards females or people from minority groups [23].

On the other hand, deployment of deep learning models in industrial settings is now a common practice [11,12,20,21]. As document image classification is now a standard component of many business process workflows, there is a dire need to enable efficient and scalable deployment of these deep learning models [2, 3,7,32]. For this purpose, many pruning strategies have been proposed in the past, which can be directly accelerated on a given hardware [11,20]. The task of a pruning algorithm is to take a large reference model and compress it such that the size of the network is minimized or the inference speed is maximized without any impact on the resulting predictions of the network. A common proxy to compute this impact is the top-1 accuracy [11,20]. To the best of the authors' knowledge, this paper is the first to analyze the performance of one such compression algorithm [26] for the task of document image classification. We show that most of the network parameters can be pruned without only a negligible effect on the usually reported performance metrics (top-1 accuracy).

Having the same top-1 accuracy results in a general false conclusion of the classifier being functionally equivalent to the uncompressed version. Hooker et al. (2020) [10] analyzed the problem in detail and showed that even though the accuracy of the classifier is retained, the two classifiers can be functionally very distinct. This is particularly concerning in cases where the uncompressed model already underwent tight scrutiny and testing. Therefore, a fair or unbiased model after pruning can disregard these attributes during the compression process. One simple but expensive solution to this problem is to ensure the scrutiny is carried out again once the model underwent pruning. Another solution is to ensure that the network shows minimum deviation from the uncompressed model during the pruning phase. Joseph et al. (2020) [12] defined a principled approach to minimize the deviations between the compressed and the uncompressed models. Even in cases where the predictions are retained, Park et al. (2020) [24] showed that the compressed models can focus on entirely different aspects of the input compared to their uncompressed counterparts.

Based on these recent findings, we also analyzed the impact of model compression for the task of document image classification. We show that direct compression results (both structured and unstructured pruning) in a large number of mismatches between the uncompressed and the compressed models. We also employ the logit-pairing loss proposed by Joseph et al. (2020) [12] to minimize these disagreements. We achieved a more than 2x reduction in the number of mismatches between the compressed and the uncompressed model using the logit-pairing loss, showing its efficacy for compressing models designed for document image classification. In particular, the contribution of this paper is three-fold:

- We employ a state-of-the-art compression technique (Rewind [26]) and show that models trained for document image classification can be successfully compressed to a range of different sparsity ratios without significantly impacting the performance of the model in both structured and unstructured pruning scenarios.
- We analyze the resulting predictions of the model to show that although the accuracy measure remains nearly constant, the prediction characteristics vary significantly between the two models i.e., the two models disagree in terms of their predictions for a number of examples.
- Finally, we show that logit-pairing can be successful in significantly reducing the number of disagreements between the two models (by more than 2x in some cases).

2   Related Work

2.1   Document Image Classification

There is rich literature on the topic of document image classification. However, we restrict ourselves to some of the most recent work in this direction. For a more comprehensive treatment of the prior work, we refer readers to the survey from Chen & Blostein (2007) [4] where they explored different features, feature representations, and classification algorithms prevalent in the past for the task of document image classification.

Shin et al. (2001) [29] defined a set of hand-crafted features which were used to cluster documents together. The selected features included percentages of textual and non-textual regions, column structures, relative font sizes, content density as well as their connected components. These features were then fed to a decision tree for classification. Following their initial work, they proposed an approach to compute geometrically invariant structural similarity along with document similarity for querying document image databases in 2006 [30]. Reddy & Govindaraju (2008) [25] used low-level pixel density information from binary images to classify documents using an ensemble of K-Means clustering-based classifiers leveraging AdaBoost. Sarkar et al. (2010) [28] proposed a new method to learn image anchor templates automatically from document images. These templates could then be used for downstream tasks, e.g., document classification or data extraction.

Kumar et al. (2012) [17] proposed the use of code-books to compute document image similarity where the document was recursively partitioned into small patches. These features were then used for the retrieval of similar documents from the database. Building on their prior work, they additionally used a random forest classifier trained in an unsupervised way on top of the computed representations to obtain document images that belonged to the same category. This enabled them to achieve unsupervised document classification [16]. The following year, they achieved state-of-the-art performance on the table and tax form retrieval tasks [18] by using the same approach as [16], but evaluated the performance in the scenario of limited training data.

After the deep learning revolution, mainly attributed to the seminal paper from Krizhevsky et al. [15] in 2012 where they introduced the famous *AlexNet* architecture, a range of deep learning-powered document classification systems were introduced. Kang et al. (2014) [13] defined one of the first uses of deep Convolutional Neural Networks (CNN) for document image classification. They achieved significant gains in performance in contrast to prior hand-coded feature engineering approaches. Afzal et al. (2015) [1] introduced *DeepDocClassifier* powered by a deep CNN (AlexNet [15] in their case), and demonstrated the potential of transfer learning where the weights of the model were initialized from the pretrained model trained on the large-scale ImageNet [27] dataset comprising of 1.28 million training images belonging to a 1000 different categories. This transformed the initial convolutional layers into generic feature extractors, achieving a significant performance boost in contrast to prior methods. Noce et al. (2016) [22] additionally included textual information extracted using an off-the-shelf OCR alongside the raw images to boost classification performance. The extracted text was fed to an NLP model for projection into the feature space. Kolsch et al. (2017) [14] used extreme learning machine on top of frozen convolutional layers initialized from pretrained AlexNet [15] model. They achieved significant gains in efficiency without any major degradation in accuracy.

Afzal et al. (2017) [2] showed that using pretrained state-of-the-art image classification networks trained on a large amount of labeled data yields significant improvements on document image classification benchmarks where they evaluated the performance of VGG, ResNet, and GoogLeNet [8,33,34]. Asim et al. (2019) [3] proposed a two-stream network taking both visual and textual inputs into the network for producing the output. In contrast to prior work, they included a feature ranking algorithm to pick the top-most features from the textual stream. Similar to [22], Ferrando et al. (2020) [7] combined predictions from OCR with that of images but leveraged the representational power of BERT [6] as the NLP model to boost performance.

Cosma et al. (2020) [5] recently analyzed the potential of self-supervised representation learning for the task of document image classification. However, their analysis was limited to some obsolete self-supervised tasks (Jigsaw puzzles) where they showed its insufficiency to learn useful representations. They also analyzed some custom variants of this jigsaw puzzle task which they found to improve performance over this trivial baseline. The authors then moved to fuse representations in a self-supervised way for both images and text using LDA and showed significant performance gains in a semi-supervised setting.

2.2   Model Compression

There has been a range of different attempts on network compression [9, 11,12,20,21,26]. These efforts can be mainly categorized into structured and unstructured pruning [11]. Unstructured pruning removes single connections from the network based on some measure of their importance. Since only a small fraction of non-useful weights are removed, there is usually a negligible loss in performance in contrast to structured pruning where whole structures are removed (usually, whole filters for convolutional networks). As the sparsity induced by structured pruning has direct implication on FLOPs, they usually result in better run-times as compared to their unstructured counterparts which are harder to accelerate [11].

Pruning of networks has been a topic of interest since their inception when LeCun et al. (1990) [19] tried to use second-order information to prune network weights in an unstructured fashion. Since then, a range of different efforts has been made. Hinton et al. (2015) [9] proposed the idea of knowledge distillation, where they distilled knowledge from a large teacher network to a smaller student network. They hypothesized that the information learned by a large teacher network is represented in its output probabilities in the form of uncertainty. Hence, they used a temperature-augmented softmax layer which diffused the probability distribution from a one-hot representation and trained the student model to mimic the probability distribution by minimizing the KL-divergence between the two. Malchanov et al. (2016) [21] used Taylor approximation to identify the influence of each weight on the resulting loss function which served as a measure of weight importance. Li et al. (2020) [20] proposed the group-sparsity algorithm where they connected filter pruning with low-rank tensor factorization in a unified framework and achieved state-of-the-art accuracy under high sparsity regime. Renda et al. (2020) [26] proposed a simple pruning algorithm called as *REWIND* which alternates between training and pruning steps where the induced sparsity is enhanced incrementally. We leverage rewind as the compression algorithm in our experiments.

2.3   Compression Identified Examples (CIEs)

Hooker et al. (2020) [10] analyzed the differences between the prediction of a compressed and the uncompressed network. They termed these examples as Compression Identified Examples (CIEs – $y'_W \neq y'_{\overline{W}}$ where y'_W represents the predictions of the uncompressed model while $y'_{\overline{W}}$ represents the predictions of the compressed model). They also analyzed the undesirable consequences of model compression, including amplification of bias and unfairness. Similarly, Parker et al. (2020) [24] showed that even in cases where the predictions of the compressed models agree with the uncompressed model, the network can focus on an entirely different set of features as compared to the uncompressed model. This has particularly large implications for model interpretability.

Joseph et al. (2020) [12] further categorized CIEs into *CIE-U* where the predictions of the uncompressed model are correct, but the predictions of the compressed model are incorrect. This is more concerning than the other case where the compressed model gets the example right. Furthermore, they also proposed a new loss formulation that was effective in reducing the number of CIEs significantly by leveraging knowledge from the uncompressed model in

---

Algorithm 1: Rewind [26] algorithm combined with our label-preserving-aware loss function.

Result: The final compressed set of parameters $(\overline{\mathcal{W}})$

Given: dataset $(\mathcal{D})$, pruning scheme (structured / unstructured), number of epochs, desired sparsity, mask M which is initially set to be all ones, uncompressed trained model weights $(\mathcal{W})$, compressed weights which are initialized to the uncompressed weights $\overline{\mathcal{W}} = \mathcal{W}$;

while *desired sparsity $\leq$ current sparsity* **do**

    M = prune($\overline{\mathcal{W}}$, scheme, current sparsity * 0.8) ;

    while *current epoch $\leq$ number of epochs* **do**

        $\overline{\mathcal{W}} = \underset{\mathbf{w}}{\arg\min}\ \mathcal{L}(\mathbf{w}, \mathcal{D}) \odot M$

    end

end

---

a logit-pairing framework. In this paper, we leverage the logit-pairing losses to analyze for the first time the impact of compression on document image classification networks.

3   Method

Let Φ be the chosen network architecture (e.g. ResNet [8], VGG [33]) without the softmax layer, σ be the softmax layer, $\mathcal{W}$ be the uncompressed set of weights, $\overline{\mathcal{W}}$ be the compressed set of weights, and $\mathcal{D}$ be the given training dataset. Almost all pruning methods attempt to minimize the cross-entropy loss after pruning the network [11, 20, 26]:

$$\mathcal{L}_{CE}(\mathbf{w}, \mathcal{D}) = \frac{1}{|\mathcal{D}|} \sum_{(\mathbf{x}, y) \in \mathcal{D}} CE(\sigma(\Phi(\mathbf{x}; \mathbf{w})), y), \qquad (1)$$

$$\overline{\mathcal{W}} = \underset{\mathbf{w}}{\arg\min}\ \mathcal{L}_{CE}(\mathbf{w}, \mathcal{D}) \qquad (2)$$

Although using such pruning strategy works fine in retaining the top-1 accuracy, they result in a large number of mismatches between the compressed and the uncompressed model (Sect. 4). Hinton et al. (2015) [9] distilled knowledge from the uncompressed model to the compressed model in the form of diffused class probabilities. This requires choosing an appropriate temperature T which is hard to tune in practice. The performance of the compressed model is particularly sensitive to the choice of this hyperparameter. Following Joseph et al. (2020) [12], we use a simple logit-pairing loss which just minimizes the difference of the logits between the two models, instead of going to the probability space. This evades the hassle of manually tuning these hyperparameters.

$$\mathcal{L}_{MSE}(\mathbf{w}, \mathcal{D}) = \frac{1}{|\mathcal{D}|} \sum_{(\mathbf{x}, y) \in \mathcal{D}} \|\Phi(\mathbf{x}; \mathbf{w})) - \Phi(\mathbf{x}; \mathcal{W}))\|^2. \qquad (3)$$

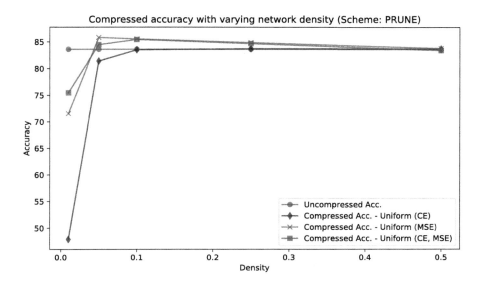

Fig. 1. Accuracy for unstructured pruning on ResNet-50 trained on the Tobacco-3482 dataset.

Minimizing the difference between the logits maximizes the agreement between the two models. However, aligning the two models when the prediction of the uncompressed model is incorrect is not really fruitful. Therefore, in order to improve the accuracy in cases where the uncompressed model got the prediction wrong, we introduce back the cross-entropy loss into the picture. Therefore, the final objective combines the two losses with a weighting factor attached to each term:

$$\mathcal{L}(\mathbf{w}, \mathcal{D}) = \alpha \cdot \mathcal{L}_{CE}(\mathbf{w}, \mathcal{D}) + \beta \cdot \mathcal{L}_{MSE}(\mathbf{w}, \mathcal{D}) \tag{4}$$

where α and β are the corresponding weightings of the loss terms for the final loss computation. Following Joseph et al. (2020) [12] where they found the simplest uniform combination to work the best, we use a uniform combination of these terms where both α and β are set to 1.0. Therefore, the final set of compressed weights are obtained as:

$$\mathcal{L}(\mathbf{w}, \mathcal{D}) = \mathcal{L}_{CE}(\mathbf{w}, \mathcal{D}) + \mathcal{L}_{MSE}(\mathbf{w}, \mathcal{D}) \tag{5}$$

$$\overline{\mathcal{W}} = \arg\min_{\mathbf{w}} \mathcal{L}(\mathbf{w}, \mathcal{D}) \tag{6}$$

As this is an iterative process, we adjust the sparsity rate by increasing the sparsity by 20% at every iteration of the update process and then retrain the model for a given number of epochs. This process is repeated until the desired level of sparsity is achieved. The final pseudo-code is given in algorithm 1. It is important to note that in contrast to Joseph et al. (2020) [12], we avoid inclusion

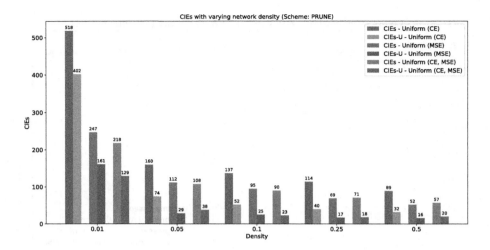

Fig. 2. Compression identified examples (CIEs) for unstructured pruning on ResNet-50 Trained on the Tobacco-3482 dataset. As CIEs represent a mismatch between the compressed and the uncompressed models, lower number of CIEs is better.

of the prediction-based cross-entropy term as cross-entropy with respect to the original labels already takes care of getting the argmax correct in cases where the logits are close to each other.

Let us now formally define what we mean by the sparsity rate. The density of the network (D) is defined as the number of non-zero weights in the compressed network $(\|\overline{\mathcal{W}}\|_0)$ in comparison to the number of non-zero weights in the uncompressed network $(\|\mathcal{W}\|_0)$.

$$D = \frac{\|\overline{\mathcal{W}}\|_0}{\|\mathcal{W}\|_0} \tag{7}$$

Sparsity (S) is simply the inverse of density. As the density is bounded between 0 and 1 in the case of compression where the size of the compressed model is reduced in comparison to the uncompressed model, sparsity can be simply computed using $1. - D$.

4   Results

We evaluate the performance and impact of model compression on a famous document image classification benchmark i.e. Tobacco-3482 dataset [2]. Tobacco-3482 dataset is comprised of 3482 images belonging to 10 different classes: advertisement, email, form, letter, memo, news, note, report, resume, and scientific publication. Since there is no predefined split of the dataset, we use randomly sampled 20% of the dataset as test set while the remaining data as train set. We define network density to be: $1.0 - sparsity$. For all our experiments, we use the ResNet-50 backbone [8] pretrained on the ImageNet dataset [27].

4.1   Unstructured Pruning

The accuracy of the model with varying levels of network densities after unstructured pruning is presented in Fig. 1. It is evident from the figure that increasing the amount of sparsity (decreasing network density) has a negligible effect on the network's performance unless the sparsity level moves down to be extremely low. We also see that in high sparsity regimes, there is a very positive impact of logit-pairing on the network's accuracy. It shows that using the proposed loss function is not only important to reduce the number of CIEs, but also positively impacts the classification accuracy.

We visualize the number of CIEs with varying levels of network density in Fig. 2. When considering low sparsity regimes, the network accuracy is already high. Therefore, there is a negligible difference in the functional form of the compressed model, resulting in a very small number of CIEs. However, as we move towards higher levels of sparsity, the trend becomes clear. The trend is especially pronounced when evaluating the network at a density of 0.01. In this case, we observe more than 2x reduction in the number of CIEs (including CIE-U) when using Uniform (CE, MSE) loss in contrast to using only the CE. The justification for the introduction of CE along with MSE is also clear from the figure where there is a positive impact of including both terms in high sparsity regimes while having a negligible difference in the low sparsity regime.

4.2   Structured (Filter) Pruning

The accuracy of the model with varying levels of network density after structured pruning is visualized in Fig. 3. One immediate impression from the figure is the fact that the impact of structured pruning is more severe in terms of accuracy of the model as compared to unstructured pruning (Fig. 1). Furthermore, the effect of using a combination of both CE and MSE on accuracy is much more pronounced in contrast to the unstructured case where the effect is only visible at the highest evaluated level of sparsity. It is interesting to note that the accuracy goes down to that of random guessing (10%) when the density of the network goes below 10%. Therefore, there is no difference in the number of CIEs or the accuracy of the network when moving between different loss formulations as the capacity of the network is insufficient to capture the relationship between the inputs and the outputs.

The number of CIEs with varying levels of network density for structured pruning is presented in Fig. 4. We see that the number of CIEs is fixed when the network density is the lowest (accuracy close to random guessing). However, once the classifier is able to regain its potential to classify the inputs, the same trend emerges where the combination of CE and MSE results in the lowest number of CIEs in contrast to using CE or MSE alone. It is however interesting to note that the trend of the reduction in the number of CIEs is also weak as compared to the one observed in the case of unstructured pruning. The reduction in the number of CIEs is less than 2x even in the best case.

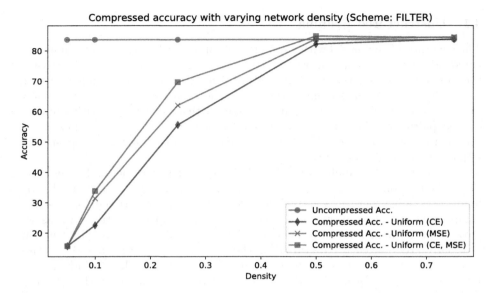

Fig. 3. Accuracy for structured (filter) pruning on ResNet-50 Trained on the Tobacco-3482 dataset.

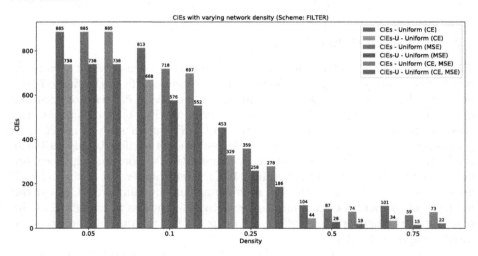

Fig. 4. Compression identified examples (CIEs) for structured (filter) pruning on ResNet-50 trained on the Tobacco-3482 dataset. As CIEs represent a mismatch between the compressed and the uncompressed models, lower number of CIEs is better.

5   Conclusion

This paper shows that it is possible to compress large CNNs trained for the task of document image classification on a range of different sparsity levels without any significant impact on the network's performance for both structured and unstructured schemes. However, our analysis showed that despite a negligible effect on metrics like top-1 accuracy, the functional outputs of the network vary significantly, specifically in high sparsity regimes. Therefore, employing label-preservation-aware loss functions provides a significant reduction in the number of mismatches (more than 2x in some cases), while also positively impacting the accuracy of the model.

Our paper focuses on a single compression algorithm i.e., rewind. However, the impact of using these loss terms also varies when considering different types of compression algorithms. Therefore, a plausible future direction is to explore the impact of different compression algorithms on the number of CIEs. Furthermore, computing mismatches is only one simple way of characterizing the differences between the compressed and the uncompressed models. More sophisticated metrics like fairness and attributions can also be taken into account along with the minimization of such metrics. We leave such exploration as part of future work.

References

1. Afzal, M.Z., et al.: Deepdocclassifier: document classification with deep convolutional neural network. In: 2015 13th International Conference on Document Analysis and Recognition (ICDAR), pp. 1111–1115 (2015). https://doi.org/10.1109/ICDAR.2015.7333933
2. Afzal, M.Z., Kölsch, A., Ahmed, S., Liwicki, M.: Cutting the error by half: Investigation of very deep cnn and advanced training strategies for document image classification. In: 2017 14th IAPR International Conference on Document Analysis and Recognition (ICDAR), vol. 1, pp. 883–888. IEEE (2017)
3. Asim, M.N., Khan, M.U.G., Malik, M.I., Razzaque, K., Dengel, A., Ahmed, S.: Two stream deep network for document image classification. In: 2019 International Conference on Document Analysis and Recognition (ICDAR), pp. 1410–1416. IEEE (2019)
4. Chen, N., Blostein, D.: A survey of document image classification: problem statement, classifier architecture and performance evaluation. Int. J. Doc. Anal. Recogn. (IJDAR) 10(1), 1–16 (2007)
5. Cosma, A., Ghidoveanu, M., Panaitescu-Liess, M., Popescu, M.: Self-supervised representation learning on document images. arXiv preprint arXiv:2004.10605 (2020)
6. Devlin, J., Chang, M.W., Lee, K., Toutanova, K.: Bert: Pre-training of deep bidirectional transformers for language understanding. arXiv preprint arXiv:1810.04805 (2018)
7. Ferrando, J., et al.: Improving accuracy and speeding up document image classification through parallel systems. In: Krzhizhanovskaya, V.V., et al. (eds.) ICCS 2020. LNCS, vol. 12138, pp. 387–400. Springer, Cham (2020). https://doi.org/10.1007/978-3-030-50417-5_29
8. He, K., Zhang, X., Ren, S., Sun, J.: Deep residual learning for image recognition. In: Proceedings of the IEEE Conference on Computer Vision and Pattern Recognition, pp. 770–778 (2016)
9. Hinton, G., Vinyals, O., Dean, J.: Distilling the knowledge in a neural network. arXiv preprint arXiv:1503.02531 (2015)

10. Hooker, S., Moorosi, N., Clark, G., Bengio, S., Denton, E.: Characterising bias in compressed models. arXiv preprint arXiv:2010.03058 (2020)
11. Joseph, V., Muralidharan, S., Garland, M.: Condensa: Programmable model compression. https://nvlabs.github.io/condensa/ (2019) Accessed 1 Jul 2019
12. Joseph, V., et al.: Reliable model compression via label-preservation-aware loss functions. arXiv preprint arXiv:2012.01604 (2020)
13. Kang, L., Kumar, J., Ye, P., Li, Y., Doermann, D.: Convolutional neural networks for document image classification. In: 2014 22nd International Conference on Pattern Recognition, pp. 3168–3172. IEEE (2014)
14. Kölsch, A., Afzal, M.Z., Ebbecke, M., Liwicki, M.: Real-time document image classification using deep cnn and extreme learning machines. In: 2017 14th IAPR International Conference on Document Analysis and Recognition (ICDAR), vol. 1, pp. 1318–1323. IEEE (2017)
15. Krizhevsky, A., Sutskever, I., Hinton, G.E.: Imagenet classification with deep convolutional neural networks. Commun. ACM **60**(6), 84–90 (2017)
16. Kumar, J., Doermann, D.: Unsupervised classification of structurally similar document images. In: 2013 12th International Conference on Document Analysis and Recognition, pp. 1225–1229. IEEE (2013)
17. Kumar, J., Ye, P., Doermann, D.: Learning document structure for retrieval and classification. In: Proceedings of the 21st International Conference on Pattern Recognition (ICPR2012), pp. 1558–1561. IEEE (2012)
18. Kumar, J., Ye, P., Doermann, D.: Structural similarity for document image classification and retrieval. Pattern Recogn. Lett. **43**, 119–126 (2014)
19. LeCun, Y., Denker, J.S., Solla, S.A.: Optimal brain damage. In: Advances in neural information processing systems, pp. 598–605 (1990)
20. Li, Y., Gu, S., Mayer, C., Gool, L.V., Timofte, R.: Group sparsity: The hinge between filter pruning and decomposition for network compression. In: Proceedings of the IEEE/CVF Conference on Computer Vision and Pattern Recognition, pp. 8018–8027 (2020)
21. Molchanov, P., Tyree, S., Karras, T., Aila, T., Kautz, J.: Pruning convolutional neural networks for resource efficient inference. arXiv preprint arXiv:1611.06440 (2016)
22. Noce, L., Gallo, I., Zamberletti, A., Calefati, A.: Embedded textual content for document image classification with convolutional neural networks. In: Proceedings of the 2016 ACM Symposium on Document Engineering, pp. 165–173 (2016)
23. Ntoutsi, E., et al.: Bias in data-driven artificial intelligence systems-an introductory survey. Wiley Interdisc. Rev. Data Min. Knowl. Discovery **10**(3), e1356 (2020)
24. Park, G., Yang, J.Y., Hwang, S.J., Yang, E.: Attribution preservation in network compression for reliable network interpretation, vol. 33 (2020)
25. Reddy, K.V.U., Govindaraju, V.: Form classification. In: Yanikoglu, B.A., Berkner, K. (eds.) Document Recognition and Retrieval XV, vol. 6815, pp. 302–307. International Society for Optics and Photonics, SPIE (2008). https://doi.org/10.1117/12.766737
26. Renda, A., Frankle, J., Carbin, M.: Comparing rewinding and fine-tuning in neural network pruning. arXiv preprint arXiv:2003.02389 (2020)
27. Russakovsky, O., et al.: ImageNet Large Scale Visual Recognition Challenge. Int. J. Comput. Vis. **115**(3), 211–252 (2015). https://doi.org/10.1007/s11263-015-0816-y
28. Sarkar, P.: Learning image anchor templates for document classification and data extraction. In: 2010 20th International Conference on Pattern Recognition, pp. 3428–3431. IEEE (2010)

29. Shin, C., Doermann, D., Rosenfeld, A.: Classification of document pages using structure-based features. Int. J. Doc. Anal. Recogn. **3**(4), 232–247 (2001)
30. Shin, C.K., Doermann, D.S.: Document image retrieval based on layout structural similarity. In: IPCV, pp. 606–612 (2006)
31. Siddiqui, S.A., Fateh, I.A., Rizvi, S.T.R., Dengel, A., Ahmed, S.: Deeptabstr: deep learning based table structure recognition. In: 2019 International Conference on Document Analysis and Recognition (ICDAR), pp. 1403–1409 (2019). https://doi.org/10.1109/ICDAR.2019.00226
32. Siddiqui, S.A., Malik, M.I., Agne, S., Dengel, A., Ahmed, S.: Decnt: deep deformable cnn for table detection. IEEE Access **6**, 74151–74161 (2018). https://doi.org/10.1109/ACCESS.2018.2880211
33. Simonyan, K., Zisserman, A.: Very deep convolutional networks for large-scale image recognition. arXiv preprint arXiv:1409.1556 (2014)
34. Szegedy, C., Ioffe, S., Vanhoucke, V., Alemi, A.: Inception-v4, inception-resnet and the impact of residual connections on learning. arXiv preprint arXiv:1602.07261 (2016)

Hierarchical and Multimodal Classification of Images from Soil Remediation Reports

Korlan Rysbayeva[✉], Romain Giot, and Nicholas Journet

LaBRI, UMR5800, Univ. Bordeaux, Bordeaux INP, CNRS, 33400 Talence, France
{korlan.rysbayeva,romain.giot,nicholas.journet}@u-bordeaux.fr

Abstract. When soil remediation specialists clean up a new site, they have a long time manually revising digital reports previously written by other experts, where they look for necessary information in accordance with similar characteristics of polluted fields. Important information lies in tables, graphs, maps, drawings and their associated captions. Therefore, experts have to be able to quickly access these content-rich elements, instead of manually scrolling through each page of entire reports. Since this information is multimodal (image and text) and follows a semantically hierarchical structure, we propose a classification algorithm that takes these two constraints into account. In contrast to existing works using either multimodal system or hierarchical classification model, we explore the combination of state-of-the-art methods from multimodal systems (image and text modalities) and hierarchical classification systems. By this combination, we tackle the constraints of our classification process: small dataset, missing modalities, noisy data, and non-English corpus. Our evaluation shows that the multimodal hierarchical system outperforms the unimodal and that the performance of multimodal system with a joint combination of hierarchical classification and flat classification on different modalities provides promising results.

Keywords: Multimodal classification · Hierarchical classification · Operational constraints

1 Introduction

Hazardous chemicals have been regularly used, utilized and spilled for decades without considering long-term issues in commercial and industrial facilities [27]. Thus, regulation services of various countries monitor the state of soil pollution for health purposes of citizens. Environmental protection agencies, companies or parties are responsible for cleaning up soil contamination. This is a laborious process which includes steps such as risk evaluation, laboratory experiments, pilot tests, field tests and further observations [10]. After performing each of

This work is supported by Abai-Verne scholarship and Innovasol Consortium.

J. Lladós et al. (Eds.): ICDAR 2021, LNCS 12821, pp. 160–175, 2021.
https://doi.org/10.1007/978-3-030-86549-8_11

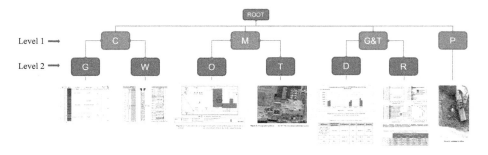

Fig. 1. *Hierarchical Tree* defined by soil remediation specialists that consist of 2-levels hierarchy. The samples illustrated for the 2nd level of classes, according to the hierarchical tree also relates to the corresponding 1st level of classes. The distribution of classes is following. **Level 1 classes:** C - cross section (505), M - maps (171), G&T - graphs and tables (416), P - photos (47). **Level 2 classes:** G - geology (195), W - well (310), O - one time (81), T - temporary (90), D - description (209), R - results (207).

these stages, specialists describe their observations and analysis of the obtained results with texts, diagrams, tables, aerial maps and photographs in a dedicated report [17] that can span from thirty pages to hundreds. Nevertheless, they are organized in similar formats and some parts may contain similar semantic hierarchical plan. Experts use experience and knowledge from previous reports to clean up a new area or re-clean a current one. Due to the problem of finding similarities between two remediation zones in terms of different characteristics from huge data source and also from specific data, it can only be personally carried out by experts. Our long-term objective is to provide a tool to assist them during structuring the huge data to search for necessary information and similarity of technical documents.

The valuable information is kept in figures and their corresponding texts. It is convenient firstly to classify the elements (figures, texts) in the document by relevant categories, which will save the time in accessing and navigating the information, instead of searching them manually. This article illustrates the first step of this objective, and presents a method for classifying the content of the different documents parts in accordance of pre-established hierarchical plan given by soil remediation specialists (Fig. 1). Our purpose is to correctly classify the data for the last level of hierarchy (2nd level). Our proposal combines state-of-the-art method for hierarchical classification [2,3,26], and multimodal classification [16,29]. We classify the data independently of each other by adding a constraint that respects the hierarchy. Moreover, to classify the data for the classes that belong to 2nd hierarchy level, which are more semantically defined, it is more convenient to additionally use text information. We are dealing with a small data source: about 1.2K images and 0.5K corresponding captions from 35 reports written by various companies with an average volume size of 50 pages. Considering that reports were provided by different experts with their inherit way of delivering information, there are constraints that includes missing modalities (some images provided without caption) and noisy data. The reports are written in French language so it is impossible to use mainstream NLP networks.

To our knowledge, several works exist on multimodal classification (*e.g.*, [14,25]) and hierarchical classification (*e.g.*, [28,31]), but there are very few works [6,23] dedicated to joint combination of multimodality and hierarchy of the classes, and even fewer works related to classification task [6]. The purpose of this article is to show the possibility of applying the combination of multimodality system and hierarchical classification on standardized data. Secondly, to show that it outperforms the individual application of multimodality system and hierarchical classification. The novelty of this work lies in comparing the multimodal hierarchical classification with multimodal classification that considers the relationship of classes as independent. Moreover, the current paper studies how the model deals with missing modality aspect and illustrates how the different fusion techniques affect the performance of hierarchical classification. Specifically, the way to fuse the modality representations affects the performance of local, global or combined approach in hierarchical classification (Sect. 2.3).

Section 2 presents the previous work. Section 3 describes the proposed methodology. Section 4 specifies the protocol of the conducted experiment and Sect. 5 contains the results of the experiment, Sect. 6 concludes.

2   Previous Work

There are few literature on MultiModal Hierarchical Classification (MMHC). We emphasize methods related to multimodal systems or hierarchical classification. Due to the nature of our application, we focus on those dedicated to classification of text and images, classification of small and imbalanced databases.

2.1   MultiModal Hierarchical Classification

Specificity of MMHC lies on both the *fusion strategy*, which belongs to modality processing and the *classification approach*, which deals with the hierarchy. The SemEval-2020 Task 8 [19] proposed three tasks: Sentiment Classification, Humor Classification and Scales of Semantic Classes for a dataset with memes of images and their corresponding short text. Das *et al.* [6] have considered these tasks as a hierarchy of classes and proposed the multi-task learning system, which combines the image feature block (ResNet [8]), text feature block (bi-LSTM [9] and GRU [4] with contextual attention) to learn all three tasks at once. They used early fusion strategy by concatenation of feature vectors to aggregate the feature block from two modalities to create task-specific features. Aggregated feature vectors passed on to smaller networks with dense layers, each one assigned to predict a sole type of fine grain sentiment label. The hierarchy dependency is managed by transfer learning of knowledge between the levels of hierarchy.

2.2   Multimodal Classification

Existing research on multimodal systems is focused on how to influence the information from multimodal feature spaces to get better performance than from their single modality counterparts. The fusion strategy can be roughly classified into *early fusion* (at feature level) and *late fusion* (at score level) [16].

Early fusion creates a joint representation of input features from multiple modalities, which is subsequently processed by a single model. The straightforward way of integrating a text with image features is a simple concatenation of their feature vectors. Wang *et al.* have used trainable CNN-RNN architecture for highlighting the meaningful text words and image regions in the Text-Image Embedding network (TieNet) [25], which is applied for classification of chest X-rays by using image and text features extracted from reports. Joint learning concatenated the two forms of representations and it uses final-fully connected layer to produce output for multi-label classification. There are also more complex fusion techniques, such as *gated summation* (using the gate value to weigh the summation of modality representations to create the fusion vector) or *bilinear transformation* (the filter that integrates the information of two vectors is concatenated with modality representations to create the fusion vector) [12,30].

Late fusion methods use a specific model on each modality and then combine the outputs to the final outcome. They integrate scores delivered by the classifiers on various features through a fixed combination rule commonly based on principle of indifference, treating all classifiers equally. The weighted averaging combines classifiers by evaluating optimal weights for each of them. Other approaches are Borda count rule, where the classifiers rank classes by giving each class the points corresponding to the number of classes ranked lower. Majority vote rule defines output by the largest number received among classifiers [11]. Separately extracting the embedding of the text and image views using pre-trained networks for classification task, Narayana *et al.* used simple weighted averaging to fuse softmax scores from image and text representations [16]. Yang *et al.* acquired the final prediction by mean-max pooling the bag-concept layer of different modalities [29]. This bag concept layer for modalities was obtained by dividing the raw articles into two modal bags - images and text paragraphs - and by calculating each modality with different networks.

2.3   Hierarchical Classification

In flat classification (FCl), the classifier is learned from labeled data instances without information about the semantic relationships between the classes. In contrast in hierarchical classification (HCl) models deal with hierarchy dependency of the labels in classification task. All existing state-of-the-art models on HCl can be divided for several approaches [1,26].

Local approach does the classification by traversing the hierarchy in top-down or bottom-up manner, learning from parent or sub-classes, accordingly. Each classifier is responsible for prediction of particular nodes or particular level of hierarchy, later combining this local prediction to generate the final classification. Fine-tuning, as a local approach, transfers the parameters from upper level to lower-level hierarchy classes while training and fine-tunes the parameters of lower levels to avoid the training from scratch. Hierarchical structure of categories for multi-label short text categorization is effectively utilized by transferring parameters of CNN trained in the upper levels to the lower levels [21]. Zhang *et al.* transferred the Ordered Neuron LSTM (ONLSTM) [20] parameters from upper level to lower level for training, and fine tuned parameters of ONLSTM between adjacent layers. They used two gates mechanism and new activation function that constructed the order and hierarchical differences among neurons. To deal with multi-label models, which suffer on categories with few training examples, Banerjee *et al.* from parent category classifiers initialized the parameters of child category and later fine-tuned the model in HTrans recursive strategy for hierarchical text classification [3].

Global approach treats the hierarchical classification as a flat multi-label classification problem, where the authors use a single classifier for all classes. *Hidden-layer initialization* approach works by initializing the final hidden layer with a layer where the co-occurrence of labels in different hierarchy is shown [1,2]. For each occurrence, the value w assigned to associated classes and value 0 assigned to the rest. The motivation for this assignment is to incline units in hidden layer by triggering only the corresponding label nodes in the output layer when they are active. *Hierarchy-aware attention mechanism* was used by Zhou *et al.* [33] to deal with the hierarchy of the classes, where the initial label embedding is directly fed into a treeLSTM [24] as input vector of specific nodes, then the output hidden state represents as the hierarchy aware label features. Furthermore, sum of product of attention value and text representation is calculated to obtain the label aligned text features. *Deep hashing* was considered using the semantic hierarchy for large-scale image retrieval problem [32]. Authors handled the hierarchy of image classes by firstly mapping the image to binary code using deep hashing model, where the class center (*e.g.*, mean value) is calculated for the lowest hierarchy level. Secondly, they updated the class center for all levels in hierarchy, which depend on the ground-level class center.

Combined approach [26] follows the hybrid method by simultaneously optimizing both local and global loss functions. The global loss tracks the dependency of labels in the hierarchy as a whole. Each local loss function enhances the propagation of gradients leading to the correct encoding of local information between classes that corresponds to the hierarchical level. Moreover, to ensure the predictions, which obey the hierarchical structure, the hierarchy violation penalty was presented. The model was trained on 21 datasets related to protein function prediction, medical images or text classification.

2.4 Discussion

Multimodal systems and hierarchical classification has been mostly studied separately using large volume of data. Even if the number of classes is large, the number of samples for the classes in deeper levels of the hierarchy is usually sufficient. Most works on text focused on the English-based corpus. Some literature related to handling the multimodal systems considers the missing modality aspect, but focused on cross-modal retrieval task [23].

3 Combined Multimodal and Hierarchical Approaches for Technical Document Content Classification

We focus on multimodal classification with hierarchy dependency of classes and study how different fusion techniques of modality representations affect the performance of hierarchical classification and show a way of improving state-of-the-art methods by using the combination of joint multimodal classification system and hierarchical classification system. Hierarchical part considers both global and local approaches [26]. In multimodal system, the concatenation (early fusion) [6] of feature representations, weighted averaging rule [16] as simple fusion rules and the mean-max pooling (late fusion) [29] are chosen as fusion techniques for text and image modalities. As this work concerns data with few training samples, where text modality can be missing, and labeling is hierarchical we revised various works of the literature tackling these constraints [11,23,26,29]. They are not directly usable in our context as they consider either hierarchy of classes [26], or the multimodality aspect [11,29] of data, but not the combination of them. Some works consider the same constraints but focuses on cross-modal retrieval task [23]. Consequently, the adaptations of state-of-the-art methods for our problem are necessary.

3.1 Proposed Architecture

The architecture of the model consists of 3 parts. (i) The raw data is processes through feature extraction stage to obtain a compact representation that contains enough information for the classification process. This representation is classified with the (ii) hierarchical and (iii) multimodal parts. Depending on the fusion technique, precisely early fusion or late fusion techniques which explained in Sect. 2.2 the hierarchical model applied after or before multimodal system respectively. This section presents feature extraction process, the hierarchical classification model and finally discusses the multimodal fusion techniques.

Feature extraction process for *Image modality* resizes the input images to 224×224 pixels (Fig. 2), then passes it to VGG16 [22] using pre-trained weights from ImageNet [7] dataset; this part of the network is kept frozen. The obtained vector goes through FC layer with 1024 dimensions and is used as input for classifier layers. *Text modality* extraction process uses bag-of-words (BoW) technique, precisely term frequency–inverse document frequency (TF-IDF) [18] to

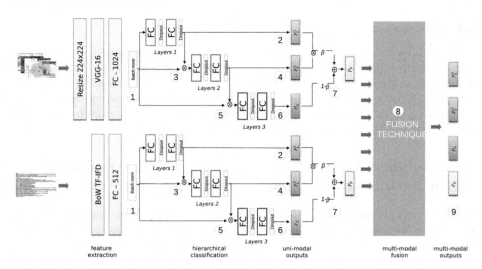

Fig. 2. The pipeline of architecture of multimodal (MM) hierarchical classification (HCl) model used during this work. The image representation passes through VGG16 pre-trained model, lately fed to the layers of hierarchical classification model ①. The text modality representations follow the same architecture except it was trained from scratch. HCl model based on the combination of local ②,④ and global approaches [26] ⑥. The hierarchy of classes handled by fine-tuning approach ③,⑤. The final score is the linear combination of two vectors from hierarchical approaches ⑦. The multimodal outputs is received after the fusion of representatives of modalities ⑨. Figure illustrates the late fusion technique of image and text modalities [16,29] ⑧; early fusion slightly simplifies the pipeline using a single branched network from ① (Sect. 3.1).

extract features from captions, as our preliminary experiments illustrated their superiority to camemBERT [13]. The vector obtained by TF-IDF has around 2500 dimensions and passed through FC layer with 512 dimensions.

The hierarchical part considers the application of the feed-forward (HMCN-F) [26] for both image and text modalities. Different combinations of layers' dimensions were tested, and finally it was decided to use 512, 256 nodes for image modality and 256, 128 nodes for text modality for each Layers 1, Layers 2 and Layers 3 (Fig. 2). The following explanation focuses on image modality, but they are similar for text modality. The feature vector of image modality is passed through Layers 1 ①. Using softmax function, the local prediction $P_{L,image}^1$ for the 4 classes of the 1st level of hierarchy is calculated ②. At the same time, the output of the last dense layer of Layers 1 is concatenated with the feature vector, which is present to Layer 2 in the architecture of the model ③. Using softmax function, the local prediction $P_{L,image}^2$ for the 6 classes of 2nd level of hierarchy is calculated ④. The last dense layer of Layer 2 is concatenated with feature vector, which was followed by passing through Layer 3 ⑤. Since global approach considers hierarchical classification as flat multi-label classification problem, 2 softmax function used for each hierarchy level to get the probability for the global prediction P_G(image) for overall 10 classes ⑥. The final prediction for 10 classes is calculated by Eq. 1, where β regulates the importance of local and global information from the class hierarchy ⑦ and $\otimes$ represents the concatenation.

$$P_{F,image} = \beta \left[P^1_{L,image} \otimes P^2_{L,image} \right] + (1 - \beta) P_{G,image} \qquad (1)$$

The multimodal system chosen fusion methods correspond to early fusion technique with 9.6M trainable parameters (*i.e.*, concatenation [6]) and late fusion technique with 6.5M trainable parameters (*i.e.*, weighted averaging [16], mean-max pooling [29]). During the experiments, each fusion technique is separately applied for the representatives of modalities. For late fusion, the fusion technique is applied for image and text representatives after being individually classified as explained in *Hierarchical part*. On the other hand, for early fusion, the fusion technique applied for feature vectors.

The weighted averaging is applied after the hierarchical model; it is the mean of two softmax tensors coming from the image and text modalities. Apart from the softmax tensors obtained for each modality $P^1_{L,image}$, $P^2_{L,image}$, $P_{G,image}$, $P_{F,image}$ and $P^1_{L,text}$, $P^2_{L,text}$, $P_{G,text}$, $P_{F,text}$, the model also calculates the corresponding 4 *multimodal* outputs, P^1_L, P^2_L, P_G and P_F ⑨. As one of the late fusion techniques, the mean-max pooling also utilizes the row-wise max pooling for the softmax tensors of image and text modalities, by firstly stacking the tensors beforehand, and consequently receives the corresponding 4 multimodal outputs. The concatenation pipeline differs from the illustration shown in Fig. 2 by transforming the fusion technique ⑧ after the feature extraction for image and text modalities respectively. Consequently, after hierarchical classification the model obtains only 4 multimodal outputs, P^1_L, P^2_L, P_G and P_F ⑨.

3.2   Loss Function

The optimizer minimizes the sum of the local (L^1_L, L^2_L) and global (L_G) loss functions with the categorical cross-entropy loss. To guarantee the consistency of hierarchical path minimizing the proposed loss function is insufficient: to penalize predictions with hierarchical violation is required [26]. The hierarchical loss appears when parent and child classes are connected. The violation of hierarchy happens when 2nd level class score is higher than the score of its parent class. Since there are more classes in 2nd level, the score is supposed to be lower for each class. The characteristic equation $x(p, c)$ is determined to define the connection between the hierarchy levels, where p indicates the class number in 1st hierarchy level, c the class number in 2nd hierarchy level. $x(p, c) = 1$, if p and c connected, $x(p, c) = 0$, if p and c is disconnected. The hierarchical violation loss is calculated for each sample i when holds the statement $Y^i_{c,2} > Y^i_{p,1}$ and exist the connection between parent and child classes:

$$L_H = \lambda \sum_i \sum_{p,c} max\{0, Y^i_{c,2} - Y^i_{p,1}\}^2 * x(p, c) \qquad (2)$$

where for sample i $Y_{c,2}^i$ represents the *child score* of class c and $Y_{p,1}^i$ represents the *parent score* of class p. $\lambda \in R$ is employed for regulating the importance of the penalty for hierarchical violations in the overall hierarchical loss function. The final loss function for whole system is optimized by

$$L_F = L_L^1 + L_L^2 + L_G + L_H \qquad (3)$$

where L_F is final loss, L_L^1, L_L^2 and L_G is the loss for local and global networks.

4   Experimental Protocol

This section gives details of the experimental procedure and describes the evaluation of interest targeted during the experiment.

4.1   Operational Dataset

The system presented in Sect. 3 was tested on a real world but private dataset. Confidentiality issues does not allow to reveal this dataset. To our knowledge, public dataset that contain multi-modal data and hierarchically defined classes does not exist. We have created our dataset by extracting images and captions from 35 reports related to soil remediation procedure. Around 500 images and their corresponding captions were automatically extracted [5] and supplemented by a manual extracting of 700 additional images with no caption. The extracted images have dimensions from 100×100 to 2000×2000 pixels, whereas the average caption length is around 44 words. Moreover, since the dataset of interest was created on the basis of reports from French companies, the considered text dataset (captions) is based on French. Soil remediation experts have labeled the images. According to Fig. 1, the dataset is spread across 2 hierarchical levels defined by the soil remediation specialists. The 1st level consists of 4 classes, the 2nd level consist of 6 classes. The samples distribution is shown in Fig. 1. To get the thorough results, the parent class *Photos*, which does not have the subclasses was transmitted for the 2nd level hierarchy, thereby the 2nd level hierarchy artificially consists of 7 classes, and global level consists of 11 classes.

Train/Validation and *Test* datasets are generated with a 10-folds stratified cross-validation that keeps the class distribution of the full dataset in each split of the data; on its turn *Train* and *Validation* dataset divided as 80% and 20%.

4.2   Hyper-parameters Configuration

Adam optimizer uses a learning rate of 10^{-3}; batches contains 16 samples; batch normalization layers are applied after first FC layer (Fig. 2); Dropout layers (0.6) are used after remaining FC layers to mitigate overfitting. Wehrmann *et al.* [26] considered two types of fusion parameters. β, which correlates the influence of local or global approach on final result (Fig. 2), and λ, which correlates the importance of hierarchical violation during calculation of loss function (Eq. 2). The value of these parameters is chosen in accordance of the best performance on the dataset of interest. For Image modality system the parameters are $\beta = 0.5$ and $\lambda = 0.1$, for Text modality system the parameters are $\beta = 0.1$ and $\lambda = 0.1$.

4.3   Evaluation Metric

Similar to the most recent works on hierarchical classification [1,21] we report weightedF1 scores for our experiments. F1 scores consider the precision (P) and recall (R) of the categories and changes between $[0,1]$, while weightedF1 computes the average of the F1 scores considering the number of instances GT_{class} of each class c.

$$weightedF_1 = \frac{\sum_c^C (F_1(c) * GT_{class}(c))}{\sum_c^C GT_{class}(c)}, F_1(c) = 2 * \frac{P(c) * R(c)}{P(c) + R(c)} \qquad (4)$$

where C is the total number of classes per hierarchy level, GT_{class} is the total number of samples per class. The weightedF1 score is calculated independently for 1st and 2nd hierarchy levels in local, global and combined approaches.

4.4   Evaluations of Interest

Several questions arise and are treated in independent evaluations.

Evaluation 1: How state-of-the-art multimodal fusion techniques perform combined with hierarchical classification for the given dataset? Three multimodal fusion techniques are combined with hierarchical classification model: weighted averaging (late fusion) weighting the modality representations equally, mean-max pooling (late fusion) and concatenation (early fusion). After receiving the results on Evaluation 1, the weighted averaging fusion of modality representation was chosen for the remaining evaluations (Sect. 5, Table 1).

Evaluation 2: Which hierarchical approach (local or global) needs to be investigated in future works? We have compared the performances of hierarchical local approach with hierarchical global approach (fine-tuning) while using the weighted averaging fusion technique for modality representations.

Evaluation 3: What is the best strategy to handle the missing modalities? Multimodal system requires the presentation of a sample of each modality, but captions maybe absent for some images. We have compared two strategies to overcome this issue: replace the missing samples with empty vectors or fallback to the monomodal system of the available modality.

Evaluation 4: How the multimodal system performs with the combination of hierarchical classification (HCl) and flat classification (FCl) models? We firstly compare the multimodal system combined with HCl and FCl systems separately, to study which modality performs better in given context, then to validate results, the multimodal system tested along with the combination of hierarchical classification and flat classification systems, taken the best of each modality performances on previous tests.

5   Results and Discussion

The results illustrated are computed globally by fusing 10 folds results (Sect. 4.1). The performance is shown for *local* and *global* approaches separately. The *final*

Table 1. WeightedF1: comparison of different fusion techniques of modality representations for hierarchical classification model. Early fusion shows better performance than the late fusion techniques. MM indicates to multimodal system.

	Late fusion						Early fusion
	Weighted averaging			Mean-max pooling			Concatenation
	Image	Text	MM	Image	Text	MM	MM
Local level1	95.88	78.68	**96.85**	95.51	80.79	41.70	**97.63***
Global level1	95.61	78.07	**96.40**	96.48	76.89	38.39	**97.54***
Final level1	96.06	78.66	**96.66**	94.53	78.45	80.27	**97.80***
Local level2	79.27	50.56	**83.26**	76.34	49.26	13.27	**83.59***
Global level2	79.05	51.25	**83.28***	72.03	26.57	19.38	**83.09**
Final level2	79.17	50.79	**83.10**	71.53	29.75	56.71	**83.95***

performance represents the performance of combined approach (Fig. 2, ⑦). The statistical significance of the results performed by Mann-Whitney rank test [15] on all folds, where the results of p-values outline superiority of one specification to another. Any sample in the dataset is assigned with one class from each level of the hierarchy. The local and global approach consist of two classifiers each. Thereby the performance is illustrated in accordance with the results obtained in Level 1 and Level 2. Using Eq. 1, the final results is supposed to be shown for 11 overall classes, but to be consistent with the illustration of results in local and global approaches, the final result is split for Level1 and Level2 categories. The early fusion (concatenation) technique creates a joint representation of image and text representations. Then the concatenated vector is used for the hierarchical classification model. Thereby the scores for Image and Text modality results is absent, and only MM classification is present.

The combination of MM and HCl systems are studied (Evaluation 1, Evaluation 2). Table 1 describes the performance of the tested multimodal fusion techniques considering hierarchical structure of labels. The concatenation fusion technique outperforms the rest probably because of the difference in number of learning weights, for Level 1 and Level 2 final scores with 97.80% and 83.95% respectively. Mann-Whitley test has confirmed that with p-value = 0.041 for Level 2. The mean-max pooling MM results shows that this fusion technique performs poorly for the tested dataset, whereas weighted averaging late fusion technique performs almost the same as early fusion technique (p-value = 0.079 for Level 1). The combination of local and global approaches works only for early fusion technique, considering the outperforming result (97.80% and 83.95%) of final scores in hierarchy levels. For the late fusion technique, the local approach performance is better for the 1st level of hierarchy. In contrast, for the 2nd level of hierarchy local and global approaches perform similarly. Furthermore, the MM outperforms the unimodality performances for both hierarchy levels, which shows that fusion of information from two modalities (Image, Text) can increase the

Table 2. WeightedF1: comparison of MM HCl system performances. *Empty vectors* corresponds to the performances of the system where the missing modality data is replaced with empty vectors and *Image fallback* corresponds to the performance of the system that fallback to the monomodal image system when text modality is unavailable.

	MM HCl (weighted averaging)		
	Unimodal	Multimodal	
	Image only	Empty vectors	Image fallback
Local level1	95.88	**96.85**	**97.68***
Global level1	95.61	**96.40**	**96.50***
Final level1	96.06	**96.66**	**97.80***
Local level2	79.27	**83.26**	**83.40***
Global level2	79.05	**83.28**	**83.69***
Final level2	79.17	**83.10**	**83.22***

results. Since the concatenation cannot show the performance of each modality separately and the mean-max pooling does not work for the dataset of interest, the weighted averaging fusion of modality representation was chosen for the remaining experiments.

The missing modality constraint is studied (Evaluation 3). Table 2 compares the performance of the image monomodal system to two variants of the multimodal system: use of an empty vector when text modality is missing or fallback to the image monomodal system otherwise. Unimodal system represents the results where the missing modality information does not influence the performance of the model. However Table 2 illustrates that the performance of the model is close for multimodal system, and unimodal performance shows worse results for both hierarchy levels (97.80% over 96.06% and 83.22% over 79.17%). This indicates that the text modality enhances the overall performance, nevertheless artificially complementing the missing text representatives with empty vectors. This ensures that empty vectors force the system to take decision with no information.

The multimodal system performance with combination of Flat Classification (FCl) and Hierarchical Classification (HCl) systems is studied (Evaluation 4). Firstly, we tested the multimodal system coupled with the HCl and FCl systems separately, to define best modality performance in each case. In the FCl system the pipeline of architecture changes by eliminating the concatenation of input vectors with the output from each network related to hierarchy approaches (Fig. 2, ③, ⑤), thereby the networks was trained independently. Secondly, the multimodal system is tested on the combination of HCl model on text modality and FCl model on image modality. Table 3 shows that the HCl model performs better on text modality (78.66% and 50.79%), FCl model performs better on image modality (96.91% and 81.33%). Moreover, the multimodal system tested

Table 3. WeightedF1: comparison of Flat classification (FCl) with Hierarchical classification (HCl) performance. The best results of each modality is represented in bold: Image modality achieves highest result under FCl system, Text under HCl system, while multimodality (MM) outperforms under combination of FCl and HCl systems.

	FCl			HCl			FCl + HCl		
	Image	Text	MM	Image	Text	MM	Image (FCl)	Text (HCl)	MM
Local level1	**96.45**	67.25	97.36*	95.88	**78.68**	96.85	96.93	67.70	**97.11**
Global level1	**96.20**	67.02	97.10	95.61	**78.07**	96.40	97.11	68.62	**97.19***
Final level1	**96.91**	67.57	97.18	96.06	**78.66**	96.66	97.19	68.09	**97.71***
Local level2	**79.43**	44.93	82.34	79.27	**50.56**	83.26	82.24	45.65	**84.61***
Global level2	**79.59**	43.81	82.16	79.05	**51.25**	83.28*	80.20	46.71	**82.97**
Final level2	**81.33**	45.09	84.90	79.17	**50.79**	83.10	82.90	46.06	**86.00***

on FCl model slightly outperforms multimodal system tested on HCl model because of simplicity of hierarchy classes with p-value = 0.018, except for Level 2 in local and global approaches (p-value = 0.104). However, the novel results were delivered by the application of multimodal system with combination of FCl and HCl models, which for modality representation overpass the performances on both modalities (97.71% and 86.00% for respective hierarchy level with (p-value = 0.014). This implies that HCl model can contribute to multimodal system performances, therefore this results acts as a foundation for perspective future research on this area. Additionally, the proposed method is competitive against humans as it completes classification in 30 min whereas the manual classification of complete dataset takes 30 h.

6 Conclusion and Perspectives

This paper proposes a multimodal and hierarchical classification framework of information contained in soil remediation reports. The main contribution of this work is the proposal of an efficient combination of state-of-the-art methods to overcome specific constraints: small dataset, missing modalities, noisy data and non-English corpus. The most relevant results are that the multimodal system enhances the performance of the unimodal systems, and regarding the hierarchical approaches, the combination of local and global approaches only works with concatenation of modality representations. For the late fusion of representations, the local approach needs to be investigated for future works. Moreover, early fusion technique for multimodal system performs better than the late fusion technique for our dataset. The multimodal system combined with hierarchical classification model performs worse than with flat classification model. However, multimodal system with hierarchical classification model for text modality and flat classification model for image modality performs better than the rest.

During this study, the trained hierarchical classification system applies both local and global hierarchical approaches, instead of having two separate networks that consider either local or global approaches, which supposedly had an impact on the obtained result. Thereby, for the future work, it is interesting to analyze the performance using these distinct networks. Moreover, in order to study whether the classifier assigns labels to classes that do not conform with the hierarchy of the categories, it is convenient to compare the post-processing label correlation [1, 2].

References

1. Aly, R., Remus, S., Biemann, C.: Hierarchical multi-label classification of text with capsule networks. In: Proceedings of the 57th Annual Meeting of the Association for Computational Linguistics: Student Research Workshop, pp. 323–330 (2019)
2. Baker, S., Korhonen, A.L.: Initializing neural networks for hierarchical multi-label text classification. In: Proceedings of the BioNLP Workshop, pp. 307–315 (2017)
3. Banerjee, S., Akkaya, C., Perez-Sorrosal, F., Tsioutsiouliklis, K.: Hierarchical transfer learning for multi-label text classification. In: Proceedings of the 57th Annual Meeting of the Association for Computational Linguistics, pp. 6295–6300 (2019)
4. Cho, K., et al.: Learning phrase representations using RNN encoder-decoder for statistical machine translation. In: Proceedings of the 2014 Conference on Empirical Methods in Natural Language Processing (EMNLP), pp. 1724–1734, October 2014
5. Clark, C., Divvala, S.: PDFFigures 2.0: mining figures from research papers. In: 2016 IEEE/ACM Joint Conference on Digital Libraries, pp. 143–152. IEEE (2016)
6. Das, S.D., Mandal, S.: Team neuro at SemEval-2020 task 8: multi-modal fine grain emotion classification of memes using multitask learning. arXiv preprint arXiv:2005.10915 (2020)
7. Deng, J., Dong, W., Socher, R., Li, L.J., Li, K., Fei-Fei, L.: ImageNet: A large-scale hierarchical image database. In: 2009 IEEE Conference on Computer Vision and Pattern Recognition, pp. 248–255. IEEE (2009)
8. He, K., Zhang, X., Ren, S., Sun, J.: Deep residual learning for image recognition. In: Proceedings of the IEEE Conference on Computer Vision and Pattern Recognition, pp. 770–778 (2016)
9. Huang, Z., Xu, W., Yu, K.: Bidirectional LSTM-CRF models for sequence tagging. arxiv 2015. arXiv preprint arXiv:1508.01991 (2015)
10. Hyman, M., Dupont, R.R.: Groundwater and soil remediation: process design and cost estimating of proven technologies, pp. 367–422. American Society of Civil Engineers (2001)
11. Kittler, J., Hatef, M., Duin, R.P., Matas, J.: On combining classifiers. IEEE Trans. Patt. Anal. Mach. Intell. **20**, 226–239 (1998)
12. Lu, D., Neves, L., Carvalho, V., Zhang, N., Ji, H.: Visual attention model for name tagging in multimodal social media. In: Proceedings of the 56th Annual Meeting of the Association for Computational Linguistics (Volume 1: Long Papers), pp. 1990–1999 (2018)
13. Martin, L., et al.: Camembert: a tasty French language model. In: Proceedings of the 58th Annual Meeting of the Association for Computational Linguistics (2020)

14. Masakuna, J.F., Utete, S.W., Kroon, S.: Performance-agnostic fusion of probabilistic classifier outputs. In: 2020 IEEE 23rd International Conference on Information Fusion (FUSION), pp. 1–8. IEEE (2020)
15. McKnight, P.E., Najab, J.: Mann-Whitney U Test, pp. 1. American Cancer Society (2010)
16. Narayana, P., Pednekar, A., Krishnamoorthy, A., Sone, K., Basu, S.: HUSE: hierarchical universal semantic embeddings. arXiv preprint arXiv:1911.05978 (2019)
17. Pastor, J., Gutiérrez-Ginés, M.J., Bartolomé, C., Hernández, A.J.: The complex nature of pollution in the capping soils of closed landfills: Case study in a mediterranean setting. In: Environmental Risk Assessment of Soil Contamination, pp. 199–223. IntechOpen, Rijeka (2014)
18. Sammut, C., Webb, G.I. (eds.): Encyclopedia of Machine Learning, chap. TF-IDF, pp. 986–987. Springer, US, Boston, MA (2010). ISBN 978-1-4899-7687-1
19. Sharma, C., et al.: SemEval-2020 Task 8: memotion analysis-the visuo-lingual Metaphor! In: Proceedings of the 14th International Workshop on Semantic Evaluation (SemEval-2020) (2020)
20. Shen, Y., Tan, S., Sordoni, A., Courville, A.: Ordered neurons: integrating tree structures into recurrent neural networks. arXiv preprint arXiv:1810.09536 (2018)
21. Shimura, K., Li, J., Fukumoto, F.: HFT-CNN: learning hierarchical category structure for multi-label short text categorization. In: Proceedings of the 2018 Conference on Empirical Methods in Natural Language Processing, pp. 811–816 (2018)
22. Simonyan, K., Zisserman, A.: Very deep convolutional networks for large-scale image recognition. Comput. Biol. Learn. Soc. 1–14 (2015)
23. Sun, C., Song, X., Feng, F., Zhao, W.X., Zhang, H., Nie, L.: Supervised hierarchical cross-modal hashing. In: Proceedings of the 42nd International ACM SIGIR on Research and Development in Information Retrieval, pp. 725–734 (2019)
24. Tai, K.S., Socher, R., Manning, C.D.: Improved semantic representations from tree-structured long short-term memory networks. In: Proceedings of the 53rd Annual Meeting of the Association for Computational Linguistics and the 7th International Joint Conference on Natural Language Processing (Volume 1: Long Papers), pp. 1556–1566 (2015)
25. Wang, X., Peng, Y., Lu, L., Lu, Z., Summers, R.M.: TieNet: text-image embedding network for common thorax disease classification and reporting in chest X-rays. In: 2018 IEEE/CVF Conference on Computer Vision and Pattern Recognition, pp. 9049–9058. IEEE (2018)
26. Wehrmann, J., Cerri, R., Barros, R.: Hierarchical multi-label classification networks. In: International Conference on Machine Learning, pp. 5075–5084 (2018)
27. Wuana, R.A., Okieimen, F.E.: Heavy metals in contaminated soils: a review of sources, chemistry, risks, and best available strategies for remediation. Heavy Metal Contamination of Water and Soil: Analysis, Assessment, and Remediation Strategies, p. 1 (2014)
28. Xue, H., Liu, C., Wan, F., Jiao, J., Ji, X., Ye, Q.: DANet: divergent activation for weakly supervised object localization. In: Proceedings of the IEEE International Conference on Computer Vision, pp. 6589–6598 (2019)
29. Yang, Y., Wu, Y.F., Zhan, D.C., Liu, Z.B., Jiang, Y.: Complex object classification: a multi-modal multi-instance multi-label deep network with optimal transport. In: Proceedings of the 24th ACM SIGKDD International Conference on Knowledge Discovery & Data Mining, pp. 2594–2603 (2018)
30. Yu, Z., Yu, J., Xiang, C., Fan, J., Tao, D.: Beyond bilinear: generalized multimodal factorized high-order pooling for visual question answering. IEEE Trans. Neural Netw. Learn. Syst. **29**(12), 5947–5959 (2018)

31. Zhang, Q., Chai, B., Song, B., Zhao, J.: A hierarchical fine-tuning based approach for multi-label text classification. In: 2020 IEEE 5th International Conference on Cloud Computing and Big Data Analytics (ICCCBDA), pp. 51–54. IEEE (2020)
32. Zhe, X., Ou-Yang, L., Chen, S., Yan, H.: Semantic hierarchy preserving deep hashing for large-scale image retrieval. arXiv preprint arXiv:1901.11259 (2019)
33. Zhou, J., et al.: Hierarchy-aware global model for hierarchical text classification. In: Proceedings of the 58th Annual Meeting of the Association for Computational Linguistics, pp. 1106–1117 (2020)

Competition and Collaboration in Document Analysis and Recognition

Daniel Lopresti[1(✉)] [iD] and George Nagy[2] [iD]

[1] Lehigh University, Bethlehem, PA 18015, USA
lopresti@cse.lehigh.edu
[2] Rensselaer Polytechnic Institute, Troy, NY 12180, USA
nagy@ecse.rpi.edu

Abstract. Over the last twenty years, competitions aimed at showcasing research on various aspects of document analysis have become a significant part of our communal activities. After a quick look at competition in general and organized competitions in other domains, we focus on the organizers' reports of the 18 competitions completed in conjunction with the 14th Conference on Document Analysis and Research, ICDAR 2017. We provide descriptive statistics on the 130 organizers of these contests, their affiliations, the 450 participants, the platforms that underlie the evaluations, and the spectrum of specified tasks. We comment on the ~ 100 citations garnered by these contests over the intervening 3.5 years. Finally, in what we consider a logical sequel, we speculate on the possibility of an alternative model of small-scale, short-range communal research based on collaboration that seems to offer benefits competitions cannot capture.

Keywords: Contests · Benchmarks · Future of document analysis · Research assessment · Performance evaluation · Technology transfer · Reproducibility

1 Introduction

We attempt to take a dispassionate look at the value and cost of competitive research in the ICDAR community. After a brief general discussion of the benefits and drawbacks of competitive research, we focus on the organizers' reports of the 18 completed competitions held in conjunction with ICDAR 2017, chosen because the intervening 3.5 years should allow sufficient time for a degree of impact to accrue. We present a snapshot of the topics, evaluation platforms, organizers, participants, and indications of the impact of these competitions. Then we take a ninety-degree turn and speculate on the possibility of collaborative research in the format of small-scale imitations of the Research Priorities and Grand Challenges set in motion by various international and national organizations, and NGOs [1–4].

Competitions for "best" solutions to preset problems are popular in computer science, software engineering and mathematics, but rare in physics, chemistry and biology. In engineering they are largely confined to student team projects like concrete canoes, matchstick bridges and solar cars. But the celebrated 2004–2005 DARPA driverless races

© Springer Nature Switzerland AG 2021
J. Lladós et al. (Eds.): ICDAR 2021, LNCS 12821, pp. 176–187, 2021.
https://doi.org/10.1007/978-3-030-86549-8_12

across the desert were strictly for grownups,[1] as were the succeeding high-stake competitions for humanoid robots and satellite launches. Amateurs, like ham radio operators, birdwatchers and wild flower enthusiasts, compete more amicably at various scales.

Among the benefits claimed for contests are directing research to important unsolved problems, promoting best solutions, calling attention to roadblocks, rewarding successful researchers, and publicizing host organizations. Putative benefits also include generating benchmark data sets and developing common ground on performance metrics. Some of the benefits purportedly hinge on the reproducibility of the methods and results of a competition. In [6], we studied the 2016–2019 ICPR and ICDAR competitions from the perspective of reproducibility, and observed that while a few are doing a good job in this regard, most fall short in significant ways that might be easy to remedy if more attention was paid to certain desirable guidelines.

The costs are less widely advertised, and may indeed be only loss of time and diversion of attention. Common evaluation metrics may reduce diversity in evaluations and discourages otherwise promising approaches, particularly when multi-dimensional metrics are arbitrarily combined into a scalar value (e.g., the F-measure). Also, once standard measures are developed, less scrupulous individuals can find ways to "game" the system. Many competitions unintentionally enable skirting ethical standards by giving entrants lengthy access to test data, ostensibly to avoid the challenges of having to run submitted code developed under complex and perhaps hard-to-reproduce software environments. Repeated competition scenarios may prove counterproductive: how many iterations does it take for reported results become too specific or lose relevance? Witness what has happened with contests based on Highleyman's data [7], UW-1 [8] and MNIST [9], which may still provide some educational value for acclimating new students, but have long-since become uninteresting from a research perspective.

2   Prior Work

There is a large body of work on competition (covert and overt) in economics, psychology, anthropology and education. As mentioned above, organized competitions abound in every sphere. Some make headlines, especially those in athletics, political elections, film, television, music, book awards, and beauty contests. In some countries, chess and Go competitions draw popular attention. Closer to our sphere are mathematics, programming and robotics contests. We do not, however, know of any other competitions in data processing, which is what document analysis research is really about.[2] The huge number of input artifacts (billions on paper, plus born-digital text and images proliferating exponentially in cellphones and in the clouds), and the infinite number of possible outputs (analyses, transformations, transcriptions and interpretations), distinguish our competitions from all others. Although several competition organizers, past and present, have written knowledgeably and thoughtfully about the benefits of competitions and the

---

[1] DARPA specifically calls out the economic advantage of offering prizes instead of directly funding research [5].

[2] The well-known examples of AI techniques programmed to beat human experts in checkers, chess, Jeopardy, and Go fall in a different category from what we are considering here.

desirable aspects of training and test data, evaluation metrics and protocols, and submission platforms, we beg the reader's indulgence to exit this Prior Work section without any references.

3   Competitions in the DAR Community

3.1   Overview

From 2003 to 2019 more than 100 competitions have been organized in conjunction with ICDAR. Participants assembled themselves into teams and registered cost-free in response to announcements posted many months earlier. Registration was typically required for access to datasets and additional guidance via email. On occasion, competitions were cancelled when there is insufficient interest expressed by the community, as we have noted elsewhere [6]. We downloaded and studied the 6- to 9-page reports from the 18 competitions completed under the aegis of ICDAR 2017 [10]. These included:

- Arabic Text Detection and Recognition in Multi-resolution Video Frames
- Baseline Detection
- Classification of Medieval Handwritings in Latin Script
- Document Image Binarization
- Handwritten Text Recognition on the READ Dataset
- Historical Document Writer Identification
- Information Extraction in Historical Handwritten Records
- Layout Analysis for Challenging Medieval Manuscripts
- Multi-font and Multi-Size Digitally Represented Arabic Text
- Page Object Detection
- Post-OCR Text Correction
- Reading Chinese Text in the Wild
- Recognition of Documents with Complex Layouts
- Recognition of Early Indian Printed Documents
- Robust Reading Challenge on COCO-Text
- Robust Reading Challenge on Multi-lingual Scene Text Detection and Script Identification
- Robust Reading Challenge on Omnidirectional Video
- Robust Reading Challenge on Text Extraction from Biomedical Literature Figures

3.2   Organization

These competitions were organized by 119 members of the ICDAR community. Ten individuals served on three or more organizing committees, and 53 on at least two. They collectively represented 33 research laboratories or universities (Fig. 1). The five most active each provided eight or more organizers to the eighteen contests. Thirteen institutions were represented by only one organizer at a single contest. We are deliberately presenting only aggregate numbers, but it is fair to say that our community includes several active clusters of competition organizers. Beyond bringing together the community at a conference, organized competitions may also provide data that organizers find useful for other purposes.

3.3 Participants

Counting the number of competitors is more difficult, because five reports do not name or count them, and other reports give only the team sobriquet for some teams. We noted similar deficiencies in some reports we tallied in our earlier paper focusing on reproducibility [6]. Using the average membership (3.3) of the 77 teams for which we have complete counts, we estimate that the 2017 ICDAR competitions attracted 430 participants (including quite a few of the organizers of other competitions) representing 130 teams. Even if there some overlap between teams in distinct competitions, the number of competitors is comparable to ICDAR attendance (data presented at the welcome session indicates there were 386 registration for the main conference). The number of teams per competition ranged from 2 to 18, and the largest team had 9 members. As one might expect, participation (Fig. 2) is far less concentrated and more geographically diverse than administration.

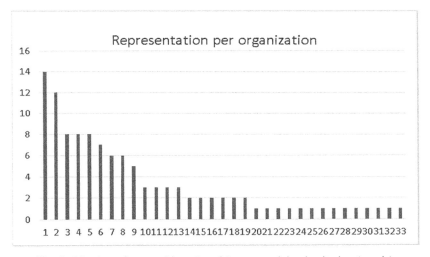

Fig. 1. Number of competitions (y-axis) vs. organizing institution (x-axis).

In terms of the distribution of organizers across contests, 89 people were involved in organizing one contest, 20 in organizing two contests, nine in organizing three contests, and one person was involved in organizing five contests.

Industrial participation was represented by teams from giants AliBaba, Google, Samsung and Tencent, and a dozen smaller and more specialized companies. The British Library and the Bibliothèque nationale de France contributed their multilingual expertise. Most of the participants were from Asian institutions in China, Vietnam, Japan and Korea. This appears largely consistent with the number of contributed papers at the main conference, which had nearly twice as many authors from China as the next highest country. A sparse sampling of the affiliations of prominent participants (and organizers) appears, roughly East-to-West, listed below.

School of ICT, Griffith University
University of Technology Sydney
National Laboratory of Pattern Recognition, Chinese Academy of Sciences
Tsinghua University, Beijing
CVPR unit, Indian Statistical Institute
National Center for Scientific Research Demokritos, Athens and Thrace
LATIS Lab, University of Sousse
National Engineering School of Sfax
Computer Vision Lab, TU Wien
DIVA group, University of Fribourg (Unifr) Fribourg
Computational Intelligence Technology Lab, University of Rostock
Paris Descartes University and Centre national de la recherche scientifique
L3i Laboratory, University of La Rochelle
University Rennes 2 and Insa Rennes,
Computer Vision Center, Universitat Autonoma de Barcelona
PRHLT research centre. Universitat Politecnica de Valencia
PRImA, University of Salford, Manchester
Brigham Young University, Provo,Utah

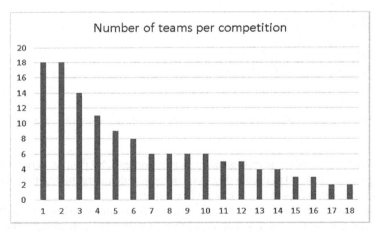

Fig. 2. Numbers of participating teams (y-axis) for each competition (x-axis).

3.4 Experimental Data

Of perennial interest is the nature of the data sets used to train and test proposed systems. According to the reports, the emphasis is on text and images that present a good variety of potential recognition problems. The resulting "convenience samples" make for interesting competitions, but they are the antithesis of the random samples necessary to predict performance on a given population of documents (such as all 19th Century conference proceedings in a national library, or an archival collection, or all Google books with a 19th Century date of publication). Random sampling is, of course, the key to inference from observations of scientific experiments [11]. We saw, however, only a few instances of small-scale random sampling (e.g. random selection of 100 pages from 10 hand-picked books). Data augmentation, the addition of labeled synthetic documents to the training set, is sometimes used to make up for the paucity of real data.

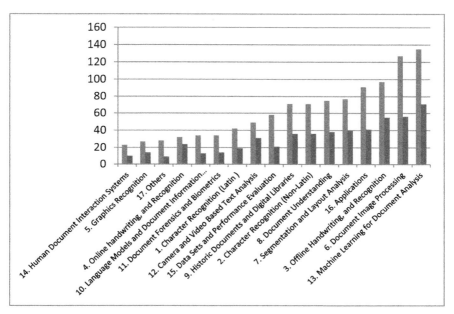

Fig. 3. Topic areas for contributed papers at the main ICDAR 2017 conference (Color figure online).

Several reports mention the desirability of larger training samples, but random sampling of a well-defined population is hampered by selective digitization and transcription of source material, which is usually driven by other priorities. For example, librarians are likely to choose what they digitize based on a document's popularity or scholarly importance. Condition of the original source may also factor into consideration: material in good condition may be preferred because it is easier and cheaper to handle during the digitization process, or material in bad condition might be preferred because documents deteriorate over time to the point where the originals can no longer be handled safely by scholars. Copyright issues may also be a factor. All of these considerations point to non-random selections made during the sampling process, which impacts the generalizability of the results.

Ten competitions featured contemporary data (including scene and video text), so we consider the remaining 8 as historical document processing (of printed, handwritten and illuminated manuscripts). Both the contemporary and the historical material spanned many languages and scripts: Arabic, Bengali, Chinese, Japanese, English, French, German and Latin. Only four of the databases were strictly English, but it is difficult to find any contemporary non-European text that does not contain any English phrases. The size of the databases varied, according to task, over almost three orders of magnitude, from under 100 pages to over 10,000 pages or cropped page images. There were contests based on cropped words, lines, and illustrations as well as on cropped page images. Four contests evaluated various tasks on scene text (from videos, cellphone images, web screenshots, and multiple cameras), and two contests tested methods on synthetic (machine-generated) text. Many contests subdivided the end-to-end pipeline, e.g., layout analysis, followed by character or word recognition. Preprocessing might involve digitization, and selecting and labeling training and validation data. There is a potential here, too, for selection bias. The variety of tasks devised by the competition organizers makes one ponder the current meaning of document.

As another representation of the research interests of the community, the distribution of submitted (blue) and accepted (red) papers for the main conference are shown in Fig. 3, reproduced from slides used for the welcome session.

3.5 Submission and Evaluation Platforms

Systems used in several competitions include Alethea from the University of Salford [12], DIVA Services from the University of Fribourg [13], ScriptNet - READ (Recognition and Enrichment of Archival Documents) of the European Union's Horizon 2000 project [14]. and RRC Annotation and Evaluation from the University of Barcelona [15]. All of these systems, which were used for two-thirds of the ICDAR 2017 competitions, were developed and used for earlier competitions. The PAGE (Page Analysis and Ground-truth Elements) Format Framework [16], also developed at Salford, is occasionally used independently of *Alethea*.

The remaining competitions used individual custom platforms programmed in python, java, or other languages, with XML, CSV, or simple text ground-truth. Most had been developed for earlier competitions. Common metrics include accuracy, precision/relevance/F-measure, edit distance, IoU (intersection over union), and various heuristic thresholds to rule out counterintuitive results. The granularity of the computation of averages is not always clearly specified.

3.6 Notes

None of the competitions addressed the question of what is to be done with residual errors and unclassified items. Some competitions, however, did suggest ways (evaluation profiles) to map their metrics into application-specific costs. It is difficult to see how the current competition paradigms could measure the human labor cost of tuning and training an existing system to a new evaluation platform and new data.

4 Journal and Conference Publications

Each report typically has one page or so summarizing each method used by the participants in a paragraph or two. Some teams submit several entries with different methods, but no method is attributed to several teams. The summaries vary greatly, even within the same competition, in the level of detail. The reports almost never disclose the email addresses of the participants.

If competitions have an impact on research, then surely this will be reflected in subsequent publications. We ascertained the number of Google Scholar citations garnered by the eighteen contemporaneous contests: 174, or 9.7 per competition (including 25 citations by the organizers, usually in reports of subsequent competitions). Our search was limited to citations of entire contests. Since the reports themselves list very few papers by the contestants, their individual or team research results will probably take longer than a three-year latency to attract wide attention.[3]

It can be noted that our data suggests not all competitions are created equal. In 2017 there was one competition that has garnered 140 + citations by April 2021, and some ICDAR competitions have gathered 600 + and 800 + citations over the years.[4] The number of citations per 2017 ICDAR contest reported at the time of our original writing (January 2021) ranged from 0 to 42. Longer retrospection will certainly give larger and more stable counts, but at the cost of shedding light only on long past activities. The skewed distributions do indeed suggest competition among the competitions. They also raise the question whether peak counts or average counts are the better measure of the value of competitions for the ICDAR community.

Finding a place to present or advertise one's work in hundreds of journals and conferences also has its competitive aspects. Some are merely financial, as in for-profit journals and conferences where the main entry barrier is a page charge or registration fee. In others, editors and referees attempt to select the contributions likely to prove most attractive to their readers and participants. Experimental reports are routinely rejected unless they can demonstrate results superior to some other experimental reports (which may be one inducement for researchers to participate in competitions).

The competition is intensified by the automation of citation counts and their application to decisions, like academic promotions and grant awards, which were never intended by the founders of scientific communication. It is a commonplace that Albert Einstein's h-index would not deserve attention without a huge adjustment for publication inflation.

We were curious about the influence of survey articles compared to the competitions. We did not find any relevant reviews published within a year of some of the competitions. Perhaps most of the competition tasks are too narrow to attract frequent review.

Citations of a competition report are just one measure of impact. The number of competitors going from the nth to the $n + 1$st is another (impact as reflected by growing community interest). We also might consider the differential citation rate for published papers related to the entrants in competition versus similar published papers by non-competitors. If by competing in a contest a researcher gets a lot more citations compared

---

[3] Systems for automatically collecting and tabulating citation counts could prove informative for a comparative here, but as with all such measures should be taken with a grain of salt. For example, a quick perusal of the results of searching "ICDAR 2017" on Microsoft Academic shows a mix with more than half of citations on the first page of results (the highest counts) going to regular papers, with a few competition reports mixed in [17].

[4] We thank the anonymous Senior PC member who provided these observations in the meta-review for our ICDAR 2021 submission.

to someone who does not compete, the payoff in terms of impact could be significant. In addition, future publications that leverage work done for a competition (the framework, data, evaluation measures) should count as a form of impact, even if the authors did not actually participate in the competition itself. Finally, it seems possible that selected results from competitions are reported in summary tables in later papers addressing the same problem and using the same data; such papers ought to cite the competition, but may instead choose to cite the scientific papers that describe the tested methods.

Still, questions remain when we ponder the impact of competitions on the trajectories of lines of research. What does it mean to suggest that one method dominates all others because it has won a particular contest? Are promising "losers" receiving due attention, or are they being shuffled off to the "scrap heap" of history, only to be rediscovered (hopefully) sometime in the future, a scenario that has played out before in pattern recognition research? Is the time and effort devoted toward developing many similarly performing methods, at the expense of leaving other territory unexplored, a good investment of the community's scarce resources?

5   Collaboration

We can speculate about a complementary model for advancing our field. What if optional research directions for our community were set each year, for overlapping two-year periods, by an IAPR committee, perhaps composed of representatives from TC-10 and TC-11? The committee (we might call this ADAR, for "Advancing Document Analysis Research") would consist of leaders in the field, such as journal editorial board members and those who have chaired or are chairing important conferences. They would select their own chairperson and maintain liaison with other relevant organizations, including funding agencies and professional societies in various countries. Their goal would be to identify a small set of agreed upon research objectives, much like what major scientific academies and funding agencies promote on a much grander scale and longer timelines [1–4].

The ADAR Committee would have two main tasks:

Task 1. Each year select five topics, suitable for experimental research, on the basis of interest level measured by recent submissions to the relevant conferences, workshops and journals. Topics may be repeated from year to year, until they reach the point of diminishing returns. We are aware that such a choice of topics is open to the objection of looking backwards rather than forwards, at the territory to be explored. We cannot, however, foresee the unforeseen, and must therefore be satisfied with innovative solutions to known problems. New problems will gather momentum through individual efforts and eventually rise to the attention of the ADAR Committee.

Task 2. Issue a Call for Participation (CfP) for each of the five topics each year. This document will give a concise description of the problem area, and add a few references to prior work and metrics. It will also set a date for the appointment of a steering committee for each topic. The steering committee will be selected from the applicants to the CfP for that topic, perhaps winnowed by some criterion for experience or a lower or higher bound for age. As a condition of appointment, each candidate will have to sign a public agreement to contribute to the Final Report due at the end of the two-year period from the appointment of the steering committee. The steering committee, in turn, will organize itself and all the participants to conduct research along the lines of the CfP and leading to the preparation of their Final Report.

The ADAR Committee will promptly submit the five yearly Calls for Participation, Lists of Participants, and Final Reports to the IAPR TC-10 and TC-11 leadership for review and posting on the respective websites, and will also request their publication in the proceedings of the next dominant DAR conference. Responsibility for the integrity of the research and the quality of the final report will rest solely with its mandatory signatories.

Credit (or blame) in the community will necessarily accrue to the ADAR Committee, the steering committees, and the participants. We believe that a good final report will be as creditable as lead authorship in a prestigious publication, and participation will be comparable to co-authorship. (Papers in experimental physics often have more than 100 authors.) A bad final report will be an albatross around the neck of its authors and participants.

Why five topics? We believe that, with a few dozen participants in each project, five is as many as can be managed by our community. Why two years? We expect that once appointed (say for overlapping five-year terms), the ADAR Committee would need about three months every year for a judicious choice of topics, and perhaps another two months for selecting the steering committees. The steering committee might need three months to set up protocols and initiate research, and three months at the end to analyze the experimental results and prepare the final report.

There is, of course, also the potential for meta-research in our proposal. The creation of new platforms for effective collaboration – for example, sharing and combining methods – would be significant contributions deserving of recognition. More attention would be aimed at the human side of the equation: the time and effort needed to develop, test, field, and maintain document analysis systems, as well as to cope with the cases they still cannot handle. Reproducibility may also be facilitated since all of the participants on a project are nominally working together employing open lines of communication.

6 Conclusions

Our snapshot of the ICDAR 2017 competitions confirms that organizers devised a variety of "challenging" tasks, constructed versatile multi-use submission and evaluation platforms, defined useful metrics, located obscure sources of digitized, transcribed and annotated "documents" spanning many centuries, scripts and languages, and attracted capable participants from much of the world. The organizers filed conscientious reports in the conference proceedings, though they differed in their emphases of different aspects of the contests and the levels of detail they disclosed varied widely.

Do the results give an accurate indication of the 2017 state of document analysis and recognition? After layout analysis and transcription, can we summarize magazine articles well enough to improve query-answer systems? Once we have located and identified all the relevant components of a technical article, can we construct an abstract more informative than the author's? Can finding and reading incidental text allow labeling photographs accurately enough to divide them into albums that make sense? Will automated analysis of old letters reveal the context of preceding and succeeding letters by the same author to the same destinataries? Will the analysis of ancient manuscripts allow confirming or contradicting current interpretation of historical events? Do the competitions point the way to the ultimate goals of DAR? What are these goals?

We saw that competition is ubiquitous and pervasive, and it surely has some merit. We listed its manifestations in the metadata generated by our research community. We tried to quantify the influence of organized research competitions on subsequent research, and compared it to the influence of journal and conference publications. We also proposed a collaborative model for experimental research different from the large-scale efforts organized by major funding agencies. We believe that organized competitions and collaborations can coexist, with some researchers more productive with one modus operandi, some with the other, and many preferring to work entirely on their own or in fluid, informal groupings. We now look forward to further joint ventures into uncharted DAR research territory.

Acknowledgements. We gratefully acknowledge the thoughtful feedback and suggestions from the anonymous reviewers as well as the cognizant Senior PC member, especially given the unconventional topic of our paper. We have incorporated several of their suggestions in this final version, and continue to ponder others. Paraphrasing one of the reviewers, our primary aim is indeed to reflect on new methods of interaction between researchers within the DAR community, to help make the community more efficient, more dynamic, more visible, and ultimately more impactful. We also thank all those who have organized competitions at ICDAR and other conferences in our field: their contributions are more significant than the recognition they receive for such efforts.

References

1. National Science Foundation: Big Ideas. https://www.nsf.gov/news/special_reports/big_ideas/
2. United Nations: Goals. https://sdgs.un.org/goals
3. Computing Research Association: Visioning. https://cra.org/ccc/visioning/visioning-activities/2018-activities/artificial-intelligence-roadmap/
4. U.K.: AI Roadmap. https://www.gov.uk/government/publications/ai-roadmap
5. Defense Advanced Research Projects Agency: Prize Challenges. https://www.darpa.mil/work-with-us/public/prizes
6. Lopresti, D., Nagy, G.: Reproducibility: evaluating the Evaluations. In: Third Workshop on Reproducible Research in Pattern Recognition (RRPR 2020), Milan, Italy (virtual), January 2021
7. Highleyman, W.: Data for character recognition studies. IEEE Trans. Electron. Comput. **12**, 135–136 (1963)
8. Liang, J., Rogers, R., Haralick, R.M., Phillips, I.: UW-ISL document image analysis toolbox: an experimental environment. In: Proceedings of the Fourth International Conference on Document Analysis and Recognition (ICDAR), Ulm (1997)
9. LeCun, Y., Cortes, C., Burges, C.J.C.: The MNIST Database. http://yann.lecun.com/exdb/mnist/
10. Competition reports. In: 14th IAPR International Conference on Document Analysis and Recognition (ICDAR), IEEE (2017)

11. Wheelan, C.: Naked Statistics: stripping the dread from the data, WW Norton & Company, New York, London (2013)
12. Clausner, C., Pletschacher, S., Antonacopoulos, A.: Aletheia - an advanced document layout and text ground-truthing system for production environments. In: Proceedings of the 11th International Conference on Document Analysis and Recognition (ICDAR2011), Beijing, China, pp. 48–52, September 2011
13. Würsch, M., Ingold, R., Liwicki, M.: DIVAServices—a RESTful web service for Document Image Analysis methods, Digital Sch. Humanit. **32**(1), i150–i156 (2017). https://doi.org/10.1093/llc/fqw051
14. Diem, M., Fiel, S., Kleber, F.: READ – Recognition and Enrichment of Archival Documents (2016). https://readcoop.eu/wpontent/uploads/2017/01/READ_D5.8_ScriptNetDataset.pdf
15. Karatzas, D., Gomez, L., Nicolaou, A., Rusinol, M.: The robust reading competition annotation and evaluation platform, In: 2018 13th IAPR International Workshop on Document Analysis Systems
16. Pletschacher, S., Antonacopoulos, A.: The PAGE (Page Analysis and Ground-truth Elements) format framework. In: Proceedings of the 20th International Conference on Pattern Recognition (ICPR2010), pp. 257-260, Istanbul, Turkey. IEEE, CS Press, 23-26 August 2010
17. Microsoft Academic search for "ICDAR 2017" on 21 February 2021. https://academic.microsoft.com/search?q=%20ICDAR%202017&f=&orderBy=0&skip=0&take=10

Handwriting Recognition

2D Self-attention Convolutional Recurrent Network for Offline Handwritten Text Recognition

Nam Tuan Ly$^{(\boxtimes)}$ (iD), Hung Tuan Nguyen (iD), and Masaki Nakagawa (iD)

Tokyo University of Agriculture and Technology, Tokyo, Japan
`nakagawa@cc.tuat.ac.jp`

Abstract. Offline handwritten text recognition is still a big challenging problem due to various backgrounds, noises, diversity of writing styles, and multiple touches between characters. In this paper, we propose a model of 2D Self-Attention Convolutional Recurrent Network (2D-SACRN) for recognizing handwritten text lines. The 2D-SACRN model consists of three main components: 1) a 2D self-attention based convolutional feature extractor that extracts a feature sequence from an input image; 2) a recurrent encoder that encodes the feature sequence into a sequence of label probabilities; and 3) a CTC-decoder that decodes the sequence of label probabilities into the final label sequence. In this model, we present a 2D self-attention mechanism in the feature extractor to capture the relationships between widely separated spatial regions in an input image. In the experiment, we evaluate the performance of the proposed model on the three datasets: IAM Handwriting, Rimes, and TUAT Kondate. The experimental results show that the proposed model achieves similar or better accuracy when compared to state-of-the-art models in all datasets.

Keywords: 2D Self-sttention · Handwritten text recognition · CNN · BLSTM · CTC

1 Introduction

Offline handwritten text recognition is an important part of handwritten document analysis problems and has been receiving much attention from numerous researchers. This task is still a big challenging problem, however, due to various backgrounds, noises, diversity of writing styles, and multiple touches between characters. Most early works on offline handwritten text recognition in western languages were often taking the segmentation-free approach based on Hidden Markov Model with features extracted from the images using a Gaussian mixture model or neural networks [1, 2]. On the other hand, for Chinese and Japanese, the segmentation-based approach was commonly applied [3, 4]. This approach segments or over-segments text into characters and fragments and then merges the fragments in the recognition state.

In recent years, many segmentation-free methods based on Deep Neural Networks (DNNs) and Connectionist Temporal Classification (CTC) [5–13] have been proposed

© Springer Nature Switzerland AG 2021
J. Lladós et al. (Eds.): ICDAR 2021, LNCS 12821, pp. 191–204, 2021.
https://doi.org/10.1007/978-3-030-86549-8_13

and proven to be powerful models for both western and oriental text recognition. The core recognition engine has been shifted from Hidden Markov Models (HMMs) to Recurrent Neural Networks (RNNs) with CTC. The principle of the CTC-based approach is to interpret the network output as a sequence of label probabilities over all labels and use an objective function to maximize the sequence probability. Convolutional Neural Networks (CNNs) are successfully employed as feature extractors in the CTC-based models [8–13]. It processes the information in a local neighborhood, so that it might not extract information from long-distance locations in an input image.

This work presents a 2D Self-Attention Convolutional Recurrent Network (2D-SACRN) model with a 2D self-attention mechanism for recognizing handwritten text lines. The proposed model consists of three main components: a feature extractor, a recurrent encoder, and a CTC-decoder. Given an input image, the feature extractor extracts a feature sequence from it by a CNN and a 2D self-attention block. The recurrent encoder encodes the feature sequence into a sequence of label probabilities by a bidirectional Long Short-Term Memory (BLSTM) network. The decoder applies a CTC to decode the sequence of label probabilities into a final label sequence. In this model, we present a 2D self-attention mechanism in the feature extractor to help the CNN to capture the relationships between widely separated spatial regions in an input image. According to our extensive experiments on the three datasets of IAM Handwriting, Rimes, and TUAT Kondate, the 2D-SACRN model achieves similar or better accuracy than the state-of-the-art models. The 2D self-attention map visualization shows that the 2D self-attention mechanism helps the feature extractor capture the relationships between widely separated spatial regions in an input image.

The rest of this paper is organized as follows: Sect. 2 describes the related work. Section 3 presents an overview of the 2D-SACRN model. Section 4 describes the datasets. Section 5 reports our experimental results and analysis. Finally, Sect. 6 draws conclusions.

2 Related Work

Early works of the CTC-based approach were introduced by A. Graves et al. [5, 6]. They combined BLSTM with CTC to recognize both online and offline handwritten English text [5]. Subsequently, they proposed Multi-Dimensional LSTM (MDLSTM) with CTC for offline handwritten Arabic text recognition [6]. Following the success of the CTC-based approach, V. Pham et al. presented an end-to-end MDLSTM with dropout followed by CTC for handwritten text recognition [7]. B. Shi et al. proposed the combination of CNN and BLSTM, followed by CTC, which called Convolutional Recurrent Neural Network (CRNN) for image-based sequence recognition [8]. Following the work of CRNN in [8], T. Bluche et al. proposed the Gated Convolutional Recurrent Neural Networks (GCRNN) for Multilingual Handwriting Recognition [9]. At the same time, N. T. Ly et al. presented the combination of pre-trained CNN and BLSTM with CTC, which is named Deep Convolutional Recurrent Network (DCRN) for recognizing offline handwritten Japanese text [10]. The DCRN model achieved better accuracy than the traditional segmentation-based method [3]. Then, they presented an end-to-end version of the DCRN model for recognizing offline handwritten Japanese text [11] and historical Japanese text [12]. J. Puigcerver et al. employed MDLSTM or CNN+LSTM, both followed by CTC for offline handwritten English and French text recognition [13].

Recently, the sequence-to-sequence (seq2seq in short) model with the attention mechanism has been successful applied to many seq2seq tasks, such as machine translation [14] and speech recognition [15]. Many segmentation-free methods have been studied based on the attention-based seq2seq model for image-based sequence recognition tasks [16–21]. J. Sueiras et al. combined the use of a horizontal sliding window and an attention-based seq2seq model for handwritten English and French text recognition [16]. T. Bluche et al. also employed an attention-based end-to-end model combined with an MDLSTM network for handwritten paragraph recognition [17, 18]. N. T. Ly et al. also proposed an attention-based seq2seq model with residual LSTM for recognizing multiple text lines in Japanese historical documents [19, 20]. Zhang et al. presented an attention-based seq2seq model with a CNN-encoder and a GRU decoder for robust text image recognition [21].

More recently, A. Vaswani et al. introduced a self-attention mechanism-based model named Transformer for seq2seq tasks [22]. The Transformer achieved state-of-the-art results in some tasks of machine translation [22] and speech recognition [23]. Following the work of Transformer, L. Kang et al. presented the CNN-Transformer model for handwritten text line recognition [24]. Meanwhile, N. T. Ly et al. presented an Attention Augmented Convolutional Recurrent Network with a self-attention mechanism for Handwritten Japanese Text Recognition [25]. These works have a limitation on capturing the spatial relation since they utilize the 1D self-attention mechanism.

3   The Proposed Method

3.1   2D Self-attention Mechanism

Convolutional Neural Networks have been proven to be compelling models and achieve state-of-the-art results in many computer vision tasks. However, convolutional operation processes the information in a local neighborhood. Thus, it is difficult to obtain information from long-distance locations. X. Wang et al. proposed the non-local operations in Non-local Neural Networks for capturing long-range dependencies in an image or video [26]. H. Zhang et al. adapted the non-local model in [26] to introduce self-attention to the GAN framework, helping both the generator and the discriminator capture the relationships between widely separated spatial regions [27]. Based on their works, in this paper, we present a 2D self-attention block in the feature extractor to help it capture the relationships between widely separated spatial regions in an input image. The architecture of the 2D self-attention block is shown in Fig. 1.

Let $X \in R^{H \times W \times C}$ denote a feature grid input to the 2D self-attention block (where H, W, and C are *height, width,* and the number of channels of the feature grid X, respectively). Firstly, the 2D self-attention block transforms the feature grid X into three feature grids: queries $Q \in R^{H \times W \times C'}$, keys $K \in R^{H \times W \times C'}$, and values $V \in R^{H \times W \times C}$ by linear projections as shown in Eq. (1):

$$Q = X \otimes W_Q \quad K = X \otimes W_K \quad V = X \otimes W_V \tag{1}$$

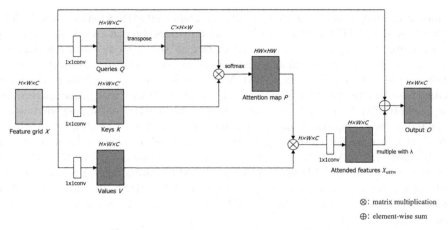

Fig. 1. Architecture of 2D self-attention block.

where the projections are the learned parameter matrices $W_Q \in \mathbb{R}^{C \times C'}, W_K \in \mathbb{R}^{C \times C'}$, and $W_V \in \mathbb{R}^{C \times C}$ with each implemented as a 1×1 convolution layer.

The 2D self-attention maps $P \in R^{H \times W \times H \times W}$ are calculated from the queries Q and the keys K as shown in Eq. (2) and Eq. (3):

$$s_{ijqk} = K_{ij} \otimes Q_{qk}^T \tag{2}$$

$$P_{ijqk} = \frac{\exp(S_{ijqk})}{\sum\limits_{q=0,k=0}^{q=H,k=W} \exp(S_{ijqk})} \tag{3}$$

where Q^T is the transpose of the queries Q and P_{ijqk} indicates how the ij^{th} location in the feature grid X attend to the qk^{th} location in the feature grid X.

Then, the attended feature grid X_{attn} are computed from the 2D self-attention maps P and the values V as shown in Eq. (4):

$$X_{attn} = (P \otimes V) \otimes W_F \tag{4}$$

where $W_F \in \mathbb{R}^{C \times C}$ is the learned parameter matrices implemented as a 1×1 convolution layer.

Finally, the output of the 2D self-attention block is calculated from the attended features X_{attn}, and the input feature grid X as follow:

$$O = X_{attn} * \lambda + X \tag{5}$$

where λ is a learnable scalar, and it is initialized as 0.

3.2 2D Self-attention Convolutional Recurrent Network

In this work, we propose a model of 2D Self-Attention Convolutional Recurrent Network (2D-SACRN) for recognizing handwritten text lines. The proposed model is composed of three main components: a feature extractor, a recurrent encoder, and a CTC-decoder, as shown in Fig. 2. They are described in the following sections.

3.2.1 Feature Extractor

In the 2D-SACRN model, we employ a CNN network followed by a 2D self-attention block to build the feature extractor. The CNN network is constructed by taking the convolutional and max-pooling layers from a standard CNN network while removing fully connected, and Softmax layers. Given an input image of size $w \times h \times c$ (where c is the color channels of image), the CNN network extracts a feature gird F of size $w' \times h' \times k$, where k is the number of feature maps in the last convolutional layer, and w' and h' depend on the w and h of input images and the number of pooling layers in the CNN network. Then, the feature grid F is fed into the 2D self-attention block to get the final attended feature grid F_{attn}. Finally, the final attended feature grid F_{attn} is unfolded to a feature sequence column by column from left to right in each feature map. The feature sequence is fed into the encoder.

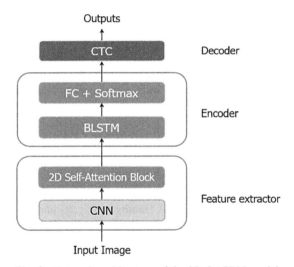

Fig. 2. Network architecture of the 2D-SACRN model.

3.2.2 Encoder

At the top of the feature extractor, the encoder encodes the feature sequence extracted from the feature extractor into a sequence of label probabilities. Mathematically, the encoder predicts label-probabilities for each feature in the feature sequence. In the 2D-SACRN model, we use a BLSTM network followed by a fully connected layer and a Softmax layer to build the encoder. The BLSTM network takes the feature sequence from the feature extractor as the input. Then, the output of the BLSTM network is fed into the fully connected layer, which converts the output feature dimension to the size of the total character set (plus 1 for CTC blank character). Finally, the Softmax layer, which is placed at the end of the encoder, generates the label probabilities at each time step.

Let $F_{seq} = (f_1, f_2 \cdots f_n)$, $E = (e_1, e_2 \cdots e_n)$ and $H = (h_1, h_2 \cdots h_n)$ denote the feature sequence, the sequence of label probabilities, and the output of the BLSTM network, respectively, where n is the number of feature vectors. Then, we have:

$$H = \text{BLSTM}\left(F_{seq}\right) \tag{6}$$

$$E = \text{Softmax}(\text{FC}(H)) \tag{7}$$

3.2.3 Decoder

At the top of the 2D-SACRN model, the decoder converts the sequence of label probabilities made by the encoder into a final label sequence. Mathematically, the decoding process is to find the final label sequence with the highest probability conditioned on the sequence of label probabilities. CTC [28] is a specific loss function designed for sequence labeling tasks where it is difficult to segment the input sequence into the final segmented sequence that exactly matches a target sequence. In this work, we employ the CTC algorithm to build the decoder to obtain the conditional probability.

The whole system is trained end-to-end using stochastic gradient descent algorithms to minimize the CTC loss. For the decoding process in the testing phase, we apply the CTC beam search with the *beamwidth* of 2 to obtain the final label sequence with the highest probability conditioned.

4 Datasets

In this paper, we conduct the experiments on the following three datasets: two widely used western handwritten datasets - IAM Handwriting [29] and Rimes [30], and one Japanese handwritten dataset - TUAT Kondate [31]. The details of them are given in the following sections.

4.1 IAM Handwriting

IAM Handwriting is an offline handwritten English text dataset compiled by the FKI-IAM Research Group. The dataset is composed of 13,353 text lines extracted from 1,539 pages of scanned handwritten English text, which were written by 657 different writers. All text lines in the IAM Handwriting are built using sentences provided by the Lancaster-Oslo/Bergen (LOB) corpus. We employ the IAM Aachen splits [32] shared by T. Bluche from RWTH Aachen University to split the dataset into three subsets: 6,482 lines (747 pages) for training, 2,915 (336 pages) lines for testing, and 976 lines (116 pages) for validation. There are 79 different characters in the dataset, including the space character.

4.2 Rimes

Rimes is a well-known handwriting French dataset compiled by A2iA's research laboratory. The dataset consists of 11,333 lines extracted from 1,500 paragraphs for training and 778 lines extracted from 100 paragraphs for testing. The original dataset does not include the validation set, so we use the lines extracted from the last 100 paragraphs of the training set as a validation set. Consequently, the Rimes dataset consists of three subsets: 10,532 lines for training, 801 lines for validation, and 778 lines for testing. There are 99 different characters in the dataset, including the space character.

4.3 TUAT Kondate

TUAT Kondate is an online handwritten database compiled by Nakagawa Lab., Tokyo University of Agri. & Tech. (TUAT). The database stores online handwritten patterns mixed of text, figures, tables, maps, diagrams and so on. It was turned to offline patterns by thickening strokes with constant width. The handwritten Japanese portion of TUAT Kondate comprises 13,685 text line images collected from 100 Japanese writers. We use the train-valid-test configuration in [11] to split the dataset into three subsets: 11,487 text line images collected from 84 writers for training; 800 text line images collected from 6 writers; and 1,398 text line images collected from 10 writers for testing. There are 3,345 different characters in the dataset.

5 Experiments

To evaluate the performance of the proposed 2D-SACRN model, we conducted experiments on the three datasets: IAM handwriting, Rimes, and TUAT Kondate. The implementation details are described in Sect. 5.1; the results of the experiments are presented in Sect. 5.2; and the visualization of the 2D self-attention map is shown in Sect. 5.3.

5.1 Implementation Details

IAM and Rimes Datasets. In the experiments on the two western datasets, the architecture of the CNN network in the feature extractor is ConvNet-1 as shown in Table 1, where 'maps', 'k', 's' and 'p' denote the number of kernels, kernel size, stride and padding size of convolutional layers, respectively. It consists of five convolutional (Conv) blocks. Each Conv block consists of one Conv layer followed by the Batch normalization [33] and the ReLU activation. To reduce overfitting, we apply dropout at the input of the last three Conv blocks (with dropout probability equal to 0.2).

At the encoder, we employ a Deep BLSTM network with 256 hidden nodes of five layers. To prevent overfitting when training the model, the dropout (dropout rate $= 0.5$) is also applied in each layer of the Deep BLSTM network. A fully connected layer and a softmax layer with the node size equal to the character set size (n $=$ 80 for IAM and 100 for Rimes) are applied after each time step of the Deep BLSTM network.

TUAT Kondate Dataset. In our experiments on the TUAT Kondate dataset, the architecture of the CNN network in the feature extractor is ConvNet-2, which consists of six Conv blocks, as shown in Table 1. The Deep BLSTM network in the encoder has three BLSTM layers with 256 hidden nodes of each layer. The other configurations are the same as the 2D-SACRN model in the experiments on the two western datasets.

Table 1. Network configurations of the CNN in the feature extractor.

Type	Configurations	
	ConvNet-1	ConvNet-2
Input	h × w image	h × w image
Conv block 1	#maps:16, k:3 × 3, s:1, p:1	#maps:16, k:3 × 3, s:1, p:1
Max-pooling1	#window:2 × 2, s:2 × 2	#window:2 × 2, s:2 × 2
Conv block 2	#maps:32, k:3 × 3, s:1, p:1	#maps:32, k:3 × 3, s:1, p:1
Max-pooling2	#window:2 × 2, s:2 × 2	#window:2 × 2, s:2 × 2
Conv block 3	#maps:48, k:3 × 3, s:1, p:1	#maps:48, k:3 × 3, s:1, p:1
Max-pooling3	#window:1 × 2, s:1 × 2	#window:2 × 2, s:2 × 2
Conv block 4	#maps:64, k:3 × 3, s:1, p:1	#maps:64, k:3 × 3, s:1, p:1
Max-pooling4	#window:2 × 1, s:2 × 1	#window:1 × 2, s:1 × 2
Conv block 5	#maps:80, k:3 × 3, s:1, p:1	#maps:80, k:3 × 3, s:1, p:1
Max-pooling5		#window:2 × 1, s:2 × 1
Conv block 6		#maps:128, k:3 × 3, s:1, p:1

5.2 Experiment Results

In order to evaluate the performance of the 2D-SACRN model, we employ the terms of Character Error Rate (CER), Word Error Rate (WER), and Sequence Error Rate (SER) that are defined as follows:

$$CER(h,S') = \frac{1}{Z} \sum_{(x,z) \in S'} ED(h(x),z) \tag{8}$$

$$WER(h,S') = \frac{1}{Z_{word}} \sum_{(x,z) \in S'} ED_{word}(h(x),z) \tag{9}$$

$$SER(h,S') = \frac{100}{|S'|} \sum_{(x,z) \in S'} \begin{cases} 0 \text{ if } h(x) = z \\ 1 \text{ otherwise} \end{cases} \tag{10}$$

where Z is the total number of target labels in S' and $ED(p, q)$ is the edit distance between two sequences p and q, while Z_{word} is the total number of words in S' and $ED_{word}(p, q)$ is the word-level edit distance between two sequences p and q.

5.2.1 English and French Text Recognition

The first experiment evaluated the performance of the 2D-SACRN model on the two western handwritten datasets: IAM Handwriting and Rimes in terms of CER and WER. To fairly compare with the previous models, we do not use any data augmentation techniques as well as linguistic context information. Table 2 shows the recognition error rates by the 2D-SACRN model and the previous models [7, 9, 13, 16, 18, 24, 34, 35] on the test set of IAM Handwriting and Rimes datasets without using the language model.

Table 2. Recognition error rates (%) on IAM and Rimes datasets.

Model	IAM		Rimes	
	CER	WER	CER	WER
CNN-1DLSTM (Moysset et al. [34])	11.52	35.64	6.14	20.11
MDLSTM (Pham et al. [7])	10.80	35.10	6.80	28.50
GNN-1DLSTM (Bluche et al. [9])*	10.17	32.88	5.75	19.74
2DLSTM (Moysset et al. [34])	8.88	29.15	4.94	16.03
2DLSTM-X2 (Moysset et al. [34])	8.86	29.31	4.80	16.42
CNN-Seq2Seq (Sueiras et al. [16])	8.80	23.80	4.80	15.90
CNN-Seq2Seq (Zhang et al. [21])	8.50	22.20	–	–
CNN-1DLSTM (Puigcerver et al. [13])	8.20	25.40	3.30	12.80
2DLSTM (Bluche et al. [18])	7.90	24.60	**2.90**	12.60
CNN-1DLSTM (Puigcerver et al. [13])*	7.73	25.22	4.39	14.05
CNN-transformers (Kang et al. [24])	7.62	24.54	–	–
Deep BLSTM+Dropout (Bluche et al. [35])	7.30	24.70	5.60	20.90
2D-SACRN (Ours)	**6.76**	**20.89**	3.43	**11.92**

*Experiments run by Moysset et al.[34]

On the IAM Handwriting dataset, the 2D-SACRN model achieved CER of 6.76% and WER of 20.89%. These results show that the 2D-SACRN model achieves the state-of-the-art accuracy and outperforms the best model in [35] by about 8% of CER and 15% of WER on the IAM Handwriting dataset. On the Rimes dataset, the 2D-SACRN model achieved CER of 3.43% and WER of 11.92%. Although its CER was considerably larger than the current state-of-the-art [18], its WER was the best.

From the above results, we conclude that the 2D-SACRN model achieves similar or better accuracy when compared to the state-of-the-art models in both IAM Handwriting and Rimes datasets.

5.2.2 Japanese Text Recognition

In the second experiment, we evaluated the performance of the 2D-SACRN model on the TUAT Kondate - offline handwritten Japanese text dataset in terms of CER and SER.

To fairly compare with the previous models, we also do not use any data augmentation techniques as well as linguistic context information. Table 3 compares the recognition error rates by the 2D-SACRN model and the previous works in [3, 10, 11, 19, 25] on the test set of TUAT Kondate without using the language model. The 2D-SACRN model achieved CER of 2.49% and SER of 12.66% on the test set of TUAT Kondate. The results imply that the 2D-SACRN model obtains the state-of-the-art results on the TUAT Kondate dataset and outperforms the best model in [25] by about 10% of CER and 25% of SER.

Table 3. Recognition error rates (%) on the test set of TUAT Kondate.

Model	Kondate	
	CER	SER
Segmentation-based method [3]	11.2	48.53
DCRN-f&s [10]	6.95	28.04
DCRN-s [10]	6.44	25.89
End-to-End DCRN [11]	3.65	17.24
Attention-based model [19]	9.17	20.10
AACRN [25]	2.73	15.74
2D-SACRN (Ours)	**2.49**	**12.66**

5.3 Effects of 2D Self-attention Mechanism

To measure the effectiveness of the 2D self-attention mechanism in the feature extractor of the 2D-SACRN, we prepared one variation, which was the same as the 2D-SACRN model except using the 2D self-attention block in the feature extractor. This variation is called 2D-SACRN_w/o_2DSelfAttn. We trained the 2D-SACRN_w/o_2DSelfAttn according to the same scheme applied to the 2D-SACRN model on the three datasets. Table 4 compares its recognition error rates with the 2D-SACRN model on the test set of the IAM Handwriting, Rimes, and TUAT Kondate datasets. In all datasets, the 2D-SACRN model slightly outperforms the 2D-SACRN_w/o_2DSelfAttn. The results imply that the 2D self-attention mechanism in the feature extractor improves the performance of the 2D-SACRN model for handwritten text recognition. On the Kondate

Table 4. Recognition error rates (%) with different feature extractors.

Model	IAM		Rimes		Kondate	
	CER	WER	CER	WER	CER	SER
2D-SACRN_w/o_2DSelfAttn	7.49	22.97	3.78	13.48	2.77	14.02
2D-SACRN	6.76	20.89	3.43	11.92	2.49	12.66

dataset, the 2D-SACRN_w/o_2DselfAttn model is slightly inferior to the SoTA model [25] but superior to the other methods, e.g., End-to-End DCRN [11] that is equivalent to the CNN-BLSTM-CTC model. Although it has similar network architecture with the End-to-End DCRN model, the dropout in some Conv layers and more hidden nodes in the BLSTM network might help improve the accuracy.

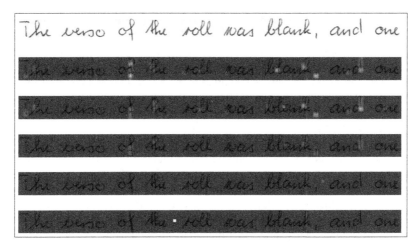

(a) A group of attention maps belonging to the first text image.

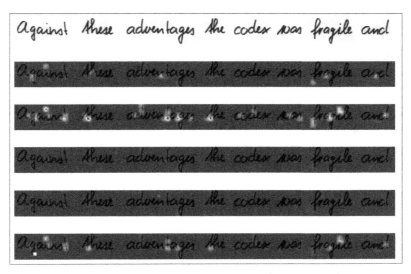

(b) A group of attention maps belonging to the second text image.

Fig. 3. The visualization of 2D self-attention maps.

5.4 Visualization of 2D Self-attention Mechanism

To verify whether the 2D self-attention helps the feature extractor to capture the relationships between widely separated spatial regions in an input image, we visualize the 2D self-attention map in 2D-SACRN for different images in the IAM Handwritten dataset. Figure 3 shows the visualization of the 2D self-attention map for two images. In each group, the top image is the original input image, while each of the other five images shows one query point with color-coded dots (blue, fuchsia, green, red, and yellow) and the 2D self-attention map for that query point. We observe that the 2D self-attention mechanism tends to focus on locations having similar texture to the query point, though these locations are far from the query point. For example, in the first group of Fig. 3(a), the blue point (top of the "f" character) attends mostly to locations around the stroke of the "f", "d", ",", and "k" characters. Besides, the fuchsia point (inside the stroke of the "a" character of the "blank") do not attend to locations around its stroke but mostly attends to the stroke of the "f", "d", and "," characters.

We also see that query points inside background regions seem not to attend mostly to any other location, such as the red and yellow points in Fig. 3(a) as well as the green and red points in Fig. 3(b). It seems because the points inside background regions do not mostly relate to any other location in the image. We also find that some query points are quite close in spatial location but have very different attention maps. For example, in the second group of Fig. 3(b), the blue point and the fuchsia point are quite close but have very different attention maps. This shows that the adjacent points may freely attend to other distant locations. These observations demonstrate that the 2D self-attention mechanism helps the feature extractor to capture the relationships between widely separated spatial regions in an input image.

6   Conclusion

In this paper, we proposed the 2D Self-Attention Convolutional Recurrent Network (2D-SACRN) model with the 2D self-attention mechanism for recognizing handwritten text lines. The 2D-SACRN model consists of three main parts: the feature extractor by Convolutional Neural Network and 2D self-attention mechanism, the encoder by bidirectional LSTM, and the decoder by Connectionist Temporal Classification. Following the experiments, the 2D-SACRN model archived the 6.76%, 3.43%, and 2.49% of character error rates on the test sets of IAM Handwriting, Rimes, and TUAT Kondate datasets, respectively. The following conclusions are drawn: 1) The 2D-SACRN model achieves similar or better accuracy when compared to the state-of-the-art models in the three datasets; 2) The 2D self-attention mechanism helps the feature extractor to capture the relationships between widely separated spatial regions and improves the performance of the 2D-SACRN model for handwritten text recognition.

In future works, we will conduct experiments of the 2D-SACRN model with other text recognition tasks such as Scene text recognition and historical text recognition. We also plan to incorporate language models into the 2D-SACRN model to improve its performance.

Acknowledgments. This research is being partially supported by the grant-in-aid for scientific research (S) 18H05221 and (A) 18H03597.

References

1. Bunke, H., Roth, M., Schukat-Talamazzini, E.G.: Off-line cursive handwriting recognition using hidden Markov models. Pattern Recogn. **28**, 1399–1413 (1995)
2. España-Boquera, S., Castro-Bleda, M.J., Gorbe-Moya, J., Zamora-Martinez, F.: Improving offline handwritten text recognition with hybrid HMM/ANN models. IEEE Trans. Pattern Anal. Mach. Intell. **33**, 767–779 (2011)
3. Nguyen, K.C., Nguyen, C.T., Nakagawa, M.: A Segmentation method of single- and multiple-touching characters in offline handwritten Japanese text recognition. IEICE Trans. Inf. Syst. **100**, 2962–2972 (2017)
4. Wang, Q.-F., Yin, F., Liu, C.-L.: Handwritten Chinese text recognition by integrating multiple contexts. IEEE Trans. Pattern Anal. Mach. Intell. **34**, 1469–1481 (2012)
5. Graves, A., Liwicki, M., Fernandez, S., Bertolami, R., Bunke, H., Schmidhuber, J.: A novel connectionist system for unconstrained handwriting recognition. IEEE Trans. Pattern Anal. Mach. Intell. **31**, 855–868 (2009)
6. Graves, A., Schmidhuber, J.: Offline handwriting recognition with multidimensional recurrent neural networks. Adv. Neural Inf. Process. Syst. **21**, 545–552 (2008)
7. Pham, V., Bluche, T., Kermorvant, C., Louradour, J.: Dropout improves recurrent neural networks for handwriting recognition. In: Proceedings of International Conference on Frontiers in Handwriting Recognition, ICFHR, pp. 285–290 (2014)
8. Shi, B., Bai, X., Yao, C.: An end-to-end trainable neural network for image-based sequence recognition and its application to scene text recognition. IEEE Trans. Pattern Anal. Mach. Intell. **39**, 2298–2304 (2017)
9. Bluche, T., Messina, R.: Gated convolutional recurrent neural networks for multilingual handwriting recognition. In: Proceedings of the International Conference on Document Analysis and Recognition, ICDAR, pp. 646–651 (2017)
10. Ly, N.-T., Nguyen, C.-T., Nguyen, K.-C., Nakagawa, M.: Deep convolutional recurrent network for segmentation-free offline handwritten Japanese text recognition. In: Proceedings of the International Conference on Document Analysis and Recognition (ICDAR), pp. 5–9 (2017)
11. Ly, N.T., Nguyen, C.T., Nakagawa, M.: Training an end-to-end model for offline handwritten Japanese text recognition by generated synthetic patterns. In: Proceedings of the International Conference on Frontiers in Handwriting Recognition (ICFHR), pp. 74–79 (2018)
12. Ly, N.-T., Nguyen, K.-C., Nguyen, C.-T., Nakagawa, M.: Recognition of anomalously deformed kana sequences in Japanese historical documents. IEICE Trans. Inf. Syst. **102**, 1554–1564 (2019)
13. Puigcerver, J.: Are Multidimensional recurrent layers really necessary for handwritten text recognition? In: Proceedings of the International Conference on Document Analysis and Recognition, ICDAR, pp. 67–72 (2017)
14. Luong, T., Pham, H., Manning, C.D.: Effective approaches to attention-based neural machine translation. In: Proceedings of the Conference on Empirical Methods in Natural Language Processing, pp. 1412–1421 (2015)
15. Bahdanau, D., Chorowski, J., Serdyuk, D., Brakel, P., Bengio, Y.: End-to-end attention-based large vocabulary speech recognition. In: Proceedings of the International Conference on Acoustics, Speech and Signal Processing (ICASSP), pp. 4945–4949 (2016)

16. Sueiras, J., Ruiz, V., Sanchez, A., Velez, J.F.: Offline continuous handwriting recognition using sequence to sequence neural networks. Neurocomputing **289**, 119–128 (2018)
17. Bluche, T., Louradour, J., Messina, R.: Scan, attend and read: end-to-end handwritten paragraph recognition with MDLSTM attention. In: Proceedings of the International Conference on Document Analysis and Recognition (ICDAR), pp. 1050–1055 (2017)
18. Bluche, T.: Joint line segmentation and transcription for end-to-end handwritten paragraph recognition. Neural Inf. Process. Syst. **29**, 838–846 (2016)
19. Ly, N.T., Nguyen, C.T., Nakagawa, M.: An attention-based end-to-end model for multiple text lines recognition in Japanese historical documents. In: Proceedings of the International Conference on Document Analysis and Recognition, ICDAR, pp. 629–634 (2019)
20. Ly, N.T., Nguyen, C.T., Nakagawa, M.: An attention-based row-column encoder-decoder model for text recognition in Japanese historical documents. Pattern Recogn. Lett. **136**, 134–141 (2020)
21. Zhang, Y., Nie, S., Liu, W., Xu, X., Zhang, D., Shen, H.T.: Sequence-to-sequence domain adaptation network for robust text image recognition. In: Proceedings of the IEEE Computer Society Conference on Computer Vision and Pattern Recognition, pp. 2735–2744 (2019)
22. Vaswani, A., et al.: Attention is all you need. In: Advances in Neural Information Processing Systems, pp. 5999–6009 (2017)
23. Dong, L., Xu, S., Xu, B.: Speech-transformer: a no-recurrence sequence-to-sequence model for speech recognition. In: Proceedings of the IEEE International Conference on Acoustics, Speech and Signal Processing - Proceedings, pp. 5884–5888 (2018)
24. Kang, L., Riba, P., Rusiñol, M., Fornés, A., Villegas, M.: Pay attention to what you read: non-recurrent handwritten text-line recognition. arXiv (2020)
25. Ly, N.T., Nguyen, C.T., Nakagawa, M.: Attention augmented convolutional recurrent network for handwritten Japanese text recognition. In: Proceedings of International Conference on Frontiers in Handwriting Recognition, ICFHR, pp. 163–168 (2020)
26. Wang, X., Girshick, R., Gupta, A., He, K.: Non-local neural networks. In: Proceedings of the IEEE Computer Society Conference on Computer Vision and Pattern Recognition, pp. 7794–7803 (2018)
27. Zhang, H., Goodfellow, I., Metaxas, D., Odena, A.: Self-attention generative adversarial networks. In: 36th International Conference on Machine Learning, ICML 2019, 2019-June, pp. 12744–12753 (2018)
28. Graves, A., Fernández, S., Gomez, F., Schmidhuber, J.: Connectionist temporal classification: labelling unsegmented sequence data with recurrent neural networks. In: Proceedings of the ACM International Conference Proceeding Series, pp. 369–376 (2006)
29. Marti, U.V., Bunke, H.: The IAM-database: an English sentence database for offline handwriting recognition. Int. J. Doc. Anal. Recognit. **5**, 39–46 (2003)
30. Grosicki, E., Carre, M., Brodin, J.-M., Geoffrois, E.: RIMES evaluation campaign for handwritten mail processing. In: Proceedings of Workshop on Frontiers in Handwriting Recognition, pp. 1–6 (2008)
31. Matsushita, T., Nakagawa, M.: A database of on-line handwritten mixed objects named "Kondate". In: Proceedings of International Conference on Frontiers in Handwriting Recognition, ICFHR, pp. 369–374 (2014)
32. IAM Aachen splits. https://www.openslr.org/56/. Accessed 19 Feb 2021
33. Ioffe, S., Szegedy, C.: Batch normalization: accelerating deep network training by reducing internal covariate shift. In: Proceedings of the International Conference on Machine Learning, ICML 2015, pp. 448–456 (2015)
34. Moysset, B., Messina, R.: Are 2D-LSTM really dead for offline text recognition? In: International Journal on Document Analysis and Recognition, pp. 193–208 (2019)
35. Bluche, T.: Deep neural networks for large vocabulary handwritten text recognition. https://tel.archives-ouvertes.fr/tel-01249405 (2015)

Handwritten Text Recognition with Convolutional Prototype Network and Most Aligned Frame Based CTC Training

Likun Gao[1,2], Heng Zhang[1], and Cheng-Lin Liu[1,2,3(✉)]

[1] National Laboratory of Pattern Recognition (NLPR), Institution of Automation, Chinese Academy of Sciences, Beijing 100190, China
{gaolikun2018,heng.zhang}@ia.ac.cn, liucl@nlpr.ia.ac.cn
[2] School of Artificial Intelligence, University of Chinese Academy of Sciences, Beijing 100049, China
[3] CAS Center for Excellence in Brain Science and Intelligence Technology, Beijing 100190, China

Abstract. End-to-end Frameworks with Connectionist Temporal Classification (CTC) have achieved great success in text recognition. Despite high accuracies with deep learning, CTC-based text recognition methods also suffer from poor alignment (character boundary positioning) in many applications. To address this issue, we propose an end-to-end text recognition method based on robust prototype learning. In the new CTC framework, we formulate the *blank* as the rejection of character classes and use the one-vs-all prototype classifier as the output layer of the convolutional neural network. For network learning, based on forced alignment between frames and character labels, the most aligned frame is up-weighted in CTC training strategy to reduce estimation errors in decoding. Experiments of handwritten text recognition on four benchmark datasets of different languages show that the proposed method consistently improves the accuracy and alignment of CTC-based text recognition baseline.

Keywords: Text recognition · Connectionist temporal classification · Convolutional prototype network · Frame alignment · Most aligned frame

1 Introduction

Text (character string) recognition, as an important sequence labeling problem, has been widely studied by researchers in industry and academia. Text recognition has potential applications in many scenarios, such as street number reading, bank checks, mail sorting, and historical documents. Due to the complexity of image layout, the diversity of handwriting styles, and the variety of image

© Springer Nature Switzerland AG 2021
J. Lladós et al. (Eds.): ICDAR 2021, LNCS 12821, pp. 205–220, 2021.
https://doi.org/10.1007/978-3-030-86549-8_14

backgrounds, text recognition is remaining a challenging task. Taking advantage of deep learning approaches, text recognition has been largely advanced in recent years. Especially, the Connectionist Temporal Classification(CTC) [8] and attention-based end-to-end frameworks, representing the state-of-the-art, have achieved superior results in many text recognition works.

Early CTC-based methods [27] used hand-crafted image features such as histogram of oriented gradient (HOG), and recurrent neural network (RNN) for context modeling. Replacing HOG with Convolutional Neural Network (CNN), Shi et al. [24] proposed an end-to-end model in scene text recognition named Convolutional Recurrent Neural Network (CRNN). Yin et al. [40] proposed a new framework using CNN for sliding-window-based classification to enable parallel training and decoding. The sliding-window-based method not only achieves better performance but also largely reduces model parameters and computation cost. As for attention-based methods, since firstly applied to scene text recognition by Shi et al. [25], this framework has been followed by many researchers. Combining with the attention mechanism, RNN integrates global information at each step and directly outputs the decoding results. In addition to the flexibility of decoding from 1D to 2D alignment, the attention framework can also memorize semantic information to improve the recognition accuracy in scene text recognition.

Despite the great success of end-to-end frameworks in text recognition, there are remaining problems. One problem is the inaccurate character position alignment, although the final recognition result (transcript) is correct. This is due to the mismatch between the confidence peak and the true position of the character in CTC-based methods [19]. For the attention-based methods, the current decoding step depends on outputs of previous steps, so once attention maps deviate from the character position, the accumulation of errors will appear [3]. Besides, the model confidence will also directly affect the recognition accuracy. With more accurate model confidence, higher model recognition performance can be reached.

To improve the character alignment and alleviate the overconfidence problem of the state-of-the-art frameworks, we propose an end-to-end text recognition method using convolutional prototype network (CPN) [37], and most aligned frame based CTC training. CPN is used to replace conventional CNN for sliding-window-based character classification based on the nearest prototype in convolutional feature space. In our prototype learning framework, the *blank* symbol can be regarded as the rejection of character classes with more reliable confidence than linear classification. In CPN training, the prototype loss (PL) loss is similar to the maximum likelihood regularization proposed in [17], which can improve the intra-class compactness in feature representation. To better differentiate between character classes and background (*blank*), we use one-vs-all prototype learning [38]. Also, to better exploit character samples in CPN training, we propose a Most Aligned Frame Selection (MAFS) based training strategy. By estimating the most aligned frames of characters in a text image, the sequence labeling problem is transformed into a character classification problem, thereby both the network training and the recognition are improved. We conducted experiments on four handwritten text datasets ORAND-CAR, CVL HDS (digit strings), IAM (English), and ICDAR-2013 (Chinese). The experimental results demonstrate the superiority of the proposed method compared with the baseline, and the benefits of both CPN and MAFS are justified.

The rest of this paper is organized as follows. Section 2 reviews some related works. Section 3 describes our proposed methods with CPN and MAFS based training. Section 4 presents our experimental results, and Sect. 5 draws concluding remarks.

2   Related Work

2.1   Text Recognition

Before the prosperity of end-to-end text recognition methods, over-segmentation-based methods with character classification and Hidden Markov Model (HMM) modeling were mostly used for Chinese and Latin handwriting [6,30]. Along with advances in deep learning, end-to-end methods have gradually become dominant in text recognition. Shi et al. [24] proposed an end-to-end RNN-CTC framework, which improves the image feature representation with CNN. This Convolutional RNN (CRNN) has gained success in various scenarios. For example, Zhan et al. [42] applied the RNN-CTC model to handwritten digit string recognition and obtained improved recognition accuracy. Ly et al. [20] used the CRNN framework in historical document recognition. Yin et al. [40] proposed a pure CNN-based model with sliding window classification and CTC decoding. With much fewer parameters and faster calculation speed than CRNN, the sliding-window-based model achieves better recognition results. As for attention-based methods, Shi et al. [25] first proposed an end-to-end framework with RNN and attention for scene text recognition. Since then, attention-based methods [4,26,32] have become popular in text recognition tasks. Recently, Bartz et al. [1] replaced the RNN-attention part with the transformer and achieved further improvements.

2.2   Prototype Learning

Prototype learning is a classical and representative method in pattern recognition. The predecessor of prototype learning is k-nearest-neighbor (KNN) classification. To reduce the storage space and computation resources of KNN, prototype reduction and learning methods (including learning vector quantization (LVQ) [11]) have been proposed. Among the methods, some designed suitable updating conditions and rules to learn the prototypes [13,15], while others treated prototypes as learnable parameters and learned the prototypes through optimizing the related loss functions [9,22,23]. A detailed review and evaluation of prototype learning methods can be found in [16]. As for text recognition, prototype learning [17,30] is also widely used before the advent of deep learning. Previous prototype-based methods mainly employ hand-crafted features before the arrival of CNN. Yang et al. [37] combined the prototype classifier with deep convolutional neural networks and proposed a convolutional prototype network (CPN) for high accuracy and robust pattern classification. They show that CPN with one-vs-all training can give better performance of open set recognition (classifying known-class samples while rejecting unknown-class samples) [38].

3   Method

We use the sliding-window model and CTC training proposed in [40] as the text recognition framework (see Fig. 1). Based on the CTC analysis, we firstly transform the CTC loss into a cross-entropy between pseudo-label distributions and probabilities output by the neural network. Then the cross-entropy loss is improved based on the CPN model and MAFS method for robust text recognition.

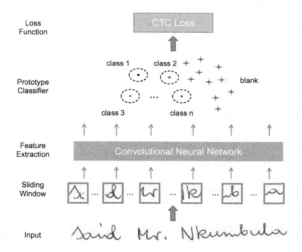

Fig. 1. An illustration of our text recognition based on convolutional prototype learning

3.1   Outline and Analysis of CTC

In the CTC recognition framework, the input is a T length sequence $Y = \{y^1, y^2, ...y^T\}$, where y^i is a L'-dimension vector from the neural network output. Except for the L characters to be recognized, there is also a *blank* class, so $L' = L + blank$.

CTC is not only a loss but also a decoding algorithm B i.e. removing the repeated labels then all blanks from the given path. By concatenating one prediction of each frame at all time-steps, a path is formed. The probability of a path is defined as

$$p(\pi|Y) = \prod_{t=1}^{T} y_{\pi_t}^t, \forall \pi \in L'^T, \tag{1}$$

where $y_{\pi_t}^t$ is the probability passing through path π at frame t. Given a sequence label l, a feasible path is defined as the path that can map onto l via B. During training, the probabilities of all feasible paths are added up as the posterior probability $P(l|Y)$ and the negative logarithm of $P(l|Y)$ is taken as the objective function:

$$p(l|Y) = \sum_{\pi \in B^{-1}(l)} p(\pi|Y). \tag{2}$$

$$Loss_{CTC} = -\log p(l|Y), \tag{3}$$

Then the loss partial differential concerning y_k^t is computed as:

$$\frac{\partial Loss_{CTC}}{\partial y_k^t} = -\frac{1}{p(l|Y)y_k^t} \sum_{\{\pi|\pi \in B^{-1}(l), \pi_t=k\}} p(\pi|Y). \tag{4}$$

We rewrite Eq.(4) as

$$\frac{\partial Loss_{CTC}}{\partial y_k^t} = -\frac{\sum_{\{\pi|\pi \in B^{-1}(l), \pi_t=k\}} p(\pi|Y)}{p(l|Y)} \frac{\partial \log y_k^t}{\partial y_k^t} = -z_k^t \frac{\partial \log y_k^t}{\partial y_k^t}, \tag{5}$$

where we regard $z_k^t = \frac{\sum_{\{\pi|\pi \in B^{-1}(l), \pi_t=k\}} p(\pi|Y)}{p(l|Y)}, k = 0, ..., L' - 1$, as the pseudo-label distribution in the CTC decoding graph at frame t. When z_k^t is regarded as a constant, we can find that Eq.(5) is a derivative form of a cross-entropy between z_k^t and y_k^t. So CTC loss is equivalent to the cross-entropy between pseudo-label distributions and classifier outputs:

$$Loss_{CE} = Loss_{CTC} = -\sum_t \sum_k z_k^t \log y_k^t. \tag{6}$$

Based on the above formulation, we can divide CTC training in each iteration into two steps (see Fig. 2): pseudo-label estimation and cross-entropy training. The first step is to estimate the pseudo-label distribution for each frame using the model output Y and the ground truth l. Secondly, update the model parameters with the cross-entropy criteria. In this step, the pseudo-label distribution plays a similar role to the one-hot label used in the classification task. Therefore, CTC loss is an alternate-updating process, where the classifier output and pseudo-label distribution interact with each other and become more and more accurate.

As we know in [7], to facilitate the calculation of loss, CTC introduces a one-way graph G based on the extended ground truth l'. For example in Fig. 2, ground truth $l = \{A, P, P\}$ and $l' = \{blank, A, blank, P, blank, P, blank\}$, and $l'(n)$ means the character of the n-th element in l'. Following this setting and the definition of pseudo-label distribution z^t, we can define raw pseudo-label

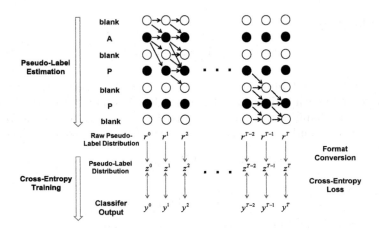

Fig. 2. CTC in two steps

distribution as $r_n^t = \frac{\sum_{\{\pi \mid \pi \in B^{-1}(l), \pi_t = l'(n)\}} p(\pi|Y)}{p(l|Y)}$, which is the probability distribution in the CTC decoding graph at frame t concerning to the n-th node. Using raw pseudo-label distribution r_n^t, we can compute pseudo-label distribution as $z_k^t = \sum_{l'(n)=k} r_n^t$. For the class k not appearing in the ground truth (except for $blank$), $z_k^t = 0$.

Equation(6) indicates that reliable confidence can improve the pseudo-label estimation, thereby making the model performance accurate. So we use CPN instead of the CNN model for a better confidence estimation. On the other hand, the most aligned frames corresponding to the ground truth are more important for training. So we can use the MAFS method to improve CTC training.

3.2 One-vs-All Prototype Learning

In the CTC recognition framework, the classifier-output confidence y_k^t is directly used for training or decoding as in Eq.(6). Although the linear classifier has achieved excellent recognition results in [40], the confidence is still not robust enough. Therefore, we use the convolutional prototype network to improve pseudo-label estimation and cross-entropy training in CTC. Prototype learning is to train one or more prototypes for each class and use the idea of template matching to classify samples. In our work, for simplicity, we use only one prototype for each character class. We assume that each character class presents a standard Gaussian distribution in the feature space, so Euclidean distance is used to describe the similarity between each sample and the prototype. But as for $blank$, a class describing all non-character samples in the CTC framework, Gaussian assumption seems unreasonable. So we choose not to learn the prototype of $blank$ but estimate its probability as the rejection of characters. Then we take inspiration from Liu's [14] work and merge multiple two-class classifiers to build One-vs-All prototype learning.

We use the prototypes of L classes to construct L two-class classifiers,

$$m_k^t = Sigmoid(-\tau(d_k^t - T_k)), \tag{7}$$

$$d_k^t = ||f(x^t) - p_k||_2, \tag{8}$$

where d_k is the Euclidean distance between the feature map $f(x^t)$ of frame t and the prototype p_k of class k. T_k is a learnable threshold and τ is a predefined temperature coefficient, which is set to 5 in our experiment. We can regard m_k^t as the confidence of a two-class classifier whether frame t belongs to class k. For the L classifiers, we follow the principle in [14] to calculate the confidence y_k^t in Eq.(6),

$$y_k^t = A^{-1} m_k \prod_{k^o \neq k} (1 - m_{k^o}^t), \qquad k \in \{k | k \neq blank\} \tag{9}$$

$$y_{blank}^t = A^{-1} \prod_{k^o} (1 - m_{k^o}^t) \tag{10}$$

$$A^t = \sum_{k \neq blank} m_k^t \prod_{k^o \neq k} (1 - m_{k^o}^t) + \prod_{k^o} (1 - m_{k^o}^t), \tag{11}$$

where A^t is a normalization factor.

In addition to optimize the cross-entropy in Eq.(6), the prototype loss is also added according to Yang et al. [37],

$$Loss = Loss_{CE} + \alpha Loss_{pl}, \tag{12}$$

where α is set to 0.01 in our experiment.

The difficulty of applying prototype learning to text recognition is that the ground truth of each frame is unknown, so it is impossible to gather samples of a certain class around the corresponding prototype. With the pseudo-label distribution z^t for weighting, this problem can be solved,

$$Loss_{pl} = \sum_t \sum_{k \neq blank} z_k^t ||f(x^t) - p_k||_2. \tag{13}$$

Based on the prototype classifier and its loss function, we can make text recognition more robust.

3.3   Most Aligned Frame Selection Based Training

As discussed at the end of Sect. 3.1 by selecting the most aligned frames with the ground truth for training, the model can converge better.

Based on the raw pseudo-label distribution r^t, the probability of frame t aligned with the c-th character in the ground truth can be computed as $F_c(t) = \frac{r_{2c+1}^t}{\sum_t r_{2c+1}^t}$, where we regard t as a random variable confirming to Gaussian distribution. Then the most aligned frame can be achieved with the expectation of t,

$$t_c = Round(\sum_t t \cdot \frac{r_{2c+1}^t}{\sum_t r_{2c+1}^t}), \tag{14}$$

where t_c means the most aligned frame of the c-th character in the ground truth and $Round(\cdot)$ is a rounding function. Here, smaller is the variance, more credible is expectation estimation. So only when the variance is smaller than a certain threshold (1 in our work), the most aligned frame of the character can be used for training. Otherwise, if the decoding result with pseudo-label distribution is consistent with the ground truth, the probability distribution r^t is confident and then the most aligned frames of this text sample can also be used for training. A schematic diagram of the method can be seen in Fig. 3.

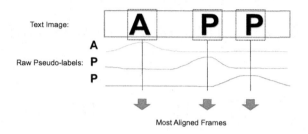

Fig. 3. An illustration of most aligned frame selection

In the training process, for high confidence of most aligned frames, we use the selected most aligned frame with the one-hot label of the corresponding class. Besides, to reduce the influence of other frames and *blank* frames, we still use the pseudo-label distribution but multiply by the weakening coefficients γ before the loss. The new CE loss function can be written as,

$$Loss_{CE} = -\gamma \sum_{(t,k)\notin\Omega} z_k^t \log y_k^t - \sum_{(t,k)\in\Omega} \log y_k^t, \tag{15}$$

where Ω indicates the set of the most aligned frames with labels. In the experiment, we choose 0.5 as the weakening coefficient γ of the non-most aligned frame.

4   Experiments

4.1   Datasets

In our experiment, four public datasets are used to evaluate our handwritten text recognition method. Two of them are handwritten digit strings named ORAND-CAR [5] and Computer Vision Lab Handwritten Digit String (CVL HDS, or CVL for short) [5], the third is an English text line dataset named IAM [21] and the last one is a Chinese handwritten dataset named ICDAR-2013 [39].

ORAND-CAR consists of 11719 images obtained from the Courtesy Amount Recognition (CAR) field of real bank checks. It can be divided into two sub-datasets CAR-A and CAR-B with different data sources. CAR-A has 2009 images for training and 3784 images for testing, while CAR-B consists of 3000 training images and 2926 testing images.

CVL HDS has been collected mostly amongst students of the Vienna University of Technology. 7960 images from 300 writers are collected, where only 1262 images are in the training set and the other 6698 images are for testing.

IAM contains unconstrained handwritten texts collected from 657 writers. As an English handwritten dataset, there are 79 categories, including numbers, letters, and punctuation. It contains 6,482 lines in the training set, 976 lines in the validation set, and 2,915 lines in the test set.

As for the Chinese handwritten dataset, we set the training set as **CASIA-HWDB** [18], which is divided into six sub-datasets. CASIA-HWDB1.0-1.2 consists of individual character samples, while CASIA-HWDB2.0-2.2 samples are handwritten text line images. There are 3118477 character images with 7356 classes and 41781 text lines with 2703 categories in the training set (816 writers out of 1020). For the test dataset **ICDAR-2013**, there are 3432 text line images. In our experiments, training sets of CASIA-HWDB are used for the model training.

All datasets are the only line labeled without character boundaries. So for the alignment experiment, we use MNIST handwritten digital dataset [12] for string synthesis and model evaluation.

4.2 Implementation Details

We use the sliding-window-based model [40] as the baseline, where the text image is divided into multiple windows equidistantly and then directly recognized by CNN. We use the same network structure as [40], but choose different sizes of windows according to different databases. In digit string recognition task, images are resized and padded to 32×256, and multi-scale windows are used with the size of 32×24, 32×28, 32×32. As for IAM dataset, image height is scaled to 32, width is scaled proportionally and multi-scale window sizes are set to 32×24, 32×32, 32×40. Models shift with step 4 in both experiments. For the Chinese handwritten dataset, we use character images in CASIA-HWDB1.0-1.2 training set to synthesize 1,250,000 text images and train the network together with the real text samples in CASIA-HWDB2.0-2.2. We scale the image to a width of 64, the multi-scale window to be 64×48, 64×64, 64×80, and the window step size to 8.

In the training process of digit string recognition, we first use the Adam optimizer [10] to train our network with a batch size of 32. The initial learning rate is 3×10^{-4}. After trained for 50 epochs, we switch to Stochastic gradient descent (SGD) with a learning rate 1×10^{-4}. After another 50 epochs, the learning rate is reduced by 0.3 times and then trained again for 50 epochs. For CVL HDS, due to the lack of training samples, a model with random initialization is not easy to converge. So we use the model trained on CAR-A for ten epochs as initialization. In the handwritten English and Chinese recognition task, we only use the Adam

Table 1. String accuracies of different models on the handwritten digital dataset.

Methods	CAR-A	CAR-B	CVL HDS
Pernambuco [5]	0.7830	0.7543	0.5860
BeiJing [5]	0.8073	0.7013	**0.8529**
FSPP [29]	0.8261	0.8332	0.7923
CRNN [24]	0.8801	0.8979	0.2601
ResNet-RNN [42]	0.8975	0.9114	0.2704
DenseNet [41]	0.9220	0.9402	0.4269
Sliding-window [40]	0.9337	0.9357	0.8010
Sliding-window + CPN	0.9430	0.9445	0.8425
Sliding-window + MAFS	0.9447	0.9425	0.8356
Sliding-window + CPN + MAFS	**0.9483**	**0.9470**	0.8512

optimizer with the initial learning rate of 3×10^{-4}, randomly initialized model can converge well.

4.3 Comparison with the State-of-the-art Methods

For handwritten digital strings, we compare our methods with state-of-the-art approaches in Table 1. On the ORAND-CAR dataset, our method achieves the best performance and can reduce the error rate by 25% in the best case. On the CVL dataset, based on careful initialization, our end-to-end framework has almost the same performance as Beijing [5] and achieves state-of-the-art recognition accuracy among the deep-learning-based methods. The Beijing method manually segments each training text image into characters for classifier training, while our method can train the model with only line labels and so is more practical for practical application.

Since the digit string datasets do not have context information, it is not suitable for attention-based methods. But for English text recognition, the attention-based model is also listed in Table 2. In the comparison, we only scale each image to a height of 32 but achieve better performance. As shown in Table 2, our proposed method has achieved the best recognition results in both character error rate (CER) and word error rate (WER). We visualize some recognition results in Fig. 4.

We also conducted experiments on the Chinese handwriting dataset, using the 5-g statistical model trained by Wu et al. [36]. The experimental results are shown in Table 3. We can also find that in large-category database, the training strategy based on the most aligned frame and the character classifier based on the convolution prototype can still improve the recognition performance of the model, which also verifies the effectiveness of our methods.

Table 2. Results of different models on the dataset IAM. (CER: character error rate; WER: word error rate.)

Methods	CER	WER
Salvador et al. [6]	9.8	22.4
Bluche [2]	7.9	24.6
Sueiras et al. [28]	8.8	23.8
Zhang et al. [43]	8.5	22.2
DAN [32]	6.4	19.6
Sliding-window	6.6	18.8
Sliding-window + CPN	6.1	18.1
Sliding-window + MAFS	6.2	17.9
Sliding-window + CPN + MAFS	**5.8**	**17.8**

Table 3. Results on the Chinese handwritten text dataset ICDAR-2013. (CR: correct rate; AR: accurate rate [35])

Methods	Without LM		With LM	
	CR	AR	CR	AR
Wu et al. [35]	87.43	86.64	-	92.61
Wang et al. [31]	90.67	88.79	95.53	94.02
Wu et al. [36]	–	–	96.32	96.20
Wang et al. [34]	89.66	–	96.47	–
Wang et al. [33]	89.12	87.00	95.42	94.83
Sliding-window	89.03	88.65	95.89	95.35
Sliding-window + MAFS	90.71	90.16	96.43	96.15
Sliding-window + CPN + MAFS	**90.92**	**90.30**	**96.64**	**96.23**

We believe that our convolutional prototype classifier is more reasonable than linear classifiers, where $blank$ is the rejection of character classes. Besides, the model can converge better due to more reliable confidence. That is why CPN can improve recognition performance in different datasets. As for our proposed MAFS, sparser pseudo-label distribution can reduce the error caused by pseudo-label estimation in the training process. Besides, most aligned frames can be paid more attention in training by reducing the impact of other frames. So our work can improve recognition performance.

4.4 Robustness and Alignment Evaluation

In this part, we analyze the effect of CPN and MAFS on model alignment and robustness. As far as we know, there is almost no evaluation standard of alignment effect in the field of text recognition, so we propose an indirect experiment

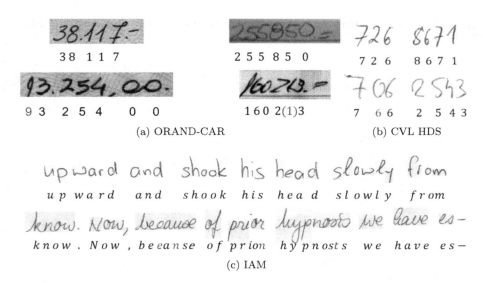

(a) ORAND-CAR   (b) CVL HDS

(c) IAM

Fig. 4. Visualization results of the datasets used in our experiments.

Table 4. Recognition rates of different models on the datasets MNIST.

	MNIST string	MNIST character
Sliding-window + CNN	0.934	0.942
Sliding-window + MAFS	0.932	0.956
Sliding-window + CPN	0.939	**0.983**

for alignment comparison. We train the model with string samples synthesized by MNIST digital images and compare the character classification accuracy on the MNIST test set. We believe the higher accuracy of character classification, the more character-aligned frames are classified correctly in sequence recognition. It also indirectly describes the alignment effect of the model. In the experiment, we randomly select samples in the MNIST dataset and splice them into strings with a length of 5 to 8. When the recognition accuracy with sequence samples is similar, character classification accuracy can be a standard for comparing model alignment effects. As shown in Table 4, with comparable recognition accuracy, CPN and MASF have higher classification accuracy than CNN, which shows that they lead to better alignment performance.

We also visualize the feature representation learned by CPN on the CAR database. Although the ground truth per frame is not available, we choose the category predicted by pseudo-label distribution for each frame as the label and draw a scatter plot. In Fig. 5, different colors represent different classes. The black dots represent the coordinate in feature dimension of prototypes, and *blank* frames are not in this figure. It can be seen from Fig. 5 that samples cluster near prototypes in the feature dimension, which proves that CPN has a robust feature representation.

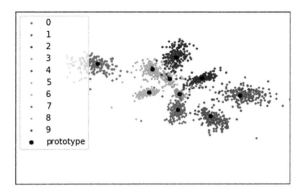

Fig. 5. Feature representation learned by CPN model on CAR (Color figure online)

5   Conclusions

In this paper, we propose a method for handwritten text recognition using convolutional prototype network for character classification and most aligned frame based CTC training. Different from previous CTC-based methods, we regard *blank* as the rejection of character classes and design a one-vs-all prototype classifier. The training strategy is based on the most aligned frame selection so as to improve the accuracy of character location and classification. Experiments on four handwritten text datasets confirm that our proposed methods can effectively improve text recognition performance, and the valuation on MNIST also verifies that CPN is beneficial for better alignment and model robustness. The proposed framework will be applied to more recognition scenarios (including scene text recognition) for further evaluation and improvement.

Acknowledgements. This work has been supported by the National Key Research and Development Program Grant 2020AAA0109702, the National Natural Science Foundation of China (NSFC) grants 61936003, 61721004.

References

1. Bartz, C., Bethge, J., Yang, H., Meinel, C.: Kiss: keeping it simple for scene text recognition. arXiv preprint arXiv:1911.08400 (2019)
2. Bluche, T.: Joint line segmentation and transcription for end-to-end handwritten paragraph recognition. arXiv preprint arXiv:1604.08352 (2016)
3. Cheng, Z., Bai, F., Xu, Y., Zheng, G., Pu, S., Zhou, S.: Focusing attention: towards accurate text recognition in natural images. In: Proceedings of the IEEE International Conference on Computer Vision, pp. 5076–5084 (2017)
4. Cheng, Z., Xu, Y., Bai, F., Niu, Y., Pu, S., Zhou, S.: Aon: towards arbitrarily-oriented text recognition. In: Proceedings of the IEEE Conference on Computer Vision and Pattern Recognition, pp. 5571–5579 (2018)

5. Diem, M., et al.: ICFHR 2014 competition on handwritten digit string recognition in challenging datasets (HDSRC 2014). In: Proceedings of the 14th International Conference on Frontiers in Handwriting Recognition, pp. 779–784. IEEE (2014)
6. Espana-Boquera, S., Castro-Bleda, M.J., Gorbe-Moya, J., Zamora-Martinez, F.: Improving offline handwritten text recognition with hybrid HMM/ANN models. IEEE Trans. Pattern Anal. Mach. Intell. **33**(4), 767–779 (2010)
7. Graves, A., Fernández, S., Gomez, F., Schmidhuber, J.: Connectionist temporal classification: labelling unsegmented sequence data with recurrent neural networks. In: Proceedings of the 23rd International Conference on Machine Learning, pp. 369–376 (2006)
8. Graves, A., Liwicki, M., Fernández, S., Bertolami, R., Bunke, H., Schmidhuber, J.: A novel connectionist system for unconstrained handwriting recognition. IEEE Trans. Pattern Anal. Mach. Intell. **31**(5), 855–868 (2008)
9. Huang, Y.S., et al.: A simulated annealing approach to construct optimized prototypes for nearest-neighbor classification. In: Proceedings of 13th International Conference on Pattern Recognition, vol. 4, pp. 483–487. IEEE (1996)
10. Kingma, D.P., Ba, J.: Adam: a method for stochastic optimization. arXiv preprint arXiv:1412.6980 (2014)
11. Kohonen, T.: The self-organizing map. Proc. IEEE **78**(9), 1464–1480 (1990)
12. LeCun, Y., Bottou, L., Bengio, Y., Haffner, P.: Gradient-based learning applied to document recognition. Proc. IEEE **86**(11), 2278–2324 (1998)
13. Lee, S.W., Song, H.H.: Optimal design of reference models for large-set handwritten character recognition. Pattern Recogn. **27**(9), 1267–1274 (1994)
14. Liu, C.L.: Classifier combination based on confidence transformation. Pattern Recogn. **38**(1), 11–28 (2005)
15. Liu, C.L., Eim, I.J., Kim, J.H.: High accuracy handwritten Chinese character recognition by improved feature matching method. In: Proceedings of the Fourth International Conference on Document Analysis and Recognition, vol. 2, pp. 1033–1037. IEEE (1997)
16. Liu, C.L., Nakagawa, M.: Evaluation of prototype learning algorithms for nearest-neighbor classifier in application to handwritten character recognition. Pattern Recogn. **34**(3), 601–615 (2001)
17. Liu, C.L., Sako, H., Fujisawa, H.: Effects of classifier structures and training regimes on integrated segmentation and recognition of handwritten numeral strings. IEEE Trans. Pattern Anal. Mach. Intell. **26**(11), 1395–1407 (2004)
18. Liu, C.L., Yin, F., Wang, D.H., Wang, Q.F.: Casia online and offline Chinese handwriting databases. In: Proceedings of the International Conference on Document Analysis and Recognition, pp. 37–41. IEEE (2011)
19. Liu, H., Jin, S., Zhang, C.: Connectionist temporal classification with maximum entropy regularization. In: Proceedings of the Advances in Neural Information Processing Systems, pp. 831–841 (2018)
20. Ly, N.T., Nguyen, K.C., Nguyen, C.T., Nakagawa, M.: Recognition of anomalously deformed kana sequences in Japanese historical documents. IEICE Trans. Inf. Syst. **102**(8), 1554–1564 (2019)
21. Marti, U.V., Bunke, H.: The IAM-database: an English sentence database for offline handwriting recognition. Int. J. Doc. Anal. Recogn. **5**(1), 39–46 (2002). https://doi.org/10.1007/s100320200071
22. Sato, A., Yamada, K.: Generalized learning vector quantization. In: Advances in Neural Information Processing Systems, pp. 423–429 (1996)

23. Sato, A., Yamada, K.: A formulation of learning vector quantization using a new misclassification measure. In: Proceedings of the Fourteenth International Conference on Pattern Recognition, vol. 1, pp. 322–325. IEEE (1998)
24. Shi, B., Bai, X., Yao, C.: An end-to-end trainable neural network for image-based sequence recognition and its application to scene text recognition. IEEE Trans. Pattern Anal. Mach. Intell. **39**(11), 2298–2304 (2016)
25. Shi, B., Wang, X., Lyu, P., Yao, C., Bai, X.: Robust scene text recognition with automatic rectification. In: Proceedings of the IEEE Conference on Computer Vision and Pattern Recognition, pp. 4168–4176 (2016)
26. Shi, B., Yang, M., Wang, X., Lyu, P., Yao, C., Bai, X.: Aster: an attentional scene text recognizer with flexible rectification. IEEE Trans. Pattern Anal. Mach. Intell. **41**(9), 2035–2048 (2018)
27. Su, Bolan, Lu, Shijian: Accurate scene text recognition based on recurrent neural network. In: Cremers, Daniel, Reid, Ian, Saito, Hideo, Yang, Ming-Hsuan. (eds.) ACCV 2014. LNCS, vol. 9003, pp. 35–48. Springer, Cham (2015). https://doi.org/10.1007/978-3-319-16865-4_3
28. Sueiras, J., Ruiz, V., Sanchez, A., Velez, J.F.: Offline continuous handwriting recognition using sequence to sequence neural networks. Neurocomputing **289**, 119–128 (2018)
29. Wang, Q., Lu, Y.: A sequence labeling convolutional network and its application to handwritten string recognition. In: Proceedings of the International Joint Conference on Artificial Intelligence, pp. 2950–2956 (2017)
30. Wang, Q.F., Yin, F., Liu, C.L.: Handwritten Chinese text recognition by integrating multiple contexts. IEEE Trans. Pattern Anal. Mach. Intell. **34**(8), 1469–1481 (2011)
31. Wang, S., Chen, L., Xu, L., Fan, W., Sun, J., Naoi, S.: Deep knowledge training and heterogeneous CNN for handwritten Chinese text recognition. In: Proceedings of the 15th International Conference on Frontiers in Handwriting Recognition, pp. 84–89. IEEE (2016)
32. Wang, T., et al.: Decoupled attention network for text recognition. Proc. AAAI Conf. Artif. Intell. **34**, 12216–12224 (2020)
33. Wang, Z.X., Wang, Q.F., Yin, F., Liu, C.L.: Weakly supervised learning for over-segmentation based handwritten Chinese text recognition. In: Proceedings of the 17th International Conference on Frontiers in Handwriting Recognition, pp. 157–162. IEEE (2020)
34. Wang, Z.R., Du, J., Wang, W.C., Zhai, J.F., Hu, J.S.: A comprehensive study of hybrid neural network hidden Markov model for offline handwritten Chinese text recognition. Proc. Int. J. Doc. Anal. Recogn. **21**(4), 241–251 (2018)
35. Wu, Y.C., Yin, F., Chen, Z., Liu, C.L.: Handwritten Chinese text recognition using separable multi-dimensional recurrent neural network. In: Proceedings of the 14th IAPR International Conference on Document Analysis and Recognition, vol. 1, pp. 79–84. IEEE (2017)
36. Wu, Y.C., Yin, F., Liu, C.L.: Improving handwritten Chinese text recognition using neural network language models and convolutional neural network shape models. Pattern Recogn. **65**, 251–264 (2017)
37. Yang, H.M., Zhang, X.Y., Yin, F., Liu, C.L.: Robust classification with convolutional prototype learning. In: Proceedings of the IEEE Conference on Computer Vision and Pattern Recognition, pp. 3474–3482 (2018)
38. Yang, H.M., Zhang, X.Y., Yin, F., Yang, Q., Liu, C.L.: Convolutional prototype network for open set recognition. IEEE Transactions on Pattern Analysis and Machine Intelligence, early access (2020)

39. Yin, F., Wang, Q.F., Zhang, X.Y., Liu, C.L.: ICDAR 2013 Chinese handwriting recognition competition. In: Proceedings of the 12th International Conference on Document Analysis and Recognition, pp. 1464–1470. IEEE (2013)
40. Yin, F., Wu, Y.C., Zhang, X.Y., Liu, C.L.: Scene text recognition with sliding convolutional character models. arXiv preprint arXiv:1709.01727 (2017)
41. Zhan, H., Lyu, S., Lu, Y.: Handwritten digit string recognition using convolutional neural network. In: Proceedings of the 24th International Conference on Pattern Recognition, pp. 3729–3734. IEEE (2018)
42. Zhan, H., Wang, Q., Lu, Y.: Handwritten digit string recognition by combination of residual network and RNN-CTC. In: Proceedings of the International Conference on Neural Information Processing, pp. 583–591. Springer (2017). https://doi.org/10.1007/978-3-319-70136-3_62
43. Zhang, Y., Nie, S., Liu, W., Xu, X., Zhang, D., Shen, H.T.: Sequence-to-sequence domain adaptation network for robust text image recognition. In: Proceedings of the IEEE/CVF Conference on Computer Vision and Pattern Recognition, pp. 2740–2749 (2019)

Online Spatio-temporal 3D Convolutional Neural Network for Early Recognition of Handwritten Gestures

William Mocaër[1,2]([envelope]) [ID], Eric Anquetil[1] [ID], and Richard Kulpa[2] [ID]

[1] University of Rennes, CNRS, IRISA, 35000 Rennes, France
eric.anquetil@irisa.fr
[2] University of Rennes, Inria, M2S, 35000 Rennes, France
{william.mocaer,richard.kulpa}@irisa.fr

Abstract. Inspired by recent spatio-temporal Convolutional Neural Networks in computer vision field, we propose OLT-C3D (Online Long-Term Convolutional 3D), a new architecture based on a 3D Convolutional Neural Network (3D CNN) to address the complex task of early recognition of 2D handwritten gestures in real time. The input signal of the gesture is translated into an image sequence along time with the trajectory history. The image sequence is passed into our 3D CNN OLT-C3D which gives a prediction at each new frame. OLT-C3D is coupled with an integrated temporal reject system to postpone the decision in time if more information is needed. Moreover our system is end-to-end trainable, OLT-C3D and the temporal reject system are jointly trained to optimize the earliness of the decision. Our approach achieves superior performances on two complementary and freely available datasets: ILGDB and MTGSetB.

Keywords: Spatio-temporal convolutional neural network · Early recognition · Handwritten gesture · Online long-term C3D · WaveNet 3D

1 Introduction

To be reactive, some applications need to know as soon as possible the intention of the user. In a tactile environment where you can zoom, scroll with direct manipulation, we should also be able in the same interactive context to do more complex actions associated with real gestures like symbols (abstract command). The coexistence of direct manipulation and abstract command is possible in interactive context only if we are able to predict very early the intention of the user, before the gesture is completed. Very few works addressed this coexistence problem, but we can find the ones of Petit and Maldivi [15] and Kurtenbach and Buxton [10]. Most existing works focus on the recognition of gestures once it is

This study is funded by the ANR within the framework of the PIA EUR DIGISPORT project (ANR-18-EURE-0022).

J. Lladós et al. (Eds.): ICDAR 2021, LNCS 12821, pp. 221–236, 2021.
https://doi.org/10.1007/978-3-030-86549-8_15

completed, only few covered the early recognition problem which is a complex new challenge for the 2D handwritten gesture recognition community, but also for the 3D human gesture recognition community.

In most of the cases, it is possible to discriminate the gesture before it is completed. We define the early recognition problem as the task to predict the class of a gesture as soon as possible. To avoid errors the system should be able to postpone the decision if more information is needed, which is linked to confidence. Early gesture recognition opens a large field of new applications with the coexistence of direct and abstract commands in the same context.

To tackle the aforementioned challenges, we propose the novel OLT-C3D network, coupled with an integrated temporal reject option system. Inspired by recent spatio-temporal CNNs [3,17] in computer vision field, we propose a new architecture based on 3D CNN. The input signal of the gesture is translated into a sequence of images along time with the cumulative trace. Then the images are passed into the 3D CNN which gives a prediction at each new frame. We provide to OLT-C3D a capacity of auto-evaluation of the prediction thanks to the temporal reject system: the prediction can be either accepted to confirm the prediction or rejected if the network needs to wait for more information to decide. The main contributions of this paper are summarized as follows:

- We designed a representation strategy to translate the online input signal in free context into an image sequence.
- We propose the OLT-C3D network, a 3D convolutional neural network built to handle 2D image along time in an online manner.
- We added a temporal reject option system to the OLT-C3D network to postpone the decision in time if more information is needed.
- The network is end-to-end trainable and do not need any post-calibration for the reject system.
- Our method achieves superior performance for the early recognition task regarding accuracy and earliness. Experiments were conducted on two freely available and complementary datasets: ILGDB [16] (mono-stroke gestures) and MTGSetB [5] (multi-touch gestures).

2   Related Works

We believe that the task of 3D gesture recognition is very close to the recognition of 2D gesture, and particularly the recognition from skeleton joints. The 3D gesture recognition community is particularly active these last years. Most work focuses on trimmed gesture recognition, some works addressed the untrimmed gesture recognition and only few tackle the early recognition problem. In 2D, the early recognition of handwritten gesture is also very few addressed. We present in this section the works related to the 3D and 2D early gesture recognition and prediction.

In 3D, some works addressed the early recognition task with template-based methods. For example, Kawashima et al. [8] proposed a method that computes the distance between input gestures and templates. The system prediction is rejected until the distance with the second most similar class template is over a threshold. Mori et al. [13] used a partial matching method to do early recognition and gesture prediction. Bloom et al. [1] also proposed an early action recognition and prediction method based on template matching using DTW. More recently, some works proposed deep learning-based methods. A system based on a recurrent 3D CNN (R3DCNN) proposed by Molchanov et al. [12] is able to do early recognition, the input video is split in clips and passed into a 3D CNN, then the output is given as input of an RNN. They used a reject system based on the confidence score emitted by the classifier and a fixed threshold. Weber et al. [19] used LSTM network with the 3D joints coordinates in input, the reject strategy consists of waiting that the system repeats the same class prediction a fixed number of consecutive frames. Boulahia et al. [2] proposed a more explicit and transparent method based on a combination of curvilinear models, they indexed the gesture completion using displacement instead of time in order to be speed-independent in the gesture representation. They used a reject system based on the confidence score given by an SVM.

Escalante et al. [6] and Liu et al. [11] addressed the task of action prediction of 3D gestures in an untrimmed stream. Both proposed a method to predict the class at any observation ratio of the gesture without any reject option. Escalante et al. method is based on naive Bayes. Liu et al. developed an architecture named SSNet based on WaveNet [14]. They used a hierarchy of stacked causal and dilated 1D convolutional layers. SSNet is able to handle a stream in real time, giving a new response to each new frame, but no reject system is incorporated to the method. Some of these methods can be adapted to the early recognition of 2D gestures.

Few works addressed the task of early recognition of **2D gestures**. Uchida et al. [18] proposed an early recognition system based on frame-classifier combination. A frame classifier at time t uses weighted combination of the previous $1 \ldots t-1$ frame classifiers. Chen et al. [4] addressed the task by using a combination of length-dependent classifiers and a system of reject based on the confidence scores of the classifiers and repetition of prediction. Recently, Yamagata et al. [20] proposed an approach to do handwriting prediction which learns the bifurcations of gestures based on an LSTM network.

The 3D gesture recognition methods particularly inspired us to develop our approach to address the early recognition of 2D gestures task.

3   Method

Firstly, inspired by trajectory-based methods [1,2,18], we propose an original **spatio-temporal representation**, representing the gesture completion in time. The online signal of the gesture is translated into an image sequence containing the trajectory, each new image of the sequence is incremented by a new piece of the trajectory. Then, the images are passed into our **OLT-C3D network**, this original network is mainly inspired by recent spatio-temporal CNNs [3,17] which have proven their abilities to learn spatio-temporal features. OLT-C3D gives a prediction at each new frame. Finally, the **temporal reject option system** is able to postpone the decision by rejecting the predictions.

3.1   Spatio-temporal Gesture Representation Strategy

In this work we present an original spatio-temporal representation of the gesture. The input signal is an online trajectory, at each instant we know the position of fingers/pen on the device. We will translate this input signal into an image sequence, each new image containing the new positions of the fingers with its previous trajectories. This representation will be the input of the spatio-temporal CNN. At each instant we will feed the CNN with the new information received in order that the input is always up to date. At the beginning, the network sees only a small piece of the gesture, at the end, it sees the full gesture trace with all the history of the gesture completion. The history is very important to see in which order the gesture is done: two gestures can have the same final shape, but are not done is the same order, this is illustrated in Fig. 1.

A naive strategy would be to feed the neural network with a new image containing the new information at each time we have new information from the device used, most of the works in 3D gesture recognition [1,11,12,17] use this strategy. Nevertheless between each time there can be only a very small amount of new information if the gesture is done slowly, or even no new information at all if the gesture is paused. By translating the input signal into a new image at each instant, this would lead to a lot of images with some duplication and a very small amount of new effective information between images. Furthermore, if we choose a larger sample rate to reduce the number of images, a gesture made slowly and the same gesture done quickly would not produce the same amount of images along time. For the quick gesture, we would not be able to recognize the gesture as soon (in terms of quantity of information) as we would be able to recognize. To tackle these problems we used, as previous studies [2,4], the quantity of information (displacement) instead of the time to quantify the effective displacement and generate a speed independent image sequence. Each new image will be incremented by the same total displacement θ. A smaller θ will lead to more images than a higher one, because the displacement between each new image will be smaller we need more images to complete the gesture. This strategy leads to three main advantages: fewer images in the sequence, more significant difference between each image and speed invariant representation. For simplicity, we use the term "frame" to designate each image incremented by a certain amount of displacement.

The network has all the history of the trace, but it cannot make the difference between a simple touch then finger up and a constant touch. We need a strategy to make this difference appear on our representation, this is necessary for some gestures in multi-touch (i.e., multiple fingers are used to make a gesture) databases like MTGSetB [5]. To this end, we add a second channel to our images, containing a "1" value if a finger is active in this coordinate at the end of the current displacement, a null value otherwise. This leads to two images: one containing the trajectory, and the other one, very sparse, containing ones where a finger is active. These two images will be used as channels by the CNN. This second channel has been used only for the multi-touch dataset MTGSetB. For mono-touch datasets such as ILGDB, it has no benefits.

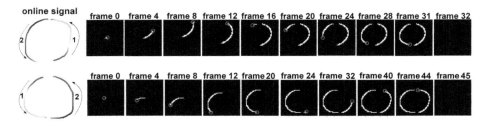

Fig. 1. Considering the final shape, two gesture classes can be involved regarding the order of the stroke. Regarding the second gesture, the order of the strokes is reversed from the first gesture. Our representation takes in consideration the order of the strokes thanks to the history. The last finger position in the segment is represented by a red pixel (surrounded by a red circle in images for visibility) in our representation. If the gesture reaches the border of the image, all the gesture is shifted in new images (this is visible in both gestures between frame 8 and 12). The end of the gesture is symbolized by a black image. (Color figure online)

Another difficulty is the *free context*: we don't know in advance what will be the size of the full gesture, so we cannot ensure that the gesture will enter correctly in the image dimension with respect to the resolution we have at the begin of the draw. One solution is to rescale the gesture at each new frame to fit the initial dimension, but we think that would break the spatio-temporality of the information that will be used by the CNN, with difficulties to perceive which new piece is just added between two frames. To keep the same scale from the begin to the end, we choose to "follow" the movement by shifting all the gesture in the opposite direction of the movement when it reaches the edge of the image. We potentially lose a piece of the gesture at each time we shift, but the network does not need it anymore because this part is still in its history.

Lastly, in a database where some gestures are subpart of others, the network needs to know when the gesture is finished. To do that, we add a black image at the end of the gestures. In real application, a strategy must be established to determine when a gesture is finished, it can be pen up from the device, explicit confirmation, time without any action... The final representation is illustrated in Fig. 1.

3.2 Online Spatio-temporal 3D CNN with Temporal Reject System

Recently, the CNNs have proved their ability on learning from time series [11,14] and image sequences [3,17]. Inspired by these approaches, we propose OLT-C3D

(Online Long-Term Convolutional 3D), a spatio-temporal 3D CNN able to treat streaming data from devices and to give a response in real time.

Online Spatio-temporal 3D CNN. We propose a new architecture mainly inspired by two networks: WaveNet [14] and C3D [17]. WaveNet is able to handle 1D temporal series in an online manner using causal and dilated convolutions. C3D is a 3D CNN able to handle 3D input: 2D images along time.

Our objective was to be able to handle 3D input in an online manner. To do that, we propose a spatio-temporal 3D CNN with causal and dilated convolutions on the temporal axis. On the spatial dimensions, the network follows standards using alternatively convolutional layer and pooling layer. An example of a 3D convolution (first layer) of OLT-C3D is provided in Fig. 2.

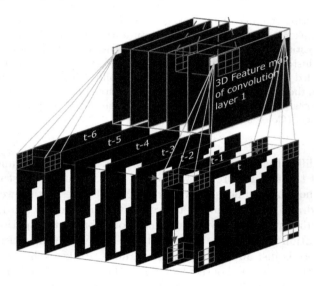

Fig. 2. Example of a spatio-temporal 3D convolution in the first layer. The filter size is 2 (time) × 3 (x axis) × 3 (y axis). The filters are applied along all the temporal axis with causal convolutions, and along spatial dimensions with classic convolutions. There is no dilatation along temporal axis for the first layer. The green filter is one of the first layer. The pink filter is one of the second layer. The filters can learn spatio-temporal patterns thanks to the 3D convolution. (Color figure online)

Our architecture is composed of 10 stacked convolutional layers. The convolutions are causal on temporal dimension. The causal convolutions let the network compute an output only from previous frames, this ensure not to use future frames for the predictions. The layers also use dilated (or "à trou") convolutions along the time axis. The dilated convolutions allow the network to grow its receptive field quickly with the layers, ensuring that all the history of the frames is used to compute the new output. These layers are divided in two blocks of 5 layers, with a dilatation rate equals to 2^i where $i \in \{0, 1, 2, 3, 4\}$ is the index of the layer in the block, the dilatation rate is 16 for last layers of blocks. Taking these dilation rates allows to increase the receptive field and to assure that all initial values given in input are taken into account. Each convolution layer is followed by a max-pooling layer applied to the two spatial dimensions, there is no pooling along the temporal dimension.

As shown in Fig. 3, the lower convolutional layer has a very small receptive field since only two frames are used to compute the output, it focuses on the last two frames. The second layer uses the result of the previous one, using indirectly four frames. The number of frames used increases with the number of layers. The top convolutional layer of the network is able to see up to 63 input frames. The receptive field must be accorded to the parameter θ defined in Sect. 3.1 and the length of gestures. With the θ we used in our experiment, 63 frames in enough to see the whole gesture completion history in most cases. These dilated convolutions allow our system to avoid the use of memory cell mechanisms like RNN used in some works [12, 19] to have a long-term memory.

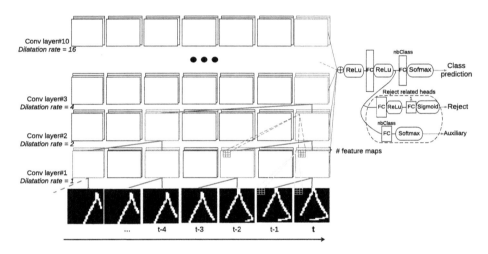

Fig. 3. The complete architecture of OLT-C3D. First, the signal is transformed into images, and fed into the network. The network makes a prediction at each instant. OLT-C3D is coupled to a temporal reject system.

The network emits a classification prediction for each new frame. To this end, the network is completed by an aggregation (average) of the feature maps from all layers corresponding (thanks to causal convolutions) to **the last frame (instant** t**)**. Because the network only uses the feature maps corresponding to the last frame, our network can handle any length of sequence and is able to give a prediction at each new frame. The aggregation is followed by a fully connected (FC) layer and then three main heads. The first head f is dedicated to the class prediction, composed of a new FC layer with the number of class neurons followed by a softmax activation function, the second head (Reject, g) and third head (Auxiliary, h) are dedicated to the reject option system. The complete architecture is provided in Fig. 3.

Temporal Rejection Option System. One main problem in early recognition is to take a decision to recognize a gesture only when the answer is sure, we want to let the classifier the possibility to postpone a decision when it estimates that it does not have enough information. For example, if we have gestures with common parts at the beginning, we want the classifier to give no response until the common part is passed. We need a mechanism in order to have a kind of confidence score, to reject or accept the current prediction.

To tackle this problem, we used SelectiveNet [7] and adapted it to a per-frame fashion. This leads us to add two new outputs: selection/reject head and auxiliary head, these two heads are included in the "Reject related heads" block shown in the Fig. 3. The reject head is composed of a FC layer with ReLu activation. Then it lasts by a FC layer with only one neuron. Sigmoid is the final activation function used in this head, we define this output as g. The goal of this output is to accept or reject the prediction. We will consider the prediction rejected if the reject head output is less than a parameter τ, and accepted if it is over. The final prediction output with respect to g is defined as:

$$(f, g_\tau)(x) = \begin{cases} f(x), & \text{if } g(x) \geq \tau \\ \text{don't know}, & \text{otherwise.} \end{cases} \tag{1}$$

We used $\tau = 0.5$ in this paper, as in SelectiveNet. Finally, the loss of the prediction head, using g, is defined as follows:

$$\mathcal{L}_{(f,g)} = \frac{1}{m} \sum_{i=1}^{m} \ell\left(f\left(x_i\right), y_i\right) g\left(x_i\right) + \lambda \Psi\left(c - \hat{\phi}\left(g\right)\right) \tag{2}$$

where c is the target coverage, λ is a hyperparameter relative to the importance of the coverage constraint and $\Psi(a) = \max(0, a)^2$. As SelectiveNet we used $\lambda = 32$. $\hat{\phi}(g)$ is the empirical coverage, i.e. the average value of $g(x)$, and ℓ is the cross entropy loss.

The auxiliary head is the same as the prediction head, but is optimized with a standard loss function $\mathcal{L}_h$, we used cross-entropy. It is used to optimize the CNN representation without focusing too much on the loss of the prediction head $\mathcal{L}_{(f,g)}$. More details about SelectiveNet can be found in the original paper [7].

Unlike in SelectiveNet where the rejection head is responsible for the rejection of a full-gesture sample, here the rejection head has to decide for each frame if the prediction is accepted. The loss has to be adapted from a per-sample fashion to a per-frame fashion. In order to do this, the coverage average $\hat{\phi}(g)$ and the loss ℓ are computed for all predictions along time. Consequently, the coverage c is more related to earliness in our case.

The final optimized loss is:

$$\mathcal{L} = \alpha \mathcal{L}_{(f,g)} + (1 - \alpha)\mathcal{L}_h \tag{3}$$

in our experiments we fixed α to 0.5 as SelectiveNet.

At each instant OLT-C3D will output one class prediction and the temporal reject system accepts or rejects the prediction if it needs more information. The network can also totally reject the gesture, even at the end, if the gesture is too close to two classes, or if it does not correspond to a known class.

4   Experimental Evaluation

We evaluate the OLT-C3D approach on two freely available datasets: ILGDB [16] which contains only **mono-stroke** gestures and MTGSetB [5] which contains **multi-touch** gestures. These two datasets are complementary in terms of gesture natures (mono/multi-stroke, mono/multi-touch), they are very interesting for early recognition experiments. We compare our scores to the state-of-the-art method on the task of early recognition [4] on these two datasets.

4.1   Network Hyperparameters and Details

The pooling size and stride of the maxpooling layers is 3 for spatial dimensions, 1 for temporal dimension (no pooling). We used a small dropout for convolutional layers of 0.1, and 0.3 for the first FC layer. ReLu is the activation function used after each convolutional layer. We optimized the network with Adam [9] with the learning rate fixed to 0.003. 85% of the training data is used for training, 15% is used as validation set. The dataset specific hyperparameters are presented in Table 1. For ILGDB, we fixed the image dimension to 30 by 30. The coordinates of the gestures are scaled by 0.2, and we augmented the training data by scaling the train gestures coordinates also by 0.3, 0.4, 0.5 and 0.6. The displacement quantity θ is fixed to 4.5 pixels (once scaled). Regarding the CNN, we found that 10 filters by layers is enough for this dataset, with 300 neurons in the FC layers. We used a batch size of 85 sequences padded with black images at the end. The target coverage c is fixed to 0.6. The hyperparameters for MTGSetB leads to a higher network (770k parameters for MTGSetB and 420k for ILGDB), this is because this dataset is more complex than ILGDB due to a higher number of gesture classes and shape varieties. These hyperparameters have been softly fine-tuned with a validation set.

Table 1. Dataset specific hyperparameters.

	Image dim.	Scale	Data aug. scale	filters	θ	FC neurons	Batch size	c
ILGDB	30×30	0.2	0.3, 0.4, 0.5, 0.6	10	4.5	300	85	0.6
MTGSetB	40×40	0.03	0.04	25	2	150	40	0.75

4.2   Measures

To evaluate the early recognition task, we use the True Acceptance Rate (TAR) which measure the accuracy of the classifier when the prediction is accepted. We also use the False Acceptance Rate (FAR) which measure the error rate when the prediction is accepted. To measure the final reject, we use the Reject Rate (RR) which is the number of samples which are totally rejected, even at the end of the sequence. Referring to the notation of the Table 2 the TAR, the FAR and RR are defined as:

$$TAR = \frac{N_A^T}{N} \; ; \; FAR = \frac{N_A^F}{N} \; ; \; RR = \frac{N_R}{N} \tag{4}$$

Regarding our network, only the classification of the first acceptation is used to compute the TAR and the FAR. If the gesture is never accepted, it is taken into account in the RR. Note that $TAR + FAR + RR = 1$. To evaluate the earliness, we used the Normalized Distance To Detection (NDtoD) which is computed as the length of the gesture made at the moment of the first acceptation over the total length of the gesture, only the accepted gestures are taken into account.

Table 2. Notations used for the measurement of reject-based systems, used in [4]. N_A^F are the training samples which are wrongly classified but accepted by the reject system while the N_A^T samples are correctly classified and accepted.

Sample set (N)	Reject option	
	Accept (N_A)	Reject (N_R)
Correctly classified (N_{cor})	True Accept (N_A^T)	False Reject (N_R^F)
Mis-classified (N_{err})	False Accept (N_A^F)	True Reject (N_R^T)

4.3 Early Recognition Results

ILGDB. The ILG database is a **mono-stroke pen-based gestures dataset** performed by 38 users. It contains 21 different gesture classes with a total of 1923 samples, 693 are used for training and 1230 for testing. The specificity of this dataset is that there are a lot of gestures which have common begins, or are subparts of other gestures. With this, the network needs to reject until the trajectory become discriminative, and it can be very late. We compared our score to the approach of Chen et al. [4] using the predefined Train/Test split furnished by the dataset. To compare fairly with them, we used the parameter t which is the number of time the same prediction class must be accepted consecutively by our reject system to be finally accepted. This is a way to reinforce the confidence, but it leads to delayed decisions and a higher reject rate, which can be less optimal than tuning the reject system. $t = 1$ is the default value for other experiments, unless explicit mention. The scores are in Table 3. We can see that for an equivalent earliness ($t = 2$ for Chen et al. and $t = 1$ for us), our network is much more accurate with 15% of better classification when the gesture is accepted, 10% less misclassification, and fewer gestures rejected. The decision is made on average at 76%, which is late but consistent with the gestures. Some gestures have common begins, for these the network needs to wait that the common part is passed. In Fig. 4a, we see that the network reject most of the prediction in early states, waiting that the gesture completion is at least 40% to begin to accept the predictions. The FAR stay low until the end. We can see a peak at 100% of the gesture completion, that is because some gestures are subpart of others, so the only way to recognize the gesture is to wait that the gesture is completely finished.

Table 3. Comparison of the OLT-C3D approach to the previous method on the datasets. TAR: True Acceptance Rate, FAR: False Acceptance Rate, RR: Reject Rate, NDtoD: Normalized Distance to Detection (earliness). t is the number of consecutive acceptation of the same class required to be finally accepted.

Dataset	t	Chen et al. [4]				OLT-C3D (ours)			
		TAR	FAR	RR	NDtoD	TAR	FAR	RR	NDtoD
ILGDB	1	30.65%	67.15%	2.20%	34.81%	**79.75%**	**16.74%**	**3.50%**	**76.81%**
	2	**64.15%**	**26.42%**	**9.43%**	**75.53%**	81.79%	14.15%	4.07%	82.56%
	3	73.98%	11.22%	14.80%	92.24%	83.25%	12.03%	4.72%	87.44%
	4	77.72%	6.26%	16.02%	97.62%	84.31%	9.51%	6.18%	91.01%
	5	77.80%	4.88%	17.32%	99.19%	85.20%	7.40%	7.40%	93.71%
	6	77.72%	4.55%	17.72%	99.68%	84.39%	6.34%	9.26%	95.54%
MTGSetB	1	**81.89%**	**14.56%**	**3.54%**	**37.04%**	89.25%	7.24%	3.51%	30.77%
	2	83.44%	10.85%	5.71%	46.82%	**90.52%**	**5.82%**	**3.66%**	**36.89%**
	3	82.38%	8.85%	8.77%	55.89%	90.94%	5.08%	3.98%	42.78%
	4	82.20%	6.06%	11.73%	66.16%	91.22%	4.25%	4.53%	48.48%
	5	80.35%	4.60%	15.05%	71.03%	91.49%	3.53%	4.98%	53.92%
	6	77.42%	3.41%	19.17%	77.54%	91.36%	2.86%	5.77%	58.92%

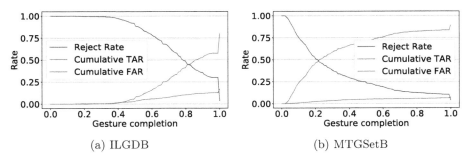

(a) ILGDB   (b) MTGSetB

Fig. 4. Behaviour of each rate (Cumulative TAR, cumulative FAR, RR) on the two datasets. The cumulative TAR at x% of completion is the number of samples accepted before x% and correctly classified over the total number of samples.

MTGSetB. The MTGSetB dataset is composed of 45 different **multi-touch gestures** regrouped into 31 rotation invariant gesture classes made by 33 users. Like ILGDB, we compared our score to Chen et al. [4]. 50% of the data are used for training. The comparison is shown in the Table 3. We can see that, for the same earliness and reject rate ($t = 1$ for Chen et al. and $t = 2$ for us), OLT-C3D has a much higher TAR (+9%), a much lower FAR (−9%). The MTGSetB dataset contains multi-touch gestures, and only few of them are subparts of others. By using the number of fingers and the initial directions, our network is able to predict the gestures very early, with few errors. In Fig. 4b, we see that the network starts to accept very early, keeping a low FAR.

Earliness Evaluation per Class. The earliness per-class is presented in Fig. 5. For ILGDB, we observe differences between classes, with earliness from an average of 60% gesture completion for the class "effect3" to 100% for the class "display1". Note that the gesture corresponding to "display1" is a subpart of two other gesture classes, so the only way to recognize the gesture correctly is to wait that the gesture is completely finished. In MTGSetB we observe a similar case with the class "A_02" which is a subpart of multiple other gesture classes. Otherwise, the gestures are accepted very early for most of the classes. The earliest predicted class is the last one, "C_04", which is accepted on average at 5% of gesture completion. The network is able to predict as early for this class because it is the only one with only one touch which begins by moving to the right, so there is enough information in the first frame in most cases.

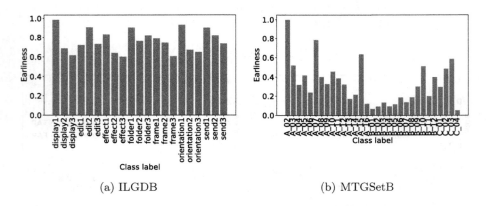

(a) ILGDB   (b) MTGSetB

Fig. 5. Earliness per class (NDtoD)

4.4 Spatio-temporal Representation Evaluation

We compare our representation discussed in Sect. 3.1 with two variants: one with only the trajectory, the other with only the finger position. The results are provided in Table 4. We see that both trajectory and finger position bring significant information to the representation. In particular, the finger position channel brings the difference between a constant touch and a touch then release.

Table 4. Comparison of the different variants of our representation on MTGSetB.

Variant	TAR	FAR	RR	NDtoD
Only trajectory (first channel)	84.1%	9.73%	6.17%	**30.71%**
Only finger position (second channel)	86.54%	10.87%	2.59%	33.28%
Both: trace and finger position	**89.25%**	**7.24%**	3.51%	**30.77%**

4.5 Qualitative Results

In the datasets, there are some gestures with common parts. The expected behavior from our network is to reject the prediction until the common part is passed. In this section, we analyze the output of our system. For example, ILGDB contains three gestures which starts like an "M" letter, the direction given after the "M" stroke is decisive. The Fig. 6 shows an example of the behavior of our network on these three labels. We see that the temporal reject system waits the decisive instant to accept the prediction. Note that this example is very representative of the behavior of the network on the "M" classes. This shows the ability of our approach to well reject in time the prediction until the common part is passed. The case of the "display1" class is also interesting, this gesture is a simple down stroke, as illustrated in Fig. 6. This gesture is a subpart of two other gestures, where one goes left and the other go right after the down stroke. The only way the network can know that it is the class "display1" is to wait until the end of the gesture (e.g., pen up for single stroke gestures). As explained in Sect. 3.1 we modeled the end of the gesture using a black image. For this gesture class the network rejects the prediction until the black image.

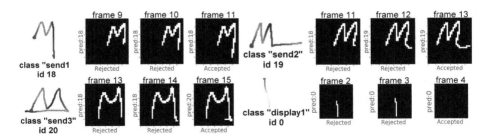

Fig. 6. Behavior on the "M" classes until the first acceptation. The system rejects predictions until the decisive instant. The class "display1" is a subpart of two other classes, the system rejects the predictions until the black frame (end of the gesture). A green label prediction on the left means a good classification. (Color figure online)

On MTGSetB, the first acceptation is made very early on average. By analyzing the number of touch and the beginning of the trajectory, the network is able to make accurate predictions and very early. An example is shown in Fig. 7. In this example, we see that the only element which makes the difference between the two gestures is the finger position channel. Without this channel, the network would not be able to discriminate these two gestures.

Fig. 7. Behavior on the classes C_01 and A_07 from MTGSetB. For the gesture on the left, the network is able to accept the prediction in the frame 1 because it can see that the finger of the left stroke has been released, and it is the only gesture which starts like that. For the gesture on the right, the finger of the left stroke is continuously pressed, that can be multiple gestures at this instant, so it rejects the predictions for these frames.

4.6   Speed Execution

Our network is able to treat streaming data at **46** frames per seconds, considering that the frame representation extraction time execution is negligible. Note that our system waits to acquire enough displacement to submit the new image to the network, in this case it should be able to respond in $\approx$**22 ms**. This is enough to be used in real-time application. Experiments were conducted on a Quadro RTX 3000. The response time can be improved using the activation sharing scheme used in SSNet [11].

5   Conclusion

In this paper, we proposed a framework composed of three main parts. First, an original **spatio-temporal representation** of the gesture, well suited to represent online gesture, regardless of its nature (mono/multi-stroke, mono/multi-touch). Then, **OLT-C3D**, an original 3D CNN able to extract spatio-temporal features in an online manner. Lastly, a **temporal reject system** to postpone the decision if necessary. Our network coupled to the temporal reject system is end-to-end trainable, and runs in real time. We showed that our method is able to make predictions very early, with very interesting performance. It opens a large field of innovative applications. Our future works will focus on an extension of this approach for early recognition of 3D gestures.

References

1. Bloom, V., Argyriou, V., Makris, D.: Linear latent low dimensional space for online early action recognition and prediction. Pattern Recogn. **72**, 532–547 (2017). https://doi.org/10.1016/j.patcog.2017.07.003
2. Boulahia, S.Y., Anquetil, E., Multon, F., Kulpa, R.: Détection précoce d'actions squelettiques 3D dans un flot non segmenté à base de modèles curvilignes. In: Reconnaissance des Formes. Image, Apprentissage et Perception, RFIAP 2018, June 2018, Paris, France, pp. 1–8 (2018)
3. Carreira, J., Zisserman, A.: Quo vadis, action recognition? A new model and the kinetics dataset. In: Proceedings of the IEEE Conference on Computer Vision and Pattern Recognition (CVPR) (2017). https://doi.org/10.1109/CVPR.2017.502
4. Chen, Z., Anquetil, E., Viard-Gaudin, C., Mouchère, H.: Early recognition of hand-written gestures based on multi-classifier reject option. In: 14th IAPR International Conference on Document Analysis and Recognition (ICDAR), vol. 01, pp. 212–217 (2017). https://doi.org/10.1109/ICDAR.2017.43
5. Chen, Z., Anquetil, E., Mouchère, H., Viard-Gaudin, C.: Recognize multi-touch gestures by graph modeling and matching. In: 17th Biennial Conference of the International Graphonomics Society. Drawing, Handwriting Processing Analysis: New Advances and Challenges. International Graphonomics Society (IGS) and Université des Antilles (UA), Pointe-a-Pitre, Guadeloupe (June 2015)
6. Escalante, H.J., Morales, E.F., Sucar, L.E.: A naïve Bayes baseline for early gesture recognition. Pattern Recogn. Lett. **73**, 91–99 (2016). https://doi.org/10.1016/j.patrec.2016.01.013
7. Geifman, Y., El-Yaniv, R.: SelectiveNet: a deep neural network with an integrated reject option. In: Chaudhuri, K., Salakhutdinov, R. (eds.) Proceedings of the 36th International Conference on Machine Learning, 09–15 June 2019, vol. 97, pp. 2151–2159. PMLR (2019)
8. Kawashima, M., Shimada, A., Nagahara, H., Taniguchi, R.: Adaptive template method for early recognition of gestures. In: 17th Korea-Japan Joint Workshop on Frontiers of Computer Vision (FCV), pp. 1–6 (2011). https://doi.org/10.1109/FCV.2011.5739719
9. Kingma, D.P., Ba, J.: Adam: a method for stochastic optimization (2017)
10. Kurtenbach, G., Buxton, W.: Issues in combining marking and direct manipulation techniques. In: Proceedings of the 4th Annual ACM Symposium on User Interface Software and Technology, UIST 1991, pp. 137–144. Association for Computing Machinery, New York (1991). https://doi.org/10.1145/120782.120797
11. Liu, J., Shahroudy, A., Wang, G., Duan, L., Kot, A.C.: Skeleton-based online action prediction using scale selection network. IEEE Trans. Pattern Anal. Mach. Intell. **42**(6), 1453–1467 (2020). https://doi.org/10.1109/TPAMI.2019.2898954
12. Molchanov, P., Yang, X., Gupta, S., Kim, K., Tyree, S., Kautz, J.: Online detection and classification of dynamic hand gestures with recurrent 3d convolutional neural network. In: The IEEE Conference on Computer Vision and Pattern Recognition (CVPR) (2016). https://doi.org/10.1109/CVPR.2016.456
13. Mori, A., Uchida, S., Kurazume, R., Taniguchi, R., Hasegawa, T., Sakoe, H.: Early recognition and prediction of gestures. In: 18th International Conference on Pattern Recognition, ICPR 2006, vol. 3, pp. 560–563 (2006). https://doi.org/10.1109/ICPR.2006.467
14. van den Oord, A., et al.: WaveNet: a generative model for raw audio. CoRR (2016)

15. Petit, E., Maldivi, C.: Unifying gestures and direct manipulation in touchscreen interfaces (December 2013)
16. Renau-Ferrer, N., Li, P., Delaye, A., Anquetil, E.: The ILGDB database of realistic pen-based gestural commands. In: Proceedings of the 21st International Conference on Pattern Recognition, ICPR 2012, pp. 3741–3744 (2012)
17. Tran, D., Bourdev, L., Fergus, R., Torresani, L., Paluri, M.: Learning spatiotemporal features with 3d convolutional networks. In: IEEE International Conference on Computer Vision (ICCV), pp. 4489–4497 (2015). https://doi.org/10.1109/ICCV.2015.510
18. Uchida, S., Amamoto, K.: Early recognition of sequential patterns by classifier combination. In: 19th International Conference on Pattern Recognition, pp. 1–4 (2008). https://doi.org/10.1109/ICPR.2008.4761137
19. Weber, M., Liwicki, M., Stricker, D., Scholzel, C., Uchida, S.: LSTM-based early recognition of motion patterns. In: 2014 22nd International Conference on Pattern Recognition, pp. 3552–3557 (2014). https://doi.org/10.1109/ICPR.2014.611
20. Yamagata, M., Hayashi, H., Uchida, S.: Handwriting prediction considering inter-class bifurcation structures. In: 17th International Conference on Frontiers in Handwriting Recognition (ICFHR), pp. 103–108 (2020). https://doi.org/10.1109/ICFHR2020.2020.00029

Mix-Up Augmentation for Oracle Character Recognition with Imbalanced Data Distribution

Jing Li[1], Qiu-Feng Wang[1(✉)], Rui Zhang[2], and Kaizhu Huang[1]

[1] Department of Intelligent Science, School of Advanced Technology,
Xi'an Jiaotong-Liverpool University, Suzhou, China
Jing.Li19@student.xjtlu.edu.cn, {Qiufeng.Wang,Kaizhu.Huang}@xjtlu.edu.cn
[2] Department of Foundational Mathematics, School of Science, Xi'an
Jiaotong-Liverpool University, Suzhou, China
Rui.Zhang02@xjtlu.edu.cn

Abstract. Oracle bone characters are probably the oldest hieroglyphs in China. It is of significant impact to recognize such characters since they can provide important clues for Chinese archaeology and philology. Automatic oracle bone character recognition however remains to be a challenging problem. In particular, due to the inherited nature, oracle characters are typically very limited and also seriously imbalanced in most available oracle datasets, which greatly hinders the research in automatic oracle bone character recognition. To alleviate this problem, we propose to design the mix-up strategy that leverages information from both majority and minority classes to augment samples of minority classes such that their boundaries can be pushed away towards majority classes. As a result, the training bias resulted from majority classes can be largely reduced. In addition, we consolidate our new framework with both the softmax loss and triplet loss on the augmented samples which proves able to improve the classification accuracy further. We conduct extensive evaluations w.r.t. both total class accuracy and average class accuracy on three benchmark datasets (i.e., Oracle-20K, Oracle-AYNU and OBC306). Experimental results show that the proposed method can result in superior performance to the comparison approaches, attaining a new state of the art in oracle bone character recognition.

Keywords: Oracle character recognition · Imbalanced data · Mix-up augmentation · Triplet loss

1 Introduction

Oracle characters carved on either animal bones or turtle shells are probably the oldest hieroglyphs in China. These characters record the history as well as the civilization during Bronze Age. Thus, it is of great significance to recognize and study these invaluable hieroglyphs which could then offer essential information for Chinese archaeology and philology. However, oracle characters can only

© Springer Nature Switzerland AG 2021
J. Lladós et al. (Eds.): ICDAR 2021, LNCS 12821, pp. 237–251, 2021.
https://doi.org/10.1007/978-3-030-86549-8_16

be recognized by a very limited number of experts manually at current, which greatly hinders the research in this area.

To tackle this problem, researchers have made various attempts in applying pattern recognition-based methods to recognize oracle characters automatically. For example, Li et al. proposed to recognize oracle inscriptions-based on graph isomorphism [9], and Guo et al. considered oracle characters as sketches and built hierarchical representations for recognition [2]. Recently, deep neural networks (DNNs) [5] have also been engaged and achieved great advances in recognizing oracle characters [6, 19]. Particularly, in [6], a large size of oracle character dataset was introduced based on which several popular deep neural networks were tested; in [19], DenseNet [4] was applied to classify oracle characters by the nearest neighbor rule with deep metric learning.

However, all these present methods appear still limited in recognizing oracle characters. On one hand, as the oldest hieroglyphs in China so far, samples of oracle characters are very scarce and difficult to be obtained; on the other hand, the available oracle characters are also seriously imbalanced in almost all the benchmark oracle datasets. To illustrate, we show the distribution of sample frequencies on the recently released dataset OBC306 [6] in Fig. 1. OBC306 contains 25,898 samples for the largest class. In sharp contrast, there are merely one or two samples for many other classes. As a consequence, DNNs trained on such kind of datasets often show inferior performance, especially on identifying minority classes. To mitigate this problem, in [19], a triplet loss-based deep metric learning strategy was proposed. Unfortunately, though obtaining good performance on the dataset Oracle-AYNU, this method appears less effective on Oracle-20K. Moreover, no evaluation results were reported on OBC306.

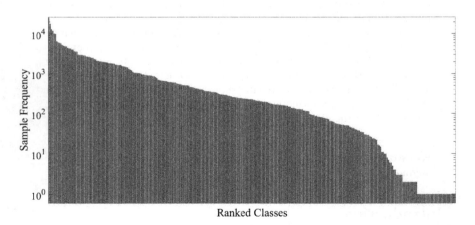

Fig. 1. The sorted number of samples per category in OBC306 dataset.

To deal with the problem caused by imbalanced data distribution, there have been various proposals in deep learning [1,7,20]. Among them, the mix-up based method has been utilized recently due to its simple but effective nature [1,15,18,20]. In [18], the authors first proposed the mix-up method to directly generate samples with linear interpolations of raw input samples and their corresponding labels. Later, Manifold Mix-up [15] was developed to implement the interpolations on randomly selected layers in the latent feature space. Recently, Remix [1] has been designed to disentangle the mixing-up weights between the features and labels, which further improved the classification performance on imbalanced datasets.

In the paper, we propose to recognize oracle characters with the mix-up augmentation to alleviate the imbalanced data issue. Specifically, we investigate both Manifold Mix-up and Remix to augment training samples in the feature space. It is found that Remix is more beneficial to oracle character recognition. Furthermore, we consolidate our framework by integrating the softmax loss and triplet loss. Namely, the softmax loss is directly calculated on the mix-up feature space, whilst the triplet loss is learned on the triplets of positive, negative and mixed features. Such strategy proves effectiveness to promote the classification accuracy. We evaluate the proposed method on three benchmark oracle datasets which are under the imbalanced problem. Experimental results demonstrate that both the mix-up strategies can promote the performance though Remix appears more effective. The results can be further improved once the triplet loss is applied. Overall, our proposed method achieves so far the best performance on the three datasets. Specifically, we improve the total class accuracies from 93.05% to **94.76%**, 77.54% to **83.65%**, and 89.81% to **91.74%** on Oracle-20K, Oracle-AYNU and OBC306, respectively. To show the effectiveness of the proposed method on minority classes, we also reported the average class accuracies, which were also improved from 90.61% to **93.50%**, 67.53% to **79.96%**, and 75.62% to **80.16%** on Oracle-20K, Oracle-AYNU and OBC306, respectively.

2   Related Work

2.1   Oracle Character Recognition

Oracle character recognition aims to classify characters from drawn or real rubbing oracle bone images. Except common issues in character recognition, e.g., large variances in the same class [2,6], most oracle datasets suffer from imbalanced and limited data problems [6,19] due to the historical archive. Namely, high-frequency characters merely belong to a few categories, while low-frequency characters account for a large proportion of entire character categories; the amount of low-frequency character is usually insufficient. A model trained on such oracle datasets could get biased towards majority characters. Additionally, oracle datasets composed of real oracle images are usually severely damaged, due to carved strokes, bone cracks, long-term burial and careless excavation. In short, the above challenges severely hinder the research on recognizing oracle characters.

Related works of oracle character recognition could be divided into two categories: traditional pattern recognition methods and deep learning methods. Traditional pattern recognition methods primarily leverage graph theory and topology. The method in [9] read oracle inscriptions based on graph isomorphism, which treated each character as an undirected graph, and endpoints and intersections as nodes. In [13], the authors converted oracle fonts to topological graphs and then recognized characters. Guo et al. proposed hierarchical representations for oracle character recognition combining Gabor-related and sparse encoder-related features [2]. However, most of the early works are complicated and rely heavily on hand-crafted feature extraction. Recently, since DNNs have achieved great success in OCR-related area [3,10,11], researchers have started to adopt DNNs to extract features for oracle character recognition [2,6,19]. For instance, in [2], Convolutional Neural Networks (CNNs) were simply combined with hierarchical representations on a small dataset Oracle-20K. Huang et al. published a large oracle dataset and provided several baselines from different CNNs [6], among which Inception-v4 [14] produces the optimal classification accuracy. In [19], the authors proposed a deep metric learning-based nearest neighbor classification method with DenseNet [4] to improve the oracle character recognition accuracy. Therefore, in our work, DenseNet [4] and Inception-v4 [14] with a softmax classifier are utilized as our baseline models. Furthermore, we propose the mix-up augmentation for oracle recognition to mitigate the imbalanced data problem.

2.2 Imbalanced Data Classification

The imbalanced data problem has attracted increasing attention since it is naturally prevalent in real world data. Most related works include two directions, i.e., re-weighing and re-sampling techniques [20]. Re-weighing aims to assign larger weights to minority categories, while re-sampling attempts to train networks on pseudo-balanced datasets through sampling. Recently, one new direction based on mix-up strategies has been widely applied, which is also viewed as a data augmentation method [1,20]. In [20], the authors verified that mix-up training is able to facilitate networks' performance on imbalanced datasets. The first work Mix-up [18] purely trained models with interpolations of samples for classification. Due to its simplicity and effectiveness, many extension methods [1,15,17] have been proposed. Manifold Mix-up [15] introduced semantic interpolations on latent space to encourage networks to generate smoother decision boundaries, leading to a higher tolerance for slightly different test examples. Cutmix [17] cut out a part of one sample at pixel-level and randomly filled with pixels from other samples in the training set. The above methods all use the same mixing-up factor for samples and labels. Considering class imbalance, Remix [1] modified the mixing formulas in [15,18], thereby balancing the generalization error between majority and minority classes.

For the imbalanced data problem in oracle character recognition, the relevant research is sparse. In [19], the authors proposed a triplet loss-based metric learning method, which aimed to reduce the class imbalance issue implicitly in oracle character recognition. This method demonstrated the improvement on the dataset Oracle-AYNU, but with a slight drop on the dataset Oracle-20K and without the evaluation on the most seriously imbalanced oracle dataset

OBC306. In this paper, we propose the mix-up augmentation to alleviate the imbalanced issue explicitly in oracle character recognition for all three benchmarks Oracle-20K, Oracle-AYNU and OBC306. Moreover, we consolidate our framework with both the softmax loss and triplet loss to promote classification performance further.

3   Methodology

The overall structure is illustrated in Fig. 2. We adopt a CNN as the backbone network. The mix-up strategy is integrated into the backbone to synthesize samples (e.g., Feature AB in Fig. 2) from original samples (e.g., Feature A and Feature B in Fig. 2). Then, the synthesized samples are utilized to calculate the softmax loss. Meanwhile, to make the extracted features in the same class more compact, we merge the triplet loss with the softmax loss into the final loss function. In the triplet loss, one triplet (i.e., one anchor, its positive and negative samples) is selected from two original samples and the synthesized sample. In the inference stage, the mix-up augmentation and Triplet branch are removed.

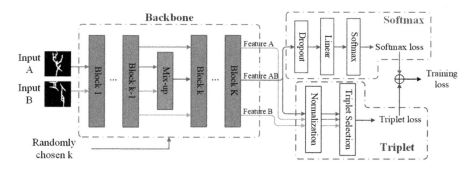

Fig. 2. The overall structure of the proposed method.

3.1   Network Architecture

To make the comparison with other reported performance fair, we adopt the DenseNet [4] for the experiments on the datasets of Oracle-20K and Oracle-AYNU and the Inception-v4 [14] for OBC306 dataset.

The utilized DenseNet consists of four dense blocks and each of which is followed by a transition layer. Each dense block has 9 layers. In each layer, there is a composite function of three continuous operations: a batch normalization layer (BN), a rectified linear unit (ReLU) and a 3×3 convolution layer. Note that internal feature maps of each layer are the concatenation of external feature maps from all preceding layers. Additionally, external feature maps of each layer contain 6 channels.

Inception-v4 consists of a stem module, inception modules (Inception-A, -B and -C) and reduction modules (Reduction-A and -B). The stem module is used to preprocess images, mainly containing multiple convolution layers and two pooling layers. After stem part, we follow the structure in [14] and utilize 4 Inception-A modules, 7 Inception-B modules and 3 Inception-C modules. Between three Inception blocks, there are reduction modules which play a role of down-sampling like transition layers in DenseNet, aiming to change the widths and heights of features. More details can be seen in [14]. After the last Inception-C, features are fed into an average pooling layer.

3.2 Mix-Up Strategy

The basic idea of mix-up strategy is straightforward, which generates new samples by the linear combination of real samples [18]. In details, two random raw images x and x' and corresponding labels y and y' are mixed linearly to generate a new image $\tilde{x}$ with a label $\tilde{y}$. The mixing process is formulated by Eq. 1 and Eq. 2.

$$\tilde{x} = \lambda x + (1 - \lambda)x', \tag{1}$$

$$\tilde{y} = \lambda y + (1 - \lambda)y', \tag{2}$$

where λ denotes the mixing factor sampled from the beta distribution $Beta(\alpha, \alpha)$. In the experiment, we set $\alpha = 2$ following [15]. y and y' are class labels with one-hot representations. An example is shown in Fig. 3(a). We can see that the mixed label $\tilde{y}$ is a continuous representation instead of the simple one-hot format (e.g., y, y').

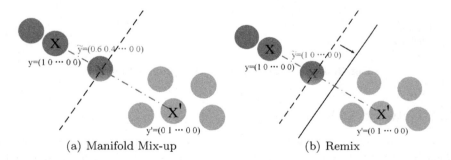

(a) Manifold Mix-up   (b) Remix

Fig. 3. An illustrated example for the visualizations of Manifold Mix-up and Remix. Blue and green cycles denote samples of minority and majority classes, respectively. The red dashed line represents all possible convex combinations of real samples x and x'. The hybrid color cycle represents synthesized sample $\tilde{x}$. Their labels are represented by y, y' and $\tilde{y}$ with one-hot or continuous values, respectively. The dashed black line is the decision boundary under Manifold Mix-up. The solid black line is the decision boundary under Remix. In the sub-figure (b), the decision boundary moves towards the majority class due to the augmented sample by remix strategy, which results in enlarging the scope of the minority class. (Color figure online)

To make the representation more flexible, we employ the mix-up strategy in the feature space, which is called by **Manifold Mix-up** [15]. Specifically, we choose k-th block in the backbone network (e.g., Block k in Fig. 2). Note that k is randomly chosen in each iteration during training. Then, the input features x and x' of this block are mixed up to generate a mixed feature $\tilde{x}$. If $k = 0$, x and x' are input images. Next, the mixed features $\tilde{x}$ are input to the following blocks.

Although Manifold Mix-up is effective for the classification [15], it does not consider the imbalanced data problem explicitly. In order to balance the data distribution, we disentangle mixing factors for features and labels in Eq. 1 and Eq. 2 to generate more minority samples. Hence, the mix-up process can be defined as

$$\tilde{x} = \lambda_x x + (1 - \lambda_x)x', \tag{3}$$

$$\tilde{y} = \lambda_y y + (1 - \lambda_y)y', \tag{4}$$

$$\lambda_y = \begin{cases} 0, & n \geq n' \quad and \quad \lambda_x < \tau \\ 1, & n \leq n' \quad and \quad 1 - \lambda_x < \tau \, , \\ \lambda_x, otherwise \end{cases} \tag{5}$$

where n and n' represent the number of samples in the classes of x and x', respectively. λ_x is the mixing factor for features that is the same as λ in Eq. 1. λ_y is the mixing factor for labels. τ controls the proportion of minority labels for mixed samples, which is set as $\tau = 0.5$ in our experiments. This method is also called by **Remix** [1]. In this way, minority labels increase virtually, in other words, data distribution becomes balanced.

An interpretation example is illustrated in Fig. 3(b). We can see that the synthesized sample close to the minority class is added to the minority class and the classification decision boundary moves towards to the majority class slightly, thereby benefiting the imbalanced data problem.

In our implementation, we use all samples in one batch for the mix-up strategy to augment minority samples. In details, the original batch contains D samples $B_0 = [x_1, x_2, ..., x_D]$. Firstly, we shuffle the samples to get a new batch $B_1 = [x_{s(1)}, x_{s(2)}, ..., x_{s(D)}]$, where $s(i)$ is the inverse shuffle function representing the original position in B_0; Secondly, we mix up samples at the same position in both batches, e.g., $B_2 = [MixUp(x_1, x_{s(1)}), MixUp(x_2, x_{s(2)}), ..., MixUp(x_D, x_{s(D)})]$ where the function $MixUp$ can be referred to Manifold Mix-up (Eq. 1 and Eq. 2) or Remix (Eq. 3, Eq. 4 and Eq. 5). Finally, the batch B_2 will be used in the training. Note that if $s(i) = i$, then this MixUp sample will be the original sample x_i. Hence, the final batch still contains N samples, which possibly includes the original samples. Compared to the original batch B_0, the samples in new batch B_2 are more diverse and enable to improve the generalization of networks.

3.3   Loss Function

In our method, we combine the softmax loss and triplet loss as the final loss function in the training

$$\mathcal{L} = \mathcal{L}_{sft} + \beta \cdot \mathcal{L}_{tri}, \tag{6}$$

where $\mathcal{L}_{sft}$ denotes the softmax loss aiming to minimize the classification errors, $\mathcal{L}_{tri}$ is the triplet loss aiming to learn a better feature representation as a regularization term and β is a trade-off parameter. In our experiments, we set $\beta = 0.001$ for Oracle-AYNU and OBC306, $\beta = 0.0001$ for Oracle-20K.

Softmax Loss. Softmax loss (a.k.a Cross-Entropy loss) has been widely used for classification and recognition tasks, which is defined by

$$\mathcal{L}_{sft} = \sum_{n=1}^{N} (-\tilde{y}_n \log(s(\tilde{x}_n))), \tag{7}$$

Here, $\tilde{x}_n$ is the n-th augmented sample, $\tilde{y}_n$ is the mixed ground-truth label of that augmented sample, N is the number of mixed training samples and $s(\cdot)$ represents the Softmax branch as shown in Fig. 2. From the Fig. 2, we can see that only mix-up samples are used in this loss function. However, the original samples are also possibly contained if the mixed two samples are the same as described in Sect. 3.2.

Triplet Loss. The goal of triplet loss is to optimize the network to extract better representations by reducing the intra-class distance and increasing the inter-class distance. It can be defined as

$$\mathcal{L}_{tri} = \sum_{i=1}^{N} [||x_i^a - x_i^p||_2^2 - ||x_i^a - x_i^n||_2^2 + \alpha_\lambda]_+. \tag{8}$$

where x_i^a represents an anchor sample, x_i^p represents a positive sample of the same category with x_i^a and x_i^n represents a negative sample of other categories. $[\cdot]_+ = max(\cdot, 0)$ and α_λ denotes a flexible λ-based margin. Following [16], $\alpha_\lambda = |2\lambda - 1|$ (Manifold Mix-up) or $\alpha_\lambda = |2\lambda_x - 1|$ (Remix). x_i^a should be closer to all x_i^p than other x_i^n with a certain margin α_λ at least. To remove the impact of the length of each sample x in the triplet loss, all samples are normalized by a L2 normalization layer firstly, i.e., $||x||_2 = 1$.

   To speed up the convergence, we use a two-stage training like [19]. In the first stage, the anchor is the mixed sample $\tilde{x}$ of two random original samples x and x'. If its mixed label $\tilde{y} = y$, then positive sample $x_p = x$ and negative sample $x_n = x'$; If $\tilde{y} = y'$, then $x_p = x'$ and $x_n = x$; otherwise, this anchor is discarded due to no samples with the same label (not a simple one-hot representation)[1]. In the second stage, we choose hard triplets to train the model. Concretely, we firstly choose samples in one batch without overlapped labels; secondly, we calculate the Euclidean distances among all samples in this batch; thirdly, for each sample x, we select the nearest one x' from other categories, which is difficult to distinguish relatively (that is why we call them hard triplets); finally, the anchor point $\tilde{x}$

---

[1] For Manifold Mix-up, the positive sample is chosen by nearest neighbour within a certain threshold since no one-hot mixed labels exist.

mixed by x and x' is obtained, the positive and negative samples are selected by the same way in the first stage. The pseudocode of the first training stage combined with the mix-up augmentation is given in Algorithm 1.

Algorithm 1. Minibatch stochastic gradient descent training of the proposed method

1: **for** number of training iterations **do**
2:     Sample a batch of images & targets with size D, $B_0 = [(x_1, y_1), ..., (x_D, y_D)]$
3:     Shuffle B_0 to get a new batch, $B_1 = [(x'_1, y'_1), ..., (x'_D, y'_D)]$
4:     Mix up B_0 and B_1 to get a mixed batch, $B_2 = [(\tilde{x}_1, \tilde{y}_1), ..., (\tilde{x}_D, \tilde{y}_D)]$
5:     Calculate L_{sft} on B_2 as defined in Eq. 7
6:     Initialize a valid batch B_{tri} for triplet loss
7:     **for** $i = 1$ to D **do**
8:         **if** $\tilde{y}_i = y_i$ **then**
9:             Anchor $x_a \leftarrow \tilde{x}_i$, positive point $x_p \leftarrow x_i$, negative point $x_n \leftarrow x'_i$
10:             Append (x_a, x_p, x_n) to B_{tri}
11:         **else if** $\tilde{y}_i = y'_i$ **then**
12:             Anchor $x_a \leftarrow \tilde{x}_i$, positive point $x_p \leftarrow x'_i$, negative point $x_n \leftarrow x_i$
13:             Append (x_a, x_p, x_n) to B_{tri}
14:         **end if**
15:     **end for**
16:     Calculate L_{tri} on B_{tri} as defined in Eq. 8
17:     Update the network by minimizing L_{sft} and L_{tri}
18: **end for**

---

4   Experiments

In this section, we firstly introduce benchmark datasets in our experiments. Then, the experimental details and evaluation metrics are given. Finally, we show the experimental results of the proposed method in details.

4.1   Datasets

As far as we known, currently there are three benchamark oracle character datasets for recognition research, including Oracle-20K [2], Oracle-AYNU [19] and OBC306 [6]. All datasets exist the issue of imbalanced data, especially OBC306. Some examples are presented in Fig. 4 and statistical information of each dataset is listed in Table 1.

Oracle-20K [2] contains 19,491 oracle character instances and 249 classes[2]. In this dataset, the largest category contains 291 instances, while the smallest

---

[2] Because this dataset cannot be downloaded officially, we collect data with slightly different with the reported numbers in [2,19], which are 20,039 instances and 261 classes.

Table 1. The statistical information of Oracle datasets.

Dataset	Training set	Test set	Overall IR	Training set IR	Class number
Oracle-20K	13,074	6,417	291:25	194:17	249
Oracle-AYNU	34,424	4,648	287:2	259:1	2,584
OBC306	232,168	77,354	25,898:2	19,424:1	277

* IR represents imbalance ratio

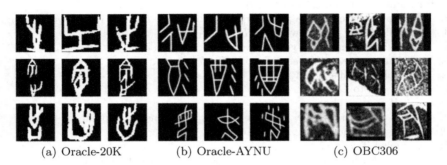

   (a) Oracle-20K   (b) Oracle-AYNU   (c) OBC306

Fig. 4. Examples from three oracle datasets. In each subfigure, images in the same row are with the same class label. All images are resized to 64 × 64 with zero paddings.

one merely contains 25 instances. In the experiments, we randomly divide the instances into training set and test set with 2:1 ratio in each class following the work [2].

Oracle-AYNU [19] was firstly evaluated by [19], which contains 2584 categories with 39072 oracle character instances in total[3]. The largest classes with 287 instances, while there are 1757 classes with fewer than 10 training instances. We follow [19] to divide training set and test set. Each class has 90% instances for training and 10% for testing randomly. If the number of instances is less than 10, the test set has one instances.

OBC306 [6] is a large oracle dataset and suffers from the severe imbalanced data issue. The official dataset contains 309,551 oracle character images with 306 classes. A small number of high-frequency characters (72 classes with more than 1000 images) accounts for 83.74% of the total number of images, but only accounts for 27.5% of the total number of categories. In addition, there is merely one image in 29 classes. Different from the work [6], we removed all classes with one image and remain 277 classes with 309,522 images in total. We randomly split the dataset into training set and test set with ratio 3:1 in each class as following [6].

4.2　Experimental Details

The proposed method is trained on end-to-end learning with Adam optimizer [8]. The learning rate is set to 0.001 at the beginning and is decayed to 0.0001 for the

---

[3] Actually, the work [19] reported 2583 categories with 39062 instances.

rest epochs, schedulers are slightly different in terms of datasets. Input images are resized to 64×64 with zero paddings. The batch size is set to 64. Note that all models in the paper are trained with the same image size and batch size. The models are implemented with the Pytorch framework on a GPU NVIDIA GTX-1080Ti.

4.3 Evaluation Metrics

The total class accuracy is widely used in classification research, which reports an overall accuracy on all classes. However, such accuracy can be dominated by the majority classes if datasets with imbalanced data distribution. Therefore, we also report the average class accuracy that calculates the accuracy on each class separately. To be specific, these two evaluation metrics are defined as

$$Total_{Acc} = \frac{1}{H} \sum_{c=1}^{C} r_c, \qquad (9)$$

$$Average_{Acc} = \frac{1}{C} \sum_{i=c}^{C} \frac{r_c}{h_c}. \qquad (10)$$

Here, H represents the total number of test images and C represents the number of classes. h_c is the number of test images of class c, r_c is the number of correctly classified images of class c.

4.4 Experiments on Mix-Up Strategy

Both Manifold Mix-up and Remix methods are evaluated and experimental results are summarized in Table 2. Note that the baselines of Oracle-20K and Oracle-AYNU are DenseNet and the baseline of OBC306 is Inception-v4.

Table 2. Comparison of classification results of different mix-up augmentation methods on Oracle-20K, Oracle-AYNU and OBC306.

Method	Oracle-20K		Oracle-AYNU		OBC306	
	Average	Total	Average	Total	Average	Total
Baseline (B)	90.61%	93.05%	67.53%	77.54%	75.62%	89.81%
B+Manifold Mix-up	91.38%	93.89%	73.91%	81.76%	76.98%	**91.65%**
B+Remix	**93.50%**	**94.64%**	**75.95%**	**82.83%**	**79.02%**	91.56%

Table 2 shows that methods with mix-up augmentation outperform the baselines in all datasets. For example, in Oracle-AYNU, Manifold Mix-up improves average and total accuracy by 6.38% and 4.22%, respectively; Remix improves average and total accuracy by 8.42% and 5.29%, respectively. It is found that the metric of average accuracy demonstrates the effectiveness of the proposed method better. Comparing the two mix-up augmentation methods, Remix performs better than Manifold Mix-up in most cases, including average and total accuracy on Oracle-20K and Oracle-AYNU and average accuracy on OBC306. The reason is that Remix explicitly generates training samples for minority classes. However, the total accuracy of Remix on OBC306 declines a little compared with that of Manifold Mix-up. Because, more samples for minority classes generated by Remix push classification boundaries to majority classes, thus damaging the accuracy of majority classes to some extent. Meanwhile, the training imbalance ratio of OBC306 is extremely large, around 19424. Therefore, a slight decrease in the accuracy of majority classes has a greater impact on the overall accuracy of OBC306, though the accuracy of minority classes increases.

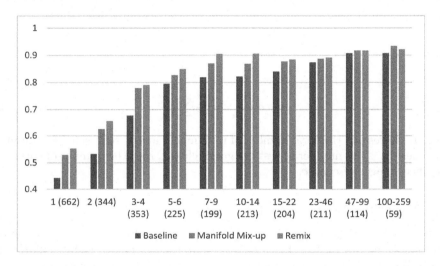

Fig. 5. Relationship between average accuracy and the number of training samples in Oracle-AYNU. The numbers in brackets represent the number of categories with the corresponding number of training samples, e.g., there are 353 classes with three or four training samples.

To show effectiveness of our models on various numbers of training data, we illustrate the relationship between average accuracy and classes sorted by the number of training data on Oracle-AYNU in Fig. 5. We can see that both mix-up augmentation methods surpass the baseline model in each range, especially in the range of less than 4 training instances (the first three ranges). For instance, for the first range, Manifold Mix-up and Remix increase the accuracies from 44.26% to 52.87% and 55.28%, respectively. It fully proves the effectiveness of our method on minority classes, even classes with one training instance. In addition, we can tell that Remix performs much better than Manifold Mix-up on most of minority classes, because Remix can augment samples for such classes, while Manifold Mix-up merely considers mixed categories (new classes). In the majority classes (e.g., the last two ranges in Fig. 5), the accuracy of Manifold Mix-up is a little better. The possible reason is that Remix sacrifices a part of the accuracy of the majority classes to facilitate the performance of minority classes. Comparing the equations of Manifold Mix-up (Eq. 1 and Eq. 2) and Remix (Eq. 3, Eq. 4 and Eq. 5), we could know that Manifold Mix-up treats every class equally, while Remix focuses more on minority classes. For example, if mixing up two samples (one from a minority class and the other from a majority class), Manifold Mix-up synthesizes a sample with one new mixed label (Eq. 2), but a sample synthesized by Remix is with high probability to be assigned to the minority class ($\lambda_y = 0$ in Eq. 5), though it contains part of features from the majority class. Therefore, in this situation, the accuracy of majority class might decrease while the accuracy of minority classes increases.

In summary, both Remix and Manifold mix-up increase the classification accuracy. However, Remix performs much better than Manifold Mix-up. Hence, we choose Remix as our data augmentation method in the following experiments.

4.5 Experiments on Triplet Loss

Table 3. Comparison of classification performance among the proposed method and other methods on Oracle-20K, Oracle-AYNU and OBC306.

Method	Oracle-20K		Oracle-AYNU		OBC306	
	Average	Total	Average	Total	Average	Total
Zhang et al. [19]	–	92.43%	–	83.37%	–	–
Zhang et al. [19]*	89.40%	92.20%	74.87%	82.15%	–	–
Huang et al. [6]†	–	–	–	–	77.64%	–
Huang et al. [6]*	–	–	–	–	75.62%	89.81%
Baseline	90.61%	93.05%	67.53%	77.54%	75.62%	89.81%
Baseline+Remix	**93.50%**	**94.64%**	75.95%	82.83%	79.02%	91.56%
Baseline+Remix+Triplet	93.43%	94.31%	**79.96%**	**83.65%**	**80.16%**	**91.74%**

* * denotes the results reproduced by us. † denotes the results on 277 classes merely while the original paper [6] reported the average accuracy as 70.28% on 306 classes (including 29 classes without training samples and with merely one sample per class in the test set). If there are no correct samples in these 29 classes, the average accuracy can be estimated by 70.28% × 306/277 = 77.64%.

Table 3 shows the experimental results of the proposed method with triplet loss. It is found that the model integrated with triplet loss improves the performance on Oracle-AYNU and OBC306 further. On the Oracle-AYNU, Remix with triplet loss method produces a large improvement on average accuracy from 75.95% to 79.96%, and total accuracy is improved from 82.83% to 83.65%. On the OBC306, Remix with triplet loss outperforms around 1.14% and 0.18% on average and total accuracy.

To compare with other oracle recognition methods, we also show the recently reported accuracies on these benchmark datasets. Note that the method in [6] adopted a pre-trained model on the ImageNet dataset [12] while ours do not use. In addition, the work in [19] was also reproduced with slightly different performances (i.e., reported 92.43% while our total accuracy is 92.20%) probably due to different experimental settings (not given in [19]). As we see in Table 3, our method outperforms the method in [19] on Oracle-AYNU by 0.28% for average accuracy. Although adding triplet loss reduces the performance of Oracle-20K slightly, the proposed method with Remix still achieves better average accuracy than the reported results from [19] by 2.21%. In OBC306, our method also surpasses pre-trained Inception-v4 [6] by 2.52%. In summary, the proposed method achieves the state-of-the-art average accuracy and total accuracy on all three oracle datasets.

5 Conclusion

In this paper, we proposed to improve oracle character recognition by integrating mix-up augmentation and triplet loss function. In the mix-up augmentation, samples of minority classes are augmented in the training, especially Remix strategy explicitly labels synthesized samples as minority classes. The triplet loss is integrated to supervise the backbone network to learn more compact representation in the same class and larger distances among different classes. We evaluated the proposed method on three benchmark oracle character datasets. The experimental results demonstrate that mix-up augmentation methods can improve the total and average class accuracies. The combination of the mix-up strategy and triplet loss improves the accuracy further and obtains the state-of-the-art performance. However, it is found that the accuracy of minority classes is still lower than that of majority classes. In the future, we will consider more strategies (e.g., generative adversarial networks) to oversample minority classes to reduce the imbalanced problem.

Acknowledgements. The work was partially supported by the following: National Natural Science Foundation of China under no.61876155 and no.61876154; Jiangsu Science and Technology Programme (Natural Science Foundation of Jiangsu Province) under no. BE2020006-4B, BK20181189, BK20181190; Key Program Special Fund in XJTLU under no. KSF-T-06, KSF-E-26, and KSF-A-10, and the open program of Henan Key Laboratory of Oracle Bone Inscription Information Processing (AnYang Normal University) under no. OIP2019H001.

References

1. Chou, H.-P., Chang, S.-C., Pan, J.-Y., Wei, W., Juan, D.-C.: Remix: rebalanced mixup. In: Bartoli, A., Fusiello, A. (eds.) ECCV 2020. LNCS, vol. 12540, pp. 95–110. Springer, Cham (2020). https://doi.org/10.1007/978-3-030-65414-6_9
2. Guo, J., Wang, C., Roman-Rangel, E., Chao, H., Rui, Y.: Building hierarchical representations for oracle character and sketch recognition. IEEE Trans. Image Process. **25**(1), 104–118 (2015)

3. Guo, Z., Xu, H., Lu, F., Wang, Q., Zhou, X., Shi, Y.: Improving irregular text recognition by integrating gabor convolutional network. In: 2019 IEEE 31st International Conference on Tools with Artificial Intelligence (ICTAI), pp. 286–293. IEEE (2019)
4. Huang, G., Liu, Z., Van Der Maaten, L., Weinberger, K.Q.: Densely connected convolutional networks. In: Proceedings of the IEEE Conference on Computer Vision and Pattern Recognition, pp. 4700–4708 (2017)
5. Huang, K., Hussain, A., Wang, Q.F., Zhang, R.: Deep Learning: Fundamentals, Theory and Applications, vol. 2. Springer, Cham (2019) https://doi.org/10.1007/978-3-030-06073-2
6. Huang, S., Wang, H., Liu, Y., Shi, X., Jin, L.: Obc306: a large-scale oracle bone character recognition dataset. In: 2019 International Conference on Document Analysis and Recognition (ICDAR), pp. 681–688. IEEE (2019)
7. Kang, B., et al.: Decoupling representation and classifier for long-tailed recognition. arXiv preprint arXiv:1910.09217 (2019)
8. Kingma, D.P., Ba, J.: Adam: a method for stochastic optimization. arXiv preprint arXiv:1412.6980 (2014)
9. Li, Q., Yang, Y., Wang, A.: Recognition of inscriptions on bones or tortoise shells based on graph isomorphism. Jisuanji Gongcheng yu Yingyong(Comput. Eng. Appl.) 47(8), 112–114 (2011)
10. Ma, M., Wang, Q.F., Huang, S., Huang, S., Goulermas, Y., Huang, K.: Residual attention-based multi-scale script identification in scene text images. Neurocomputing 421, 222–233 (2021)
11. Qian, Z., Huang, K., Wang, Q.F., Xiao, J., Zhang, R.: Generative adversarial classifier for handwriting characters super-resolution. Pattern Recogn. 107, 107453 (2020)
12. Russakovsky, O., et al.: Imagenet large scale visual recognition challenge. Int. J. Comput. Vis. 115(3), 211–252 (2015)
13. Shaotong, G.: Identification of oracle-bone script fonts based on topological registration. Comput. Digit. Eng 44(10), 2001–2006 (2016)
14. Szegedy, C., Ioffe, S., Vanhoucke, V., Alemi, A.: Inception-v4, inception-resnet and the impact of residual connections on learning. In: Proceedings of the AAAI Conference on Artificial Intelligence, vol. 31 (2017)
15. Verma, V., et al.: Manifold mixup: Better representations by interpolating hidden states. In: International Conference on Machine Learning, pp. 6438–6447. PMLR (2019)
16. Xu, M., Zhang, J., Ni, B., Li, T., Wang, C., Tian, Q., Zhang, W.: Adversarial domain adaptation with domain mixup. In: Proceedings of the AAAI Conference on Artificial Intelligence, vol. 34, pp. 6502–6509 (2020)
17. Yun, S., Han, D., Oh, S.J., Chun, S., Choe, J., Yoo, Y.: Cutmix: regularization strategy to train strong classifiers with localizable features. In: Proceedings of the IEEE/CVF International Conference on Computer Vision, pp. 6023–6032 (2019)
18. Zhang, H., Cisse, M., Dauphin, Y.N., Lopez-Paz, D.: mixup: Beyond empirical risk minimization. arXiv preprint arXiv:1710.09412 (2017)
19. Zhang, Y.K., Zhang, H., Liu, Y.G., Yang, Q., Liu, C.L.: Oracle character recognition by nearest neighbor classification with deep metric learning. In: 2019 International Conference on Document Analysis and Recognition (ICDAR), pp. 309–314. IEEE (2019)
20. Zhang, Y., Wei, X.S., Zhou, B., Wu, J.: Bag of tricks for long-tailed visual recognition with deep convolutional neural networks (2021)

Radical Composition Network for Chinese Character Generation

Mobai Xue[1], Jun Du[1,2(✉)], Jianshu Zhang[1], Zi-Rui Wang[1,3], Bin Wang[4], and Bo Ren[4]

[1] University of Science and Technology of China, Hefei, China
{xmb15,xysszjs,cs211}@mail.ustc.edu.cn, jundu@ustc.edu.cn
[2] Guangdong Artificial Intelligence and Digital Economy Laboratory (Pazhou Lab), Guangzhou, China
[3] Chongqing University of Posts and Telecommunications, Chongqing, China
wangzr@cqupt.edu.cn
[4] Youtu Lab, Tencent, Shenzhen, China
{bingolwang,timren}@tencent.com

Abstract. Recently, the generation of Chinese characters attracts many researchers. Many excellent works only focus on Chinese font transformation, which is, Chinese character can be transformed into another font style. However, there is no research to generate a Chinese character of new category, which is an interesting and important topic. This paper introduces a radical combination network, called RCN, to generate new Chinese character categories by integrating radicals according to the caption which describes the radicals and spatial relationship between them. The proposed RCN first splits the caption into pieces. A self-recurrent network is employed as an encoder, aiming at integrating these caption pieces and pictures of radicals into a vector. Then a vector which represents font/writing style is connected with the output from encoder. Finally a decoder, based on deconvolution network, using the vector to synthesize the picture of a Chinese character. The key idea of the proposed approach is to treat a Chinese character as a composition of radicals rather than a single character class, which makes the machine play the role of Cangjie who invents Chinese characters in ancient legend. As a kind of important resource, the generated characters can be reused in recognition tasks.

Keywords: Radical analysis · Chinese character generation · Data augmentation

1 Introduction

Chinese character generation is a challenging problem due to the large number of existing Chinese characters which is still increasing. Meanwhile, significant differences are observed between countless font/writing styles. The generation of chinese characters with novel classes and font/writing styles is a complex task.

© Springer Nature Switzerland AG 2021
J. Lladós et al. (Eds.): ICDAR 2021, LNCS 12821, pp. 252–267, 2021.
https://doi.org/10.1007/978-3-030-86549-8_17

the recognition of RAN

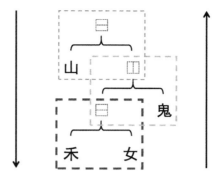

the generation of RCN

Fig. 1. The radical analysis by RAN and radical combination by RCN for the same character. The red, blue and green rectangles in RAN represent subtrees of recognition output. And in RCN, they represent the inputs of each step. (Color figure online)

In recent years, some teams investigate the font conversion of existing Chinese characters and have achieved success. [2] transfers the style of given glyph to the contents of unseen ones, capturing highly stylize fonts found in the real-world such as those on movie posters or infographics. [5] realizes the artistic style transfer of pictures and it can be used on the conversion of font style. [20] has achieved excellent results from font style transfer conditional adversarial networks based on encoder-decoder model. However, these algorithms can only generate Chinese characters with existent classes and have no ability to generated Chinese characters with novel classes. Moreover, these algorithms treat each Chinese character as a whole without considering the internal structure and basic components among Chinese characters.

In this paper, we propose a novel Chinese character generation network, called radical combination network (RCN), for generating characters of new categories. To create a new Chinese character, choosing a coding method that can be understood by computer is necessary. A number, a vector or a picture, neither of them is a good choice. This coding method needs to describe the internal details of Chinese characters, so that a new character can also be encoded in this way. A neural network named RAN [24] proposed a radical-based method to describe Chinese character. It treats a character as a composition of basic structural components, called radicals. The recognition results of RAN include radicals and spatial structures between them, which is a suitable encoding method for character generation. And only about 500 radicals [13] is adequate to describe more than 20,000 Chinese characters, which could reduce the cost of the project. Therefore, we design the RCN network according to the captions provided by RAN. RCN integrates radicals one by one at each step and finally outputs a picture of a Chinese character. Figure 1 shows the radical analysis by RAN and radical combination by RCN for the same character. RAN treats a Chinese character as a tree, in which each leaf node represents a radical and each parent node represents the spatial structure of the two children. RAN recognizes a character following the tree in Fig. 1 from top to bottom, and RCN combines the radicals from bottom to top to realize the generation of a character. At step, radicals or the output from previous are integrated together and finally output a complete character.

It is clear to see that RCN attempts to imitate the development of Chinese characters. Chinese characters developed from simple ones to complex ones. Single-component characters were created first, then two-component characters were made up by two radicals, and then multi-component radicals appeared. People added a radical on an existing Chinese character to create new characters, according to the meaning, pronunciation or shape of radicals. RCN is designed to follow this characteristic.

Regarding to the network architecture, the proposed RCN is based on encoder-decoder model. A densely connected convolutional networks (DenseNet) [7,8,11,22] is employed to pre-process the pictures of radicals. We employ the network in work [8] and modify the parameters, so that the network can achieve better results in our tasks. We propose a tree-structured recurrent neural network (TRN) as the encoder to integrate the radicals. Refer to generative adversarial nets (GAN) [6,14,16], the decoder consists of several deconvolution networks [4,15]. Due to different mission objectives from GAN, we abandoned the discriminator and retained the generator.

As the new characters can be created, RCN provides a novel method on data augmentation. Although these generation samples may be blurry and unsightly for human vision, they can provide a great deal of information to recognition network.

The main contributions of this study are summarized as follows:

- We propose RCN for the Chinese character generation which is not dependent image input and realizes the transfer from caption to image.
- Based on radicals, the proposed method can create new Chinese characters. RCN can generated Chinese characters with novel classes and font/writing styles by the novel captions and mixing up font vectors.
- We experimentally demonstrate how the generated data improve the recognition rate, especially for unseen characters in printed/handwritten.
- RCN can increase the diversity of data sets and reduce the cost of collection and annotation effectively.

2   Related Works

2.1   Recursive Neural Network

In the past few years, lots of efforts have been made for recursive neural network. Tree-LSTM [19] proposes a generalization of LSTMs to tree-structured network topologies. Improved tree-LSTM networks [1,25] propose effective mothods for

Natural Language Processing. Recursive neural networks [12,18] perform well on sentiment classification at the sentence level and phrase level. Refer to the above works, we modify the framework to adapt to Chinese character generation. The proposed encoder TRN is an improved recurrent neural network and designed to matching the input caption, which can receive three input vectors.

2.2 Data Augmentation

Traditional methods of data augmentation adopt image processing on training set upon the retaining the captions. The easiest and most common method of data augmentation label-preserving transformations [10], including translation, horizontal mapping and changing the value of RGB channels. SMOTE [3] uses random interpolation on training set to synthesis data. Mixup [23] adopts linear interpolation to mix up two samples as a new sample. Compared with traditional methods, RCN can create samples with new captions, which greatly improve the diversity.

3   Radical Analysis and Caption Segmentation

RAN proposes a method to captioning Chinese characters based on detailed analysis of Chinese radicals and structures. It decomposes Chinese characters into corresponding captions, and we employ the captions to generate Chinese characters. As for spatial structures among radicals, [24] shows ten common structures in Fig. 2, where "a" represents a left-right structure, "d" represents a top-bottom structure, "stl" represents a top-left-surround structure, "str" represents a top-right-surround structure, "sbl" represents a bottom-left-surround structure, "sl" represents a left-surround structure, "sb" represents a bottom-surround structure, "st" represents a top-surround structure, "s" represents a surround structure and "w" represents a within structure. They use a pair of braces to constrain a single structure in character captions. In the existing Chinese characters, these ten structures are not uniform distribution. "a" and "d" appear more often.

Before these captions are sent into RCN, the complete captions need to be split to pieces, as shown in the Fig. 2. Each piece includes 3 symbols and the first is a spatial symbol which describes the spatial structure between the other two symbols. The second and the third symbol could be radical characters or filled character "#". A radical character represents an existing radical and filled character "#" represents a subtree of the caption. These caption pieces describe parts of characters, which are the inputs of RCN.

4   Network Architecture of RCN

RCN based on encoder-decoder model is shown in Fig. 3 and it is a reverse form of RAN. First, the caption of the character is clustered into several pieces

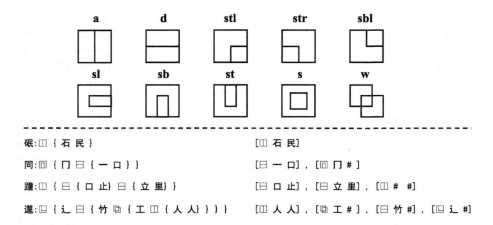

Fig. 2. Graphical representation of ten common spatial structures between Chinese radicals and several examples of captions from sequence to pieces.

according to structural relationship and the encoder of RCN chronologically fuses them into high-level representations with a self-recurrent mechanism. Then, a vector which represents font/writing style is connected with the output from encoder. Finally, the decoder of RCN with deconvolution neural network utilizes the representation at the end to generate a Chinese character picture.

4.1 Data Processing

The description of a captation can form a tree perfectly. Naturally, a caption is expanded into piece sequence C according to tree depth-first traversal order, which is described in Sect. 3. Each piece $c_i \in C$ includes 3 basic components c_{i0}, c_{i1} and c_{i2}.

$$C = \{c_1, c_2, ..., c_T\} , \ c_i \in \mathbb{R}^3 \tag{1}$$

where T is the length of C.

We employ three learnable codebooks (G_s, G_r, G_f) to represent the spatial structure relationship, radical skeleton and font/writing style.

4.2 TRN Encoder

Encoder is a tree-structured recurrent neural network, called TRN. TRN consists of basic units, TRN cell, which is multiply used in the process of generation. In this section, we focus on the architecture of the encoder: generation process of TRN and the self-recurrent structure of TRN and the details of TRN cell.

Overall Architecture of TRN. The structure of TRN is not fixed, which depends on the specific caption. So TRN can be applied to tree-structured caption with variable depth. Figure 4 shows an example of encoding captions, which

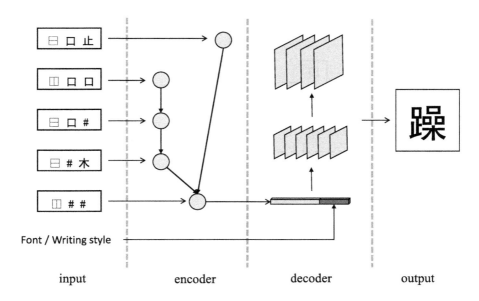

Fig. 3. The overall architecture of RCN for Chinese character generation. We employ a tree-structured encoder and each circle represents an encoder cell.

is shown in two different ways: the tree-structured following the logical structure and the sequence-structure following the timing sequence.

The input caption is expanded into piece sequence C according to tree depth-first traversal order, which is described in Sect. 3. Each piece $c_i \in C$ includes 3 basic components c_{i0}, c_{i1} and c_{i2}. At each step, TRN cell generates a vector according to the local description of the caption c_i. As shown in Figs. 4 the tree-structured shows the structure of a character intuitively, which is consist with human cognition. However, the sequence-structure is used in code implementation. The input caption is treated as a sequence following the depth-first traversal order of the tree.

Tree-Structured Recurrent Network of TRN. In this section, we introduce how the tree-structured recurrent of TRN is implemented. Compared with traditional sequence recurrent network, tree-structured recurrent network calls the output across several steps, not limited to the previous step. In actual processing, we need memory module (a LIFO stack) S to store the outputs O_t from each step t.

At each step t, we input the local description of the caption c_i into TRN cell and get the output O_t. Send O_t into the memory module S at step t, and take O_t out when it is called in further step. The memory module ensures the output O_t can be called correctly in subsequent operations. At step T (the length of

tree-structure

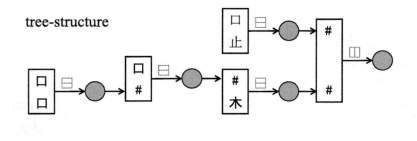

sequence-structure

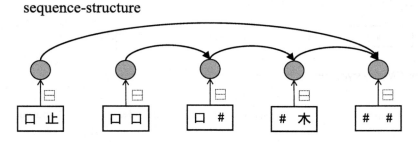

Fig. 4. The architecture of encoder TRN and each circle represents a TRN cell. The tree-structure follows the logical structure and the sequence-structure follows the timing sequence.

C), the caption is completely encoded. There must be only one item (O_T) in the memory module S and the output is sent to decoder to generate a character.

It is easier to understand TRN by analogy with RNN, whose extended structure is sequence. TRN and RNN both are recurrent neural network, and the outputs are both obtained by iteration. But they update the intermediate results in different ways. At each time step t, the hidden state h_t of the RNN is updated by

$$h_t = f_{\text{RNN}}(x_t, h_{t-1}) \tag{2}$$

where f_{RNN} is a non-linear activation function and x_t is the input vector. For TRN, the interim output O_t is updated by Eq. 3.

$$O_t = f_{\text{TRN}}(c_t, S) \tag{3}$$

f_{RNN} only calls h_{t-1} at step t, while f_{TRN} may call O_{t-n} where $0 < n < t$. The two child nodes of the parent node may all be subtrees, which causes the difference between tree-structured and sequence-structure. As shown in Fig. 4, at the final step, f_{TRN} calls the output from step 3 and step 0.

TRN Cell. The processing of TRN encoding is realized by running TRN cell several times. TRN cell processes input caption c_t into high dimensional representation. Figure 5 shows the architecture of TRN cell.

At each step t, input caption piece c_t includes c_{t0}, c_{t1} and c_{t2}. They are transformed into high-dimensional representations through codebooks G_s and G_r, which represent the spatial relationship and radical skeleton. We acquire the quantized representation of spatial character c_{t0} from codebook G_s and c_{t0} is represented as a l-dimensional vector e_{t0}. Since c_{t1} and c_{t2} could be radical characters or filled characters, we need to judge how the corresponding vectors e_{t1} and e_{t2} is obtained.

$$e_{t2} = \begin{cases} G_s[c_{t2}] & c_{t2} \text{ is radical character} \\ s_{-1} & c_{t2} \text{ is filled character} \end{cases}$$

$$e_{t1} = \begin{cases} G_s[c_{t1}] & c_{t1} \text{ is radical character} \\ s_{-1} & c_{t1} \text{ is filled character \& } c_{t2} \text{ is radical character} \\ s_{-2} & c_{t1} \text{ and } c_{t2} \text{ are both filled character} \end{cases} \tag{4}$$

where $G_s[c_{t1}], bmG_s[c_{t2}], e_{t1}, e_{t2} \in \mathbb{R}^L$. s_{-1} and s_{-2} represent the last and the second to last item in memory module S.

Then, concate e_{t0}, e_{t1} and e_{t2}.

$$e_t = \{e_{t0}, e_{t1}, e_{t2}\} \tag{5}$$

where $e_t \in \mathbb{R}^{2L+l}$. We use a feed forward network to change the dimension to L, which ensures that the format of output O_t is match when it is called. The feed forward network consists of 3 linear transformations with 2 ReLU activation in between.

$$\begin{aligned} O_t &= \text{FFN}_0(e_t) \\ &= \max(\max(0, e_t W_0 + b_0)W_1 + b_1)W_2 + b_2 \end{aligned} \tag{6}$$

where $W_0 \in \mathbb{R}^{2L+l \times 2L}, W_1 \in \mathbb{R}^{2L \times L}, W_2 \in \mathbb{R}^{L \times L}$.

4.3 Decoder

The decoder aims to generate a picture of the target Chinese character, which is a grayscale image. First, the font/writing style is transformed into high-dimensional representations through codebooks G_f and O^* is the concatenation of the font/writing style vector and the output O_T from encoder. A feed forward network is used to enhance the fitting ability of the decoder and further fuse the skeleton information and font style. The feed forward network consists of 2 linear transformations with a ReLU activation in between.

$$\text{FFN}_1(O^*) = \max(0, e_t W_3 + b_3)W_4 + b_4 \tag{7}$$

where $W_3 \in \mathbb{R}^{L \times 2L}, W_4 \in \mathbb{R}^{2L \times L}$. Then, the decoder consists of four deconvolution layers [4] and decodes O^* into an array $\hat{Y}$, where $O^* \in \mathbb{R}^{L \times 1 \times 1}$ and $\hat{Y} \in \mathbb{R}^{1 \times H \times W}$. During decoding, the number of channels is reduced from L to 1,

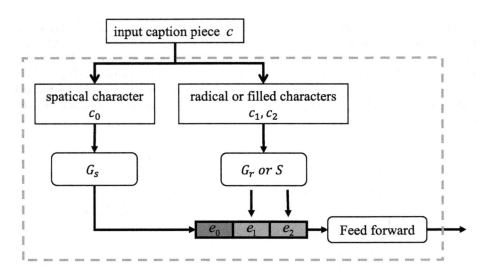

Fig. 5. The architecture of TRN cell. The spatial characters and radical characters are transformed into high-dimensional representations through codebooks. The filled character represents the item in memory module S

and the size is expanded from 1×1 to $H \times W$. Finally, we use contrast stretching operation for $\hat{Y}$ and achieve normalization.

$$\bar{Y} = \frac{\hat{Y} - \min(\hat{Y})}{\max(\hat{Y}) - \min(\hat{Y})} \tag{8}$$

RCN can generate novel font/writing style by weighted addition on font/writing style vectors and a novel caption can be transformed into a novel Chinese character. These new samples can increase the diversity of data sets and reduce the cost of collection and annotation effectively.

4.4   Perceptual Loss

We employ perceptual loss function [9] that measure high-level perceptual and semantic differences between generated sample and ground truth. Rather than encouraging the pixels of the generated image to exactly match the pixels of the target image, we instead encourage them to have similar feature representations as computed by the loss network φ. Let $\varphi_j(x)$ be the activations of the jth layer of the network φ when processing the image; if jth layer is a convolutional layer then $\varphi_j(x)$ will be a feature map of shape $C_j \times H_j \times W_j$. The feature reconstruction loss is the (squared, normalized) Euclidean distance between feature representations:

$$l_j = \frac{\left\| \varphi_j(Y) - \varphi_j(\bar{Y}) \right\|}{C_j H_j W_j} \tag{9}$$

5 Experiment

The innovation of our work is that Chinese characters are generated by using the defined radicals. The generated Chinese characters should be similar to printed style or handwritten style. Therefore, RCN creates new Chinese characters and also provides a novel method for data augmentation. In this section, we introduce several experiments to show the effect of RCN. Section 5.1 explains the network parameters and describes the division of database. Section 5.2 shows some generated samples of RCN and analyses the restrictions on random captions. Section 5.3 illustrates how generated samples can improve the recognition rate of RAN on unseen characters. In Sect. 5.4, we verify the effectiveness of this method on handwritten characters and show the improvement of the recognition.

5.1 Experimental Parameters and Database

In TRN cell, dimension l of vectors in codebook G_s is set to 256 and dimension L of vectors in codebook G_r is set to 1024. The decoder consists of four deconvolution layers of kernel size 4×4 with stride 2 and padding 1. The dimension of vectors in codebook G_f is set to 1024. The dimensions of output vectors from each layer are 256, 64, 8 and 1, respectively. In all our experiments, the loss network φ is the 16-layer VGG network [17] and the pre-trained model is directly acquired from torchvision package.

As for the recognition model, RAN has the same configuration as [21]. A DenseNet is employed as the encoder. The initially convolutional layer comprises 48 convolutions of size 7×7 with stride 2 and each block comprises 10 bottleneck layers and growth rate is set to 24. The decoder is two unidirectional layers with 256 GRU units.

We prepare the printed and handwritten data sets and design related experiments to verify the effectiveness of our method in simple and complex scenes. In order to generate new Chinese characters, we prepare 20,000 random captions of nonexistent Chinese characters. We introduce the details and partitions of the data in the following.

Printed Database. In the experiments on printed characters, we choose 27,533 Chinese characters in 5 font styles (Hei, Song, Kai, Deng, Fangsong) as database and divide them into training set and testing. There are 114,665 printed characters in 5 font styles in training set, which contains 22,933 classes. And there are 20,860 printed characters in testing set, which contains 4,172 classes that do not appear in training set.

Handwritten Database. The handwritten database is the CASIA database including HWDB1.0, 1.1 and 1.2. We use HWDB1.0 and 1.1 set as training set which contains 3,755 classes, 720 writing styles and 2,674,784 handwritten

Chinese characters. HWDB1.2 set is employed as testing set and we choose 3,277 unseen classes for evaluating the performance on unseen Chinese characters. We make sure all radicals of these characters are covered in 3,755 training classes. Note that the Chinese characters in HWDB1.2 set are not common and usually have more complicated radical composition.

Random Caption. The network generates the specified characters according to the description of the input caption. It is feasible to generate novel/nonexistent Chinese characters by inputting corresponding captions. However, the caption cannot be completely random due to the limitation of the default composition rules of Chinese character structure. A radical has given positions in character, which is determined by the shape of the radical. For example, it's unusual to observe a radical with vertical bar shape appears in a top-bottom structure.

Therefore, we need to count the frequency of the combination of each radical and spatial structure. It is necessary to avoid the combination that does not appear in the training set when generating random captions. We generate a random caption from top to bottom according to the tree structure. First, we select a spatial structure as the parent node randomly. Then, the radicals that appear in such structure are selected as the corresponding child nodes. In order to generate tree structure annotated with different depths, we replace radical by using a subtree with probability 0.2 and continue to select child nodes until the tree structure is complete.

5.2 Experiments on Generation of RCN

In this section, we show some generated results of RCN. In the experiments on printed characters, we generate new Chinese characters with different font styles, and verify the influence of training set capacity on the generation effect. In the experiments on handwritten characters, we expand the categories and writing style of the database.

In the experiments on printed characters, we use the printed training set mentioned in Sect. 5.1 and increase the training set from 4,586 to 22,933 with 20% stride to observe the change of generation effect. The printed testing set is employed to validate the effect on unseen characters. As shown in the top half of the Fig. 6, the generated results get clearer and the details become smoother with the increasing of training set. We display some results of testing set in the middle part of Fig. 6 and some results of random caption input in the lower part.

In the experiments on handwritten characters, we use the HWDB training set and testing set mentioned in Sect. 5.1. As shown in the top half of the Fig. 7, we generate mixed-up writing style characters that are mentioned in Sect. 4.3. We display some characters with testing classes in the lower part.

Fig. 6. The generated result of printed Chinese characters. The first line shows 6 groups of generated results with the increasing training data, the obtained characters become more and more clearly. The middle part shows some results of testing set and the lower part shows some results of random caption inputs.

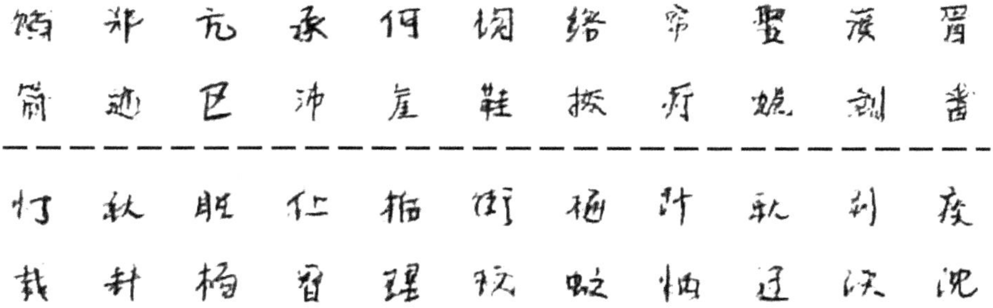

Fig. 7. The generated result of handwritten Chinese characters. The top half part shows mixed-up writing style characters, and the lower part shows the characters with testing classes.

In this experiment, we find that captions contain semantic information which can be regarded as sentences. We consider that semantic information in captions can not be intuitively understood. Based on the observation of the experimental results, the generated results created by testing captions perform much better than the ones created by random captions. It can be explained via analogy with a translation work. If a verb in a sentence is replaced by a noun, the sentence may be translated into an incorrect one. Despite the grammatical constraints, a sentence made up of random words may probably become an outlier for translation network. It is the same reason to RCN, an unsuitable radical in a caption may lead to a blurry output, although we create the captions according to the restriction of radicals' shape. However, the characters in the testing set are

Table 1. Result of experiment on printed characters.

Training data			Accuracy					
Training	Testing	Random	Hei	Song	FangSong	Deng	Kai	Avg
114,665	-	-	88.68	88.56	87.19	87.17	86.80	87.79
114,665	20,860	-	92.97	92.84	91.91	91.89	90.00	92.03
114,665	20,860	20,000	93.81	93.93	92.94	92.78	90.11	92.83
114,665	20,860	40,000	95.16	95.04	94.01	93.95	91.23	93.99
114,665	20,860	80,000	95.00	94.90	93.99	94.20	92.46	**94.22**

existing ones and the captions contain correct semantic information. These captions are fitter to the training set and radicals in the generated results are clearer and more distinguishable.

We also observe that the performance of generated results can not be completely random. In the one hand, captions are limited by the default rules of existent Chinese character structure, which is mentioned in Sect. 4.3. On the other hand, a radical may have various performances, which depends on its position in character. A radical may perform differently in different positions of a character. It may be a smaller one at the left in a left-right structure, which is learned from existing Chinese characters in the training set. These constraints make generated results fit to the human cognition.

5.3   Data Augmentation Experiments on Printed Characters

In this section, we use the newly created classes as data augmentation and show how the generated results improve the recognition rate on unseen characters. In this experiment, we use the printed training set to train the recognition model RAN and observe the increase of recognition rate by adding new samples. We add 20,860 generated samples with testing class and 80,000 with random classes in 5 font styles. The printed testing set is used for evaluating the performance on unseen characters.

The recognition rate of each font and the whole test set are shown in Table 1. The part of training data displays the number of samples. 'Training' represents the printed training set; 'Testing' represents the generated samples with testing classes and 'Random' represents the generated samples with random classes. With the addition of the generated samples of testing classes, the accuracy increases from 87.79% to 92.03%. And the accuracy increases to 94.22% further with the generated samples of random classes increase from 20,000 to 80,000.

5.4   Data Augmentation Experiments on Handwritten Characters

In this section, we use the newly created classes and writing styles to train the recognition model, and show the improvement of recognition rate on characters with unseen classes and writing style. In this experiment, we use the handwritten

Table 2. Result of experiment on handwritten characters.

Training data				Accuracy
Raw	Classes	Styles	Classes+Styles	
2,674,784	-	-	-	40.82 [21]
2,674,784	1,440,000	-	-	43.23
2,674,784	2,880,000	-	-	44.17
2,674,784	5,760,000	-	-	44.89
2,674,784	-	1,126,500	-	42.50
2,674,784	-	2,253,000	-	43.22
2,674,784	-	4,506,000	-	43.66
2,674,784	5,760,000	4,506,000	-	45.42
2,674,784	5,760,000	4,506,000	9,600,000	**46.12**

training set to train the recognition model RAN and the handwritten testing set is used for evaluation. We use the generated samples to train the network and observe the impact of novel classes and writing styles on the recognition effect.

We reproduce the experimental result in [21] as a baseline. As shown in Table 2, we expand the training data on classes and writing styles. 'Raw' represents the handwritten training set mentioned in Sect. 5.1; 'Classes' represents the generated samples with novel classes that are not shown in the training set; 'Styles' represents the samples with mixed-up writing styles and 'Classes+Styles' represents the samples with novel classes and writing styles at the same time. We observe that the newly created classes are more helpful in recognizing characters with unseen classes and unseen writing styles. When all the generated samples are used to train the recognition model, the recognition rate increases from 40.82% to 46.12%.

6 Conclusion

In this paper, we introduce a novel model named radical combination network for Chinese character generation. We show the function of the generation for Chinese characters with novel classes and font/writing styles by RCN, and the effect of data augmentation on recognizing unseen Chinese characters. In future work, we plan to improve the quality of generated images and apply this work to the generation of natural scene text with complex background. We also will employ dual learning and jointly optimize the generation and recognition tasks for better performance.

Acknowledgements. This work was supported in part by the MOE-Microsoft Key Laboratory of USTC, and Youtu Lab of Tencent.

References

1. Ahmed, M., Samee, M.R., Mercer, R.E.: Improving tree-LSTM with tree attention. In: 2019 IEEE 13th International Conference on Semantic Computing (ICSC), pp. 247–254. IEEE (2019)
2. Azadi, S., Fisher, M., Kim, V.G., Wang, Z., Shechtman, E., Darrell, T.: Multi-content gan for few-shot font style transfer. In: Proceedings of the IEEE Conference on Computer Vision and Pattern Recognition, pp. 7564–7573 (2018)
3. Chawla, N.V., Bowyer, K.W., Hall, L.O., Kegelmeyer, W.P.: SMOTE: synthetic minority over-sampling technique. J. Artif. Intell. Res. **16**, 321–357 (2002)
4. Cho, S., Wang, J., Lee, S.: Handling outliers in non-blind image deconvolution. In: 2011 International Conference on Computer Vision, pp. 495–502. IEEE (2011)
5. Gatys, L.A., Ecker, A.S., Bethge, M.: Image style transfer using convolutional neural networks. In: Proceedings of the IEEE Conference on Computer Vision and Pattern Recognition, pp. 2414–2423 (2016)
6. Goodfellow, I., et al.: Generative adversarial nets. In: Advances in Neural Information Processing Systems, pp. 2672–2680 (2014)
7. He, K., Zhang, X., Ren, S., Sun, J.: Deep residual learning for image recognition. In: Proceedings of the IEEE Conference on Computer Vision and Pattern Recognition, pp. 770–778 (2016)
8. Huang, G., Liu, Z., Van Der Maaten, L., Weinberger, K.Q.: Densely connected convolutional networks. In: Proceedings of the IEEE Conference on Computer Vision and Pattern Recognition, pp. 4700–4708 (2017)
9. Johnson, J., Alahi, A., Fei-Fei, L.: Perceptual losses for real-time style transfer and super-resolution. In: Leibe, B., Matas, J., Sebe, N., Welling, M. (eds.) ECCV 2016. LNCS, vol. 9906, pp. 694–711. Springer, Cham (2016). https://doi.org/10.1007/978-3-319-46475-6_43
10. Krizhevsky, A., Sutskever, I., Hinton, G.E.: ImageNet classification with deep convolutional neural networks. In: Advances in Neural Information Processing Systems, pp. 1097–1105 (2012)
11. Larsson, G., Maire, M., Shakhnarovich, G.: FractalNet: ultra-deep neural networks without residuals. arXiv preprint arXiv:1605.07648 (2016)
12. Li, J., Luong, M., Jurafsky, D., Hovy, E.: When are tree structures necessary for deep learning of representations. Artificial Intelligence. arXiv (2015)
13. Li, X., Zhang, X.: The writing order of modern Chinese character components. J. Modernization Chin. Lang. Educ. **2**, 26–41 (2013)
14. Mirza, M., Osindero, S.: Conditional generative adversarial nets. arXiv preprint arXiv:1411.1784 (2014)
15. Noh, H., Hong, S., Han, B.: Learning deconvolution network for semantic segmentation (2015)
16. Radford, A., Metz, L., Chintala, S.: Unsupervised representation learning with deep convolutional generative adversarial networks. arXiv preprint arXiv:1511.06434 (2015)
17. Simonyan, K., Zisserman, A.: Very deep convolutional networks for large-scale image recognition. Computer Science (2014)
18. Socher, R., et al.: Recursive deep models for semantic compositionality over a sentiment treebank. In: Proceedings of the 2013 Conference on Empirical Methods in Natural Language Processing, pp. 1631–1642 (2013)
19. Tai, K.S., Socher, R., Manning, C.D.: Improved semantic representations from tree-structured long short-term memory networks. Computation and Language. arXiv (2015)

20. Tian, Y.: zi2zi: Master Chinese calligraphy with conditional adversarial networks (2017)
21. Wang, W., Zhang, J., Du, J., Wang, Z.R., Zhu, Y.: DenseRAN for offline handwritten Chinese character recognition. In: 2018 16th International Conference on Frontiers in Handwriting Recognition (ICFHR), pp. 104–109. IEEE (2018)
22. Xie, S., Girshick, R., Dollár, P., Tu, Z., He, K.: Aggregated residual transformations for deep neural networks. In: Proceedings of the IEEE Conference on Computer Vision and Pattern Recognition, pp. 1492–1500 (2017)
23. Zhang, H., Cisse, M., Dauphin, Y.N., Lopez-Paz, D.: mixup: beyond empirical risk minimization. arXiv preprint arXiv:1710.09412 (2017)
24. Zhang, J., Zhu, Y., Du, J., Dai, L.: Radical analysis network for zero-shot learning in printed Chinese character recognition. In: 2018 IEEE International Conference on Multimedia and Expo (ICME), pp. 1–6. IEEE (2018)
25. Zhu, X., Sobihani, P., Guo, H.: Long short-term memory over recursive structures. In: International Conference on Machine Learning, pp. 1604–1612. PMLR (2015)

SmartPatch: Improving Handwritten Word Imitation with Patch Discriminators

Alexander Mattick, Martin Mayr[(✉)], Mathias Seuret, Andreas Maier, and Vincent Christlein

Pattern Recognition Lab, Friedrich-Alexander University Erlangen, Nürnberg, Germany
{alex.mattick,martin.mayr,mathias.seuret,andreas.maier, vincent.christlein}@fau.de

Abstract. As of recent generative adversarial networks have allowed for big leaps in the realism of generated images in diverse domains, not the least of which being handwritten text generation. The generation of realistic-looking handwritten text is important because it can be used for data augmentation in handwritten text recognition (HTR) systems or human-computer interaction. We propose SmartPatch, a new technique increasing the performance of current state-of-the-art methods by augmenting the training feedback with a tailored solution to mitigate pen-level artifacts. We combine the well-known patch loss with information gathered from the parallel trained handwritten text recognition system and the separate characters of the word. This leads to a more enhanced local discriminator and results in more realistic and higher-quality generated handwritten words.

Keywords: Offline handwriting generation · Generative adversarial networks · Patch discriminator

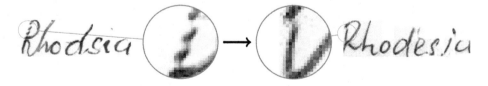

Fig. 1. Reduction of pen-level artifacts with tailored patch discriminator.

1 Introduction

Automatically generating handwritten text is an active field of research due to its high complexity and due to multiple possible use cases. Especially for

© Springer Nature Switzerland AG 2021
J. Lladós et al. (Eds.): ICDAR 2021, LNCS 12821, pp. 268–283, 2021.
https://doi.org/10.1007/978-3-030-86549-8_18

handwritten text recognition (HTR) of historical documents, the possibility to produce extra writer-specific synthetic training data could be a game-changer, because manual transcriptions of a sufficient amount of epoch and document-specific data can be very cost- and time-intensive for research projects dealing with ancient documents. Synthetically produced data that are based on the type of document and styles of the different writers can drastically improve the performance of HTR systems.

Algorithms for handwriting generation can be divided into two different categories: online generation, in which the dataset contains strokes as a time-series, and offline generation, in which only the result of the text is accessible. While online generation has the opportunity to exploit the sequential nature of real handwritten texts, it has the drawback of the complicated assessment of data. In contrast, there are quite many ways to get images of text lines for an offline generation.

The state-of-the-art offline handwriting generation model on word-level, known as GANwriting [13] suffers from unrealistic artifacts, see Fig. 1. This greatly reduces the degree of authenticity of the generated handwriting and often allows quick identification. In this paper, we extend their approach by adding a patch level discriminator to mitigate often produced artifacts. We are going to introduce three different variations: a naive version (NaivePatch) that follows a sliding-window approach, a centered version (CenteredPatch) which uses the attention window of the jointly-trained recognition system, and a smart version (SmartPatch) that additionally inputs the expected character into the patch discriminator. In particular, we make the following contributions:

(1) We extend the current state-of-the-art method for handwritten word imitation [13] with an extra lightweight patch-based discriminator. We derive this additional discriminator from a naive version, via a patch-centered version that incorporates attention masks from the handwritten word recognition system, and finally, propose a new smart discriminator that also uses a patch-centered version of the word images in combination with knowledge of the characters' positions and the recognizer's predictions.[1]

(2) We further show that HTR can robustly validate the synthetically produced outputs.

(3) Finally, we thoroughly evaluate our results compared to other methods and also to genuine word images through qualitative and quantitative measurements.

This work is structured as follows. First, we give an overview of current related work in Sect. 2. The methodology is outlined in Sect. 3. We describe and analyze GANwriting's method [13], which serves as the basis for our approach. We present our extension to improve robustness against pen-level artifacts. In Sect. 4, we evaluate our approach and compare the produced outputs against other methods and real data. First, we compare the FID score of the approaches. Second, we measure the readability with the usage of a sequence-to-sequence handwritten word recognition system. And third, a user study is conducted to

---

[1] Code: https://github.com/MattAlexMiracle/SmartPatch

assess which data is looking most realistic to human subjects. Finally, we discuss our results in Sect. 5 and conclude the paper with Sect. 6.

2   Related Work

Previous approaches to generating human handwriting can be divided by their data used for training the models: online and offline handwriting.

Online Handwriting Synthesis. The first method which achieved convincing-looking outputs was the method of Alex Graves [9]. This approach is based on Long Short-Term Memory (LSTM) cells to predict new positions given the previous ones. Also, skip connections and an attention mechanism were introduced to improve the results. The main problem of this method is the often changing writing style towards the end of the sequence. Chung et al. [5] overcome this problem by using Conditional Variational RNNs (CVRNNs). Deepwriting [2] separately processed *content* and *style*. *Content* in this case describes the information/words to be written, while *style* describes writer-specific glyph attributes like slant, cursive vs. non-cursive writing, and variance between glyphs. The disadvantage of this method is the necessary extra information in the online handwriting training data. Beginning and ending tokens of every single character have to be added to the data to produce persistent results. Aksan et al. [1] introduced Stochastic Temporal CNNs replacing the CVRNNs in the architecture to further reduce inconsistency in generating realistic-looking handwriting. Another approach for generating online handwriting [3] uses the theory of kinematics to generate parameters for a sigma-lognormal model. These parameters may later be distorted to produce new synthetic samples. The problem of all these methods is the crucial need for online handwriting data for training and also for inference. Mayr et al. [16] solved this for inference by converting offline data into online handwriting, processing the online data, and going back to the spatial domain. Still online data is necessary for training the writer style transfer. Another disadvantage of this method is the lack of an end-to-end approach, changing one stage in the pipeline leads to the retraining of all following stages.

Offline Handwriting Synthesis. Currently, many offline handwriting generation methods make use of the generative adversarial network paradigm [8] to simultaneously generate either words or entire lines. ScrabbleGAN [7] uses stencil-like overlapping generators to build long words with, while consistent, random style. A big advantage of ScrabbleGAN is the use of unsupervised learning which allows for many more real-world use cases due to the increase of available training data. GANwriting [13] can be seen as development on top of ScrabbleGAN without the stencil-like generators, but with the ability to select specific styles based on author examples. Davis et al. [6] present further development by using a mixture of GAN and autoencoder training to generate entire lines of handwritten text based on the extracted style. This approach shares some similarities to ours in the sense that they also use the recognizer's attention window to localize characters within the image, though they use this for style extraction rather than

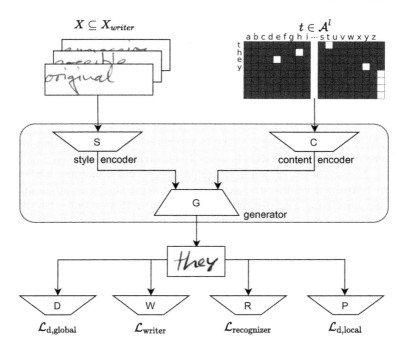

Fig. 2. Basic architecture sketch. Given an input image and new text, a GAN-based approach with additional loss terms generates the text with the style of a selected writer.

improved discriminators. For convenience, we further call the method of Davis et al. lineGen.

3   Methodology

3.1   Background

Figure 2 depicts our architecture. In large parts, we follow the work of GANwriting [13] but additionally introduce a local patch loss. It consists of six interconnected sub-models. The style encoder encodes several example words obtained by an author into a corresponding style-latent space. The content encoder transforms a one-hot encoded character-vector into two content vectors $g_1(c)$ and $g_2(c)$ where $g_1(c)$ works on an individual character level and is concatenated with the style-vector. Conversely, $g_2(c)$ encodes the entire input string into a set of vectors, which later serves as input for the adaptive instance normalization [11] layers in the decoder. The concatenated content-style vector is then used in the fully-convolutional generator. This generator consists of several AdaIn layers [11] which injects the global content vector $g_2(c)$ to different up-sampling stages. The original learning objective contains three units: (1) A writer classifier that tries to predict the style encoder's input scribe, (2) a handwritten text recognizer

Fig. 3. Frequently appearing artifacts in the outputs of GANwriting.

(specifically [17]) consisting of a VGG-19 backbone [19], an attention-layer, and a two-layered bidirectional GRU [4], and (3) a discriminator model. The text and writer classifier are both trained in a standard supervised manner, while the discriminator is trained against the generator following the generative-adversarial network [8] paradigm: The discriminator is trained to maximize the difference between real and fake samples, while the generator is trained to minimize it. The total generator loss then consists of a sum of the losses of text, writer, and discriminator network [13]:

$$\min_{G,W,R} \max_{D} \mathcal{L}(G, D, W, R) = \mathcal{L}_\mathrm{d}(G, D) + \mathcal{L}_\mathrm{w}(G, W) + \mathcal{L}_\mathrm{r}(G, R), \qquad (1)$$

where G is the text generated by the generator, D is real data, W is the real writer and R the real text input into the encoder. $\mathcal{L}_\mathrm{d}, \mathcal{L}_\mathrm{w}, \mathcal{L}_\mathrm{r}$ are the discriminator, writer, and recognizer loss respectively.

Even though GANwriting produces sufficiently accurate words at a glance, closer inspection of the generated words shows some typical artifacts visualized in Fig. 3. The left image shows noticeable vertical banding artifacts for the letter "D". It is quite unlikely to see such a pattern in real images because humans tend to produce continuous strokes in their handwriting. Another prominent artifact is depicted on the right-hand side. There the letter "e" in the middle of the word almost vanishes. By contrast, the strokes in real handwriting have regular intensities throughout the word, except the pen runs out of ink which could be the case for the last letters, but not only for one letter in the middle of the word. We hypothesize that the flaws in the pen-level happen due to the global discriminator loss not being fine-grained enough to discourage locally bad behavior. Therefore, we introduce additional local losses (Fig. 5) to explicitly discourage unrealistic line appearance.

3.2   Local Discriminator Design

Naive Patch Discriminator (NaivePatch). We first determine the baseline-efficacy of using local discriminators, by naively splitting the image into overlapping square patches of size height × height. We use a step size of $\frac{\text{height}}{2}$ making sure that there's considerable overlap between patches as to not generate hard borders between patch zones. A similar strategy in the generator has already been proven to be advantageous by ScrabbleGAN [7] which learned individual overlapping "stencil" generators for each glyph. The expected local loss is then added to Eq. (1). Let $q \in P_\mathrm{real}$, $p \in P_\mathrm{fake}$ be the set of patches stemming from image D,G, resulting in the new loss term

 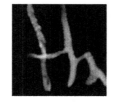 

Patch centered at "e"   Patch centered at "H"   Patch centered at "u"

Fig. 4. Centered patches around letters, automatically found using text recognition module's attention windows.

$$\min_{G,W,R} \max_{D} \mathcal{L}(G, D, W, R) = \mathcal{L}_{\text{d,glob}}(G, D) + \mathcal{L}_{\text{w}}(G, W) + \mathcal{L}_r(G, R) + \mathcal{L}_{\text{d,loc}}(p, q).$$

(2)

The local discriminator model was chosen as small as possible to benchmark the addition of local discriminators, rather than simply the scaling of GANwriting with additional parameters. Specifically, we selected the discriminator design introduced in the Pix2Pix architecture [12]. The receptive field was chosen to be 70×70, as this was the smallest size that entirely covers the patch. This discriminator design shows to be the right size to work with our small patches without over-parameterizing the discriminator. In contrast to PatchGAN [12] we use this discriminator only on the small patches extracted from the image, rather than convolving the discriminator across the whole image like PatchGAN. The discriminator loss function is the standard saturating loss presented in [8] applied to the patches $p \in P_{\text{fake}}, q \in P_{\text{real}}$ resulting in

$$\mathcal{L}_{\text{d,loc}}(p, q) = \mathbb{E}[\log D_{\text{loc}}(q)] + \mathbb{E}[1 - \log D_{\text{loc}}(p)].$$

(3)

Centered Patch Discriminator (CenteredPatch). We further explore the idea of using the text recognition system trained alongside the generator as a useful prior for locating sensible patches in the samples: As part of the sequence-to-sequence model used for digit recognition, an attention map is generated across the latent space of the generator [17]. Since the encoder up to this point is fully convolutional, we can assume similar locality in embedding as in feature space. This means that by up-sampling the attention-vector to the full width of the word image, we obtain windows centered on the associated character. This also allows for a flexible number of patches per image. We found linear up-sampling and the use of the maximum of the attention window as the center led to the most accurate location around specific letters (see Fig. 4). Simple nearest-neighbor up-sampling or matching the relative position of the maximum value in the attention window and the width of the image did not produce consistent results, presumably because different letters set their maximal attention-window at slightly different points relative to the letter's center. Preliminary tests with adapting the patch size depending on character proved very unstable in early training and have not been analyzed further. The loss function, patch size, and model are the same as in NaivePatch to maintain comparability between approaches.

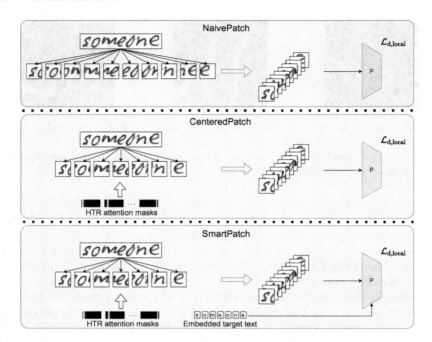

Fig. 5. Architectural overview of our methods: All discriminator architectures share the same Pix2Pix discriminator design [12]. In the case of the smart patch discriminator, we use a linear layer to project the encoded text into a compact latent-space and inject this as additional channels before the discriminator's second-to-last layer. The latter two approaches use the recognition module's up-sampled attention masks for accurately localizing the center of patches.

Smart Patch Discriminator (SmartPatch). The centered patch discriminator (CenteredPatch) is extended further by injecting the label of the expected character into the patch. This is done by first projecting the one-hot encoded character vector into a compact latent space $C_{enc} \in \mathbb{R}^8$ and then appending the vector as extra channels into the CNN's penultimate layer. In this way, we minimize the architectural differences between the NaivePatch, CenteredPatch and SmartPatch.

$$\mathcal{L}_{d,loc}(p, q, c_{real}, c_{fake}) = \mathbb{E}[\log D_{loc}(q, c_{real})] - \mathbb{E}[1 - \log D_{loc}(p, c_{fake})] \quad (4)$$

where $c_{real} \in C_{enc}$ and $c_{fake} \in C_{enc}$ are the character-labels for the fake and real image, respectively. In the case of generated images, we also use the actual character that is supposed to be at the given location, rather than the prediction supplied by the recognition module. This stabilizes training especially for writers with bad handwriting where the recognition module doesn't produce good predictions.

4   Evaluation

We compare the quality of the generated words using three evaluation metrics. First, we make use of the Fréchet Inception Distance (FID) [10] to measure the

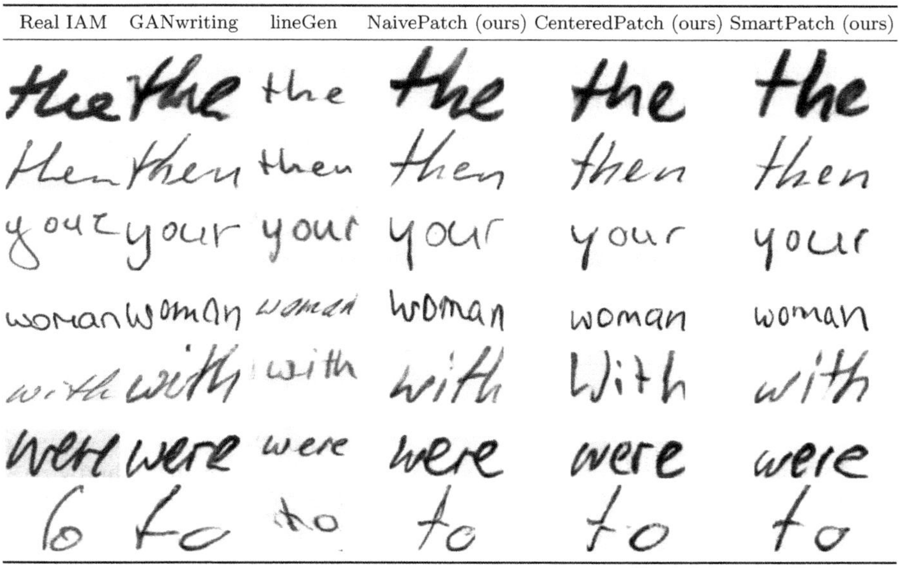

Real IAM	GANwriting	lineGen	NaivePatch (ours)	CenteredPatch (ours)	SmartPatch (ours)

Fig. 6. Randomly chosen outputs of different words. For each row the priming image and the content is the same. More examples can be found in Appendix A.

feature similarity. This metric is widely used in the field of generation tasks in combination with GANs. Second, we measure the legibility of our synthetically created data with a word-based HTR system trained on real data. Third, we conduct a user study to see if we can robustly forge handwriting that humans don't recognize the difference.

For testing, we re-generate data of the IAM dataset [15] using the RWTH Aachen test split. Some random examples can be seen in Fig. 6. For the *style decoder*, we make sure to shuffle each writer's sample images after each generated word to facilitate the generation of different instances of words with the same content. In this way, we avoid creating the image for the same writer and text-piece multiple times and to give us a more accurate assessment of the mean performance of each of the approaches. Aside from removing horizontal whitespaces from small words (as is done in the IAM dataset itself), we perform no additional augmentation or alteration of the GAN outputs.

4.1 FID Score

Table 1 shows the FID score between the real data and the synthetic samples. Empirically, all our three proposed methods produce a lower FID score than GANwriting [13] with the naive approach ranking best. However, note that these values should be taken with a grain of salt: FID is computed using an Inception Net [20] trained on ImageNet [18]. Empirically, this produces sound embeddings

Table 1. Fréchet Inception Distance (FID) computed between real data and the synthetic data. Also, Character-Error-Rate (CER) and loss of the handwritten text recognition system (HTR loss). All approaches using local discriminators seem to perform significantly better than the baseline. We also want to highlight the lower run-to-run variance for the patch discriminators.

Architecture	FID	CER	HTR loss
Real IAM	–	6.27 ± 0.00	23.68 ± 0.00
GANwriting[13]	51.16	4.68 ± 0.54	17.02 ± 0.17
NaivePatch (ours)	**40.33**	4.32 ± 0.33	13.00 ± 0.12
CenteredPatch (ours)	42.05	4.18 ± 0.24	13.41 ± 0.05
SmartPatch (ours)	49.00	**4.04** ± 0.18	**12.54** ± 0.06

for generative models trained on ImageNet or similar datasets, but not necessarily for domains far away from ImageNet, such as monochromatic handwritten text lines. This has also been noted by Davis *et al.* [6]. There is also the problem of how the conditional generation capabilities of handwriting generation is used: Should you use random content? Should you choose random noise for the style? We opted to use the same "synthetic IAM" approach as used later in the HTR tests as this should provide a lower bound on unconditional generation as well (with random content and text the distributions will diverge more). In general, we mainly report FID for comparability with GANwriting [13].

4.2   Handwritten Text Recognition

We evaluate the legibility of the words by training a state-of-the-art handwritten text recognition model [14] on the IAM dataset [15] and by performing evaluations on synthetic data from the baseline, as well as on all our three patch discriminator outputs. For training the text recognition model, we use the RWTH Aachen partition and remove all words containing non-Latin symbols (such as "!.?)") and all words with more than ten characters, as neither the baseline nor our approaches were trained using long words or special characters.

We chose the HTR system by Kang *et al.* [14] due to its high performance and because it is different from the one used in the adversarial loss of the generation systems. In this way, we account for potential biases introduced in the generator's architecture.

We observe that approaches augmenting the global loss with local discriminators produce more (machine)-legible content, with more sophisticated approaches further reducing the CER shown in Table 1. We especially want to highlight the stability of the CenteredPatch, and SmartPatch: We observe significantly less variance across evaluation runs, which lends credence to our claim that locally attending discriminators to symbols allows for the reduction of extreme failure cases (this can also be observed when visually inspecting the samples).

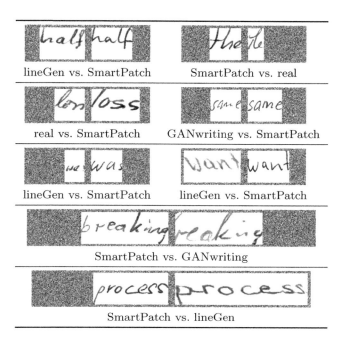

Fig. 7. Examples of the user study including real data and synthetic data produced by GANwriting, lineGen, and our approach SmartPatch.

4.3   User Study

In order to properly assess the quality of our handwriting system, we conducted a user study[2] comparing the produced synthetic words of GANwriting [13], handwritten line generation [6], and our best working method SmartPatch against each other. We also compare each approach against real images of the test set of IAM [15]. Example samples can be seen in Fig. 7.

Participants were instructed to determine the more realistic sample, given two images from different sources. The samples were created by sampling uniformly at random from the synthetic IAM dataset described in Sect. 4.2, and matching pairs of samples with the same content and writer. For line generation (lineGen), we used the code and models provided by [6] and generated individual words from the IAM dataset. Since it was not trained on individual words, but rather on entire lines, the comparison between GANwriting and our method is not perfectly fair, but we choose to include it in our benchmark as it is another strong baseline to compare against. All samples were normalized and aligned to a common baseline, with the background being uniform noise to cancel out any decision-impacting bias, such as contrast.

In total, the study consists of six sets with twenty images each, comparing the configurations SmartPatch vs. GANwriting, SmartPatch vs. lineGen,

---

[2] https://forms.gle/TNoZvxihJNUJiV1b9

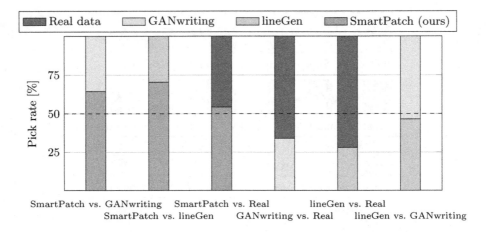

Fig. 8. Pick rate in the user study. Each bar represents one comparison.

SmartPatch vs. Real, Real vs. GANwriting, Real vs. lineGen and GANwriting vs. lineGen.

Figure 8 shows the pick rate for each comparison in percent. The pick rate is defined as the percentage one image was chosen over its counterpart based on the total amount of choices. Our SmartPatch method (■■) outperforms both lineGen (■■) and GANwriting (□□) with pick rates of 70.5% and 64.5%. Also, in 54.4% of the cases our results were more often chosen than the actual IAM images (■■). However, real data is favored with 66.0% over GANwriting and 71.9% over lineGen. In a direct comparing of the latter two methods, GANwriting is slightly preferred by 53.5%.

Figure 9 shows the different picking rates for different word lengths for each method. We categorized the words into small, medium, and large. Small words contain one to three characters, medium words contain four to five characters and large words contain seven to 10 characters. For small words, our method has the highest pick rate of 70%, even more than real images. Small words from GANwriting and lineGen are only chosen by 42% and 28%, respectively. For medium words, GANwriting improves to almost 50%, but lineGen stays nearly the same. Our approach performs equally well for medium words than real words, but for large words, the pick rate further increases for real data and our method is just above 50%. GANwriting's pick rate decreases quite drastically to under 25% for large words whereas lineGen advances to almost 50%. Also note, that for words with ten characters lineGen is picked in 60% of the cases. Our method performs the best for words of length three with a pick rate of 72%.

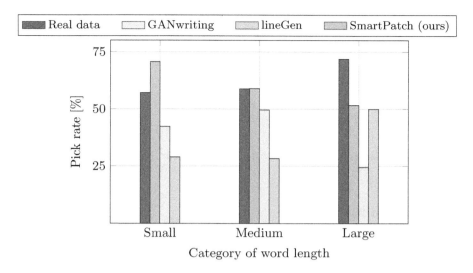

Fig. 9. Pick rate relative to word lengths. Small words have a length between 1 and 3 characters, medium words between 4 and 6 characters and large words between 7 and 10 characters.

5   Discussion

Having a higher pick rate than real images especially for small words is a very interesting outcome of the user study. Presumably, the reason is that participants value readability significantly when comparing the "realness" of samples. Especially small, hastily written words like "the" and "to" tend to be smudged by human writers, making them difficult to read individually. Our method penalizes unreadable characters explicitly, leading to more readable characters for small words (see Fig. 7, top right). This effect might have been enhanced by the fact that subjects had only a single word to analyze and were explicitly asked to distinguish real from machine-generated examples. Additionally, small artifacts introduced during the scanning of IAM samples or malfunctioning pens may also introduce artifacts that participants interpret as generation errors (see Fig. 7, second row, first image). In these cases, the small inaccuracies in stroke consistency are interpreted not as errors in scanning or a dried-out pen, but rather as very weak versions of the grave errors witnessed e.g., Fig. 3. The evaluation of character recognition rate with the HTR system supports our thesis that GANwriting and our approach are focused on producing readable outputs. Further, this questions the conducted user studies of all handwriting generation methods to evaluate whether the generated handwriting images are looking real or not. Also only using HTR systems to grade will not be enough of a quantitative measurement if the improvements in this field further continue this fast. Like Davis et al. [6] already note, we also doubt the validity of reporting the FID score as a good rating for different handwriting generation tasks.

Fig. 10. Comparisions between GANwriting (left) and our SmartPatch method (right).

6   Conclusion

We introduced a new lightweight local patch discriminator that significantly improves the state-of-the-art system for handwritten word generation. Specifically, our method increases the verisimilitude of generated characters, as seen in Fig. 10, by explicitly attending to characters that create more convincing outputs compared to the current state-of-the-art. This extra feedback module was developed especially for this use case but shows the great improvement of incorporating prior knowledge into the training process. We performed a user study that further shows the use of our method for generating handwritten words and that it produces more realistic text than other state-of-art methods.

Appendix

A   Further Examples are shown in Fig. 11

Word	Real IAM	GANwriting	lineGen	NaivePatch (ours)	CenteredPatch (ours)	SmartPatch (ours)
sweet	sweet	sweet	sweet	sweet	sweet	sweet
severe	severe	severe	severely	severe	severe	severe
someone	someone	someone	someone	someone	someone	someone
reading	reading	reading	reading	reading	reading	reading
of	of	of	of	of	of	of
lower	lower	lower	lower	lower	lower	lower
it	it	it	it	it	it	it
inevitable	inevitable	inevitable	inevitable	inevitab	inevitable	inevitabl
home	home	home	home	home	home	home
for	for	for	for	for	for	for
do	do	do	door	do	do	do
clutch	clutch	clutch	clutch	clutch	clutch	clutch
animals	animals	animals	animals	animals	animals	animals
and	and	and	and	and	and	and

Fig. 11. Comparison of randomly chosen outputs. For each row the priming image and the content is the same.

References

1. Aksan, E., Hilliges, O.: STCN: stochastic temporal convolutional networks. In: International Conference on Learning Representations (2019). https://openreview. net/forum?id=HkzSQhCcK7

2. Aksan, E., Pece, F., Hilliges, O.: DeepWriting: making digital ink editable via deep generative modeling, pp. 1–14. Association for Computing Machinery, New York (2018). https://doi.org/10.1145/3173574.3173779

3. Bhattacharya, U., Plamondon, R., Dutta Chowdhury, S., Goyal, P., Parui, S.K.: A sigma-lognormal model-based approach to generating large synthetic online handwriting sample databases. Int. J. Doc. Anal. Recogn. (IJDAR) **20**(3), 155–171 (2017). https://doi.org/10.1007/s10032-017-0287-5

4. Cho, K., van Merriënboer, B., Bahdanau, D., Bengio, Y.: On the properties of neural machine translation: encoder-decoder approaches. In: Proceedings of SSST-8, 8th Workshop on Syntax, Semantics and Structure in Statistical Translation, Doha, Qatar, pp. 103–111. Association for Computational Linguistics (October 2014). https://doi.org/10.3115/v1/W14-4012

5. Chung, J., Kastner, K., Dinh, L., Goel, K., Courville, A.C., Bengio, Y.: A recurrent latent variable model for sequential data. In: Cortes, C., Lawrence, N.D., Lee, D.D., Sugiyama, M., Garnett, R. (eds.) Advances in Neural Information Processing Systems 28, pp. 2980–2988. Curran Associates, Inc. (2015). https://proceedings. neurips.cc/paper/2015/file/b618c3210e934362ac261db280128c22-Paper.pdf
6. Davis, B., Tensmeyer, C., Price, B., Wigington, C., Morse, B., Jain, R.: Text and style conditioned GAN for generation of offline handwriting lines. In: British Machine Vision Conference (BMVC) (2020). https://www.bmvc2020-conference. com/assets/papers/0815.pdf
7. Fogel, S., Averbuch-Elor, H., Cohen, S., Mazor, S., Litman, R.: ScrabbleGAN: semi-supervised varying length handwritten text generation. In: 2020 IEEE/CVF Conference on Computer Vision and Pattern Recognition (CVPR), pp. 4323–4332 (2020). https://doi.org/10.1109/CVPR42600.2020.00438
8. Goodfellow, I., et al.: Generative adversarial nets. In: Ghahramani, Z., Welling, M., Cortes, C., Lawrence, N., Weinberger, K.Q. (eds.) Advances in Neural Information Processing Systems, vol. 27. Curran Associates, Inc. (2014). https://proceedings. neurips.cc/paper/2014/file/5ca3e9b122f61f8f06494c97b1afccf3-Paper.pdf
9. Graves, A.: Generating sequences with recurrent neural networks. arXiv:1308.0850 [cs] (June 2014)
10. Heusel, M., Ramsauer, H., Unterthiner, T., Nessler, B., Hochreiter, S.: GANs trained by a two time-scale update rule converge to a local Nash equilibrium. In: Guyon, I., et al. (eds.) Advances in Neural Information Processing Systems, vol. 30. Curran Associates, Inc. (2017). https://proceedings.neurips.cc/paper/2017/ file/8a1d694707eb0fefe65871369074926d-Paper.pdf
11. Huang, X., Belongie, S.: Arbitrary style transfer in real-time with adaptive instance normalization. In: 2017 IEEE International Conference on Computer Vision (ICCV), pp. 1510–1519 (2017). https://doi.org/10.1109/ICCV.2017.167
12. Isola, P., Zhu, J., Zhou, T., Efros, A.A.: Image-to-image translation with conditional adversarial networks. In: 2017 IEEE Conference on Computer Vision and Pattern Recognition (CVPR), pp. 5967–5976 (2017). https://doi.org/10.1109/ CVPR.2017.632
13. Kang, L., Riba, P., Wang, Y., Rusiñol, M., Fornés, A., Villegas, M.: GANwriting: content-conditioned generation of styled handwritten word images. In: Vedaldi, A., Bischof, H., Brox, T., Frahm, J.-M. (eds.) ECCV 2020. LNCS, vol. 12368, pp. 273–289. Springer, Cham (2020). https://doi.org/10.1007/978-3-030-58592-1_17
14. Kang, L., Toledo, J.I., Riba, P., Villegas, M., Fornés, A., Rusiñol, M.: Convolve, attend and spell: an attention-based sequence-to-sequence model for handwritten word recognition. In: Brox, T., Bruhn, A., Fritz, M. (eds.) GCPR 2018. LNCS, vol. 11269, pp. 459–472. Springer, Cham (2019). https://doi.org/10.1007/978-3-030-12939-2_32
15. Marti, U.V., Bunke, H.: The IAN-database: an English sentence database for offline handwriting recognition. Int. J. Doc. Anal. Recogn. 5(1), 39–46 (2002). https:// doi.org/10.1007/s100320200071
16. Mayr, M., Stumpf, M., Nicolaou, A., Seuret, M., Maier, A., Christlein, V.: Spatio-temporal handwriting imitation. In: Bartoli, A., Fusiello, A. (eds.) ECCV 2020. LNCS, vol. 12539, pp. 528–543. Springer, Cham (2020). https://doi.org/10.1007/ 978-3-030-68238-5_38
17. Michael, J., Labahn, R., Grüning, T., Zöllner, J.: Evaluating sequence-to-sequence models for handwritten text recognition. In: 2019 International Conference on Document Analysis and Recognition (ICDAR), pp. 1286–1293 (2019). https://doi.org/ 10.1109/ICDAR.2019.00208

18. Russakovsky, O., et al.: ImageNet large scale visual recognition challenge. Int. J. Comput. Vis. **115**(3), 211–252 (2015). https://doi.org/10.1007/s11263-015-0816-y
19. Simonyan, K., Zisserman, A.: Very deep convolutional networks for large-scale image recognition (2015). https://arxiv.org/abs/1409.1556
20. Szegedy, C., Vanhoucke, V., Ioffe, S., Shlens, J., Wojna, Z.: Rethinking the inception architecture for computer vision. In: 2016 IEEE Conference on Computer Vision and Pattern Recognition (CVPR), pp. 2818–2826 (2016). https://doi.org/10.1109/CVPR.2016.308

Scene Text Detection and Recognition

Reciprocal Feature Learning via Explicit and Implicit Tasks in Scene Text Recognition

Hui Jiang[1], Yunlu Xu[1], Zhanzhan Cheng[1,2(✉)], Shiliang Pu[1], Yi Niu[1],
Wenqi Ren[1], Fei Wu[1], and Wenming Tan[1]

[1] Hikvision Research Institute, Hangzhou, China
chengzhanzhan@hikvision.com
[2] Zhejiang University, Hangzhou, China

Abstract. Text recognition is a popular topic for its broad applications. In this work, we excavate the implicit task, character counting within the traditional text recognition, without additional labor annotation cost. The implicit task plays as an auxiliary branch for complementing the sequential recognition. We design a two-branch reciprocal feature learning framework in order to adequately utilize the features from both the tasks. Through exploiting the complementary effect between explicit and implicit tasks, the feature is reliably enhanced. Extensive experiments on 7 benchmarks show the advantages of the proposed methods in both text recognition and the new-built character counting tasks. In addition, it is convenient yet effective to equip with variable networks and tasks. We offer abundant ablation studies, generalizing experiments with deeper understanding on the tasks. Code is available.

Keywords: Scene text recognition · Character counting

1 Introduction

Scene text recognition (STR) has attracted great attention in recent years, whose goal is to translate a cropped scene text image into a string. Thanks to the rapid development of deep learning, modern STR methods have achieved remarkable advancement on public benchmarks [1]. As the high precision acquired on most of the commonly used datasets, *e.g.*, above 90% or even 95% accuracy on IIIT5K [21], SVT [34], ICDAR 2003/2013 [16], recent works tend to more specific problems instead of general representative ability, including irregular text [28], low-quality imaging [22], attention drift [4,6,35], efficiency [36] and *etc*. An emerging trend is to enhance the networks with additional branches [4,6,33,39], generating a tighter constraint on the original sequential task. It can be roughly grouped in two categories: (1) Approaches introducing *additional annotations or tasks*

H. Jiang and Y. Xu—Contributed equally. Supported by National Key R&D Program of China (Grant No. 2018YFC0831601).

J. Lladós et al. (Eds.): ICDAR 2021, LNCS 12821, pp. 287–303, 2021.
https://doi.org/10.1007/978-3-030-86549-8_19

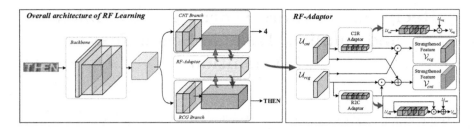

Fig. 1. Overall architecture of the proposed network. Given an image input, the shared backbone extracts visual features, which then are fed into the RCG and CNT branch. The two branches are trained by respective supervision, and intertwined and mutually assisted through a novel RF-Adaptor.

assisting the sole sequential recognition and (2) Approaches *exploiting the original tasks* limited to the word-level annotations.

In the first group, using *additional information from another task or detailed supervision* is intuitive and usually ensures the boosting performance. For instance, in FAN [4] and DSAN [6], character-level position was imported for finer constraints on word recognition. SEED [23] brought pre-trained language model supervised from wording embedding in Natural Language Processing (NLP) [2] field to compensate the visual features. However, all the above approaches bring in extra annotation cost, enlarging from word-level labels to character-level or needing source from other domains, *e.g.*, language model.

Another group *exploiting original tasks and supervision* without introducing extra supervision or tasks is meanwhile sprouting, such as multi-task learning [17] of variable networks (*e.g.*, CTC-based [7] and Attn-based [20][1] models) sharing partial features. Though expecting the joint learning of two branches for supplementary features on both sides, the improvement is not always satisfying [4,12]. As claimed in [12], CTC branch is promoted under the guidance of Attn, which turns into the main contribution in GTC approach that gets a more efficient and effective CTC model. Conversely, the Attn branch cannot be likewise promoted with CTC, since the feature learning in Attn is harmed by the joint partner CTC loss. In essence, the multi-task training seems like a trade-off between the features from the two branches. We attribute it to the similartaxis of dual-branch rooting in the same supervision, the original strong branch can not be enhanced any further. *In another word, the limitation of the multi-task learning in STR (without extra annotations) lies in the onefold word-level supervision.* So far, what we are wondering is *'Can a strong recognizor be promoted with another task without extra annotation cost?'*. Observing there is hierarchically implicit information in text itself, including tokens, semantics, text-length and the like, apart from text strings. Obviously, the implicit labels above cannot

---

[1] CTC is short for Connectionist Temporal Classification, and Attn is short for Attention Mechanism in the whole manuscript, both of which are mainstreaming approaches in sequence-to-sequence problems.

cover complete information in the string, but we make a conjecture that it leads the network to learn its typical features differing from the usual text labels. Regarding not bringing extra annotations or resources(*e.g.*, tokens or semantics in text relies on extra corpus), we take the text-length as the additional supervision, generating a new task named *character counting*. The total character numbers can be directly obtained from known words, not requiring any labor cost.

In this paper, we propose a novel text recognition framework, called Reciprocal Feature Learning (RF-L). It consists of two tasks. The main task is a recognizor (denoted as RCG) outputting a target string. And we also exploit the implicit labels to generate an auxiliary task named *counting* (denoted as CNT) to predict the occurrence number of characters, as shown in Fig. 1. The supervision of CNT can be inferred from RCG, but the training goals have some difference. The feature learned in CNT focuses more on the category-unaware occurrence of characters, while the RCG puts effort on discrepancy between types of classes and even the End-Of-Sentence (EOS) signals. Intuitively, we hope to utilize the discrepancy and build relation between the two branches with separate tasks. In this work, we propose a well-designed module called Reciprocal Feature Adaptor (RF-Adaptor). It transfers the complementary data from one branch to the other without any extra annotations, playing a role of assembling features and adapting to the task. Note that the RF-Adaptor allows bi-directional data flowing between both the tasks. The main contributions can be summarized as:

(1) We dig the implicit task in the traditional STR, *i.e.,* character counting, without any extra annotation cost. The counting network supervised by new exploited labels can be regarded as an auxiliary part in addition to the recognition task, facilitating positive outcomes. Also, we offer a strong baseline for the sole newly-built character counting task based on the existing STR datasets.
(2) We propose a multi-task learning framework called RF-L for STR through exploiting the complementary effect between two different tasks, word recognition and counting respectively. And the two tasks are learned in their own branch, in interaction to the other, via a simple yet effective RF-Adaptor module.
(3) The proposed method achieves impressive improvements in multiple benchmarks, not only in STR tasks but also in counting. The auxiliary network and the adaptor can be easily integrated into deep neural network with any other scene text recognition method, which boosts the single task via the proposed RF-L framework as verified in extensive experiments.

2   Related Works

2.1   Scene Text Recognition

Scene text recognition is an important challenge in computer vision and various methods have been proposed. Conventional methods were based on hand-crafted features including sliding window methods [34], strokelet generation [37], histogram of oriented gradients descriptors [30] and *etc*. Recent years, deep neural network has dominated the area. Approaches can be grouped in *segmentation-free* and *segmentation-based* methods.

In the first *segmentation-free* category, Shi et al. [27] have brought recurrent neural networks for the varying sequence length, and the CTC-based recognition poses a milestone in OCR community, not only STR, but also the handwriting [7] or related tasks [14]. Later works [4,28] applied attention mechanism in STR towards higher level of accurate text recognition. Recently, some specific modules appeared like transformation [28], high-resolution pluggable module [22], advanced attention [35] and enhanced semantic modeling [38]. Although focused on different modules or handling with variable problems, their main body can be roughly classified into CTC-based [27,32] and Attn-based methods [4,28,35,38]. *Segmentation-based* approaches are becoming another new hotspot [18,31]. They are usually more flexible than sequential decoders in the recognition of irregular text with orientation or curvature, and offer finer constraints on features, due to the detailed supervision. However, these methods remain limited because they can not work with only sequence-level labels.

2.2 Auxiliary Supervision and Tasks in STR

As modern networks [1,28] have acquired high precision on most of the commonly-used datasets, *e.g.*, above 90% or even 95% accuracy on IIIT5K [21], SVT [34], ICDAR 2003/2013 [16], it is usual practice to induce auxiliary supervision with new losses [4,6,33,39] appending to the sequential recognition network. The key of promoting results comes from not only the well-designed module in assistance with the main task but also the detailed information from additional expensive annotation. For instance, [4] handled with the attention drift through an additional focusing attention network aligning the attention center with character-level position annotations. [39] designed dual-branch mutual supervision from order segmentation and character segmentation tasks. [6] enhanced the encoding features through an auxiliary branch training the character-level classifier with extra supervision. [33] utilized mutual learning of several typical recognition networks, including CTC-based [27], Attn-based [28] and segmentation-based [18] methods. Though effective through auxiliary branches, all these above require the character-level supervision. Limited by the word-level labels, multi-branch assistance meets much difficulty. Early work [17] in speech intuitively trained tasks simultaneously by CTC and Attn-based loss but it does not work well in STR as claimed [4,12]. Thus, GTC [12] proposed to enhance CTC-based branch under the guidance from the Attn-based branch, but the Attn-based branch would not be reinforced through the same solution. Among these works [12,17], mechanized application of shared backbones with multiple losses is adopted, while the relation between the branches, the explicit and implicit tasks, has not been fully exploited.

3 Methodology

3.1 Overall Architecture

An overview of the proposed architecture is shown in Fig. 1, an end-to-end trainable Reciprocal Feature Learning (RF-L) network. It aims at exploiting the information from the given word-level annotations and strengthening the feature between different tasks by explicit and implicit annotations. The network can be divided into four parts: the backbone network, two task branches including text recognition and character counting, and reciprocal feature adaptor interwinding the two branches. Given an input image, the shared backbone is used to extract visual features from the input images. Then the character counting branch and text recognition branch utilize the encoded feature for respective tasks simultaneously. The relation between the two branches is built upon a well-designed module called Reciprocal Feature Adaptor (RF-Adaptor). It transfers useful messages from one branch to another, replenishing features in a single task with the high-related partner. Note that the RF-Adaptor allows bi-directional data flowing and synchronously reinforces the targeted tasks on both sides.

3.2 Backbone

The backbone is shared by the two targeted tasks, where any existing feature extractor can be applied in this module. In these work, we follow the widely-used encoders [1] including VGG [29]-based [27], ResNet [10]-based backbone [4]. The feature module (noted as $F(\cdot)$) encodes the input image x with scale $H \times W$ into feature map V, where C, H_{out}, W_{out} denote the output channels, output height and width size.

$$V = F(x), V \in R^{C \times H_{out} \times W_{out}} \tag{1}$$

Note that we share shallow layers of backbones, *i.e.*, the stage-1 and stage-2 in the ResNet-based encoder and the convolution layer ranging from 1 to 5 out of 7 in the VGG-based encoder. The remained higher layer of features are learned in each branch respectively. We use ResNet [4] if not specified in experiments.

3.3 Character Counting Branch

The character counting (short in CNT) branch is responsible for predicting the total character occurrence numbers. Given the encoded features from the shared backbone, the CNT branch outputs the text length. It's worth mentioning that the supervision for the task can be directly computed from the word-level recognition labels and thus does not require any extra annotation cost. Similar to [36] who predicts the character category-aware occurrence for each text, our prediction of total character counting (a scalar) can be realized through two methods, regression-based prediction with Mean Square Error (MSE) loss, and classification-based with Cross-Entropy loss, respectively as show in Eq. 2.

$$L_{cnt} = \begin{cases} MSE(\hat{y}_{cnt}, y_{cnt}), & if\ Regression \\ CrossEntropy(\hat{y}_{cnt}, y_{cnt}), & if\ Classification \end{cases} \tag{2}$$

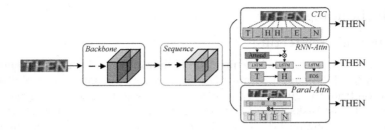

Fig. 2. The segmentation-free text recognition networks.

where $\hat{y}_{cnt}$ and y_{cnt} are the predicted scalars and the groundtruth (*i.e., the text-length*, like 3 for 'cnt') respectively. Besides, regarding the unbalanced of character occurring, we count the ratio α of character appearing numbers, and use $(1 - \alpha)$ as factors to re-weight the loss in training (denoted as *w. Class Balance*), and it works well in both functions in Sect. 4.2.

3.4   Text Recognition Branch

The text recognition (short in RCG) branch is typically sequential prediction which can directly use any existing recognition networks [1, 27, 28, 35].

As shown in Fig. 2, the existing segmentation-free recognition models can be mainly grouped into CTC-based [27, 32] or Attn-based methods [1, 28, 35]. The latter group usually has relatively superior performance while the former is more efficient and friendly to real applications. Specifically, the group of Attn-based decoders include the dominating RNN-based Attn [1, 28], the emerging Parallel Attn [35, 38] (short in Paral-Attn), and Transformer [26]. As framework of Transformer-based encoder-decoder [26] is largely different from the other CTC or Attn-based pipelines, we do not discuss in this paper. Therefore, in this work, we reproduce the CTC [27], Bilstm-Attn [1] (*w.o.* TPS transformation), and Paral-Attn [35] models as three representatives among most existing networks. Note that we are focusing on the feature learning in the recognition networks, so we adopt mainstreaming models only with the main recognizor *without* any pluggable functional modules (*e.g.,* TPS-based transformation [28], high-resolution unit [22], semantic enhancement [23, 38], position clues [39]). The loss function L_{rcg} in RCG branch is the standard CTC loss or the Cross-Entropy (CE) loss for Attn-based methods.

3.5   Reciprocal Feature Adaptor

Instead of the existing works limited by the similartaxis of dual-branch (*e.g.,* CTC and Attn) features [4, 12], which results from the same supervising labels in word-level STR, we dig the implicit task and labels (*i.e.,* character counting). We arrange it as an additional branch to train with the recognizor simultaneously, as shown in Fig. 1. And we are motivated to uncover and thus build the relationship between the two branches. We believe that a good dual-task mutual assistance majors in two abilities: (1) Offering complementary message from one branch to the other. (2) Adapting feature from one task to the other. Therefore, we propose a module called Reciprocal Feature Adaptor (RF-Adaptor) in order to realize the mutual feature assistance. Specifically, given the separate features from two branches $\mathcal{U}_{cnt}$ and $\mathcal{U}_{rcg}$ as input, the mutual-enhanced features $\mathcal{V}_{cnt}$ and $\mathcal{V}_{rcg}$ via bidirectional adaptors can be formulated as:

$$\mathcal{V}_* = \mathcal{U}_{cnt} \diamond \mathcal{U}_{rcg} \tag{3}$$

where $\diamond$ is the binary operator with two-side inputs, which includes the usual approaches like *element-wise multiplication* (noted in $\odot$), *element-wise addition* (noted in $\oplus$) and *concatenation* (noted in $\copyright$), as shown in Fig. 3. In detail, CNT seems inherently contained in RCG, and its supervision offers much less information. Therefore, RCG can supplement CNT with abundant informative details. On the other hand, not requiring discriminating the characters, the feature learning in CNT focuses more on the text itself, and thus can obtain more purified representation and not be bothered by the confusing patterns. For convenience, procedure of generating the enhanced RCG features $\mathcal{V}_{rcg}$ is called *CNT-to-RCG* (short in *C2R*). Conversely, producing $\mathcal{V}_{cnt}$ is *RCG-to-CNT* (short in *R2C*).

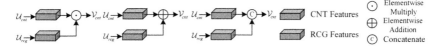

Fig. 3. Fusion of the two-branch features in the proposed RF-Adaptor. The figure shows the RCG-to-CNT one-way fusion, and in turn the CNT-to-RCG is similar.

- In *R2C*, as the partner $\mathcal{U}_{rcg}$ contains more information than $\mathcal{U}_{cnt}$, it can replenish the $\mathcal{U}_{cnt}$ a lot via manipulating $\oplus$ on the two features.
- In *C2R*, $\mathcal{U}_{rcg}$ itself has rich information and $\mathcal{U}_{cnt}$ has more purified features, so CNT plays a role of feature selector or weighting factor for $\mathcal{U}_{rcg}$. The selector is implemented via a learnable gate, which is designed to suppress the text-unrelated noise and thus refine $\mathcal{U}_{rcg}$ via the $\odot$ operation.

Note the direct concatenation $\copyright$ regards the two input equally and thus is not recommended for our asymmetry (with unequal information) dual branches, so we do not apply it in our RF-Adaptor. Rigorous ablations of the three fusion operations are shown later in Table 2 with explanations in Sect. 4.4.

Upon the chosen *fusion* operation $\oplus$ in *R2C* and $\odot$ in *C2R*, and not satisfied by the direct combination on the two-branch features, we resort to some lightweight modules, which we name them Feature Enhancement (short in *FE*) modules (formulated in $\mathcal{F}(\cdot)$), to dispose of the input $\mathcal{U}_{rcg}$ and $\mathcal{U}_{cnt}$ for better providing the supplementary information to the other. Furthermore, using $\mathcal{I}(\cdot)$ to represent identity mapping, we specify Eq. 3 into

$$\begin{aligned} \mathcal{V}_{cnt} &= \mathcal{I}(\mathcal{U}_{cnt}) \oplus \mathcal{F}_c(\mathcal{U}_{rcg}) \\ \mathcal{V}_{rcg} &= \mathcal{F}_r(\mathcal{U}_{cnt}) \odot \mathcal{I}(\mathcal{U}_{rcg}), \end{aligned} \tag{4}$$

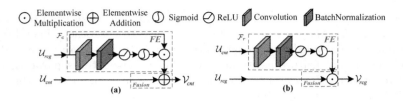

Fig. 4. The structures for the *FE* blocks. (a) is in *R2C* and (b) is in *C2R*.

where $\mathcal{F}_c(\cdot)$ and $\mathcal{F}_r(\cdot)$ are both *FE* modules in *C2R* and *R2C* respectively. Inspired by the convolutional attention-based modules [3,11] which can be easily integrated into the CNNs for feature refinement, we would like to apply the idea into our setups for a better adaptation on features from one branch to the other. Differently, instead of self-enhancement as the existing methods [3,11] with single input source, we have the dual-branch features from different tasks. Intuitively, the *FE* modules are designed in two separate ways as details in the following.

- $\mathcal{F}_c(\cdot)$ is the self-reinforced features through multiplying the importance learnt on the context using one 1×1 convolution, Normalization, one ReLU and a sigmoid function sequentially, as Fig. 4(a). The function outputs an enhanced RCG features $\mathcal{F}_c(\mathcal{U}_{rcg})$ which is expected to suit the CNT branch via end-to-end adaptive learning. In the feedforward of *R2C*, producing $\mathcal{V}_{cnt}$ can be represented as the original $\mathcal{U}_{cnt}$, supplemented by useful information from $\mathcal{U}_{rcg}$ via a refinement module on $\mathcal{U}_{rcg}$ itself.
- $\mathcal{F}_r(\cdot)$ is rescaling the feature on each channel per coordinate as Fig. 4(b), *i.e.*, using features from $\mathcal{U}_{cnt}$ as soft-attention on the $\mathcal{U}_{rcg}$ to produce an enhanced RCG feature output $\mathcal{V}_{rcg}$. In this case, the $\mathcal{F}_r(\cdot)$ is similarly composed of one 1×1 convolution, Normalization, ReLU and a sigmoid function orderly. It functions on the $\mathcal{U}_{cnt}$ to compute the importance for each channel and position in the original input RCG features $\mathcal{U}_{rcg}$, which can be seen as using the additional $\mathcal{U}_{cnt}$ for re-weighting the $\mathcal{U}_{rcg}$. More details and understanding of the proposed RF-Adaptor are studied and shown in Sect. 4.4.

3.6 Optimization

The supervision is working on each branch respectively. The loss function of the overall Reciprocal Feature Learning (RF-L) framework can be formulated by sum of the losses from two branches.

$$L = L_{cnt} + \lambda L_{rcg} \tag{5}$$

where λ is 1.0 by default. L_{rcg} and L_{cnt} are introduced in Sect. 3.3 and 3.4. The training can be in an end-to-end approach without any pre-training or stage-wise strategies. Note that we do not adjust the parameters λ for a delicate balance, while we merely expect to affirm the mutual assistance on both branches.

4   Experiments

4.1   Datasets

Following existing benchmarks [1], we train the models on 2 synthetic datasets *MJSynth* (MJ) [13] and *SynthText* (ST) [8] using word-level annotations. Note that models are only trained with the above 2 datasets and directly evaluated in the following 7 public testsets without any finetuning stage or tricks.

IIIT5K(IIIT) [21] contains 3,000 images of scene texts and born-digital images for evaluation, which are collected from the websites. **SVT** [34] has 647 images cropped from Google Street View for evaluation. **ICDAR2003/2013/2015 (IC03/IC13/IC15)** [15,16,19] offer respectively 860, 867, 1811 images for evaluation, which were created for the Robust Reading Competition in the International Conference on Document Analysis and Recognition (ICDAR) 2003/2013/2015. **SVTP** [24] contains 645 images for evaluation and the images are collected from Google Street View. Many of the images in dataset are heavily distorted. **CUTE80(CT)** [25] contains 288 cropped images for evaluation. Most of them are curved text images.

4.2   Tasks and Metrics

Scene Text Recognition. Text recognition is a popular task in the deep learning community. Following the existing [1], we regard it as a sequential recognition problem and evaluate the overall performance on the word accuracy.

Character Counting. Few work uses scene text images for other tasks. Noting that Xie et al. [36] solved the STR problem from a new perspective which is analogical to our counting task, but they finally transfer the learned feature into sequential outputs as common STR task. One more step, we explicitly build the task called character counting using the existing STR public datasets. Annotations for the counting can be directly transferred from the word-level labels, so it does not require extra labor. Given the category-unaware predictions and the counting labels c_i for image i, the RMSE and relRMSE [36] are calculated by

$$RMSE = \sqrt{\frac{1}{N}\sum_{i=1}^{N}(\widehat{c}_i - c_i)^2}$$
$$relRMSE = \sqrt{\frac{1}{N}\sum_{i=1}^{N}\frac{(\widehat{c}_i - c_i)^2}{c_i + 1}}. \tag{6}$$

For convenience, we also evaluate the character counting on the ratio of correctly predicted images in ablations.

4.3   Implementation Details

All images are resized to 32×100 before inputting to the feature extractor and encoded to the feature map of 1×26. The parameters are initialized using He's

Table 1. Comparison of counting results based on Cross-Entropy and regression-based methods, evaluated on the ratio of correctly predicted images (%).

Methods	w.o. Class Balance		w. Class Balance	
	Regular (see Footnote 2)	Irregular	Regular	Irregular
CE	89.5	78.5	93.2	83.5
Regression	93.3	82.3	**94.6**	**84.5**

Table 2. Performance of feature fusing in *C2R* and *R2C*. The CNT-to-RCG is evaluated on word accuracy (%) the same with the RCG task, while the RCG-to-CNT is evaluated on the ratio (%) of correctly predicted images on the CNT task (similarly in Table 3 and Table 4). Finally, *C2R* uses $\odot$ (*multiplication*) and *R2C* uses $\oplus$ (*addition*).

Methods	CNT-to-RCG		RCG-to-CNT	
	Regular	Irregular	Regular	Irregular
Baseline	77.4	53.2	84.3	69.7
$\odot$ Multiplication	**79.4(+2.0%)**	**54.3(+1.1%)**	88.2(+3.9)	74.6(+4.9)
$\oplus$ Addition	79.0(+1.6)	**54.3(+1.1%)**	**88.5(+4.2%)**	**76.6(+6.9%)**
ⓒ Concatenation	78.6(+1.2)	53.1(-0.1)	88.6(+4.3)	75.1(+5.4)

method [9] if not specified. AdaDelta optimizer is adopted with the initial learning rate 1. Batch size is set to 128. The details of backbone are introduced in Sect. 3.1. For text recognition, the hidden dimension in RNN and Attn is 512. And the *FE* channel dimension in RF-Adaptor is 512. For quick ablation, we conduct extensive ablations partially on 10% MJSynth (MJ) [13] using a simplified Paral-Attn model, where the modules in DAN [35] are replaced with linear mapping for Attn decoder. And the remaining results are following the standard setups [1]. All the comparisons are only with word-level annotations.

4.4   Ablation Studies

Design of CNT Network. We formulate the counting problem as a total number of all-class characters with loss functions designed in two ways, the regression and the Cross-Entropy (CE) loss. We find it performs better in *regression-based* methods *w. Class Balance* as shown in Table 1, solely in counting tasks.[2]

Design of RF-Adaptor. The design for the cross-branch message learner is focused on *combining two-branch messages*, and *suiting to the other task* as illustrated in Sect. 3.5. We divide it into 2 steps orderly: (1) Feature Enhancement (*FE*): the attention-based enhancement, (2) Fusion: the fusion operation. We do

---

[2] *Regular* and *Irregular* represent mean accuracy under 4 regular datasets (*i.e.*, IIIT5K, SVT and IC03/13) and 3 irregular datasets (*i.e.*, IC05, SVTP and CT).

Table 3. Effect of different components in RF-adaptor.

Methods	Modules		CNT-to-RCG		RCG-to-CNT	
	Fusion	FE	Regular	Irregular	Regular	Irregular
Baseline			77.4	53.2	84.3	69.7
RF-Adaptor *w.o.FE*	✓		79.4(+2.0)	54.3(+1.1)	88.5(+4.2)	76.6(+6.9)
RF-Adaptor	✓	✓	**79.6(+2.2)**	**55.4(+2.2)**	**90.4(+6.1)**	**77.0(+7.3)**

Table 4. Comparison with SOTA recognition models with their training settings. We only report the results using MJ and ST without extra data or annotations. The best accuracy is in **bold** and the second one is denoted in <u>underline</u>.

Methods	Year	Training data	Benchmark							Avg. Acc	
			IIIT	SVT	IC03	IC13	IC15	SVTP	CT	Regular	Irregular
CRNN [27]	2016	MJ	78.2	80.8	89.4	-	-	-	-	-	-
AON [5]	2018	MJ+ST	87.0	82.8	91.5	-	-	73.0	76.8	-	-
NRTR [26]	2018	MJ+ST	90.1	**91.5**	94.7	-	79.4	**86.6**	80.9	-	<u>82.3</u>
ASTER [28]	2019	MJ+ST	93.4	89.5	94.5	91.8	-	78.5	79.5	92.3	-
TPS-Bilstm-Attn [1]	2019	MJ+ST	87.9	87.5	<u>94.9</u>	93.6	77.6	79.2	74.0	91.0	76.9
AutoSTR [40]*	2020	MJ+ST	<u>94.7</u>	<u>90.9</u>	93.3	<u>94.2</u>	<u>81.8</u>	81.7	-	**93.2**	-
RobustScanner [39]†	2020	MJ+ST	**95.3**	88.1	-	-	-	79.5	**90.3**	-	-
Bilstm-Attn [1] (see Footnote 3)	2019	MJ+ST	93.7	89.0	92.3	93.2	79.3	81.2	80.6	92.1	80.4
Bilstm-Attn *w.* RF-L	-	MJ+ST	94.1	88.6	<u>94.9</u>	**94.5**	**82.4**	82.0	82.6	<u>93.0(+0.9)</u>	**82.4(+2.0)**
DAN [35] (see Footnote 4)	2020	MJ+ST	93.4	87.5	94.2	93.2	75.6	80.9	78.0	92.1	78.2
DAN *w.* RF-L	-	MJ+ST	94.0	87.7	93.6	93.5	76.7	<u>84.7</u>	77.8	92.2(+0.1)	79.7(+1.5)

ablation on each point independently for clear explanation on each module in the proposed RF-Adaptor.

- We first discuss the *fusion* procedure ($\diamond$ in Eq. 3), as it is *necessary* though is the last step in all the three. We compare among the approaches including concatenation, bit-wise multiplication or sum, shown in Fig. 3 and Table 2.
- The proposed *FE* modules are inspired from the channel attention [3] and broaden it into the two-branch input networks. Comparing RF-Adaptor and RF-Adaptor *w.o. FE* in Table 3, it shows the effectiveness of the *FE* module not only in *C2R* but also in *R2C* direction.

4.5   Comparison with State-of-the-Art Approaches

Scene Text Recognition. In order to verify the generalization of the proposed approach, we evaluate our proposed RF-L in form of two networks based on Attention, the strong baseline Bilstm-Attn [1] and DAN [35] based on the well-designed decoupled attention.   For fair comparison, we train the baseline using the released code[3] and resource[4] from authors under the same

---

[3] We use the code at https://github.com/clovaai/deep-text-recognition-benchmark.

[4] We use the code and pretrained parameters released at https://github.com/Wang-Tianwei/Decoupled-attention-network.

Samples
w.o RF-L
w. RF-L
vll
evil
gujara
gujarat
ugh
laugh
souris
solaris
Samples
w.o RF-L
w. RF-L
pipinang
ppinang
alibabal
alibaba
chance
change
refore
before

Fig. 5. Visualizations on STR tasks in comparison with the model *w.* and *w.o RF-L*.

Table 5. Performance of character counting with existing methods in evaluation of RMSE/relRMSE (smaller is better). The best results for each test set are in **bold**.

Methods	ACE [36] (see Footnote 5)		ACE w. RF-L		CNT		CNT w. RF-L	
	RMSE	relRMSE	RMSE	relRMSE	RMSE	relRMSE	RMSE	relRMSE
IIIT	0.477	0.169	0.323	0.133	0.300	0.128	**0.272**	**0.115**
SVT	0.963	0.361	0.890	0.326	0.455	0.165	**0.455**	**0.164**
IC03	0.555	0.206	0.509	0.192	0.372	0.147	**0.352**	**0.138**
IC13	0.518	0.193	0.502	0.188	0.275	0.107	**0.268**	**0.106**
IC15	0.889	0.364	0.896	0.361	0.614	0.261	**0.604**	**0.256**
SVTP	1.389	0.499	1.414	0.514	**0.724**	0.258	0.747	**0.256**
CT	1.001	0.443	1.200	0.442	0.854	0.420	**0.835**	**0.368**

experiment setups, without additional dataset, augmentation, rectification, stage-wise finetuning or other training tricks. As Bilstm-Attn is the solely recognition networks, we supplement the task with CNT branch as auxiliary via RF-L. The comparisons between the SOTA approaches are reported in Table 4. It shows that the strong baseline can be further promoted through multi-task without extra annotation. Bilstm-Attn w. RF-L obtains the obvious improvement on the overall 7 benchmarks, especially over 2% above the original on IC03, IC15 and CT. Visualizations are as Fig. 5. It can be seen that the missing or the redundant characters can be effectively corrected via the RF-L, *e.g.*, the word 'evil', 'alibaba'. Besides, the images with distraction like 'change' can be easier to recognize through the enhanced features. It is similar when equipped to DAN where the promotion is also obvious compared to the original one. Note that our results are only slightly overwhelmed on 2 out of 7 datasets by recent AutoSTR [40] (with $*$) and RobustScanner [39] (with $\dagger$). The former applies backbones specific for each dataset via automated network search, which relies on validation data in all 7 benchmarks. The latter uses specific pre-processing on the input and evaluate on the best of 3-sibling image group.

Character Counting. In scene text images, character counting task has not been introduced. We first build the benchmark using the same public 7 test datasets as the traditional STR tasks. Similarly, we also use the 2 synthetic datasets for training and evaluate directly (without finetuning) on these

Table 6. Performance of RCG tasks and CNT tasks, evaluated on word recognition accuracy (%) and characters counting accuracy of total images (%) respectively.

Methods	Branch		Direction		Benchmark							Avg.
	RCG	CNT	C2R	R2C	IIIT	SVT	IC03	IC13	IC15	SVTP	CT	Acc
RCG	✓				90.0	82.8	87.6	89.0	72.4	71.0	73.3	81.3
RCG w. CNT (JT-L)	✓	✓			89.6	83.9	92.6	91.7	72.6	74.0	78.1	82.4(+1.1)
RCG w. Fixed CNT (RF-L)	✓	✓	✓		90.2	86.7	92.2	91.6	73.2	76.0	79.5	82.8(+1.5)
RCG w. CNT (Unidirectional RF-L)	✓	✓	✓		90.7	86.6	92.6	91.2	73.2	76.0	80.2	82.9(+1.7)
RCG w. CNT (Bidirectional RF-L)	✓	✓	✓	✓	90.3	85.8	92.2	93.0	73.8	75.8	77.8	83.3(+2.0)
CNT		✓			92.5	93.0	96.3	95.6	84.2	85.0	85.8	89.4
CNT w. RCG (JT-L)	✓	✓			93.0	94.3	96.2	96.1	84.9	86.4	83.7	89.8(+0.4)
CNT w. Fixed RCG (RF-L)	✓	✓		✓	91.6	92.9	96.5	96.0	86.0	87.3	87.2	89.9(+0.5)
CNT w. RCG (Unidirectional RF-L)	✓	✓		✓	92.6	93.5	96.6	95.2	86.0	86.7	89.6	90.0(+0.6)
CNT w. RCG (Bidirectional RF-L)	✓	✓	✓	✓	93.5	94.0	96.7	95.7	85.5	86.7	88.9	90.3(+0.9)

non-homologous test sets. Among existing works in STR research field, no former results have been reported according to this task. Only the approach [36] has something related, partly belonging to the scope the counting tasks but differing in the evaluation. Therefore, we re-implement the ACE [36] with the release[5] and train models using the same backbone and training setups with our proposed approach. Table 5 shows the comparison, which verifies the effectiveness of our proposed CNT branch and the promotion for CNT task from the other (*i.e.*, *RCG*) branch through the proposed RF-Learning.

5 Discussions

5.1 Effectiveness of RF-Learning Framework

Mutual Promotion of Explicit and Implicit Tasks. In Table 6, given the explicit text recognition task and the implicit character counting task, we intuitively use the shared features for joint learning (denoted in *JT-L*) as a stronger baseline than the single RCG or CNT. Comparing *RCG* and all kinds of *RCG w. CNT*, the assistance from CNT is obvious, where the gain on average accuracy ranges from +1.1% (*JT-L*) to +2.0% (*Bidirectional RF-L*). It is similar in turn for the RCG's effects on the CNT task. Even the simple *JT-L* with shared backbone can help a lot. We attribute it to the additional supervision though implicit labels contained in the main task from a different optimization goal. Conversely, the assistance from RCG to CNT tasks is similar.

---

[5] As no reported character counting results, we re-implement the model using the released code at https://github.com/summerlvsong/Aggregation-Cross-Entropy.

Table 7. Performance of recognition tasks with the proposed CNT branch through reciprocal-feature learning with different reproduced recognition mainstreaming networks, including variable backbones and decoders.

Encoder	Decoder	w. CNT (RF-L)	IIIT	SVT	IC03	IC13	IC15	SVTP	CT	Avg. Gain
VGG	Bilstm-Attn		91.2	85.5	92.6	92.1	77.5	77.7	73.6	
VGG	Bilstm-Attn	✓	91.8	86.9	92.9	92.9	78.0	78.9	74.7	+0.9
ResNet	Bilstm-Attn		93.7	89.0	92.3	93.2	79.3	81.2	80.6	
ResNet	Bilstm-Attn	✓	94.1	88.6	94.9	94.5	82.4	82.0	82.6	+1.4
ResNet	CTC		91.7	85.8	91.5	91.7	74.1	73.2	76.7	
ResNet	CTC	✓	92.1	86.9	92.1	92.4	76.5	75.8	78.9	+1.5

Table 8. Performance of the proposed framework adapted to ACE in evaluation of both word recognition accuracy and character counting performance evaluated in RMSE.

Methods	Auxiliary		RCG accuracy (%)				CNT RMSE			
	CNT	RCG	IIIT	SVT	IC03	IC15	IIIT	SVT	IC03	IC15
ACE			87.5	81.8	89.9	67.5	0.477	0.963	0.555	0.889
w. RCG (RF-L)		✓	**88.4**	**83.8**	90.2	70.0	**0.323**	0.890	0.518	0.896
w. CNT (RF-L)	✓		**88.4**	83.6	**90.3**	**70.1**	0.327	**0.886**	**0.514**	**0.884**

Advantage of RF-Adaptor Module. We have chosen the RF-Adaptor structure in Sect. 4.4. Here we compare different training strategies using the RF-Adaptor. We compare 3 approaches as (1) **Fixed CNT(RF-L)**: the fusion (in RF-Adaptor) is with pre-fixed encoded features from the trained independent CNT network (in the preceding order), without end-to-end adaptive adjusting procedure, (2) **Unidirectional RF-L**: only CNT-to-RCG one-pass message flow is applied and (3) **Bidirectional RF-L**: the end-to-end bidirectional interactive training with the two branches is carried on. All of the three types achieve better results than the *CNT(JT-L)*, which verifies the design of the RF-Adaptor module is truly beneficial in combining the two-branch features. Among these, the *Bidirectional RF-L* performs above the others by a large margin, around 0.9% average gain on 7 benchmarks above the traditional joint learning and 2.0% gain compared to the original single RCG. Similar effect can be also found in the CNT task. The structure design, the bidirectional flowing and the end-to-end dual-task optimization all contribute to the effectiveness of the RF-L.

5.2 Generalization of the Proposed Approach

Backbones and Decoders. We excavate exhaustively the mutual relation between the explicit (text recognition) task and the commonly-ignored implicit (character counting) task. We regard the proposed RF-L a general mechanism, and a plug-and-in tools for different settings. For the network-unawareness, we supplement CNT branch with different kinds of backbones, including VGG [27], and ResNet [4], and different decoders including CTC [7], RNN-based Attn [28], shown in Table 7. All these variable networks, regardless of changing backbones or decoders conformably show the effectiveness and good adaptation of RF-L.

ACE-Related Task Formulations. We go deeper into the proposed framework with a far more interesting experiment. The aforementioned are from the perspective of recognition or the new-proposed character counting tasks, but we observe an intermediate zone where ACE [36] solved the sequential recognition task. Compared to the normal RCG tasks, it optimizes the loss function ignoring the ordering of character occurrence and serialize the output to the final string. While compared to CNT tasks, it pays more attention on category-aware character appearance. It also means, we can regard the ACE-based network in the perspective of auxiliary assistance to both RCG and CNT tasks. Applied in our proposed RF-L framework, we use it as the RCG branch assisted by our CNT branch, and conversely implement it as the CNT as auxiliary branch to our major RCG task. Table 8 shows that despite the role the ACE network plays in our framework, it can be enhanced, so does the other concurrent branch.

6   Conclusion

In this paper, we excavate the implicit tasks in the traditional STR, and design a two-branch reciprocal feature learning framework. Through exploiting the complementary effect between explicit and implicit tasks, *i.e.*, text recognition and character counting, the feature is reliably enhanced. Extensive experiments show the effectiveness of the proposed methods both in STR tasks and the counting tasks, and the reciprocal feature learning framework is easy to equip with variable networks and tasks. We offer abundant ablation studies, generalizing experiments with deeper understanding on the tasks.

References

1. Baek, J., et al.: What is wrong with scene text recognition model comparisons? Dataset and model analysis. In: ICCV, pp. 4714–4722 (2019)
2. Bojanowski, P., Grave, E., Joulin, A., Mikolov, T.: Enriching word vectors with subword information. Trans. Assoc. Comput. Linguist. **5**, 135–146 (2017)
3. Cao, Y., Xu, J., Lin, S., Wei, F., Hu, H.: GCNet: non-local networks meet squeeze-excitation networks and beyond. In: ICCV Workshops, pp. 1971–1980 (2019)
4. Cheng, Z., Bai, F., Xu, Y., Zheng, G., Pu, S., Zhou, S.: Focusing attention: towards accurate text recognition in natural images. In: ICCV, pp. 5076–5084 (2017)
5. Cheng, Z., Xu, Y., Bai, F., Niu, Y., Pu, S., Zhou, S.: AON: towards arbitrarily-oriented text recognition. In: CVPR, pp. 5571–5579 (2018)
6. Gao, Y., Huang, Z., Dai, Y., Xu, C., et al.: DSAN: double supervised network with attention mechanism for scene text recognition. In: VCIP, pp. 1–4 (2019)
7. Graves, A., et al.: Connectionist temporal classification: labelling unsegmented sequence data with recurrent neural networks. In: ICML, pp. 369–376 (2006)
8. Gupta, A., Vedaldi, A., Zisserman, A.: Synthetic data for text localisation in natural images. In: CVPR, pp. 2315–2324 (2016)
9. He, K., Zhang, X., Ren, S., Jian, S.: Delving deep into rectifiers: surpassing human-level performance on ImageNet classification. In: ICCV, pp. 1026–1034 (2015)
10. He, K., Zhang, X., Ren, S., Sun, J.: Deep residual learning for image recognition. In: CVPR, pp. 770–778 (2016)

11. Hu, J., Shen, L., Sun, G.: Squeeze-and-excitation networks. In: CVPR, pp. 7132–7141 (2018)
12. Hu, W., Cai, X., Hou, J., Yi, S., Lin, Z.: GTC: guided training of CTC towards efficient and accurate scene text recognition. In: AAAI, pp. 11005–11012 (2020)
13. Jaderberg, M., Simonyan, K., Vedaldi, A., et al.: Synthetic data and artificial neural networks for natural scene text recognition. CoRR abs/1406.2227 (2014)
14. Kang, L., Riba, P., Rusiñol, M., et al.: Pay attention to what you read: non-recurrent handwritten text-line recognition. CoRR abs/2005.13044 (2020)
15. Karatzas, D., et al.: ICDAR 2015 competition on robust reading. In: ICDAR, pp. 1156–1160. IEEE (2015)
16. Karatzas, D., et al.: ICDAR 2013 robust reading competition. In: ICDAR, pp. 1484–1493. IEEE (2013)
17. Kim, S., Hori, T., Watanabe, S.: Joint CTC-attention based end-to-end speech recognition using multi-task learning. In: ICASSP, pp. 4835–4839 (2017)
18. Liao, M., et al.: Scene text recognition from two-dimensional perspective. In: AAAI, pp. 8714–8721 (2019)
19. Lucas, S.M., Panaretos, A., Sosa, L., Tang, A., Wong, S., Young, R.: ICDAR 2003 robust reading competitions. In: ICDAR, pp. 682–687 (2003)
20. Luong, T., Pham, H., Manning, C.D.: Effective approaches to attention-based neural machine translation. In: EMNLP, pp. 1412–1421 (2015)
21. Mishra, A., Alahari, K., Jawahar, C.: Scene text recognition using higher order language priors. In: BMVC, pp. 1–11 (2012)
22. Mou, Y., et al.: PlugNet: degradation aware scene text recognition supervised by a pluggable super-resolution unit. In: Vedaldi, A., Bischof, H., Brox, T., Frahm, J.-M. (eds.) ECCV 2020. LNCS, vol. 12360, pp. 158–174. Springer, Cham (2020). https://doi.org/10.1007/978-3-030-58555-6_10
23. Qiao, Z., Zhou, Y., Yang, D., Zhou, Y., Wang. W.: SEED: semantics enhanced encoder-decoder framework for scene text recognition. In: CVPR, pp. 13525–13534 (2020)
24. Quy Phan, T., Shivakumara, P., Tian, S., Lim Tan, C.: Recognizing text with perspective distortion in natural scenes. In: ICCV, pp. 569–576 (2013)
25. Risnumawan, A., Shivakumara, P., Chan, C.S., Tan, C.L.: A robust arbitrary text detection system for natural scene images. Exp. Syst. Appl. **41**, 8027–8048 (2014)
26. Sheng, F., Chen, Z., Xu, B.: NRTR: a no-recurrence sequence-to-sequence model for scene text recognition. In: ICDAR, pp. 781–786 (2019)
27. Shi, B., Bai, X.: An end-to-end trainable neural network for image-based sequence recognition and its application to STR. TPAMI **39**(11), 2298–2304 (2016)
28. Shi, B., Yang, M., Wang, X., Lyu, P., Yao, C., Bai, X.: ASTER: an attentional scene text recognizer with flexible rectification. TPAMI **41**(9), 2035–2048 (2019)
29. Simonyan, K., Zisserman, A.: Very deep convolutional networks for large-scale image recognition. In: ICLR, pp. 770–778 (2015)
30. Su, B., Lu, S.: Accurate scene text recognition based on recurrent neural network. In: ACCV, pp. 35–48 (2014)
31. Wan, Z., He, M., Chen, H., Bai, X., Yao, C.: TextScanner: reading characters in order for robust scene text recognition. In: AAAI, pp. 12120–12127 (2020)
32. Wan, Z., Xie, F., Liu, Y., Bai, X., Yao, C.: 2D-CTC for scene text recognition. CoRR abs/1907.09705 (2019)
33. Wan, Z., Zhang, J., Zhang, L., Luo, J., Yao, C.: On vocabulary reliance in scene text recognition. In: CVPR, pp. 11422–11431 (2020)
34. Wang, K., Babenko, B., Belongie, S.: End-to-end scene text recognition. In: ICCV, pp. 1457–1464 (2011)

35. Wang, T., et al.: Decoupled attention network for text recognition. In: AAAI, pp. 12216–12224 (2020)
36. Xie, Z., Huang, Y., Zhu, Y., Jin, L., Liu, Y., Xie, L.: Aggregation cross-entropy for sequence recognition. In: CVPR, pp. 6538–6547 (2019)
37. Yao, C., Bai, X., Liu, W.: Strokelets: a learned multi-scale mid-level representation for scene text recognition. IEEE Trans. Image Process. **25**, 2789–2802 (2016)
38. Yu, D., Li, X., Zhang, C., Han, J., Ding, E.: Towards accurate scene text recognition with semantic reasoning networks. In: CVPR, pp. 12110–12119 (2020)
39. Yue, X., Kuang, Z., Lin, C., Sun, H., Zhang, W.: RobustScanner: dynamically enhancing positional clues for robust text recognition. In: Vedaldi, A., Bischof, H., Brox, T., Frahm, J.-M. (eds.) ECCV 2020. LNCS, vol. 12364, pp. 135–151. Springer, Cham (2020). https://doi.org/10.1007/978-3-030-58529-7_9
40. Zhang, H., Yao, Q., Yang, M., Xu, Y., Bai, X.: AutoSTR: efficient backbone search for scene text recognition. In: Vedaldi, A., Bischof, H., Brox, T., Frahm, J.-M. (eds.) ECCV 2020. LNCS, vol. 12369, pp. 751–767. Springer, Cham (2020). https://doi.org/10.1007/978-3-030-58586-0_44

Text Detection by Jointly Learning Character and Word Regions

Deyang Wu, Xingfei Hu, Zhaozhi Xie, Haiyan Li, Usman Ali,
and Hongtao Lu[✉]

Department of Computer Science and Engineering, Shanghai Jiao Tong University,
Shanghai, China
{wudeyang,huxingfei,xiezhzh,lihaiyan_2016,usmanali,htlu}@sjtu.edu.cn

Abstract. Text detection in natural scenes has developed significantly in recent years. Segmentation-based methods are widely used for text detection because they are robust to detect text of any shape. However, most state-of-the-art works are limited to word/line level detection as character-level data annotation is too expensive. Considering the close connection between characters and words, we propose a detector containing four different headers: Gaussian map, offset map, mask map, and centerline map, to obtain word-level and character-level prediction results simultaneously. Besides, we design a weakly supervised method to fully use the word-level labels of the real dataset to generate character-level pseudo-labels for training. We perform rigorous experiments on multiple benchmark datasets. Results demonstrate that our method achieves state-of-the-art results. Specifically, we achieve an F-measure of 85.2 on the dataset CTW1500, which is 1.3% higher than the state-of-the-art methods.

Keywords: Scene text detection · Weakly supervised learning · Character detection

1 Introduction

Text detection in natural scenes is a crucial issue in the field of computer vision. Enabling a computer to accurately and efficiently extract text information from images is of great significance to application scenarios such as automatic driving, map navigation, and image search. Given the appearance variability of text in natural scenes such as perspective distortion, complicated image background, uneven lighting, and low resolution, etc., it is still a challenge to detect text in natural scenes efficiently.

This paper is supported by NSFC (No. 61772330, 61876109), Shanghai Municipal Science and Technology Major Project (2021SHZDZX0102), the Shanghai Key Laboratory of Crime Scene Evidence (No. 2017XCWZK01), and the Interdisciplinary Program of Shanghai Jiao Tong University (No. YG2019QNA09).
H. Lu—Also with the MoE Key Lab of Artificial Intelligence, AI Institute, Shanghai Jiao Tong University.

J. Lladós et al. (Eds.): ICDAR 2021, LNCS 12821, pp. 304–318, 2021.
https://doi.org/10.1007/978-3-030-86549-8_20

In recent years, with the development of deep learning and computer vision, great progress has been made in text detection. One of the ideas is to treat the text directly as a special object. Inspired by general object detection algorithms such as SSD [16], Faster-RCNN [25], these methods [11,12,17,36,39] are designed by modifying the region proposal network and bounding box regression of general detectors to localize text instances directly. Such methods have achieved good results on multiple datasets. However, their performance is undermined by standard convolution's receptive field when encountering long text. Besides, due to anchor shape limitation, this type of method performs poorly on complex text shapes such as curved text.

It is observed that the text itself is composed of some similar components, which require a smaller receptive field than the entire text. Compared to detecting the entire text, some methods [20,26,30,38] only predict sub-text components and then aggregate them into a text instance. As pixels can be regarded as the most fine-grained regions, some scholars have also proposed using segmentation to classify each pixel [4,13,31–33,35,40]. By their nature, these types of component-based methods significantly reduce the dependence on the CNN receptive field and can detect text of any shape.

However, due to the high cost of labeling, most of the existing datasets for text detection do not provide character-level labeling. Both regression-based methods and segmentation-based methods can only give detection results at the word or line level. Existing methods cannot get accurate alignments between feature areas and targets when encountering low-quality images, which will lead to poor recognition results [2]. Because characters and words are closely connected, we envision that the joint supervision of both character-level and word-level detection will improve the overall performance. Therefore, to locate text regions in the image, we design a model comprising a character detection module based on Gaussian map and offset map, and a word detection module consisting of mask map and centerline map. The two components share the same backbone network. Different supervision signals enable the entire network to learn the information of text areas from different granularities. To provide character-level supervision for a dataset that only provides word-level annotations such as Total-Text [3], we design a weakly-supervised approach for pseudo label generation. Different from other methods like WeText [29], WordSup [8], CRAFT [1], our method does not need to iterate over the entire dataset to generate pseudo-labels. Instead, we use a single image for finetuning with a lower learning rate(LR). Because character-level modules (Gaussian map and offset map) and word-level modules(mask map and centerline map) share the same feature extraction net, our detector can quickly generate high-quality character-level pseudo-labels for training under the supervision of the word-level label of a single image. Our main contributions are three-fold.

1. We design a novel framework that jointly trains character and word detection modules by exploiting the relationship between characters and text regions.

2. Based on our network structure, we propose a fast weakly-supervised method for generating character-level pseudo-labels. Our method only needs a single image and word-level annotations to generate pseudo-labels quickly.
3. We conduct an extensive set of experiments on multiple datasets to prove the effectiveness of our framework. Our model outperforms existing methods on word-level or line-level prediction and additionally provides character-level detection results as well.

2   Related Work

Before the advent of deep learning, most works mainly use hand-designed features such as MSER [21], SWT [5] to extract the area containing the text.

In recent years, most of the methods are based on deep learning. Many models draw inspiration from general object detection [16,25] and semantic segmentation [7,14]. According to different method types, text detection in natural scenes can be divided into instance-level methods and component-level methods.

Instance-level methods regard words as a special object and use general object detection methods to generate rectangular boxes as the detection result. Among them, TextBoxes [12] drew on the idea of SSD [16], predicted text areas in parallel at different scales, and redesigned the size and number of anchors at different scales to better adapt to text lines of different sizes and aspect ratios. Textboxes++ [11] and DMPNet [17] directly regressed the four vertices of the text area. Textboxes++ [11] also proposed a cascaded Non-Maximum Suppression(NMS) to speed up the network. EAST [39] further simplified whole pipelines, predicting the bounding box's position directly. SPCNet [36] introduced a Re-Score module to reduce false-positive predictions. These types of methods have achieved good results on the corresponding datasets. However, limited by CNN's receptive field and the possible geometric and perspective deformation of the text in the image, instance-level methods do not perform well in detecting larger font sizes and long sentences. Besides, bounded by the algorithm's design, these methods performed poorly on the curved text.

Component-level methods decompose the entire text line into characters, text blocks, and even more fine-grained pixels. Therefore, these types of methods can naturally process text in any direction and shape. CTPN [30] used fixed-width anchors to detect text components, then applied bidirectional LSTM [27] network module to learn the context information of the text sequence. Inspired by CTPN [30] and SSD [16], SegLink [26] utilized pre-set different sizes of anchors to predict text regions on feature maps of different scales and got 8 Links to connect subregions of different layers in different directions. Zhang et al. [38] further utilized Graph Convolutional Network to connect different components. For pixel-level detection, PixelLink [4] learned the connection relationship of different pixels to determine the text instance. Wu et al. [35] made three classification predictions for each pixel: edge, text, and background. SAE [31] learned a high-dimensional vector for each pixel to distinguish the adhesion problem of different instances in the segmentation map. PSE [32] learned text areas of different size ratios and then used a breadth-first algorithm to aggregate different scales' predictions into the final detection result. PAN [33] adopted high-dimensional vectors similar to SAE [31] to distinguish different text instances. DB [13] proposed differentiable binarization to distinguish the boundary part of the text better. TextMountain [40] introduced a new representation method to define the center and border probability accurately. In character-level detection, CRAFT [1] applied Gaussian heat maps to learn the positions of characters and the connection relationship between them.

Weakly and Semi-supervised Methods. The basic idea of these methods is to initialize a model T, then apply rules to generate character-level pseudo-labels based on the results produced by T. The pseudo labels are used as additional supervision to refine the character-detector. Among them, WeText [29] used word-level annotations and a fixed threshold to filter results, WordSup [8] assumed that texts are straight lines and used an eigenvalue-based metric to measure their straightness. CRAFT [1] utilized the length of the ground truth to evaluate the quality of pseudo-labels.

3   Methodology

Figure 1 shows the overall architecture of our model. First, the image is passed through our backbone network, comprising a convolutional neural network (CNN) with feature pyramid network (FPN) [14] to extract fused feature map F. F is then fed to the character detection module and word detection module for generating word-level and character-level prediction results. The supervision for character detection is provided using the Gaussian map (G) and offset map (O), while the word detection is supervised using mask map (M) and centerline map (C). In the following subsections, we will first elaborate on our character and word detection modules, followed by a detailed description of our weakly supervised learning method.

3.1   Character Module

The character detection module learns a Gaussian map and an offset map to detect individual characters within text regions.

Gaussian Map. Inspired by the ideas of G-RMI [23], TextMountain [40], and CRAFT [1], we believe that for a single character, the amount of information contained in the middle region is greater than along the edges. Therefore, it is more reasonable to learn the character's center point than the entire character area. However, it is difficult to define the center point of a character. Instead, we use the Gaussian map to represent the character area. For a square of length H, whose center point is P_{center}, the standard Gaussian generation process is as

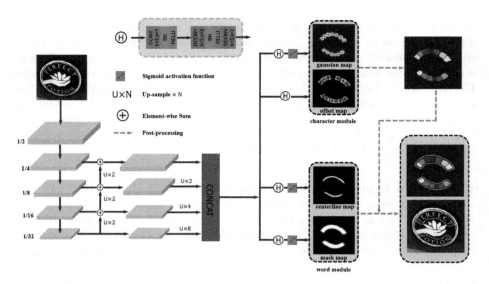

Fig. 1. The overall architecture of our model. The features from the backbone network are enhanced by FPN. After that, Gaussian map, offset map, mask map, and centerline map are generated.

follows. First, we generate a value for each $P_{x,y}$ according to Eq. 2. The value of v determines the variance of the Gaussian distribution (in the experiment, we set it to 3.34), and d in Eq. 1 represents the Euclidean distance of the point $P_{x,y}$ from P_{center}. Then we apply min-max normalization to the Gaussian map. Since characters have variable shapes (as shown in Fig. 2), we make corresponding geometric transformations to the standard Gaussian diagram to adapt to different characters. In the inference process, we use the Gaussian map to locate a single character's central area with a threshold (0.7), thereby distinguishing different characters.

$$d = \sqrt{(x - x_{center})^2 + (y - y_{center})^2}, \tag{1}$$

$$G_{(x,y)} = e^{-\frac{1}{2}\left(\frac{v \times d}{H}\right)^2} \tag{2}$$

Offset Map. Considering that it is difficult to specify a threshold on the Gaussian map to get the character's bounding box, we propose an offset map to help the network complete this task. For each pixel in the character area, the value of the offset ($\Delta x, \Delta y$) is generated, where Δx and Δy respectively represent the horizontal and vertical distance from the center of the character, as shown in Fig. 2. In the inference stage, we first use the Gaussian map to get the center area of the character and then use Δx and Δy of each pixel to determine which character it belongs to. If the calculated position is not in the character's central area, we consider the pixel belongs to the background. Since the calculation can be parallel, this module will not slow down the overall inference speed.

To train our character detection module, we optimize both the Gaussian map and the offset map using MSE loss:

$$L_g = \frac{1}{N} \sum_{i=1}^{N} (g_i - \hat{g}_i)^2,$$

(3)

$$L_o = \frac{1}{2N} \sum_{i=1}^{N} (ox_i - o\hat{x}_i)^2 + (oy_i - o\hat{y}_i)^2,$$

(4)

where L_g and L_o represent the loss of the Gaussian map and the loss of the offset map, respectively. g and $\hat{g}$ are the ground truth and the predicted value of the Gaussian map, respectively. Similarly, ox and oy refer to the ground truth values for the offset map, while $o\hat{x}$ and $o\hat{y}$ are the corresponding predicted values.

3.2   Word Module

Due to the good performance of semantic segmentation methods, the mask map is widely used in text detection [2,32,33]. We have also followed this excellent idea as the mask map can suppress false positive predictions [36]. However, simply applying the segmentation method to the text detection field will cause unnecessary adhesion. Therefore, we learn a centerline map [20] to refine the word-level instance distinction. In the inference stage, the mask map is used to eliminate the noise from other feature maps; the centerline is used to connect the character instances to generate word-level or line-level results.

For the centerline map and mask map, we find that only using dice loss [22] would make the network difficult to converge. Hence we add focal loss [15] to assist network training.

$$L_{dice}(Q, P) = 1 - \frac{2\sum_{x,y} P_{x,y} Q_{x,y}}{\sum_{x,y} P_{x,y}^2 + \sum_{x,y} Q_{x,y}^2},$$

(5)

$$L_{focal}(Q, P) = -\theta \sum_{x,y} (1 - P_{x,y})^\epsilon Q_{x,y} log(Q_{x,y})$$
$$- (1 - \theta) \sum_{x,y} P_{x,y}^\epsilon (1 - Q_{x,y}) log(1 - Q_{x,y}),$$

(6)

where L_{dice} is the dice loss and L_{focal} is the focal loss. Q and P represent ground truth and prediction respectively.

$$L_M = L_{focal}(M, \hat{M}) + L_{dice}(M, \hat{M}),$$

(7)

$$L_C = L_{focal}(C, \hat{C}) + L_{dice}(C, \hat{C}),$$

(8)

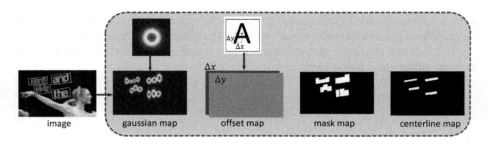

Fig. 2. Label generation. The red and the green quadrilaterals in the image represent the annotation of word-level and character-level, respectively. (Color figure online)

where L_M represents the mask loss and L_C represent the centerline loss. M and C denote the ground truth of the mask map and the centerline map respectively, while $\hat{C}$ and $\hat{M}$ are the corresponding predicted values. Based on empirical analysis, we set θ to 0.75, and ϵ to 2.0, respectively.

3.3 Label Generation on SynthText

Given a text image in SynthText [6], character-level and word-level labels are provided in the form of a set of polygons. We use character-level polygon labels to generate both the Gaussian map and the offset map, and word-level labels to generate the mask map and the centerline map, as shown in Fig. 2.

For each character polygon, we first generate a standard Gaussian distribution heat map, according to Eq. 2. Then we make a geometric transformation of the Gaussian heat map to adapt to different text shapes in the image, then we get the ground truth of Gaussian map. For the ground truth of offset map, we first find the geometric center point P_{center} based on the Gaussian map, and then calculate the horizontal and vertical distances of each pixel in the polygon from P_{center}.

The mask map is defined as a binary mask where the pixels within the polygon are labeled as 1. For the centerline map, we first use the polygon corresponding to a word to generate the start and the end of the word. Our method draws on TextSnake [20], which uses the cosine value of the unit vector corresponding to the polygon edge as the basis for determining whether an edge belongs to the start or the end. We then take 30% of the average width of the polygon as the centerline's width and shrink it at both the start and the end to avoid possible overlap with the neighboring words.

3.4 Weakly-Supervised Learning

Most of the existing datasets for text detection do not provide character-level label information. Therefore, we design a weakly-supervised method to generate character-level pseudo-labels. For a single image, even if the character module

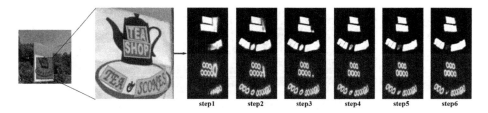

Fig. 3. Pseudo label generation using our weakly-supervised method. For each unit in step 1–step 6, the upper one represents the mask map, and the lower one represents the Gaussian map.

has no supervision signal, the quality of character prediction will be improved under the supervision of the word module.

First, we use character-level labels and word-level labels to pretrain our network on the synthetic dataset to get initial model T. We then freeze the character module (Gaussian map and offset map) of T and use only single-image word-level labels as the supervision signal to train the word module as well as the backbone network on respective dataset. We set a small LR and train only for a few epochs (we set the LR value to 5e−4 and the number of epochs to 9). The specific process and visualization results are shown in Fig. 3.

When our model trains in a weakly-supervised manner, we use the number of characters provided by the dataset to filter the generated labels to ensure the quality of pseudo-labels. Assuming that the number of ground truth characters in a specific word region is n_g, and the number of generated Gaussian kernels based on pseudo-labels is n_p. Empirically, we only use the pseudo-labels of the region for training when $|n_g - n_p|/n_g$ is in the range of (0.7, 1.3). After obtaining the character-level pseudo-labels, we train our model on the real dataset, just like on the synthetic dataset.

During inference, we first use the predicted mask map with a threshold (0.5) to filter out noise from all maps (G, O, M, C). Then we use a threshold (0.7) on G to get the central region of character instance (G_c). The offset map is used to distinguish which character the pixel belongs to in the region M. Finally, we connect the characters using C and get the final detection result.

The overall loss function is designed as a weighted sum of the Gaussian map loss L_g, the offset map loss L_o, the mask map loss L_m, and centerline map loss L_c:

$$L = \alpha \times L_g + \beta \times L_o + \gamma \times L_m + \omega \times L_c, \tag{9}$$

in our experiments, we set α, β, γ, and ω to 50, 0.001, 1, 1, respectively.

4   Experiments

We evaluate our method on ICDAR 2015, Total-Text, and CTW1500. Experiments demonstrate that, after integrating the character module and the word

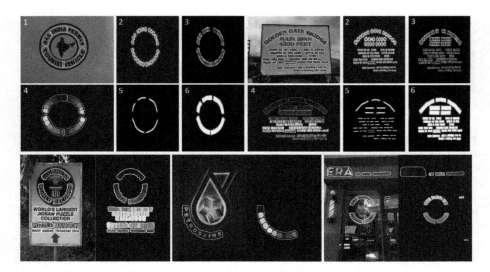

Fig. 4. Some visualization results of our detector. For each unit in the first line, 1, 2, 3, 4, 5, and 6 represent the original image, Gaussian map, offset map, final detection result, centerline map, and mask map, respectively.

module, our model obtains remarkable performance and outperforms state-of-the-art methods.

4.1 Datasets

SynthText [6] is a synthetic dataset proposed by Gupta et al. in 2016 comprising 800k images. The dataset is generated by carefully adding word instances in 8000 selected images of natural scenes. Both character-level and word-level labels are available for the text. Following other works [13,33], we use this dataset only to pre-train our model.

ICDAR2015 [9] contains 1000 training and 500 test images taken by *Google Glass* with a resolution of $720*1080$, where many images are blurry. The dataset only provides quadrilateral labeling boxes for English text, while areas containing other languages are set as 'do not care'. Only word-level labels are provided for this dataset.

Total-Text [3] focuses on images containing text areas of different shapes, including horizontal, tilted, and curved text. It is divided into 1255 training and 300 test images. Total-Text's labeling is word-level.

CTW1500 [19] focuses on curved text, including 1000 training images and 500 test images. The labeling of the dataset is line-level.

4.2 Implementation Details

We initialize our backbone ResNet50 network with weights of a pre-trained ImageNet classification model. Then, we train the whole model on SynthText for

2 epochs, learning initial features for character and word-level detection. After that, we finetune our model on the corresponding dataset for 800 epochs. Input image size is set to 640 × 640. The 'do not care' areas in all datasets are ignored during training. Specifically, 'do not care' area is multiplied by 0 before calculating loss. In actual training, we adopt the warmup strategy to ensure faster convergence.

Adam optimizer is used in the training process [10]. Data augmentation includes random rotation (−10 to 10°), random crop (640 × 640), random horizontal flip, and color change (Gaussian blur, sharpening, saturation, contrast adjustment). Furthermore, we adopt online hard example mining (OHEM) [28] to solve the class imbalance problem. Following previous methods, we report precision (P), recall (R), and F-measure (F) as evaluation metrics.

4.3   Ablation Study

We conduct rigorous ablation experiments on Total-Text. The experiments show the effectiveness of the proposed weakly supervised module and highlight the importance of jointly learning character and work-level detection. The detailed experimental results are shown in Table 1.

Table 1. Detection results with different settings. Iterations represent the number of steps to finetune the network on a single image for pseudo-labels generation.

Method	Iterations	Total-text		
		P(%)	R(%)	F(%)
Baseline	-	70.9	62.7	66.5
Baseline+Char	0	74.6	77.9	76.2
Baseline+Char	3	84.9	85.4	85.1
Baseline+Char	6	84.5	85.9	85.2
Baseline+Char	9	85.0	87.2	86.1

Baseline. We use the word module as our baseline. Specifically, we use a threshold (0.5) to filter out the mask map's negative samples directly and then use each of the connected components as a text instance. Our baseline achieves an F-measure of 66.5%.

Character-Word Joint Module. Our model has achieved a great improvement compared to directly using the word module. Even if we do not iteratively improve the label quality on a single image, the overall model still improves by 9.7%. Our model raised F-measure from 66.5% to 86.1% compared to the baseline after improving the quality of character-level labels. This proves the effectiveness of our proposed character-word joint module.

Weakly-Supervised Module. To verify our weakly-supervised module's effectiveness, we conducted different iterations on a single image to generate character-level pseudo-labels. Compared to methods that do not use weak supervision to improve character labels' quality, we use the 3rd, 6th, and 9th iterations of pseudo-labels to train the network. In the end, the F-measure increased by 8.9%, 9.0%, 9.9%, respectively. Our experiments also show that a few iterations will produce great improvements, but too many iterations will lead to worse results.

4.4   Comparison with Previous Methods

We compare our method with several state-of-the-art works on three datasets, including two benchmarks for curved text and one benchmark for multi-direction and blur text. The experimental results are shown in Table 2. Some qualitative results are visualized in Fig. 4.

Table 2. Comparative results on Total-Text, CTW1500, and ICDAR2015. * denotes that the results are collected from Liu et al. [19]. 'P', 'R', and 'F' refer to precision, recall, and F-measure, respectively.

Method	Total-Text			CTW1500			ICDAR2015		
	P(%)	R(%)	F(%)	P(%)	R(%)	F(%)	P(%)	R(%)	F(%)
CTPN(2016) [30]	-	-	-	60.4*	53.8*	56.9*	74.2	51.6	60.9
TextSnake(2018) [20]	82.7	74.5	78.4	67.9	85.3	75.6	84.9	80.4	82.6
PSE(2019) [32]	-	-	-	84.8	79.7	82.2	86.9	84.5	85.7
CRAFT(2019) [1]	87.6	79.9	83.6	86.0	81.1	83.5	89.8	84.3	86.9
SAE(2019) [31]	-	-	-	82.7	77.8	80.1	85.1	84.5	84.8
TextField (2019) [37]	81.2	79.9	80.6	83.0	79.8	81.4	84.3	83.9	84.1
Mask TTD(2019) [18]	79.1	74.5	76.7	79.7	79.0	79.4	86.6	**87.6**	**87.1**
DB(2020) [13]	87.1	82.5	84.7	86.9	80.2	83.4	88.2	82.7	85.4
TextPerceptron(2020) [24]	**88.1**	78.9	83.3	**88.7**	78.2	83.1	**91.6**	81.8	86.4
ContourNet(2020) [34]	86.9	83.9	85.4	83.7	84.1	83.9	87.6	86.1	86.9
Our method	85.0	**87.2**	**86.1**	86.2	**84.3**	**85.2**	85.8	87.2	86.5

We verify the accuracy and robustness of our method on curved text datasets Total-Text and CTW1500. For Total-Text, we achieve P: 85.0%, R: 87.2%, and F: 86.1%. Our recall value of 87.2% exceeds the best method [34] by a margin of 3.3%. Overall, our approach shows well-balanced results on P and R. Despite a higher recall rate, the accuracy of detection stays very high, resulting in the highest F-measure of 86.1%. Further inspection reveals that Total-Text contains more challenging samples, resulting in a generally lower recall rate of other methods, which indicates that our method has obvious advantages under challenging conditions. The ground truth for CTW1500 is provided as line-level annotations.

Fig. 5. Images under different brightness. Among the images on the left, the top one is our detection result; the middle one is the image after enhancing the brightness; the image below is the ground truth, where red polygon represents 'do not care'. The image on the right represents the situation of different brightness. γ is set to 0.04, 0.1, 0.4, 2.5, 5.0, 10.0, respectively.

Compared with word-level annotated datasets, it has a very large aspect ratio, making it a more challenging dataset. Our proposed method achieves P: 86.2%, R: 84.3% and F: 85.2% on CTW1500, outperforming the best method [34] by 1.3% on F-measure. This shows that our method not only performs well on word-level datasets, but also on more challenging line-level datasets. For ICDAR15, we also achieved competitive results.

4.5 Robustness Analysis

To further verify the generalization performance of our model, we directly take the model trained on Total-Text and test its performance on ICDAR2015. Our detector achieved an F-measure of 81.4%, which proves that our method generalizes well to unseen data.

Table 3. Detection results under different brightness conditions. The first column shows various values of γ in Eq. 10 to modify brightness of the image.

γ	DB [13]			Our method		
	P (%)	R (%)	F (%)	P (%)	R (%)	F (%)
0.04	79.7	29.9	43.5	68.0	60.0	63.8
0.1	84.3	58.6	69.1	78.4	74.3	76.3
0.4	87.3	77.9	82.3	83.4	82.6	83.0
2.5	87.3	79.6	83.3	84.1	83.5	83.8
5.0	86.8	68.1	76.3	81.9	76.8	79.3
10.0	82.6	44.4	57.8	68.5	55.0	61.0

We further test our model under challenging scenarios by taking an incorrectly annotated image from the Total-Text test set. As depicted in the left image of Fig. 5, the text is visible only after adjusting the brightness, which is otherwise hidden in the dark image. Furthermore, we adjust the brightness of images according to Eq. 10, where I represents the pixel's value (In the experiment, γ is set to 0.04, 0.1, 0.4, 2.5, 5.0, 10.0 respectively). We compared our detector with DB [13], and the results are shown in Table 3. The results show that our method has better performance when facing very low and very high-brightness images.

$$O = \left(\frac{I}{255}\right)^{\gamma} \times 255 \tag{10}$$

5   Conclusion

In this paper, we considered the inherent relationship between characters and words. Based on this, we proposed a new model to make full use of synthetic datasets and real datasets to generate character-level and word-level predictions. We also introduced a novel weakly-supervised method to generate the character's pseudo label. The effectiveness of our method has been demonstrated on several public benchmarks. In the meantime, our model showed good generalization performance. In future work, we consider extending our method to an end-to-end text recognition system.

References

1. Baek, Y., Lee, B., Han, D., Yun, S., Lee, H.: Character region awareness for text detection. In: Proceedings of the IEEE Conference on Computer Vision and Pattern Recognition, pp. 9365–9374 (2019)
2. Cheng, Z., Bai, F., Xu, Y., Zheng, G., Pu, S., Zhou, S.: Focusing attention: towards accurate text recognition in natural images. In: Proceedings of the IEEE International Conference on Computer Vision, pp. 5076–5084 (2017)
3. Ch'ng, C.K., Chan, C.S.: Total-text: a comprehensive dataset for scene text detection and recognition. In: 2017 14th IAPR International Conference on Document Analysis and Recognition (ICDAR), vol. 1, pp. 935–942. IEEE (2017)
4. Deng, D., Liu, H., Li, X., Cai, D.: PixelLink: detecting scene text via instance segmentation. In: Proceedings of the AAAI Conference on Artificial Intelligence, vol. 32 (2018)
5. Epshtein, B., Ofek, E., Wexler, Y.: Detecting text in natural scenes with stroke width transform. In: 2010 IEEE Computer Society Conference on Computer Vision and Pattern Recognition, pp. 2963–2970. IEEE (2010)
6. Gupta, A., Vedaldi, A., Zisserman, A.: Synthetic data for text localisation in natural images. In: Proceedings of the IEEE Conference on Computer Vision and Pattern Recognition, pp. 2315–2324 (2016)
7. He, K., Gkioxari, G., Dollár, P., Girshick, R.: Mask R-CNN. In: Proceedings of the IEEE International Conference on Computer Vision, pp. 2961–2969 (2017)
8. Hu, H., Zhang, C., Luo, Y., Wang, Y., Han, J., Ding, E.: WordSup: exploiting word annotations for character based text detection. In: Proceedings of the IEEE International Conference on Computer Vision, pp. 4940–4949 (2017)
9. Karatzas, D., et al.: ICDAR 2015 competition on robust reading. In: 2015 13th International Conference on Document Analysis and Recognition (ICDAR), pp. 1156–1160. IEEE (2015)

10. Kingma, D.P., Ba, J.: Adam: a method for stochastic optimization. arXiv preprint arXiv:1412.6980 (2014)
11. Liao, M., Shi, B., Bai, X.: TextBoxes++: a single-shot oriented scene text detector. IEEE Trans. Image Process. **27**(8), 3676–3690 (2018)
12. Liao, M., Shi, B., Bai, X., Wang, X., Liu, W.: TextBoxes: a fast text detector with a single deep neural network. In: Proceedings of the AAAI Conference on Artificial Intelligence, vol. 31 (2017)
13. Liao, M., Wan, Z., Yao, C., Chen, K., Bai, X.: Real-time scene text detection with differentiable binarization. In: AAAI, pp. 11474–11481 (2020)
14. Lin, T.Y., Dollár, P., Girshick, R., He, K., Hariharan, B., Belongie, S.: Feature pyramid networks for object detection. In: Proceedings of the IEEE Conference on Computer Vision and Pattern Recognition, pp. 2117–2125 (2017)
15. Lin, T.Y., Goyal, P., Girshick, R., He, K., Dollár, P.: Focal loss for dense object detection. In: Proceedings of the IEEE International Conference on Computer Vision, pp. 2980–2988 (2017)
16. Liu, W., et al.: SSD: single shot multibox detector. In: Leibe, B., Matas, J., Sebe, N., Welling, M. (eds.) ECCV 2016. LNCS, vol. 9905, pp. 21–37. Springer, Cham (2016). https://doi.org/10.1007/978-3-319-46448-0_2
17. Liu, Y., Jin, L.: Deep matching prior network: toward tighter multi-oriented text detection. In: Proceedings of the IEEE Conference on Computer Vision and Pattern Recognition, pp. 1962–1969 (2017)
18. Liu, Y., Jin, L., Fang, C.: Arbitrarily shaped scene text detection with a mask tightness text detector. IEEE Trans. Image Process. **29**, 2918–2930 (2019)
19. Liu, Y., Jin, L., Zhang, S., Luo, C., Zhang, S.: Curved scene text detection via transverse and longitudinal sequence connection. Pattern Recogn. **90**, 337–345 (2019)
20. Long, S., Ruan, J., Zhang, W., He, X., Wu, W., Yao, C.: TextSnake: a flexible representation for detecting text of arbitrary shapes. In: Ferrari, V., Hebert, M., Sminchisescu, C., Weiss, Y. (eds.) ECCV 2018. LNCS, vol. 11206, pp. 19–35. Springer, Cham (2018). https://doi.org/10.1007/978-3-030-01216-8_2
21. Matas, J., Chum, O., Urban, M., Pajdla, T.: Robust wide-baseline stereo from maximally stable extremal regions. Image Vis. Comput. **22**(10), 761–767 (2004)
22. Milletari, F., Navab, N., Ahmadi, S.A.: V-Net: fully convolutional neural networks for volumetric medical image segmentation. In: 2016 4th International Conference on 3D Vision (3DV), pp. 565–571. IEEE (2016)
23. Papandreou, G., et al.: Towards accurate multi-person pose estimation in the wild. In: Proceedings of the IEEE Conference on Computer Vision and Pattern Recognition, pp. 4903–4911 (2017)
24. Qiao, L., et al.: Text perceptron: towards end-to-end arbitrary-shaped text spotting. Proc. AAAI Conf. Artif. Intell. **34**, 11899–11907 (2020)
25. Ren, S., He, K., Girshick, R., Sun, J.: Faster R-CNN: towards real-time object detection with region proposal networks. IEEE Trans. Pattern Anal. Mach. Intell. **39**(6), 1137–1149 (2016)
26. Shi, B., Bai, X., Belongie, S.: Detecting oriented text in natural images by linking segments. In: Proceedings of the IEEE Conference on Computer Vision and Pattern Recognition, pp. 2550–2558 (2017)
27. Shi, X., Chen, Z., Wang, H., Yeung, D.Y., Wong, W.K., Woo, W.: Convolutional LSTM network: a machine learning approach for precipitation nowcasting. In: Advances in Neural Information Processing Systems 28, pp. 802–810 (2015)

28. Shrivastava, A., Gupta, A., Girshick, R.: Training region-based object detectors with online hard example mining. In: Proceedings of the IEEE Conference on Computer Vision and Pattern Recognition, pp. 761–769 (2016)
29. Tian, S., Lu, S., Li, C.: WeText: scene text detection under weak supervision. In: Proceedings of the IEEE International Conference on Computer Vision, pp. 1492–1500 (2017)
30. Tian, Z., Huang, W., He, T., He, P., Qiao, Yu.: Detecting text in natural image with connectionist text proposal network. In: Leibe, B., Matas, J., Sebe, N., Welling, M. (eds.) ECCV 2016. LNCS, vol. 9912, pp. 56–72. Springer, Cham (2016). https://doi.org/10.1007/978-3-319-46484-8_4
31. Tian, Z., et al.: Learning shape-aware embedding for scene text detection. In: Proceedings of the IEEE Conference on Computer Vision and Pattern Recognition, pp. 4234–4243 (2019)
32. Wang, W., et al.: Shape robust text detection with progressive scale expansion network. In: Proceedings of the IEEE Conference on Computer Vision and Pattern Recognition, pp. 9336–9345 (2019)
33. Wang, W., et al.: Efficient and accurate arbitrary-shaped text detection with pixel aggregation network. In: Proceedings of the IEEE International Conference on Computer Vision, pp. 8440–8449 (2019)
34. Wang, Y., Xie, H., Zha, Z.J., Xing, M., Fu, Z., Zhang, Y.: ContourNet: taking a further step toward accurate arbitrary-shaped scene text detection. In: Proceedings of the IEEE/CVF Conference on Computer Vision and Pattern Recognition, pp. 11753–11762 (2020)
35. Wu, Y., Natarajan, P.: Self-organized text detection with minimal post-processing via border learning. In: Proceedings of the IEEE International Conference on Computer Vision, pp. 5000–5009 (2017)
36. Xie, E., Zang, Y., Shao, S., Yu, G., Yao, C., Li, G.: Scene text detection with supervised pyramid context network. Proc. AAAI Conf. Artif. Intell. **33**, 9038–9045 (2019)
37. Xu, Y., Wang, Y., Zhou, W., Wang, Y., Yang, Z., Bai, X.: TextField: learning a deep direction field for irregular scene text detection. IEEE Trans. Image Process. **28**(11), 5566–5579 (2019)
38. Zhang, S.X., et al.: Deep relational reasoning graph network for arbitrary shape text detection. In: Proceedings of the IEEE/CVF Conference on Computer Vision and Pattern Recognition, pp. 9699–9708 (2020)
39. Zhou, X., et al.: EAST: an efficient and accurate scene text detector. In: Proceedings of the IEEE conference on Computer Vision and Pattern Recognition, pp. 5551–5560 (2017)
40. Zhu, Y., Du, J.: TextMountain: accurate scene text detection via instance segmentation. Pattern Recogn. **110**, 107336 (2020)

Vision Transformer for Fast and Efficient Scene Text Recognition

Rowel Atienza[✉][iD]

Electrical and Electronics Engineering Institute, University of the Philippines,
Quezon City, Philippines
rowel@eee.upd.edu.ph

Abstract. Scene text recognition (STR) enables computers to read text in natural scenes such as object labels, road signs and instructions. STR helps machines perform informed decisions such as what object to pick, which direction to go, and what is the next step of action. In the body of work on STR, the focus has always been on recognition accuracy. There is little emphasis placed on speed and computational efficiency which are equally important especially for energy-constrained mobile machines. In this paper we propose ViTSTR, an STR with a simple single stage model architecture built on a compute and parameter efficient vision transformer (ViT). On a comparable strong baseline method such as TRBA with accuracy of 84.3%, our small ViTSTR achieves a competitive accuracy of 82.6% (84.2% with data augmentation) at 2.4× speed up, using only 43.4% of the number of parameters and 42.2% FLOPS. The tiny version of ViTSTR achieves 80.3% accuracy (82.1% with data augmentation), at 2.5× the speed, requiring only 10.9% of the number of parameters and 11.9% FLOPS. With data augmentation, our base ViTSTR outperforms TRBA at 85.2% accuracy (83.7% without augmentation) at 2.3× the speed but requires 73.2% more parameters and 61.5% more FLOPS. In terms of trade-offs, nearly all ViTSTR configurations are at or near the frontiers to maximize accuracy, speed and computational efficiency all at the same time.

Keywords: Scene text recognition · Transformer · Data augmentation

1 Introduction

STR plays a vital role for machines to understand the human environment. We invented text to convey information through labels, signs, instructions and announcements. Therefore, for a computer to take advantage of this visual cue, it must also understand text in natural scenes. For instance, a "Push" signage on a door tells a robot to push it to open. In the kitchen, a label with "Sugar" means that the container has sugar in it. A wearable system that can read "50" or "FIFTY" on a paper bill can greatly enhance the lives of visually impaired people.

© Springer Nature Switzerland AG 2021
J. Lladós et al. (Eds.): ICDAR 2021, LNCS 12821, pp. 319–334, 2021.
https://doi.org/10.1007/978-3-030-86549-8_21

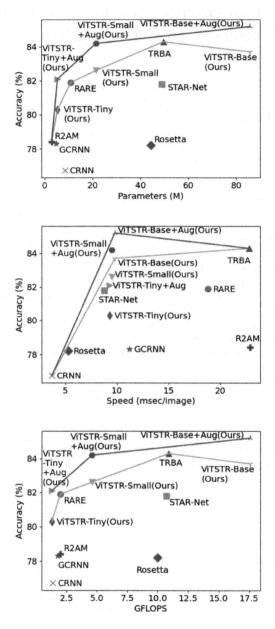

Fig. 1. Trade-offs between accuracy vs number of parameters, speed and computational load (FLOPS). +Aug uses data augmentation. Almost all versions of ViTSTR are at or near the frontiers to maximize the performance on all metrics. The slope of the line is the accuracy gain as the number of parameters, speed or FLOPS increases. The steeper the slope, the better. Teal line includes ViTSTR with data augmentation.

STR is related but different from the more developed field of Optical Character Recognition (OCR). In OCR, symbols on a printed front facing document are detected and recognized. In a way, OCR operates in a more structured setting. Meanwhile, the objective of STR is to recognize symbols in varied unconstrained settings such as walls, signboards, product labels, road signs, markers, etc. Therefore, the inputs have many degrees of variation in font style, orientation, shape, size, color, texture and illumination. The inputs are also subject to camera sensor orientation, location and imperfections causing image blur, pixelation, noise, and geometric and radial distortions. Weather disturbances such as glare, shadow, rain, snow and frost can also greatly affect the performance of STR.

In the body of work on STR, the emphasis has always been on accuracy with little attention paid to speed and computing requirements. In this work, we attempt to put balance on accuracy, speed and efficiency. Accuracy refers to the correctness of recognized text. Speed is measured by how many text images are processed per unit time. Efficiency can be approximated by the number of parameters and computations (e.g. FLOPS) required to process one image. The number of parameters reflects the memory requirements while FLOPS estimates the number of instructions needed to complete a task. An ideal STR is accurate and fast while requiring only little computing resources.

In the quest to beat the SOTA, most models are zeroing on accuracy with inadequate discussion on the trade off. In order to instill balance on the importance of accuracy, speed and efficiency, we propose to take advantage of the simplicity and efficiency of vision transformers (ViT) [7] such as Data-efficient image Transformer (DeiT) [34]. ViT demonstrated that SOTA results in ImageNet [28] recognition can be achieved using a transformer [35] encoder only. ViT inherited all the properties of a transformer including its speed and computational efficiency. Using the model weights of DeiT which is simply a ViT trained by knowledge distillation [13] for better performance, we built an STR that can be trained end-to-end. This resulted to a simple single stage model architecture that is able to maximize accuracy, speed and computational performance. The tiny version of our ViTSTR achieves 80.3% accuracy (82.1% with data augmentation), is fast at 9.3 msec/image, with a small footprint of 5.4M parameters and requires much less computations at 1.3 Giga FLOPS. The small version of ViTSTR achieves a higher accuracy of 82.6% (84.2% with data augmentation), is also fast at 9.5 msec/image while requiring 21.5M parameters and 4.6 Giga FLOPS. With data augmentation, the base version of ViTSTR achieves 85.2% accuracy (83.7% no augmentation) at 9.8 msec/image but requires 85.8M parameters and 17.6 Giga FLOPS. We adopted the reference *tiny*, *small* and *base* to indicate which ViT/DeiT transformer encoder was used in ViTSTR. As shown in Fig. 1, almost all versions of our proposed ViTSTR are at or near the frontiers of accuracy vs speed, memory, and computational load indicating optimal trade-offs. To encourage reproducibility, the code of ViTSTR is available at https://github.com/roatienza/deep-text-recognition-benchmark.

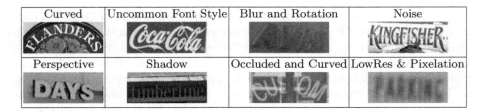

Curved	Uncommon Font Style	Blur and Rotation	Noise
Perspective	Shadow	Occluded and Curved	LowRes & Pixelation

Fig. 2. Different variations of text encountered in natural scenes

2   Related Work

For machines, reading text in the human environment is a challenging task due to different possible appearances of symbols. Figure 2 shows examples of text in the wild affected by curvature, font style, blur, rotation, noise, geometry, illumination, occlusion and resolution. There are many other factors that could affect text images such as weather condition, camera sensor imperfection, motion, lighting, etc.

Reading text in natural scenes generally requires two stages: 1) text detection and 2) text recognition. Detection determines the bounding box of the region where text can be found. Once the region is known, text recognition reads the symbols in the image. Ideally, a method is able to do both at the same time. However, the performance of SOTA end-to-end text reading models is still far from modern-day OCR systems and remains an open problem [5]. In this work, our focus is on text recognition of 96 Latin characters (i.e. 0–9, a-Z, etc.).

STR identifies each character of a text in an image in the correct sequence. Unlike object recognition where usually there is only one category of object, there may be zero or more characters for a given text image. Thus, STR models are more complex. Similar to many vision problems, early methods [24,38] used hand-crafted features resulting to poor performance. Deep learning has dramatically advanced the field of STR. In 2019, Baek *et al.* [1] presented a framework that models the design patterns of modern STR. Figure 3 shows the four stages or modules of STR. Broadly speaking, even recently proposed methods such as transformer-based models, No-Recurrence sequence-to-sequence Text Recognizer (NRTR) [29] and Self-Attention Text Recognition Network (SATRN) [18] can fit into **Rectification-Feature Extraction (Backbone)-Sequence Modelling-Prediction** framework.

The Rectification stage removes the distortion from the word image so that the text is horizontal or normalized. This makes it easier for Feature Extraction (Backbone) module to determine invariant features. Thin-Plate-Spline (TPS) [3] models the distortion by finding and correcting fiducial points. RARE (Robust-text recognizer with Automatic REctification) [31], STAR-Net (SpaTial Attention Residue Network) [21], and TRBA (TPS-ResNet-BiLSTM-Attention) [1] use TPS. ESIR (End-to-end trainable Scene text Recognition) [41] employs an iterative rectification network that significantly boosts the performance of text recognition models. In some cases, no rectification is employed such as in CRNN

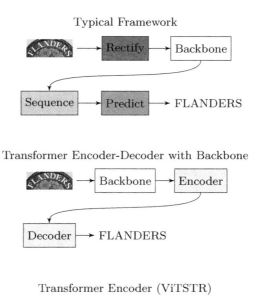

Fig. 3. STR design patterns. Our proposed model, ViTSTR, has the simplest architecture with just one stage.

(Convolutional Recurrent Neural Network) [30], R2AM (Recursive Recurrent neural networks with Attention Modeling) [17], GCRNN (Gated Recurrent Convolution Neural Network) [36] and Rosetta [4].

The role of Feature Extraction (Backbone) stage is to automatically determine the invariant features of each character symbol. STR uses the same feature extractors in object recognition tasks such as VGG [32], ResNet [11], and a variant of CNN called RCNN [17]. Rosetta, STAR-Net and TRBA use ResNet. RARE and CRNN extract features using VGG. R2AM and GCRNN build on RCNN. Transformer-based models NRTR and SATRN use customized CNN blocks to extract features for transformer encoder-decoder text recognition.

Since STR is a multi-class sequence prediction, there is a need to remember long-term dependency. The role of Sequence modelling such as BiLSTM is to make a consistent context between the current character features and the past/future characters features. CRNN, GRCNN, RARE, STAR-Net and TRBA use BiLSTM. Other models such as Rosetta and R2AM do not employ sequence modelling to speed up prediction.

The Prediction stage examines the features resulting from the Backbone or Sequence modelling to arrive at a sequence of characters prediction. CTC (Connectionist Temporal Classification) [8] maximizes the likelihood of an output sequence by efficiently summing over all possible input-output sequence alignments [5].

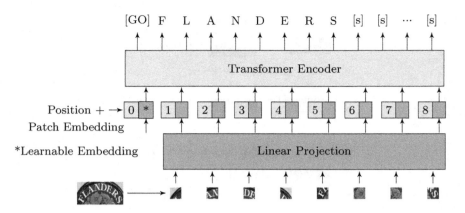

Fig. 4. Network architecture of ViTSTR. An input image is first converted into patches. The patches are converted into 1D vector embeddings (flattened 2D patches). As input to the encoder, a learnable patch embedding is added together with a position encoding for each embedding. The network is trained end-to-end to predict a sequence of characters. [GO] is a pre-defined start of sequence symbol while [s] represents a space or end of a character sequence.

Alternative to CTC is Attention Mechanism [2] that learns the alignment between the image features and symbols. CRNN, GRCNN, Rosetta and STAR-Net use CTC. R2AM, RARE and TRBA are Attention-based.

Like in natural language processing (NLP), transformers overcome sequence modelling and prediction by doing parallel self-attention and prediction. This resulted to a fast and efficient model. As shown in Fig. 3, current transformer-based STR models still require a Backbone and a Transformer Encoder-Decoder. Recently, ViT [7] proved that it is possible to beat the performance of deep networks such as ResNet [11] and EfficientNet [33] on ImageNet1k [28] classification by using the transformer encoder only but pre-training it on very large datasets such as ImageNet21k and JFT-300M. DeiT [34] demonstrated that ViT does not need a large dataset and can even achieve better results but it must be trained using knowledge distillation [13]. ViT, using pre-trained weights of DeiT, is the basis of our proposed fast and efficient STR called ViTSTR. As shown in Fig. 3, ViTSTR is a very simple model with just one stage that can easily halve the number of parameters and FLOPS of a transformer-based STR.

3   Vision Transformer for STR

Figure 4 shows the model architecture of ViTSTR in detail. The only difference between ViT and ViTSTR is the prediction head. Instead of single object-class recognition, ViTSTR must identify multiple characters with the correct sequence order and length. The prediction is done in parallel.

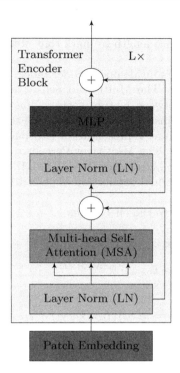

Fig. 5. A transformer encoder is a stack of L identical encoder blocks.

The ViT model architecture is similar to the original transformer by Vaswani *et al.* [35]. The difference is only the encoder part is utilized. The original transformer was designed for NLP tasks. Instead of word embeddings, each input image $\mathbf{x} \in \mathbb{R}^{H \times W \times C}$ is reshaped into a sequence of flattened 2D patches $\mathbf{x}^p \in \mathbb{R}^{N \times P^2 C}$. The image dimension is $H \times W$ with C channels while the patch dimension is $P \times P$. The resulting patch sequence length is N. The transformer encoder uses a constant width D for embedding and features in all its layers. To match this size, each flattened patch is converted to an embedding of size D via linear projection. This is shown as small boxes with teal color in Fig. 4.

A learnable class embedding of the same dimension D is prepended with the sequence. A unique position encoding of the same dimension D is added to each embedding. The resulting vector sum is the input to the encoder. In ViTSTR, a learnable position encoding is used.

In the original ViT, the output vector corresponding to the learnable class embedding is used for object category prediction. In ViTSTR, this corresponds to the [GO] token. Furthermore, instead of just extracting one output vector, we extract multiple feature vectors from the encoder. The number is equal to the maximum length of text in our dataset plus two for the [GO] and [s] tokens. We use the [GO] token to mark the beginning of the text prediction and [s] to indicate the end or a space. [s] is repeated at the end of each text prediction up to the maximum sequence length to mark that nothing follows after the text characters.

Figure 5 shows the layers inside one encoder block. Every input goes through Layer Normalization (LN). The Multi-head Self-Attention layer (MSA) determines the relationships between feature vectors. Vaswani *et al.* [35] found out that using multiple heads instead of just one allows the model to jointly attend to information from different representation subspaces at different positions. The number of heads is H. The Multilayer Perceptron (MLP) performs feature extraction. Its input is also layer normalized. The MLP is made of 2 layers with GELU activation [12]. Residual connection is placed between the output of LN and MSA/MLP.

In summary, the input to the encoder is:

$$\mathbf{z}_0 = [\mathbf{x}_{class}; \mathbf{x}_p^1 \mathbf{E}; \mathbf{x}_p^2 \mathbf{E}; ...; \mathbf{x}_p^N \mathbf{E}] + \mathbf{E}_{pos}, \tag{1}$$

where $\mathbf{E} \in \mathbb{R}^{P^2 C \times D}$ and $\mathbf{E}_{pos} \in \mathbb{R}^{(N+1) \times D}$.
The output of MSA block is:

$$\mathbf{z}_l' = MSA(LN(\mathbf{z}_{l-1})) + \mathbf{z}_{l-1}, \tag{2}$$

for $l = 1...L$. L is the depth or the number of encoder blocks. A transformer encoder is made of a stack of L encoder blocks.

The output of the MLP block is:

$$\mathbf{z}_l = MLP(LN(\mathbf{z}_l')) + \mathbf{z}_l', \tag{3}$$

for $l = 1...L$.

Finally, the head is made of a sequence of linear projections forming the word prediction:

$$\mathbf{y}_i = Linear(\mathbf{z}_L^i), \tag{4}$$

for $i = 1...S$. S is the maximum text length plus two for [GO] and [s] tokens. Table 1 summarizes the ViTSTR configurations.

Table 1. ViTSTR configurations

ViTSTR Version	Patch size P	Depth L	Embedding size D	No. of Heads H	Seq. length S
Tiny	16	12	192	3	27
Small	16	12	384	6	27
Base	16	12	768	12	27

4   Experimental Results and Discussion

In order to evaluate different strong baseline STR methods, we used the framework developed by Baek *et al.* [1]. A unified framework is important in order to arrive at a fair evaluation of different models. A unified framework ensures consistent train and test conditions are used in the evaluation. Following discussion describes the train and test datasets which have been the point of contention in performance comparisons. Using different train and test datasets can heavily tilt in favor or against a certain performance reporting.

After discussing the train and test datasets, we present the evaluation and analysis across different models using the unified framework.

4.1   Train Dataset

Fig. 6. Samples from datasets with synthetic images.

Due to the lack of a big dataset of real data, the practice in STR model training is to use synthetic data. Two popular datasets are used: 1) MJSynth (MJ) [14] or also known as Synth90k and 2) SynthText (ST) [9].

MJSynth (MJ) is a synthetically generated dataset made of 8.9M realistically looking words images. MJSynth was designed to have 3 layers: 1) background, 2) foreground and 3) optional shadow/border. It uses 1,400 different fonts. The font kerning, weight, underline and other properties are varied. MJSynth also utilizes different background effects, border/shadow rendering, base coloring, projective distortion, natural image blending and noise.

SynthText (ST) is another synthetically generated dataset made of 5.5M word images. SynthText was generated by blending synthetic text on natural images. It uses the scene geometry, texture, and surface normal to naturally blend and distort a text rendering on the surface of an object within the image. Similar to MJSynth, SynthText uses random fonts for its text. The word images were cropped from the natural images embedded with synthetic text.

In the STR framework, each dataset contributes 50% to the total train dataset. Combining 100% of both datasets resulted to performance deterioration [1]. Figure 6 shows sample images from MJ and ST.

4.2 Test Dataset

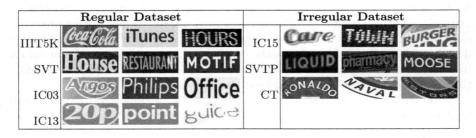

Fig. 7. Samples from datasets with real images.

The test dataset is made of several small publicly available STR datasets of text in natural images. These datasets are generally group into two: 1) Regular and 2) Irregular.

The regular datasets have text images that are frontal, horizontal and have minimal amount of distortion. IIIT5K-Words [23], Street View Text (SVT) [37], ICDAR2003 (IC03) [22] and ICDAR2013 (IC13) [16] are considered regular datasets. Meanwhile, irregular datasets contain text with challenging appearances such curved, vertical, perspective, low-resolution or distorted. ICDAR2015 (IC15) [15], SVT Perspective (SVTP) [25] and CUTE80 (CT) [27] belong to irregular datasets. Figure 7 shows samples from regular and irregular datasets. For both datasets, only the test splits are used for the evaluation.

Table 2. Train conditions

Train dataset: 50%MJ + 50%ST	Batch size: 192
Epochs: 300	Parameter initialization: He [10]
Optimizer: Adadelta [40]	Learning rate: 1.0
Adadelta ρ: 0.95	Adadelta ϵ: $1e^{-8}$
Loss: Cross-Entropy/CTC	Gradient clipping: 5.0
Image size: 100×32	Channels: 1 (grayscale)

Regular Dataset

- **IIIT5K** contains 3,000 images for testing. The images are mostly from street scenes such as sign board, brand logo, house number or street sign.
- **SVT** has 647 images for testing. The text images are cropped from Google Street View images.

- **IC03** contains 1,110 test images from ICDAR2003 Robust Reading Competition. Images were captured from natural scenes. After removing words that are less than 3 characters in length, the result is 860 images. However, 7 additional images were found to be missing. Hence, the framework also contains the 867 test images version.
- **IC13** is an extension of IC03 and shares similar images. IC13 was created for the ICDAR2013 Robust Reading Competition. In the literature and in the framework, two versions of the test dataset are used: 1) 857 and 2) 1,015.

Irregular Dataset

- **IC15** has text images for the ICDAR2015 Robust Reading Competition. Many images are blurry, noisy, rotated, and sometimes of low-resolution since these were captured using Google Glasses with the wearer undergoing unconstrained motion. Two versions are used in the literature and in the framework: 1) 1,811 and 2) 2,077 images. The 2,077 version contains rotated, vertical, perspective-shifted and curved images.
- **SVTP** has 645 test images from Google Street View. Most are images of business signage.
- **CT** focuses on curved text images captured from shirts and product logos. The dataset has 288 images.

4.3 Experimental Setup

The recommended training configurations in the framework are listed in Table 2. We reproduced the results of several strong baseline models: CRNN, R2AM, GCRNN, Rosetta, RARE, STAR-Net and TRBA for a fair comparison with ViT-STR. We trained all models for at least 5 times using different random seeds. The best performing weights on the test datasets are saved to get the mean evaluation scores.

For ViTSTR, we used the same train configurations except that the input is resized to 224×224 to match the dimension of the pre-trained DeiT [34]. The pre-trained weights file of DeiT is automatically downloaded before training ViTSTR. ViTSTR can be trained end-to-end with no parameters frozen.

Tables 3 and 4 show the performance scores of different models. We report the accuracy, speed, number of parameters and FLOPS to get the overall picture of trade-offs as shown in Fig. 1. For accuracy, we follow the framework evaluation protocol in most STR models of case sensitive training and case insensitive evaluation. For speed, the reported numbers are based on model run time on a 2080Ti GPU. Unlike in other model benchmarks such as in [19,20], we do not rotate vertical text images (e.g. Table 5 IC15) before evaluation.

4.4 Data Augmentation

Using a recipe of data augmentation specifically targeted for STR can significantly boost the accuracy of ViTSTR. In Fig. 8, we can see how different data

Table 3. Model accuracy. Bold: highest for all, Underscore: highest no augmentation.

Model	IIIT	SVT	IC03		IC13		IC15		SVTP	CT	Acc. %	Std.
	3000	647	860	867	857	1015	1811	2077	645	288		
CRNN [30]	81.8	80.1	91.7	91.5	89.4	88.4	65.3	60.4	65.9	61.5	76.7	0.3
R2AM [17]	83.1	80.9	91.6	91.2	90.1	88.1	68.5	63.3	70.4	64.6	78.4	0.9
GCRNN [36]	82.9	81.1	92.7	92.3	90.0	88.4	68.1	62.9	68.5	65.5	78.3	0.1
Rosetta [4]	82.5	82.8	92.6	91.8	90.3	88.7	68.1	62.9	70.3	65.5	78.4	0.4
RARE [31]	86.0	85.4	93.5	93.4	92.3	91.0	73.9	68.3	75.4	71.0	82.1	0.3
STAR-Net [21]	85.2	84.7	93.4	93.0	91.2	90.5	74.5	68.7	74.7	69.2	81.8	0.1
TRBA [1]	<u>87.8</u>	<u>87.6</u>	<u>94.5</u>	<u>94.2</u>	**93.4**	<u>92.1</u>	<u>77.4</u>	<u>71.7</u>	78.1	<u>75.2</u>	<u>84.3</u>	0.1
ViTSTR-Tiny	83.7	83.2	92.8	92.5	90.8	89.3	72.0	66.4	74.5	65.0	80.3	0.2
ViTSTR-Tiny+Aug	85.1	85.0	93.4	93.2	90.9	89.7	74.7	68.9	78.3	74.2	82.1	0.1
ViTSTR-Small	85.6	85.3	93.9	93.6	91.7	90.6	75.3	69.5	78.1	71.3	82.6	0.3
ViTSTR-Small+Aug	86.6	87.3	94.2	94.2	92.1	91.2	77.9	71.7	81.4	77.9	84.2	0.1
ViTSTR-Base	86.9	87.2	93.8	93.4	92.1	91.3	76.8	71.1	<u>80.0</u>	74.7	83.7	0.1
ViTSTR-Base+Aug	**88.4**	**87.7**	**94.7**	**94.3**	93.2	**92.4**	**78.5**	**72.6**	**81.8**	**81.3**	**85.2**	0.1

Table 4. Model accuracy, speed, and computational requirements on a 2080Ti GPU.

Model	Accuracy %	Speed msec/image	Parameters 1×10^6	FLOPS 1×10^9
CRNN [30]	76.7	3.7	8.5	1.4
R2AM [17]	78.4	22.9	2.9	2.0
GRCNN [36]	78.3	11.2	4.8	1.8
Rosetta [4]	78.4	5.3	44.3	10.1
RARE [31]	82.1	18.8	10.8	2.0
STAR-Net [21]	81.8	8.8	48.9	10.7
TRBA [1]	84.3	22.8	49.6	10.9
ViTSTR-Tiny	80.3	9.3	5.4	1.3
ViTSTR-Tiny+Aug	82.1	9.3	5.4	1.3
ViTSTR-Small	82.6	9.5	21.5	4.6
ViTSTR-Small+Aug	84.2	9.5	21.5	4.6
ViTSTR-Base	83.7	9.8	85.8	17.6
ViTSTR-Base+Aug	85.2	9.8	85.8	17.6

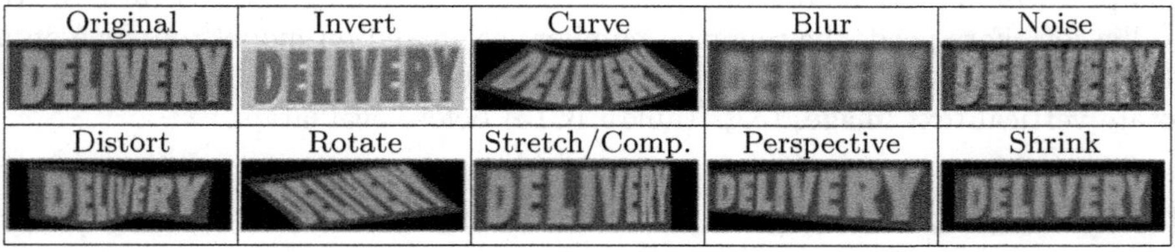

Fig. 8. Illustration of data augmented text images designed for STR.

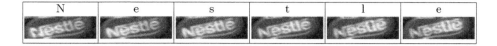

N	e	s	t	l	e

Fig. 9. ViTSTR attention as it reads out **Nestle** text image.

augmentations alter the image but not the meaning of text within. Table 3 shows that applying RandAugment [6] on different image transformations such as inversion, curving, blur, noise, distortion, rotation, stretching/compressing, perspective, and shrinking improved the generalization of ViTSTR-Tiny by +1.8%, ViTSTR-Small by +1.6% and ViTSTR-Base by 1.5%. The biggest increase in accuracy is on irregular datasets such as CT (+9.2% tiny, +6.6% small and base), SVTP (+3.8% tiny, +3.3% small, +1.8% base), IC15 1,811 (+2.7% tiny, +2.6% small, +1.7% base) and IC15 2,077 (+2.5% tiny, +2.2% small, +1.5% base).

Table 5. ViTSTR sample failed prediction from each test dataset. From first to last row: input image, ground truth, prediction, dataset. Wrong symbol prediction in underline.

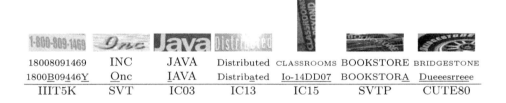

18008091469	INC	JAVA	Distributed	CLASSROOMS	BOOKSTORE	BRIDGESTONE
1800B09446Y	Onc	IAVA	Distribated	Io-14DD07	BOOKSTORA	Dueeesrreee
IIIT5K	SVT	IC03	IC13	IC15	SVTP	CUTE80

4.5 Attention

Figure 9 shows the attention map of ViTSTR as it reads out a text image. While the attention is properly focused on each character, ViTSTR also pays attention to neighboring characters. Perhaps, a context is placed during individual symbol prediction.

4.6 Performance Penalty

Every time a stage in an STR model is added, there is a gain in accuracy but at a cost of slower speed and bigger computational requirements. For example, RARE↪TRBA increases the accuracy by 2.2% but requires 38.8M more parameters and slows down the task completion by 4 msec/image. Replacing the CTC stage by Attention like in STAR-Net↪TRBA significantly slows down the computation from 8.8 msec/image to 22.8 msec/image to gain an additional

2.5% in accuracy. In fact, the slowdown due to change from CTC to Attention is > 10× as compared to adding BiLSTM or TPS in the pipeline. In ViTSTR, the transition from tiny to small version requires an increase in embedding size and number of heads. No additional stage is necessary. The performance penalty to gain 2.3% in accuracy is increase in number of parameters by 16.1M. From tiny to base, the performance penalty to gain 3.4% in accuracy is additional 80.4M parameters. In both cases, the speed barely changed since we use the same parallel tensor dot product, softmax and addition operations in MLP and MSA layers of the transformer encoder. Only the tensor dimension is increased resulting to a minimal 0.2 to 0.3 msec/image slowdown in task completion. Unlike in multi-stage STR, an additional module requires additional sequential layers of forward propagation which can not be parallelized resulting into a significant performance penalty.

4.7   Failure Cases

Table 5 shows sample failed predictions by ViTSTR-Small from each test dataset. The main causes of wrong prediction are confusion between similar symbols (e.g. 8 and B, J and I), scripted font (e.g. I in Inc), glare on a character, vertical text, heavily curved text image and partially occluded symbol. Note that in some of these cases, even a human reader can easily make a mistake. However, humans use semantics to resolve ambiguities. Semantics has been used in recent STR methods [26, 39].

5   Conclusion

ViTSTR is a simple single stage model architecture that emphasizes balance in accuracy, speed and computational requirements. With data augmentation targeted for STR, ViTSTR can significantly increase the accuracy especially for irregular datasets. When scaled up, ViTSTR stays at the frontiers to balance accuracy, speed and computational requirements.

Acknowledgements. This work was funded by the University of the Philippines ECWRG 2019–2020. GPU machines have been supported by CHED-PCARI AIRSCAN Project and Samsung R&D PH. Special thanks to the people of Computer Networks Laboratory: Roel Ocampo, Vladimir Zurbano, Lope Beltran II, and John Robert Mendoza, who worked tirelessly during the pandemic to ensure that our network and servers are continuously running.

References

1. Baek, J., et al.: What is wrong with scene text recognition model comparisons? dataset and model analysis. In: ICCV, pp. 4715–4723 (2019)
2. Bahdanau, D., Cho, K., Bengio, Y.: Neural machine translation by jointly learning to align and translate. arXiv preprint arXiv:1409.0473 (2014)

3. Bookstein, F.L.: Principal warps: thin-plate splines and the decomposition of deformations. Trans. Pattern Anal. Mach. Intell. **11**(6), 567–585 (1989)

4. Borisyuk, F., Gordo, A., Sivakumar, V.: Rosetta: large scale system for text detection and recognition in images. In: International Conference on Knowledge Discovery & Data Mining, pp. 71–79 (2018)

5. Chen, X., Jin, L., Zhu, Y., Luo, C., Wang, T.: Text recognition in the wild: A survey. arXiv preprint arXiv:2005.03492 (2020)

6. Cubuk, E.D., Zoph, B., Shlens, J., Le, Q.V.: Randaugment: practical automated data augmentation with a reduced search space. In: Proceedings of the IEEE/CVF Conference on Computer Vision and Pattern Recognition Workshops, pp. 702–703 (2020)

7. Dosovitskiy, A., et al.: An image is worth 16x16 words: transformers for image recognition at scale. In: ICLR (2020)

8. Graves, A., Fernández, S., Gomez, F., Schmidhuber, J.: Connectionist temporal classification: labelling unsegmented sequence data with recurrent neural networks. In: ICML, pp. 369–376 (2006)

9. Gupta, A., Vedaldi, A., Zisserman, A.: Synthetic data for text localisation in natural images. In: CVPR, pp. 2315–2324 (2016)

10. He, K., Zhang, X., Ren, S., Sun, J.: Delving deep into rectifiers: surpassing human-level performance on imagenet classification. In: ICCV, pp. 1026–1034 (2015)

11. He, K., Zhang, X., Ren, S., Sun, J.: Deep residual learning for image recognition. In: CVPR, pp. 770–778 (2016)

12. Hendrycks, D., Gimpel, K.: Gaussian error linear units (gelus). arXiv preprint arXiv:1606.08415 (2016)

13. Hinton, G., Vinyals, O., Dean, J.: Distilling the knowledge in a neural network. arXiv preprint arXiv:1503.02531 (2015)

14. Jaderberg, M., Simonyan, K., Vedaldi, A., Zisserman, A.: Synthetic data and artificial neural networks for natural scene text recognition. In: NIPS Workshop on Deep Learning (2014)

15. Karatzas, D., et al.: Icdar 2015 competition on robust reading. In: ICDAR, pp. 1156–1160. IEEE (2015)

16. Karatzas, D., et al.: Icdar 2013 robust reading competition. In: ICDAR, pp. 1484–1493. IEEE (2013)

17. Lee, C.Y., Osindero, S.: Recursive recurrent nets with attention modeling for ocr in the wild. In: CVPR, pp. 2231–2239 (2016)

18. Lee, J., Park, S., Baek, J., Oh, S.J., Kim, S., Lee, H.: On recognizing texts of arbitrary shapes with 2d self-attention. In: CVPR Workshops, pp. 546–547 (2020)

19. Li, H., Wang, P., Shen, C., Zhang, G.: Show, attend and read: a simple and strong baseline for irregular text recognition. AAAI **33**, 8610–8617 (2019)

20. Litman, R., Anschel, O., Tsiper, S., Litman, R., Mazor, S., Manmatha, R.: Scatter: selective context attentional scene text recognizer. In: CVPR, pp. 11962–11972 (2020)

21. Liu, W., Chen, C., Wong, K.Y.K., Su, Z., Han, J.: Star-net: a spatial attention residue network for scene text recognition. In: BMVC, vol. 2, p. 7 (2016)

22. Lucas, S.M., et al.: Icdar 2003 robust reading competitions: entries, results, and future directions. Int. J. Doc. Anal. Recogn. **7**(2–3), 105–122 (2005)

23. Mishra, A., Alahari, K., Jawahar, C.: Scene text recognition using higher order language priors. In: BMVC. BMVA (2012)

24. Neumann, L., Matas, J.: Real-time scene text localization and recognition. In: CVPR, pp. 3538–3545. IEEE (2012)

25. Phan, T.Q., Shivakumara, P., Tian, S., Tan, C.L.: Recognizing text with perspective distortion in natural scenes. In: ICCV, pp. 569–576 (2013)
26. Qiao, Z., Zhou, Y., Yang, D., Zhou, Y., Wang, W.: Seed: semantics enhanced encoder-decoder framework for scene text recognition. In: CVPR, pp. 13528–13537 (2020)
27. Risnumawan, A., Shivakumara, P., Chan, C.S., Tan, C.L.: A robust arbitrary text detection system for natural scene images. Expert Syst. Appl. **41**(18), 8027–8048 (2014)
28. Russakovsky, O., et al.: Imagenet large scale visual recognition challenge. Int. J. Comput. Vision **115**(3), 211–252 (2015)
29. Sheng, F., Chen, Z., Xu, B.: Nrtr: a no-recurrence sequence-to-sequence model for scene text recognition. In: ICDAR, pp. 781–786. IEEE (2019)
30. Shi, B., Bai, X., Yao, C.: An end-to-end trainable neural network for image-based sequence recognition and its application to scene text recognition. Trans. Pattern Anal. Mach. Intell. **39**(11), 2298–2304 (2016)
31. Shi, B., Wang, X., Lyu, P., Yao, C., Bai, X.: Robust scene text recognition with automatic rectification. In: CVPR, pp. 4168–4176 (2016)
32. Simonyan, K., Zisserman, A.: Very deep convolutional networks for large-scale image recognition. In: ICLR (2015)
33. Tan, M., Le, Q.: Efficientnet: rethinking model scaling for convolutional neural networks. In: ICML, pp. 6105–6114. PMLR (2019)
34. Touvron, H., Cord, M., Douze, M., Massa, F., Sablayrolles, A., Jégou, H.: Training data-efficient image transformers & distillation through attention. arXiv preprint arXiv:2012.12877 (2020)
35. Vaswani, A., et al.: Attention is all you need. In: NeuRIPS, pp. 6000–6010 (2017)
36. Wang, J., Hu, X.: Gated recurrent convolution neural network for ocr. In: NeuRIPS, pp. 334–343 (2017)
37. Wang, K., Babenko, B., Belongie, S.: End-to-end scene text recognition. In: ICCV, pp. 1457–1464. IEEE (2011)
38. Yao, C., Bai, X., Liu, W.: A unified framework for multioriented text detection and recognition. Trans. Image Process. **23**(11), 4737–4749 (2014)
39. Yu, D., et al.: Towards accurate scene text recognition with semantic reasoning networks. In: CVPR, pp. 12113–12122 (2020)
40. Zeiler, M.D.: Adadelta: an adaptive learning rate method. arXiv preprint arXiv:1212.5701 (2012)
41. Zhan, F., Lu, S.: Esir: end-to-end scene text recognition via iterative image rectification. In: CVPR, pp. 2059–2068 (2019)

Look, Read and Ask: Learning to Ask Questions by Reading Text in Images

Soumya Jahagirdar[1], Shankar Gangisetty[1(✉)], and Anand Mishra[2]

[1] KLE Technological University, Hubballi, India
{01fe17bcs212,shankar}@kletech.ac.in
[2] Vision, Language, and Learning Group (VL2G), Indian Institue of Technology
Jodhpur, Jodhpur, Rajasthan, India
mishra@iitj.ac.in

Abstract. We present a novel problem of text-based visual question generation or TextVQG in short. Given the recent growing interest of the document image analysis community in combining text understanding with conversational artificial intelligence, e.g., text-based visual question answering, TextVQG becomes an important task. TextVQG aims to generate a natural language question for a given input image and an automatically extracted text also known as OCR token from it such that the OCR token is an answer to the generated question. TextVQG is an essential ability for a conversational agent. However, it is challenging as it requires an in-depth understanding of the scene and the ability to semantically bridge the visual content with the text present in the image. To address TextVQG, we present an OCR-consistent visual question generation model that Looks into the visual content, Reads the scene text, and Asks a relevant and meaningful natural language question. We refer to our proposed model as OLRA. We perform an extensive evaluation of OLRA on two public benchmarks and compare them against baselines. Our model – OLRA automatically generates questions similar to the public text-based visual question answering datasets that were curated manually. Moreover, we 'significantly' outperform baseline approaches on the performance measures popularly used in text generation literature.

Keywords: Visual question generation (VQG) · Conversational AI · Visual question answering (VQA)

1 Introduction

> "To seek truth requires one to ask the right questions."
>
> *Suzy Kassem*

Developing agents that can communicate with a human has been an active area of research in artificial intelligence (AI). The document image analysis community has also started showing interest in this problem lately as evident from the

J. Lladós et al. (Eds.): ICDAR 2021, LNCS 12821, pp. 335–349, 2021.
https://doi.org/10.1007/978-3-030-86549-8_22

Input   **Output**

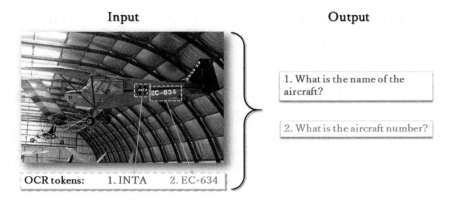

Fig. 1. **TextVQG**. We introduce a novel problem of visual question generation by leveraging text in the image. Given an image and an OCR token automatically extracted from it, our goal is to generate a meaningful natural language question whose answer is the OCR token.

efforts of the community on text-based visual question answering [4, 25, 28, 39], and ICDAR 2019 robust reading challenge on scene text visual question answering [1]. Visual question answering (VQA) is only one desired characteristic of a conversational agent where the agent answers a natural language question about the image. An intelligent agent should also have the ability to ask meaningful and relevant questions with respect to its current visual perception. Given an image, generating meaningful questions, also known as visual question generation (VQG) is an essential component of a conversational agent [30]. VQG is a precursor of visual dialogue systems and might help in building large-scale VQA datasets automatically. To the best of our knowledge, the document image analysis community has not yet looked into the important problem of VQG leveraging text. In this work, we fill up this gap prevailing in the literature by introducing a novel problem of text-based visual question generation or TextVQG in short.

The TextVQG has the following goal: given a natural image containing text, and an OCR token extracted from it, generate a natural language question whose answer is the OCR token. This problem is challenging because it requires in-depth semantic interpretation of the text as well as the visual content to generate meaningful and relevant questions with respect to the image. VQG has been a well-explored area in vision and language community [13, 17, 30, 43]. However, these works often ignore the text appearing in the image, and only restrict themselves to the visual content while generating questions. It should be noted that text in the image helps in asking semantically meaningful questions connecting visual and textual content. For example, consider the image shown in Fig. 1. Given an image of an aircraft and *INTA* and *EC-634* words written on it, we aim to automatically generate questions such as "What is the name of the aircraft?" and "What is the aircraft number?".

Baselines: Motivated by the baselines presented in the VQG literature [30], a few of the plausible approaches to address TextVQG are as follows: (i) **maximum-entropy language model (MELM):** which uses the extracted OCR tokens based on their confidence scores along with detected object to generate question, (ii) **seq2seq model:** which first generates a caption for the input image, and then, this caption is fed into a seq2seq model to generate a question, and (iii) **GRNN model:** where the CNN feature of the input image is passed to a gated recurrent unit (GRU) to generate questions. We empirically observe that these methods often fall short in performance primarily due to their inability (a) to semantically interpret the text and visual content jointly, and (b) to establish consistency between generated question and the OCR token which is supposed to be the answer. To overcome these shortcomings, we propose a novel TextVQG model as described below.

Our Approach. To encode visual content, scene text, its position and bringing consistency between generated question and OCR token; we propose an OCR-consistent visual question generation model that Looks into the visual content, Reads the scene text, and Asks a relevant and meaningful natural language question. We refer to this architecture for TextVQG as OLRA. OLRA begins by representing visual features using pretrained CNN, extracted OCR token using FastText [5] followed by an LSTM, and positions of extracted OCR using positional representations. Further, these representations are fused using a multimodal fusion scheme. The joint representation is further passed to a one-layered LSTM-based module and a maximum likelihood estimation-based loss is computed between generated and reference question. Moreover, to ensure that the generated question and corresponding OCR tokens are consistent with each other, we add an OCR token reconstruction loss which is computed by taking l_2-loss between the original OCR token feature and the representation obtained after passing joint feature to a multi-layer perception. The proposed model OLRA is trained in a multi-task learning paradigm where a weighted combination of the OCR-token reconstruction and question generation loss is minimized to generate a meaningful question. We evaluate the performance of OLRA on TextVQG, and compare against the baselines. OLRA significantly outperforms baseline methods on two public benchmarks, namely, ST-VQA [4] and TextVQA [39].

Contributions of this Paper. The major contributions of this paper are two folds:

1. We draw the attention of the document image analysis community to the problem of visual question generation by leveraging text in the image. We refer to this problem as TextVQG. TextVQG is an important and unexplored problem in the literature with potential downstream applications in building visual dialogue systems and augmenting training sets of text-based visual question answering models. We firmly believe that our work will boost ongoing research efforts [4,25,37,39] in the broader area of conversational AI and text understanding.

2. We propose OLRA – an OCR-consistent visual question generation model that looks into the visual content, reads the text, and asks a meaningful and relevant question for addressing TextVQG. OLRA automatically generates questions similar to the datasets that are curated manually. Our model viz. OLRA significantly outperforms the baselines and achieves a BLEU score of 0.47 and 0.40 on ST-VQA [4] and TextVQA [39] datasets respectively.

2   Related Work

The performance of scene text recognition and understanding has significantly improved over the last decade [12,21,27,31,36,42]. It has also started influencing other computer vision areas such as scene understanding [15], cross-modal image retrieval [24], image captioning [37], and Visual question answering (VQA) [4,25,39]. Among these, text-based VQA works [4,25,39] in the literature can be considered one of the major steps by document image analysis community towards conversational AI. In the follow-up sections, we shall first review the text-based VQA followed by VQG which is the primary focus of this work.

2.1   Text-Based Visual Question Answering

Traditionally, VQA works in the literature focus only on the visual content [2,14,32], and ironically fall short in answering the questions that require reading the text in the image. Keeping the importance of text in the images for answering visual questions, researchers have started focusing on text-based VQA [4,20,28,39]. Among these works, authors in [39] use top-down and bottom-up attention on text and visual objects to select an answer from the OCR token or a vocabulary. The scene text VQA model [4] aims to answer the questions by performing reasoning over the text present in the natural scene. In OCR-VQA [28], OCR is used to read the text in the book cover images, and a baseline VQA model is proposed for answering questions enquiring about author name, book title, book genre, publisher name, etc. Typically, these methods are based on a convolutional neural network (CNN) to perform the visual content analysis, a state-of-the-art text recognition engine for detecting and reading text, and a long short-term memory (LSTM) network to encode the questions. More recently, T-VQA [20] presents a progressive attention module and a multimodal reasoning graph for reading and reasoning. Transformer and graph neural network-based models have also started gaining popularity for addressing text-based VQA [9,11]. Another direction of work in text-based visual question answering is textKVQA [38] where authors propose a VQA model that reads textual content in the image, connects it with a knowledge graph and performs reasoning using a gated graph neural network to arrive at an accurate answer.

2.2   Visual Question Generation

VQG is a dual task of VQA and is essential for building visual dialogue systems [8,13,17,19,29,30,43,44]. Moreover, the ability to generate relevant question to the image is a core to in-depth visual understanding. Question generation from images as well as from raw text have been a well-studied problem in the literature [10,13,17,22,35,43]. Automatic question generation techniques in NLP have also enabled chat-bots [10,22,35].

Among visual question generation works, [30], focused on generating natural and engaging questions from the image, and provided with three distinct datasets, each covering object to event-centric images. Authors in [30] proposed three generative models for tackling the task of VQG which we believe are the baselines for any novel VQG tasks, and we adopt these models for TextVQG and compare it with the proposed model. Fan et al. [8] generates diverse questions of different types [8], such as, when, what, where, which and how questions. In [17], goal-driven variational auto-encoder model is used to generate questions by maximizing the mutual information between visual content as well as the expected category of the answer. Authors in [19] posed VQG as a dual task to VQA and jointly addressed both the task.

These works in the literature restrict their scope to asking questions only about visual content and ignore any scene text present in the image. We argue that conversational agents must have the ability to ask questions by semantically bridging text in the image with visual content. Despite its fundamental and applied importance, the problem of TextVQG - text-based visual question generation has not been looked into in the literature. We fill this gap in the literature through our work.

3   Proposed Model: OLRA

Given a natural image containing text and an OCR token extracted from the image, TextVQG aims to generate a natural language question such that the answer is the OCR token. To solve this problem, the proposed method should be able to successfully recognize text and visual content in the image, semantically bridge the textual and visual content and generate meaningful questions. For text detection and recognition, we rely on one of the successful modern scene text detection and recognition methods [3,36].[1] Further, the pre-trained CNN is used for computing the visual features and an LSTM is used to generate a question. The overall architecture of the proposed model viz. OLRA is illustrated in Fig. 2.

OLRA has the following three modules:

(i) Look and Read Module: In this module, we extract convolutional features from the given input image I. We use pre-trained CNN ResNet-50 which gives us

---

[1] For one of the datasets namely TextVQA, OCR-tokens extracted from Rosetta [6] are provided with the dataset.

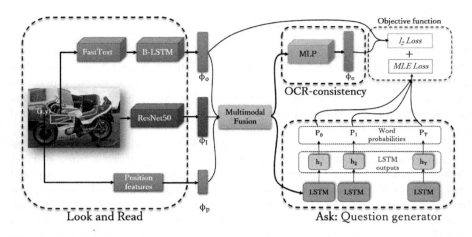

Fig. 2. OLRA. Our proposed visual question generator architecture viz. OLRA has three modules, namely, (i) Look and Read, (ii) OCR-consistency and (iii) Ask module. Note: since OCR tokens can be more than a single word, we pass their FastText embeddings to a B-LSTM. Please refer to Sect. 3 for more details. [**Best viewed in color**]. (Color figure online)

512-dimensional features ϕ_I. Further, since our objective is to generate questions by relating visual content and text appearing in the image, we use [3,36] for detecting and recognizing text. We prefer to use [3,36] for text detection and recognition due to its empirical performance. However, any other scene text recognition module can be plugged in here. Once we detect and recognize the text, we obtain its FastText [5] embedding. Since OCR tokens can be proper nouns or noisy, and therefore, can be out of vocabulary; we prefer FastText over Word2Vec [26] or Glove [33] word embeddings. The FastText embedding of OCR token is fed to a bi-directional long short term memory (B-LSTM) to obtain a 512-dimensional OCR-token feature or ϕ_o.

Further, positions of OCR can play a vital role in visual question generation. For example as illustrated in Fig. 3, in a sports scene, the number appearing in a sportsman's jersey and an advertisement board will require us to generate questions such as "What is the jersey number of the player who is pitching the ball?" and "What is the number written on advertisement board?" respectively. In order to use OCR token positions in our framework, we use 8 features i.e., topleft-x, topleft-y, width, height, rotation, yaw, roll and pitch of the bounding box as ϕ_p.[2]

Once these three features, i.e., ϕ_o: OCR features, ϕ_I: image features and ϕ_p: positions features are obtained, our next task is to learn joint representation by judiciously combining them. To this end, we use a multi-layer perceptron $\mathcal{F}$ to obtain joint representation Ψ as follows:

---

[2] When we use CRAFT for text detection, we use only first four positional features i.e., topleft-x, topleft-y, width and height of the bounding box.

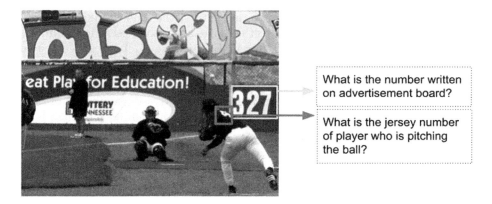

Fig. 3. Importance of positions of OCR tokens for question generation. Numbers detected in different positions in the image i.e., number 4 and 327 in this example demands generating different questions.

$$\Psi = \mathcal{F}(W\phi). \tag{1}$$

Here, W is a learnable weight matrix and $\phi = [\phi_o; \phi_I; \phi_p]$ is concatenation of features. The joint feature obtained at this module Ψ is a 512-dimensional vector that is fed to OCR token-reconstruction and question generator as described next.

(ii) OCR-Consistency Module: After learning the joint representation of both image and OCR features, the model should be able to generate questions given specific tokens. To ensure this consistency, it is important that joint representation learned after fusion (Ψ) preserves the OCR representation (ϕ_o). We reconstruct $\hat{\phi}_o$ by passing joint feature Φ to a multi-layer perceptron to obtain the reconstructed answer. We minimize the l_2-loss between the original answer and the reconstructed answer representations, i.e.,

$$\mathcal{L}_a = ||\phi_o - \hat{\phi}_o||_2^2. \tag{2}$$

(iii) Ask Module: Our final module generates the relevant questions to the image. We use a decoder LSTM to generate the question q' from combined space as explained in the previous section. The minimization of maximum-likelihood error (MLE) between the generated question and the true question in training set makes the model learn to generate appropriate questions. The 512-dimensional joint feature vector serves as the initial state to an LSTM. We produce the output question one word at a time using the LSTM. This happens until we encounter an end-of-sentence token. At each step during decoding, we feed in an embedding of the previously predicted word and predict the next word. The loss between generated and true questions guides the generator to generate better questions. The obtained joint feature Ψ from the previous module acts as an input to the question generator. This question generator captures the time-varying character-istics and outputs questions related to the image and text present in the image. As mentioned above, we minimize the MLE objective $\mathcal{L}_q$ between generated question q' and ground truth question q.

Training: OLRA is trained in a multitask learning paradigm for two tasks, i.e., OCR-token reconstruction and question generation. We use training set of text-based visual question answering datasets to train the model by combining both MLE loss $\mathcal{L}_q$ and OCR-token reconstruction loss $\mathcal{L}_a$, as follows:

$$\mathcal{L} = \mathcal{L}_q + \lambda\mathcal{L}_a, \tag{3}$$

where λ is a hyperparameter which controls relative importance in optimizing these two losses, and ensures that the generated questions have better semantic structure with respect to the textual content along with the scene of the image.

4   Experiments and Results

In this section, we experimentally validate our proposed model for TextVQG. We first discuss the datasets and performance measures in Sect. 4.1 and Sect. 4.2 respectively. We then describe the baseline models for TextVQG in Sect. 4.3 followed by implementation details in Sect. 4.4. The quantitative and qualita-tive experimental results of the proposed model and the comparative analysis is provided in Sect. 4.5.

4.1   Datasets

As this is the first work towards a visual question generation model that can read the text in the image, there is no dedicated dataset available for this task. We, therefore, make use of the two popular text-based VQA datasets, namely, TextVQA [39] and ST-VQA [4]. Note that unlike these datasets where origi-nally the task is to answer a question about the image, our aim is to generate questions. In other words, given an image and an OCR token (often answers in these datasets), our proposed method learns to automatically generate questions similar to these manually curated datasets. A brief statistical description related to these datasets are as follows:

1. **ST-VQA** [4]: consists of $23,038$ images with $31,791$ question-answer pairs obtained from different public datasets. A total of $16,063$ images with $22,162$ questions and $2,834$ images with $3,910$ questions are used for training and testing respectively, considering only the question-answer pairs having OCR tokens as their answers.
2. **TextVQA** [39]: comprises of $28,418$ images obtained from openImages [16] with $45,336$ questions. It also provides OCR tokens extracted from Rosetta [6]. We use $21,953$ images with $25,786$ questions and $3,166$ images with $3,702$ questions in all as train and test set respectively, considering only the question-answer pairs having OCR tokens as their answers.

4.2   Performance Metrics

Following the text-generation literature [7], we use popular evaluation measures such as bilingual evaluation understudy (BLEU), recall-oriented understudy for gisting evaluation - longest common sub-sequence (ROUGE-L), and metric for evaluation of translation with explicit ordering (METEOR). The BLEU score compares n-grams of the generated question with the n-grams of the reference question and counts the number of matches. The ROUGE-L metric indicates similarity between two sequences based on the length of the longest common sub-sequence even though the sequences are not contiguous. The METEOR is based on the harmonic mean of uni-gram precision and recall and is considered a better performance measure in the text generation literature [18]. Higher values of all these performance measures imply better matching of generated questions with the reference questions.

4.3   Baseline Models

Inspired by the VQG models in literature [30], we present three baseline models by adopting them for TextVQG task.

Maximum Entropy Language Model (MELM). Here, a set of OCR tokens extracted from the image along with a set of detected objects using Faster-RCNN [34] are fed to a maximum entropy language model to generate a question.

Seq2seq Model. In this baseline, we first obtain a caption of the image using a method proposed in [23]. Then, the generated caption and the identified OCR tokens from the images are passed to a seq2seq model [40]. In seq2seq model, the encoder contains an embedding layer followed by an LSTM layer and for decoding we use an LSTM layer followed by a dense layer. The seq2seq model is trained for question generation tasks using the train set of datasets described earlier.

Gated Recurrent Neural Network (GRNN). In this model, we obtain visual features using InceptionV3 [41]. This yields a feature vector of $1 \times 1 \times 4096$ dimensions that are then passed to a gated recurrent unit (GRU) to generate a question. We train GRU for question generation by keeping all the layers of InceptionV3 unchanged.

4.4   Implementation Details

We train our proposed network using the Adam optimizer with a learning rate of $1e-4$, batch size of 32, and a decay rate of 0.05. The value of λ in Eq. 3 is set to 0.001. The maximum length of the generated questions is set to 20. We train the model for 10 epochs. In multi-modal fusion, a two-layer attention network is used with feature sizes of 1032 and 512 respectively. The model is trained on a single NVIDIA Quadro P5000.

Table 1. Comparison of OLRA with baselines. We observe that the proposed model viz. OLRA clearly outperforms baselines on both the datasets for TextVQG.

Method	ST-VQA [4] dataset			TextVQA [39] dataset		
	BLEU	METEOR	ROUGE-L	BLEU	METEOR	ROUGE-L
MELM	0.34	0.12	0.31	0.30	0.11	0.30
Seq2seq	0.29	0.12	0.30	0.27	0.11	0.29
GRNN	0.36	0.12	0.32	0.33	0.12	0.30
Ours (OLRA)	**0.47**	**0.17**	**0.46**	**0.40**	**0.14**	**0.40**

4.5 Results and Discussions

Quantitative Analysis. We evaluate the performance of our proposed model i.e., OLRA, and compare it against three baseline approaches, namely MELM, seq2seq, and GRNN on ST-VQA and TextVQA datasets. This comparative result is shown in Table 1 using performance measures discussed in Sect. 4.2. Note that higher values for all these popularly used performance measure is considered superior.

While experimenting with MELM baseline, we observe that MELM trained on both OCR and FasterRCNN features performed better with an increase in the BLEU score by 0.04 over text only MELM and by 0.02 over object only MELM. However, among the three baseline approaches, GRNN generates comparatively better questions. Our proposed model i.e., OLRA significantly outperforms all the baseline models. For example, on ST-VQA dataset, OLRA improves BLEU score by 0.11, METEOR score by 0.05, and ROUGE-L score by 0.14 as compared to the most competitive baseline i.e., GRNN. It should be noted that under these performance measures these gains are considered significant [7]. We observe similar performance improvement on TextVQA dataset as well.

Qualitative Analysis. We perform a detailed qualitative analysis of the baselines as well as our proposed model. We first show a comparison of generated questions using all the three baselines versus OLRA in Fig. 4. We observe that the baselines are capable of generating linguistically meaningful questions. However, they do not fulfill the sole purpose of **Look**, **Read** and **Ask**-based question generation. For the first example in Fig. 4 the expected question is "What does the street sign say?", the baseline approaches and the proposed model generate nearly the same question. But, as the complexity of the scene increases, the baseline models fail to generate appropriate questions and the proposed model due to its well-designed look, read, ask, and OCR-consistency modules, generates better questions. For example, in the last column of Fig. 4, MELM generates "What is the brand?" but it fails to specify the kind of the product in the scene. Whereas the proposed model generates "What is the brand of the milk?" which is very close to the target question.

Further, more results of our model are shown in Fig. 5. Here, we represent OCR token using green bounding boxes, correctly generated questions in blue

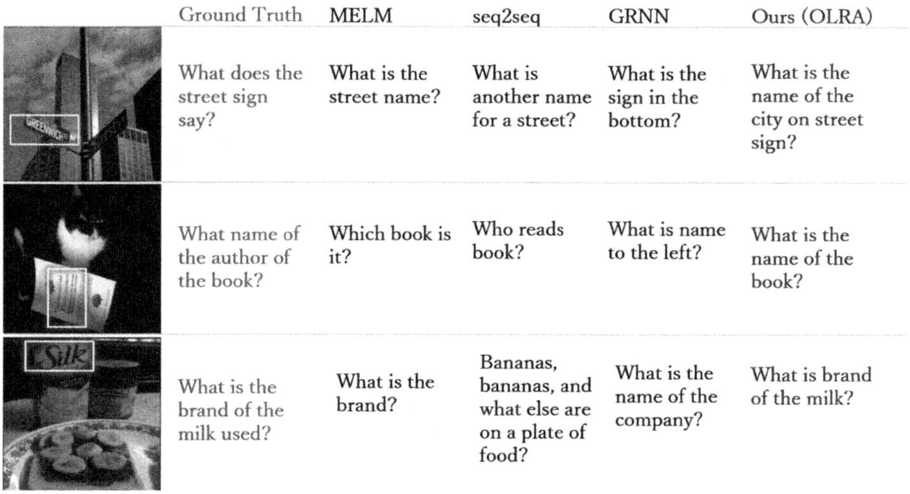

	Ground Truth	MELM	seq2seq	GRNN	Ours (OLRA)
	What does the street sign say?	What is the street name?	What is another name for a street?	What is the sign in the bottom?	What is the name of the city on street sign?
	What name of the author of the book?	Which book is it?	Who reads book?	What is name to the left?	What is the name of the book?
	What is the brand of the milk used?	What is the brand?	Bananas, bananas, and what else are on a plate of food?	What is the name of the company?	What is brand of the milk?

Fig. 4. Qualitative comparison. Visual question generation by MELM, seq2seq, and GRNN baselines, and ours on a set of images from ST-VQA dataset.

Table 2. Ablation study. BLEU scores analysis (i) with inclusion of positional information and (ii) with n-word answers on both the datasets.

OLRA	ST-VQA [4] dataset	TextVQA [39] dataset
w/o position	0.44	0.39
w/position	0.45	0.39
w/position and OCR-consistency	0.47	0.40
w/1-word answers	0.48	0.41
w/2-word answers	0.47	0.39
w/3-word answers	0.46	0.40

color text, and incorrectly generated questions in red color text. Consider the example of image of a train in Fig. 5. Here, our method successfully generates the question "What is the number on the front of the train?". Similar such example question generations can be seen in Fig. 5.

The failure of our model pronounced when either OCR-token is misinterpreted (for example, the word "derp" on the computer screen is misinterpreted as computer monitor brand, and the word "TELEPHONE" on a telephone booth is misinterpreted as company name) or there is a need of generating questions whose answer is not an OCR token, for example: "Are there any numbers on the identification label?".

Ablation Study: We perform two ablation studies to demonstrate: (i) utility of the positional information and (ii) model's capability to generate those questions which has multi-word answers.

OCR: 222003
Gen-Q: What is the
number on the front of
the train?
GT-Q: What six
numbers are visible ?

OCR: the pickerel
Gen-Q: What is the name
of the store on the left?
GT-Q: What is the name of
the inn?

OCR: derp
Gen-Q: What is the brand of the
computer monitor?
GT-Q: What does the computer
screen on the right say?

ST-VQA

OCR: calais 5
Gen-Q: What does
the sign say?
GT-Q: What is
written on the sign?

OCR: cocio
Gen-Q: What is the brand of the
beer ?
GT-Q: What is the brand of the
drink ?

OCR: Ana
Gen-Q: What is the
name of the airline?
GT-Q: What is the
airline of the blue plane?

OCR: Fique de bem
Gen-Q: What is the name of the
book?
GT-Q: What book is that?

OCR: rpm
Gen-Q: What is the name of the
team?
GT-Q: What initials are on the
back of the shirt worn by the
person in the middle?

OCR: telephone
Gen-Q: What is
the name of the
company?
GT-Q: What does
the gray sign say?

TextVQA

OCR: Samsung
Gen-Q: What is the name of
the company?
GT-Q: What is the brand of the
phone?

OCR: Yournamehere
Gen-Q: What is the name of
the team on man's shirt?
GT-Q: What is the name on
the back of the jerseys?

OCR: no
Gen-Q: What is the
license plate number?
GT-Q: Are there any
numbers on the
identification label?

Fig. 5. Samples of generated questions on ST-VQA and TextVQA datasets with OCR tokens shown in green bounding box. Correctly and incorrectly generated questions are shown in blue and red colors respectively. Gen-Q and GT-Q denote generated question from proposed method and ground-truth question respectively. [**Best viewed in color**]. (Color figure online)

Model without positional information considers image features ϕ_I + token features ϕ_o, and with positional information considers image features ϕ_I + token features ϕ_o + positional information ϕ_p to generate questions. We observe that model with positional information and OCR-consistency i.e., our full model (OLRA) as shown in Table 2 enhances the BLEU score and quality of generated question over other models on both ST-VQA and TextVQA datasets.

Further, in Table 2, we also show OLRA's performance on generating those questions which has 1-word, 2-word and 3-word answers. Based on statistical analysis with respect to ST-VQA test set, there are, 69% 1-word, 19.6% 2-word, 7% 3-word, and 4.4% above 3-word question-answer pairs. While in TextVQA testset, there are, 63.7% 1-word, 21.3% 2-word, 8% 3-word and 7% above 3-word answers. We observe that OLRA BLEU score is nearly the same and the model performs equally well in all the three cases (see Table 2). This ablation study suggest that our model is capable of generating even those questions which has two or three-length word as answer.

5   Conclusions

We introduced the novel task of 'Text-based Visual Question Generation', where given an image containing text, the system is tasked with asking an appropriate question with respect to the OCR token. We proposed OLRA – an OCR-consistent visual question generation model to ask meaningful and relevant visual questions. OLRA outperformed three baseline approaches on two public benchmarks. As the first work towards developing a visual question generation model that can read, we restrict our scope to generating simple questions whose answer is the OCR token itself. Generating complex questions that require deeper semantic and commonsense reasoning, improving text-based VQA models by augmenting data automatically generated using TextVQG models, are few tasks that we leave as future works.

We firmly believe that the captivating novel task and the benchmarks presented in this work will encourage researchers to develop better TextVQG models, and thereby gravitate ongoing research efforts of the document image analysis community towards conversational AI.

References

1. ICDAR 2019 Robust Reading Challenge on Scene Text Visual Question Answering. https://rrc.cvc.uab.es/?ch=11. Accessed 01 Feb 2021
2. Antol, S., et al.: VQA: visual question answering. In: ICCV (2015)
3. Baek, Y., Lee, B., Han, D., Yun, S., Lee, H.: Character region awareness for text detection. In: CVPR (2019)
4. Biten, A.F., et al.: Scene text visual question answering. In: ICCV (2019)
5. Bojanowski, P., Grave, E., Joulin, A., Mikolov, T.: Enriching word vectors with subword information. Trans. Assoc. Comput. Linguistics 5, 135–146 (2017)
6. Borisyuk, F., Gordo, A., Sivakumar, V.: Rosetta: large scale system for text detection and recognition in images. In: KDD (2018)

7. Celikyilmaz, A., Clark, E., Gao, J.: Evaluation of text generation: a survey. CoRR abs/2006.14799 (2020)
8. Fan, Z., Wei, Z., Li, P., Lan, Y., Huang, X.: A question type driven framework to diversify visual question generation. In: IJCAI (2018)
9. Gao, D., Li, K., Wang, R., Shan, S., Chen, X.: Multi-modal graph neural network for joint reasoning on vision and scene text. In: CVPR (2020)
10. Gülçehre, Ç., Dutil, F., Trischler, A., Bengio, Y.: Plan, attend, generate: Planning for sequence-to-sequence models. In: NIPS (2017)
11. Hu, R., Singh, A., Darrell, T., Rohrbach, M.: Iterative answer prediction with pointer-augmented multimodal transformers for textvqa. In: CVPR (2020)
12. Jaderberg, M., Simonyan, K., Vedaldi, A., Zisserman, A.: Reading text in the wild with convolutional neural networks. Int. J. Comput. Vision **116**(1), 1–20 (2016)
13. Jain, U., Zhang, Z., Schwing, A.G.: Creativity: generating diverse questions using variational autoencoders. In: CVPR (2017)
14. Johnson, J., Hariharan, B., van der Maaten, L., Fei-Fei, L., Zitnick, C.L., Girshick, R.B.: CLEVR: a diagnostic dataset for compositional language and elementary visual reasoning. In: CVPR (2017)
15. Karaoglu, S., Tao, R., Gevers, T., Smeulders, A.W.M.: Words matter: scene text for image classification and retrieval. IEEE Trans. Multimed. **19**(5), 1063–1076 (2017)
16. Krasin, I., et al.: Openimages: a public dataset for large-scale multi-label and multi-class image classification. Dataset available from https://github.com/openimages (2017)
17. Krishna, R., Bernstein, M., Fei-Fei, L.: Information maximizing visual question generation. In: CVPR (2019)
18. Lavie, A., Agarwal, A.: METEOR: an automatic metric for MT evaluation with high levels of correlation with human judgments. In: WMT@ACL (2007)
19. Li, Y., et al.: Visual question generation as dual task of visual question answering. In: CVPR (2018)
20. Liu, F., Xu, G., Wu, Q., Du, Q., Jia, W., Tan, M.: Cascade reasoning network for text-based visual question answering. In: ACM Multimedia (2020)
21. Long, S., He, X., Yao, C.: Scene text detection and recognition: the deep learning era. Int. J. Comput. Vision **129**(1), 161–184 (2021)
22. Lopez, L.E., Cruz, D.K., Cruz, J.C.B., Cheng, C.: Transformer-based end-to-end question generation. CoRR abs/2005.01107 (2020)
23. Luo, R., Price, B.L., Cohen, S., Shakhnarovich, G.: Discriminability objective for training descriptive captions. In: CVPR (2018)
24. Mafla, A., de Rezende, R.S., Gómez, L., Larlus, D., Karatzas, D.: Stacmr: scene-text aware cross-modal retrieval. CoRR abs/2012.04329 (2020)
25. Mathew, M., Karatzas, D., Manmatha, R., Jawahar, C.V.: DocVQA: a dataset for VQA on document images. In: WACV (2021)
26. Mikolov, T., Yih, W., Zweig, G.: Linguistic regularities in continuous space word representations. In: HLT-NAACL (2013)
27. Mishra, A., Alahari, K., Jawahar, C.V.: Scene text recognition using higher order language priors. In: BMVC (2012)
28. Mishra, A., Shekhar, S., Singh, A.K., Chakraborty, A.: OCR-VQA: visual question answering by reading text in images. In: ICDAR (2019)
29. Misra, I., Girshick, R.B., Fergus, R., Hebert, M., Gupta, A., van der Maaten, L.: Learning by asking questions. In: CVPR (2018)
30. Mostafazadeh, N., Misra, I., Devlin, J., Mitchell, M., He, X., Vanderwende, L.: Generating natural questions about an image. In: ACL (2016)

31. Neumann, L., Matas, J.: Real-time lexicon-free scene text localization and recognition. IEEE Trans. Pattern Anal. Mach. Intell. **38**(9), 1872–1885 (2016)
32. Patro, B.N., Kurmi, V.K., Kumar, S., Namboodiri, V.P.: Deep bayesian network for visual question generation. In: WACV (2020)
33. Pennington, J., Socher, R., Manning, C.D.: Glove: global vectors for word representation. In: EMNLP (2014)
34. Ren, S., He, K., Girshick, R.B., Sun, J.: Faster R-CNN: towards real-time object detection with region proposal networks. IEEE Trans. Pattern Anal. Mach. Intell. **39**(6), 1137–1149 (2017)
35. Serban, I.V., et al.: Generating factoid questions with recurrent neural networks: the 30m factoid question-answer corpus. In: ACL (2016)
36. Shi, B., Bai, X., Yao, C.: An end-to-end trainable neural network for image-based sequence recognition and its application to scene text recognition. IEEE Trans. Pattern Anal. Mach. Intell. **39**(11), 2298–2304 (2017)
37. Sidorov, O., Hu, R., Rohrbach, M., Singh, A.: Textcaps: a dataset for image captioning with reading comprehension. In: ECCV (2020)
38. Singh, A.K., Mishra, A., Shekhar, S., Chakraborty, A.: From strings to things: Knowledge-enabled VQA model that can read and reason. In: ICCV (2019)
39. Singh, A., et al.: Towards VQA models that can read. In: CVPR (2019)
40. Sutskever, I., Vinyals, O., Le, Q.V.: Sequence to sequence learning with neural networks. In: NIPS (2014)
41. Szegedy, C., Vanhoucke, V., Ioffe, S., Shlens, J., Wojna, Z.: Rethinking the inception architecture for computer vision. In: CVPR (2016)
42. Wang, K., Babenko, B., Belongie, S.J.: End-to-end scene text recognition. In: ICCV (2011)
43. Yang, J., Lu, J., Lee, S., Batra, D., Parikh, D.: Visual curiosity: Learning to ask questions to learn visual recognition. In: CoRL. Proceedings of Machine Learning Research (2018)
44. Zhang, S., Qu, L., You, S., Yang, Z., Zhang, J.: Automatic generation of grounded visual questions. In: IJCAI (2017)

CATNet: Scene Text Recognition Guided by Concatenating Augmented Text Features

Ziyin Zhang, Lemeng Pan$^{(\boxtimes)}$, Lin Du, Qingrui Li, and Ning Lu

AI Application Research Center, Huawei Technologies, Shenzhen, China
{zhangziyin1,panlemeng,dulin09,liqingrui,luning12}@huawei.com

Abstract. In this paper, we propose an end-to-end trainable text recognition model that consists of an auxiliary augmentation module and a text recognizer. Well-established traditional methods in computer vision (such as binarization, sharpening and morphological operations) play critical roles in image preprocessing. These operations are proven to be particularly useful for downstream computer vision tasks. In order to achieve better results, case-by-case hyperparameter adjustment is often required. Inspired by traditional CV methods, we propose an auxiliary network to mimic traditional CV operations. The auxiliary network acts like an image preprocessing module to extract rich augmented features from the input image to ease the downstream recognition difficulty. We studied three types of augmentation modules with parameters that can be learned directly via gradient back-propagation. This way, our method combines traditional CV techniques and deep neural network by joint learning. The proposed method is extensively tested on major benchmark datasets to show that it can boost the performance of the recognizers, especially for degraded text images in various challenging conditions.

Keywords: Scene text recognition · Feature learning · Hybrid techniques

1 Introduction

In recent years, deep neural networks gained tremendous success in the field of computer vision (CV). Krizhevsky et al. [12] achieved massive success in the imagenet competition in 2012 with Alexnet. Deep Learning (DL) was reignited and has dominated the field ever since due to a substantially improved performance compared to traditional methods. DL based methods are data-driven and rely heavily on immense computing power. Rapid progression in hardware device capabilities, including memory capacity, computing power, optics, camera lenses, and internet availability laid a solid foundation for the eruption of DL methods. As deep learning based methods develop, scene text recognition (STR) has also made significant progress in the meantime. Long et al. [17] provided a survey that summarized major advancements of scene text detection and recognition methods in the deep learning era.

© Springer Nature Switzerland AG 2021
J. Lladós et al. (Eds.): ICDAR 2021, LNCS 12821, pp. 350–365, 2021.
https://doi.org/10.1007/978-3-030-86549-8_23

Traditional text recognition methods follow the so-called bottom-up paradigm. Firstly, text are to be detected by the sliding window method using hand-crafted features such as connected components [24], or histogram of oriented gradients (HOG) descriptor [4] to separate characters from backgrounds. Then recognition of individual characters is solved as a typical classification problem. In the deep learning era, the top-down approach that attempts to solve the recognition problem using the whole text line image became the mainstream. Jaderberg et al. [10] first introduced a CNN-based method that regards each word as an object. Thus, the text recognition problem is converted to a classification problem by training a classifier with a large number of classes. Shi et al. [26] suggested a model that feeds the output of the CNN backbone to a recurrent neural network to recognize text images of varying length. Connectionist Temporal Classification (CTC) [6] is adopted to translate the per-frame predictions into the final text sequence. Sequence to sequence with attention mechanism was another technique that was employed to improve character sequence prediction. The attention mechanism performs soft feature selection and helps the model to capture implicit character-level language statistics. Lee et al. [13] first proposed a recursive recurrent network with attention mechanism for scene text recognition. Yang et al. [29] extend the attention mechanism to select local 2D features when decoding individual characters to address the irregular text recognition problem.

The rise of DL dramatically improves the accuracy of a series of difficult CV problems, including image classification, object detection, and scene text recognition(STR). Meanwhile, questions such as: "does DL make traditional Computer Vision (CV) techniques obsolete", "has DL superseded traditional computer vision", "is there still a need to adopt traditional CV techniques when DL seems to be so effective" have been brought up in the community [23] recently. In this article, we propose method Con**CAT**enating Net (CATNet) to tackle hard scene text cases including images in low-contrast, low-resolution, and uneven illumination. The key idea of our proposed method is to fuse rich text features as input to ease the recognition difficulty for the downstream network. Well-established traditional methods, such as adaptive thresholding, unsharp masking and morphological operations, can produce new features that can benefit succeeding tasks. In the pre-DL era, binarization and mathematical morphology contribute a wide range of operators to image processing. The operators are proven to be particularly useful for downstream computer vision tasks including edge detection, noise removal, image enhancement and image segmentation. For example, the Sobel kernel is one of the feature extraction methods which is used for edge detection. However, all of the methods mentioned above involve hyperparameters which should be adjusted case-by-case. It is difficult to use a few sets of predefined parameters to accommodate a wide range of scenarios. Inspired by traditional CV methods, we propose an auxiliary network that incorporates three types of augmentation modules based on traditional CV techniques. Traditional methods are modualarized with parameters that can be learned directly via gradient back-propagation. Each augmentation module produces a 1-channel feature map and multiple augmentation modules can be used in parallel to generate multiple features. The augmented features are then concatenated to the input image to form a larger tensor with rich features. This augmented tensor becomes the new input of the downstream text recognition network. The architecture of our proposed method is shown in Fig. 1.

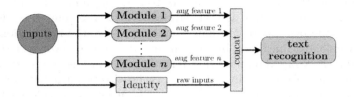

Fig. 1. Overview of the proposed frame work. Augmentation modules are designed to mimic traditional CV methods. The auxiliary modules extract richer features from the input image through traditional CV operations. Each module produces a 1-channel feature map. These maps are then concatenated to the input image to form a larger tensor with rich features. The augmented tensor becomes the new input of the downstream text recognition network. The framework is end-to-end trainable.

Is there still a need for an additional network to extract features from input images while kernels in convolutional networks basically do the same thing? In other words, is the auxiliary augmentation network redundant? Hornik [8] rigorously established that standard multilayer feedforward network with as few as one hidden layer are capable of approximating any Borel measurable function to any desired degree of accuracy, provided sufficiently many hidden units are available. In other words, multilayer feedforward networks are a class of universal approximators. Goodfellow et al. [5] pointed out that although a feedforward network with a single layer is sufficient to represent any function, but the layer may be infeasibly large and may fail to learn and generalize correctly. Although deep neural networks have the ability to approximate universal functions, it could be still beneficial to introduce an auxiliary network to lower the resources required, aide the network to learn the complex mapping with existing optimization algorithms, and making the learning processes more efficient. By exploiting well-developed traditional CV operations, useful features become available to the network in the early training stage and guide the network convergence throughout the entire optimization process.

Our model was developed in the process of solving real challenges encountered in the text recognition task in the manufacturing setting. For example, for text recognition task in the industrial setting, specular light, paint colors, diverse fonts/text appearances, glares, and low-contrast images make it challenging for text recognizers to accurately decode the serial numbers of product parts. With the help of augmentation modules, the text recognizer could achieve significant improvements on degraded text images in challenging manufacturing settings.

We contribute in the following way: 1) We provide an end-to-end trainable scene text recognizer with pluggable augmentation modules, which serve as an image preprocessor to boost recognition accuracy. We conducted extensive experiments on various benchmark STR dataset to show that the proposed method significantly boost the performance of text recognizer, especially for low-quality/low-contrast images. 2) We propose a method that combines traditional CV operations with deep learning based method. We show that two types of methods do not conflict with each other. Instead, synergy between traditional CV methods and DL methods could boost the performance of text recognizers. 3) We cleaned the COCO-text dataset for our experiment, and we will open-source the cleaned version of the COCO-text dataset to the community for future research. A private dataset contains 1000 text images from the manufacturing assembly line will also be released.

2   Related Work

For scene text recognition task, irregularly shaped text, blurring/degradation, styled fonts, images with low contrast/brightness could pose challenges and decrease the recognition accuracy. Various attempts has been made to ease the difficulty of the downstream recognition task.

RARE [25] and ASTER [25] proposed a rectification network that goes before the recognition network to handle irregular text in various shapes. The rectification network rectifies an input image with a predicted Thin-Plate-Spline (TPS) [2] transformation. The spatial transformation is modeled by Spatial Transformer Network (STN) [11]. Note that the whole model could be trained end-to-end, and the spatial transformation is learned implicitly to enhance the global text recognition accuracy.

Luo et al. [18] also pointed out that recognition models remained inadequately strong to handle multiple disturbances from various challenges and proposed a multi-object rectification network (MORN) that rectifies images to reduce the difficulty of the recognition. The text image is first decomposed into multiple small parts, then the offset is predicted for each part. Finally, the original image is sampled to obtain the rectified image, which is easier to recognize.

Wang et al. [28] suggest that the performance of text recognizers drops sharply when the input image is blurred or in low resolution. The authors proposed TextSR that combines a super-resolution network and a text recognition network. A novel Text Perceptual Loss (TPL) is proposed so that text recognizer could guide the generator to produce clear text for easier recognition.

Mou et al. [22], proposed a pluggable super-resolution (PSU) for auxiliary training to assist the recognition network. The PSU is designed to build super-resolution images given degraded images, and it shares the same CNN backbone with the recognition branch. Such network architecture is expected to help the backbone learn how to add high-resolution information on top of features extracted from degradation images.

Luo et al. [19] proposed to use an agent network to generate more effective training samples for the recognition network. By controlling a set of fiducial points, the agent network learns geometric transformations based on the recognition network's output. The agent network and the recognition network are jointly trained. Although the augmentation is learnable, a stand-alone agent network is trained to generate "harder" examples to train the recognizer.

Base on the related work we summarized above, the performance of text recognizers can be boosted by: 1) rectification of irregular text shape, 2) super resolution of degraded text images; 3) online data augmentation that generates hard training samples for the text recognizer. Influenced by related methods, we designed augmentation modules to mimic traditional CV operations, which incorporate prior knowledge as part of the input for the recognition network. The augmentation module extracts features that were long-tested and well-proven to be effective so that the difficulty of the recognition task is reduced.

3   Methodology

The proposed model comprises two parts, the auxiliary augmentation network, and the text recognition network. Three types of augmentation modules are studied in our research. In the following sections, we first describe the differentiable adaptive thresholding module (DAT) in Subsect. 3.1; differentiable unsharp masking is introduced in Subsect. 3.2; the morphological denoise module is covered in Subsect. 3.3. Then, we present the module composition and configuration conventions in Subsect. 3.4. Recognition network is described in Subsect. 3.5.

3.1   Differentiable Adaptive Thresholding (DAT)

Thresholding or binarization is a simple yet effective way to partition an image into foreground and background. The objective of binarization is to modify an image's representation into another representation that is easier to process. While thresholding applies a fixed global threshold value on the whole image, adaptive thresholding, on the other hand, determines pixel-wise threshold values based on a small region around each pixel. Therefore the output is more reasonable if there is a noticeable illumination gradient in the image. Local threshold values are usually calculated using either the mean of neighboring pixels or the weighted mean of neighboring pixels using a 2D Gaussian kernel. Two hyperparameters are needed in adaptive thresholding: 1) window size or block size, for the size of the square neighbor region where the local threshold value comes from; 2) offset, a global constant subtracted from the computed means.

Specific window size and offset are needed for a given image to generate a clear binarized outcome. We propose differentiable adaptive thresholding (DAT) to achieve this. It takes a resized image and predicts a suitable window size and offset for the original image. Note that no matter what averaging method is adopted to compute the weighted mean, it essentially produces a low-pass filter with different frequency responses for the corresponding averaging method. From this perspective of view, DAT does not directly predict a value for window size. Instead, it predicts a weight vector used to compute a weighted sum from a set of base kernels. The base kernels are a series of averaging filters with the same length but different window sizes. They share the same center. Zeroes are filled so that different-sized filters are of the same length. For example, if a base kernel set has a window size ranged from 2 to 5, length of 7, then the four base kernels are defined in Table 1:

Table 1. Illustration of base kernels

Base	Example
Base kernel 1	[0, 0, 0, 0.5, 0.5, 0, 0]
Base kernel 2	[0, 0, 0.3333, 0.3333, 0.3333, 0, 0]
Base kernel 3	[0, 0, 0.25, 0.25, 0.25, 0.25, 0]
Base kernel 4	[0, 0.2, 0.2, 0.2, 0.2, 0.2, 0]

DAT adopts a light-weight CNN (named AugCNN in this paper) followed by two fully connected layers with softmax activation to produce weights for each base kernel. Since base kernels are all 1D averaging kernels and weights sum up to 1, the outcome kernel is also a 1D kernel with weights sum up to 1. Therefore, different weights lead to different frequency responses. 2D filtering can be conducted by applying the outcome kernel to the input image along the x-axis and y-axis successively. This process is efficient even if the length is large. DAT predicts the offset value through another two fully connected layers sharing the same CNN. Hyperbolic tangent activation is used to bound the output and a scale factor is multiplied to obtain the final offset value. To make thresholding differentiable, we adopt differentiable binarization (DB) proposed by Liao et al. [16], which is defined as:

$$\hat{B}_{i,j} = \mathrm{sigmoid}(k(P_{i,j} - T)) \tag{1}$$

where $\hat{B}$ is the feature map after DB, T is the threshold predicted by DAT. Moreover, k is the amplifying factor.

In practice, the two fully connected layers simultaneously predict logits of kernel weights and threshold, and thus its output dimension is (kernel number + 1), We let the window size of the base kernel set to range from 3 to 27; we further set the length to 31 and the amplifying factor k to 50. The initialization of the last fully connected layer is described as follows. The weights corresponding to the threshold are initialized as zeroes, and the bias vector is initialized as a zero vector so that the initial value of the threshold in the forward pass is zero. Kernel weights are initialized using Xavier Initialization.

The scheme described above only outputs a 1-channel feature map. We now extends the method to allow output of arbitrary number of channels to be produced. We only increase the number of output channels of the last fully connected layer to $m \times$ (kernel number + 1), then the final output would consist m channels. The scheme of DAT is shown in Fig. 2.

3.2 Differentiable Unsharp Masking (USM)

Unsharp masking (USM) is a technique used to increase image sharpness. Unlike binarization, unsharp masking does not distinguish foreground and background.

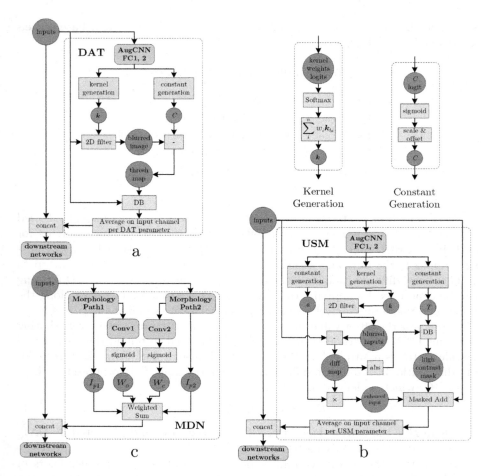

Fig. 2. Illustration of the a) Differentiable Adaptive Thresholding (DAT) module and kernel generation and costant generation; b) differentiable Unsharp Masking (USM) Module and c) Morphological De-Noise (MDN) module

Instead, USM aims to keep most details of original images. The high-frequency parts of the original image are the sharp details, which can be computed by the difference between the original image and its blurred version. To sharpen is to enhance the sharp details, therefore these details are amplified and added back to the original image:

$$I_{\text{sharpened}} = I_{\text{original}} + a(I_{\text{original}} - I_{\text{blurred}})$$

where a denotes the amount or magnitude to sharp.

This operation may incur some undesired noticeable outcome edge, or introduce image noise. However, in digital unsharp masking, undesired effects can be reduced by a high-contrast mask. The mask, produced by a given "high contrast" threshold, selects those pixels to be sharpened. Therefore, three hyperparameters are involved in unsharp masking. They are kernel size for Gaussian blur, threshold T for picking high-contrast value and amount a for the magnitude to sharp. We propose the differentiable unsharp masking module (USM) to learn these parameters automatically. Learning kernel size and threshold utilizes the same mechanism as DAT. Hence, USM also needs an AugCNN followed by two fully connected layers. As for the amount parameter a, an additional output channel is added in the last fully connected layers. Sigmoid activation is then applied, a scale factor is finally multiplied, and thereby the amount can be obtained. Empirically, the two fully connected layers simultaneously predict logits of kernel weights, the threshold, and the amount. We let the window size of the base kernel to range from 3 to 27. We further set the length to 31, the amplifying factor k to 50 and the scale factor for computing amount to 10. The same initialization method as of DAT is used to initialize kernel weights and the threshold. The initialization of the amount a is in the same as threshold, except for its corresponding bias term. The value of this bias is the logit of the initial value of amount (which is set to 1.5). The scheme of USM is described in Fig. 2.

3.3 Morphological De-noise Module (MDN)

Morphological operations are powerful tools to solve problems like shape detection and noise removal [20]. Dilation and Erosion are the most basic morphological operations. Dilation gradually expands the boundaries of foreground objects in an image, while erosion gradually erodes away the boundary regions. The value and number of pixels to expand/thin the boundaries are based upon the pixels on the image and the structuring elements (SE) used as a morphological kernel. Let $I \in \mathbb{R}^{m \times n}$ be the input gray-scale image, $W \in \mathbb{R}^{p \times q}$ be the structuring elements, then 2D Dilation ($\oplus$) and 2D Erosion ($\ominus$) on I is defined as the following [21]:

$$(I \oplus W_d)(x, y) = \max_{i \in S_1, j \in S_2} (I(x + i, y + j) + W_d(i, j))$$
$$(I \ominus W_e)(x, y) = \min_{i \in S_1, j \in S_2} (I(x - i, y - j) - W_e(i, j))$$

$$(2)$$

where $S_1 = [-\lfloor \frac{p}{2} \rfloor, \ldots, 0, \ldots, \lfloor \frac{p}{2} \rfloor]$ and $S_2 = [-\lfloor \frac{q}{2} \rfloor, \ldots, 0, \ldots, \lfloor \frac{q}{2} \rfloor]$, W_d and W_e are dilation and erosion SE or kernels, respectively. Note that W_d and W_e can be optimized by gradient back-propagation, and the optimal parameter values are learned implicitly by maximizing the global text recognition accuracy.

Morphological Opening and Closing are commonly used to remove noise. Morphological Opening is defined as an Erosion operation followed by a Dilation operation, and Closing is defined as a dilation operation followed by an Erosion operation. We adopt the architecture of smaller MorphoN from Mondal et al. [20] as our Morphological De-noise Module (MDN). It contains two paths, and each path is a sequence of alternate morphological operations, mimicking Opening and Closing operations. Traditionally, an Opening or Closing operation uses the same structuring element (or kernel) for both dilation and erosion operations, but MDN is bounded by no such constraint. Thus, the SE associated with each

Erosion and Dilation module could be different. Unlike [20], we use kernel size 3×3 instead of 8×8 for all Dilation, Erosion, and Convolution layers. The two paths of MDN are shown in Table 2:

Table 2. Description of the Morphological De-noise Module (MDN)

Path1-Conv	$1D_{3 \times 3} - 1D_{3 \times 3} - 1D_{3 \times 3} - 1E_{3 \times 3} - 1E_{3 \times 3} - 1D_{3 \times 3} - 1D_{3 \times 3} - 1E_{3 \times 3}$
	$-1E_{3 \times 3} - 1E_{3 \times 3} - 2@3 \times 3 - \tanh -2@3 \times 3 - \tanh -1@3 \times 3 - \text{sigmoid}$
Path2-Conv	$1E_{3 \times 3} - 1E_{3 \times 3} - 1E_{3 \times 3} - 1D_{3 \times 3} - 1D_{3 \times 3} - 1E_{3 \times 3} - 1E_{3 \times 3} - 1D_{3 \times 3}$
	$-1D_{3 \times 3} - 1D_{3 \times 3} - 2@3 \times 3 - \tanh -2@3 \times 3 - \tanh -1@3 \times 3 - \text{sigmoid}$

Both paths output a morphological manipulated map (I_{p1}, I_{p2}) and a weight map (W_o, W_c). The final output I_{out} is the weighted sum from I_{p1} and I_{p2} as shown the following formula:

$$I_{out} = \frac{W_o \odot I_{p1} + W_c \odot I_{p2}}{W_o + W_c} \tag{3}$$

where $\odot$ means pixel-wise multiplication.

The MDN module is illustrated in Fig. 2.

3.4   Module Composition and Configuration Conventions

Conventions. The proposed DAT, USM and MDN modules can either be used alone or in combination. When used in combination, modules can be connected in either sequential or parallel fashion. We propose the convention of module composition. We let "n[Module]" indicate n simple parallel modules. For example, "3DAT" means that three feature maps obtained by three parallel adaptive thresholding filters are concatenated to the raw input image. We let "([Module1]-[Module2]-...-[ModuleN])" to indicate a sequentially connected composite module. For example, "(1DAT-1MDN)" means that one DAT module is used, followed by an MDN, and the resulting feature map from MDN is attached to the raw input image. Equivalent notation "1(DAT-MDN)" can also be used if each type of module only produces one output channel. Furthermore, "Module1, Module2, ..." denotes several composite modules that are parallel to each other. Table 3 shows some examples of our configuration conventions.

Table 3. Examples of Configuration Conventions of Modules

Configuration	Meaning	Output channels
1DAT, 1MDN	One DAT and one MDN are used	2
3DAT, 1(MDN-DAT)	A three-channel DAT and one composite module from one MDN followed by one DAT are used	4
1DAT, 1(DAT-MDN), 1USM	One DAT, one composite module from one DAT followed by one MDN, and one USM are used	3

Shared AugCNN. As mentioned above, both DAT and USM need an AugCNN followed by two fully connected layers to predict proper augmentation hyperparameters. In practice, they share the same AugCNN followed by distinct fully connected branches. No matter how many DATs or USMs are used (unless neither is used), only one AugCNN is needed in the network architecture. Figure 3 shows a complete example of our conventions:

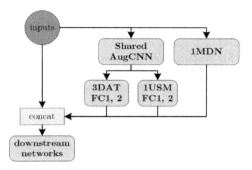

Fig. 3. Demonstration of "3DAT,1USM,1MDN" model

3.5 Recognition Network

The output of the augmentation network can be fed to any recognition networks such as: CRNN [26], ASTER [27], and transformer-based recognizer. We adopt CRNN (ResNet18+BiLSTM+CTC) as our baseline recognition network.

4 Datasets

4.1 Synthetic Datasets

Labeling scene text images is costly, time-consuming, and labor-intensive. Therefore, synthetic data is widely used in the field of STR. SynthText (ST) and MJSynth (MJ) are the two most popular synthetic datasets to train scene text recognizers. There are approximately 9 million and 8 million synthetic text images in ST and MJ, respectively. Our models are trained only on ST and MJ.

4.2 Real-World Datasets

We evaluate our methods on widely used real-world benchmarks. When evaluating, all non-alphanumeric characters are not accounted for, and all letters are case-insensitive. Baek et al. [1] thoroughly discussed the specifics of benchmark datasets in STR. We briefly describe the datasets we used to evaluate the performance of our method:

IIIT5K contains 2000 labeled word images for training and 3000 for testing.

ICDAR 2013 (IC13) contains 848 labeled word images for training and 1095 for testing.

ICDAR 2015 (IC15) contains 4468 labeled word images for training and 2077 for testing. Since they are cropped from incidental scene images captured under random perspective, some text images are oriented, perspective, or curved.

Street View Text Perspective (SVTP) contains 639 labeled word images, all for testing. They are cropped from side-view snapshots in Google Street View and thus suffering from severe perspective distortion.

CUTE80 (CT80) contains 288 curved labeled word images, all for testing.

COCO-Text (COCO-T) contains 42618 text images for training, 9896 for validation, and 9838 for testing. We cleaned COCO-T's training and validation patches. Also, the original testset images does not come with associated labels. We add text labels to the images for the testset. There are several issues with the original COCO-T labels. About 13% images are in low resolution or heavily occluded and therefore not readable to professional human labelers. Some image patches contain vertical text or contain no text at all. We illustrate some problematic examples in COCO-T in Fig. 4. When labeling, we dropped all patches that are not recognizable or contain no characters. We tag all recognizable patches with one of four following rotation angles (clockwise): 0, 90, 180, and 270°. When evaluating on COCO-T, images without a non-zero rotation tag will be rotated for the text recognizer to process properly. After label cleaning, 37710 valid patches from training, 8638 valid patches from validation and 7859 valid patches from testset are kept. Our method is evaluated on the test set only. We will open-source a cleaned version of COCO-T for future researchers.

Fig. 4. Problematic image samples from COCO-T dataset

4.3   Private Dataset

Manufacture Assembly Line Text (MALT) is a self-built dataset that contains images taken by industrial cameras from manufacturing assembly lines for electronics. These texts were formed by different crafts, such as printing, laser engraving, plastic injection, embossing, thus incurring various texts appearances. Some of them also suffer from degradations like blurriness, over-exposure, oxidation or corrosion, which further challenge the OCR system. We split the MALT dataset into training set, validation set and test set. The test set contains 1000 images, and it will be open-sourced for future research. Before evaluating, we fine-tune our models on a mixed set consisting of the MALT training set, 50k random samples from MJ, and 50k random samples from ST, for 10k steps at a learning rate of 0.001. The final evaluation is conducted on the MALT test set.

5   Experiments

5.1   Implementation Details

Networks. As our methods are pluggable, the network can be trained end-to-end no matter what text recognizer is used. In the ablation study, we experiment on CRNN $ResNet18 + BiLSTM + CTC$, which adopts CTC loss. ResNet-18 [7] is used as the backbone network.

Optimization. We conduct all experiments using the ADADELTA optimizer with learning rate of 1.0, training for 1 million steps. Learning rate decays at 400k, 600k and 800k step at a decay rate of $0.01^{1/3}$. The batch size is set to 128. In each batch, samples are randomly drawn and the number of samples from MJ and ST are equal. All images are resized to (32, 140). Each experiment is conducted on a single NVIDIA Tesla V100 GPU.

5.2   Experiment Results

We train a CRNN model $ResNet18 + BiLSTM + CTC$ from scratch as our baseline model. We then compare the performance of our proposed methods with the baseline model. No lexicon is used in the evaluation on any benchmark dataset. We will use configuration conventions described in Subsect. 3.4 to denote each experiment. Recognition word accuracy (in percentages) is the evaluation metric. Recognition results are summarized in Table 4.

Table 4. Scene text recognition performance among various widely used benchmarks. Config1: [1USM,1(MDN-DAT)]+CRNN, Config2: [1USM,1(DAT-MDN)]+CRNN; DAT: Differentiable Adaptive Thresholding, USM: Differentiable Unsharp Masking, MDN: Morphological De-noise Module

Methods	Data	IIT5K	IC03		IC13		IC15		SP	CUTE80	COCO-T	MALT
		3000	860	867	857	1015	1811	2077	645	288	7859	1000
CRNN [26]	90K	78.2	89.4	-	-	86.7	-	-	-	-	-	-
RARE [25]	90K	81.9	90.1	-	-	88.6	-	-	-	71.8	59.2	-
FAN [3]	90K, ST, C	87.4	-	94.2	-	93.3	70.6	-	-	-	-	-
ASTER [27]	90K, ST	93.4	94.5	-	-	91.8	-	76.1	-	-	-	-
SAR [15]	90K, ST, SA	91.5	-	-	-	91.0	-	69.2	-	-	-	-
SATRN [14]	90K, ST	92.8	-	96.7	-	94.1	-	79.0	86.5	-	-	-
CRNN(Baseline)	90K, ST	91.3	90.7	-	-	90.7	-	70.6	71.3	73.2	61.7	63.4
1DAT+CRNN	90K, ST	92.6	91.5	-	-	90.6	-	72.5	72.4	76.4	62.8	69.5
1USM+CRNN	90K, ST	91.1	90.0	-	-	89.9	-	71.4	69.8	76.7	62.0	64.9
1MDN+CRNN	90K, ST	92.0	90.7	-	-	90.0	-	72.3	70.5	72.9	62.4	77.1
1(DAT-MDN)+CRNN	90K, ST	91.9	90.7	-	-	90.6	-	73.1	71.3	75.7	63.1	71.0
1(MDN-DAT)+CRNN	90K, ST	91.3	90.1	-	-	89.9	-	72.0	72.1	75.0	62.3	74.5
Config1	90K, ST	93.0	90.8	-	-	90.6	-	72.9	75.0	72.7	63.6	78.6
Config2	90K, ST	92.2	90.7	-	-	90.0	-	72.9	76.4	73.0	63.2	75.1

Table 4 summarizes the recognition results among 7 widely used datasets and our private dataset MALT. We conduct a series of experiments with different combinations of augmentation modules. We compare our method with the baseline model on benchmark datasets to demonstrate the effectiveness. We first tried combining five types of one-channel augmentation module with a CRNN text recognizer. The DAT module is designed to enhance blurred edges and correct low-contrast images. However more noise might be introduced as a side effect. The USM module is expected to retain more details. The MDN module is designed to remove small particle-like noises. Sequentially connected modules DAT-MDN and MDN-DAT are expected to function similar to DAT, but output features with less induced noise. With one-channel augmentation module, text recognizer shows moderate improvements on benchmark datasets with more degraded images. We further explored more complex combinations of augmentation modules. Experiment results show that many different combinations of augmentation modules achieved significant improvement in most of the cases. Moreover, our methods performed particularly well on datasets with a large amount of hard-to-read images in various degraded conditions (such as IC15, COCO-T, and MALT). We list top two configurations to demonstrate the effectiveness of our method. Both Config1 and Config2 outperformed the CRNN baseline model in most benchmark datasets by a large margin. DAT, USM, and MDN are designed to obtain more robust feature maps by mimicking traditional CV operations. Based on experiment results, we can see that the proposed modules indeed improves the text recognition baseline models.

Note that the augmentation modules are plugable, which makes it possible to combine augmentation modules with any kind of text recognizer. In future research, the effectiveness of proposed augmentation modules will be verified on more recent state-of-the-art text recognizers.

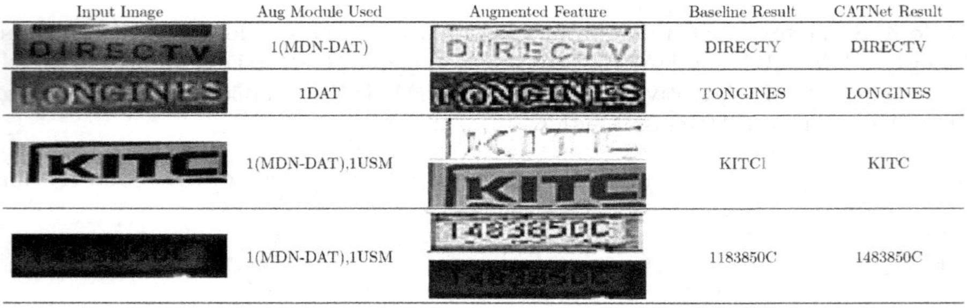

Input Image	Aug Module Used	Augmented Feature	Baseline Result	CATNet Result
	1(MDN-DAT)		DIRECTY	DIRECTV
	1DAT		TONGINES	LONGINES
	1(MDN-DAT),1USM		KITCl	KITC
	1(MDN-DAT),1USM		1183850C	1483850C

Fig. 5. Augmented features and recognition results by [26] and by CATNet

We illustrate some hard examples that were originally recognized incorrectly by the baseline model. By adding different combinations of augmentation modules, hard examples could be recognized correctly. Feature maps generated by different augmentation modules are visualized in Fig. 5. Visualization results indicate that the auxiliary module has improved feature quality and reduced the difficulty of the recognition network for challenging cases. The recognition performance is improved as a result.

6 Conclusions

In this paper, we propose an end-to-end trainable text recognition model that consists of an augmentation module and text recognizer. The augmentation module fuses feature obtained from morphological operations. The augmentation module aims to extract rich features from various morphological operations and fuse the features with the raw input image to ease the difficulty of the downstream recognition task. Effective transformations are learned implicitly by enhancing the global text recognition accuracy. This way, our method bridges the gap between the traditional CV techniques and deep neural network by joint learning. The proposed method is simple yet effective. It can automatically adapt to general text recognition tasks without any manual modification. Extensive experiments show that our method boosts the performance of the recognizers, especially for degraded text images in challenging conditions. We would like to explore a more systematic way to determine and utilize useful augmented features in the future. The "Squeeze-and-Excitation" block proposed by Hu et al. [9] could be a possible solution. Extending our method to solve low-quality images other computer vision tasks such as multiple object detection and recognition is also another interesting future direction.

References

1. Baek, J., Kim, G., Lee, J., Park, S., Han, D., Yun, S.: What is wrong with scene text recognition model comparisons? Dataset and model analysis. In: Proceedings of the IEEE International Conference on Computer Vision, pp. 4715–4723, (2019)
2. Bookstein, F.L.: Principal warps: thin-plate splines and the decomposition of deformations. IEEE Trans. Pattern Anal. Mach. Intell. **11**(6), 567–585 (1989)
3. Cheng, Z., Bai, F., Xu, Y., Zheng, G., Pu, S., Zhou, S.: Focusing attention: towards accurate text recognition in natural images. In: Proceedings of International Conference on Computer Vision (ICCV), pp. 5086–5094 (2017)
4. Dalal, N., Triggs, B.: Histograms of oriented gradients for human detection. In: Proceedings of Computer Vision and Pattern Recognition (CVPR) (2005)
5. Goodfellow, I., Bengio, Y., Courville, A.: Deep Learning. MIT Press (2016). http://www.deeplearningbook.org
6. Graves, A., Fernández, S., Gomez, F., Schmidhuber, J.: Connectionist temporal classification: labelling unsegmented sequence data with recurrent neural networks. In: ICML, pp. 369–376 (2006)
7. He, K., Zhang, X., Ren, S., Sun, J.: Deep residual learning for image recognition. In: CVPR, pp. 770–778 (2016)
8. Hornik, K., Stinchcombe, M., White, H.: Multilayer feedforward networks are universal approximators. Neural Netw. **2**(5), 359–366 (1989)
9. Hu, J., Shen, L., Sun, G.: Squeeze-and-excitation networks. In: CVPR (2018)

10. Jaderberg, M., Simonyan, K., Vedaldi, A., Zisserman, A.: Reading text in the wild with convolutional neural networks. IJCV **116**(1), 1–20 (2016)
11. Jaderberg, M., Simonyan, K., Zisserman, A., Kavukcuoglu, K.: Spatial transformer networks. In: NIPS (2015)
12. Krizhevsky, A., Sutskever, I., Hinton, G.E.: ImageNet classification with deep convolutional neural networks. In: Proceedings of the 25th International Conference on Neural Information Processing Systems, NIPS 2012, vol. 1, pp. 1097–1105 (2012)
13. Lee, C.-Y., Osindero, S.: Recursive recurrent nets with attention modeling for OCR in the wild. In: Proceedings of Computer Vision and Pattern Recognition CVPR, pp. 2231–2239 (2016)
14. Lee, J, Park, S., Baek, J., Oh, S.J., Kim, S., Lee, H.: On recognizing texts of arbitrary shapes with 2d self-attention. In: Proceedings of the IEEE/CVF Conference on Computer Vision and Pattern Recognition Workshops, pp. 546–547 (2020)
15. Li, H., Wang, P., Shen, C., Zhang, G.: Show, attend and read: a simple and strong baseline for irregular text recognition. Proc. AAAI Conf. Artif. Intell. **33**, 8610–8617 (2019)
16. Liao, M., Wan, Z., Yao, C., Chen, K., Bai, X.: Real-time scene text detection with differentiable binarization. In: 34th AAAI Conference on Artificial Intelligence (2020)
17. Long, S., He, X., Yao, C.: Scene text detection and recognition: the deep learning era. CoRR abs/1811.04256 (2018)
18. Luo, C., Jin, L., Sun, Z.: MORAN: a multiobject rectified attention network for scene text recognition. Pattern Recogn. **90**, 109–118 (2019)
19. Luo, C., Zhu, Y., Jin, L., Wang, Y.: Learn to augment: joint data augmentation and network optimization for text recognition. In: Proceedings of the IEEE/CVF Conference on Computer Vision and Pattern Recognition, pp. 13746–13755 (2020)
20. Mondal, R., Purkait, P., Santra, S., Chanda, B.: Morphological networks for image de-raining. In: Couprie, M., Cousty, J., Kenmochi, Y., Mustafa, N. (eds.) DGCI 2019. LNCS, vol. 11414, pp. 262–275. Springer, Cham (2019). https://doi.org/10.1007/978-3-030-14085-4_21
21. Mondal, R., Santra, S., Chanda, B.: Dense morphological network: an universal function approximator. arXiv preprint arXiv:1901.00109 (2019)
22. Mou, Y., et al.: PlugNet: degradation aware scene text recognition supervised by a pluggable super-resolution unit. In: Vedaldi, A., Bischof, H., Brox, T., Frahm, J.-M. (eds.) ECCV 2020. LNCS, vol. 12360, pp. 158–174. Springer, Cham (2020). https://doi.org/10.1007/978-3-030-58555-6_10
23. Nash, W., Drummond, T., Birbilis, N.: A review of deep learning in the study of materials degradation. npj Mater. Degrad. **2**, 37 (2018). https://doi.org/10.1038/s41529-018-0058-x
24. Neumann, L., Matas, J.: Real-time scene text localization and recognition. In: Proceedings of the CVPR, pp. 3538–3545. IEEE (2012)
25. Shi, B., Wang, X., Lyu, P., Yao, C., Bai, X.: Robust scene text recognition with automatic rectification. In: Proceedings of Computer Vision and Pattern Recognition (CVPR), pp. 4168–4176 (2016)
26. Shi, B., Bai, X., Yao, C.: An end-to-end trainable neural network for image-based sequence recognition and its application to scene text recognition. TPAMI **39**, 2298–2304 (2017)
27. Shi, B., Yang, M., Wang, X., Lyu, P., Bai, X., Yao, C.: ASTER: an attentional scene text recognizer with flexible rectification. IEEE Trans. Pattern Anal. Mach. Intell. **31**(11), 855–868 (2018)

28. Wang, W., et al.: TextSR: contentaware text super-resolution guided by recognition. arXiv:1909.07113 (2019)
29. Yang, X., He, D., Zhou, Z., Kifer, D., Giles, C.L.: Learning to read irregular text with attention mechanisms. In: Proceedings of International Joint Conference on Artificial Intelligence (IJCAI), pp. 3280–3286 (2017)

Explore Hierarchical Relations Reasoning and Global Information Aggregation

Lei Li[1], Chun Yuan[2(✉)], and Kai Fan[3]

[1] Department of Computer Science and Technology, Tsinghua University,
Beijing, China
lei-li18@mails.tsinghua.edu.cn
[2] Tsinghua Shenzhen International Graduate School, Beijing, China
yuanc@sz.tsinghua.edu.cn
[3] Alibaba DAMO Academy, Hangzhou, China

Abstract. Existing Graph Convolution Networks (GCNs) mainly explore the depth structure with focusing on aggregating information from the local 1-hop neighbor, leading to the common issues of over-smoothing and gradient vanishing, and further hurting the model performance on downstream tasks. To alleviate these deficiencies, we tentatively explore the width of the graph structure and propose Wider Graph Convolution Network (WGCN), targeting to reason the hierarchical relations and aggregate the global information from multi-hops dilated neighbor in a wide but shallow framework. Meanwhile, Dynamic Graph Convolution (DGC) is adopted via the masking and attention mechanisms to distinguish and re-weight informative neighborhood nodes, for stabilizing the model optimization. We demonstrate the effectiveness of WGCN on three popular tasks, including Document classification, Scene text detection and Point cloud segmentation. On the basis of achieving state-of-the-art performance among presented tasks, we verify that exploration through width expansion of GCNs is more effective than stacking more layers for model improvements.

Keywords: Graph Convolution Networks · Dynamic Graph Convolution · Document classification · Scene text detection

1 Introduction

Graph Convolution Networks (GCNs) have revolutionized various applications due to their well-suit for handling graph data [27], such as documents [28], texts [30], and point clouds [12]. Existing GCNs are multilayer neural networks that operate directly on the graph structure and generate the node representations from the most relevant portions in the graph, which is commonly known

Supported by NSFC project Grant No. U1833101, SZSTI under Grant No. JCYJ20190809172201639, the Joint Research Center of Tencent and Tsinghua, and National Key R&D Program of China (No. 2020AAA0108303).

J. Lladós et al. (Eds.): ICDAR 2021, LNCS 12821, pp. 366–381, 2021.
https://doi.org/10.1007/978-3-030-86549-8_24

as aggregating the local 1-hop neighbor information [17]. In order to capture the global topology structure in the graph and reason the relations of nodes from the larger receptive field, one usually stacks more graph convolution layers [12,28].

However, the design patterns of current GCNs have several deficiencies. First, information contained from the local 1-hop neighbor is likely to be incomplete and not optimal [5], limiting the performance on downstream tasks. Second, some works [17] raised that with the model depth increased, the contribution of distant nodes to information aggregation decreased exponentially, forcing the model to focus more on local rather than global structure. Third, the nodes readily converge to similar representations to trigger over-smoothing problem [10] in training the deep GCNs, increasing the difficulty of the convergence in model optimization due to gradient vanishing problem.

To alleviate the above defects, most recent works have proposed to improve the aggregation fashion of nodes, or enlarge the receptive field to involve more nodes for the aggregation, like GraphSAGE [6] and ResGCN [12]. However, both methods insufficiently explored the width of GCNs and still needed to stack a quite number of layers for performance improvement. Besides, fixed receptive field of the models is not conducive to perform effectively relational reasoning and information aggregation in the graph.

In this work, we intensively explore the width of GCNs, and empirically prove that the width plays a more important and effective role than the depth for performance improvement. In addition, Dynamic Graph Convolution is proposed to adaptively change the receptive field from multi-hops dilated neighbor, which is efficiently for hierarchical reasoning and global aggregation. **We summarize our contributions as follows: (i)** We propose Wider Graph Convolution Network (WGCN), a wide but shallow framework, where the hierarchical relations reasoning and the global information aggregation are performed on multi-hops dilated neighbor. **(ii)** We introduce dynamic graph convolution (DGC) to adaptively change the receptive field in the graph. This module can help to distinguish and re-weight informative neighborhood nodes through masking and attention mechanisms, while stabilizing the training process. **(iii)** We conduct empirical evaluation on three tasks, including Document classification, Scene text detection and Point cloud segmentation. The results verify that WGCN achieves significant improvements by effectively exploiting the width of GCNs.

2   Related Work

Graph Convolution Networks. Graph Convolution Networks (GCNs) are typically divided into two categories: spectral-based and spatial-based approaches. Spectral-based approaches define the graph convolution operators through polynomials of graph Laplacian, which is equivalent to conducting spectral filtering in the graph spectral domain. Defferrard et al. proposed ChebNet [2] that approximated the filter by Chebyshev polynomials of the diagonal matrix of eigenvalues. Kipf et al. [10] introduced the first-order approximation of Cheb-Net to achieve its impressive performance in classification tasks. Spatial-based

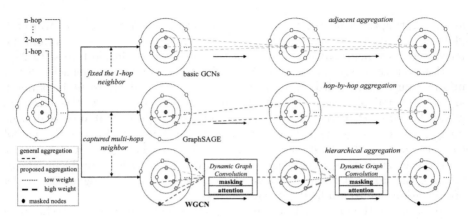

Fig. 1. Schematic illustration of different aggregation steps.

approaches usually construct the graph convolution operators by aggregating and transforming the nodes of the 1-hop neighbor. GraphSAGE [6] obtained multi-hops neighbor through hop-by-hop aggregation from multiple graph convolution layers. In LGCN [4], a fixed number of neighborhood nodes are selected for each target node, and a learning graph convolution layer is proposed to realize regular convolution of graphs. However, they did not fully explore the width of GCNs for hierarchical reasoning of multi-hops neighbor. Figure 1 illustrates the main difference of information aggregation steps among the basic GCNs [2,10], GraphSAGE [6], and our WGCN.

Dynamic Graph Convolution. Recent works [12,20,25], generally called Dynamic Graph Convolution, have been proposed to demonstrate the effectiveness of learning better node representations, by applying adaptive operators in the graph. Compared to the GCNs that performed with the fixed graph convolution operators, the graph structure of DGC based methods are allowed to be variable in each layer. For instance, Edge-Conditioned Convolution (ECC) [20] introduced dynamic edge-conditional filters, whose weights are conditioned on edge labels and dynamically generated for the target node. EdgeConv [25] proposed to dynamically change the neighbors in the current feature space, for reconstructing the graph after every graph convolution layer (GCL). For comparison, we introduce DGC via the masking and attention mechanisms to efficiently perform hierarchical relations reasoning from multi-hops neighbor in the graph.

Application of GCNs on Downstream Tasks. New generalizations and definitions have been rapidly developed to handle the complexity of the graph-based data on many downstream tasks. For instance, Text GCN [28] regarded the documents and words as undifferentiated nodes to construct the corpus graph and adopted the basic GCNs [10] for Document classification. GraphText [30] applied the modified GCNs to bridge the segmentation-based text proposal model and the relational reasoning network for Scene text detection. ResGCN [12] introduced the *Dilated k-NN* to promote the representational capacity of GCNs, increasing the receptive field of nodes for Point cloud segmentation. In contrast, we perform extensive experiments to verify that the proposed WGCN establishes the new state-of-the-art standards on all mentioned tasks.

3   Our Proposal

In this section, we first elaborate the distinct definitions of multi-hops dilated neighbor in the graph. We then introduce the details of our novel Wider Graph Convolution Network and Dynamic Graph Convolution.

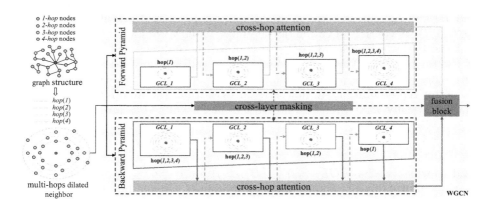

Fig. 2. The pipeline of the proposed Wider Graph Convolution Network.

3.1   Definitions of Multi-hops Dilated Neighbor

Let $G = (\nu, \varepsilon)$ denotes the graph where ν is the set of unordered nodes and ε is the set of undirected edges, while $e_{i,j} \in \varepsilon$ implies that v_i and v_j are connected to each other in this graph. For any two nodes v_i and v_j: (1) if there is a directly connected edge $e_{i,j}$ in G, then v_j is included in the 1-hop neighbor of v_i; (2) if it takes at least n edges from v_i to v_j, then v_j is included in the n-hop neighbor of v_i; (3) if the node has at least 2-hops neighbor, for instance, containing both the 1- and 2-hop neighbor, which will be collectively referred to as multi-hops dilated neighbor. Particularly, we introduce a new concept **node-wise multi-hops dilated neighbor** $N_k^{(H,S)}(v_i)$. Without loss of generality, we define:

Definition. *Let* $N_k^{(H,S)}(v_i)$ *denotes the multi-hops dilated neighbor of* v_i *with* $H = \{h_1 = n_1, h_2 = n_2, ..., h_k = n_k\}$ *and* $S = \{h_1 = s_1, h_2 = s_2, ..., h_k = s_k\}$ *in the k-th layer, then the whole receptive field* $RF = \{n_1 + n_2 + ... + n_k\}$ *will* be captured by sampling a total number of neighborhood nodes up to $SN = \{s_1 + s_2 + ... + s_k\}$.

H means the whole contained hop scales, and S indicates the number of sampled neighborhood nodes at each hop scale. For example, setting $H = \{h_1 = 1, h_2 = 2\}$ and $S = \{h_1 = 8, h_2 = 4\}$ means that we build multi-hops dilated neighbor by sampling 8 and 4 neighborhood nodes at the 1- and 2-hop neighbor, respectively. We perform random sampling of neighborhood nodes at each hop scale for the sake of computational simplicity.

3.2   Wider Graph Convolution Network

An example of the proposed WGCN with four layers is given in Fig. 2, which is built under the following settings: $(H_1 = \{1\}, S_1 = \{8\})$, $(H_2 = \{1, 2\}, S_2 = \{8, 8\})$, $(H_3 = \{1, 2, 3\}, S_3 = \{8, 8, 4\})$, and $(H_4 = \{1, 2, 3, 4\}, S_4 = \{8, 8, 4, 4\})$, and H_i and S_i represent the obtained multi-hops dilated neighbor in the i-th GCL. WGCN consists of two major modules, namely the forward pyramid (FP) and the backward pyramid (BP), to capture the bidirectional representations of the target node. The bidirectional setting will not only make the architecture more compatible with the following DGC operators, but also demonstrate its better representations in downstream tasks than the unidirectional setting. Mathematically, a K-layer FP and BP can be formulated as follows:

$$FP = \left\{ N_1^{(H_1, S_1)}(v_i), .., N_K^{(H_K, S_K)}(v_i) \right\} BP = \left\{ N_K^{(H_K, S_K)}(v_i), .., N_1^{(H_1, S_1)}(v_i) \right\}. \quad (1)$$

And in which we have $(H_C, S_C) \subseteq (H_D, S_D)$ for $1 \leq C \leq D \leq K$. It implies that the multi-hops dilated neighbor captured in H_C should be contained in H_D, and the corresponding sampled nodes in S_C are included in S_D as well. Hierarchical relations reasoning between nodes across multi-hops dilated neighbor is performed by sequentially increasing one hop scale at each GCL of the proposed FP and decreasing one hop scale of BP. For each layer in the basic GCNs, the key is to learn an aggregation function g as well as an update function f to generate the target node v_i's representations. Previous implementations comprehensively consider the own features x_i of the target node and the features of all its neighborhood nodes $\aleph(v_i)$, which is commonly denoted as the 1-hop neighbor in previous settings. Similarly, the corresponding convolutional operators for multi-hops dilated neighbor in our WGCN could be formulated as:

$$G_{k+1}(v_i) = f_k(x_i, \ g_k(\aleph(v_i))) = f_k\left(x_i, \ g_k\left(\aleph_k^{(H, \ S)}(v_i)\right)\right), \quad (2)$$

where $\aleph_k^{(H, S)}(v_i)$ denotes the features of the multi-hops dilated neighbor in the k-th layer. Instead of using the local 1-hop neighbor which may be incomplete and not optimal, the proposed WGCN is adept in aggregating global information with the larger receptive field. Besides, this aggregation procedure is efficiently balanced by the contributions from not only close but also distant neighborhood nodes, which is achieved by the introduced DGC and is detailed below.

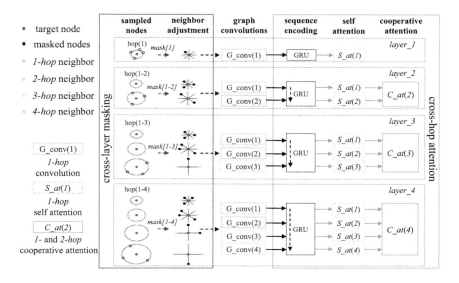

Fig. 3. Schematic illustration of introduced Dynamic Graph Convolution.

3.3 Dynamic Graph Convolution

Previous works [6,10] perform information aggregation of the local 1-hop neighbor, resulting in not only the contribution of distant nodes decreasing exponentially as the model goes deeper, but also the performance degradation caused by irrelevant or corrupted neighborhood nodes. In contrast, Dynamic Graph Convolution is introduced in order to adaptively adjust the graph structure by distinguishing and re-weighting informative nodes through the captured multi-hops dilated neighbor. Specifically, cross-layer masking mechanism is built across multiple graph convolution layers of proposed WGCN, for rewarding relevant and informative neighborhood nodes while penalizing leftovers. Meanwhile, cross-hop attention mechanism is proposed across the whole hop scales in each GCL, for balancing both the close and distant nodes within aggregated multi-hops dilated neighbor. The whole process is shown in Fig. 3.

Cross-Layer Masking. For the target node v_i, we create the soft mask on each hop scale by two weight matrices $M_F \in R^{H \times S}$ and $M_B \in R^{H \times S}$, which are initialized and fed to the two pyramids of WGCN, respectively. H is the whole hop scales contained in each pyramid and S is the total number of sampled neighborhood nodes. Given the h-th ($1 \leq h \leq H$) hop scale in the k-th ($1 \leq k \leq K$) layer of the K-layer FP, the features of sampled nodes and M_F are first adopted to predict the soft mask $m_k^h(v_i)$ as the following,

$$m_k^h(v_i) = \sigma \left(MaxPool \left(\aleph_k^{(n_h, \, s_h)}(v_i) \cdot M_F(h) \cdot MeanPool(\aleph_k^{(n_h, \, s_h)}(v_i)) \right) \right). \quad (3)$$

Where $M_F(h)$ denotes the h-th row of M_F, "." indicates the matrix multiplication, and σ is the sigmoid function. The obtained $m_k^h(v_i)$ serves as the information gatekeeper, which will be multiplied by the h-hop neighbor features and processed by a linear transformer to get the final output:

$$\tilde{o}_k^h(v_i) = \aleph_k^{(n_h, \, s_h)}(v_i) \otimes m_k^h(v_i), \quad o_k^h(v_i) = W_k \cdot \tilde{o}_k^h(v_i) + b_W, \quad (4)$$

where $\otimes$ denotes the element-wise product operation. In this way, the information aggregation at the h-th hop scale is restricted to a dynamic sub-part of the whole graph. The informative nodes at each hop scale will be rewarded and the leftovers will be penalized, making the information aggregation more discriminative. Furthermore, this mechanism is conductive to eliminate irrelevant nodes, resulting in a lighter architecture and stabilizing the training process.

Cross-Hop Attention. We introduce cross-hop attention mechanism to re-weight the importance of each node across the captured whole hop scales in each layer. This module mainly contains a sequence encoder, an intra-hop self-attention unit, and an inter-hop cooperative attention unit. Given the k-th layer of each pyramid in WGCN, we follow the ascending order of the hop scale to input the features of neighborhood nodes into the encoder. Formally, for each hop $h = 1, 2, ...$, we have $g_k^h(v_i) = GRU_k\left(\tanh\left(G_k \cdot o_k^{\tilde{h}}(v_i) + b_G\right)\right)$, where $\tilde{h}$ is the set of hop scales from the 1-hop to the h-th hop, and G_k is the weight matrix and b_G is the bias. By inferring the feature representations of the contained multi-hops dilated neighbor as the sequence encoding process, the features distribution and local structure from the neighbor within lower hop scale are helpful to supervise the neighbor within higher hop scale. Therefore, nodes from distant neighbor could be efficiently aggregated while maintaining the structure stability of the whole multi-hops dilated neighbor during training. To re-weight the importance of nodes at each hop scale, the outputs of the sequence encoder is fed to a self-attention unit as follows,

$$S_{at}(h) = \bigcap_{j=1}^{s_h}(a_k^{h, \, j}(v_i) \otimes o_k^{h, \, j}(v_i)), \quad a_k^{h, \, j}(v_i) = \frac{e^{g_k^{h, \, j}(v_i)}}{\sum_{j=1}^{s_h} e^{g_k^{h, \, j}(v_i)}}. \quad (5)$$

Where $\bigcap$ denotes the concatenation operators. Considering the interaction of neighborhood nodes from different hop scales, a cooperative attention unit is then adopted to re-weight across the whole hop scales for balancing both the close and distant nodes. We introduce the cooperative attention learning process for $h \in [1, H-1]$, which is designed to measure the attention coefficients of the captured receptive field based on adjacent hop scales,

$$C_{at}(h, \, h+1) = softmax\left(W(h) \cdot S_{at}(h+1) \oplus W(h)^\top \cdot S_{at}(h)\right), \quad (6)$$
$$W(h) = C_{at}(h-1, \, h) \cdot S_{at}(h+1)^\top, \quad W(1) = S_{at}(1) \cdot S_{at}(2)^\top.$$

Where $\oplus$ represents the concatenation of feature vectors. The co-attention coefficients are obtained through the weighted calculation of the adjacent two hop

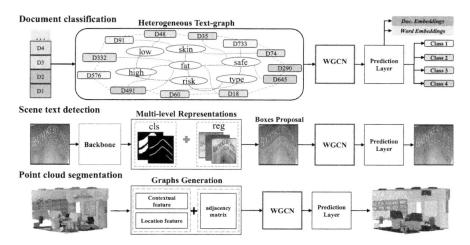

Fig. 4. Apply our WGCN to different tasks.

Table 1. Different configurations of our proposal.

Model	Nodes	Dilated neighbor	Depth	Width
$WGCN_{4L\,4W}$	(16, 8, 8, 4)	hop(1,1–2,1–3,1–4)	4	4
-M / -A / -M-A	(16, 8, 8, 4)	hop(1,1–2,1–3,1–4)	4	4
-BP / -F$_{usion}$	(16, 8, 8, 4)	hop(1,1–2,1–3,1–4)	4	4
$WGCN_{4L\,3W}$	**(16, 8, 8)**	**hop(1,1–2,1–3,1–3)**	4	**3**
$WGCN_{4L\,5W}$	**(16, 8, 8, 4, 4)**	**hop(1,1–2,1–4,1–5)**	4	**5**
$WGCN_{3L\,4W}$	(16, 8, 8, 4)	**hop(1,1–2,1–4)**	**3**	4
$WGCN_{5L\,4W}$	(16, 8, 8, 4)	**hop(1,1–2,1–3,1–4,1–4)**	**5**	4

scales, while the weight $W(h)$ is pre-computed by comprehensively considering the hierarchical information of the $(h-1)$-, h- and $(h+1)$-th hop scales. Thus, the co-attention output $C_{at}(H-1,H)$ of the k-th layer is obtained and finally passed through the multi-layer perceptron (MLP) with residual connections between consecutive layers: $\tilde{x}_{k+1}(v_i) = MLP(C_{at}(H-1,H)) + x_k(v_i)$. When iterating the above process until the last layer of WGCN, the row-wise concatenation of the output from both pyramids is performed to produce the representations of the target node, which will be sent to the subsequent prediction layer according to applications, according to the following formulation $x_{k+1}(v_i) = (MaxPooling(M_F \oplus M_B)) \cdot (\tilde{x}_{k+1}^{FP}(v_i) \oplus \tilde{x}_{k+1}^{BP}(v_i))$.

4   Experiments

We conduct extensive experiments on three tasks namely Document classification, Scene text detection and Point cloud segmentation. Figure 4 illustrates the basic rationale of our WGCN applied in relevant tasks.

Configurations of WGCN. Different configurations of proposal are illustrated in Tabel 1. For a reference model called $WGCN_{4L4W}$, it is built with 4 layers deep and 4 hop scales wide. The hop(1,1–2,1–3,1–4) combined with (16, 8, 8, 4) denotes the first layer of FP includes only the 1-hop neighbor while the last layer contains the 1-, 2-, 3-, and 4-hop dilated neighbor. Besides, the number of sampled neighborhood nodes at 1-, 2-, 3- and 4-hop are 16, 8, 8 and 4, respectively. -M, -A, -BP and -F_{usion} denote to remove cross-layer masking, cross-hop attention, the backward pyramid (BP), and the feature fusion of FP and BP, separately.

4.1   Document Classification

Implementation Details. We evaluate on three popular corpora [28]: *Ohsumed* (Ohs.), *20-Newsgroups* (20NG) and *Movie Review* (MR). The proposed WGCN (a reference model called $WGCN_{4L4W}$) and its variants are compared with several strong baselines including: (1) *traditional method* (TF-IDF+LR); (2) *context-embedding based methods* (LSTM [15], fastText [9], and LEAM [22]); (3) *graph based methods* (PTE [21], Graph-CNN(-C) with Chebyshev filter [2], and Text GCN [28]).

For this task, the text-graph is first constructed to make use of the document (or word) nodes from multi-hops dilated neighbor for hierarchical relations reasoning. The number of nodes in the text-graph is equal to the total number of input documents and unique words in the built vocabulary, while all the node features are initialized randomly with the standard normal distribution. The popular Simhash algorithm is adopted to compute the similarity of two documents by calculating the Hamming distance between document pairs for building the *document-document* edges. Besides, the *document-word* edges are built according to the term frequency-inverse document frequency (TF-IDF). Similar to [28], the sliding windows with fixed size on all documents are utilized to capture the global word co-occurrence information and point-wise mutual information (PMI), which are employed to measure weights of the *word-word* edges (see visualisations in the top part of Fig. 4). The classification loss is generally computed as the cross entropy L_{ce} between the predicted per-class probabilities and the one-hot encoded ground truth. Besides, we introduce the loss L_h for the captured multi-hops dilated neighbor, $L_h = \lambda_1||M_F||_2^2 + \lambda_2||M_B||_2^2 + \lambda_3 \frac{1}{||M_F| - |M_B||}$, where λ_1 and λ_2 are used to force M_F and M_B to be sparse, so that only part of the captured nodes are distinguished for aggregation. λ_3 is adopted to guide the forward pyramid and the backward pyramid to generate distinct node representations as well as graph structure distributions, thus avoiding the overfitting. The final loss for this task is computed as the sum: $L_{cls} = L_h + L_{ce}$. In the experiments, M_F and M_B are randomly initialized with standard normal distribution, while λ_1, λ_2 and λ_3 are set to 0.001, 0.001 and 0.002 respectively. The threshold of text similarity is set to 0.5. The size of sliding window of *word-word* edges is fixed at 20. The hidden layer size is 200. The dropout rate is 0.5. 10% of the training set are held-out for validation. We trained our models for a maximum of 200 epochs using Adam optimizer with the same initial learning rate (0.001) and with early stop strategy of patience epochs $= 10$. We adopted the same hyper-parameter settings for other tasks if no special instruction. We repeat each experiment 20 times to report the mean and the standard deviation of classification accuracy.

Table 2. We compare our WGCN to several algorithms on three popular corpora. The top two accuracy (%) values on the verification set are highlighted in each corpus.

Model	Ohsumed	20NG	MR
TF-IDF+LR	54.35 ± 0.00	82.54 ± 0.00	75.82 ± 0.00
LSTM [15]	42.17 ± 1.09	63.98 ± 1.48	76.18 ± 0.35
fastText [9]	54.33 ± 0.51	71.25 ± 0.36	73.95 ± 0.16
LEAM [22]	57.71 ± 0.82	79.60 ± 0.22	76.91 ± 0.48
PTE [21]	51.89 ± 0.33	75.32 ± 0.31	72.52 ± 0.40
Graph-CNN(-C) [2]	63.52 ± 0.58	81.57 ± 0.30	77.04 ± 0.25
Text-GCN [28]	68.36 ± 0.58	86.34 ± 0.12	76.74 ± 0.24
$\boldsymbol{WGCN_{4L\,4W}}$	$\mathbf{70.34 \pm 0.31}$	$\mathbf{88.51 \pm 0.21}$	$\mathbf{79.16 \pm 0.16}$
-M	68.59 ± 0.57	87.10 ± 0.37	77.42 ± 0.39
-A	69.15 ± 0.39	88.05 ± 0.24	78.84 ± 0.21
-M-A	68.15 ± 0.54	86.38 ± 0.42	76.93 ± 0.35
-BP	68.87 ± 0.36	84.38 ± 0.39	76.05 ± 0.30
-F$_{usion}$	69.27 ± 0.29	88.03 ± 0.30	78.58 ± 0.27
(Width) $\boldsymbol{3W}$	69.29 ± 0.38	87.93 ± 0.32	78.80 ± 0.28
(Width) $\boldsymbol{5W}$	$\mathbf{70.47 \pm 0.35}$	$\mathbf{88.16 \pm 0.21}$	$\mathbf{79.31 \pm 0.26}$
(Depth) $\boldsymbol{3L}$	69.15 ± 0.38	87.56 ± 0.27	78.31 ± 0.24
(Depth) $\boldsymbol{5L}$	69.04 ± 0.27	87.71 ± 0.35	78.37 ± 0.25

Experimental Results. Table 2 illustrates consistent superior performance (higher accuracy and smaller standard deviation) of WGCN. In ablation test, the accuracy of the corresponding variant models ($WGCN_{4L4W}$ -A and $WGCN_{4L4W}$ -F$_{usion}$) decrease but they still exceed the compared algorithms. Besides, the bidirectionally and introduced Dynamic Graph Convolution are critical for the performance promotion, when they are removed from the model, the performance (of $WGCN_{4L4W}$ -BP and $WGCN_{4L4W}$ -M-A) drops significantly. It shows that employing bidirectional representation learning instead of unidirectional way and adaptively adjusting the graph structure are contribute to perform hierarchical relations reasoning, and thus making the aggregation of the global information from captured receptive field more effectively.

We further provide the incremental experiments to explore the contributions of width expansion in bottom four rows of Table 2. By comparing $WGCN_{4L5W}$ and $WGCN_{5L4W}$, we find that explore through width expansion of GCNs is more effective than stacking more layers for accuracy improvement.

Table 3. Ablation study for the weight ratios (WR) and the weight variances (WR) from the last layer of the forward pyramid in $WGCN_{4L4W}$. WR-1(%) denotes the WR of the 1-hop neighbor, and WV-1(%) denotes the WV of the 1-hop neighbor.

Method	Dateset	WR-1	WR-2	WR-3	WR-4	Accuracy
WGCN w/o GRU	Ohsumed	71.9	23.4	3.9	0.8	68.53 ± 0.59
WGCN	Ohsumed	65.7	16.9	**11.1**	**6.3**	$\mathbf{70.34 \pm 0.31}$
WGCN w/o GRU	MR	75.6	22.8	1.3	0.3	78.03 ± 0.28
WGCN	MR	70.4	13.1	**8.6**	**7.9**	$\mathbf{79.16 \pm 0.16}$
Method	Dateset	WV-1	WV-2	WV-3	WV-4	Accuracy
WGCN w/o GRU	Ohsumed	0.0035	0.0093	0.0062	0.0001	68.53 ± 0.59
WGCN	Ohsumed	0.1827	0.0149	**0.0641**	**0.0597**	$\mathbf{70.34 \pm 0.31}$
WGCN w/o GRU	MR	0.0418	0.0027	0.0330	0.0012	78.03 ± 0.28
WGCN	MR	0.2663	0.0755	**0.1343**	**0.0881**	$\mathbf{79.16 \pm 0.16}$

Furthermore, we introduce the GRU unit as the sequence encoder in DGC for its effective role in maintaining the structure stability of the whole multi-hops dilated neighbor, while balancing the contribution of both the close and the distant nodes. For verification, we introduce the weight ratio (WR) and the weight variance (WV) of sampled nodes in each hop scale as the indicator to verify the effectiveness of the introduced unit. The weight ratio measures the relative importance of each hop scale, by calculating the proportion of the sum weights from current hop scale in the whole dilated neighbor. The weight variance estimates the difference among neighborhood nodes at the specific hop scale. We adopted the last layer of the FP as well as the first layer of the BP in WGCN to calculate the above indicators. Table 3 illustrates that the GRU unit is helpful to significantly improve the weighting ratio at the higher hop scale (especially at the 3- or the 4-hop neighbor), demonstrating that our proposal is able to capture the correlation and the dependence between neighborhood nodes from the longer distance. Besides, the GRU unit endows each hop scale the larger weighted variance, making our WGCN more discriminative to pay more attention to the informative nodes from captured multi-hops dilated neighbor.

Receptive Field Analysis. We perform the experiments to explore the captured receptive field of our proposal ($WGCN_{4L4W}$) and compare with ResGCN [12]. For the 4-layer ResGCN, 16 (the same as in its original implementation) or 36 (for fair comparison) neighborhood nodes were sampled for each target node from the 1-hop neighbor. For each target node, we take the average number of neighborhood nodes actually involved in the aggregation according to the masked results in DGC of each layer as the receptive field captured by this layer. According to Table 4, we conclude that proposed WGCN obtains the larger average receptive field. Especially in the last layer, the corresponding receptive field captured by WGCN (22.8) significantly surpasses ResGCN (1) (16). While

ResGCN (2) receives the receptive field of 36 in each GCL, its performance drops sharply due to the introduction of irrelevant or corrupted nodes. Combined with the introduced masking and attention mechanisms, the average of captured receptive field across all layers of our WGCN is restricted to a reasonable value (17.9), making our proposal aggregate more informative nodes with the larger receptive field. Finally, with few extra computational complexity, our proposal establishes the new state-of-the-art results on this challenging dataset.

Table 4. The average captured receptive field (RF) of each algorithm, followed by a detailed analysis of each layer. CN: captured neighbor.

Method	Nodes	CN	Avg. RF	Acc.
ResGCN(1) [12]	16	1-hop	16	62.53 ± 0.75
ResGCN(2) [12]	36	1-hop	36	51.65 ± 6.28
WGCN	**36**	**1∼4-hop**	**17.9**	**70.34 ± 0.31**
Method	1st layer	2nd layer	3rd layer	4th layer
ResGCN(1) [12]	16	16	16	16
ResGCN(2) [12]	36	36	36	36
WGCN	**12.6**	**15.7**	**20.3**	**22.8**

4.2 Scene Text Detection

We evaluate the proposed method on three standard benchmarks: *CTW-1500*, *ICDAR2015*, and *MSRA-TD500*. The proposed method is compared with several popular text detectors including SegLink [19], EAST [31], TextSnake [16], PixelLink [3], PSENet [23], CRAFT [1], PAN [24], ContourNet [26] and Grapg-Text [30]. The pipeline of this task is illustrated in the middle pannel of Fig. 4, including two main modules namely text region prediction network and graph reasoning network. In the first module, we start to apply the VGG-16 equipped with FPN [14] as backbone to predict the classification confidence of the text region, while performing the regressions of the rotation angle and the center line of text instances. A series of candidate boxes will be extracted along the direction perpendicular to the text writing, and the local graphs are then roughly established to link between different boxes (text nodes). In the second module, our proposal (a reference model called $WGCN_{4L4W}$) is adopted to perform deep relations reasoning and infer the likelihood of linkages between the target node and its multi-hops dilated neighbor. According to the results, text nodes will be grouped into arbitrary-shaped text region.

Curved Text Detection. We evaluate the performance for detecting close or arbitrary-shaped texts on the curved benchmarking dataset *CTW-1500*. As shown in Table 5, the proposed method achieves recall, precision and H-mean rate of 85.4%, 87.1% and 86.2%, significantly outperforming the counterparts by

Table 5. Experimental results on *CTW-1500*, *ICDAR 2015*, and *MSRA-TD500*. The best score is highlighted in bold. R: Recall, P: Precision, H: Hmean.

Method	CTW-1500			ICDAR 2015			MSRATD500		
	R	P	H	R	P	H	R	P	H
SegLink [19]	40.0	42.3	40.8	–	–	–	70.0	86.0	77.0
EAST [31]	49.1	78.7	60.4	78.3	83.3	80.7	61.6	81.7	70.2
TextSnake [16]	85.3	67.9	75.6	84.9	80.4	82.6	73.9	83.2	78.3
PixelLink [3]	–	–	–	82.0	85.5	83.7	–	–	–
PSENet [23]	79.7	84.8	82.2	84.5	86.9	85.7	–	–	–
CRAFT [1]	81.1	86.0	83.5	–	–	–	78.2	88.2	82.9
PAN [24]	81.2	86.4	83.7	–	–	–	83.8	84.4	84.1
ContourNet [26]	84.1	83.7	83.9	**86.1**	87.6	86.9	–	–	–
GraphText [30]	–	–	–	84.7	88.5	86.6	82.3	88.1	85.1
WGCN	**85.4**	**87.1**	**86.2**	85.8	**89.3**	**87.5**	**84.2**	**88.5**	**86.3**

a large margin. Benefiting from the introduced multi-hops dilated neighbor for hierarchical relations reasoning, our method achieves promising results on representing arbitrary-shaped texts especially with varying degrees of curvature.

Multi-oriented Text Detection. We evaluate our method on *ICDAR 2015* to validate its ability for detecting multi-oriented texts. As shown in Table 5, our model shows advantages in both precision and Hmean. The relative improvement of precision and Hmean score reaches 7.2% and 8.4% compared to the baseline model called EAST [31], while achieving the promotion of 1.9% and 0.7% compared to the leader method namely ContourNet [26].

Multi-language Text Detection. To test the robustness of WGCN to multiple languages with long texts, we evaluate our method on the *MSRA-TD500* benchmark, which are listed in Table 5. Our method shows consistent performance advantages in recall, precision and Hmean. Compared to GraphText [30] which introduced the basic GCNs to aggregate the local 1-hop neighbor for linkage predictions, WGCN performs global aggregation from captured multi-hops dilated neighbor which is conducive to detecting long texts. Qualitative results shown in Fig. 5 can also demonstrate the effectiveness of proposed algorithm.

4.3   Point Cloud Segmentation

We evaluate on the Stanford 3D semantic parsing dataset, including 3D scans from Matterport scanners in 6 areas of 271 rooms. We compare our WGCN with baseline methods including PointNet++ [18], 3DRNN+CF [29], PointCNN [13], ResGCN-28 [12] and SSP+SPG [11]. The results of semantic segmentation are reported on widely used metrics: overall accuracy (*OA*), mean perclass accuracy

| Input | Boxes Generation | Linkage Prediction |

Fig. 5. Qualitative results on benchmark datasets of our proposal.

Table 6. Comparison of our WGCN with other algorithms on the S3DIS dataset.

Methods	OA	mAcc	mIOU	Methods	mPrec	mRec
PointNet++ [18]	81.0	67.1	53.2	OccuSeg [7]	72.8	60.3
3DRNN+CF [29]	86.9	73.6	56.3	PointGroup [8]	69.6	69.2
PointCNN [13]	88.1	75.6	65.4	$WGCN_{4L4W}$ **-M-A**	70.5	67.9
ResGCN-28 [12]	85.9	–	60.0	$WGCN_{4L4W}$ **-BP**	72.3	68.4
SSP+SPG [11]	87.9	78.3	68.4	$WGCN_{4L4W}$ **-F**$_{usion}$	72.7	68.2
$WGCN_{4L4W}$	87.6	79.8	68.8	$WGCN_{4L4W}$	72.4	**69.8**
$WGCN_{4L5W}$	**89.8**	**80.5**	**69.3**	$WGCN_{4L5W}$	**73.1**	69.7

($mAcc$) and mean per-class intersection-over-union ($mIOU$). The overall pipeline is depicted in the bottom part of Fig. 4. As the input features of cloud points provide 3D spatial coordinates, we adopt the pre-defined distance metric as [12] to search the corresponding multi-hops dilated neighbor.

The main results are revealed in Table 6, indicating that our WGCN achieves very competitive results on all metrics. Particularly, our 4-layer reference model ($WGCN_{4L4W}$) beats a 28-layer ResGCN [12]. Besides, we conducted empirical evaluations between our WGCN and some methods including the current leaders (e.g. OccuSeg [7], PointGroup [8]) on the S3DIS dataset. Following these latest methods, we employed the 6-fold cross validation and used the mean precision (mPrec)/mean recall (mRec) with an IoU threshold 0.5 to perform evaluations. As shown in Table 6, our proposals achieve competitive performance in terms of both mPrec and mRec, although OccuSeg [7] and PointGroup [8] are specifically designed for 3D instance segmentation tasks.

5   Conclusion

In this paper, we propose a wide but shallow learning framework, namely Wider Graph Convolution Network, to capture the multi-hops dilated neighbor in the graph. By constructing Dynamic Graph Convolution through masking and attention mechanisms, the feature representations of nodes as well as the graph structure are adaptively adjusted at each layer for hierarchical relations reasoning and global information aggregation. Experiments demonstrate the effectiveness of our proposal on several popular tasks.

References

1. Baek, Y., Lee, B., Han, D., Yun, S., Lee, H.: Character region awareness for text detection. In: Proceedings of the IEEE Conference on Computer Vision and Pattern Recognition, pp. 9365–9374 (2019)
2. Defferrard, M., Bresson, X., Vandergheynst, P.: Convolutional neural networks on graphs with fast localized spectral filtering. In: NeurIPS, pp. 3844–3852 (2016)
3. Deng, D., Liu, H., Li, X., Cai, D.: Pixellink: detecting scene text via instance segmentation. In: Proceedings of the AAAI Conference on Artificial Intelligence, vol. 32 (2018)
4. Gao, H., Wang, Z., Ji, S.: Large-scale learnable graph convolutional networks. In: Proceedings of the 24th ACM SIGKDD International Conference on Knowledge Discovery & Data Mining, pp. 1416–1424 (2018)
5. Gong, L., Cheng, Q.: Exploiting edge features for graph neural networks. In: Proceedings of the IEEE Conference on Computer Vision and Pattern Recognition, pp. 9211–9219 (2019)
6. Hamilton, W., Ying, Z., Leskovec, J.: Inductive representation learning on large graphs. In: NeurIPS, pp. 1024–1034 (2017)
7. Han, L., Zheng, T., Xu, L., Fang, L.: Occuseg: occupancy-aware 3d instance segmentation. In: Proceedings of the IEEE/CVF Conference on Computer Vision and Pattern Recognition, pp. 2940–2949 (2020),
8. Jiang, L., Zhao, H., Shi, S., Liu, S., Fu, C.W., Jia, J.: Pointgroup: dual-set point grouping for 3d instance segmentation. In: Proceedings of the IEEE/CVF Conference on Computer Vision and Pattern Recognition, pp. 4867–4876 (2020)
9. Joulin, A., Grave, E., Bojanowski, P., Mikolov, T.: Bag of tricks for efficient text classification. arXiv preprint arXiv:1607.01759 (2016)
10. Kipf, T.N., Welling, M.: Semi-supervised classification with graph convolutional networks. arXiv preprint arXiv:1609.02907 (2016)
11. Landrieu, L., Boussaha, M.: Point cloud oversegmentation with graph-structured deep metric learning. arXiv preprint arXiv:1904.02113 (2019)
12. Li, G., Muller, M., Thabet, A., Ghanem, B.: Deepgcns: can gcns go as deep as cnns? In: Proceedings of the IEEE International Conference on Computer Vision, pp. 9267–9276 (2019)
13. Li, Y., Bu, R., Sun, M., Wu, W., Di, X., Chen, B.: Pointcnn: convolution on x-transformed points. In: NeurIPS, pp. 820–830 (2018)
14. Lin, T.Y., Dollár, P., Girshick, R., He, K., Hariharan, B., Belongie, S.: Feature pyramid networks for object detection. In: Proceedings of the IEEE Conference on Computer Vision and Pattern Recognition pp. 2117–2125 (2017)

15. Liu, P., Qiu, X., Huang, X.: Recurrent neural network for text classification with multi-task learning. arXiv preprint arXiv:1605.05101 (2016)
16. Long, S., Ruan, J., Zhang, W., He, X., Wu, W., Yao, C.: Textsnake: a flexible representation for detecting text of arbitrary shapes. In: Proceedings of the European Conference on Computer Vision (ECCV), pp. 20–36 (2018)
17. Nathani, D., Chauhan, J., Sharma, C., Kaul, M.: Learning attention-based embeddings for relation prediction in knowledge graphs. arXiv:1906.01195 (2019)
18. Qi, C.R., Yi, L., Su, H., Guibas, L.J.: Pointnet++: deep hierarchical feature learning on point sets in a metric space. In: Advances in Neural Information Processing Systems, pp. 5099–5108 (2017)
19. Shi, B., Bai, X., Belongie, S.: Detecting oriented text in natural images by linking segments. In: Proceedings of the IEEE Conference on Computer Vision and Pattern Recognition, pp. 2550–2558 (2017)
20. Simonovsky, M., Komodakis, N.: Dynamic edge-conditioned filters in convolutional neural networks on graphs. In: Proceedings of the IEEE Conference on Computer Vision and Pattern Recognition, pp. 3693–3702 (2017)
21. Tang, J., Qu, M., Mei, Q.: Pte: predictive text embedding through large-scale heterogeneous text networks. In: KDD, pp. 1165–1174. ACM (2015)
22. Wang, G., et al.: Joint embedding of words and labels for text classification. arXiv preprint arXiv:1805.04174 (2018)
23. Wang, W., et al.: Shape robust text detection with progressive scale expansion network. In: Proceedings of the IEEE Conference on Computer Vision and Pattern Recognition, pp. 9336–9345 (2019)
24. Wang, W., et al.: Efficient and accurate arbitrary-shaped text detection with pixel aggregation network. In: Proceedings of the IEEE International Conference on Computer Vision, pp. 8440–8449 (2019)
25. Wang, Y., Sun, Y., Liu, Z., Sarma, S.E., Bronstein, M.M., Solomon, J.M.: Dynamic graph cnn for learning on point clouds. ACM Trans. Graphics (TOG) **38**(5), 1–12 (2019)
26. Wang, Y., Xie, H., Zha, Z.J., Xing, M., Fu, Z., Zhang, Y.: Contournet: taking a further step toward accurate arbitrary-shaped scene text detection. In: Proceedings of the IEEE/CVF Conference on Computer Vision and Pattern Recognition, pp. 11753–11762 (2020)
27. Wu, Z., Pan, S., Chen, F., Long, G., Zhang, C., Philip, S.Y.: A comprehensive survey on graph neural networks. IEEE Trans. Neural Netw. Learn. Syst. (2020)
28. Yao, L., Mao, C., Luo, Y.: Graph convolutional networks for text classification. In: Proceedings of the AAAI Conference on Artificial Intelligence, vol. 33, pp. 7370–7377 (2019)
29. Ye, X., Li, J., Huang, H., Du, L., Zhang, X.: 3D recurrent neural networks with context fusion for point cloud semantic segmentation. In: Proceedings of the European Conference on Computer Vision (ECCV), pp. 403–417 (2018)
30. Zhang, S.X., et al.: Deep relational reasoning graph network for arbitrary shape text detection. In: Proceedings of the IEEE/CVF Conference on Computer Vision and Pattern Recognition, pp. 9699–9708 (2020)
31. Zhou, X., et al.: East: an efficient and accurate scene text detector. In: Proceedings of the IEEE Conference on Computer Vision and Pattern Recognition, pp. 5551–5560 (2017)

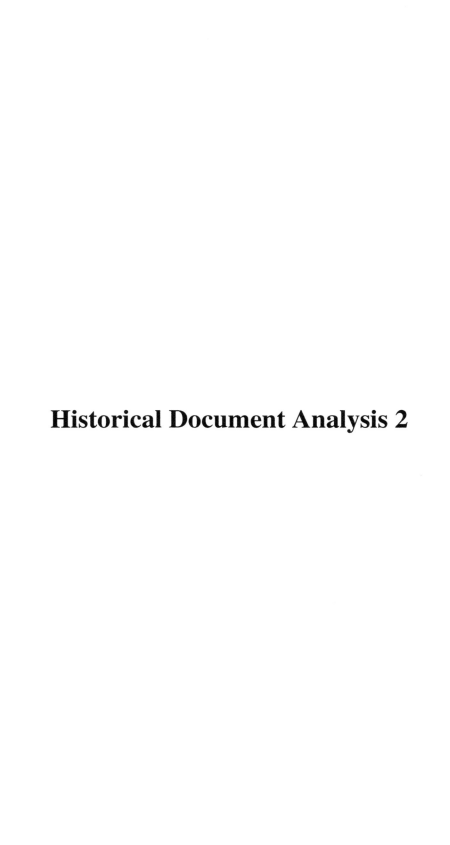

Historical Document Analysis 2

One-Model Ensemble-Learning for Text Recognition of Historical Printings

Christoph Wick[1]([✉])[ID] and Christian Reul[2][ID]

[1] Planet AI GmbH, Warnowufer 60, 18057 Rostock, Germany
`christoph.wick@planet-ai.de`
[2] University of Würzburg, Am Hubland, 97074 Würzburg, Germany
`christian.reul@uni-wuerzburg.de`

Abstract. In this paper, we propose a novel method for Automatic Text Recognition (ATR) on early printed books. Our approach significantly reduces the Character Error Rates (CERs) for book-specific training when only a few lines of Ground Truth (GT) are available and considerably outperforms previous methods. An ensemble of models is trained simultaneously by optimising each one independently but also with respect to a fused output obtained by averaging the individual confidence matrices. Various experiments on five early printed books show that this approach already outperforms the current state-of-the-art by up to 20% and 10% on average. Replacing the averaging of the confidence matrices during prediction with a confidence-based voting boosts our results by an additional 8% leading to a total average improvement of about 17%.

Keywords: Optical Character Recognition · Ensemble learning · Historical document analysis · Early printed books

1 Introduction

While Automatic Text Recognition (ATR) of printed modern fonts can be considered as solved, ATR of the earliest historical printings dating back to the 15th century still represents a challenge due to the high variability among different typefaces but also degradation. Currently, it is still necessary to train book-specific models to reliably achieve Character Error Rates (CERs) close to 1% or lower on this material (see, e.g., [9,17]). These networks can be trained by Open Source ATR-engines (e.g., Calamari [21], Kraken [6], OCRopus3 [3], or Tesseract [16]) which most commonly provide a combination of Convolutional Neural Network (CNN) and Long-Short-Term-Memory-Cell (LSTM) networks. As default, these are trained using the Connectionist Temporal Classification (CTC)-algorithm [5] which automatically learns the alignment of a line image and its corresponding transcription, facilitating Ground Truth (GT) production.

A key ingredient in order to obtain highly performant models is the amount of available training GT. Since these transcriptions mostly have to be created manually which is a cumbersome and time-consuming task, it is mandatory to develop training procedures that optimise the performance for limited GT,

© Springer Nature Switzerland AG 2021
J. Lladós et al. (Eds.): ICDAR 2021, LNCS 12821, pp. 385–399, 2021.
https://doi.org/10.1007/978-3-030-86549-8_25

e.g., just a few hundred lines. Several techniques have already been proposed and are widely used: Data augmentation is a very common one because it is a simple yet effective approach to artificially increase the size of the dataset. A related technique in ATR is to synthetically render new lines from available computer fonts. For historical printings, however, this is challenging since suitable book-specific fonts are usually not available and their creation is very tedious. Naturally, starting the training from an existing model which was trained on other (historical) data leads to better results because the network has already learned useful features [11]. Another technique which was proposed by Reul et al. [10] is cross-fold-training with confidence-voting which led to a significant performance gain. The idea is to train an ensemble of several models and then assemble the output by using the confidence values of the individual voters. The current drawback of this method is that the decoded sequences of the voters have to be aligned before voting can kick in. In this step, several ambiguities can occur which currently rely on strict heuristics to be resolved. The reason is that the underlying CTC-algorithm learns an arbitrary alignment since only the order of characters but not the actual position of a character within the line is respected. In practice, the predicted location highly correlates with the actual character position in the image, but variations by a few pixels make a position wise voting, e.g., by averaging the confidences, impossible.

In this paper, we tackle this shortcoming by proposing a novel approach to train an ensemble of voters combined into one single model which enables us to directly average the confidence matrices of the individual voters. While each voter is trained analogously with an individual CTC-loss, we add an additional CTC-loss on the combined output, thereby forcing the voters to align their predictions. We show on five different historical books and with a varying number of training lines that our approach outperforms the previous method by about 10% on average, and over 20% when using only 100 lines of GT. In general, the method trains on the complete GT, that is, splitting the limited GT into separate training and validation sets is not required. Moreover, our method is simple to implement and can straightforwardly be integrated into existing ATR engines.

The remainder of this paper is structured as follows: First, we introduce and discuss related work regarding ensemble learning in general and in the context of Optical Character Recognition (OCR). The following section presents our proposed approach condisting of the training. validation, and prediction procedures. After describing our datasets we evaluate and discuss the results of several different experiments before concluding the paper.

2   Related Work

Ensemble learning (see e.g., [15] for a recent review) is a common technique to obtain performance boosts in many state-of-the-art supervised machine learning approaches such as Deep Learning. The fundamental idea is to train multiple diverse yet performant voters whose predictions are combined to obtain a stronger, less erroneous outcome. Sagi et al. [15] differentiate between several approaches to train such an ensemble: input manipulation, varying the learning algorithm, dataset partitioning, output manipulation, and hybrid approaches. To fuse the output, there are typically two options: weighting the output or incorporating a meta model that learns how to combine the output.

In the following, we list several publications dealing with ensemble learning in the context of OCR whereby the traditional approach to fuse multiple outputs is voting. Diverse ensembles can either be achieved by transcribing the same text with different OCR engines (e.g., [1,2,12]) or by manipulating the data: Reul et al. [10] split the data into a n-cross-fold whereby each voter of the ensemble is trained on $n-1$ parts and validated on the remaining one. Voting in the former examples is typically performed using the ISRI tools by Rice and Nartker [14] who provide their so-called Longest Common Substring (LCS) algorithm [13] which enables to align several sequences of possibly different length. The final output for individual characters is determined by majority voting wile ties are broken using heuristics. Reul et al. [10] improved the LCS algorithm by including the character confidences: First, the decoded paths of each voter are aligned, then, instead of majority voting, the confidences of the individual characters, both the actual output as well as the alternatives, of each match are accumulated and the maximum is returned. This resulted in a performance boost of up to 50% and more on early printed books compared to a single model. However, confidence-voting is only possible if the characters of each sequences within a match have the same length which enforces the application of heuristics in any other case.

The overall problematic of the spiky timings in the posterior confidences of a CTC model was already investigated in the related field of speech recognition. For example, Kurata and Audhkhasi [8] proposed a two stage approach to fuse the output of multiple CTC by first training a guiding model that forces other models to have the same alignment. Instead, we introduce a one stage approach to learn the alignment among the models.

3   Proposed Method

In this section, we describe our proposed methodology to initialise an ensemble of n individual models (voters) which is then jointly trained. First, the CNN/LSTM-network architecture of a voter is introduced, then we outline the prediction, training, and validation procedure (see Fig. 1).

3.1   Network Architecture

The CNN/LSTM network architecture of each individual model M_i is composed of two convolutional and max-pooling layers and a subsequent bidirectional LSTM-layer whose output is obtained by concatenating each output of the forward and backward LSTMs (see Fig. 2). Next, a dropout layer with a

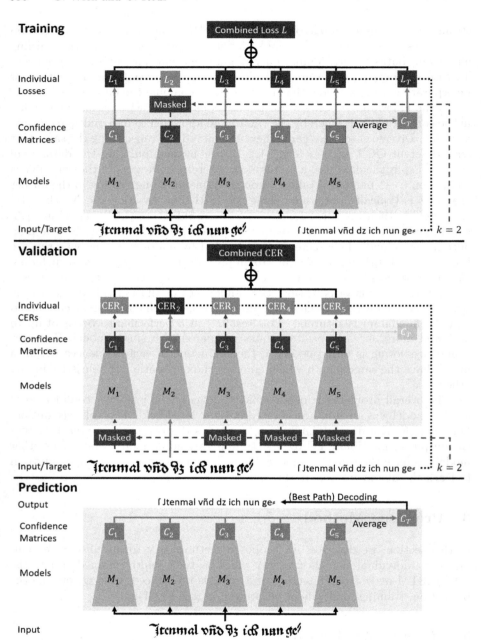

Fig. 1. Workflow consisting of training, validation, and prediction.

Fig. 2. The arcitecture of the CNN/LSTM-network used in our experiments. The sub- and superscript in $C/P_{k_y \times k_x}^{s_y \times s_x}$ describe the size of the kernel k and strides s along both dimensions (width x, height y). The depth of the feature map is given in square brackets, that is, the number of filters for a convolutional layer. The outputs of the bidirectional LSTM with 200 hidden nodes are concatenated.

dropout ratio of 0.5 and the final dense layer are appended yielding the confidence matrix C_i. The output dimension of C_i equates the (subsampled) sequence length times the number of characters in the alphabet $|\mathcal{A}|$ plus one (the blank label required for the CTC-algorithm).

The full network (see Fig. 1) comprises n independent CNN/LSTM subgraphs and a combined confidence matrix C_T which is obtained by averaging all C_i. Having the same input and output specifications as an individual CNN/LSTM network, it can be utilised alike.

3.2 Prediction Procedure

The prediction (see bottom of Fig. 1) of an image line is straightforward: After each sub model M_i processed the image independently to obtain a confidence matrix C_i, the average C_T is decoded. In our case, we utilize CTC best-path decoding, but any other more sophisticated approach such as beam-search could be applied instead.

3.3   Training Procedure and Masking

Since averaging the results of an ensemble performs best if the voters are individually strong but diverse, we adopt the cross-fold-training mechanism as proposed by Reul et al. [10]: The complete training data is split evenly into n distinct folds whereby each voter only sees $n-1$ partitions during training using the left out part for validation (see next Section). We realise this setup by randomly assigning a number $k = 1, ..., n$ to each sample of the dataset. During training, the voter with the corresponding k is masked out, that is, its probability matrix is multiplied by zero, consequently not contributing to the averaged final output (see Fig. 1). We examine two further masking strategies for comparison: The *no masking* approach does not mask a sample during training at all, whereas the *random masking* strategy masks a random voter independent of k. In these two cases, each individual model sees all lines during training whereby *random masking* introduces a higher variance during training because skipping some samples leads to a different actual order of samples for each voter.

We compute a traditional CTC-loss L_i for each of the individual voters as well as one for the combined model L_T. The combined loss L is obtained by summing up each of the $n + 1$-losses (see Fig. 1): $L = L_T + \sum_{i=1}^{n} L_i$. While ensuring that each voter learns a robust model on its own this approach also enforces the alignment of the outputs of all voters. In fact, preliminary experiments showed poor results when setting $L = L_T$.

3.4   Validation Procedure

During training, we compute the loss and CER after each epoch by presenting a sample exclusively to its corresponding voter, i.e. the one that has not seen this sample during training, masking out any other voter (see middle of Fig. 1). The total CER is the average of all individual CERs. This enables tracking the validation error which determines the best ensemble and when to stop training using early stopping. Both the *random* and *no* masking approach possibly see every instance of the dataset during training which is why no real validation can be performed. Instead, we apply the same scheme as for the *fold* masking approach, where each of the n voters uses a fixed portion of $1/n$ of the training data for validation. Naturally, this leads to overfitting.

4   Datasets

In this section, we introduce our dataset comprising five different early printed books. Figure 3 shows an example line for each book whereas Table 1 lists the training and evaluation split, the number of total characters, the average number of characters per line, and the alphabet size $|\mathcal{A}|$. The books were selected from the *Kallimachos* sub corpus of the publicly available[1] *GT4HistOCR* dataset published by Springmann et al. [18] because of their varying difficulty in terms of typography and quality.

---

[1] https://zenodo.org/record/1344132.

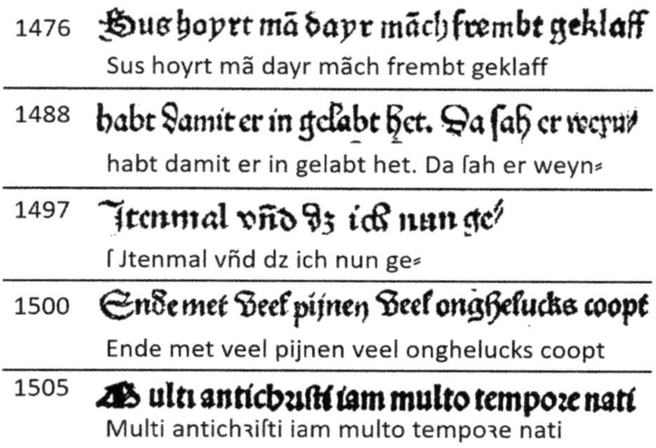

1476	$\mathfrak{Hus\ hoyrt\ m\tilde{a}\ dayr\ m\tilde{a}ch\ frembt\ geklaff}$
	Sus hoyrt mã dayr mãch frembt geklaff

1488	habt Şamit er in geſabt ḥet. Şa ſaḥ er weyn⸗
	habt damit er in gelabt het. Da ſah er weyn⸗

1497	ʃJtenmal vñd Ꝫꝫ ich nun ge⸗
	ʃ Jtenmal vñd dz ich nun ge⸗

1500	Ende met Ꝟeel pijnen Ꝟeel ongheſucks coopt
	Ende met veel pijnen veel onghelucks coopt

1505	Ꜳ ulti antichꝛiſti iam multo tempoꝛe nati
	Multi antichꝛiſti iam multo tempoꝛe nati

Fig. 3. An example line and its corresponding GT from each of the five books.

Table 1. Overview of the number of training and evaluation lines of the five books used in our experiments.

Year	(Short) Title	GT lines			Chars		
		Total	Train	Eval.	Total	Per line	$|\mathcal{A}|$
1476	Historij	3,160	1,000	2,160	103,665	32.8	64
1488	Der Heiligen Leben	4,178	1,000	3,178	187,385	44.9	66
1497	Cirurgia	3,476	1,000	2,476	119,859	34.5	73
1500	Der narrenscip	2,500	1,000	1,500	126,513	50.6	63
1505	Nauis stultifera	4,713	1,000	3,713	299,244	63.5	80

5   Experiments

All experiments were performed on a system comprising a NVIDIA GeForce GTX 1080 Ti GPU and an Intel Core i7-6850K CPU. Since the Open Source ATR engine Calamari already provides an implementation of the confidence-voting approach of Reul et al. we used it to compute our reference values and added our proposed method[2]. Consequently, our proposed and the reference approach both rely on the same pre- and postprocessing: all images are rescaled to a fixed line height of 48px and centre normalised, repeated spaces in the GT text are unified, and spaces at the beginning and the end of a line are trimmed. Throughout our experiments, we used the ADAM optimiser [7], gradient clipping of 5, a constant learning rate of 0.001, and a batch size of 5. To determine the best model and to automatically stop training when no further significant improvements are expected, we utilized early stopping: After each epoch, we apply the current model to the validation set, calculate the CER, and either

---

[2] See our implementation at https://github.com/Calamari-OCR/calamari/blob/master/calamari_ocr/ocr/model/ensemblegraph.py.

Table 2. Comparison of the CER of voting five individual models (Vot.) and training an ensemble as one. The improvements (Imp.) are listed for training on a different number of lines for five different historical books. All values are given in percent.

Lines	100			250			500			1,000		
Method	Vot.	One	Imp.	Vot.	One	Imp.	Vot.	One	Imp.	Vot.	One	Imp.
1476	5.13	4.87	5.07	2.32	2.19	5.60	1.42	1.33	6.34	0.98	0.92	6.12
1488	3.51	3.43	2.28	1.46	1.23	15.75	0.83	0.73	12.05	0.51	0.46	9.80
1497	7.38	6.61	10.43	3.36	3.02	10.12	1.84	1.49	19.02	1.07	0.96	10.28
1500	3.33	2.85	14.41	1.86	1.70	8.60	1.24	1.25	−0.81	1.01	0.93	7.92
1505	4.95	3.89	21.41	2.66	2.16	18.80	1.67	1.53	8.38	1.32	1.21	8.33
Avg.	10.72			11.78			9.00			8.49		

keep the previously determined best model or replace it by the new one. If the best model could not be improved on for five epochs the training is stopped.

In the first experiment we examine the influence of the number of available GT lines for training and compare our method to the state-of-the-art. Then, we vary the masking method and the number of folds. Finally, we measure and compare the amount of time required for training and prediction. We exclude experiments that show that an ensemble approach performs significantly better than a single model because this was already verified thoroughly by Reul et al. [10] against who we compare.

5.1 Influence of the Number of Training Lines

Table 2 shows the obtained CERs when varying the number of training lines for each of the five books. We used 100, 250, 500, and 1,000 lines since this roughly depicts typical steps in an *iterative training approach*[3] as described in [9]. Our proposed approach outperforms the original cross-fold-training and confidence-voting approach in almost every experiment by reducing the error by up to 21.41% and by about 10% on average. There is only one case in which our model performed worse but only by less than 1%. The performance gap increases if only a few lines are available for training.

A general observation is that the influence of the number of training lines, alphabet size $|\mathcal{A}|$, and average characters per line (see Table 1) can not be esti-

---

[3] Since it is usually unclear how many lines of GT are required to achieve a certain CER and the transcriptions effort correlates with the amount of errors within the ATR result, it is usually advantageous to perform the GT production iteratively: Starting from an often quite erroneous output of an existing mixed model only a minimal amount of GT (for example 100 lines) is produced and used to train a first book-specific model. In most cases, applying this model to unseen data (for example 150 further lines) already results in a significantly better ATR output which can be corrected much faster than before. After training another model these steps are repeated until a satisfactory CER is reached or the whole book is transcribed.

Table 3. Comparison of different masking approaches (see Sect. 3.3) with $n = 5$: *no*, *random*, and *fold masking*. We list the CERs in percent for all five books and 100, 250, 500, and 1,000 lines in the training data set. The last row is the average across all books.

Lines	100			250			500			1,000		
Mask	No	Rnd.	Fold	No	Rnd.	Fold	No	Rnd.	Fold	No	Rnd.	Fold
1476	4.81	4.59	4.87	2.13	1.98	2.19	1.49	1.61	1.33	0.98	0.91	0.92
1488	3.16	3.29	3.43	1.11	1.32	1.23	0.74	0.74	0.73	0.45	0.45	0.46
1497	7.70	7.92	6.61	3.09	3.01	3.02	1.72	1.72	1.49	1.02	1.22	0.96
1500	2.87	2.75	2.85	1.64	1.62	1.70	1.17	1.30	1.25	0.97	1.02	0.93
1505	3.83	4.26	3.89	2.21	2.31	2.16	1.52	1.54	1.53	1.26	1.22	1.21
Avg.	4.45	4.56	4.33	2.04	2.05	2.06	1.32	1.38	1.27	0.94	0.96	0.90

mated reliably. For example, book 1505 performs the worst by some distance when trained with 1,000 lines despite having the highest average number of characters per line. A probable explanation is that it has the largest $|\mathcal{A}|$ and consequently there are more rare characters that have to be learned and might be underrepresented in the training set. However, book 1488 performs best even though book 1500 has a smaller $|\mathcal{A}|$ and saw more characters during training due to a higher number of chars per line. The primary reason for this observations is the different quality of the material which is influenced, for instance, by the typeface or by degradation.

5.2 Influence of the Masking Method

This section evaluates the three applied masking approaches: *no masking*, *random masking*, and *fold masking* (see Sect. 3.3). Table 3 lists the results for all five books and 100, 250, 500, and 1,000 training lines, as well as the averages per line and method. The results reveal that all approaches achieve quite similar results, but fold masking performs best for every number of lines on average. This had to be expected since only in this case a clean stopping criteria for early stopping can be found which prevents obtaining a severely overfitted model. Furthermore, since each model does not see a (disctinct) fifth of the training data, the models are more diverse, improving the effectiveness of voting.

Interestingly, even though random noise induces more variance during training, using no mask at all yields slightly better results on average. Our explanation is that if the model is overfitting anyway, it is better to use as much data as possible instead of skipping instances randomly.

5.3 Influence of the Number of Folds

Table 4 lists the CERs obtained by varying the number of voters within the model to 3, 5, and 10. Since we use the *fold masking* approach, the actual size

Table 4. Comparison of varying the number of folds n (see Sect. 3.3). We list the CERs in percent for all five books and 100, 250, 500, and 1,000 lines in the training data set, respectively, as well as the average across all books. The last row show the average training times in minutes and seconds.

Lines	100			250			500			1,000		
Folds	3	5	10	3	5	10	3	5	10	3	5	10
1476	5.16	4.87	4.96	1.91	2.19	2.11	1.39	1.33	1.36	0.93	0.92	0.88
1488	3.31	3.43	3.33	1.36	1.23	1.30	0.74	0.73	0.72	0.47	0.46	0.49
1497	7.47	6.61	6.77	2.93	3.02	2.92	1.74	1.49	1.67	0.96	0.96	1.72
1500	2.72	2.85	2.79	1.84	1.70	1.75	1.27	1.25	1.66	1.03	0.93	1.22
1505	4.17	3.89	4.54	2.21	2.16	2.71	1.57	1.53	1.56	1.23	1.21	1.30
Avg.	4.57	4.33	4.48	2.05	2.06	2.16	1.34	1.27	1.39	0.92	0.90	1.12
Time	02:16	03:17	05:47	04:08	06:08	09:38	07:17	10:38	17:24	14:31	19:01	30:16

of the training dataset for each of the three models is 67%, 80%, and 90% of all lines, respectively. The remainders form the validation set. On average and independent of the number of lines, using five folds led to the best results. For only a small number of lines, the three fold setup leads to worse models because fewer lines are seen per model during training (67% instead of 80%). On the contrary, ten folds seem to use a too small portion of lines for validation (10% instead of 20%) promoting less generalised and consequently less effective voters. Therefore, this can also be considered as very similar to training without masking which manifests in similar average performances (CER of 4.57 vs. 4.45, see Table 3). The last row of Table 4 confirms that, as expected, increasing the number of folds also results in an increase of training time. Note that while the time per iteration doubles from 5 to 10 folds (not shown), the total training time is not increasing proportionally due to early stopping and only differs by a factor of 1.5 (about 10 min) for training a model using 1,000 lines of GT. In general, the trade-off between training time and accuracy encourages to consistently split into five folds by default.

5.4   Cross-Fold-Voting Using Trained One-Model Ensemble

In this experiment, we split our trained One-model into its five individual models and vote their outputs by the original confidence-voting approach of Reul et al. This enables us to compare the performance of the voting algorithm and our method that simply averages the five probability matrices. Table 5 lists the results.

On average, combining confidence-voting with our training methodology (OneV) significantly outperforms the default approach of averaging the confidence matrices (One) by about 8% on average. While this improves the effectiveness of the method even further we, unfortunately, cannot give a founded explanation for this behaviour, yet. The differences between confidence-voting

Table 5. Comparing our default One-model approach (One) of fusing the output of different voters by averaging their probability matrix outputs, and splitting the model into separate voters and applying the confidence-voting algorithm (OneV) of Reul et al. The improvement (Imp.) and all CERs are given in percent.

Lines	100			250			500			1,000		
Method	One	OneV	Imp.	One	OneV	Imp.	One	OneV	Imp.	One	OneV	Imp.
1476	4.87	4.42	9.24	2.19	2.10	4.11	1.33	1.23	7.52	0.92	0.90	2.17
1488	3.43	3.04	11.37	1.23	1.13	8.13	0.73	0.67	8.22	0.46	0.41	10.87
1497	6.61	6.04	8.62	3.02	2.60	13.91	1.49	1.38	7.38	0.96	0.88	8.33
1500	2.85	2.59	9.12	1.70	1.48	12.94	1.25	1.19	4.80	0.93	0.88	5.38
1505	3.89	3.69	5.14	2.16	1.97	8.80	1.53	1.44	5.88	1.21	1.17	3.31
Avg.	4.33	3.96	8.64	2.08	1.86	9.90	1.27	1.18	6.64	0.90	0.85	5.36

and averaging the probability matrices have to be located directly within the voting mechanism. Confidence-voting first aligns the decoded sequences and then drops matches (substrings) of different lengths. We expect that this procedure removes uncertain voters that could disturb the prediction of the more confident ones. To verify this statement, we provide an error analysis in the next section. In summary, these experiments reveal that the strength of the proposed methodology does not rely on the confidence-voting but rather on the training procedure which enforces an alignment. Consequently, models trained in this fashion further benefit from a more sophisticated voting, e.g., the confidence-voting. Naturally, this voting is also slower than averaging during prediction (see Sect. 5.6).

5.5 Error Analysis

To get a better idea of the achieved improvement's nature, we compared the confusion tables of all methods using the output for 250 training lines over all books since it produced the biggest improvement. Because of the known vulnerability of the alignment step during prediction of the cross-fold-voting method, we differentiate between error types where the corresponding GT and prediction have different lengths (insertions and deletions) and the ones with equal length (confusions). Note that the amount of evaluation data varies among the books and has not been normalised for this experiment and that we counted only the frequency of each error notwithstanding its length. Hence the difference in terms of overall improvement compared to Tables 2 and 5.

Table 6 shows the results. The relative error values confirm the widespread assumption that deletion present the biggest source of error when it comes to the ATR of historical printings, being responsible for about half of all occurring errors. Confusions make up for about one third of errors while deletions are only responsible for a little more than 13%. When comparing the proposed One-model approach (One) to the original cross-fold method (Vot.) it becomes apparent that the former is significantly more effective when dealing with confusions, insertions,

Table 6. Comparison of the errors produced by the original cross-fold- training and confidence-voting method (Vot.), the proposed One-Model (One) approach, and its combination (training as One-model, then split and predict as cross-fold using confidence-voting OneV.) using the 250 line output averaged over all books. For each method the absolute and relative portion of confusions (Conf.), insertions (Ins.), and deletions (Del.) is given. The last three rows compare the results of the approaches and show the relative improvement (Impr.) for all categories. One/Vot. thereby denotes the achieved improvement of One over Vot. and so on.

	Method	Absolute errors				Relative errors			
		Conf.	Ins.	Del.	All	Conf.	Ins.	Del.	All
Errors	Vot.	4,777	1,745	6,571	13,093	35.5%	13.3%	50.2%	100%
	One	3,285	1,399	5,887	10,571	31.1%	13.2%	55.7%	100%
	OneV.	3,456	1,324	4,762	9,542	36.2%	13.9%	49.9%	100%
Impr.	One/Vot.	31.2%	19.8%	10.4%	19.3%				
	OneV./One	−5.2%	5.4%	19.1%	9.7%				
	OneV./Vot.	27.7%	24.1%	27.5%	27.1%				

Table 7. Comparison of prediction and training times of a single voter, confidence-voting of five models, and our proposed One-model approach. Training times are the total times in seconds for each approach and number of lines in the training set. Prediction times including the speed up (SU) of using a GPU are listed in iterations per second (it/s).

	Training [s]				Predict [it/s]		
	100	250	500	1,000	GPU	CPU	SU
Single voter	31	47	67	105	154	105	1.5
Voted models	156	237	336	524	46	26	1.8
One-model	142	284	514	928	119	38	3.1

and deletions (improvements of 31.2%, 19.8%, and 10.4% over the cross-fold method, respectively). Interestingly, combining our approach with confidence-voting (OneV) over One mainly lowered deletions with 19.1%. Combined with a slightly better performance when it comes to insertions (5.4%), this adds up to a significant overall performance gain of almost 10% despite a notable deterioration when dealing with confusions (−5.2%). We think that this observation confirms our explanation of the previous section why OneV is better than One: for every detected confusion, insertion, or deletion, voting drops the individual results with the greatest difference compared to the other ones even though they might have a high confidence. The error analysis shows that this considerably influences deletions which represent the main source of errors.

As shown before, the combined approach (OneV) provides the best overall result achieving similar and significant improved values for all error types. The One-model approach training decreases primarily confusion errors while confidence-voting improves deletions with a minor effect on confusions.

5.6   Timing

In this section, we measure the prediction and training times of our proposed method and the approach of Reul et al. [10] which is to train five models separately and vote their predictions. All experiments were performed on book 1476 using five folds. First, we measure the total training time for varying numbers of training lines (see Table 7). We include the training time of a single voter by averaging the five individual voting models and the accumulated total training time which consequently is five times the mean. For our One-model approach, we simple measure the complete training time. The application of early stopping defines the stopping criteria of the training and thus the actual training times.

As expected, the training times rise with an increasing number of lines. However, compared to training on 100 lines, the amount of time required by the approach of Reul et al. increases by a factor of 1.5, 2.2 and 3.4, for 250, 500, and 1,000 lines, respectively, while our training times increase by 2.0, 3.6, and 6.5. Therefore, our model performs better (see Sect. 5.1) but requires more time (about factor 2) for training. The reason is early stopping: since in our approach, the validation accuracy is computed by averaging the five different branches (see Fig. 1), any improvement of a single voter is sufficient to continue training.

To compare the prediction times, we measure the number of samples (iterations per second) that can be processed by each pipeline, including the times for sample loading, preparation, network interference, decoding, and voting. For comparison, we add the times of a single voter. The main outcome is that our approach runs significantly faster during prediction. Naturally, a single voter is considerably faster on the CPU but only marginally when using a GPU. The reason is that the pre- and postprocessing on the CPU can run in parallel with the inference of the model on the GPU. Since the GPU capacity is not yet fully utilised, the difference between the single voter and the One-model is small since the five voters can be run in parallel.

6   Conclusion and Future Work

The experiments showed that already the One-model approach clearly and consistently outperforms existing Open Source methods for ensemble-based ATR on historical printings by about 10% on average. Only in a single one out of 20 experiments our proposed approach performed worse but by less than 1% of relative CER. The additional combination with confidence-voting resulted in an even higher total improvement of about 17%. The error analysis revealed that the One-model reduces confusion errors by about one third, the confidence-voting additionally reduces mainly deletions resulting in an improvement of 27% of all errors accumulated across all five books. While, as expected, the prediction times are the same, the training times are higher by up to a factor of two for

1,000 training lines when using a single GPU[4]. In practice, this increase can be neglected because the trade-off between a higher time for training which runs fully automatically and a lower CER, and thus manually spent correction time, is highly satisfying. In conclusion, we recommend adopting our method for training book-specific models on early printed books because the implementation is straightforward and promises a significantly increased performance.

While this paper mainly focused on the application to book-specific training, voting ensembles have been shown to be highly effective when using them in scenarios dealing with *mixed models* which can be applied out of the box without further production of and training on book-specific GT. Consequently, experiments towards this direction represent the next logical step. Furthermore, we aim to adapt our approach for different domains in which GT is scarce: An obvious field is Handwritten Text Recognition (HTR) (see, e.g., [19]) but also historical music recognition (see, e.g., [4,20]). Especially the latter is of interest because the GT production is even more tedious than for ATR.

To reduce the CER even further, experiments using pretrained weights and data augmentation are scheduled. Combining our proposed method with these established accuracy-improving techiques will likely improve the results even further since we expect the effects to behave orthogonally. Finally, after significantly improving the training step of ensembles we clearly see room for further developments regarding the applied alignment and confidence-voting algorithm since its heuristic-based handling of diagreements of different lenghts clearly is far from optimal. However, due to the peculiarities of the blank character this is far from trivial and represents a challenging task for the future.

References

1. Al Azawi, M., Liwicki, M., Breuel, T.: Combination of multiple aligned recognition outputs using WFST and LSTM. In: 2015 13th International Conference on Document Analysis and Recognition (ICDAR), pp. 31–35. IEEE (2015)
2. Boschetti, F., Romanello, M., Babeu, A., Bamman, D., Crane, G.: Improving OCR accuracy for classical critical editions. In: Research and Advanced Technology for Digital Libraries, pp. 156–167 (2009)
3. Breuel, T.: High performance text recognition using a hybrid convolutional-LSTM implementation. In: 14th IAPR International Conference on Document Analysis and Recognition (ICDAR), pp. 11–16. IEEE (2017)
4. Calvo-Zaragoza, J., Toselli, A.H., Vidal, E.: Handwritten music recognition for mensural notation with convolutional recurrent neural networks. Pattern Recogn. Lett. (2019)
5. Graves, A., Fernández, S., Gomez, F., Schmidhuber, J.: Connectionist temporal classification: labelling unsegmented sequence data with recurrent neural networks. In: Proceedings of the 23rd International Conference on Machine learning, pp. 369–376. ACM (2006)

---

[4] Since the traditional cross-fold training approach consists of n, usually five, independent training sub processes it is possible to minimise the training duration by running these processes in parallel if several, ideally n, GPUs are available. However, we think that the presence of, at most, a single GPU should be considered the default case.

6. Kiessling, B.: Kraken - an Universal Text Recognizer for the Humanities. DH 2019 Digital Humanities (2019)
7. Kingma, D., Ba, J.: Adam: a method for stochastic optimization. In: ICLR (2014)
8. Kurata, G., Audhkhasi, K.: Guiding ctc posterior spike timings for improved posterior fusion and knowledge distillation. arXiv preprint arXiv:1904.08311 (2019)
9. Reul, C., et al.: Ocr4all–an open-source tool providing a (semi-) automatic ocr workflow for historical printings. App. Sci. **9**(22), 4853 (2019)
10. Reul, C., Springmann, U., Wick, C., Puppe, F.: Improving OCR accuracy on early printed books by utilizing cross fold training and voting. In: 2018 13th IAPR International Workshop on Document Analysis Systems (DAS). pp. 423–428. IEEE (2018). https://ieeexplore.ieee.org/document/8395233
11. Reul, C., Wick, C., Springmann, U., Puppe, F.: Transfer learning for OCRopus model training on early printed books. 027.7 J. Libr. Cult. **5**(1), 38–51 (2017). http://dx.doi.org/10.12685/027.7-5-1-169
12. Rice, S.V., Jenkins, F.R., Nartker, T.A.: The fifth annual test of OCR accuracy. Information Science Research Institute (1996)
13. Rice, S.V., Kanai, J., Nartker, T.A.: An algorithm for matching OCR-generated text strings. Int. J. Pattern Recogn. Artif. Intell. **8**(05), 1259–1268 (1994)
14. Rice, S.V., Nartker, T.A.: The ISRI analytic tools for OCR evaluation. UNLV/Information Science Research Institute, TR-96-02 (1996)
15. Sagi, O., Rokach, L.: Ensemble learning: a survey. WIREs Data Mining Knowl. Disc. **8**(4), e1249 (2018). https://doi.org/10.1002/widm.1249
16. Smith, R.: An overview of the Tesseract OCR engine. In: Ninth International Conference on Document Analysis and Recognition (ICDAR 2007), vol. 2, pp. 629–633. IEEE (2007)
17. Springmann, U., Lüdeling, A.: OCR of historical printings with an application to building diachronic corpora: a case study using the RIDGES herbal corpus. Digital Human. Q. **11**(2) (2017). http://www.digitalhumanities.org/dhq/vol/11/2/000288/000288.html
18. Springmann, U., Reul, C., Dipper, S., Baiter, J.: Ground truth for training ocr engines on historical documents in german fraktur and early modern latin. JLCL Spec. Issue Autom. Text Layout Recogn. **33**(1), 97–114 (2018). https://jlcl.org/content/2-allissues/2-heft1-2018/jlcl-2018-1.pdf
19. Sánchez, J.A., Romero, V., Toselli, A.H., Villegas, M., Vidal, E.: Icdar 2017 competition on handwritten text recognition on the read dataset. In: 2017 14th IAPR International Conference on Document Analysis and Recognition (ICDAR), vol. 01, pp. 1383–1388 (2017). https://doi.org/10.1109/ICDAR.2017.226
20. Wick, C., Puppe, F.: Experiments and detailed error-analysis of automatic square notation transcription of medieval music manuscripts using CNN/LSTM-networks and a neume dictionary. J. New Music Res., 1–19 (2021)
21. Wick, C., Reul, C., Puppe, F.: Calamari - a high-performance tensorflow-based deep learning package for optical character recognition. Digital Human. Q. **14**(1) (2020)

On the Use of Attention in Deep Learning Based Denoising Method for Ancient Cham Inscription Images

Tien-Nam Nguyen[1]([✉])(iD), Jean-Christophe Burie[1](iD), Thi-Lan Le[2](iD), and Anne-Valerie Schweyer[3](iD)

[1] Laboratoire Informatique Image Interaction (L3i) La Rochelle University, Avenue Michel Crépeau, 17042 La Rochelle Cedex 1, France
{tnguye28,jcburie}@univ-lr.fr

[2] School of Electronics and Telecommunications, Hanoi University of Science and Technology, Hanoi, Vietnam
lan.lethi1@hust.edu.vn

[3] Centre Asie du Sud-Est (CASE), CNRS, Paris, France
anne-valerie.schweyer@cnrs.fr

Abstract. Image denoising is one of the most important steps in the document image analysis pipeline thanks to its good effect into the rest of the workflow. However, the noise in historical documents is totally different from the common noise present in other classical problems of image processing. It is particularly the case of the image of Cham inscriptions obtained by the stamping of ancient stele. In this paper, we leverage the advantage of deep learning to adapt with these noisy conditions. The proposed network follows an encoder-decoder structure by combining convolution/deconvolution operators with symmetrical skip connections and residual blocks for improving reconstructed image. Furthermore, global attention fusion is proposed to learn the relevant regions in the image. Our experiments demonstrate the proposed method can't only remove unwanted parts in the image, but also enhance the visual quality for the Cham inscriptions.

Keywords: Document image analysis · Historical document · Image denoising · Attention · Cham inscription

1 Introduction

Exploring cultural heritage has attracted many researchers these last decades. Historical handwritten documents are important evidence in order to understand historical events and especially the ones of extinct civilizations. The Cham inscriptions are written from the Cham language system, which has been used from the very early centuries AD in Champa (nowadays Vietnam coastal areas) and some nearby areas. The descendants of the Cham population represent one part of the community in Southeast Asia. Nowadays, the Cham inscriptions are

© Springer Nature Switzerland AG 2021
J. Lladós et al. (Eds.): ICDAR 2021, LNCS 12821, pp. 400–415, 2021.
https://doi.org/10.1007/978-3-030-86549-8_26

mainly carved on steles of stone. Over time, the aging and climatic conditions have damaged the characters and created bumps. Many unwanted parts or gaps appeared on the stones making the visual quality of the image degraded significantly. The readability has become a real challenge for archaeologists, historians, as well as for people curious about Cham culture. The preservation of this cultural heritage is an important problem that needs to be considered. Similar research works to preserve palm leaf manuscripts have been done in [3].

In Document Image Analysis (DIA), a binarization process is usually used as a pre-processing step before applying the Optical Character Recognition (OCR) step. However, as mentioned above, these inscriptions have been damaged. Moreover, the image of the inscriptions, we work with, are obtained by a stamping process which may also create noise. Thus, traditional image binarization can not work well on these types of inscriptions due to the various degradation. Hence, an adaptive approach to remove noise and improve the visual quality of these inscriptions is needed before analyzing the content.

With the tremendous performance in many computer vision tasks, deep neural networks (DNNs) have demonstrated the ability in not only traditional tasks but also on complex learning tasks. Despite the outcome from DNNs, applying the latest techniques to the document historical problems have not been studied enough yet. Image denoising task could be considered as an image translation task that maps one image from the noisy domain to the cleaned domain [17]. Inspired by this idea, the supervised [1,2] or unsupervised [4,5] approach have shown promising results in image denoising problems.

Attention is proposed as an auxiliary module to make model robust to salient information rather than learning insignificant background parts in the image. Attention has obtained successful results in natural language processing problems such as [20,21] and is gradually used in computer vision problems [22]. Specifically, at different scales, an image contains different information. At lower scale it reflects general information while it contains detailed information at higher scale [30]. To efficiently leverage this information, we propose global attention fusion which accumulates attention from different scales to enhance denoised image quality.

Our main contributions are briefly detailed as follows: First, to the best of our knowledge, this is the first time attention module is embedded into image denoising method on historical documents. Second, we proposed an adaptive way to use the attention module in the training model and the global attention fusion module in cooperation with the loss function providing higher both qualitative and quantitative results. Finally, we tested the effectiveness of the proposed method by comparing it with different denoising methods of the literature on the Cham inscription dataset.

The rest of the paper is organized as follows. In Sect. 2, we present a brief overview of related works on image denoising, especially for historical documents, based on both traditional and deep learning approaches. In Sect. 3, we introduced the Cham inscription dataset. In Sect. 4, the details of the proposed approach are presented. The experimental results and comparison with traditional approaches are described in Sect. 5. Finally, a conclusion with some future research directions are given in Sect. 6.

2   Related Work

Image Denoising. Usually, traditional methods have been applied for image denoising problems based on the characteristics of noise in the corrupted image. Such methods can be categorized as spatial filters [6–8] or transform domain [13,14,16]. The denoising in spatial domain is based on the observation that noise appears at high frequency and low pass filter is adapted to eliminate this noise. However, these filter-based methods are not robust as each type of noise requires distinct filter kernel size. Another approaches are based on the pioneer work of non-local means [9]. The denoised image is estimated by weighting with similar patches on different locations of image. To handle with the blurry results, some suitable regularization methods can be used to enhance the quality of the denoised image such as: total variation regularization [10], sparse prior [11], low rank prior [12]. In the transform domain, image will be converted to a new domain where the characteristics of the noisy part are different with the clean part. Fourier transform, Wavelet transform [13,14,16] are common transform domain approaches. The main disadvantage of transformation-based approach is choosing the kind of transform or wavelet bases which are suitable for data. Besides the model driven methods, data driven methods are conventional approaches which are based on statistical representation of the data. These methods try to represent a set of images into sub-components with some prior assumption then remove redundancy in the representation which is estimated the noise present in the image. Independent Component Analysis [34] and Principal Component Analysis [36] are widely represented for these methods. The BM3D algorithm [15] is an impressive work which leverages the advantage of spatial domain and the transform domain. In general, these traditional approaches depend on pre-defined rules, which reduce the robustness to different types of noises. Although these methods have demonstrated successful results, especially BM3D that achieves very promising results in comparison with deep learning based approaches, the main problem is that all the methods work under the assumption that noise is an additive white Gaussian noise with a given standard deviation. However, in real problems, this requirement is not adaptable, noise needs to be represented by a more complex function. Then most of the research shifted to deep learning based approaches. One of the first works using deep learning approach is proposed in [17]. They used a multi-layer perceptron (MLP) to learn the mapping directly from noisy to clean image. Based on this idea, many works have shown competitive results with very deep learning-based approaches such as: [1,2,37]. In the work of [1], they used a deep convolutional network made of two parts : an encoder for learning clean parts in the image and a decoder to reconstruct the original shape of the image from them. Instead of directly mapping from noisy image to noisy-free image, [2] proposed a model for leaning the noisy space in the image, the noisy-free image can be obtained

Fig. 1. Sample inscriptions in the Cham dataset

by subtracting the noisy image and the noise extracted from the model. Related to our work, [18] and [19] applied generative modeling, successfully used in the context of historical handwritten document analysis. However, the mentioned approaches are equally considering the role of each pixel, the outcome results will not clearly distinguish between background (noise) and foreground pixels (characters). This hence lowers the qualitative and quantitative results. To resolve this issue, we adapted the attention module.

Attention. Based on the principle of human vision, attention mechanism is proposed to be robust to the important parts of the image instead of learning irrelevant parts by assigning higher weight to the useful region. Attention can directly be integrated as a component in the model. In general, attention can be split into two types: self-attention [23–25] which is computed from only one input feature and general attention [26] which is computed from two or more input features. Depending on the problems and the architecture of models, the appropriate attention is selected.

3   Dataset

We now present the Cham inscription dataset used in our research works. A preliminary introduction about Cham language can be found at [32]. Since no dataset and no ground truth were available, our first contribution was to build a dataset and the corresponding ground truth. The images of Cham inscriptions have been obtained by a stamping process. This work has been done in Vietnam by archaeologists during excavations in the field or by curators in museums, when the steles were deposited there. The stamping process allows to copy the inscriptions carved on the stone (ancient stele) on a large sheet of paper. However, all gaps and bumps are also "copied". The dimensions of the images are variable from 1000 to 6000 pixels either height or width. The sheets of paper are then scanned to obtain digital images. Our dataset, at the moment, consists of 100 "documents" (duplicates of ancient stone steles). Besides the limitation of available ground truth, the challenges of these inscriptions also come from the range of date. The inscriptions that have been collected fall within a wide chronological range from the 6th to the 15th century AD and are written either in Cham or in Sanskrit. The use of each of these two languages leads to a slightly different writing system, which must be considered for the text analysis but they can be processed in the same way for the denoising task. In addition, the degradation of many inscriptions poses an additional challenge for working on Cham inscriptions. As we can see in Fig. 1, there are many parts of the text that may seem undesirable. By simple observation, it is often impossible to tell whether certain strokes on the stone form part of a letter or a group of letters or are noise. The annotation has been done by a linguistic expert. A raster graphic editor has been used to remove all pixels of noise and correct the missing pixels of text. As the cleaning task is time-consuming and to avoid processing large-size documents, each "document" image has been split in text line images. Our dataset consists of 190 text line images that have been cleaned one by one by the expert. So, the ground truth consists of a set of binary images without any noise and with cleaned characters (according to the expert knowledge).

4   Proposed Approach

4.1   Architecture

As shown in Fig. 2, the proposed model consists of two main components: a baseline encoder-decoder model and the attention module.

Baseline Model: The baseline of the proposed model is inspired by Unet [27] that follows the encoder-decoder architecture. Input image is first processed by a consecutive Convolution-BatchNorm-Relu layer to the 1×1 size at the bottle layer. Then from the output of bottleneck layer, reconstructed image is achieved by doing the series Deconvolution-BatchNorm-Relu (up-sampling step). However, the up-sampling step is done from the bottleneck layer where the size of the data is small, much of the information will be lost in the reconstructed images. Instead of down-sampling the input image to the 1×1 size, the input image is down-sampled twice before going through a sequence of Resnet block [29] in the middle (the blue blocks in Fig. 2). This type of architecture is similar to the coarse-to-fine generator in [28].

Attention: In order to reduce the redundancy of the information when using skip connection [1] between down sampling steps and up sampling steps, we adapt the attention gate introduced in [26]. This module selects appropriate information instead of keeping all information from the down-sampling step. The integration of the attention gate in skip connection can be explained as below.

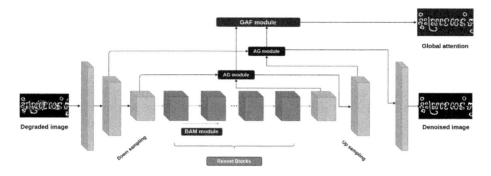

Fig. 2. Architecture of the proposed model.

First, an attention map is generated from F_{D2} and F_{U2}. This attention map has high value at the similar feature between F_{D2} and F_{U2}. The modification features F'_{D2} is computed through multiplying F_{D2} with the attention map.

$$F'_{D2} = F_{D2} * \sigma(W_n * (\omega((W_d * F_{D2} + W_u * F_{U2})))) \tag{1}$$

where F_{D2}, F_{U2} are the features at the down sampling step and up sampling step respectively, W_d, W_u and W_n are the parameters of convolution layers, ω is Relu activation function, σ is the sigmoid activation function. After that, we concatenate F'_{D2} and F_{U2} features as common skip connection then upsample it to F_{U1}.

$$F_{U1} = W_t * (concat([F_{U2}, F_{D2'}])) \tag{2}$$

where W_t are parameters of deconvolution layer. This process also repeats at the next upper scale. The details of module are represented in Fig. 3.

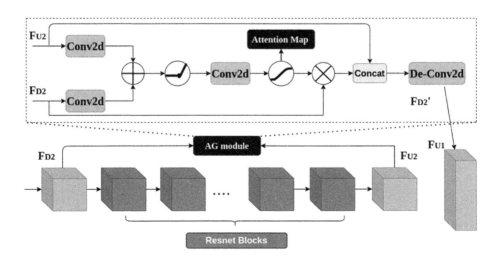

Fig. 3. Integration of attention gate in skip connection [26]

For the sequence Resnet block in the middle of our architecture, we modified the normal connection between Resnet block by adding the Bottleneck Attention Module (BAM) [24] after every Resnet block to improve the separation of the features at low-level such as: gaps or strokes on the surface and gradually focus on the exact target at a high-level of semantic (characters) by integrating attention from both channel and spatial information. After the sequence Resnet block + BAM (see in Fig. 4), the model aims to highlight salient features of character regions while deducting the features of non-relevant regions. Shortly, refined features by BAM can be described as:

$$F' = F + F * \sigma(M_c(F) + M_s(F)) \tag{3}$$

where F is the output feature of the previous Resnet block. M_c, M_s are the channel attention and spatial attention, respectively. More details of each channel attention and spatial attention can be found in the [24].

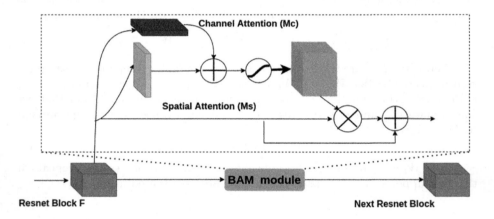

Fig. 4. Refined features by BAM [24]

Global Attention Fusion: Due to the difficulty when reading Cham inscriptions, it is more important to keep and enhance the quality of the characters than removing only the noise. Therefore, we proposed a global attention fusion module by integrating attention at multiple scales to help the model focus more on the pixels of the characters. This module works as below. First, we concatenate the attention map from different scales of the attention gate module.

$$C(A) = concat(A_1, resize(A_2)) \tag{4}$$

where A_i be the attention map generated from the attention gate at different scales, i is the scale of input image. The concatenated attention is then followed by a deconvolution layer to generate global attention map which has

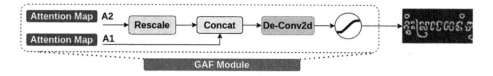

Fig. 5. Global attention fusion module

the same shape of input image then sigmoid activation function to normalize the coefficients of them to the range (0,1).

$$C_s = \sigma(W_A * C(A)) \tag{5}$$

where W_A is the parameters of deconvolution layer, σ is the sigmoid activation function. (See Fig. 5). These coefficients of global attention map (C_s) represent the confidence of the generated pixels for the character regions in the image. As we can see in Fig. 6, at the beginning, only simple patterns which model easily to recognize are noises or characters, will correctly reconstructed with the high confidence score. After some epochs, this score gradually increases to reveal difficult patterns of the characters which have lower scores at the first epochs.

| Epoch 1 | Epoch 25 | Epoch 100 |

Fig. 6. Evolution of confidence score. Bright pixels have a high score, dark ones have a low score.

4.2 Objective Function

The training objective function combines two main functions. The first function, weighted L1 Loss (weight reconstruction loss), helps to reconstruct the denoised image of important regions closest to the ground truth images.

$$L_1^w = ||C_s * \hat{y} - y||_1 \tag{6}$$

where C_s is the confidence score output from GAF module. The second one is the perceptual loss L_p [30] which forces a generated image from the model

Fig. 7. Attention map with and without constraint on the C_s. We can see that in the middle of the figure, when the model is trained without constraint, character regions and empty regions have high coefficients. In the right image, when using constraint, the model aims to reveal only character regions, the other regions have a low score.

to have a perceptual feature similar to the one of the ground truth images. Perceptual features are intermediate features extracted from different layers of a pre-trained model such as VGG, Inception or Resnet.

$$L_p = \sum_{i=1}^{n} \frac{1}{C_i * H_i * W_i} \sum_{h=1}^{H_i} \sum_{w=1}^{W_i} \sum_{c=1}^{C_i} ||F_{h,w,c}^i(\hat{y}) - F_{h,w,c}^i(y)||_1 \qquad (7)$$

where $F_{h,w,c}^i$ is output feature from list n features named as 'relu1-2', 'relu2-2', 'relu3-3', 'relu4-3' of VGG network.

We found that if we used C_s in the Eq. 6 without constraint, the coefficients of C_s have high values in regions almost empty because for those regions, the model is easy to reconstruct. After that, the optimization for these regions will not change. (See Fig. 7). To avoid that we simply used the mean of C_s as an additional constraint that helps C_s to consider only the character regions instead of the empty regions.

Overall the objective function can be described as:

$$L = \lambda_{l_1} * L_1^w + \lambda_{l_{cs}} * mean(C_s) + \lambda_{l_p} * L_p \qquad (8)$$

where λ_{l_1}, $\lambda_{l_{cs}}$ and λ_{l_p} are the weights of the weight reconstruction loss (L_1^w), mean value final attention map (C_s) and perceptual loss (L_p) in the total loss, respectively. Each weight indicates the contribution of each component in the total loss function. The higher the weight is, the greater the contribution of the components is. The model has to balance both pixel level and high features level based on different weights of each loss.

5   Experiments

Dataset: We evaluated the proposed model on the dataset presented in Sect. 3. This dataset consists of images of Cham inscriptions written in Cham or Sanskrit. Due to the different size of each inscription, we normalized the size of the training images. Instead of resizing the whole text line image which leads to distortion, we simply cropped the original image into sub-images which have a size of 256×512 pixels. 140 text line images were split into approximate 2500 images as training data while 50 other text line images were used as testing data. Furthermore, due to the limitation available data, augmentation strategy was also studied. We used some data augmentation below: random rotation (some text lines are not horizontal), random erasing, similar mosaic augmentation [31] (combining images by cutting parts from some regions and pasting them onto the augmented image).

Training Setting: We used the Adam optimizer with initial learning rate 0.0002. The weight λ_{l_1}, $\lambda_{l_{cs}}$ and λ_{l_p} were determined experimentally 1, 0.1, 2, respectively. Further studies will be done to evaluate the influence of these weights on the results but they are not in the scope of this paper. All models in our experiments were trained from scratch with the same the number of training data. The experiments were done on 2080Ti GPU 12 GB memory.

Evaluation Metrics: In order to evaluate the denoising performance, we used two metrics: Peak Signal-to-Noise Ratio (PSNR) and Structural Similarity (SSIM). PSNR is a quality metric which measures how the denoised image is close to the ground truth image while SSIM metric measures the similarity structure between two images. The highest PSNR and SSIM values indicate the better results.

5.1 Ablation Study

Table 1. Ablation study on the sub-component in our model

Experiment	Condition	PSNR	SSIM
Exp 1	Baseline model	15.45 ± 2.65	0.898 ± 0.05
Exp 2	Baseline model + Attention gate	15.57 ± 2.62	0.900 ± 0.049
Exp 3	Baseline model + BAM module	15.86 ± 2.76	0.904 ± 0.048
Exp 4	**Proposed model**	**15.98 ± 2.82**	**0.905 ± 0.048**

Table 1 shows the ablation study on the effects of each component in the proposed model which goes from baseline model, BAM module, AG module, and the global attention fusion module. On the SSIM metric, the results are similar because on text line images, the structure of the image is simple, and all methods can simply preserve the original structure. On the PSNR metric, we can see that the use of the BAM module and AG module have slightly improved results. The proposed method can achieve better results than each module combined separately and shows an improvement in comparison to the baseline model.

5.2 Comparison with Different Approaches

Table 2 presents a comparison of our approach with other methods of the literature applied on our dataset. We split these methods into two main groups. In

Table 2. Quantitative results on our dataset

Method	Avg-PSNR	Avg-SSIM
Original	11.93	0.651
TV [10]	12.68	0.567
NLM [9]	12.01	0.489
BM3D [15]	11.64	0.623
Ostu [38]	10.96	0.713
Sauvola [40]	10.44	0.689
Niblack [39]	10.59	0.694
NMF [35]	12.63	0.749
FastICA [34]	12.69	0.754
Unet [27]	15.54	0.900
Pix2pix [33]	14.05	0.875
Pix2pixHD [28]	13.26	0.855
Proposed method	**15.98**	**0.905**

the first group, the input image is directly processed without any training step. The second group consists of methods with a training step integrating knowledge from both original (degraded image) and clean image. The first group gathers traditional denoising methods (TV [10], NLM [9], BM3D [15]) and binarization methods (Otsu [38], Sauvola [40], Niblack [39]). The traditional methods, in average, PSNR is improved but the value of the SSIM metric is lower than

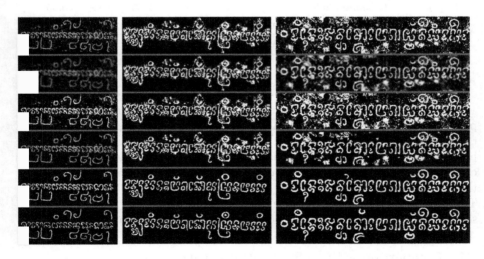

Fig. 8. Some qualitative results on cham inscriptions with **low** degradation. From top to bottom: Original image, NLM, Otsu, FastICA, Proposed method, Ground truth.

the original images. This can be explained by the fact that these methods are efficient on regions where the size of the noise is relatively small, but it leads the output image blurrier, the SSIM score is lower than the original input image. The results with binarization methods show that these approaches are not adapted since they don't reduce the noise (lower PSNR than original image) even if they enhance the global visual quality of the image providing a better SSIM than original image. Qualitative results on images with different methods on the low and high degradation are shown respectively in Fig. 8 and Fig. 9.

In the second group, both NMF [35] and FastICA [34] methods improve both the PSNR and SSIM metrics but we observed that the denoising ability is very limited when the patterns of noise are similar to some parts of the characters affecting the readability. The last methods of this second group consist of deep learning-based approaches. All methods have significantly boosted both the quantitative (PSNR and SSIM metrics) and qualitative results. The proposed method achieves better results on both metrics. Besides enhancing denoising results compared to the other methods, our approach gave better results because it generates pixels with a high confidence value in the foreground and thus provides a better visual quality.

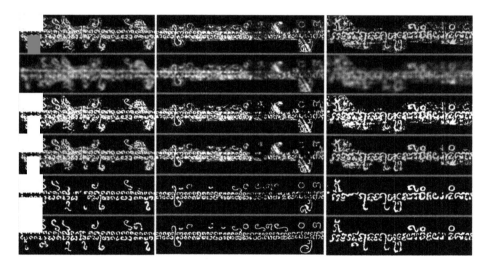

Fig. 9. Some qualitative results on cham inscriptions with **high** degradation. From top to bottom: Original image, NLM, Otsu, FastICA, Proposed method, Ground truth.

5.3 Qualitative Results

If the proposed method improves statistically the quality of the image, the readability is also an important parameter for the next step : the recognition process. In order to evaluate the relevance of the approach we asked an expert in Cham language to estimate the performance of our method. We asked her to analyse qualitatively two different criteria: performance on noise removal and character readability. For each criterion, we defined four level assessments: very bad, bad, normal, and good, corresponding respectively to the 1, 2, 3, 4 score. The qualitative evaluation is obtained by computing, for each experiment detailed in Table 1, the average score for each criterion. However, instead of evaluating the results on the whole testing set, it was split into three subsets depending of the historical period : 7th-9th century (5 images), 10th-12th century (19 images) and 13th-15th century (26 images). The quality of the inscriptions depends on their age due to the fact that the damages are more important on the older stones. The separation of images into chronological categories has been motivated by the evolution of writing, starting from often irregular and less codified scripts, passing through the blossoming and mastery of characters and ending with less refined scripts and less distinguished characters, creating more confusion for deciphering. So, we created 3 categories by this separation. Table 3 presents the qualitative results. For the noise removal criterion, we observe that the integration of attention module (AG or BAM) into the baseline model improves significantly the results compared to the simple baseline model. For character readability criterion, the proposed method outperforms the other approaches. This qualitative evaluation shows the effect of the GAF module by encouraging the model to generate character pixel with higher confidence value. The improvement of contrast between foreground and background part as well as the quality of the characters, it increases the readability of the Cham inscription.

Table 3. Average score of the qualitative evaluation on our testing dataset split in 3 sets depending of the historical period

Aspects	Condition	7th–9th	10th–12th	13th–15th
Noise removal	Baseline	3 ± 0	2.45 ± 0.6	1.53 ± 0.58
	Baseline + BAM	3 ± 0	2.8 ± 0.52	2.31 ± 0.68
	Baseline + AG	3 ± 0	2.6 ± 0.5	1.85 ± 0.61
	Proposed method	$\mathbf{3.4 \pm 0.55}$	$\mathbf{2.84 \pm 0.66}$	$\mathbf{2.65 \pm 0.77}$
Character readability	Baseline	2.2 ± 0.45	1.94 ± 0.41	1.34 ± 0.41
	Baseline + BAM	3 ± 0.7	2.05 ± 0.4	1.88 ± 0.41
	Baseline + AG	3.2 ± 0.45	2.15 ± 0.5	1.96 ± 0.41
	Proposed method	$\mathbf{3.4 \pm 0.55}$	$\mathbf{2.52 \pm 0.77}$	$\mathbf{2.34 \pm 0.41}$

6   Conclusion

In this work, we present an approach based on attention model for denoising old Cham inscription images. We have first introduced a new dataset for the image denoising problem of Cham inscriptions. We have detailed several experiments and analysed the benefits as well the disadvantages of each method. The quantitative and qualitative evaluations show that the proposed approach improves the quality of the Cham inscription in terms of noise removal and character readability. However, a room of improvement is certainly possible by using more data for the training step in order to consider more styles of degradation. So, we envisage to increase the size of the dataset to improve our model. In future work, we also plan to tackle the next step: the recognition of Cham Inscriptions.

Acknowledgment. This work is supported by the French National Research Agency (ANR) in the framework of the ChAMDOC Project, n°ANR-19-CE27-0018-02.

References

1. Mao, X.J., Shen, C., Yang, Y.B.: Image restoration using very deep convolutional encoder-decoder networks with symmetric skip connections. In: Proceedings of the 30th International Conference on Neural Information Processing Systems (2016)
2. Zhang, K., Zuo, W., Chen, Y., Meng, D., Zhang, L.: Beyond a gaussian denoiser: residual learning of deep cnn for image denoising. IEEE Trans. Image Process **26**(7), 3142–3155 (2017)
3. Kesiman, M.W.A., et al.: Benchmarking of document image analysis tasks for palm leaf manuscripts from southeast Asia. J. Imaging **4**(2), 43 (2018)
4. Lehtinen, J., et al.: Noise2noise: Learning image restoration without clean data. In: International Conference on Machine Learning (2018)
5. Krull, A., Buchholz, T.O., Jug, F.: Noise2void-learning denoising from single noisy images. In: IEEE Conference on Computer Vision and Pattern Recognition (2019)
6. Pitas, I., Venetsanopoulos, A.N.: Nonlinear Digital Filters: Principles and Applications, vol. 84. Springer, New York (2013) https://doi.org/10.1007/978-1-4757-6017-0
7. Wiener, N.: Extrapolation, Interpolation, and Smoothing of Stationary time Series: with Engineering Applications. MIT Press, Cambridge (1950)
8. Tomasi, C., Manduchi, R.: Bilateral filtering for gray and color images. In: Sixth International Conference on Computer Vision, pp. 839–846. IEEE (1998)
9. Buades, A., Coll, B., Morel, J.M.: A non-local algorithm for image denoising. In: IEEE Conference on Computer Vision and Pattern Recognition, pp. 60–65 (2005)
10. Rudin, L.I., Osher, S., Fatemi, E.: Nonlinear total variation based noise removal algorithms. Phys. D Nonlinear Phenom. **60**(1–4), 259–268 (1992)
11. Elad, M., Aharon, M.: Image denoising via sparse and redundant representations over learned dictionaries. IEEE Trans. Image Process. **15**(12), 3736–3745 (2006)
12. Dong, W., Shi, G., Li, X.: Nonlocal image restoration with bilateral variance estimation: a low-rank approach. IEEE Trans. Image Process. **22**(2), 700–711 (2012)
13. Choi, H., Baraniuk, R.: Analysis of wavelet-domain wiener filters. In: Proceedings of the IEEE-SP International Symposium on Time-Frequency and Time-Scale Analysis, pp. 613–616 (1998)
14. Ram, I., Elad, M., Cohen, I.: Generalized tree-based wavelet transform. IEEE Trans. Signal Process. **59**(9), 4199–4209 (2011)
15. Dabov, K., Foi, A., Katkovnik, V., Egiazarian, K.: Image denoising by sparse 3-d transform-domain collaborative filtering. IEEE Trans. Image Process. **16**(8), 2080–2095 (2007)
16. Portilla, J., Strela, V., Wainwright, M.J., Simoncelli, E.P.: Image denoising using scale mixtures of gaussians in the wavelet domain. IEEE Trans. Image Process. **12**(11), 1338–1351 (2003)

17. Burger, H.C., Schuler, C.J., Harmeling, S.: Image denoising: Can plain neural networks compete with bm3d? In: 2012 IEEE Conference on Computer Vision and Pattern Recognition, pp. 2392–2399 (2012)

18. Dumpala, V., Kurupathi, S.R., Bukhari, S.S., Dengel, A.: Removal of historical document degradations using conditional gans. In: ICPRAM (2019)

19. Souibgui, M.A., Kessentini, Y.: De-gan: a conditional generative adversarial network for document enhancement. In: IEEE Transactions on PAMI (2020)

20. Luong, T., Pham, H., Manning, C.D.: Effective approaches to attention-based neural machine translation. In: Proceedings of the 2015 Conference on Empirical Methods in Natural Language Processing, pp. 1412–1421. Lisbon, Portugal (September 2015)

21. Vaswani, A., et al.: Attention is all you need. In: Proceedings of the 31th International Conference on Neural Information Processing Systems (2017)

22. Carion, N., Massa, F., Synnaeve, G., Usunier, N., Kirillov, A., Zagoruyko, S.: End-to-end object detection with transformers. In: Vedaldi, A., Bischof, H., Brox, T., Frahm, J.-M. (eds.) ECCV 2020. LNCS, vol. 12346, pp. 213–229. Springer, Cham (2020). https://doi.org/10.1007/978-3-030-58452-8_13

23. Woo, S., Park, J., Lee, J.Y., Kweon, I.S.: Cbam: convolutional block attention module. In: Proceedings of the European Conference on Computer Vision, pp. 3–19 (2018)

24. Park, J., Woo, S., Lee, J.Y., Kweon, I.S.: Bam: bottleneck attention module. In: British Machine Vision Conference (2018)

25. Zhang, H., Goodfellow, I., Metaxas, D., Odena, A.: Self-attention generative adversarial networks. In: International Conference on Machine Learning, pp. 7354–7363. PMLR (2019)

26. Schlemper, J., et al.: Attention gated networks: Learning to leverage salient regions in medical images. Med. Image Anal. **53**, 197–207 (2019)

27. Ronneberger, O., Fischer, P., Brox, T.: U-net: convolutional networks for biomedical image segmentation. In: Navab, N., Hornegger, J., Wells, W.M., Frangi, A.F. (eds.) MICCAI 2015. LNCS, vol. 9351, pp. 234–241. Springer, Cham (2015). https://doi.org/10.1007/978-3-319-24574-4_28

28. Wang, T.C., Liu, M.Y., Zhu, J.Y., Tao, A., Kautz, J., Catanzaro, B.: High-resolution image synthesis and semantic manipulation with conditional gans. In: Proceedings IEEE Conference on Computer Vision and Pattern Recognition (2018)

29. He, K., Zhang, X., Ren, S., Sun, J.: Deep residual learning for image recognition. In: Proceedings of the IEEE Conference on Computer Vision and Pattern Recognition, pp. 770–778 (2016)

30. Johnson, J., Alahi, A., Fei-Fei, L.: Perceptual losses for real-time style transfer and super-resolution. In: European Conference on Computer Vision, pp. 694–711 (2016)

31. Bochkovskiy, A., Wang, C.Y., Liao, H.Y.M.: Yolov4: Optimal speed and accuracy of object detection. arXiv preprint arXiv:2004.10934 (2020)

32. Nguyen, M.T., Shweyer, A.V., Le, T.L., Tran, T.H., Vu, H.: Preliminary results on ancient cham glyph recognition from cham inscription images. In: 2019 International Conference on Multimedia Analysis and Pattern Recognition (MAPR), pp. 1–6. IEEE (2019)

33. Isola, P., Zhu, J.Y., Zhou, T., Efros, A.A.: Image-to-image translation with conditional adversarial networks. In: Proceedings of the IEEE Conference on Computer Vision and Pattern Recognition, pp. 1125–1134 (2017)

34. Hyvarinen, A., Hoyer, P., Oja, E.: Sparse code shrinkage: Denoising by nonlinear maximum likelihood estimation. Adv. Neural Inf. Process. Syst. **11**, 473–479 (1999)
35. Févotte, C., Idier, J.: Algorithms for nonnegative matrix factorization with the beta-divergence. Neural Comput. **23**(9), 2421–2456 (2011)
36. Deledalle, C.A., Salmon, J., Dalalyan, A.S., et al.: Image denoising with patch based pca: local versus global. BMVC **81**, 425–455 (2011)
37. Zhang, K., Zuo, W., Zhang, L.: Ffdnet: toward a fast and flexible solution for cnn based image denoising. IEEE Trans. Image Process **27**(9), 4608–4622 (2018)
38. Otsu, N.: A threshold selection method from gray-level histograms. IEEE Trans. Syst. Man Cybern. **9**(1), 62–66 (1979)
39. Niblack, W.: An Introduction to Digital Image Processing. Strandberg Publishing Company, Birkeroed (1985)
40. Sauvola, J., Pietikainen, M.: Adaptive document image binarization. Pattern Recogn. **33**(2), 225–236 (2000)

Visual FUDGE: Form Understanding via Dynamic Graph Editing

Brian Davis[1]([✉]), Bryan Morse[1], Brian Price[2], Chris Tensmeyer[2], and Curtis Wiginton[2]

[1] Brigham Young University, Provo, USA
{briandavis,morse}@byu.edu
[2] Adobe Research, San Jose, USA
{bprice,tensmeye,wigingto}@adobe.com

Abstract. We address the problem of *form understanding*: finding text entities and the relationships/links between them in form images. The proposed FUDGE model formulates this problem on a graph of text elements (the vertices) and uses a Graph Convolutional Network to predict changes to the graph. The initial vertices are detected text lines and do not necessarily correspond to the final text entities, which can span multiple lines. Also, initial edges contain many false-positive relationships. FUDGE edits the graph structure by combining text segments (graph vertices) and pruning edges in an iterative fashion to obtain the final text entities and relationships. While recent work in this area has focused on leveraging large-scale pre-trained Language Models (LM), FUDGE achieves the same level of entity linking performance on the FUNSD dataset by learning only visual features from the (small) provided training set. FUDGE can be applied on forms where text recognition is difficult (e.g. degraded or historical forms) and on forms in resource-poor languages where pre-training such LMs is challenging. FUDGE is state-of-the-art on the historical NAF dataset.

Keywords: Form understanding · Relationship detection · Entity linking

1 Introduction

Paper forms are a convenient way to collect and organize information, and it is often advantageous to digitize such information for fast retrieval and processing. While OCR and handwriting recognition (HWR) methods can extract raw text from a form image, we aim to understand the layout and relationships among the text elements (e.g., that "Name:" is associated with "*Lily Johnson*"). The term *form understanding* was recently coined [10] as the task of extracting the full structure of information from a form image. We define the task as: given a form image, identify the semantic text entities and the relationships between them *with no prior form template*. This work focuses on finding relationships among

© Springer Nature Switzerland AG 2021
J. Lladós et al. (Eds.): ICDAR 2021, LNCS 12821, pp. 416–431, 2021.
https://doi.org/10.1007/978-3-030-86549-8_27

text entities, although this requires first detecting, segmenting, and classifying text into entities. After entities are predicted, the relationship detection or entity linking task is simply predicting which text entities have a semantic relationship.

Most recent works on form understanding rely primarily on large-scale pre-trained Language Models (LMs) [8,18,21], which in turn have a dependency on accurate text detection and recognition. Such approaches may perform poorly in domains where OCR/HWR results are poor or on languages with limited resources to train such LMs. (LayoutLM [21] is pre-trained on 11 million documents.) OCR/HWR often struggle on damaged or degraded historical documents and on document images captured inexpertly with a smartphone.

In contrast, we present a purely visual solution, improving on our previous visual form understanding method in Davis et al. [5]. Given forms in an unfamiliar language, humans can generally infer the text entities and their relationships using layout cues and prior experience, which we aim to approximate. Our approach doesn't require language information and could be applied to visually similar languages (e.g., those sharing the same script), possibly without fine-tuning. In this work we attempt to show that a visual model trained on a small dataset without language information is, on several tasks, comparable to methods that rely on large amounts of pre-trained language information.

Similar to some prior works [4,5,18], we model forms as a graph, where text segments are the vertices and pairwise text relationships are edges. In [5] we scored pairwise heuristic relationship proposals independently using visual features and applied global optimization as a post-processing step to find a globally coherent set of edges. We improve upon this by making our model more end-to-end trainable and by jointly predicting relationships with Graph Convolutional Network (GCN). Additionally, [5] was unable to predict text entities that span multiple text lines, a problem solved with our dynamic graph editing. An alternative formulation to solve form understanding visually is to treat it as a pixel labeling problem, as in Sarkar et al. [16]. However, it is not clear from [16] how to infer form structure (bounding boxes and relationships) from pixel predictions, and the proposed (even dilated) CNN model could have difficulty modeling relationships between spatially distant elements. Instead we use a GCN that directly predicts the form structure and does not need to rely on limited receptive fields to propagate information spatially.

Our proposed FUDGE (Form Understanding via Dynamic Graph Editing) model is a multi-step process involving text line detection, relationship proposals, graph editing to group text lines into coherent text entities and prune edges, leading to relationship prediction (Fig. 1). We use GCNs so that semantic and relationship predictions can be jointly predicted. We initialize the graph vertices with detected text lines (visual detection of semantically grouped words is hard). However, the relationships of interest are between text entities, which can be multiple text lines. FUDGE is unique from other GCNs as we dynamically edit the graph during inference to predict which vertices should be merged into a single vertex. This groups text lines into text entities and corrects over-segmented initial text detections.

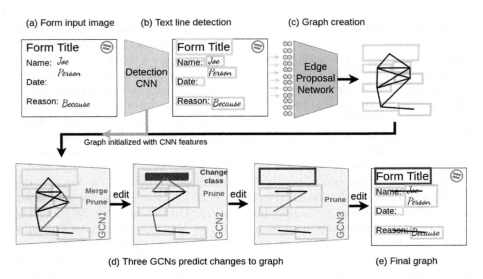

Fig. 1. Overview of our model, FUDGE. From a form image (a) text line detection (b) is performed. Then (c) an edge proposal score is computed for each possible edge. After thresholding the scores, the remaining edges form the graph. The graph is initialized with spatial features and features from the CNN detector. A series of GCNs are run (d), each predicting edits to the graph (pruning irrelevant edges, grouping text lines into single entities, correcting oversegmented lines). The final graph (e) is the text entities and their relationships.

In summary, our primary contributions are:

- a GCN that jointly models form elements in the context of their neighbors,
- an iterative method to allow the GCN to merge vertices and prune edges,
- extensive experiments on the FUNDS [10] and NAF [5] datasets which validate that FUDGE performs comparably to methods that pre-train on millions of images.

Our code is available at https://github.com/herobd/FUDGE.

2   Prior Work

Automated form processing has been of interest for decades, but generally template-based solutions have been used to extract information from a few known form layouts. Recently, the idea of form understanding [10] has become an area of interest. This has evolved from extraction of information in a template-free setting [12] to capturing all the information a form contains. Recent methods have leveraged the astounding progress of large pre-trained language models [6], focusing on language rather than visual elements to understand forms.

DocStruct [18] is a language-focused approach for form relationship detection. They blend three feature sources: semantic/language from BERT [6], spatial text bounding box coordinates, and visual infromation extracted from pixels. Their ablation results [18] show that the language and spatial features are the most valuable, which is reasonable given that the visual features are extracted only from the text areas and thus ignore most of the visual context around the relationships, especially distant ones.

BROS [8] is an unpublished method that builds on the ideas of LayoutLM [21], which is primarily a BERT model [6] with additional spatial and visual features appended to the word embeddings. Similar to DocStruct, LayoutLM only extracts visual features immediately around the text instances. BROS [8] adds better spatial representations and awareness to the model in addition to better pre-training tasks, but do not improve its use of visual features compared to LayoutLM. BROS solves both the text entity extraction (detection) task and entity linking (relationship detection). Entity extraction is done in two steps: start-of-entity detection and semantic classification, and next-token classification which collects the other tokens composing the entity. Relationship detection is done by computing an embedding for each start-of-entity token and doing a dot product multiplication across all pairs of start-of-entity token embeddings, giving the relationship scores. Form2Seq [1] is similar to LayoutLM, but uses LSTMs instead of Transformers and omits visual features entirely.

The unpublished LayoutLMv2 [20] is a Transformer based model that uses a much stronger visual component, adding visual tokens to the input and vision-centric pre-training tasks. For forms, LayoutLMv2 has only been evaluated on the entity detection and not on relationship detection. Another example of a good blend of visual and language information is used by both Attend, Copy, Parse [12] and Yang et al. [22]. Both encode the text as dense vectors and append these vectors to the document image pixels at the corresponding spatial locations. A CNN can then perform the final task, which is information extraction in [12] and semantic segmentation in [22].

In contrast, Davis et al. [5] and Sarkar et al. [16] propose language-agnostic models. In [5] we focused on the relationship detection problem and used a CNN backbone to detect text lines. Relationship candidates were found using a line-of-sight heuristic and each candidate was scored using a visual-spatial classifier. However, each candidate was scored independently, and a separate postprocessing optimization step was needed to resolve global consistency issues. This latter step also required predicting the number-of-neighbors for each text line, which may not be accurate on more difficult datasets. Sarkar et al. [16] focus on extracting the structure of forms but treat it as a semantic segmentation (pixel labeling) problem. They use a U-Net architecture, and at the lowest resolution include dilated convolutions to allow information to transfer long distances. Sarkar et al. [16] predicts all levels of the document hierarchy in parallel, making it quite efficient.

Aggarwal et al. [2] offers an approach that is architecturally like a language-based approach but uses contextual pooling of CNN features like [5]. They determine a context window by identifying a neighborhood of form elements and use a CNN to extract image features from this context window. They also extract language features (using a non-pre-trained LSTM), which are combined with the visual features for the final decisions. However, the neighborhood is found using

k-nearest neighbors with a distance metric which could be sensitive to cluttered or sparse forms and long distance relationships.

GCNs have been applied to other structured document tasks, such as table extraction [13,15]. Carbonell et al. [4] use a GCN to solve the entity grouping, labeling and linking tasks on forms. They use word embedding and bounding box information as the initial node features and they do not include any visual features. They use k-nearest neighbors to establish edges. In our method, we update our graph for further GCN processing, particularly grouping entities together. Carbonell et al. [4] predict the entity grouping from a GCN and then sum the features of the resulting groups. Rather than processing more with a GCN, they predict entity class and linking from these features. Unlike most other non-visual methods, Carbonell et al. [4] do not use any pre-training.

3   FUDGE Architecture

FUDGE is based on Davis et al. [5]. We use the same detection CNN backbone and likewise propose relationships, extract local features around each proposed relationship, and then classify the relationships. FUDGE differs in three important ways:

– In [5] we used line-of-sight to propose relationships, which can cause errors due to false positive detections and form layouts that don't conform to the line-of-sight assumption. FUDGE instead learns an edge proposal.
– Instead of predicting each relationship in isolation, we put the features into a graph convolutional network (GCN) so that a joint decision can be made. This also allows semantic labels for the text entities to be predicted jointly with the relationships, as they are very related tasks.
– We allow several iterative edits to the graph, which are predicted by the GCN. Text lines are grouped into single text entities, and oversegmented text lines are corrected, by aggregating groups of nodes into new single nodes. Spurious edges are pruned.

Figure 1 shows an overview of FUDGE. We now go into each component in detail.

3.1   Text Line Detection

We use the same text line detector as [5], only ours does not predict the number of neighbors. This detector is a fully-convolutional network with wide horizontally strided convolutions and a YOLO predictor head [14]. We threshold predictions at 0.5 confidence. The detector is both pre-trained and fine-tuned during training of the relationship detection. The detection makes an auxiliary class prediction; the final text entity class prediction is made by the GCN.

3.2   Edge Proposal

While line-of-sight is effective for the simplified NAF dataset contributed by Davis et al. [5], it is brittle and doesn't apply to all cases. FUDGE instead learns an edge proposal network using a simple two-layer linear network with ReLU activation in between. It receives features from each possible pair of detected text lines and then predicts the likelihood of an edge (they are either oversegmented, part of the same entity, or parts of entities that have a relationship). Half of the relationships with the highest scores (maximum of 900) are used to build the initial undirected graph. The features are: difference of x and y position for all corresponding corners and the center of the boxes, height and width of both boxes, L2 distance of all corresponding corners, normalized x and y position for both bounding boxes, whether there is a line of sight between the boxes (computed as in [5]), and the detection confidences and class predictions for both boxes. We predict both permutations of the pair orders and average them.

3.3   Feature Extraction

While the relationship proposal step gives us the initial graph structure, we also need to initialize the graph with features. We use a GCN architecture with features on both the nodes and the edges (described in Sect. 3.4). We use two types of features: spatial features, similar to those used in edge proposal, and visual features from two layers of the detection CNN (high- and low-level features). We perform an ROIAlign [7] for a context window, and then a small CNN processes those features, eventually pooling to a single vector. This is almost identical to the features extraction in Davis et al. [5], only differing in resolution, padding, and the CNN hyper-parameters.

For nodes, the context window is the bounding box surrounding the text line(s) composing the entity, padded by 20 pixels (image space) on all sides. The ROIAlign pools the features to 10×10 resolution and the two feature layers are appended. Two mask layers are appended to these features: one of all detected text boxes, and the other of just the text boxes belonging to this entity (these are from the same window as the ROIAlign). These are passed to a small CNN which ends with global pooling (exact network in Table 1). The bounding box surrounding all the text lines of the entity is used to compute additional spatial features: detection confidence, normalized height, normalized width, and class prediction; these are appended to the global pooled features from the CNN.

For edges, the context window encompasses all the text lines composing the two nodes, padded by 20 pixels. The ROIAlign pools the features into 16×16 resolution (larger than the resolution than for nodes, as more detail exists in these windows). Appended to the CNN features are three mask layers: all detected text boxes, and one for each of the nodes containing all of the text boxes for the entity. These are passed to a small CNN which ends with global pooling (exact network in Table 1). The two bounding boxes surrounding all the text lines of each entity are used to compute the spatial features: normalized height of both entities, normalized width of both boxes, the class predictions of each entity, and distance between the corner points of the two entities (top-left to top-left, etc.). These are appended to the features from the CNN.

These node and edge features are passed through a single linear node or edge transition layer to form the initial features of the graph.

Table 1. Architecture of feature extraction CNNs for nodes and edges.

Node (starts at 10 × 10, 320 channels)	Edge (starts at 16 × 16, 320 channels)
Depth-wise seperable 3 × 3 conv, 64	Depth-wise seperable 3 × 3 conv, 128
Depth-wise seperable 3 × 3 conv, 64	Depth-wise seperable 3 × 3 conv, 128
Max pool 2 × 2	Max pool 2 × 2
Depth-wise seperable 3 × 3 conv, 128	Depth-wise seperable 3 × 3 conv, 256
3 × 3 conv, 256	Depth-wise seperable 3 × 3 conv, 256
Global average pooling	Max pool, 2 × 2
	3 × 3 conv, 256
	Global average pooling

We ROIAlign high- and low-level features from the CNN (second-to-last conv layer and first pool layer) as high-level features generally contain the more interesting and descriptive information but may leave out certain low-level features that were irrelevant for the detection task.

3.4   Iterated Graph Editing with GCN

We use a series of three GCNs, each performing the same iterated predictions. We apply the first GCN to the initial graph and then use its predictions to update the graph structure and features. The next GCN is then applied to the updated graph, which is updated again, and so forth. The final graph contains the (predicted) text entities and relationships. This process is seen both at the bottom of Fig. 1 and as actual predictions in Fig. 2.

Each GCN in composed of several *GN blocks*, as outlined in Sect. 3.2 of Battaglia et al. [3], without the global attributes and using attention to aggregate edge features for the node update. The GN block first updates the edge features from their two node features and then updates node features from their edge features. The GN block is directed, so we duplicate edge features to create edges in both directions. At the end of a GCN, the predictions and features for the two directions are averaged for each edge.

Our **GN block edge update** concatenates the current edge features with its two nodes' features. These are passed to a two-layer fully connected ReLU network. The output is summed with the previous features (residual) to produce the new edge features.

Our **GN block edge aggregation** (for a given node) applies multi-head attention [17] using the node's features as the query and its edges' features as the keys and values. We use 4 heads.

Our **GN block node update** first appends the aggregation of the edge features with the node's current features. This is passed through a two-layer fully connected ReLU network. The output is summed with the previous features (residual) to produce the new node features.

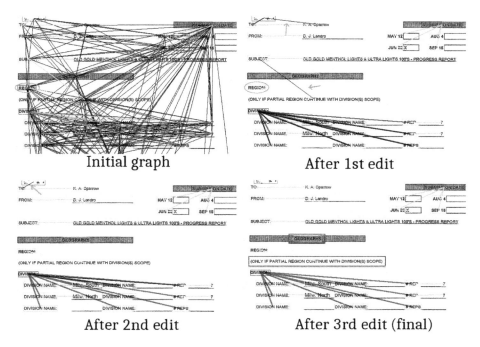

Initial graph

After 1st edit

After 2nd edit

After 3rd edit (final)

Fig. 2. Example of iterative graph edits on a FUNSD image. Blue, cyan, yellow, and magenta boxes indicate predicted header, question, answer, and other entities. Red boxes are missed entities. Green, yellow, and red lines indicate true-positive, false-positive, and false-negative relationship predictions. Orange marks draw attention to edits. (Color figure online)

All the GCNs use an input and hidden size of 256 (all linear layers are 256 to 256). The first two GCNs have 7 layers, the last has 4; this is based on the observation that most decisions are made earlier on, and so more power is given to the earlier GCNs.

Each node of the GCN predicts the semantic class for that text entity (or incomplete entity). Each edge predicts four things: (1) whether the edge should be *pruned*, (2) whether the entities should be *grouped* into a single entity, (3) if these are oversegmented text lines and should be *merged* (corrected), and (4) if this is a true *relationship*. We threshold each of these predictions and update the graph accordingly. The final graph update uses the relationship prediction to prune edges (so remaining edges are relationships).

If two or more nodes are to be grouped or merged, their features are averaged and their edges are aggregated. Any resulting duplicate edges have their features averaged. If two nodes are to be merged to fix an oversegmentation, the bounding box for the text line is replaced by the one encompassing both, and their class predictions are averaged.

The features introduced with the initial graph are based on the original bounding boxes, which are potentially modified during a group or merge edit. We reintroduce the initial features again at the start of each GCN. We reuse the same initial features (before the transition layer) for unmodified nodes and edges, and compute new ones for the modified nodes and their edges. These are appended to the final features of the previous GCN and passed through a single linear node or edge transition layer to become the features of the new graph (each GCN has its own transition layers).

As an ablation, we also show results of FUDGE without a GCN, where the GCN is replaced by two fully-connected networks which predict from the initial edge and node features respectively. The graph is still updated in the same way, it is just the GCN being replaced by individual fully-connected networks. To compensate for the lack of complexity, the non-GCN network's fully-connected networks have double the number of hidden channels as the GCN.

The thresholds used to determine if an edit will be made are different on each iteration and were heuristically chosen. The specific thresholds are: 1st edit {merge: 0.8, group: 0.95, prune: 0.9}, 2nd edit {merge: 0.9, group: 0.9, prune: 0.8}, 3rd edit {merge: 0.9, group: 0.6, prune: 0.5}. Merge thresholds are initially lower, since we want merges to occur first. Grouping is a higher level decision and so its threshold is higher initially. The prune decision is kept relatively high until the final edit as it's generally desirable to keep edges around.

We use GroupNorm [19] and Dropout in all fully connected and convolutional networks.

4   Training

We train the detector first and then train the other components while continuing to fine-tune the detector. The detection losses are based on YOLO [14] and are identical to [5]. The edge proposals, the GCN edge predictions and the GCN node (class) predictions are all supervised by binary cross-entropy losses.

When computing the GCN losses, we align the predicted graph to the ground truth (GT) by assigning each predicted text line to a GT text line. From these the proper edge GT can be determined. We assign predicted and GT text lines by thresholding (at 0.4) a modified IOU which optimally clips the GT bounding boxes horizontally to align them with the predictions; this allows the correct assignment of oversegmented predictions. If multiple text line predictions are assigned to the same GT text line bounding box, the edges between their nodes them are given a merge GT. Any edges between nodes with predicted text lines assigned to GT bounding boxes that are part of the same GT text entity are given a grouping GT. Any edges between two nodes with text line predictions that are assigned to GT text lines which are part of two GT entities with a GT relationship between them are given a relationship GT. Any edges which don't have either a merge, group, or relationship GT are given GT to be pruned. For the edge proposal GT, the prune GT is computed for all possible edges. Nodes are given the GT class of their assigned GT text entities.

Because of memory restrictions we cannot train on an entire form image. Instead, we sample a window of size 800×800 for the FUNSD dataset and 600×1400 for the NAF dataset. We use a batch size of 1. We also randomly rescale images as a form of data augmentation. The scale is uniformly sampled from a range of 80%–120% and preserves aspect ratios.

We train using the AdamW optimizer [11] with a weight decay of 0.01. The detection-only pre-training uses a learning rate of 0.01, and increases the learning rate from zero for the first 1000 iterations, training a total of 250,000 iterations. The full training uses a learning rate of 0.001, increasing the learning rate from zero for the first 1000 iterations. The detector's pre-trained weights are frozen for the first 2000 iterations. In half of the iterations (randomly assigned) during training, we create the initial graph using GT text line bounding boxes. At 590,000 iterations, we drop the learning range by a factor of 0.1 over 10,000 iterations. We then apply stochastic weight averaging (SWA) [9] for an additional 100,000 iterations, averaging at every iteration.

5 Datasets

Form understanding is a growing area of research, but there are only limited results on public datasets available with which to compare. There are two large public datasets of form images annotated for our relationship detection task: the FUNSD dataset [10] and the NAF dataset [5]. We did all development on the training and validation sets only.

The **FUNSD dataset** [10] contains 199 low resolution scans of modern forms. The FUNSD dataset has 50 images as a test set; we divide the training images into a 120 image training set and a 19 image validation set. The forms mostly contain printed text, though some handwriting is present. The images of the FUNSD dataset are relatively clean, though low resolution. The FUNSD dataset is labeled with word bounding boxes with the corresponding transcription, grouping of words into semantic or text entities (one or more lines of text) with a label (header, question, answer, or other), and relations between the text entities. We preprocess the data to group the words into text lines.

The **NAF dataset** [5] contains images of historical forms, 77 test set images, 75 validation set images, and 708 training set images. The images are high resolution, but the documents have a good deal of noise. Most of the forms are filled in by hand and some forms are entirely handwritten. The dataset is labeled with text line bounding boxes with two labels (preprinted text, input text) and the relationships between the text lines. Unlike the FUNSD dataset, there isn't a notion of text entities, rather all the lines which would compose a text entity merely have relationships connecting them. Additionally, the text transcription is unavailable for the NAF dataset, meaning it cannot be used by methods that rely on language. We resize these images to 0.52 their original size.

In Davis et al. [5] we only evaluated on a subset of the NAF dataset, the forms which do not contain tables or fill-in-the-blank prose (e.g. "I ____, on the _ day of ____, do hereby...". In this work we use the full dataset, although we ignore

Table 2. Relationship detection (entity linking) on the NAF dataset

Full NAF, all relationships			
Method	Recall	Precision	F1
Davis et al. [5]	54.62	45.53	49.60
FUDGE no GCN	61.23	51.96	56.21
FUDGE	59.92	54.92	**57.31**
Simple subset of NAF, key-value relationships only (averaged per document as in [5])			
Method	Recall	Precision	F1
Davis et al. [5]	59.9	65.4	60.7
FUDGE	63.6	73.2	**66.0**

tables; tables are not annotated and the models learn to ignore them. In [5] we only detected relationships between preprinted and input text (key-value), not relationships between text lines of the same semantic class (which would indicate being the same text entities). Here we evaluate using all relationships.

6 Evaluation and Results

Qualitative results can be seen in Fig. 3. Quantitatively, we evaluate performance using the relationship and entity detection (micro) recall, precision, and F1 score (F-measure). For a correct text entity detection, the prediction must have a bounding box overlapping (with at least 0.5 IOU) each of the text lines making up a ground truth text entity, contain no additional text lines, and have the correct semantic label (class). We adopt the method of scoring of relationships from [8]; for a predicted relationship to be correct, the two predicted entities it is between must contain at least the first text line (word in [8]) and correct class of two ground truth entities with a relationship. For the NAF dataset, each text line is its own text entity.

In Davis et al. [5] we introduced the NAF dataset, but only evaluated on a simplified subset of the data and only on key-value relationships. In this work we use all the images and relationships of the NAF dataset. We retrain [5] on this larger and harder dataset. We train [5] for 600,000 iterations and use SWA for 100,000 iterations just as we trained FUDGE. We previously [5] used far fewer iterations as it as on a much smaller dataset. The inclusion of SWA to [5] makes the comparison more fair as SWA provides a significant boost to performance. We also train FUDGE on the simplified NAF dataset. The relationship detection accuracy for both the full and simple subset of the NAF dataset are reported in Table 2. We also report the text detection accuracy in Table 3 for the full NAF. As can be seen in Table 2, FUDGE significantly outperforms [5] in relationship detection for both the simplified and full NAF dataset. FUDGE performs similarly to [5] on text line detection, which is reasonable as they share the same text detection backbone.

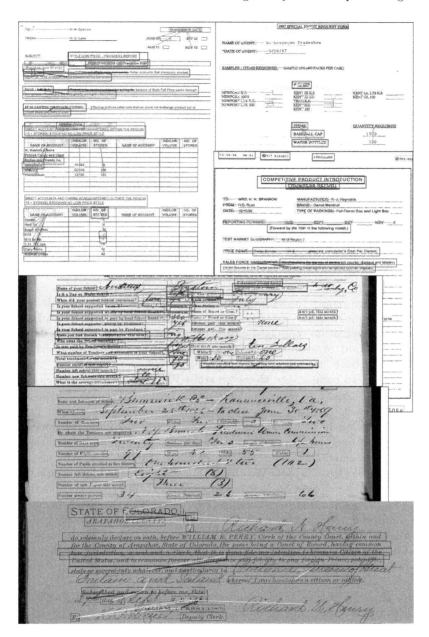

Fig. 3. Example results of FUDGE's final decisions on crops of three FUNSD images and three NAF images. Blue, cyan, yellow, and magenta boxes indicate predicted header, question, answer, and other entities for the FUNSD images. Red boxes indicate a missed entity (correct class is colored on bottom of red box). Green, yellow, and red lines indicate true-positive, false-positive, and false-negative relationship predictions. (Color figure online)

Table 3. Text line detection on the full NAF dataset

Method	Recall	Precision	F1
Davis et al. [5]	83.23	66.20	73.75
FUDGE no GCN	81.34	69.45	74.93
FUDGE	80.16	68.22	73.71

Table 4. Relationship detection (entity linking) on FUNSD dataset

Method	GT OCR info	# Params	Recall	Precision	F1
Carbonell et al. [4]	Word boxes + Transcription	201M	–	–	39
LayoutLM (reimpl.) [8]	Word boxes + Transcription	–	41.29	44.45	42.81
BROS [8]	Word boxes + Transcription	138M	64.30	69.86	**66.96**
Word-FUDGE	Word boxes	17M	62.88	70.79	**66.60**
FUDGE no GCN	None	12M	50.81	54.18	52.41
FUDGE	None	17M	54.04	59.49	56.62

In Table 4 we report the same relationship detection metrics for the FUNSD dataset. We compare against Carbonell et al. [4] and the unpublished BROS [8] method. Davis et al. [5] cannot be directly applied to the FUNSD dataset as it lacks a method of grouping the text lines into text entities. Carbonell et al. and BROS both use the dataset provided OCR word bounding boxes and transcriptions. To compare to these we train Word-FUDGE, which is FUDGE trained always using the provided OCR word boxes (as opposed to using text lines or predicted boxes). It sees the word boxes as oversegmented lines and learns to merge them into text lines. Word-FUDGE matches the performance of BROS, a method that is pre-trained on over ten million additional document images before being fine-tuned on the FUNSD training set, whereas we use *only* the FUNSD training set. We think this shows that while the relationship detection problem can be solved with a language-centered approach like BROS, it requires far more data than a visual approach to reach the same performance. We expect a superior approach would combine strong visual features with pre-trained language information. The non-GCN version of FUDGE performs almost as well as the GCN version (see also Table 2), indicating that predicting in context is either not very necessary or that FUDGE is unable to learn to use the GCN effectively.

Xu et al. [21] first presented a semantic labeling task for the FUNSD dataset, which is to predict the semantic entities with their labels, given the word bounding boxes and their transcription. We compare our text entity detection against various other methods in Table 5, both normally and using ground truth word bounding boxes. Our model is outperformed in this metric by the language-centered approaches. This is understandable; while understanding the layout and having vision helps for this task, it isn't as essential if the language is understood well enough. FUDGE with and without the GCN perform the same, which is surprising as we would expect the context provided by the GCN would improve class predictions.

Table 5. Text entity detection/Semantic labeling on the FUNSD dataset

Method	GT OCR info	# Params	Recall	Precision	F1
Carbonell et al. [4]	Word boxes + Transcription	201M	–	–	64
LayoutLM$_{BASE}$ [21]	Word boxes + Transcription	113M	75.97	81.55	78.66
LayoutLM$_{LARGE}$ [21]	Word boxes + Transcription	343M	75.96	82.19	78.95
BROS [8]	Word boxes + Transcription	138M	80.56	81.88	81.21
LayoutLMv2$_{BASE}$ [20]	Word boxes + Transcription	200M	80.29	85.39	82.76
LayoutLMv2$_{LARGE}$ [20]	Word boxes + Transcription	426M	83.24	85.19	**84.20**
Word-FUDGE	Word boxes	17M	70.75	76.74	73.63
FUDGE no GCN	None	12M	63.64	66.57	65.07
FUDGE	None	17M	64.90	68.23	66.52

Table 6. Hit@1 on FUNSD dataset with GT text entities

Method	Hit@1
DocStruct without visual features [18]	55.94
DocStruct [18]	58.19
FUDGE no GCN	66.48
FUDGE	**68.28**

DocStruct [18] evaluated the relationship detection problem as a retrieval problem, where a query child must retrieve its parent (answers retrieve questions, questions retrieve headers). This view of the problem doesn't account for predicting which nodes have parents in the first place. We compare results on the Hit@1 metric, a measure of how often, for each child query, the parent is correctly returned as the most confident result. DocStruct [18] uses the ground truth OCR text boxes and transcriptions, and also uses the ground truth grouping of text entities ("text fragment" in [18]). We also use this information for comparison by forcing FUDGE to make the correct text grouping in its first graph edit step, and preventing any further grouping. Our results are compared in Table 6. FUDGE significantly outperforms DocStruct [18] with and without the GCN. However, we don't feel this metric demonstrates general performance as it relies on ground truth text entity grouping and does not measure the ability to detect if a relationship does not exist for a query.

While the previous results have validated the use of the GCN for relationship detection, we also perform an ablation experiment exploring other aspects of FUDGE: the number of graph edit steps, our edge proposal network compared to the line-of-sight proposal used in [5], and the impact of being able to correct text line detections. The results are presented in Table 7. Using graph editing and our improved edge proposal improve overall performance. In particular, having at least one intermediate edit step improves entity detection. The merging of oversegmented text lines makes only a little improvement, which is to be expected given that the text line detection makes few errors on the FUNSD dataset. Because we did not perform an exhaustive hyper-parameter search for our primary model, some different choices in the number of GCN layers leads to slightly better results than our primary model.

Table 7. Ablation models' relationship and entity detection on FUNSD dataset

GCN layers	Edit iters	Edge proposal	Allow merges	Relationship detection			Entity detection		
				Recall	Precision	F1	Recall	Precision	F1
10	1	Network	Yes	59.98	45.84	51.96	66.58	63.02	64.75
8,8	2	Network	Yes	56.45	59.22	**57.80**	62.58	66.35	64.41
3,3,3	3	Network	Yes	54.85	59.50	57.08	65.44	69.14	**67.24**
3,3,3	3	Line-of-sight	Yes	48.14	63.89	54.90	65.39	68.92	67.11
3,3,3	3	Network	No	55.73	57.97	56.83	66.99	68.10	**67.44**
7,7,4	3	Network	Yes	54.04	59.49	56.62	64.90	68.23	66.52

Looking at the relationship detection errors made by FUDGE in detail, it can be seen that the majority of false negative relationships are actually caused by poor entity detection, which is why the performance increases so dramatically with the use of ground truth bounding boxes.

7   Conclusion

We present FUDGE, a visual approach to form understanding that uses a predicted form graph and edits it to reflect the true structure of the document. The graph is created from detected text lines (vertices) and the initial edges are proposed by a simple network. In three iterations, GCNs predict whether to combine vertices to group text elements into single entities or to prune edges.

FUDGE uses no language information, but it performs similarly to methods using large pre-trained language models. We believe this demonstrates that most form understanding solutions do not put enough emphasis on visual information.

References

1. Aggarwal, M., Gupta, H., Sarkar, M., Krishnamurthy, B.: Form2Seq : A framework for higher-order form structure extraction. In: Conference on Empirical Methods in Natural Language Processing (EMNLP) (2020)
2. Aggarwal, M., Sarkar, M., Gupta, H., Krishnamurthy, B.: Multi-modal association based grouping for form structure extraction. In: IEEE/CVF Winter Conference on Applications of Computer Vision (WACV) (2020)
3. Battaglia, P., et al.: Relational inductive biases, deep learning, and graph networks. arXiv (2018). https://arxiv.org/pdf/1806.01261.pdf
4. Carbonell, M., Riba, P., Villegas, M., Fornés, A., Lladós, J.: Named entity recognition and relation extraction with graph neural networks in semi structured documents. In: 25th International Conference on Pattern Recognition (ICPR) (2020)
5. Davis, B., Morse, B., Cohen, S., Price, B., Tensmeyer, C.: Deep visual template-free form parsing. In: International Conference on Document Analysis and Recognition (ICDAR) (2019)

6. Devlin, J., Chang, M.W., Lee, K., Toutanova, K.: BERT: Pre-training of deep bidirectional transformers for language understanding. In: Conference of the North American Chapter of the Association for Computational Linguistics: Human Language Technologies (NAACL-HLT) (2019)
7. He, K., Gkioxari, G., Dollár, P., Girshick, R.: Mask R-CNN. In: International Conference on Computer Vision (ICCV), pp. 2961–2969 (2017)
8. Hong, T., Kim, D., Ji, M., Hwang, W., Nam, D., Park, S.: BROS: A pre-trained language model for understanding texts in document (2021). https://openreview. net/forum?id=punMXQEsPr0
9. Izmailov, P., Podoprikhin, D., Garipov, T., Vetrov, D., Wilson, A.: Averaging weights leads to wider optima and better generalization. In: 34th Conference on Uncertainty in Artificial Intelligence (UAI) (2018)
10. Jaume, G., Kemal Ekenel, H., Thiran, J.: FUNSD: a dataset for form understanding in noisy scanned documents. In: International Conference on Document Analysis and Recognition Workshops (ICDARW) (2019)
11. Loshchilov, I., Hutter, F.: Decoupled weight decay regularization. In: International Conference on Learning Representations (ICLR) (2019)
12. Palm, R.B., Laws, F., Winther, O.: Attend, copy, parse end-to-end information extraction from documents. In: International Conference on Document Analysis and Recognition (ICDAR) (2019)
13. Qasim, S.R., Mahmood, H., Shafait, F.: Rethinking table recognition using graph neural networks. In: International Conference on Document Analysis and Recognition (ICDAR) (2019)
14. Redmon, J., Farhadi, A.: YOLOv3: An incremental improvement. arXiv preprint arXiv:1804.02767 (2018)
15. Riba, P., Dutta, A., Goldmann, L., Fornés, A., Ramos, O., Lladós, J.: Table detection in invoice documents by graph neural networks. In: International Conference on Document Analysis and Recognition (ICDAR) (2019)
16. Sarkar, M., Aggarwal, M., Jain, A., Gupta, H., Krishnamurthy, B.: Document structure extraction using prior based high resolution hierarchical semantic segmentation. In: European Conference on Computer Vision (ECCV) (2020)
17. Vaswani, A., et al.: Attention is all you need. In: 31st Conference on Neural Information Processing Systems (NIPS) (2017)
18. Wang, Z., Zhan, M., Liu, X., Liang, D.: DocStruct: a multimodal method to extract hierarchy structure in document for general form understanding. In: Findings of the Association for Computational Linguistics: EMNLP (2020)
19. Wu, Y., He, K.: Group normalization. In: European Conference on Computer Vision (ECCV), pp. 3–19 (2018)
20. Xu, Y., et al.: LayoutLMv2: Multi-modal pre-training for visually-rich document understanding. In: 59th Annual Meeting of the Association for Computational Linguistics (ACL) (2021)
21. Xu, Y., Li, M., Cui, L., Huang, S., Wei, F., Zhou, M.: LayoutLM: pre-training of text and layout for document image understanding. In: International Conference on Knowledge Discovery & Data Mining (KDD) (2020)
22. Yang, X., Yumer, E., Asente, P., Kraley, M., Kifer, D., Lee Giles, C.: Learning to extract semantic structure from documents using multimodal fully convolutional neural networks. In: Conference on Computer Vision and Pattern Recognition (CVPR) (2017)

Annotation-Free Character Detection in Historical Vietnamese Stele Images

Anna Scius-Bertrand[1,2]([✉]), Michael Jungo[1], Beat Wolf[1], Andreas Fischer[1,3], and Marc Bui[2]

[1] iCoSys, University of Applied Sciences and Arts Western Switzerland, Sierre, Switzerland
{annascius-bertrand,michaeljungo,beatwolf,andreasfischer}@hefr.ch
[2] Ecole Pratique des Hautes Etudes, PSL, Paris, France
{annascius-bertrand,marcbui}@ephe.sorbonne.fr
[3] DIVA, University of Fribourg, Fribourg, Switzerland

Abstract. Images of Historical Vietnamese stone engravings provide historians with a unique opportunity to study the past of the country. However, due to the large heterogeneity of thousands of images regarding both the text foreground and the stone background, it is difficult to use automatic document analysis methods for supporting manual examination, especially with a view to the labeling effort needed for training machine learning systems. In this paper, we present a method for finding the location of Chu Nom characters in the main text of the steles without the need of any human annotation. Using self-calibration, fully convolutional object detection methods trained on printed characters are successfully adapted to the handwritten image collection. The achieved detection results are promising for subsequent document analysis tasks, such as keyword spotting or transcription.

Keywords: Object detection · Self-calibration · FCOS · YOLO · Historical vietnamese steles · Chu Nom Characters

1 Introduction

Centuries-old stone engravings in man-sized steles in Vietnam carry invaluable information for historians, who are interested to learn more about the life of common people in the villages [14]. Over the past decades, thousands of steles have been copied to paper, then photographed, thus creating a comprehensive collection of digital documents [15,16] that facilitate the examination of the engravings, as currently being done in the ERC project VIETNAMICA[1]. However, due to the sheer size of the document collection, which contains tens of thousands of images, manual inspection is difficult and time-consuming.

Automatic document image analysis would be of tremendous help for the historians. Even if an automatic transcription of the engraved Chu Nom characters

---

[1] https://vietnamica.hypotheses.org.

© Springer Nature Switzerland AG 2021
J. Lladós et al. (Eds.): ICDAR 2021, LNCS 12821, pp. 432–447, 2021.
https://doi.org/10.1007/978-3-030-86549-8_28

is not immediately feasible, knowing more about the general layout structure, number and size of the characters, or spotting particular keywords, can be helpful for browsing and exploring the image collection.

Over the past decade, methods for historical document analysis have seen a strong progress [6], driven in large part by the advent of robust convolutional approaches to image enhancement [21], layout analysis [2], keyword spotting [22], and automatic transcription [3], to name just a few. However, unlike earlier more heuristic methods to solve these problems, one of the main constraints is the need for human-annotated samples to drive the machine learning process.

In the case of historical Vietnamese steles, a large amount of labeled samples would be needed to cover all the different layouts, character shapes, text resolutions, stone backgrounds, fissures, ornaments, etc. that are different from one stele to another. For obtaining an impression of the variety of steles, we refer the reader to Fig. 7 in the experimental section. Annotations are difficult to obtain, especially for rare Nom characters that require expert knowledge.

The ideal properties of a document analysis method in this context would be that it avoids preprocessing, does not rely on an explicit segmentation prior to recognition, and does not require any human annotation. In this paper, we explore the feasibility of using fully convolutional object detection methods to achieve these goals, aiming at transferring knowledge from printed Nom characters directly to the stone engravings.

Such an approach is encouraged by the findings of Yang et al. [27], showing that object detection can be used successfully to detect Chinese characters in historical documents. Furthermore, specifically targeting Nom characters, Nguyen et al. [13] report strong results for character extraction and recognition in historical manuscripts using convolutional neural networks, i.e. a U-Net architecture [19] to perform semantic segmentation, followed by a watershed segmentation of character regions, and finally a coarse-to-fine recognition with different convolutional models. An interesting aspect of their work, which relates to our goals, is that they were able to train the U-Net in large parts with printed characters, which were then fine-tuned on real manuscript pages. In previous work, a U-Net architecture has also been applied by Scius-Bertrand et al. [20] in the context of Vietnamese steles, together with a seam carving algorithm, to first detect pixels belonging to the main text and then segmenting them reliably into columns. However, further denoising, detection, and recognition algorithms would be needed to locate individual characters.

In the present paper, we investigate object detection methods trained on printed Nom characters to find the location of main text characters on the steles directly. Without using human annotations, we control the behavior of the convolutional neural networks by means of unsupervised layout analysis and a self-calibration step, for allowing the network to take the different stele backgrounds into account[2]. The proposed approach is studied for two object detection methods, namely YOLO [17] and FCOS [24], on a dataset of 2,036 stele images

---

[2] Readers interested in the source code and the dataset are referred to our GitHub repository https://github.com/asciusb/annotationfree.

434   A. Scius-Bertrand et al.

containing hundreds of thousands of characters. 65 stele images have been manually annotated for the purpose of performance evaluation.

In the following, we describe the dataset, provide more details about object detection, introduce the proposed layout analysis and self-calibration methods, present the experimental results, and draw some conclusions.

2   Dataset of Historical Vietnamese Steles

The collection of stele inscriptions includes about 40,000 unique copies collected from 1910 to 1954 by the French School of the Far East (EFEO) and, since 1995, by the Han-Nom Institute [14]. The majority of the steles are from the 17th to 19th century. The steles were mainly located in the north of Vietnam, where production and trade made the farmers rich enough to have steles, such as in the rice plains and along river banks. In terms of content, 90% of the steles are at the village or quarter level. They are often internal village or family documents. The steles are a mine of information on the social, cultural, economic, religious and linguistic life of the villages.

Reading this corpus involves many challenges. As the steles are often outside, they have suffered damage due to the weather or the deterioration of the stone (impact, cracks, etc.). The layout of the steles is also very diverse. The borders are more or less wide and more or less ornate. The title can be engraved or can appear in relief. The text columns are irregular and can be divided into two columns. All these elements make automation more complex.

To test our system, we had access to 2,036 photographs of stele stamps used in the VIETNAMICA project. A stamping is a faithful reproduction of a stele thanks to a roll of ink passed over the whole of a sheet glued together with banana juice. The engraved characters are white and those in relief are black. The only preprocessing in this paper consists of inverting the color of the steles, such that the characters of the main text appear in black. For the training data we have generated synthetic documents inspired by the steles, namely text in Chu Nom organised in columns. For printing Nom characters, we rely on a font, which was kindly provided by the Vietnamese Nom Preservation Foundation[3]. The annotated test set of 65 stele images contains on average 375 characters per stele, with a minimum of 20 characters and a maximum of 883.

3   Convolutional Character Detection

3.1   You Only Look Once (YOLO)

YOLO [17] (You Only Look Once) was the pioneer of one-stage object detection by showing competitive results with a single convolutional neural network. Its simplicity and efficiency made it one of the most popular object detection models and inspired further improvements of YOLO itself, by addressing its shortcomings, as well as other one-stage object detectors.

---

[3] http://www.nomfoundation.org.

Backbone   Feature Pyramid / Neck   Head
(a)   (b)   (c)

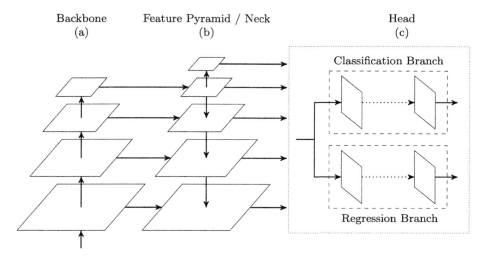

Fig. 1. One-stage object detection models are comprised of three major parts: (a) the backbone extracts features from the input image, (b) a feature pyramid combines the extracted feature maps to enrich the features across different scales, which are then processed individually by the head (c), where two separate branches are used for the classification and the bounding box regression respectively. The specific implementation for all three parts can be freely chosen and some models may include additional modifications, but the overall architecture remains.

The model has been improved iteratively with rather small changes and in the latest version, YOLOv4 [1], the focus put on optimizing the current architecture by comparing various existing building blocks in order to find the best possible configuration. As a result, YOLOv4 consists of the following: A variation of the Cross Stage Partial Network (CSPNet) [25] is used as the backbone, a modified Path Aggregation Network (PAN) [12] as the neck and the existing anchor based head of YOLOv3 [5]. Figure 1 illustrates these system components.

We have decided to use YOLOv5 [8]. Contrary to its name, it is not the successor of YOLOv4, but rather a port of YOLOv3 to PyTorch, which has been improved independently from YOLOv4, but then also adopted most improvements introduced by YOLOv4.

3.2 Fully Convolutional One-Stage Object Detection (FCOS)

FCOS [24] is a fully convolutional one-stage and anchor-free object detector. The primary difference of the underlying method compared to YOLO is the omission of anchors, which serve as initial bounding boxes that are further refined in order to get tightly fitting bounding boxes around the objects (Fig. 2a). In contrast to this, FCOS uses a single point to span up a bounding box by predicting the distances towards each of the four sides (Fig. 2b).

(a) YOLO   (b) FCOS

Fig. 2. Difference in bounding box prediction: (a) YOLO refines a given anchor box (b) FCOS predicts the distances to all four sides from a point

The backbone is most commonly any variation of ResNet [7], but any desired CNN can be used in its place. We have decided to use EfficientNet [23] because it has shown to provide strong results with great performance. One key change to the EfficientNet model was replacing Batch Normalization with Group Normalization [26] because of having to use smaller batch sizes, due to needing larger input images for this task, where Batch Normalization is not as effective.

A particularly important aspect of the architecture for the bounding box localization in FCOS is the Feature Pyramid Network (FPN) [9], which generates multiple feature maps at different scales, five to be exact, where later feature maps flow back into earlier stages. This feature pyramid results in semantically stronger feature maps across all levels, making the detection more robust to various scales.

Another aspect that made one-stage and anchor-free object detectors more viable, was the selection of loss functions. The Focal Loss [10] is used as the classification loss to address the imbalance between foreground and background. As one-stage detectors create dense predictions, i.e. they predict every single point in a feature map, there are disproportionally more points that are considered background than foreground. Having such a massive imbalance, the model may easily be swayed towards predicting background when even the smallest uncertainty exists, as it is the much more likely outcome.

$$\mathcal{L}_{cls} = FL(p, y) = \begin{cases} -\alpha(1-p)^{\gamma}\log(p) & \text{if } y = 1 \text{ (foreground)} \\ -(1-\alpha)p^{\gamma}\log(1-p) & \text{otherwise (background)} \end{cases} \tag{1}$$

The Focal Loss (Eq. 1) is an extension of the cross entropy loss by adding the factors $(1-p)^{\gamma}$ and p^{γ} respectively, which are designed to reduce the loss for easily classified examples, putting more emphasis on misclassified ones, when $\gamma > 0$. Setting $\gamma = 0$ makes the Focal Loss equivalent to the α-balanced cross entropy loss. $\alpha \in [0, 1]$ is a weighting factor to also address the class imbalance, but only to balance the importance of positive/negative examples, whereas the Focal Loss also addresses the importance between easy/hard examples.

For the bounding box regression, the Generalized Intersection over Union Loss (GIoU) [18] is used. The regular Intersection over Union (IoU) is not well suited as a loss, since non-overlapping boxes always have an IoU of zero, regardless of how far apart they are, making the optimization infeasible. This shortcoming is addressed by the GIoU with the introduction of an additional term:

$$\mathcal{L}_{reg} = 1 - GIoU = 1 - \left(\frac{|A \cap B|}{|A \cup B|} - \frac{|C \setminus (A \cup B)|}{|C|} \right) \tag{2}$$

where A and B are two bounding boxes and C is the smallest enclosing bounding box, i.e. the smallest bounding box to contain both A and B. The goal is to have an enclosing bounding box C with the smallest possible area. In other words, the empty area between the A and B should be minimized.

Similar to the classification imbalance, multiple points may be used as the starting point for the bounding box regression, where some of them can be harmful for the model. In order to alleviate that problem, the Adaptive Training Sample Selection (ATSS) [28] has been employed, where only the $k = 9$ best candidates are selected per FPN level for each ground truth bounding box, based on the centerness, because starting points closer to the center of the bounding box are favorable and produce the best results.

4   Self-calibration

In this section, we introduce the proposed calibration method, which allows to adapt a character detection system from printed characters to the handwritten image collection. An overview of the method is provided first, followed by a description of the individual components.

4.1   Method Overview

Figure 3 provides an overview of the self-calibration. First, a comprehensive Chu Nom font (see Sect. 2) and a generic background are used to create an initial training dataset. Afterwards, a convolutional character detection system (see Sect. 3) is trained and applied to the entire dataset of Vietnamese steles. The character detection results, which are far from being perfect before calibration, are then processed by a layout analysis method that aims to form consistent main text columns. If a main text area can be identified, it is filled with a homogeneous non-text area to create a synthetic background image. The obtained background images are then provided as input to the page synthesis for creating more realistic pages, together with an estimated number of columns and characters.

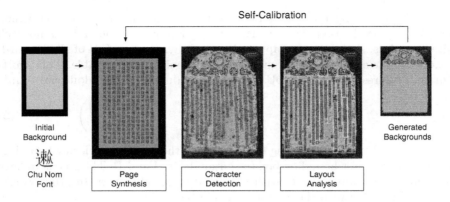

Fig. 3. Method overview

4.2 Page Synthesis

In order to generate synthetic training pages, random Chu Nom characters are printed in columns on a page background. The characters are selected either uniformly from the entire set of characters in the font, or according to a specific Chu Nom text, respecting the character frequencies observed in the text.

For the initial page synthesis, a simple background with gray text background and black border is used. The number of columns and characters per column are varied randomly to obtain different character sizes. When placing characters in a column, the width of the characters is fixed and the padding between the characters is varied randomly. For the second page synthesis, the layout analysis method generates hundreds of stele backgrounds. Characters are printed on the main text area according to the estimated number of columns and characters per column.

After placing the characters on the background, several random image distortions are applied to support the training of the character detection system. Distortions include blur, changes in brightness, translation, as well as salt and pepper noise.

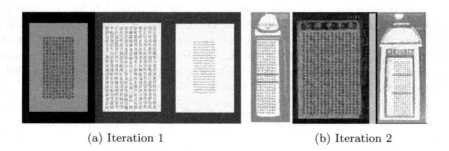

(a) Iteration 1   (b) Iteration 2

Fig. 4. Page synthesis

Figure 4 illustrates generated training samples during the first iteration with the initial background and during the second iteration with different stele backgrounds. Note that even though the main text area is not always well detected, the generated stele backgrounds allow the character detection system to include relevant background patterns.

4.3   Character Detection

The generated training samples contain bounding box coordinates of each character as well as class labels, such that character detection can be conjointly trained for localization and classification (see Sect. 3).

However, due to computational constraints, we have clustered the tens of thousands of Chu Nom characters into 500 categories, in order to allow the neural networks to train with reasonable memory requirements.

Fig. 5. Character clusters

For clustering, we create binary images of the printed characters with a width of 50 pixels and extract 100 grid features $(x_1, \ldots, x_{100})$ using a regular 10×10 grid, where x_i is the number of black pixels in the respective cell. Afterwards, k-means clustering is applied for forming k character classes. For the case $k = 1$, all characters belong to the same class and the character detection system is optimized for localization only. Figure 5 illustrates seven exemplary clusters for $k = 500$.

4.4   Layout Analysis

The goal of layout analysis is to identify the main text region of a stele, the number of columns, and the number of characters per column based on the character detection results. If these estimates can be done with a high confidence, a homogeneous text background pattern is identified and used to fill the main text region. The generated stele background images are then used for page synthesis.

Avoiding the need for manual annotation, layout analysis relies on unsupervised clustering. It aims to form consistent columns based on two basic assumptions:

1. Within the same stele the main text characters have a similar size.
2. Within the same column characters are relatively close together.

Algorithm 1. Column Detection

Require:
    $\mathcal{D}$: set of detected characters
    $(\epsilon = 2, k = 3)$: DBSCAN parameters
    $\tau_s = 0.3$: maximum size deviation
    $\tau_x = 0.1$: maximum horizontal deviation
Ensure:
    $\mathcal{C}$: set of main text columns
1: $m \leftarrow$ character with the median area in $\mathcal{D}$
2: $\mathcal{D} \leftarrow \mathcal{D} - \{c \in \mathcal{D} : \frac{|c.width - m.width|}{m.width} > \tau_s \vee \frac{|c.height - m.height|}{m.height} > \tau_s\}$
3: $\mathcal{C} \leftarrow DBSCAN(\mathcal{D}, \epsilon, k, d_{m,\tau_x}(.,.))$

We use DBSCAN [4] to perform the column clustering. This algorithms groups together points that have at least k neighbors within distance ϵ. Algorithm 1 details the proposed column detection method. First, the median character m is determined among all detected bounding boxes with respect to its area (line 1). Afterwards, characters whose width or height deviate more than τ_s from the median character are discarded, following the assumption that main text characters have a similar size (line 2). Finally, column clusters are formed using DBSCAN with parameters ϵ and k and a special column metric d_{m,τ_x} (line 3). The column metric is defined as

$$d_{m,\tau_x}(c_1, c_2) = \begin{cases} \infty, & \text{if } \frac{|c_1.x - c_2.x|}{m.width} > \tau_x \\ \frac{|c_1.y - c_2.y|}{m.height}, & \text{otherwise} \end{cases}$$

where $c.x$ and $c.y$ are the bounding box center coordinates of character c. That is, characters can only be assigned to the same column if their horizontal position does not deviate more than τ_x.

Note that because the self-calibration does not have access to annotated training samples, reasonable defaults have to be chosen for the parameters. With the suggested defaults in Algorithm 1, main text characters must not deviate more than $\tau_s = 30\%$ in width or height from the median character. They are grouped into columns if their horizontal deviation is less than $\tau_x = 10\%$ and there are at least $k = 3$ other characters above or below within the range of $\epsilon = 2$ character heights, following the assumption that characters in the same column are relatively close together.

By using the median character as a reference, the proposed column detection method is scale-invariant. This is important for the dataset of Vietnamese steles, where the character size varies significantly from one stele to another.

The final step of layout analysis is to identify a homogeneous non-text patch and use it to fill the main text region, i.e. the bounding box of all detected columns, such that the engraved characters are erased and an empty background is obtained. For finding such a patch, we systematically scan areas above, below, and between columns, convert them to grayscale, and choose the area with minimum standard deviation of its grayscale values.

The empty background image is then provided to the page synthesis method for the next self-calibration step, together with the bounding box of the main text, the number of detected columns, and the maximum number of characters per column. In order to increase the quality of the results, background images are used only if there are at least 100 characters in the main text.

Figure 6 illustrates two layout analysis examples. Red boxes indicate characters that are too small or too big, blue boxes are outliers with respect to the column clustering, green boxes are part of a column, yellow boxes are drawn around columns and the main text area, and the cyan box indicates the homogeneous non-text patch. Note that although the character detection quality is rather low during the first iteration, the layout analysis still detects useful main text areas. Examples of resulting synthetic pages are illustrated in Fig. 4.

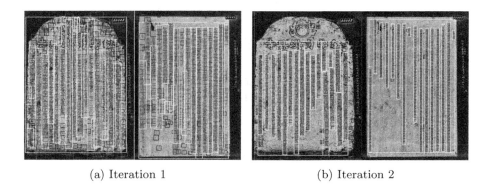

(a) Iteration 1   (b) Iteration 2

Fig. 6. Layout analysis (Color figure online)

5   Experimental Evaluation

To test the possibilities and limitations of annotation-free character detection, we have conducted a series of experiments on the dataset of 2,036 stele images with both YOLO and FCOS. In the following, we describe the experimental setup and evaluation measures in more detail before presenting the results.

5.1   Setup

The neural networks are trained on synthetic data and then applied to the 2,036 real stele images without manual interaction. The proposed self-calibration is performed without using human annotations, neither for fine-tuning the neural networks nor for optimizing meta-parameters. To test the performance of the method, a subset of 65 stele images has been manually annotated with character bounding boxes. This test set consists of two parts.

- **TEST-A.** In a first step, 10 stele images have been annotated that show some diversity in layout but contain main text characters that are relatively well-readable for the human eye.
- **TEST-B.** In a second step, to increase the difficulty, we have randomly selected 55 other samples[4]. Some of them are barely readable.

We consider several experimental setups that aim to investigate key aspects of the annotation-free character detection method, including the number of character classes C for object detection, the size of the dictionary D for page synthesis, the size of the synthetic training set, and the number of training epochs. Starting from a baseline setup, only one parameter is changed at a time.

- **Baseline-10K.** Object detection is only trained for bounding box regression, not for classification ($C = 1$), all characters of the font are used for page synthesis ($D = 26,969$), and the network is trained one epoch on 10,000 synthetic pages.
- **C-500.** Use k-means to cluster the characters into 500 classes ($C = 500$).
- **D-4855.** Consider a reduced set of frequent characters that appear in the famous Tale of Kieu ($D = 4,855$).
- **Baseline-30K.** Use a training set of 30,000 synthetic training pages.
- **Fully-Trained.** Train the object detection network until convergence, typically 10–15 epochs.

For each setup, we evaluate the character detection performance before layout analysis (**Raw detection**), after layout analysis (**Layout analysis**), and after self-calibration (**Calibrated**). The general configuration of the two object detection methods is described in Sect. 3. For YOLO, we consider a YOLOv5m model that has been pretrained on the COCO dataset [11]. For FCOS, we consider an EfficientNet-B4 backbone, which is trained from scratch using Adam optimization. Training one epoch on 30,000 synthetic pages with a batch size of 8 and two Titan RTX cards takes about 80 min.

5.2   Evaluation Measures

For evaluating the detection results, we first compute an optimal assignment between the detected character boxes and the ground truth annotations, using the intersection over union (IoU) $\frac{|A \cap B|}{|A \cup B|}$ as the matching cost and solving a linear sum assignment problem.

Several metrics are then computed from the optimal assignment, including the mean IoU, precision, recall, F_1 score, and the *character detection accuracy* (CDA), which we define as

$$CDA = \frac{N - S - D - I}{N} \tag{3}$$

---

[4] More specifically, three random selections have been performed. 11 samples have been selected among already transcribed steles, 22 samples from the dataset used in previous work [20], and 22 samples from the rest of the dataset.

where N is the number of characters in the ground truth, S is the number of substitution errors, D is the number of deletions (characters that are not retrieved), and I is the number of insertions (false positives of the detection system). Matching characters are considered as correct if their IoU is greater than 0.5, otherwise they are counted as substitution errors. Note that this definition is similar to the character recognition accuracy for optical character recognition. Because the number of insertion errors is not limited, it can also be negative.

5.3 Results

Table 1 reports character detection results for the initial experiment with 10'000 synthetic pages, training the detection systems for one epoch. Besides the raw detection output, we also evaluate the characters that are considered as main text by the layout analysis. In this weakly trained scenario, FCOS performs better than YOLO. While the increase in characters (C-500) or the focus on more frequent characters (D-4855) does not have a strong effect, the self-calibration clearly helps to adapt to the steles. The 87.1% character detection accuracy correspond to a remarkable increase of +35.8% when compared with the baseline. Layout analysis improves the results as long as the networks are not calibrated but slightly decreases the performance for calibrated models.

Table 2 reports the outcome of detection systems that are trained more extensively on 30,000 synthetic pages. In this setup, YOLO achieves quite a remarkable performance of 85.3% after layout analysis even without calibration, i.e. only trained with printed characters on the simple gray background with black border. After calibration, both YOLO and FCOS achieve a character detection accuracy of 87.3%, which is similar to the best FCOS result in Table 1. The comparison with a fully trained network indicates that FCOS overfits with more training epochs. YOLO, on the other hand, does not overfit, which we assume is due to the positive effect of pretraining on COCO.

Table 3 shows the results for the more difficult test set TEST-B. Even though FCOS has a very low detection accuracy at the beginning (due to a large number of insertion errors; see Eq. 3), the layout analysis can make sense of the results and produces useful stele backgrounds for self-calibration, which increases the final performance to 60.7%. For this more difficult test set, YOLO proves more robust and achieves a significantly better result of 75.3%.

In all cases, the raw output of the calibrated system, trained until convergence on the synthetic pages and adapted one epoch on the synthetic stele backgrounds emerges as the most stable configuration. Table 4 summarizes the different evaluation measures for TEST-A in this configuration, showing that the character detection accuracy (CDA) is closely related to the recall of the system.

Figure 7 illustrates exemplary detection results. In general, the system is able to detect characters of different sizes, even for styles that differ strongly from the printed characters used for training (see for example the fourth stele from the left in Fig. 7). Errors are often observed at the border of the main text region, or due to variations in the layout, e.g. when a single column is continued with two smaller columns below.

Table 1. Character detection accuracy on the initial test dataset (TEST-A) using 10,000 synthetic training pages. The baseline is compared with an extended number of classes (C-500), a reduced number of Nom characters (D-4855), and a self-calibrated system. The best result is highlighted in bold font.

		Baseline-10K	C-500	D-4855	Calibrated
FCOS	Raw detection	51.3	34.2	56.7	**87.1**
	Layout analysis	55.2	53.6	59.5	85.0
YOLO	Raw detection	29.4	25.3	25.2	**35.8**
	Layout analysis	34.8	29.6	34.0	33.9

Table 2. Character detection accuracy on the initial test dataset (TEST-A) using 30,000 synthetic training pages. The baseline is compared with a fully trained and calibrated system.

		Baseline-30K	Fully-Trained	Calibrated
FCOS	Raw detection	52.7	48.8	**87.3**
	Layout analysis	59.6	52.9	84.7
YOLO	Raw detection	64.4	72.4	**87.3**
	Layout analysis	66.2	85.3	86.8

Table 3. Character detection accuracy on the extended test dataset (TEST-B) using 30,000 synthetic training pages. The baseline is compared with a fully trained and calibrated system.

		Baseline-30K	Fully-Trained	Calibrated
FCOS	Raw detection	1.1	3.4	**60.7**
	Layout analysis	24.9	22.9	59.9
YOLO	Raw detection	30.2	57.7	**75.3**
	Layout analysis	44.8	68.1	71.6

Table 4. Performance evaluation of the fully trained and calibrated system for TEST-A using 30,000 synthetic training pages.

	CDA	IoU	Precision	Recall	F1 Score
FCOS	87.3	65.1	98.9	87.3	92.6
YOLO	87.3	66.8	96.3	88.9	92.3

Fig. 7. Exemplary character detection results. The ground truth is marked in blue, detected characters in green, deletion errors (missing characters) in red, and insertion errors (false positives) in magenta. (Color figure online)

6   Conclusions

The experimental results indicate a quite surprising ability of the convolutional object detection methods to transfer knowledge from a printed Nom font to the heterogeneous images of stone engravings. The proposed annotation-free approach is expected to be very useful for an initial exploration of the document collection, providing immediate access to hundreds of thousands of character images, predominant layouts, estimated number of characters per stele, etc. – meta-information that is obtained "for free" without manually labeling stele images beforehand.

Regarding the performance of the system, there is clearly a margin of improvement. In addition to the suggested background calibration, it would be interesting to include a text foreground calibration as well, e.g. by means of confidently extracted character images in a self-training scenario, or by using neural networks to generate characters in a similar style. Furthermore, the detected character locations may be used for keyword spotting and, eventually, for automatic transcription. Finally, it might be rewarding to explore the concept for other scripts and languages.

Acknowledgements. This work has been supported by the Swiss Hasler Foundation (project 20008). It has also received funding from the European Research Council (ERC) under the European Union's Horizon 2020 research and innovation programme (grant agreement No. 833933 - VIETNAMICA).

We would like to thank Bélinda Hakkar, Marine Scius-Bertrand, Jean-Michel Nafziger, René Boutin, Morgane Vannier, Delphine Mamie and Tobias Widmer for annotating bounding boxes during more than hundred hours to create the ground truth of the test set.

References

1. Bochkovskiy, A., Wang, C.Y., Liao, H.Y.M.: Yolov4: Optimal speed and accuracy of object detection. arXiv:2004.10934 (2020)

2. Borges Oliveira, D.A., Viana, M.P.: Fast CNN-based document layout analysis. In: Proceedings International Conference on Computer Vision Workshops (ICCVW), pp. 1173–1180 (2017)
3. Clanuwat, T., Lamb, A., Kitamoto, A.: KuroNet: Pre-modern Japanese Kuzushiji character recognition with deep learning. In: Proceedings 15th International Conference on Document Analysis and Recognition (ICDAR), pp. 607–614 (2019)
4. Ester, M., Kriegel, H.P., Sander, J., Xu, X.: A density-based algorithm for discovering clusters in large spatial databases with noise. In: Proceedings 2nd International Conference on Knowledge Discovery and Data Mining, pp. 226–231 (1996)
5. Farhadi, A., Redmon, J.: Yolov3: An incremental improvement. arXiv:1804.02767 (2018)
6. Fischer, A., Liwicki, M., Ingold, R. (eds.): Handwritten historical document analysis, recognition, and retrieval – State of the art and future trends. World Scientific (2020)
7. He, K., Zhang, X., Ren, S., Sun, J.: Deep residual learning for image recognition. In: Proceedings International Conference on Computer Vision and Pattern Recognition (CVPR), pp. 770–778 (2016)
8. Jocher, G., et al.: ultralytics/yolov5: v4.0 - nn.SiLU() activations, Weights & Biases logging, PyTorch Hub integration (2021). https://doi.org/10.5281/ZENODO.4418161
9. Lin, T.Y., Dollár, P., Girshick, R., He, K., Hariharan, B., Belongie, S.: Feature pyramid networks for object detection. In: Proceedings International Conference on Computer Vision and Pattern Recognition (CVPR). pp. 2117–2125 (2017)
10. Lin, T.Y., Goyal, P., Girshick, R., He, K., Dollár, P.: Focal loss for dense object detection. In: Proceedings International Conference on Computer Vision (ICCV), pp. 2980–2988 (2017)
11. Lin, T.Y., et al.: Microsoft COCO: common objects in context. In: Proceedings 13th European Conference on Computer Vision (ECCV), pp. 740–755 (2014)
12. Liu, S., Qi, L., Qin, H., Shi, J., Jia, J.: Path aggregation network for instance segmentation. In: Proceedings International Conference on Computer Vision and Pattern Recognition (CVPR), pp. 8759–8768 (2018)
13. Nguyen, K.C., Nguyen, C.T., Nakagawa, M.: Nom document digitalization by deep convolution neural networks. Pattern Recogn. Lett. 133, 8–16 (2020)
14. Papin, P.: Aperçu sur le programme "Publication de l'inventaire et du corpus complet des inscriptions sur stèles du Viêt-Nam". Bull. de l'École Française d'Extrême-Orient 90(1), 465–472 (2003)
15. Papin, P., Manh, T.K., Nguyên, N.V.: Corpus des inscriptions anciennes du Vietnam. EPHE, EFEO, Institut Han-Nôm (2005–2013)
16. Papin, P., Manh, T.K., Nguyên, N.V.: Catalogue des inscriptions du Viêt-Nam. EPHE, EFEO, Institut Han-Nôm (2007–2012)
17. Redmon, J., Divvala, S., Girshick, R., Farhadi, A.: You only look once: unified, real-time object detection. In: Proceedings International Conference on Computer Vision and Pattern Recognition (CVPR), pp. 779–788 (2016)
18. Rezatofighi, H., Tsoi, N., Gwak, J., Sadeghian, A., Reid, I., Savarese, S.: Generalized intersection over union: a metric and a loss for bounding box regression. In: Proceedings of International Conference on Computer Vision and Pattern Recognition (CVPR), pp. 658–666 (2019)
19. Ronneberger, O., Fischer, P., Brox, T.: U-Net: convolutional networks for biomedical image segmentation. In: Proceedings International Conference on Medical Image Computing and Computer-Assisted Intervention (MICCAI), pp. 234–241 (2015)

20. Scius-Bertrand, A., Voegtlin, L., Alberti, M., Fischer, A., Bui, M.: Layout analysis and text column segmentation for historical Vietnamese steles. In: Proceedings 5th International Workshop on Historical Document Imaging and Processing (HIP), pp. 84–89 (2019)
21. Stewart, S., Barrett, B.: Document image page segmentation and character recognition as semantic segmentation. In: Proceedings 4th International Workshop on Historical Document Imaging and Processing (HIP), pp. 101–106 (2017)
22. Sudholt, S., Fink, G.A.: PHOCNet: a deep convolutional neural network for word spotting in handwritten documents. In: Proceedings 15th International Conference on Frontiers in Handwriting Recognition (ICFHR), pp. 277–282 (2016)
23. Tan, M., Le, Q.: EfficientNet: rethinking model scaling for convolutional neural networks. In: International Conference on Machine Learning (ICML), pp. 6105–6114 (2019)
24. Tian, Z., Shen, C., Chen, H., He, T.: FCOS: fully convolutional one-stage object detection. In: Proceedings of International Conference on Computer Vision (ICCV), pp. 9627–9636 (2019)
25. Wang, C.Y., Liao, H.Y.M., Wu, Y.H., Chen, P.Y., Hsieh, J.W., Yeh, I.H.: CSPNet: a new backbone that can enhance learning capability of CNN. In: Proceedings International Conference on Computer Vision and Pattern Recognition Workshops (CVPRW), pp. 390–391 (2020)
26. Wu, Y., He, K.: Group normalization. In: Proceedings European Conference on Computer Vision (ECCV), pp. 3–19 (2018)
27. Yang, H., Jin, L., Huang, W., Yang, Z., Lai, S., Sun, J.: Dense and tight detection of Chinese characters in historical documents: datasets and a recognition guided detector. IEEE Access 6, 30174–30183 (2018)
28. Zhang, S., Chi, C., Yao, Y., Lei, Z., Li, S.Z.: Bridging the gap between anchor-based and anchor-free detection via adaptive training sample selection. In: Proceedings International Conference on Computer Vision and Pattern Recognition (CVPR), pp. 9759–9768 (2020)

Document Image Processing

DocReader: Bounding-Box Free Training of a Document Information Extraction Model

Shachar Klaiman[1(✉)] and Marius Lehne[2(✉)]

[1] SAP AI, Dietmar-Hopp-Allee 16, 69190 Walldorf, Germany
`shachar.klaiman@sap.com`
[2] SAP AI, Münzstraße 15, 10178 Berlin, Germany
`marius.lehne@sap.com`

Abstract. Information extraction from documents is a ubiquitous first step in many business applications. During this step, the entries of various fields must first be read from the images of scanned documents before being further processed and inserted into the corresponding databases. While many different methods have been developed over the past years in order to automate the above extraction step, they all share the requirement of bounding-box or text segment annotations of their training documents. In this work we present DocReader, an end-to-end neural-network-based information extraction solution which can be trained using *solely* the images and the target values that need to be read. The DocReader can thus leverage existing historical extraction data, completely eliminating the need for any additional annotations beyond what is naturally available in existing human-operated service centres. We demonstrate that the DocReader can reach and surpass other methods which require bounding-boxes for training, as well as provide a clear path for continual learning during its deployment in production.

Keywords: Document information extraction · Deep learning · OCR · Attention · RNN

1 Introduction

Information extraction from documents is an indispensable task in many scenarios. The information extracted can vary depending on the down stream task, but normally includes *global* document information which is expected to appear once on the document, e.g., document date, recipient name, etc., and *tabular* information which can be in the form of an actual table or an itemized list in the document. In the context of business documents, where we will focus our discussion on, global information is also referred to as header fields whereas tabular information is often referred to as line-items. There is a myriad of different document templates which normally prevents one, in all but very narrow applications, from easily developing a rule-based extraction solution. In some

© Springer Nature Switzerland AG 2021
J. Lladós et al. (Eds.): ICDAR 2021, LNCS 12821, pp. 451–465, 2021.
https://doi.org/10.1007/978-3-030-86549-8_29

large-scale applications, centralized spend management for example, one could even reach the extreme but realistic situation where most templates are seen only once.

The majority of current state-of-the-art automated information extraction solutions are based on an initial optical character recognition (OCR) step. As such they follow the generic two-step pipeline: first extract all the text from the document and only then localize and retrieve the requested information from the previously extracted text [1,2]. There have been multiple approaches in the literature for solving the localization step. While some initial approaches were "pure" natural language processing (NLP) approaches, e.g., [3], studies demonstrated the importance of including also the positional information of the text on the image. This led to a different set of models incorporating methods from both computer vision (CV) and NLP, see for example [4]. The described two-step setup, inevitably requires position-labels, i.e., bounding-box annotations, for training the models.

Training the localization models requires annotating tens-of-thousands of records, leading to substantial costs as well as a significant delay in training before the annotation work is over. The training dataset contains a large set of pairs of documents and labels. The labels which are constructed using the annotation results are a collection of the fields which we want to extract from the document, i.e., the various textual elements we wish to extract from the document, as well as bounding boxes for each of these fields, i.e., the coordinates of a rectangle surrounding each of the text snippets we would like to extract from the document. Yet another unavoidable complexity with the annotation process, lies in the need to select samples from the historical database to be annotated. Bad sampling leads to multiple annotation rounds which can further delay achieving the needed model performance. The sampling is usually needed due to the shear size of the historical database in many cases.

In this work we present a different approach to information extraction models. Our approach is motivated through the vast amounts of historical manually extracted data already available wherever a human-based extraction task is being performed. In contrast to the training data used for the localization models discussed above, this historical data does not contain any bounding box information. Thus, we only have access to the document images and the extracted key-value pairs. We therefore need to design an information extraction model which can be trained without any bounding-box annotations. This led us to the design of the DocReader model which is trained using only the *weakly* annotated data, i.e., training data consisting of only images and key-value string pairs, but can reach and surpass the performance of models using *strongly* annotated data. The DocReader model demonstrates how a data-driven approach, where the model is designed based on the available *very large* training data rather than preparing the training data to fit the model, can alleviate many of the challenges in acquiring training data while still providing state-of-the-art performance on the given task.

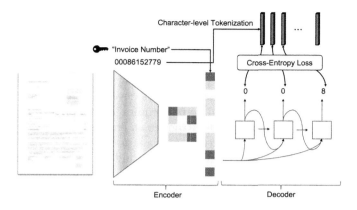

Fig. 1. Overview of the complete DocumentReader model. A document and an extraction key are provided as an input. The key is used to condition the attention layer. The image is first processed by an encoder. The resulting feature map is passed together with the key into an RNN-based decoder. The cross-entropy loss is computed from character based tokenization of the ground truth and the projected output of the RNN.

2   Related Work

The standard approach to information extraction is a two stage process, that requires an initial OCR step followed by a second information localization step. Localization can be performed on the extracted text sequence, by training a NER model [5]. With such an approach however spatial information is lost. In recent publications different methodologies are proposed to incorporate this spatial or relational information to improve extraction performance. In NER a simple way is to add coordinates to the word tokens [6] or augment the word tokens with positional embeddings [2]. In [7] the authors augment each text token with graph embeddings that represents the context of that token. In [8] the authors propose a system that identifies a set of candidates for a query based on a pretrained generic NER model. Scoring of the candidates is performed by an embedding of its neighborhood. The flexibility of this approach is limited by the availability of an external NER model that has a schema that generalizes the schema of the queries.

 CharGrid [4] combines extracted text with methodologies from computer vision. The extraction task is formulated here as an instance segmentation task with a two-dimensional grid of encoded characters as input. The predicted segmentation masks and bounding boxes are overlaid with the word-boxes of the OCR in order to retrieve the relevant text from the invoice. BertGrid [9] extended the CharGrid input concept by using a grid of contextualized embeddings as an input instead of encoding single character tokens.

All of the above systems rely on an external OCR solution. This means that the error of the extractions is bounded by the OCR error, meaning that even a perfect extraction model would still be limited by the OCR accuracy. Additionally, strong annotations such as token label or bounding boxes are required.

Another relevant line of research deals with the OCR in the wild task. Here one wishes to extract the text from an image, but the text is normally very short and embedded in some other part of the image. Since the task requires the extraction of all the text on the image, models developed to solve this task can be trained with only text labels and do not require location information through bounding boxes. The authors in [10] introduce an end-to-end OCR system based on an attention mechanisms to extract relevant text from an image. Similarly, in [11], the authors use spatial transformers to find and extract from text from relevant regions. While these models can be trained without bounding-box information, they have been so far limited to extracted all the text from the image and as such could not be directly ported to the document information extraction task.

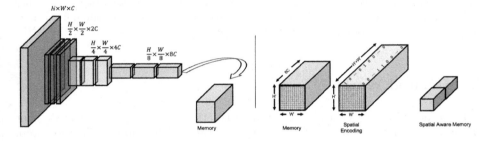

Fig. 2. Left: structure of the encoder model. An input image is passed through several convolution blocks. Right: spatial augmentation of the memory. Coordinate positions are separately encoded as one-hot vectors and concatenated to the memory on the channel axis.

3   Method

3.1   Model Architecture

We propose an end-to-end neural network architecture that generates a specific text from an input image specified by an extraction key, e.g. invoice number on an invoice. This architecture, as sketched in Fig. 1, follows an encoder-decoder pattern, where the feature map coming out of the encoder is conditioned on the provided extraction-key.

The input to the model is an image of height H and width W. It is passed into an encoder with a VGG-inspired [12] structure identical to [13]. A sketch of the encoder is shown in Fig. 2 on the left. It consists out of an initial convolution layer with a base channel size of C followed by three convolution blocks. Each block is comprised of a sequence of three convolution layer. The first layer in each of the first three blocks are convolutions with a stride of 2. Hence, the resolution after the first three blocks is reduced by a factor of 8. The number of channels doubles with each block. The convolutions in block 3 are dilated with a factor of 2. Every convolution layer is followed by a dropout layer, a batch-normalization layer, and a ReLu activation function. The output of the encoder is a dense intermediate representation of the input image with the dimensions $(H', W', 8C)$, which is also often referred to as the *memory*. This memory is then used by the decoder to generate the output sequence.

Before passing the memory into the decoder we fuse the output of a spatial-encoder along the channel axis of the memory, thereby adding positional information to each of the features in the memory. The spatial encoder could be a straight-forward one-hot encoding of the height and width of the memory, as depicted in Fig. 2 on the right, and first presented in [10], a fixed one-dimensional encoding [14] or a trainable positional embedding. In our model we have experimented with all of the above options and concluded that the one-hot positional encoding provides the best compromise between accuracy and model size. As a result we obtain a spatially augmented memory.

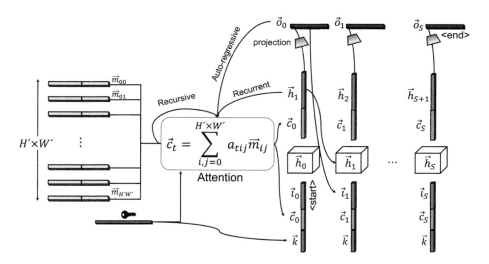

Fig. 3. Network structure of the decoder. An attention layer computes of the memory and the key a context vector. In each RNN step the cell receives the key, context vector and the previous character. The state of the RNN is concatenated with the context vector and passed into a projection layer that outputs a character.

The decoder, see Fig. 3, is based on a recurrent neural network coupled with an attention layer. It receives two inputs. The first input is the spatially augmented memory. The second input is the key that determines the field of interest to be extract from the document, e.g., invoice number. This key is encoded using either a one-hot or a trainable embedding. For the recurrent layer in the DocReader we chose a LSTM [15] and for the attention layer we use an *augmented* and *conditioned* version of the sum-attention layer [16].

Documents normally contain many potential relevant text sections in various degrees of readability. Localizing and recognizing the correct information on the document is thus a key challenge in the information extraction task. On a coarser level the correct location of the information of interest needs to be identified. On a more detailed level the correct sequence characters need to be pinpointed in each step. To tackle these challenges, we augment the base sum-attention layer with additional inputs as depicted in Fig. 3 as explained in detail below.

The attention score is computed at each decoding step for each spatial element in the memory, represented by the vector m_{ij}. We additionally condition the attention layer on the input key by adding the vector representation of the key k as an additional attribute to the scoring function. Making the model aware of the key allows us to train a model on multiple keys simultaneously. We further introduce two additional modifications to the attention layer for supporting the character by character decoding by providing additional information about the previous decoding step. We include the previously predicted character o_{t-1} into the scoring function. We refer to this attribute of the attention as auto-regressive. We also include the previous step's "flattened" attention weights, a_{t-1}, as an additional parameter to the scoring function. With this additional information, the attention layer can use the attended location of previous decoding step to assist it in deciding where to focus next. We call this mechanism recursive attention. This idea is similar to a coverage vector that has been used in neural machine translation and was introduced in [17].

In summary, we provide the scoring function with the memory m_{ij} and the current LSTM cell state h, the embedded key k, the previous attention weights a_{t-1} and the previous predicted character o_{t-1}. The scoring function then reads:

$$f\left(m, k, a, b, c\right) = \tanh\left(W_m m + W_k k + W_a a + W_b b + W_c c\right) \qquad (1)$$

$$a_{tij} = softmax\left(v^t f\left(m_{ij}, k, h_{t-1}, o_{t-1}, a_{t-1}\right)\right) \qquad (2)$$

The context vector is then calculated from the memory in the following way:

$$c_t = \sum_{i,j}^{H' \times W'} a_{tij} m_{ij} \qquad (3)$$

The attention layer outputs the context vector for the next step as well as the attention weights themselves. The resulting context vector is then concatenated to the character embedding l_{t-1} of the input at time $t-1$ and the key embedding to form the LSTM cell input.

$$h_t = \text{RNNStep}\left(h_{t-1}, l_{t-1}, c_t, k\right) \qquad (4)$$

The LSTM cell output at each step is concatenated with the step's context-vector and fed into a softmax-layer which projects the input of the layer into the prediction probabilities over the different characters in the vocabulary.

3.2   Training

As an end-to-end solution, one would expect to train the DocReader model from scratch. Indeed, for other models with similar architectures, e.g., attentionOCR [10], the authors have shown that given sufficient training data, one can train the entire network from scratch. In contrast to the "OCR in the wild" task studied in [10], we focus here on document information extraction. The main difference is that we are not interested in all the text in the image but rather only a very specific segment of text from the image. This makes learning from scratch substantially more complicated. Characters from the expected string could appear in multiple places on the document and the task of initially localizing the information on the image is much harder. Furthermore, given a randomly initialized encoder, the feature map going into the decoder has yet to learn the concept of characters which means we are trying to simultaneously learn to "see", "find", and "read" which is expectedly hard. We emphasize, that the above challenges in the document extraction task compared to the "OCR in the wild" task apply in general and are not limited to the difficulty of training the model from scratch. The model now extracts only a specific text element from the image and not the entire image text making the *signal* the model needs to learn extremely sparse in comparison.

The above challenges, led us to consider transfer-learning as a starting point for training the DocReader. Since the input to the DocReader is an image, it is natural to consider any state-of-the-art pretrained image network. This approach was used in multiple previous studies [10,18]. Since the DocReader not only needs to distinguish textual from non textual areas in the image but rather also distinguish between different texts in the image, a generic image classifier would not be a good choice here. Therefore, we chose here, to use the encoder of a pretrained OCR model instead. Specifically, we use the encoder from the recently published OCR model: CharGrid-OCR [13]. The usage of a pretrained OCR model, provides an excellent starting point for the DocReader, since the feature map obtained from the decoder already allows the decoder to distinguish between different characters in the image. We note in passing, that our choice to facilitate the training using a pretrained OCR model does not in principle exclude the possibility of training the DocReader from scratch given enough training data and computational resources.

In practice, the DocReader is trained using a two phase procedure. During the first phase the weights of the encoder, which are taken from the pretrained OCR model, are frozen. The second phase commences after convergence is reached in the first training phase. In the second phase all weights in the model are trainable. As explained above, the decoder must "see", "find", and "read" from the encoder output. By first freezing the encoder, we allow the decoder to concentrate on the latter two actions instead of trying to optimize all three at once. Our experiments show that without this initial training phase, the decoder tends to quickly overfit the training data, completely bypassing the attention module. In other words, the decoder "memorizes" rather than "reads". As we shall explicitly demonstrate in a later section, the results of the DocReader after only the first training phase are still very poor. Thus, we cannot leave the encoder fixed throughout and competitive results can only be achieved by fine-tuning the entire network.

Normally, one wishes to extract multiple, possibly very different, fields from the input document. With respect to the DocReader model's training, this raises the question whether one should train multiple single-key models or rather a single multi-key model. The DocReader's built-in key-conditioning allows one to do both. In practice we observe that multi-key models tend to perform better than many single-key models. The improvement likely stems from better training of the encoder as the multiple-keys, which comprise of different characters and appear on different locations in the image, allow for a better coverage of the feature map being trained. Nevertheless, one must also consider the issue of data sampling and training time when choosing how many keys to simultaneously train on. Since the DocReader extracts in every step the value for a single-key, the number of records in the training set scales linearly with the number of keys being trained on, because we add the same invoice with a different label to the training set. Furthermore, some of the keys do not appear on all documents which means the model needs to learn to predict a missing value for certain document and key pairs. We eventually divide the keys into groups having similar missing-value support in the training set. This allowed us to avoid predicting to many false-negatives on keys which are almost always present but avoid the need to exclude more rare fields.

4   Information Extraction from Invoices

4.1   Data

Our training dataset consists of $1.5M$ scanned single-page invoices. The invoices are in different languages and the majority of invoice templates appear only a single time in the dataset. Each invoice image is accompanied with the human extracted values (strings) for a set of fields, keys, for example, invoice number, invoice date, etc. Our test set which comprises of 1172 documents from 12 countries, contains also multi-page documents. By excluding overlapping vendors between the training and test sets, we make sure that the test set does not contain any of the invoice templates from the training set. We note that the above strict separation of templates between training and test set produces a very stringent performance measure, but provides a very realistic measure of the generalization capabilities of the trained model.

In order to compare with other state-of-the-art approaches which require bounding-box annotations for training, we also fully annotated a training set of $40k$ scanned invoices, i.e., key-value strings and bounding-box annotations. The invoices in this set are in multiple languages with the very large variety in templates. We shall refer to this training set as the strongly-annotated training set in the following. Also the vendors of the strongly-annotated set are excluded from the above test set.

Both of the above datasets are proprietary in-house datasets. Therefore, in order to visualize the DocReader's extractions we use documents from the invoice category of the RVL-CDIP test set [19].

4.2 Implementation Details

As described above, we reuse a part of the encoder of the pretrained CharGrid-OCR model [13] as the encoder of the DocReader. Here, however, lies one of the big challenges in training the DocReader without any positional information, i.e., bounding boxes. The OCR model is usually trained on crops from the full, 300 dpi, document. Since the OCR training data contains positional information, cropping is not an issue. When we now want to use the same encoder for the DocReader, however, we no longer have any positional information and therefore we cannot crop the training images. In other words, we must feed the DocReader the entire image of the document. As a result of this, we must reduce the resolution of the scanned image, and thus pretrain the OCR model on the reduced resolution as well. In practice we found that an input resolution of $(832, 640)$ works very well.

We choose the base channel number C to be 32. For the convolutional layer in the encoder we use 3×3 kernel. The dropout layer in the encoder have a dropout probability of 0.1 during training. The LSTM cell has a size of 256 and the attention layer has a dimensionality of 256. The token vocabulary encompasses ASCII characters including punctuation characters[1]. We also add special tokens for start of sequence, padding, end of sequence and out of vocabulary. For training we use a batch size of 16.

The ground truth is provided as a sequence of characters. We add a start token at the beginning of the training and an end token at the end of the sequence. We observe that adding several warm-up tokens at the beginning of the sequence can facilitate the decoding process. This provides the attention layer some extra steps to determine the correct location of the text to be extracted. During training we make use of teacher forcing. For inference we feed the output of the previous step into the decoder. We stop decoding if either a stop token was encountered or the maximum sequence length has been reached. We compute a cross-entropy loss for each character individually and sum the loss over the sequence.

Special care must be taken when choosing the learning rate for training the DocReader. As explained in Sect. 3.1, the DocReader model is constructed from 3 modules: the encoder, the attention, and the decoder. If the learning rate is too high, the attention module will simply be ignored and the decoder will memorize the input samples and will not generalize. One can see this very clearly from the attention maps where the attention in such a case is spread over uniformly over the entire image instead of targeting the desired key. For our training we use the Adam optimizer and a learning rate of $2 \cdot 10^{-4}$. We note in passing that automated *optimal* learning rate algorithms such as that detailed in [20] fail to

---

[1] This is equivalent with the set of characters returned by string.printable in Python.

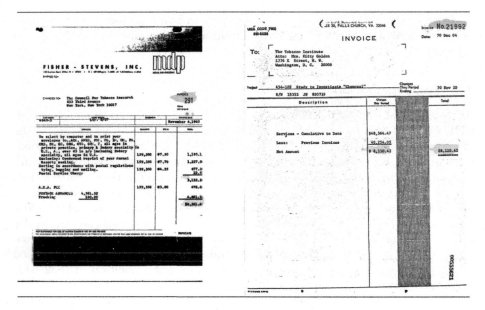

Fig. 4. DocReader attention maps for the predictions of invoice number (orange) and invoice amount (turquoise) on the two invoice scans taken from the invoice category of the RVL-CDIP test set [19]. When multiple instance of a field appear on the invoice, the DocReader attends to all instance simultaneously. (Color figure online)

resolve this type of dependency since the training loss will continue to go down even if the attention module is *bypassed*.

4.3   Experiments and Results

We trained and evaluated our model on the extraction task of 6 different fields from invoice scans. Motivated by the ratio of documents where the different fields occurs we trained 3 different models. The first was trained to extract the fields: invoice number, invoice date, total amount, and currency. The second model was trained on extracting the purchase order number (PO) from invoices. The third model was trained to extract the tax amount from the invoice.

Training is performed in the two phase training procedure as described in Sect. 3.2. The model performance during training time on the evaluation set is shown in Fig. 5. We observe an increase of model performance in the first phase that levels off at at quite low value. Hence, using only the features from the frozen pre-trained encoder is not enough to perform well on this task. Fine-tuning the whole model enables us then to achieve significantly better results.

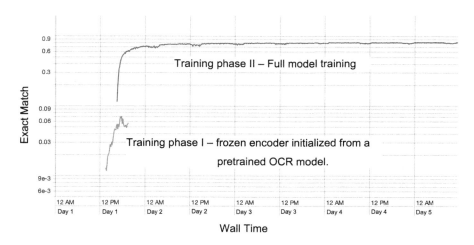

Fig. 5. Exact string match metric on the evaluation set during the two phase training regimen as described in Sect. 3.2. During the first phase here shown in blue the encoder weights are initialized from a pre-trained OCR model [13], but remain frozen. In the second phase the training is continued on all weights of the model. (Color figure online)

The model's performance was evaluated using an exact string match for the predicted values against the ground truth values. Fields that are not present on a document are represented by empty strings. The metric therefore also covers cases where the model falsely extracts a value for a non-existing fields or where the model falsely does not return a value for an existing field, i.e., false negatives and false positives. We chose this stricter metric since it guarantees automation of the extraction process. It is worth keeping in mind that, in contrast to the DocReader, models which copy the predicted value from an OCR output depend on the correctness of the OCR which cannot be controlled by the extraction model.

In Table 1 we compare the DocReader's performance on all the extracted fields. We compare the performance to a CharGrid model [4] which was trained on the $40k$ fully annotated set, see Sect. 4.1. The CharGrid model was previously shown to provide state-of-the-art results on invoice information extraction. The DocReader model show comparable results on the invoice date, invoice amount, and purchase order number fields, and substantial improvement on invoice number, tax amount, and currency.

Figure 4 depicts the attention maps for the invoice number and invoice amount predictions of the DocReader model. The attention maps are created by summing the attention maps of all single character prediction steps and then resizing the attention map to the original image size. The two invoices shown are taken from the invoice category of the RVL-CDIP test set [19]. The invoices

Table 1. Exact string match of the predicted and ground-truth strings on the test set defined in Sect. 4.1. We compare the DocReader trained on 1.5M invoices to the state-of-the-art CharGrid trained on 40k invoices [4] on the extraction task of 6 different fields from invoice scans. For predicting the invoice currency, the displayed CharGrid accuracy (*) is the combined accuracy of CharGrid and a separately trained OCR-text currency Classifier. See the text for more details. For four of the six fields, we also show the performance of the DocReader model when trained using only the training set used to train the CharGrid model (DocReader(40k)).

Model	Number	Date	Amount	PO	Tax	Currency
CharGrid	0.72	0.82	0.87	**0.82**	0.47	0.82*
DocReader	**0.79**	**0.83**	**0.90**	0.79	**0.80**	**0.97**
DocReader(40k)	0.58	0.78	0.84	–	–	0.89

in this test set are of rather poor scanning quality and appear rather different than the invoices the model was trained on. Still, the DocReader succeeds in extracting the chosen fields accurately. It is interesting to note, that in cases of multiple instance of the field on the invoice, the DocReader simultaneously reads from all the instances.

Compared to OCR-based extraction models, the DocReader can also predict *global* fields which cannot always be found in the text on the invoice itself. One such field, for example, is the invoice currency. In many invoices there is no currency symbol at all. Even when a currency symbol is on the invoice, it is often not sufficient to make a concrete prediction. Consider a US invoice where the amount has the $ symbol next to it. One cannot infer that the currency is USD from the currency symbol. Instead, one would have to also examine the invoice address fields to determine the actual invoice currency. Since the DocReader is not restricted to localize first and then copy the value from the OCR output, it can directly infer the currency from the invoice image. As depicted in Fig. 6, in cases where the currency symbol is on the invoice, the DocReader reads the symbol directly whereas whenever the currency symbol is insufficient to make a currency prediction, the model attends the address field as well in order to determine the correct currency. Our previous state-of-the-art solution for currency employed a combination of the OCR-based CharGrid model with a fallback on a OCR-text-based currency classifier in the case the CharGrid could not retrieve any decisive currency prediction. As Table 1 clearly shows, the DocReader offers a very significant improvement on the performance on this *global* field.

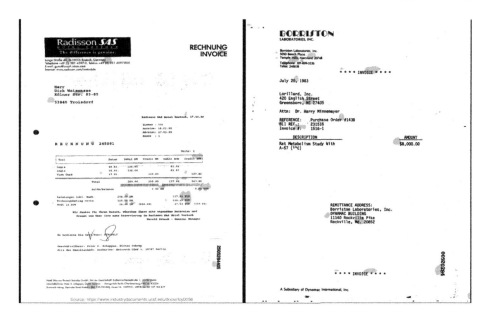

Fig. 6. DocReader attention maps for the predictions of invoice currency. If the currency symbol/identifier is directly on the invoice, the DocReader attends to that identifier directly. In cases where the symbol is not on the invoice or where the symbol is not sufficient to make a currency prediction, e.g., the $ sign is not sufficient to distinguish between USD and AUD, the DocReader infers the currency from the address section of the invoice as a human extractor would. The two invoices were taken from the invoice category of the RVL-CDIP test set [19].

5   Discussion

The results of the experiments described in Sect. 4, demonstrate how in situations where we have a very large amount of *weakly* annotated historical data, the DocReader model can yield state-of-the-art results while freeing us from the need to additionally annotate any data. As can be expected (see Table 1), without access to a substantial amount of historical data, methods based on positional annotations, i.e., *strong* annotations, have a clear advantage over the DocReader.

One point to keep in mind when using the historical extraction data directly, is that the model is only able to learn the extraction pattern which already exists in the data. Thus, one cannot define a set of desired extraction rules which other approaches would give as their annotation guidelines. Instead, we can at most try and restrict certain rules by preprocessing the values but in principle the model learns to read the extraction pattern set by the humans performing these task before. This could also be perceived as an inherit advantage of the method since we do not need to understand the extraction logic before hand, the model will learn this logic by itself. It does though require that previous extractions have been made in a consistent manner.

We note that the DocReader model can be thought of as a very generic targeted OCR model. The flexible conditioning mechanism we incorporated directly into the attention module allows one to choose essentially any consistently appearing information on an image and read it out. The model also provides automatic post-processing so long as the post-processed values are consistent across the training set. Thus one can, for example, directly extract the date in ISO format even when the training set is composed from documents from many countries where the date format on the documents themselves can be very different.

Since we do use a trained OCR model to initialize the DocReader encoder, it is fair to ask whether we are indeed completely free of positional annotations as these are used in the training of the OCR model. As we mention in Sect. 3.2, this initialization of the model might not be strictly necessary [10]. Furthermore, the annotations for training the OCR model are generated and require no manual annotations as explained in [13]. This means that even with the DocReader initialization from the OCR model, no additional manual positional annotations are required at all.

6   Conclusion

The DocReader model, presented in this work shows how a data-driven model design can lead to substantially improved performance on the document information extraction task. By using only the *available* historical extraction data, the DocReader mitigates the challenges and costs which come with specialized data annotations. Our end-to-end approach allows us to go beyond the standard extraction tasks of identifying the correct content and reading the corresponding value. The DocReader model can also directly solve image classification tasks, e.g., predicting invoice currency, making use of very diverse information available on the document image such as the document template and the interplay between different fields.

References

1. Zhang, P., et al.: TRIE: end-to-end text reading and information extraction for document understanding. In: Proceedings of the 28th ACM International Conference on Multimedia, pp. 1413–1422 (2020)
2. Xu, Y., Li, M., Cui, L., Huang, S., Wei, F., Zhou, M.: Layoutlm: pre-training of text and layout for document image understanding. In: Proceedings of the 26th ACM SIGKDD International Conference on Knowledge Discovery and Data Mining, pp. 1192–1200 (2020)
3. Lample, G., Ballesteros, M., Subramanian, S., Kawakami, K., Dyer, C.: Neural architectures for named entity recognition. arXiv preprint arXiv:1603.01360 (2016)
4. Katti, A.R., et al.: Chargrid: towards understanding 2D documents. In: EMNLP (2018)
5. Graliński, F., et al.: Kleister: a novel task for information extraction involving long documents with complex layout. arXiv preprint arXiv:2003.02356 (2020)

6. Sage, C., Aussem, A., Eglin, V., Elghazel, H., Espinas, J.: End-to-end extraction of structured information from business documents with pointer-generator networks. In: Proceedings of the Fourth Workshop on Structured Prediction for NLP, pp. 43–52 (2020)
7. Liu, X., Gao, F., Zhang, Q., Zhao, H.: Graph convolution for multimodal information extraction from visually rich documents. In: Proceedings of the 2019 Conference of the North American Chapter of the Association for Computational Linguistics: Human Language Technologies, Volume 2 (Industry Papers), pp. 32–39 (2019)
8. Majumder, B.P., Potti, N., Tata, S., Wendt, J.B., Zhao, Q., Najork, M.: Representation learning for information extraction from form-like documents. In: Proceedings of the 58th Annual Meeting of the Association for Computational Linguistics, pp. 6495–6504 (2020)
9. Denk, T.I., Reisswig, C.: BERTgrid: contextualized embedding for 2D document representation and understanding. In: Workshop on Document Intelligence at NeurIPS 2019 (2019)
10. Wojna, Z., et al.: Attention-based extraction of structured information from street view imagery. In: 2017 14th IAPR International Conference on Document Analysis and Recognition (ICDAR), vol. 1, pp. 844–850. IEEE (2017)
11. Bartz, C., Yang, H., Meinel, C.: SEE: towards semi-supervised end-to-end scene text recognition. In: Proceedings of the AAAI Conference on Artificial Intelligence, vol. 32 (2018)
12. Simonyan, K., Zisserman, A.: Very deep convolutional networks for large-scale image recognition. arXiv preprint arXiv:1409.1556 (2014)
13. Reisswig, C., Katti, A.R., Spinaci, M., Höhne, J.: Chargrid-OCR: end-to-end trainable optical character recognition through semantic segmentation and object detection. In: Workshop on Document Intelligence at NeurIPS 2019 (2019)
14. Vaswani, A., et al.: Attention is all you need. In: NIPS (2017)
15. Hochreiter, S., Schmidhuber, J.: Long short-term memory. Neural Comput. **9**(8), 1735–1780 (1997)
16. Bahdanau, D., Cho, K., Bengio, Y.: Neural machine translation by jointly learning to align and translate. arXiv preprint arXiv:1409.0473 (2014)
17. Tu, Z., Lu, Z., Liu, Y., Liu, X., Li, H.: Modeling coverage for neural machine translation. arXiv preprint arXiv:1601.04811 (2016)
18. Schwarcz, S., Gorban, A., Lee, D.S., Gibert, X.: Adapting style and content for attended text sequence recognition. In: IEEE Winter Conference on Applications of Computer Vision (WACV) (2020)
19. Harley, A.W., Ufkes, A., Derpanis, K.G.: Evaluation of deep convolutional nets for document image classification and retrieval. In: International Conference on Document Analysis and Recognition (ICDAR), pp. 991–995 (2015)
20. Smith, L.N.: Cyclical learning rates for training neural networks. In: 2017 IEEE Winter Conference on Applications of Computer Vision (WACV), pp. 464–472. IEEE (2017)

Document Dewarping with Control Points

Guo-Wang Xie[1,2], Fei Yin[2], Xu-Yao Zhang[1,2], and Cheng-Lin Liu[1,2,3(✉)]

[1] School of Artificial Intelligence, University of Chinese Academy of Sciences,
Beijing 100049, People's Republic of China
xieguowang2018@ia.ac.cn

[2] National Laboratory of Pattern Recognition, Institute of Automation of Chinese
Academy of Sciences, 95 Zhongguancun East Road,
Beijing 100190, People's Republic of China
{fyin,xyz,liucl}@nlpr.ia.ac.cn

[3] CAS Center for Excellence of Brain Science and Intelligence Technology,
Beijing, People's Republic of China

Abstract. Document images are now widely captured by handheld devices such as mobile phones. The OCR performance on these images are largely affected due to geometric distortion of the document paper, diverse camera positions and complex backgrounds. In this paper, we propose a simple yet effective approach to rectify distorted document image by estimating control points and reference points. After that, we use interpolation method between control points and reference points to convert sparse mappings to backward mapping, and remap the original distorted document image to the rectified image. Furthermore, control points are controllable to facilitate interaction or subsequent adjustment. We can flexibly select post-processing methods and the number of vertices according to different application scenarios. Experiments show that our approach can rectify document images with various distortion types, and yield state-of-the-art performance on real-world dataset. This paper also provides a training dataset based on control points for document dewarping. Both the code and the dataset are released at https://github.com/gwxie/Document-Dewarping-with-Control-Points.

Keywords: Dewarping document image · Control points · Deep learning

1 Introduction

Document image has become very common and important in our daily life because of their convenience in archiving, retrieving and sharing valuable information. Unlike the controllable operating environment of the scanner, camera-captured document images often suffer from distortions and background, due to physical deformation of the paper, shooting environment and camera positions. The above factors will significantly increase the difficulty of information extraction and content analysis. For reducing the influence of distortion in document image processing, many dewarping approaches have been proposed in the literature.

© Springer Nature Switzerland AG 2021
J. Lladós et al. (Eds.): ICDAR 2021, LNCS 12821, pp. 466–480, 2021.
https://doi.org/10.1007/978-3-030-86549-8_30

Traditional approaches [1,16,18] usually require complex process, external conditions or strong assumptions to construct 2D or 3D rectification models by extracting the hand-crafted features of document images. To improve generalization ability in difficult scenarios, some deep learning based approaches have been proposed and promising performance in rectification is obtained. As shown in Fig. 1(a), a widely-used approach is to exploit the Encoder-Decoder architecture as a generic feature extractor to predict some pixel-wise information, such as forward mapping (each cell represents the coordinates of the pixels in the dewarped output image, and the pixels correspond to pixels in the warped input image) [8,9,21] in Fig. 1(b) and dewarping image (image-image translation) [12] or backward mapping (each cell represents the coordinates of the pixels in the warped input image) [3,5,10] in Fig. 1(c). Although the Encoder-Decoder architecture has achieved satisfying performance, further research is needed for more flexible and lightweight approaches.

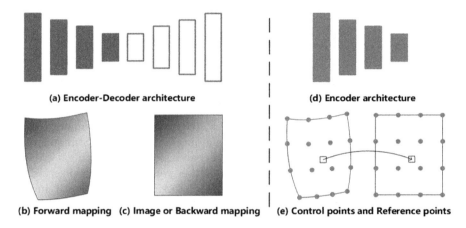

(a) Encoder-Decoder architecture

(d) Encoder architecture

(b) Forward mapping (c) Image or Backward mapping

(e) Control points and Reference points

Fig. 1. Dewarping architecture. (a) the Encoder-Decoder architecture is used as a generic feature extractor to predict the expression of the pixel-wise such as (b) forward mapping and (c) image or backward mapping in recent rectifying systems. Our proposed sparse points is based on rectification which only exploits (d) Encoder architecture to predict (e) control points and reference points so as to achieves similar effects as (a) but in a more flexible and practical way.

This paper proposes a novel approach to rectify distorted document image and remove background finely. As shown in Fig. 1(d), we take advantage of the Encoder architecture for extracting semantic information from image automatically, which is used to predict control points and reference points in Fig. 1(e). The control points and reference points are composed of the same number of vertices and describe the shape of the document in the image before and after rectifying, respectively. Then, we use interpolation method between control points and reference points to convert sparse mappings to backward mapping, and remap the original distorted document image to the rectified image. The control points are flexible and controllable, which facilitates the interaction with people to adjust the sub-optimal points. Experiments show that the method based on control points can rectify various deformed document images, and yield state-of-the-art performance on real-world dataset. Furthermore, our approach can be edited multiple times to improves its practicability when the correction effect is not satisfied, which alleviates the disadvantages of weak operability in end-to-end methods. We can flexibly select post-processing methods and the number of vertices according to different application scenarios. Compared to the pixel-wise regression, control points is more practical and efficient. To inspire future researches on this direction, we also provide a new dataset based on control points for document dewarping.

2   Related Works

In recent years, researchers have investigated a variety of approaches to rectify distorted document. We give a brief overview of these methods from the perspectives of handcrafted features and learning-based features.

Handcrafted Features-Based Rectification. Prior to the prevalence of deep learning, most approaches constructed 2D and 3D rectification models by extracting hand-crafted features of document images. Some of these approaches utilized visual cues of the document image to reconstruct the document surface, such as text lines [6,15,16], illumination/shading [2,18,23] etc. Stamatopoulos et al. [15] rectified the document image in a coarse scale by detecting words and text lines, and used baseline correction to further refine and normalize individual words. Tian et al. [16] estimated the 2D distortion grid by identifying and tracing text lines, and then estimated the 3D deformation from the 2D distortion grid. Liu et al. [6] estimated baselines' shape and characters' slant angles, and then exploited thinplate splines to recover the distorted image to be flat. Wada et al. [18] and Courteille et al. [2] employed the technique of shape from shading (SfS) and restored the distorted image based on the reconstructed 3D shape. Similarly, Zhang et al. [23] proposed a generic SfS method considering the perspective projection model and various lighting conditions, and then mapped the 3D surface back to a plane. Moreover, there were many approaches that use geometric properties or multi-view to rectify distorted document. Brown et al. [1] used the 2-D boundary of the imaged material to correct geometric distortion. He et al. [4] extracted page boundary curves to reconstruct the 3D surface. Tsoi et al. [17] utilized the boundary information from multiple views of the document image to recover geometric distortions. For the existence of simple skew, binder curl, and folded deformation of the document, handcrafted features-based rectification demonstrated good performance. However, these methods are difficult to be employed in dealing with distorted documents captured from natural scenes due to their complicated geometric distortion and changeable external conditions.

Learned Features-Based Rectification. With the progress of deep learning research, a lot of works exploited the features learned from document image to recover geometric distortions. Ramanna et al. [12] synthesized dewarped image by the deep learning network of image-image translation (pix2pixhd). Ma et al. [9] created a large-scale synthetic dataset by warping non-distorted document images and proposed a stacked U-Net to predict the forward mapping for the warping. Liu et al. [8] adopted adversarial network to predict a dense unwarping grid at multiple resolutions in a coarse-to-fine fashion. In order to improve generalization in real-world images, many approaches [3,5,10] focused on generating more realistic training dataset which has a more similar distribution to the real-world image. Das and Ma et al. [3] and Markovitz et al. [10] exploited multiple ground-truth annotations in both 2D and 3D domain to predict the backward mapping of a warped document image. Li et al. [5] proposed patch-based learning approach and stitch the patch results into the rectified document. Although higher-quality training dataset and richer ground-truth annotations make it easier for the model to learn useful features, it also increases the difficulty for engineers to build the datasets and also the models. In order to better tradeoff between model complexity and rectification performance, Xie et al. [21] proposed a novel framework to estimate pixel-wise displacements and foreground/background classification. Compared to prior approaches, [21] achieved better performance with various distorted document images, but there is still room for improvement in the computational complexity and post-processing steps.

3   Approach

3.1   Definition

Previous studies treated the geometric rectification task as a dense grid prediction problem, which take a 2D image as input and output a forward mappings (each grid represents the coordinates of the pixels in the dewarped ouput image, and the pixels correspond to pixels in the warped input image) or backward mappings (each grid represents the coordinates of the pixels in the warped input image). Our method simplifies this process to directly predict the sparse mappings, and then uses interpolation to convert it into a dense backward mapping. To facilitate explanation, we define the following concepts:

Vertex represents the coordinates of a point in document image. In this paper, we can move the vertices by changing the coordinates.

Control Points consist of a set of vertices. As shown in Fig. 2 (c), control points are distributed on the distorted image to describe the geometric deformation of the document.

Reference Points consists of the same number of vertices as the control points. As shown in Fig. 2 (d), the reference point describes the regular shape. Document Dewarping could be realized using unwarping grid by matching the control points and the reference points.

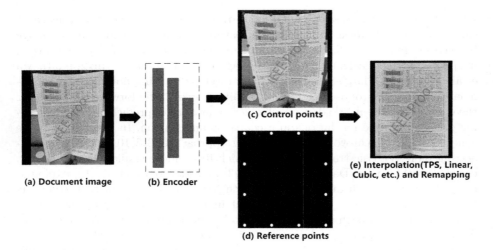

Fig. 2. Dewarping pipeline. (a) input deformed document image. (b) Encoder architecture for extracting semantic information which will be exploited to predict (c) control points and (d) reference points. Then, we convert sparse mappings to dense mapping and get the rectified image by (e) interpolation method and remapping respectively. Only 12 control points are used in this pipeline.

3.2   Dewarping Process

Figure 2 illustrates the process in our work. First, an image of a deformed document is fed into network to obtain two output branches. Our approach adopts the Encoder architecture as feature extractor, and then exploits the learned feature to predict control points and reference points respectively in a multi-task manner. Second, as shown in Fig. 1(e), we construct the rectified grid by moving the control points to the position of the reference points and converting it to pixel-wise location mapping. In order to move the position of the control points and convert sparse mappings to dense mapping, we employ interpolation method [11] (TPS, Linear, Cubic, etc.) between control points and reference points. After that, pixels are extracted from one place in the original distorted document image and mapped to another position of the rectified image. Compared with previous methods based on DNNs, our approach is simple and easy to implement.

3.3   Network Architecture

As shown in Fig. 2 and Fig. 3, our approach takes the image of deformed document as an input and predicts control points $\mathbb{R}^{31 \times 31 \times 2}$ and reference points $\mathbb{R}^{31 \times 31 \times 2}$. Control points consist of 31×31 coordinates to match the same number of reference points so as to construct the rectified grid. Since the reference points are composed of a regular grid, they can be constructed by the intervals of points between the horizontal and vertical directions $\binom{v'}{h'}$.

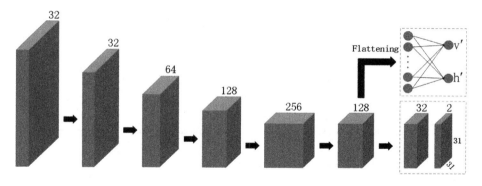

Fig. 3. Network architecture. An encoder extracts image features and sends them to two branches. The upper branch is a fully connected neural network which predicts the interval between reference points. The lower branch is a two-layer convolutional network to predict control points.

In our network, the first two layers of encoder use two convolutional layers with the strides of 2 and 3 × 3 kernels. Inspired by the architecture from [21], we use the same structure in our encoder architecture, including the dilated residual block and the spatial pyramid with stacked dilated convolution. After each convolution, the Batch Normalization and ReLU are applied. Then, we flatten the features of the last layer and feed them into the fully connected network to predict the intervals $\binom{v'}{h'}$ between reference points. Simultaneously, we use two-layer convolutional network to predict control points $\mathbb{R}^{31 \times 31 \times 2}$, which apply Batch Normalization and PReLU after the first convolution.

3.4   Training Loss Functions

We train our models in a supervised manner by using synthetic reference points and control points as the ground-truth. Training loss functions are composed of two parts. One is used to regress the position of the point, and the other is the interval between two points in the horizontal and vertical directions.

The Smooth L1 loss [7,13] is used for position regression on control points, which is less sensitive to outliers. It is defined as:

$$z_i = \begin{cases} 0.5 \left(p_i - \hat{p}_i\right)^2, & \text{if } |p_i - \hat{p}_i| < 1 \\ |p_i - \hat{p}_i| - 0.5, & \text{otherwise} \end{cases} \tag{1}$$

$$L_{smoothL1} = \frac{1}{N_c} \sum_{i}^{N_c} z_i, \tag{2}$$

N_c is the number of control points, p_i and $\hat{p}_i$ respectively denote the ground-truth and predicted the position in 2D grid.

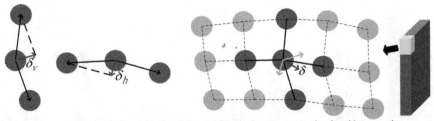

(a) Displacement trend in vertical and Horizontal   **(b) Displacement trend at local intersection**

Fig. 4. Differential Coordinates. We use differential coordinates as an alternative representation of vertex coordinates. δ_v and δ_h respectively represent the displacement trend of the center point in the horizontal and vertical. δ represents the local positional correlation between the intersection and its connection points

Different from facial keypoints detection, the content layout of document images is irregular. Although the Smooth L1 loss guide the model on how to place each vertex in an approximate position, it is difficult to represent the relationship of a point relative to its neighbors or local detail at each surface point. To make a more fault-tolerant model to better describe the shape, we use differential coordinates as an alternative representation for the center coordinates. As shown in Fig. 4, δ_v and δ_h respectively represent the displacement trend of the center point in the horizontal and vertical, which helps to maintain correlation in the corresponding direction. Similarly, δ represents the local positional correlation between the intersection and two directions, which can be defined as :

$$\delta = \sum_{j=1}^{k}(p_j - p_i),\tag{3}$$

k is the number of elements, p_i is the intersection. Inspired by [21], we use a constraint to expect the predicted displacement trend at local intersection to be as close as possible to the ground-truth. The displacement trend represents the relative relationship between a local region and its central point. We formulate L_c function as follows:

$$L_c = \frac{1}{N_c}\sum_{i}^{N_c}\left(\delta_i - \hat{\delta}_i\right)^2\tag{4}$$

The predictions of control points are not independent of each other by using correlation constraints between vertices. In Fig. 4 (b), we associate 4 directly connected vertices and set $k = 5$, which constrains the relationship between a point and its neighbors. We can constrain vertices in wide-area by expanding the correlation points in all four directions simultaneously. As shown in Fig. 5, we associate four vertices in each direction, which allows the model better learn the shape of the document image.

Fig. 5. Wide-Area Correlation. To make the model better describe the shape, we adjust the number of constrained related vertices. The first group associates 4 neighbor vertices and sets k for 5. The second group is the center point connected directly with 4 vertices in each direction and sets k for 17.

We use L1 loss to perform interval regression on reference points, which is defined as:

$$L_r = \frac{1}{N_r} \sum_{i}^{N_r} \|d_i - \hat{d}_i\|_1, \tag{5}$$

where $N_r = 2$, d represents the interval between two points in the horizontal or vertical directions. The final loss is defined as a linear combination of different losses:

$$L = L_{smoothL1} + \alpha L_c + \beta L_r, \tag{6}$$

where α and β are weights associated to L_c and L_r.

3.5 Interactivity

Control points are flexible and controllable, which facilitates the interaction with people to change the resolution of the rectified image, choose the number of vertices and adjust the sub-optimal vertices. Furthermore, our method can also be used to produce annotations for distorted document image.

Resize the Resolution of the Rectified Image. The smaller input image requires less computing, but some information maybe lost or unreadable. For facilitating implementation, we use smaller image as input to get control points and adjust these as follows:

$$O_{new} = F\left(\mathcal{M}/R_{old} * R_{new}; I_{new}\right), \tag{7}$$

where $\mathcal{M} \in \mathbb{R}^{31 \times 31 \times 2}$ is original control points, R_{old} and R_{new} respectively denote original image resolution and adjusted resolution, F is the interpolation and remapping, I_{new} and O_{new} are the distorted image and the rectified image with new-resolution. In this way, we can freely change resolution of the rectified image.

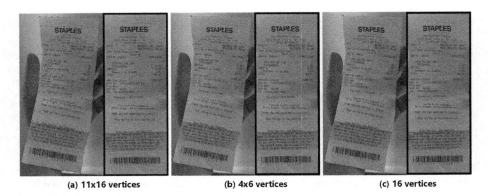

(a) 11x16 vertices   (b) 4x6 vertices   (c) 16 vertices

Fig. 6. Sparse vertices. We can choose the number of vertices according to step length in Table 1. For simple deformations, a small number of control points can be used to achieve similar results.

Choose the Number of Vertices. Control points consist of 31×31 coordinates, which can rectify a variety of distorted images. When using TPS interpolation in Fig. 2 (e), more vertices can take more deformation information into account, but more calculation time is needed. As shown in Fig. 6, we can get similar results with fewer control points. For datasets of different difficulty, we can select the appropriate number of vertices according to step length in Table 1. When we rectify simpler distorted document image in Fig. 2, internal vertices can also be omitted and the dewarping could be realized quickly with only 12 control points.

Table 1. Select vertices by step length. The number of **vertices** corresponding to the **step** length. When we use a small number of vertices, the speed of calculation can be further improved in the post-processing steps.

Step	1	2	3	5	6	10	15	30
Vertices	31	16	11	7	6	4	3	2

Adjust the Sub-optimal Vertices. Although generating more realistic training dataset can improve generalization in real-world images, it is difficult to ensure that all images can be well rectified with limited data. In addition to exploring more excellent and robust models, manual fine-tuning of the sub-optimal vertices is a more intuitive method. As shown in Fig. 7, we drag control point to new locations or edit Laplacian mesh [14] to change the mesh's shape. Like [19], a similar approach can also be used to automatically adjust the control points by iteratively optimizing the shallow neural network.

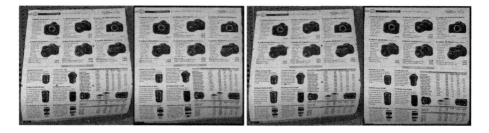

Fig. 7. Move the vertices. The first group is the predicted initial position (4×4 vertices). The second group is the adjusted position. By adjusting the coordinates of the sub-optimal vertices, we can get a better rectified image.

Distorted Document Annotation via Control Points. Our method can be used as a semi-automatic annotation tool for distorted document image. Regardless of whether the rectifying systems shown in Fig. 1 use forward mapping, backward mapping or control points as the middleware to rectify image, these methods face the problem of difficulty in annotation. Existing methods and recent advances focuse on generating more realistic training dataset to improve generalization in real-world images, but it is difficult to rely on rendering engines to simulate the real environment. Our method provides a feasible way for distorted document annotation. Due to the flexibility and controllability of control points, it is convenient to interact with people and adjust the sub-optimal points. Satisfactory control points can also get forward mapping and backward mapping by interpolation.

4   Experiments

We synthesize 30K distorted document image for training. The network is trained with Adam optimizer and hyperparameters are set as $\alpha = 0.1$ and $\beta = 0.01$. We set the batch size as 32 and learning rate as $2e-4$ which was reduced by a factor of 0.5 after each 40 epochs.

4.1   Datasets

Inspired by [9, 21], we synthesize 30K distorted document image in 2D mesh. The scanned document such as receipts, papers, etc. using two functions proposed by [1] to change the distortion type, such as folds and curves. We augment the synthetic images by adding random shadows, affine transformation, gaussian blur, background textures, jitter in the HSV color space and resize them into 992×992 (keeping the aspect ratio and zooming in or out along the longest side, then filling zero for padding). Meanwhile, we perform the same geometric transformation on a sparse grid $\mathbb{R}^{61 \times 61 \times 2}$ to get the control points. As shown in Fig. 8, a sparse grid is a set of points with intervals and evenly distributed on the scanned document. In this way, the ground-truth of control points and reference points can be obtained respectively. We can directly predict the coordinates of the control points. Similarly, we can also convert them into the position offset of the corresponding point, which is used to supervise the offset of the control point and then convert it into coordinates. In the experiment, these two groups of supervision methods have similar effects. In addition, the number of vertices in control points can be selected according to requirements. As shown in Table 2, we can choose different step length to change the density of the grid and the length of the interval. In our work, the step size is set to 2.

Table 2. Select the sparseness of vertices in the grid through different step length.

Step	1	2	3	4	5	6	10	12	15	20	30	60
Vertices	61	31	21	16	13	11	7	6	5	4	3	2

   (a) Scan image   (b) Reference points   (c) Distorted image   (d) Control points   (e) Synthetic data

Fig. 8. Synthesize distorted document image. We uniformly sample a set of (b) reference points on (a) scanned document image, and then perform geometric deformation on them to get (c) distorted image and (d) control points. (e) Synthetic data consists of distorted image, reference points and control points.

4.2 Results

We train network on synthetic data and test it on the benchmark dataset used by Ma et al. [9] which has various real-world distorted document images. Compared with previous method, control points rectify distortions while removing background. In addition, the final backward mapping obtained by interpolation from sparse control points can effectively avoid local pixel jitter. As shown in Fig. 9, the pixel-wise regression methods [3,9] pay attention to the global and local relations, which often ignore the correction of document edge area. Although the multi-task method [21] reduces edge blur by adding foreground segmentation tasks, the edge of most rectified image was still not neat enough. Our proposal addresses the difficulties in better balancing global and local. Control points focus on the global and interpolation method refines the local. These make our method simpler and more flexible while ensuring the correction effect.

Fig. 9. Results on the Ma et al. [9] **benchmark dataset.** Col 1: Original distorted images, Col 2: Results of Ma et al. [9], Col 3: Results of Das and Ma et al. [3], Col 4: Results of Xie et al. [21], Col 5: Position of control points, Col 6: Results of our method. In the first two Row, our method uses 11×16 vertices to rectify distortion. The last three Row use 31×31 vertices.

We use Multi-Scale Structural Similarity (MS-SSIM) [20] and Local Distortion (LD) [22] for quantitative evaluation. MS-SSIM evaluates the global similarity between the rectified image and scanned image in multi-scale. LD computes the local metric by using SIFT flow between the rectified image and the ground truth image. The quantitative comparisons between MS-SSIM and LD are shown

Table 3. A comparison of different real-world distorted document images was made on the Ma et al. [9] benchmark dataset.

Method	MS-SSIM ↑	LD ↓
Ma et al. [9]	0.41	14.08
Liu et al. [8]	0.4491	12.06
Das and Ma et al. [3]	0.4692	8.98
Xie et al. [21]	0.4361	**8.50**
Our	**0.4769**	9.03

in Table 3. Our method demonstrates state-of-the-art performance in the quantitative metric of global similarity, and slightly weaker than the best method in local metric.

Table 4. Compare interpolation method and number of control points on the Ma et al. [9] benchmark dataset.

Interpolation	Vertices	MS-SSIM ↑	LD ↓	Time
TPS	11×16	**0.4769**	**9.03**	330 ms
TPS	4×6	0.4694	9.19	61 ms
TPS	16	0.4638	9.58	46 ms
Linear	31×31	0.4757	9.08	**25 ms**
Linear	16×16	0.4757	9.09	25ms

To trade-off between computational complexity and rectification performance, we compare different interpolation method and the number of vertices as shown in Table 4. Thin plate splines (TPS) are a spline-based technique for data interpolation and smoothing. When we use TPS interpolation, dense control points can better rectify complex distortions, but it also increases the amount of calculation. In contrast, linear interpolation requires less calculation, but the effect is slightly worse. We run our network and post-processing on a NVIDIA TITAN X GPU which processes 2 input images per batch and Intel(R) Xeon(R) CPU E5-2650 v4 which rectifies distorted image by using Linear interpolation in multiprocessing, respectively. Our implementation takes around 0.025 s to process a 992×992 image.

5   Conclusion

In this paper, we propose a novel approach using control points to rectifying distorted document image. Different from pixel-wise regression, our method exploits the encoder architecture to predict control points and reference points, so that the complexity of the neural network is reduced greatly. Control points facilitates the interaction with people to change the resolution of the rectified image, choose the number of vertices and adjusts the sub-optimal vertices, which are more controllable and practical. Furthermore, our method is more flexible in selecting post-processing methods and the number of vertices. The control points can also be used as a preprocessing step to realize semi-automatic distorted document image annotation with some further control point editing by humans. Although our approach has better tradeoff between computational complexity and rectification performance, further research is needed to explore more lightweight and efficient models.

Acknowledgements. This work has been supported by the National Key Research and Development Program Grant 2020AAA0109702, the National Natural Science Foundation of China (NSFC) grants 61733007, 61721004.

References

1. Brown, M.S., Tsoi, Y.C.: Geometric and shading correction for images of printed materials using boundary. IEEE Trans. Image Process. **15**(6), 1544–1554 (2006)
2. Courteille, F., Crouzil, A., Durou, J.D., Gurdjos, P.: Shape from shading for the digitization of curved documents. Mach. Vision Appl. **18**(5), 301–316 (2007)
3. Das, S., Ma, K., Shu, Z., Samaras, D., Shilkrot, R.: DewarpNet: single-image document unwarping with stacked 3D and 2D regression networks. In: Proceedings of the IEEE International Conference on Computer Vision, pp. 131–140 (2019)
4. He, Y., Pan, P., Xie, S., Sun, J., Naoi, S.: A book dewarping system by boundary-based 3D surface reconstruction. In: 2013 12th International Conference on Document Analysis and Recognition, pp. 403–407. IEEE (2013)
5. Li, X., Zhang, B., Liao, J., Sander, P.V.: Document rectification and illumination correction using a patch-based CNN. ACM Trans. Graphics **38**(6), 1–11 (2019)
6. Liu, C., Zhang, Y., Wang, B., Ding, X.: Restoring camera-captured distorted document images. Int. J. Doc. Anal. Recogn. **18**(2), 111–124 (2015)
7. Liu, W., et al.: SSD: single shot multibox detector. In: Leibe, B., Matas, J., Sebe, N., Welling, M. (eds.) ECCV 2016. LNCS, vol. 9905, pp. 21–37. Springer, Cham (2016). https://doi.org/10.1007/978-3-319-46448-0_2
8. Liu, X., Meng, G., Fan, B., Xiang, S., Pan, C.: Geometric rectification of document images using adversarial gated unwarping network. Pattern Recogn. **108**, 107576 (2020)
9. Ma, K., Shu, Z., Bai, X., Wang, J., Samaras, D.: DocUNet: document image unwarping via a stacked U-Net. In: Proceedings of the IEEE Conference on Computer Vision and Pattern Recognition, pp. 4700–4709 (2018)
10. Markovitz, A., Lavi, I., Perel, O., Mazor, S., Litman, R.: Can you read me now? Content aware rectification using angle supervision. In: Vedaldi, A., Bischof, H., Brox, T., Frahm, J.-M. (eds.) ECCV 2020. LNCS, vol. 12357, pp. 208–223. Springer, Cham (2020). https://doi.org/10.1007/978-3-030-58610-2_13
11. Meijering, E.: A chronology of interpolation: from ancient astronomy to modern signal and image processing. Proc. IEEE **90**(3), 319–342 (2002)
12. Ramanna, V., Bukhari, S.S., Dengel, A.: Document image dewarping using deep learning. In: International Conference on Pattern Recognition Applications and Methods (2019)

13. Ren, S., He, K., Girshick, R., Sun, J.: Faster R-CNN: towards real-time object detection with region proposal networks. In: Neural Information Processing Systems (2015)
14. Sorkine, O.: Laplacian mesh processing. In: Eurographics (STARs), p. 29 (2005)
15. Stamatopoulos, N., Gatos, B., Pratikakis, I., Perantonis, S.J.: Goal-oriented rectification of camera-based document images. IEEE Trans. Image Process. **20**(4), 910–920 (2010)
16. Tian, Y., Narasimhan, S.G.: Rectification and 3D reconstruction of curved document images. In: Proceedings of the IEEE Conference on Computer Vision and Pattern Recognition, pp. 377–384. IEEE (2011)
17. Tsoi, Y.C., Brown, M.S.: Multi-view document rectification using boundary. In: Proceedings of the IEEE Conference on Computer Vision and Pattern Recognition, pp. 1–8. IEEE (2007)
18. Wada, T., Ukida, H., Matsuyama, T.: Shape from shading with interreflections under a proximal light source: Distortion-free copying of an unfolded book. Int. J. Comput. Vision **24**(2), 125–135 (1997)
19. Wang, N., Zhang, Y., Li, Z., Fu, Y., Liu, W., Jiang, Y.-G.: Pixel2Mesh: generating 3D mesh models from single RGB images. In: Ferrari, V., Hebert, M., Sminchisescu, C., Weiss, Y. (eds.) ECCV 2018. LNCS, vol. 11215, pp. 55–71. Springer, Cham (2018). https://doi.org/10.1007/978-3-030-01252-6_4
20. Wang, Z., Simoncelli, E.P., Bovik, A.C.: Multiscale structural similarity for image quality assessment. In: The Thrity-Seventh Asilomar Conference on Signals, Systems & Computers, 2003, vol. 2, pp. 1398–1402. IEEE (2003)
21. Xie, G.-W., Yin, F., Zhang, X.-Y., Liu, C.-L.: Dewarping document image by displacement flow estimation with fully convolutional network. In: Bai, X., Karatzas, D., Lopresti, D. (eds.) DAS 2020. LNCS, vol. 12116, pp. 131–144. Springer, Cham (2020). https://doi.org/10.1007/978-3-030-57058-3_10
22. You, S., Matsushita, Y., Sinha, S., Bou, Y., Ikeuchi, K.: Multiview rectification of folded documents. IEEE Trans. Pattern Anal. Mach. Intell. **40**(2), 505–511 (2017)
23. Zhang, L., Yip, A.M., Brown, M.S., Tan, C.L.: A unified framework for document restoration using inpainting and shape-from-shading. Pattern Recogn. **42**(11), 2961–2978 (2009)

Unknown-Box Approximation to Improve Optical Character Recognition Performance

Ayantha Randika[1]([✉]), Nilanjan Ray[1], Xiao Xiao[2], and Allegra Latimer[2]

[1] University of Alberta, 116 St and 85 Ave, Edmonton, AB, Canada
{ponnampe,nray1}@ualberta.ca
[2] Intuit Inc., 2700 Coast Ave, Mountain View, CA, USA
{Xiao_Xiao,Allegra_Latimer}@intuit.com

Abstract. Optical character recognition (OCR) is a widely used pattern recognition application in numerous domains. There are several feature-rich, general-purpose OCR solutions available for consumers, which can provide moderate to excellent accuracy levels. However, accuracy can diminish with difficult and uncommon document domains. Preprocessing of document images can be used to minimize the effect of domain shift. In this paper, a novel approach is presented for creating a customized preprocessor for a given OCR engine. Unlike the previous OCR agnostic preprocessing techniques, the proposed approach approximates the gradient of a particular OCR engine to train a preprocessor module. Experiments with two datasets and two OCR engines show that the presented preprocessor is able to improve the accuracy of the OCR up to 46% from the baseline by applying pixel-level manipulations to the document image. The implementation of the proposed method and the enhanced public datasets are available for download (https://github.com/paarandika/Gradient-Approx-to-improve-OCR).

Keywords: OCR · Gradient approximation · Preprocessing · Optical character recognition

1 Introduction

Optical Character Recognition (OCR) is the process of extracting printed or handwritten text from an image into a computer understandable form. Recent OCR engines based on deep learning have achieved significant improvements in both accuracy, and efficiency [5]. Thanks to these improvements, several commercial OCR solutions have been developed to automate document handling tasks in various fields, including cloud-based OCR services from big names in cloud technologies. This new generation of OCR solutions come with all the benefits of the Software as a Service (SaaS) delivery model where the consumer does not have to think about the hardware and maintenance.

One drawback of commercial OCR solutions is that they are often general-purpose by design, while different domains may have different document types

© Springer Nature Switzerland AG 2021
J. Lladós et al. (Eds.): ICDAR 2021, LNCS 12821, pp. 481–496, 2021.
https://doi.org/10.1007/978-3-030-86549-8_31

with unique aberrations and degradations, which can hinder OCR performance. While it is theoretically possible to either train an open-source OCR or to build a new OCR engine from scratch to accommodate the domain shift, these solutions are not realistic in practice for OCR users in the industry, given the amount of resources required for training and the potential degradation of efficiency. Therefore, the more viable approach is to enhance a commercial OCR solution with pre- or post-processing to improve accuracy.

The output of OCR is a text string and can be post-processed using natural language-based techniques [34] or by leveraging outputs from multiple OCR engines [27]. This is beyond the scope of this work. Instead, we focus on preprocessing to enhance input quality prior to the OCR, which usually occurs in the image domain, though earlier hardware-based preprocessors have worked in the signal domain [10,36]. Image preprocessing for OCR includes image binarization, background elimination, noise removal, illumination correction and correcting geometric deformations, all of which aim to produce a simpler and more uniform image that reduces the burden on the OCR engine [31].

In this work, we propose a preprocessing solution which performs pixel-level manipulations that can be tweaked to accommodate any OCR engine, including commercial 'unknown-box' OCRs, in which the components are not open-sourced or known to us. To demonstrate the ability of our preprocessor, we use Point of Sales (POS) receipts, which often have poor printing and ink quality, and therefore renders a challenging task for OCR software [12]. Additionally, we include the VGG synthetic word dataset [14] to establish the generality of our solution. Section 2 discusses existing preprocessing techniques used to improve OCR performance. Section 3 presents our approach to the problem, and Sect. 4 contains the implementation and evaluation of our method. Section 5 discusses the experiment results, and Sect. 6 concludes the article with future work.

2 Background

Image binarization, which converts a grayscale image into black and white, is one of the most widely used preprocessing techniques for OCR since the 1980s. Otsu [24] is a popular binarization method, which uses the grayscale histogram to find a threshold value. Object Attribute Thresholding [19] uses grayscale and run-length histograms for unconstrained document images with complex backgrounds. O'Gorman proposed a global thresholding method using local connectivity details [23]. Similarly, a five-step document binarization method [21] focuses on connectivity issues in characters. Sauvola and Pietikäinen proposed an adaptive binarization method [31] which classifies the image into subcomponents and assigns different threshold values to pixels based on the component types. The double thresholding method developed by Chen et al. uses edge information to generate a binary image [4]. Some more recent binarization techniques incorporate deep learning. For example, DeepOtsu [9] applies Convolutional Neural Networks (CNN) iteratively, while Vo et al. [35] adopt a CNN-based hierarchical architecture.

Skeletonization is another popular preprocessing technique [15], which aims to reduce the dimensions of an object [29]. In the context of characters, skeletonization reduces the stroke thickness to 1-D curves. Similar to binarization, there are several existing methods for Skeletonization [29], some of which are specifically designed for OCR [15]. Other preprocessing approaches exist in addition to binarization and skeletonization. Bieniecki et al. proposed methods to correct geometrical deformations in document images [2]. An independent component analysis-based method has been developed for handheld images of inscriptions [6]. Harraj and Raissouni combined different image enhancement techniques with Otsu binarization [8]. Deep learning-based Super-Resolution (SR) is employed in [16] and [26]. Sporici et al. presented a CNN-based preprocessing method where convolution kernels are generated using Reinforcement Learning (RL) [33].

Different preprocessing methods address different shortcomings in incoming images for OCR engines. Binarization and skeletonization focus on creating simpler images, while techniques such as super-resolution attempt to add more details. Geometrical correction methods create uniform characters in shape and orientation. The shared goal of these different methods is to produce output images which the OCR engine is more comfortable or familiar with. However, OCR engines have different underlying architectures and mechanisms, which may lead to different expectations for incoming images or tolerance for various defects. Therefore, we hypothesize that a preprocessor optimized for a given OCR engine would be able to produce a better approximation of the optimal images expected by this specific engine, leading to higher OCR accuracy compared to traditional generic preprocessing methods.

3 Proposed Method

The straightforward way to optimize the parameters of a preprocessor, which manipulates images to approximate the optimal input distribution for the OCR, is to use the optimal input distribution as ground truth. However, such intermediate training data are rarely available. Another way of optimizing the parameters of the preprocessor is by calculating the gradient of the OCR error and propagating it to the preprocessor using the backpropagation algorithm and updating the preprocessor parameters to minimize the error. However, the problem with this approach is that the internal mechanisms of the proprietary OCR engines cannot be accessed. Further, OCR engines may contain non-differentiable components for which gradient computation or estimation may not be possible. Therefore, the OCR engine needs to be treated as an unknown-box in the training pipeline.

Since there is no direct way of calculating the gradient of an unknown-box, to utilize the backpropagation algorithm to train the preprocessor component, the gradient of the error produced by the input needs to be estimated. A popular method for estimating the gradient is using the Score-Function Estimator (SFE) (1) [20]:

$$\nabla_\theta \mathbb{E}_{x \sim p_\theta}[f(x)] = \mathbb{E}_{x \sim p_\theta}[f(x)\nabla_\theta \log p_\theta(x)], \tag{1}$$

Algorithm 1: Gradient approximation by NN

---

input: $\sigma, S, \{Training\ Images, Ground\ Truths\}$
for $I, p_{gt} \in \{Training\ Images,\ Ground\ Truths\}$ **do**
$\quad$ $g = Preprocessor(I)$;
$\quad$ intialize s to 0;
$\quad$ **while** $s < S$ **do**
$\qquad$ sample $\epsilon_s \sim \mathcal{N}(0, \sigma)$;
$\qquad$ $\mathcal{M}_s = \mathcal{M}(Approximator(g + \epsilon_s), OCR(g + \epsilon_s))$;
$\qquad$ $s = s + 1$;
$\quad$ $min_\phi \sum_s \mathcal{M}_s$;
$\quad$ $min_\psi \mathcal{Q}(Approximator(g), p_{gt})$;

---

where $f(x)$ is the unknown-box function and p_θ is the input distribution parameterized by θ. In RL, (1) was developed into the REINFORCE algorithm [37]. Even though this estimator is unbiased, it can have a high variance, especially in higher dimensions [20]. By re-parameterizing x as $x = \theta + \sigma\epsilon$ where $\epsilon \sim \mathcal{N}(0, \mathcal{I})$, the score function can be written in the following format [30]:

$$\nabla_\theta \mathbb{E}_{\epsilon \sim \mathcal{N}(0, \mathcal{I})}[f(\theta + \sigma\epsilon)] = \frac{1}{\sigma} \mathbb{E}_{\epsilon \sim \mathcal{N}(0, \mathcal{I})}[f(\theta + \sigma\epsilon)\epsilon]. \tag{2}$$

In the context of medical image analysis, Nguyen and Ray proposed EDPCNN [22] based on the universal function and gradient approximation property of Neural Networks (NN) [11]. EDPCNN estimates the gradient of a non-differentiable dynamic programming (DP) algorithm by training a NN as an approximation for the DP algorithm and uses the approximated gradients to train the learning-based component in the processing pipeline. During inference, the approximating NN is removed, and only the non-differentiable original function is used. Jacovi et al. also proposed the same method under the name 'Estinet' and demonstrated its performance with tasks such as answering 'greater than or less than' questions written in natural language [13].

In this work, we use the online algorithm proposed by [22] to estimate the gradient of the OCR component. Our approach is detailed in Algorithm 1, which consists of two loops. In the 'inner loop', noise is added to 'jitter' the input to the OCR, then the error $\mathcal{M}$ between OCR and the approximator is accumulated as $\sum_s \mathcal{M}_s$. The 'outer loop' optimizes approximator parameters (ϕ) by minimizing the accumulated error $\sum_s \mathcal{M}_s$ and freezing the parameters (ψ) of the preprocessor. The other minimization in the 'outer loop' optimizes the parameters (ψ) of the preprocessor model while the approximator parameters (ϕ) are frozen. For this second minimization, the error $\mathcal{Q}$ is calculated by comparing the approximator output with the ground truth. Note that it is an alternating optimization between the preprocessor and the approximating NN. In its basic components, this algorithm is similar to the DDPG [18] algorithm in RL.

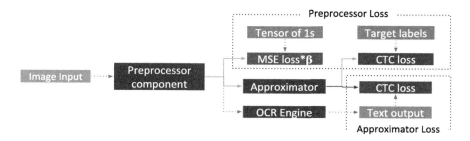

Fig. 1. Overview of the proposed training pipeline. The yellow and purple arrows represent the computation paths equipped with backpropagation of preprocessor loss and approximator loss, respectively. (Color figure online)

4 Implementation and Experiments

4.1 Implementation

NN-based Approximation Method The overview of the proposed training pipeline based on Algorithm 1 is depicted in Fig. 1. The loss function used for optimization of approximator parameters is Connectionist Temporal Classification (CTC) loss [7], which appears as $\mathcal{M}$ in Algorithm 1. In addition, mean square error (MSE) loss is calculated by comparing the preprocessor output with a 2-dimensional tensor of ones: $J_{m \times n}$ where n and m are the dimensions of the input image. In this context, it represents a completely white image. Sum of the CTC loss and the MSE loss is used as the loss function (3) to optimize the preprocessor parameters (ψ) in Algorithm 1.

$$\mathcal{Q} = CTC(Approximator(g), p_{gt}) + \beta * MSE(g, J_{m \times n}). \tag{3}$$

In (3), $g = Preprocessor(Image)$ and p_{gt} is the associated ground truth text for the input $Image$. MSE loss component in the loss function nudges the preprocessor to produce a white image. A completely white image implies no output or incorrect output from the approximator, which increases CTC error. We hypothesise that this composite loss function will reduce background clutter while preserving the characters. β acts as a hyperparameter to control the effect of MSE loss.

The architecture of the preprocessor component is based on the U-Net [28] architecture. We use the U-Net variation [3] with added batch normalization layers. The number of input and output channels is changed to one since we work with greyscale images. The sigmoid function is used as the final activation function to maintain output values in the range $[0, 1]$. Convolutional Recurrent Neural Network (CRNN) [32] model is used as the approximator to simulate the text recognition of OCR. CRNN model is a simple model, which can avoid gradient vanishing problems when training end-to-end with the preprocessor. An OCR contains different components for text detection, segmentation and recognition. However, CRNN only supports text recognition.

Algorithm 2: Gradient approximation by SFE

Function GetGradients(*Image*, p_{gt}, σ, n):

$\quad g = Preprocessor(Image)$;

$\quad$ sample $\epsilon_1, \ldots \epsilon_n \sim \mathcal{N}(0, \mathcal{I})$;

$\quad \epsilon_{n+i} = -\epsilon_i$ for $i = 1, \ldots n$;

$\quad \mathcal{L}_i = \mathcal{L}(OCR(g + \sigma\epsilon_i), p_{gt})$ for $i = 1, \ldots 2n$;

$\quad gradient = \frac{1}{2n\sigma} \sum\limits_{i=1}^{2n} \mathcal{L}_i\epsilon_i$;

$\quad$ **return** *gradient*;

SFE-Based Method. Since the proposed method is based on gradient estimation, we implement the SFE method (2) in Algorithm 2 that can be viewed as a gradient estimation alternative. Unlike the probability distribution output of CRNN in the NN-based method, OCR outputs a text string. CTC loss is not intended for direct string comparison, and there is no need for a differentiable loss function in SFE; hence, Levenshtein distance [17] is used as the loss function, represented by $\mathcal{L}$ in Algorithm 2. This loss is similar to (4), an evaluation metric discussed in Sect. 4.2, except for the multiplication by a constant. n perturbations of ϵ are sampled from the normal distribution, and the 'mirrored sampling' is used to reduce the variance [30]. The generated $2n$ samples are sent to the OCR, and the resulting text is evaluated to produce the final loss. The same U-Net model is utilized as the preprocessor, and an adaptation of (3) is used as the compound loss to optimize it by replacing the CTC loss component with Levenshtein distance. The gradient of the Levenshtein distance component of the compound loss is calculated with Algorithm 2.

4.2   Datasets and Evaluation

Document samples from two different domains are used to evaluate the preprocessor. The dataset referred to as 'POS dataset' consists of POS receipt images from three datasets: ICDAR SOIR competition dataset [12], Findit fraud detection dataset [1] and CORD dataset[25]. Since the Findit dataset is not intended as an OCR dataset, ground truth text and word bounding boxes are generated using Google Cloud Vision API[1] in combination with manual corrections. Additionally, it contains a fraction of the ICDAR SOIR competition dataset and the entire CORD dataset. Dataset images are patches extracted from POS receipts resized to have a width of 500 pixels and a maximum of 400 pixels height. The complete POS dataset contains 3676 train, 424 validation and 417 test images with approximately 90k text areas. The dataset referred to as the 'VGG' dataset is created by randomly selecting a subset of sample images from the VGG synthetic word dataset [14]. The VGG dataset contains 50k train, 5k validation and 5k test images, each containing a word.

---

[1] https://cloud.google.com/vision/.

Two free and opensource OCR engines: Tesseract[2], a popular well established opensource OCR engine and EasyOCR[3], a newer opensource OCR engine are used to train the preprocessor. Throughout the study, both OCR engines are treated as unknown-box components. Two metrics are used to measure the OCR performance variation with preprocessing. Since word-level ground truth values are available for both datasets, word-level accuracy is measured by the percentage of words matched with the ground truth. As a character-level measurement, Levenshtein distance [17] based Character Error Rate (CER) is used. CER is defined as:

$$CER = 100 * (i + s + d)/m, \qquad (4)$$

where i is the number of insertions, s is the number of substitutions, and d is the number of deletions done to the prediction to get the ground truth text. m is the number of characters in the ground truth. The OCR engine's performance is measured with original images to establish a baseline, and then the OCR engine is rerun with preprocessed images, and the performance is measured again.

4.3   Training Details

A set of preprocessors are trained for fifty epochs according to the Algorithm 1. Learning rates used for preprocessor and approximator are 5×10^{-5} and 10^{-4} respectively. The approximator is pre-trained with the dataset for fifty epochs before inserting it into the training pipeline to avoid the cold-start problem. Adam optimizer is used in both training paradigms. We maintained $\beta = 1$ in (3) and $S = 2$ in Algorithm 1. σ in Algorithm 1 is randomly selected from a uniform distribution containing 0, 0.01, 0.02, 0.03, 0.04 and 0.05. Small standard deviation values for noise are used since the images are represented by tensors in the range $[0.0, 1.0]$. POS dataset images are padded to feed into U-Net. To feed into CRNN, words are cropped using bounding-box values and padded to the size of 128×32 pixels. Since each VGG sample only contains a single word, the size of 128×32 pixels is used for both components. On average, it took $19\,h$ to train a preprocessor with an Nvidia RTX 2080 GPU, while EasyOCR was running on a different GPU.

A different set of preprocessors (referred to as SFE-preprocessor here onwards) are trained with Algorithm 2 using Adam optimizer and a learning rate of 5×10^{-5}. To avoid mirrored gradients cancelling each other at the start, the preprocessor is first trained to output the same image as the input. OCR engine output does not vary much with the perturbations of ϵ if the σ is too small. If the σ is too large, every perturbation produces empty text. Both of these scenarios lead to 0 gradients. By trial and error, a constant σ of 0.05 is used. The speed bottleneck of the pipeline is at the OCR. To reduce the training time, $n = 5$ is used. Similar to the previous training scenario, we maintained $\beta = 1$ (refer to (3)), and the same image processing techniques are used for both

---

[2] https://github.com/tesseract-ocr/tesseract.
[3] https://github.com/JaidedAI/EasyOCR.

datasets. On average, it took 79 h to train an SFE-preprocessor with an Nvidia RTX 2080 GPU while EasyOCR was running on a different GPU.

5   Results and Discussion

Results in Table 1 and Table 2 show that the performances of both OCR engines are improved by preprocessing, and NN approximation-based preprocessors outperform SFE-preprocessors. To some extent, the poor performance of the SFE-preprocessor might be attributed to the small number of perturbations. However, when considering computational time, using large n appears unpragmatic. In both cases, gradient approximation has proved to work, and it appears that the NN approximation-based method handles the image domain better than SFE.

Figure 2 and Fig. 3 depict sample output images produced by the NN approximation based preprocessors. It can be observed that shadows, complex backgrounds and noise are suppressed to improve the contrast of the text. This provides a clear advantage to the OCR engine. The bleaching of darker text in images can be the effect of MSE loss. Preprocessors have introduced mutations to the characters' shapes, which is more significant in the POS dataset than in the VGG dataset. In the POS dataset, characters have gained more fluid and continuous strokes, especially low-resolution characters printed with visible 'dots'. Additionally, the 'Tesseract preprocessor' was able to do some level of skew corrections to the text in VGG images. Based on the accuracy improvements and reduction of CER, it can be concluded that these mutations provide extra guidance in character recognition.

Our approach does not require clean document images as labels. The preprocessor is trained directly with the text output omitting the need for intermediate clean data. Due to the lack of clean ground truth images for our dataset, to compare with learning-based methods, we used pre-trained weights. In Table 3, five binarization methods and one SR method are compared against our preprocessor on the POS dataset. Vo [35], DeepOtsu [9] and robin[4] are originally trained with high resolutions images. Therefore, to mitigate the effect of low-resolution, POS dataset images are enlarged by a factor of 2 before binarization and reduced back to the original size before presenting to the OCR engine. Similarly, with SR method [26], the images are enlarged by a factor of 2 and presented the same enlarged images to the OCR. According to the results, our method has outperformed all six methods compared. Sauvola, robin, DeepOtsu and SR methods were able to increase the Tesseract accuracy, and the SR method shows the largest improvement. With robin, CER has increased despite the slight accuracy gain. Only the SR method improved EasyOCR accuracy. Furthermore, severe loss of details can be observed in the outputs produced by these methods (Fig. 4).

---

[4] https://github.com/masyagin1998/robin.

Table 1. Accuracy and CER before and after proposed NN approximation-based pre-processing.

OCR used for training and testing	Dataset	Without preprocessing		With preprocessing			
		Accuracy ↑	CER ↓	Accuracy ↑	CER ↓	Accuracy gain	CER reduction
Tesseract	POS	54.51%	26.33	83.36%	8.68	28.86%	17.66
Tesseract	VGG	18.52%	64.40	64.94%	14.70	46.42%	49.70
EasyOCR	POS	29.69%	44.27	67.97%	16.46	38.27%	27.81
EasyOCR	VGG	44.80%	26.90	57.48%	17.15	12.68%	9.75

Table 2. Proposed NN approximation-based vs. SFE-based preprocessors.

OCR used for training and testing	Dataset	NN-based preprocessing		SFE-based preprocessing	
		Accuracy ↑	CER ↓	Accuracy ↑	CER ↓
Tesseract	POS	83.36%	8.68	69.17%	16.62
Tesseract	VGG	64.94%	14.70	27.76%	52.98
EasyOCR	POS	67.97%	16.46	46.63%	28.13
EasyOCR	VGG	57.48%	17.15	47.02%	24.69

Table 3. OCR engine accuracy on POS dataset: comparison with other preprocessing methods.

Preprocessing method	Tesseract		EasyOCR	
	Accuracy ↑	CER ↓	Accuracy ↑	CER ↓
OCR *(no preprocessing)*	54.51%	26.33	29.69%	44.27
Otsu [24]	50.98%	29.84	16.96%	52.30
Sauvola [31]	55.39%	25.20	20.19%	48.49
Vo [35]	51.72%	31.81	21.95%	50.07
robin	57.18%	28.45	27.59%	43.55
DeepOtsu [9]	62.47%	21.88	26.33%	42.63
SR [26]	67.13%	15.90	37.51%	31.11
Ours	**83.36%**	**8.68**	**67.97%**	**16.46**

Table 4. OCR accuracy when trained and tested with different engines.

Dataset	OCR used for training	OCR used for testing	Test accuracy ↑	Test CER ↓
POS dataset	Tesseract	EasyOCR	40.44%	31.28
VGG dataset	Tesseract	EasyOCR	47.14%	21.77
POS dataset	EasyOCR	Tesseract	60.94%	21.81
VGG dataset	EasyOCR	Tesseract	21.64%	51.96

Fig. 2. Sample inputs and outputs from the VGG test dataset. Column 1: input images, Column 2: image output produced by the model trained with Tesseract, Column 3: image output produced by the model trained with EasyOCR.

To test the effect of individualized preprocessing, preprocessors trained for one OCR engine is tested with a different OCR engine (Table 4). After the preprocessing, the OCR engine yields better accuracy than the baseline in the test. However, the accuracy gain is lower than the accuracy obtained by the same OCR engine the preprocessor has trained with. Therefore, it is reasonable to assume that preprocessing has added OCR engine specific artifacts or mutations to the image to improve recognition. This behaviour confirms that different OCR engines expect inputs to be optimized differently, thus individualized preprocessing serves better than generic preprocessing. Additionally, Fig. 5 shows that the CRNN model has been able to well approximate the recognition capability of the OCR with different dataset sizes. However, given that the CRNN model can only recognize text but does not have text detection capabilities, it cannot fully replace the OCR engine.

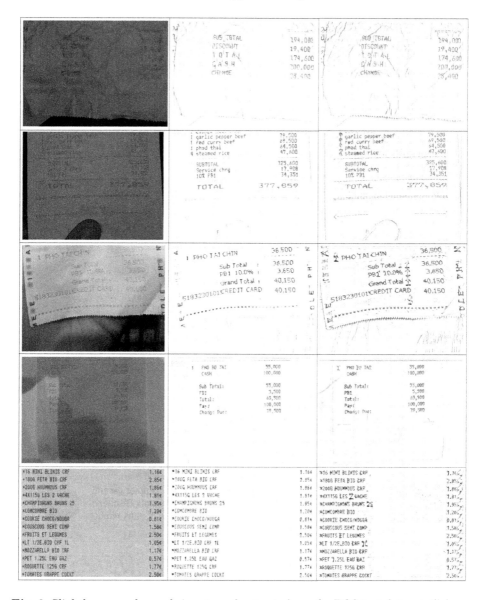

Fig. 3. Slightly cropped sample inputs and outputs from the POS test dataset. Column one contains the sample input images. Columns two and three contain outputs produced by the proposed NN-based method trained with Tesseract and EasyOCR, respectively.

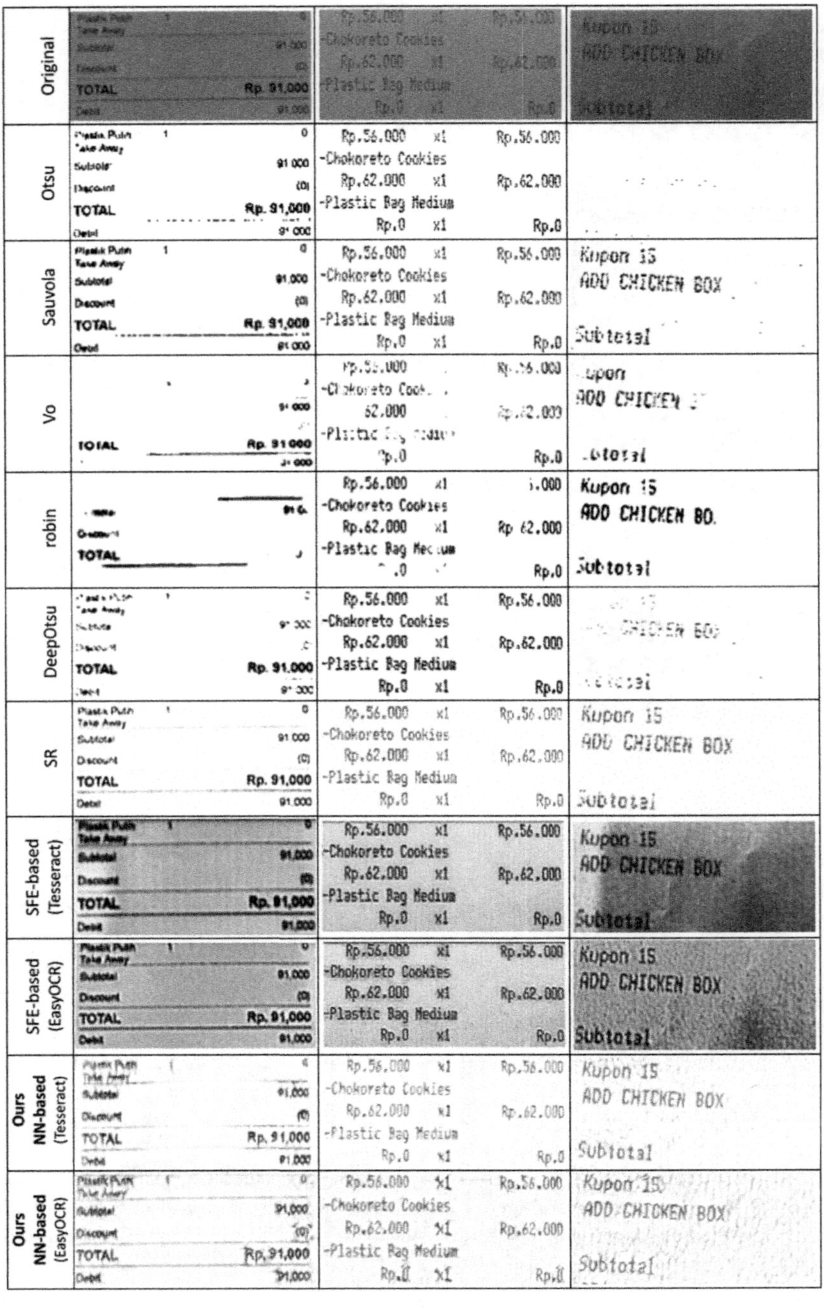

Fig. 4. Cropped input (POS test dataset) and output images from methods considered in Table 3 and our models. The images in three columns have average CER reductions of 6.5, 61 and 73.5, respectively, from left to right. CER is based on our NN-based models.

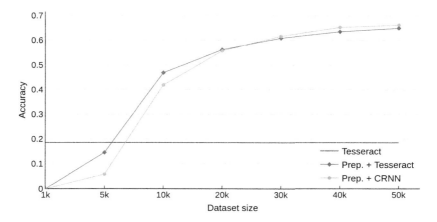

Fig. 5. The test set accuracy of OCR engine and CRNN model with different training sizes of VGG dataset. Tesseract accuracy without preprocessing is added for reference.

6 Conclusion

In this study, we have proposed a novel method to create an individualized preprocessor to improve the accuracy of existing OCR solutions. Two different approaches to the proposed method were tested. One of them, the NN approximation-based training pipeline, was able to improve the performance of two OCR engines on two different datasets, which demonstrates the power and versatility of the proposed preprocessing method based on gradient approximation for OCR tasks. The recognized downside of this approach is the added complexity in the training regimen and the increased number of hyperparameters, and the need for ground truth text with bounding boxes. Different approximation models can be considered to reduce the need for text bounding boxes in the ground truth as future work. Furthermore, different preprocessor architectures can be investigated in the same setting to see if they yield better results.

Acknowledgements. Authors acknowledge funding support from Intuit Inc. and the University of Alberta, Canada.

References

1. Artaud, C., Sidère, N., Doucet, A., Ogier, J., Yooz, V.P.D.: Find it! fraud detection contest report. In: ICPR, pp. 13–18 (2018). https://doi.org/10.1109/ICPR.2018. 8545428
2. Bieniecki, W., Grabowski, S., Rozenberg, W.: Image preprocessing for improving OCR accuracy. In: 2007 International Conference on Perspective Technologies and Methods in MEMS Design, pp. 75–80 (2007). https://doi.org/10.1109/ MEMSTECH.2007.4283429

3. Buda, M., Saha, A., Mazurowski, M.A.: Association of genomic subtypes of lower-grade gliomas with shape features automatically extracted by a deep learning algorithm. Comput. Biol. Med. **109**, 218–225 (2019). https://doi.org/10.1016/j.compbiomed.2019.05.002

4. Chen, Q., Sun, Q.S, Heng, P.A., Shen Xia, D.: A double-threshold image binarization method based on edge detector. Patt. Recogn. **41**, 1254–1267 (2008). https://doi.org/10.1016/j.patcog.2007.09.007

5. Chen, Y., Shao, Y.: Scene text recognition based on deep learning: a brief survey. In: ICCSN (2019)

6. Garain, U., Jain, A., Maity, A., Chanda, B.: Machine reading of camera-held low quality text images: an ICA-based image enhancement approach for improving OCR accuracy. In: ICPR, pp. 1–4 (2008). https://doi.org/10.1109/ICPR.2008.4761840

7. Graves, A., Fernández, S., Gomez, F., Schmidhuber, J.: Connectionist temporal classification: labelling unsegmented sequence data with recurrent neural networks. In: ACM International Conference Proceeding Series, vol. 148, pp. 369–376. ACM Press (2006). https://doi.org/10.1145/1143844.1143891

8. Harraj, A.E., Raissouni, N.: OCR accuracy improvement on document images through a novel pre-processing approach. Signal Image Process. Int. J. **6**, 01–18 (2015). https://doi.org/10.5121/sipij.2015.6401

9. He, S., Schomaker, L.: DeepOtsu: document enhancement and binarization using iterative deep learning. Patt. Recogn. **91**, 379–390 (2019). https://doi.org/10.1016/j.patcog.2019.01.025

10. Hicks, J.R., Eby Jr, J.C.: Signal processing techniques in commercially available high-speed optical character reading equipment. In: Tao, T.F. (ed.) Real-Time Signal Processing II, vol. 0180, pp. 205–216. SPIE (1979). https://doi.org/10.1117/12.957332

11. Hornik, K., Stinchcombe, M.B., White, H.: Universal approximation of an unknown mapping and its derivatives using multilayer feedforward networks. Neural Netw. **3**, 551–560 (1990)

12. Huang, Z., et al.: ICDAR 2019 competition on scanned receipt OCR and information extraction. In: ICDAR, pp. 1516–1520 (2019). https://doi.org/10.1109/ICDAR.2019.00244

13. Jacovi, A., et al.: Neural network gradient-based learning of black-box function interfaces. In: ICLR. arXiv (2019)

14. Jaderberg, M., Simonyan, K., Vedaldi, A., Zisserman, A.: Synthetic data and artificial neural networks for natural scene text recognition. In: Workshop on Deep Learning, NIPS (2014)

15. Lam, L., Suen, C.Y.: An evaluation of parallel thinning algorithms for character recognition. IEEE Trans. Pattern Anal. Mach. Intell. **17**, 914–919 (1995). https://doi.org/10.1109/34.406659

16. Lat, A., Jawahar, C.V.: Enhancing OCR accuracy with super resolution. In: ICPR, pp. 3162–3167 (2018). https://doi.org/10.1109/ICPR.2018.8545609

17. Levenshtein, V.I.: Binary codes capable of correcting deletions, insertions, and reversals. In: Soviet Physics. Doklady, vol. 10, pp. 707–710 (1965)

18. Lillicrap, T.P., et al.: Continuous control with deep reinforcement learning. arXiv preprint arXiv:1509.02971 (2015)

19. Liu, Y., Feinrich, R., Srihari, S.N.: An object attribute thresholding algorithm for document image binarization. In: ICDAR, pp. 278–281 (1993). https://doi.org/10.1109/ICDAR.1993.395732

20. Mohamed, S., Rosca, M., Figurnov, M., Mnih, A.: Monte Carlo gradient estimation in machine learning. J. Mach. Learn. Res. **21**, 1–63 (2019). http://arxiv.org/abs/1906.10652

21. Chang, M.S., Kang, S.M., Rho, W.S., Kim, H.G., Kim, D.J.: Improved binarization algorithm for document image by histogram and edge detection. In: ICDAR, vol. 2, pp. 636–639 (1995). https://doi.org/10.1109/ICDAR.1995.601976

22. Nguyen, N.M., Ray, N.: End-to-end learning of convolutional neural net and dynamic programming for left ventricle segmentation. In: Proceedings of Machine Learning Research, vol. 121, pp. 555–569. PMLR (2020)

23. Ogorman, L.: Binarization and multithresholding of document images using connectivity. CVGIP: Graph. Mod. Image Process. **56**, 494–506 (1994). https://doi.org/10.1006/cgip.1994.1044

24. Otsu, N.: A threshold selection method from gray-level histograms. IEEE Trans. Syst. Man Cybern. **9**, 62–66 (1979)

25. Park, S., et al.: CORD: a consolidated receipt dataset for post-OCR parsing. In: Workshop on Document Intelligence at NeurIPS 2019 (2019). https://openreview.net/forum?id=SJl3z659UH

26. Peng, X., Wang, C.: Building super-resolution image generator for OCR accuracy improvement. In: Bai, X., Karatzas, D., Lopresti, D. (eds.) DAS 2020. LNCS, vol. 12116, pp. 145–160. Springer, Cham (2020). https://doi.org/10.1007/978-3-030-57058-3_11

27. Reul, C., Springmann, U., Wick, C., Puppe, F.: Improving OCR accuracy on early printed books by utilizing cross fold training and voting. In: 2018 13th IAPR International Workshop on Document Analysis Systems, pp. 423–428 (2018).https://doi.org/10.1109/DAS.2018.30

28. Ronneberger, O., Fischer, P., Brox, T.: U-Net: convolutional networks for biomedical image segmentation. In: Navab, N., Hornegger, J., Wells, W.M., Frangi, A.F. (eds.) MICCAI 2015. LNCS, vol. 9351, pp. 234–241. Springer, Cham (2015). https://doi.org/10.1007/978-3-319-24574-4_28

29. Saha, P.K., Borgefors, G., di Baja, G.S.: Chapter 1 - skeletonization and its applications - a review. In: Saha, P.K., Borgefors, G., di Baja, G.S. (eds.) Skeletonization, pp. 3–42. Academic Press (2017). https://doi.org/10.1016/B978-0-08-101291-8.00002-X

30. Salimans, T., Ho, J., Chen, X., Sidor, S., Sutskever, I.: Evolution strategies as a scalable alternative to reinforcement learning (2017)

31. Sauvola, J., Pietikäinen, M.: Adaptive document image binarization. Patt. Recogn. **33**, 225–236 (2000). https://doi.org/10.1016/S0031-3203(99)00055-2

32. Shi, B., Bai, X., Yao, C.: An end-to-end trainable neural network for image-based sequence recognition and its application to scene text recognition. IEEE Trans. Patt. Anal. Mach. Intell. **39**, 2298–2304 (2017). https://doi.org/10.1109/TPAMI.2016.2646371

33. Sporici, D., Cuşnir, E., Boiangiu, C.A.: Improving the accuracy of tesseract 4.0 OCR engine using convolution-based preprocessing. Symmetry **12**, 715 (2020). https://doi.org/10.3390/SYM12050715

34. Thompson, P., McNaught, J., Ananiadou, S.: Customised OCR correction for historical medical text. In: 2015 Digital Heritage, vol. 1, pp. 35–42 (2015). https://doi.org/10.1109/DigitalHeritage.2015.7413829

35. Vo, Q.N., Kim, S.H., Yang, H.J., Lee, G.: Binarization of degraded document images based on hierarchical deep supervised network. Patt. Recogn. **74**, 568–586 (2018).https://doi.org/10.1016/j.patcog.2017.08.025

36. White, J.M., Rohrer, G.D.: Image thresholding for optical character recognition and other applications requiring character image extraction. IBM J. Res. Dev. **27**, 400–411 (1983). https://doi.org/10.1147/rd.274.0400
37. Williams, R.J.: Simple statistical gradient-following algorithms for connectionist reinforcement learning. Mach. Learn. **8**, 229–256 (1992). https://doi.org/10.1007/bf00992696

Document Domain Randomization for Deep Learning Document Layout Extraction

Meng Ling[1(✉)], Jian Chen[1], Torsten Möller[2], Petra Isenberg[3], Tobias Isenberg[3], Michael Sedlmair[4], Robert S. Laramee[5], Han-Wei Shen[1], Jian Wu[6], and C. Lee Giles[7]

[1] The Ohio State University, Columbus, USA
{ling.253,chen.8028,shen.94}@osu.edu
[2] University of Vienna, Wien, Austria
torsten.moeller@univie.ac.at
[3] Université Paris-Saclay, CNRS, Inria, LISN, Gif-sur-Yvette, France
{petra.isenberg,tobias.isenberg}@inria.fr
[4] University of Stuttgart, Stuttgart, Germany
michael.sedlmair@visus.uni-stuttgart.de
[5] University of Nottingham, Nottingham, UK
robert.laramee@nottingham.ac.uk
[6] Old Dominion University, Norfolk, USA
jwu@cs.odu.edu
[7] The Pennsylvania State University, State College, USA
clg20@psu.edu

Abstract. We present **d**ocument **d**omain **r**andomization (DDR), the first successful transfer of CNNs trained only on graphically rendered pseudo-paper pages to real-world document segmentation. DDR renders pseudo-document pages by modeling randomized textual and non-textual contents of interest, with user-defined layout and font styles to support joint learning of fine-grained classes. We demonstrate competitive results using our DDR approach to extract nine document classes from the benchmark CS-150 and papers published in two domains, namely annual meetings of Association for Computational Linguistics (ACL) and IEEE Visualization (VIS). We compare DDR to conditions of *style mismatch*, fewer or more *noisy* samples that are more easily obtained in the real world. We show that high-fidelity semantic information is not necessary to label semantic classes but style mismatch between train and test can lower model accuracy. Using smaller training samples had a slightly detrimental effect. Finally, network models still achieved high test accuracy when correct labels are diluted towards confusing labels; this behavior hold across several classes.

Keywords: Document domain randomization · Document layout · Deep neural network · Behavior analysis · Evaluation

1 Introduction

Fast, low-cost production of consistent and accurate training data enables us to use deep convolutional neural networks (CNN) to downstream document understanding

ⓒ Springer Nature Switzerland AG 2021
J. Lladós et al. (Eds.): ICDAR 2021, LNCS 12821, pp. 497–513, 2021.
https://doi.org/10.1007/978-3-030-86549-8_32

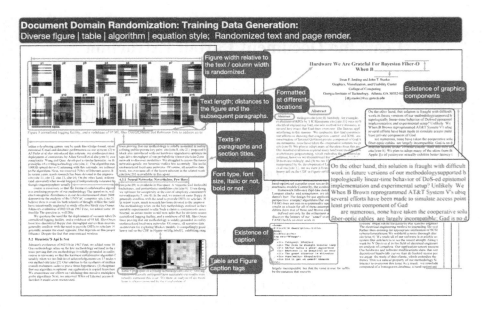

Fig. 1. Illustration of our document domain randomization (DDR) approach. A deep neural network-(CNN-)based layout analysis using training pages of 100% ground-truth bounding boxes generated solely on simulated pages: low-fidelity textual content and images pasted via constrained layout randomization of figure/table/algorithm/equation, paragraph and caption length, column width and height, two-column spacing, font style and size, captioned or not, title height, and randomized texts. Nine classes are used in the real document layout analysis with no additional training data: *abstract*, *algorithm*, *author*, *body-text*, *caption*, *equation*, *figure*, *table*, and *title*. Here the colored texts illustrate the semantic information; all text in the training data is black. (Color figure online)

[12,36,41,42]. However, carefully annotated data are difficult to obtain, especially for document layout tasks with large numbers of labels (time-consuming annotation) or with fine-grained classes (skilled annotation). In the scholarly document genre, a variety of document formats may not be attainable at scale thus causing imbalanced samples, since authors do not always follow section and format rules [9,27]. Different communities (e.g., computational linguistics vs. machine learning, or computer science vs. biology) use different structural and semantic organizations of sections and subsections. This diversity forces CNN paradigms (e.g., [35,42]) to use millions of training samples, sometimes with significant amounts of noise and unreliable annotation.

To overcome these training data production challenges, instead of the time-consuming manual annotating of real paper pages to curate training data, we generate pseudo-pages by randomizing page appearance and semantic content to be the "surrogate" of training data. We denote this as *document domain randomization* (*DDR*) (Fig. 1). DDR uses simulation-based training document generation, akin to domain randomization (DR) in robotics [19,33,39,40] and computer vision [14,28]. We randomize layout and font styles and semantics through graphical depictions in our page generator. The idea is that with enough page appearance randomization, the real page would appear to the model as just another variant. Since we know the bounding-box locations while rendering the training data, we can theoretically produce any number of highly accurate (100%) training samples following the test data styles. A key question is what styles and semantics can be randomized to let the models learn the essential features of interest on pseudo-pages so as to achieve comparable results for label detection in real article pages.

We address this question and study the behavior of DDR under numerous attribution settings to help guide the training data preparation. Our contributions are that we:

- **Create DDR—a simple, fast, and effective training page preparation method to significantly lower the cost of training data preparation.** We demonstrate that DDR achieves competitive performance on the commonly used benchmark CS-150 [10], ACL300 of Association for Computational Linguistics (ACL), and VIS300 of IEEE visualization (VIS) on extracting nine classes.
- **Cover real-world page styles using randomization to produce training samples that infer real-world document structures.** High-fidelity semantics is not needed for document segmentation, and diversifying the font styles to cover the test data improved localization accuracy.
- **Show that limiting the number of available training samples can lower detection accuracy.** We reduced the training samples by half each time and showed that accuracy drops at about the same rate for all classes.
- **Validated that CNN models remained reasonably accurate after training on noisy class labels of composed paper pages.** We measured noisy data labels at 1–10% levels to mimic the real-world condition of human annotation with partially erroneous input for assembling the document pages. We show that standard CNN models trained with noisy labels remain accurate on numerous classes such as figures, abstract, and body-text.

2   Related Work

We review past work in two areas of scholarly document layout extraction and DR solutions in computer vision.

2.1   Document Parts and Layout Analysis

PDF documents dominate scholarly publications. Recognizing the layout of this unstructured digital form is crucial in down-stream document understanding tasks [5,12,17,27,36]. Pioneering work in training data production has accelerated CNN-based document analysis and has achieved considerable real-world impact in digital libraries, such as CiteSeer[x] [5], Microsoft Academic [36], Google Scholar [13], Semantic Scholar [26], and IBM Science Summarizer [9]. In consequence, almost all existing solutions attempt to produce high-fidelity realistic pages with the correct semantics and figures, typically by annotating existing publications, notably using crowd-sourced [11] and smart annotation [20] or decoding markup languages [2,11,22,27,34,35,42]. Our

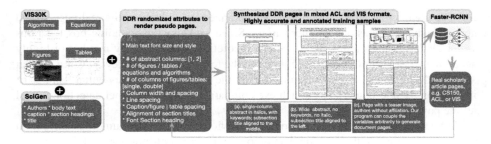

Fig. 2. DDR render-to-real workflow. Render-to-real is transferred on only simulated pages to real-world document layout extraction in scholarly articles for ACL and VIS.

solution instead uses rendering-to-real pseudo pages for segmentation by leveraging randomized page attributes and pseudo-texts for automatic and highly accurate training data production.

Other techniques manipulate pixels to synthesize document pages. He et al. [18] assumed that text styles and fonts within a document were similar or follow similar rules. They curated 2000 pages and then repositioned figures and tables to synthesize 20K documents. Yang et al. [41] synthesized documents through an encoder-decoder network itself to utilize both *appearance* (to distinguish text from figures, tables, and line segments) and *semantics* (e.g., paragraphs and captions). Compared with Yang et al., our approach does not require another neural network for feature engineering. Ling and Chen [24] also used a rendering solution and they randomized figure and table positions for extracting those two categories. Our work broadens this approach by randomizing many document structural parts to acquire both structural and semantic labels.

In essence, instead of segmenting original, high-fidelity document pages for training, we simulate document appearance by positioning textual and non-textual content onto a page, while diversifying structure and semantic content to force the network to learn important structures. Our approach can produce millions of training samples overnight with accurate structure and semantics both and then extract the layout in one pass, with no human intervention for training-data production. Our assumption is that, if models utilize textures and shape for their decisions [16], these models may well be able to distinguish among figures, tables, and text.

2.2 Bridging the Reality Gap in Domain Randomization

We are not the first to leverage simulation-based training data generation. Chatzimparm-pas et al. [6] provided an excellent review of leveraging graphical methods to generate simulated data for training-data generation in vision science. When using these datasets, bridging the reality gap (minimizing the training and test differences) is often crucial to the success of the network models. Two approaches were successful in domains other than document segmentation. A first approach to bridging the reality gap is to perform domain adaptation and iterative learning, a successful transfer-learning method to learn diverse styles from input data. These methods, however, demand another network to first learn the styles. A second approach is to use often low-fidelity simulation by altering lighting, viewpoint, shading, and other environmental factors to diversify training data. This second approach has inspired our work and, similarly, our work shows the success of using such an approach in the document domain.

3   Document Domain Randomization

Given a document, our goal with DDR is to accurately recognize document parts by making examples available at the training stage by diversifying a distinct set of appearance variables. We view synthetic datasets and training data generation from a computer graphics perspective, and use a two-step procedure of modeling and rendering by randomizing their input in the document space:

- We use **modeling** to create the semantic textual and non-textual content (Fig. 2).

  - **Algorithms, figures, tables, and equations.** In the examples in this paper, we rely on the VIS30K dataset [7,8] for this purpose.
  - **Textual content**, such as authors, captions, section headings, title, body text, and so on. We use randomized yet meaningful text [38] for this purpose.

- With **rendering** we manage the visual look of the paper (Fig. 1). We use:

  - a diverse set of other-than-body-text components (figures, tables, algorithms, and equations) randomly chosen from the input images;
  - distances between captions and figures;
  - distances between two columns in double-column articles;
  - target-adjusted font style and size;
  - target-adjusted paper size and text alignment;
  - varying locations of graphical components (figures, tables) and textual content.

Modeling Choices. In the modeling phase, we had the option of using content from publicly available datasets, e.g., Battle et al.'s [3] large Beagle collection of SVG figures, Borkin et al.'s [4] infographics, He et al.'s [18] many charts, and Li and Chen's scientific visualization figures [23], not to mention many vision databases [21,37]. We did not use these sources since each of them covers only a single facet of the rich scholarly article genre and, since these images are often modern, they do not contain images from scanned documents and thus could potentially bias CNN's classification accuracy. Here, we chose VIS30K [7,8], a comprehensive collection of images including tables, figures, algorithms, and equations. The figures in VIS30K contain not only charts and tables but also spatial data and photos. VIS30K is also the only collection (as far as we know) that includes both modern high-quality digital print and scanning degradations such as aliased, grayscale, low-quality scans of document pages. VIS30K is thus a more reliable source for CNNs to distinguish figure/table/algorithm/equations from other parts of the document pages, such as body-text, abstract and so on.

We used the semantically meaningful textual content of SciGen [38] to produce texts. We only detect the bounding boxes of the body-text and do not train models for

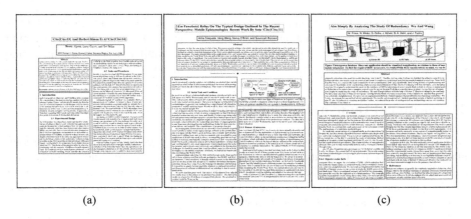

(a)   (b)   (c)

Fig. 3. Synthesized DDR pages in mixed ACL and VIS formats. Ground-truth labels and bounding boxes are produced automatically. Left: single-column abstract in italics, with keywords; subsection title centered. Middle: wide abstract, no keywords, no italic, subsection title left-aligned, Right: page with teaser image, authors without affiliations. Our program can couple the variables arbitrarily to generate document pages. (Color figure online)

As a result, we know the token-level semantic content of these pages. Sentences in paragraphs are coherent. Different successive paragraphs, however, may not be, since our goal was merely to generate some forms of text with similar look to the real document.

Rendering Choices. As Clark and Divvala rightly point out, font style influences prediction accuracy [11]. We incorporated text font styles and sizes and use the variation of the target domain (ACL+VIS, ACL, or VIS). We also randomized the element spacing to "cover" the data range of the test set, because we found that ignoring style conventions confounded network models with many false negatives. We arranged a random number of figures, tables, algorithms, and equations onto a paper page and used randomized text for title, abstract, and figure and table captions (Fig. 2)

We show some selected results in Fig. 3. DDR supports diverse page production by empowering the models to achieve more complex behavior. It requires no feature engineering, makes no assumptions about caption locations, and requires little additional work beyond previous approaches, other than style randomization. This approach also allows us to create 100% accurate ground-truth labels quickly in any predefined randomization style, because, theoretically, users can modify pages to minimize the reality gap between DDR pages and the target domain of use. DDR also requires neither decoding of markup languages, e.g., XML, or managing of document generation engines, e.g., LATEX, nor curation.

4   Evaluation of DDR

In this section we outline the core elements of our empirical setup and procedure to study DDR behaviors. Extensive details to facilitate replication are provided in the Supplemental Materials online. We also release all prediction results (see our Reproducibility statement in Sect. 5)

- **Goal 1. Benchmark and page style** (Sect. 4.1): We benchmark DDR on the classical CS-150 dataset, and two new datasets of different domains: computational linguistics (ACL300) and visualization (VIS300). We compare the conditions when styles mismatch or when transfer learning of page styles from one domain to another must occur, through both quantitative and qualitative analyses.
- **Goal 2. Label noise and training sample reduction** (Sect. 4.2): In two experiments, we assess the sensitivity of the CNNs to DDR data. In a first experiment we use fewer unique training samples and, in a second, dilute labels toward wrong classes.

Synthetic Data Format. All training images for this research were generated synthetically. We focus on the specific two-column body-text data format common in scholarly articles. This focus does not limit our work since DDR enables us to produce data from any paper style. Limiting the style, however, allows us to focus on the specific parametric space in our appearance randomization. By including semantic information, we showcase DDR's ability to localize token-level semantics as a stepping-stone to general-purpose training data production, covering both semantics and structure.

CNN Architecture. In all experiments, we use the Faster-RCNN architecture [31] implemented in tensorpack [1] due to its success in structural analyses for table detection in PubLayNet [42]. The input is images of the DDR generated paper pages. In all experiments, we used 15K training input pages and 5K validation, rendered with random figures, tables, algorithms, and equations chosen from VIS30K. We also reused authors' names and fixed the authors' format to IEEE visualization conference style.

Input, Output, and Measurement Metric. Our detection task seeks CNNs to output the bounding box locations and class labels of nine types: abstract, algorithm, author, body-text, caption, equation, figure, table, and title. To measure model performance, we followed Clark and Divvala's [11] evaluation metrics. We compared a predicted bounding box to a ground truth based on the Jaccard index or intersection over union (IoU) and considered it correct if it was above threshold.

We used four metrics (accuracy, recall, F1, and mean average precision (mAP)) to evaluate CNNs' performance in model comparisons, and the preferred ones are often chosen based on the object categories and goals of the experiment. For example, **precision and recall.** *Precision = true positives/(true positives + false positives))* and *Recall = true positives/true positives + false negatives*. Precision helps when the cost of the false positives is high. Recall is often useful when the cost of false negatives is high. **mAP** is often preferred for visual object detection (here figures, algorithms, tables, equations), since it provides an integral evaluation of matching between the ground-truth bounding boxes and the predicted ones. The higher the score, the more accurate the model is for its task. **F1** is more frequently used in text detection. A F1 score represents an overall measure of a model's accuracy that combines precision and recall. A higher F1 means that the model generates few false positives and few false negatives, and can identify real class while keeping distraction low. Here, *F1 = 2 × (precision × recall)/(precision + recall)*.

We report mAP scores in the main text because they are comprehensive measures suitable. to visual components of interest. In making comparisons with other studies for test on CS-150x, we show three scores precision, recall, and F1 because other studies [10] did so. All scores are released for all study conditions in this work.

4.1   Study I: Benchmark Performance in a Broad and Two Specialized Domains

Preparation of Test Data. We evaluated our DDR-based approach by training CNNs to detect nine classes of textual and non-textual content. We had two hypotheses:

- H1. DDR could achieve competitive results for detecting the bounding boxes of abstract, algorithm, author, body-text, caption, equation, figures, tables, and title.
- H2. Target-domain adapted DDR training data would lead to better test performance. In other words, train-test discrepancies would lower the performance.

We collected three test datasets (Table 1). The first CS-150x used all 716 double-column pages from the 1176 CS-150 pages [10]. CS-150 had diverse styles collected from several computer science conferences. Two additional domain-specific sets were chosen based on our own interests and familiarity: ACL300 had 300 randomly sampled articles (or 2508 pages) from the 55,759 papers scraped from the ACL anthology website; VIS300 contains about 10% (or 2619 pages) of the document pages in randomly partitioned articles from 26,350 VIS paper pages of the past 30 years in Chen et al. [8]. Using these two specialized domains lets us test H2 to measure the effect of using images generated in one domain to test on another when the reality gap could be large. Ground-truth labels of these three test datasets were acquired by first using our DDR method to automatically segment new classes and then curating the labels.

Table 1. Three test datasets.

Name	Source	Page count
CS-150x	CS-150	716
ACL300	ACL anthology	2508
VIS300	IEEE	2619

DDR-Based CS-150 Stylized Train and Tested on CS-150x. We generated CS-150x-style using DDR and tested it using CS-150x of two document classes, *figure* and *table*. While we could have trained and tested on all nine classes, we think any comparisons would need to be fair [15]. Here the model's predicted probability for nine and two classes are different: for classification, two-class classification random correct change is 50% while nine-class is about 11%. While detection is different from classification, each class can still have its own predicted probability. We thus followed the original CS-150 work of Clark and Divvala [10] in detecting figures and tables.

Table 2 shows the evaluation results for localizing figures and tables, demonstrating that our results from synthetic papers are compatible to those trained to detect figure and table classes. Compared to Clark and Divvala's PDFFigures [10], our method had a slightly lower precision (false-positives) but increased recall (false negatives) for both figure and table detection. Our F1 score for table detection is higher and remains competitive for figure detection.

Table 2. Precision (P), recall (R), and F1 scores on figure (f) and table (t) extractions. All extractors extracted two class labels (figure and table) except the two models in Katona [20], which were trained on eight classes.

Extractor	P_f	R_f	$F1_f$	P_t	R_t	$F1_t$
PDFFigures [10]	0.957	0.915	0.936	0.952	0.927	0.939
Praczyk and Nogueras-Iso [30]	0.624	0.500	0.555	0.429	0.363	0.393
Katona [20] U-Net*	0.718	0.412	0.276	0.610	0.439	0.510
Katona [20] SegNet*	0.766	0.706	0.735	0.774	0.512	0.616
DDR-(CS-150x) (ours)	**0.893**	**0.941**	**0.916**	**0.933**	**0.952**	**0.943**

Understanding Style Mismatch in DDR-Based Simulated Training Data. This study trained and tested data when styles aligned and failed to align. The test data were real-document pages of ACL300 and VIS300 with nine document class labels shown in Fig. 2. Three DDR-stylized training cohorts were:

- **DDR-(ACL+VIS):** DDR randomized to both ACL and VIS rendering style.
- **DDR-(ACL):** DDR randomized to ACL rendering style.
- **DDR-(VIS):** DDR randomized to VIS rendering style.

These three training and two test data yielded six train-test pairs: training CNNs on DDR-(ACL+VIS), DDR-ACL, and DDL-VIS and testing on ACL300 and VIS300, for the task of locating bounding boxes for the nine categories from each real-paper page in two test sets. Transfer learning then must occur when train and test styles do not match, such as models tested on VIS300 for ACL-styled training (DDR-(ACL)), and vice versa.

Real Document Detection Accuracy. Figure 4 summarizes the performance results of our models in six experiments of all pairs of training CNNs on DDR-(ACL+VIS), DDR-ACL, and DDL-VIS and testing on ACL300 and VIS300 to locate bounding boxes from each paper page in the nine categories.

Both hypotheses H1 and H2 were supported. Our approach achieved competitive mAP scores on each dataset for both figures and tables (average 89% on ACL300 and 98% on VIS300 for figures and 94% on both ACL300 and VIS300 for tables). We also see high mAP scores on the textual information such as *abstract, author, caption, equation*, and *title*. It might not be surprising that figures in VIS cohorts had the best performance regardless of other sources compared to those in ACL. This supports the idea that figure style influences the results. Also, models trained on mismatched styles

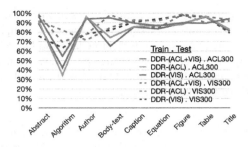

Fig. 4. Benchmark performance of DDR in six experiments. Three DDR training data (DDR customized to be inclusive (ACL+VIS), target-adapted to ACL or VIS, or not) and two test datasets (ACL300 or VIS300) for extracting bounding boxes of nine classes. Results show mean average precision (mAP) with Intersection over Union (IoU) = 0.8. In general, DDRs that are more inclusive (ACL+VIS) or target-adapted were more accurate than those not. (Color figure online)

(train on DDR-ACL and test on VIS, or train on DDR-VIS and test on ACL) in general are less accurate (the gray lines) in Fig. 4 compared to the matched (the blue lines) or more diverse ones (the red lines).

Error Analysis of Text Labels. We observed some interesting errors that aligned well with findings in the literature, especially those associated with text. Text extraction was often considered a significant source of error [11] and appeared so in our prediction results compared to other graphical forms in our study (Fig. 5). We tried to use GROBID [27], ParsCit, and Poppler [29] and all three tools failed to parse our cohorts, implying that these errors stemmed from text formats unsupported by these popular tools.

As we remarked that more accurate font-style matching would be important to localize bounding boxes accurately, especially when some of the classes may share similar textures and shapes crucial to CNNs' decisions [16]. The first evidence is that algorithm is lowest accuracy text category (ACL300: 34% and VIS300: 42%). Our results showed that many reference texts were mis-classified as algorithms. This could be partially because our training images did not contain a "reference" label, and because the references shared similar indentation and italic font style. This is also evidenced by additional qualitative error analysis of text display in Fig. 6. Some classes can easily fool CNNs when they shared fonts. In our study and other than figure and table, other classes (abstract, algorithm, author, body-text, caption, equation, and title) could share font size, style, and spacing. Many ACL300 papers had the same title and subsection font and this introduced errors in title detection. Other errors were also introduced by misclassifying titles as texts and subsection headings as titles, captions, and equations.

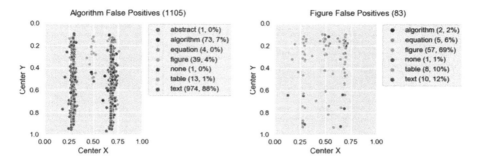

Fig. 5. Error distribution by categories: algorithm and figure. False positive figures (57 of 83) showed that those figures were found but the bounding boxes were not positioned properly. 974 among 1,105 false positive algorithms were mostly text (88%).

Error Correction. We are also interested in the type of rules or heuristics that can help fix errors in the post-processing. Here we summarize data using two *modes* of prediction errors on all data points of the nine categories in ACL300 and VIS300. The first kind of heuristics is rules that are almost impossible to violate: e.g., there will always be an abstract on the first page with title and authors (*page order heuristic*). Title will always appear in the top 30% of the first page, at least in our test corpus (*positioning heuristic*). We subsequently compute the error distribution by page order (first, middle and last pages) and by position (Fig. 7). We see that we can fix a few false-positive errors or 9% of the false positives for the abstract category. Similarly, we found that a few abstracts could be fixed by page order (i.e., appeared on the first page) and about another 30% fixed by position (i.e., appeared on the top half of the page.) Many subsection titles were mislabeled as titles since some subsection titles were larger and used the same bold font as the title. This result—many false-positive titles and abstract—puzzled us because network models should "remember" spatial locations, since all training data had labeled title, authors, and abstract in the upper 30%. One explanation is that within the text categories, our models may not be able to identify text labeling in a large font as a title or section heading as explained in Yang et al. [41].

4.2   Study II: Labeling Noises and Training Sample Reduction

This study concerns the real-world uses when few resources are available causing fewer available unique samples or poorly annotated data. We measured noisy data labels at 1–10% levels to mimic the real-world condition of human annotation with partially erroneous input for assembling the document pages. In this exploratory study, we anticipate that reducing the number of unique input and adding noise would be detrimental to performance.

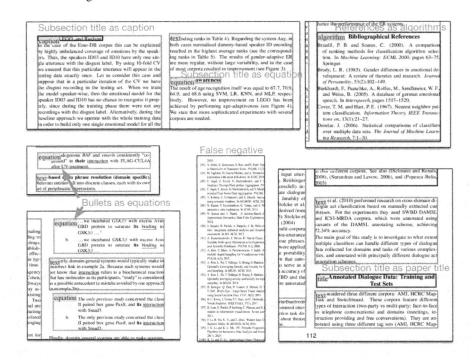

Fig. 6. Some DDR model prediction errors.

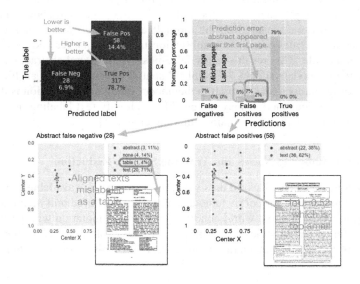

Fig. 7. DDR errors in abstract (train: DDR-(ACL), test: ACL300).

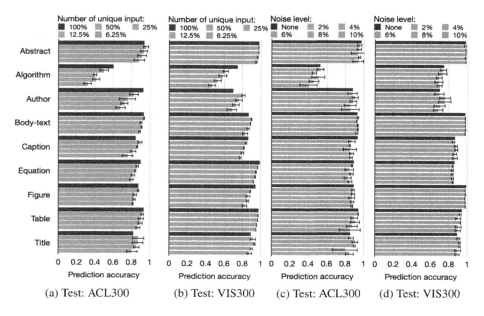

Fig. 8. DDR robustness (train: DDR-(ACL+VIS); test: ACL300 and VIS300). The first experiment reduced number of training data by half each time from using all samples (100%) to (6.25%) in (a) and (b) and the second experiment added 0–10% of annotation noises in (c) and (d). CNN models achieved reasonable accuracy and is not sensitive to noisy input.

Training Sample Reduction. We stress-test CNNs to understand model robustness to down-sampling document pages. Our DDR modeling attempts to cover the data range appearing in test. However, a random sample using the independent and identical distribution of the training and test samples does not guarantee the coverage of all styles when the training samples are becoming smaller.

Here, we reduced the number of samples from DDR-(ACL+VIS) by half each time, at 50% (7500 pages), 25% (3750 pages), 12.5% (1875 pages), and 6.25% (938 pages) downsampling levels, and tested on ACL300 and VIS300. Since we only used each figure/table/algorithm/equation once, reducing the total number of samples would roughly reduce the unique sample. Figure 8(a)–(b) showed the CNN accuracy by the number of unique training samples. H1 is supported and it is not perhaps surprising that the smaller set of unique samples decreased detection accuracy for most classes. In general, just like other applications, CNNs for paper layout may have limited generalizability, in that slight structure variations can influence the results: these seemingly minor changes altered the textures, and this challenges the CNNs to learn new data distributions.

Labeling Noise. This study involves observing the performance of DDR training samples on CNN on random 0–10% noise to the eight of the nine classes other than body-text. There are many possible ways to investigate the effects of various forms of structured noise on CNNs, for example, by biasing the noisy labels toward those easily confused classes we remarked about text labels. Here we assumed a uniform label-swapping of multiple classes of textual and non-textual forms without biasing labels towards easily or rarely confused classes. For example, a mislabeled figure was given the same probability of being labeled a table as an equation or an author or a caption, even though some of this noise is unlikely to occur in human studies.

Figure 8(c)–(d) show performance results when labels were diluted in the training sets of DDR-(ACL+VIS). H2 is supported. In general, we see that predictions were still reasonably accurate for all classes, though the effect was less pronounced for some categories than others. Also, models trained with DDR have demonstrated relatively robust to noises. Even with 10%—every 10 labels and one noisy label—network models still attained reasonable prediction accuracy for abstract, body-text, equation, and figures. Our result partially align with findings of Rolnick et al. [32], in that models were reasonably accurate (>80% prediction accuracy) to sampling noise. Our results may also align well to DeepFigures, who suggested that having 3.2% errors of their 5.5-million labels might not affect performance.

5   Conclusion and Future Work

We addressed the challenging problem of scalable trainable data production of text that would be robust enough for use in many application domains. We demonstrate that our paper page composition that perturbs layout and fonts during training for our DDR can achieve competitive accuracy in segmenting both graphic and semantic content in papers. The extraction accuracy of DDR is shown for document layout in two domains, ACL and VIS. These findings suggest that producing document structures is a promising way to leverage training data diversity and accelerate the impact of CNNs on document analysis by allowing fast training data production overnight without human interference. Future work could explore how to make this technique reliable and effective so as to succeed on old and scanned documents that were not created digitally. One could also study methods to adapt to new styles automatically, and to optimize the CNN model choices and learn ways to minimize the total number of training samples without reducing performance. Finally, we suggest that DDR seems to be a promising research direction toward bridging the reality gaps between training and test data for understanding document text in segmentation tasks.

Reproducibility. We released additional materials to provide exhaustive experimental details, randomized paper style variables we have controlled, the source code, our CNN models, and their prediction errors (http://bit.ly/3qQ7k2A). The data collections (ACL300, VIS300, CS-150x, and their meta-data containing nine classes) is on IEEE dataport [25].

Acknowledgements. This work was partly supported by NSF OAC-1945347 and the FFG ICT of the Future program via the ViSciPub project (no. 867378).

References

1. Github: Tensorpack Faster R-CNN (February 2021). https://github.com/tensorpack/tensorpack/tree/master/examples/FasterRCNN

2. Arif, S., Shafait, F.: Table detection in document images using foreground and background features. In: Proceedings of the DICTA, pp. 245–252. IEEE, Piscataway (2018). https://doi.org/10.1109/DICTA.2018.8615795

3. Battle, L., Duan, P., Miranda, Z., Mukusheva, D., Chang, R., Stonebraker, M.: Beagle: automated extraction and interpretation of visualizations from the web. In: Proceedings of the CHI, pp. 594:1–594:8. ACM, New York (2018). https://doi.org/10.1145/3173574.3174168

4. Borkin, M.A., et al.: What makes a visualization memorable? IEEE Trans. Vis. Comput. Graph. **19**(12), 2306–2315 (2013). https://doi.org/10.1109/TVCG.2013.234

5. Caragea, C., et al.: CiteSeer^x: a scholarly big dataset. In: de Rijke, M., et al. (eds.) ECIR 2014. LNCS, vol. 8416, pp. 311–322. Springer, Cham (2014). https://doi.org/10.1007/978-3-319-06028-6_26

6. Chatzimparmpas, A., Jusufi, I.: The state of the art in enhancing trust in machine learning models with the use of visualizations. Comput. Graph. Forum **39**(3), 713–756 (2020). https://doi.org/10.1111/cgf.14034

7. Chen, J., et al.: IEEE VIS figures and tables image dataset. IEEE Dataport (2020). https://doi.org/10.21227/4hy6-vh52. https://visimagenavigator.github.io/

8. Chen, J., et al.: VIS30K: a collection of figures and tables from IEEE visualization conference publications. IEEE Trans. Vis. Comput. Graph. **27**, 3826–3833 (2021). https://doi.org/10.1109/TVCG.2021.3054916

9. Choudhury, S.R., Mitra, P., Giles, C.L.: Automatic extraction of figures from scholarly documents. In: Proceedings of the DocEng, pp. 47–50. ACM, New York (2015). https://doi.org/10.1145/2682571.2797085

10. Clark, C., Divvala, S.: Looking beyond text: Extracting figures, tables and captions from computer science papers. In: Workshops at the 29th AAAI Conference on Artificial Intelligence (2015). https://aaai.org/ocs/index.php/WS/AAAIW15/paper/view/10092

11. Clark, C., Divvala, S.: PDFFigures 2.0: mining figures from research papers. In: Proceedings of the JCDL, pp. 143–152. ACM, New York (2016). https://doi.org/10.1145/2910896.2910904

12. Davila, K., Setlur, S., Doermann, D., Bhargava, U.K., Govindaraju, V.: Chart mining: a survey of methods for automated chart analysis. IEEE Trans. Pattern Anal. Mach. Intell. **43** (2021, to appear). https://doi.org/10.1109/TPAMI.2020.2992028

13. Dong, X., et al.: Knowledge vault: a web-scale approach to probabilistic knowledge fusion. In: Proceedings of the KDD, pp. 601–610. ACM, New York (2014). https://doi.org/10.1145/2623330.2623623

14. Dosovitskiy, A., et al.: FlowNet: learning optical flow with convolutional networks. In: Proceedings of the ICCV, pp. 2758–2766. IEEE, Los Alamitos (2015). https://doi.org/10.1109/ICCV.2015.316

15. Funke, C.M., Borowski, J., Stosio, K., Brendel, W., Wallis, T.S., Bethge, M.: Five points to check when comparing visual perception in humans and machines. J. Vis. **21**(3), 1–23 (2021). https://doi.org/10.1167/jov.21.3.16

16. Geirhos, R., Rubisch, P., Michaelis, C., Bethge, M., Wichmann, F.A., Brendel, W.: ImageNet-trained CNNs are biased towards texture; increasing shape bias improves accuracy and robustness (2018). https://arxiv.org/abs/1811.12231

17. Giles, C.L., Bollacker, K.D., Lawrence, S.: CiteSeer: an automatic citation indexing system. In: Proceedings of the DL, pp. 89–98. ACM, New York (1998). https://doi.org/10.1145/276675.276685

18. He, D., Cohen, S., Price, B., Kifer, D., Giles, C.L.: Multi-scale multi-task FCN for semantic page segmentation and table detection. In: Proceedings of the ICDAR, pp. 254–261. IEEE, Los Alamitos (2017). https://doi.org/10.1109/ICDAR.2017.50

19. James, S., Johns, E.: 3D simulation for robot arm control with deep Q-learning (2016). https://arxiv.org/abs/1609.03759

20. Katona, G.: Component Extraction from Scientific Publications using Convolutional Neural Networks. Master's thesis, Computer Science Department, University of Vienna, Austria (2019)
21. Krishna, R., et al.: Visual genome: connecting language and vision using crowdsourced dense image annotations. Int. J. Comput. Vis. **123**(1), 32–73 (2017). https://doi.org/10.1007/s11263-016-0981-7
22. Li, M., et al.: DocBank: a benchmark dataset for document layout analysis. In: Proceedings of the COLING, pp. 949–960. ICCL, Praha, Czech Republic (2020). https://doi.org/10.18653/v1/2020.coling-main.82
23. Li, R., Chen, J.: Toward a deep understanding of what makes a scientific visualization memorable. In: Proceedings of the SciVis, pp. 26–31. IEEE, Los Alamitos (2018). https://doi.org/10.1109/SciVis.2018.8823764
24. Ling, M., Chen, J.: DeepPaperComposer: a simple solution for training data preparation for parsing research papers. In: Proceedings of the EMNLP/Scholarly Document Processing, pp. 91–96. ACL, Stroudsburg (2020). https://doi.org/10.18653/v1/2020.sdp-1.10
25. Ling, M., et al.: Three benchmark datasets for scholarly article layout analysis. IEEE Dataport (2020). https://doi.org/10.21227/326q-bf39
26. Lo, K., Wang, L.L., Neumann, M., Kinney, R., Weld, D.S.: S2ORC: the semantic scholar open research corpus. In: Proceedings of the ACL, pp. 4969–4983. ACL, Stroudsburg (2020). https://doi.org/10.18653/v1/2020.acl-main.447
27. Lopez, P.: GROBID: combining automatic bibliographic data recognition and term extraction for scholarship publications. In: Agosti, M., Borbinha, J., Kapidakis, S., Papatheodorou, C., Tsakonas, G. (eds.) ECDL 2009. LNCS, vol. 5714, pp. 473–474. Springer, Heidelberg (2009). https://doi.org/10.1007/978-3-642-04346-8_62
28. Mayer, N., et al.: A large dataset to train convolutional networks for disparity, optical flow, and scene flow estimation. In: Proceedings of the CVPR, pp. 4040–4048. IEEE, Los Alamitos (2016). https://doi.org/10.1109/CVPR.2016.438
29. Poppler: Poppler. Dataset and online search (2014). https://poppler.freedesktop.org/
30. Praczyk, P., Nogueras-Iso, J.: A semantic approach for the annotation of figures: application to high-energy physics. In: Garoufallou, E., Greenberg, J. (eds.) MTSR 2013. CCIS, vol. 390, pp. 302–314. Springer, Cham (2013). https://doi.org/10.1007/978-3-319-03437-9_30
31. Ren, S., He, K., Girshick, R., Sun, J.: Faster R-CNN: towards real-time object detection with region proposal networks. IEEE Trans. Pattern Anal. Mach. Intell. **39**(6), 1137–1149 (2017). https://doi.org/10.1109/TPAMI.2016.2577031
32. Rolnick, D., Veit, A., Belongie, S., Shavit, N.: Deep learning is robust to massive label noise. arXiv preprint arXiv:1705.10694 (2017)
33. Sadeghi, F., Levine, S.: CAD^{2}RL: real single-image flight without a single real image. In: Proceedings of the RSS, pp. 34:1–34:10. RSS Foundation (2017). https://doi.org/10.15607/RSS.2017.XIII.034
34. Siegel, N., Horvitz, Z., Levin, R., Divvala, S., Farhadi, A.: FigureSeer: parsing result-figures in research papers. In: Leibe, B., Matas, J., Sebe, N., Welling, M. (eds.) ECCV 2016. LNCS, vol. 9911, pp. 664–680. Springer, Cham (2016). https://doi.org/10.1007/978-3-319-46478-7_41
35. Siegel, N., Lourie, N., Power, R., Ammar, W.: Extracting scientific figures with distantly supervised neural networks. In: Proceedings of the JCDL, pp. 223–232. ACM, New York (2018). https://doi.org/10.1145/3197026.3197040
36. Sinha, A., et al.: An overview of Microsoft Academic Service (MAS) and applications. In: Proceedings of the WWW, pp. 243–246. ACM, New York (2015). https://doi.org/10.1145/2740908.2742839

37. Song, S., Lichtenberg, S.P., Xiao, J.: SUN RGB-D: a RGB-D scene understanding bench-mark suite. In: Proceedings of the CVPR, pp. 567–576. IEEE, Los Alamitos (2015). https://doi.org/10.1109/CVPR.2015.7298655

38. Stribling, J., Krohn, M., Aguayo, D.: SCIgen - an automatic CS paper generator (2005). Online tool: https://pdos.csail.mit.edu/archive/scigen/

39. Tobin, J., Fong, R., Ray, A., Schneider, J., Zaremba, W., Abbeel, P.: Domain randomization for transferring deep neural networks from simulation to the real world. In: Proceedings of the IROS, pp. 23–30. IEEE, Piscataway (2017). https://doi.org/10.1109/IROS.2017.8202133

40. Tremblay, J., et al.: Training deep networks with synthetic data: bridging the reality gap by domain randomization. In: Proceedings of the CVPRW, pp. 969–977. IEEE, Los Alamitos (2018). https://doi.org/10.1109/CVPRW.2018.00143

41. Yang, X., Yumer, E., Asente, P., Kraley, M., Kifer, D., Lee Giles, C.: Learning to extract semantic structure from documents using multimodal fully convolutional neural networks. In: Proceedings of the CVPR, pp. 5315–5324. IEEE, Los Alamitos (2017). https://doi.org/10.1109/CVPR.2017.462

42. Zhong, X., Tang, J., Yepes, A.J.: PubLayNet: largest dataset ever for document layout analysis. In: Proceedings of the ICDAR, pp. 1015–1022. IEEE, Los Alamitos (2019). https://doi.org/10.1109/ICDAR.2019.00166

NLP for Document Understanding

Distilling the Documents for Relation Extraction by Topic Segmentation

Minghui Wang[1], Ping Xue[1], Ying Li[2,3]([⊠]), and Zhonghai Wu[2,3]

[1] School of Software and Microelectronics, Peking University, Beijing, China
{minghui_wang,xue.ping,li.ying}@pku.edu.cn
[2] National Engineering Center of Software Engineering, Peking University,
Beijing, China
[3] Key Lab of High Confidence Software Technologies (MOE), Peking University,
Beijing, China
zhwu@ss.pku.edu.cn

Abstract. Sentence-level relation extraction (RE) is always hard to identify the relations across sentences. Thus, researchers are now moving to the document-level RE. Since a document can be regarded as a long sequence connected by multiple sentences, document-level RE obtains cross-sentence relations by reasoning and aggregating information between entity mentions. However, document-level RE faces the low-efficiency issue. Intuitively, encoding the whole document is undoubtedly a more expensive action than encoding a single sentence. And we notice that in the process of relation extraction, documents always contain much irrelevant content, which not only wastes encoding time, but also brings potential noise information. Based on this observation, we propose a novel framework to identify and discard such irrelevant content. It integrates the topic segmentation to distill the document into topically coherent segments with concise and comprehensive information of entities. To further utilize these segments, we propose a topic enhancement module in the framework to enhance the predicted relations. The experimental results on two datasets indicate that our framework outperforms all the baseline models. Furthermore, our framework can increase the speed by 3.06 times and reduce the memory consumption by 2.90 times while maximizing the F1 score.

Keywords: Relation extraction · Topic segmentation · Bi-Attention mechanism

1 Introduction

Relation Extraction (RE) aims at identifying semantic relations between named entities in text, which plays an important role in many applications such as knowledge base population [11] and question answering [3].

Most prior RE works concentrate on the relations within a single sentence [21,23], known as the sentence-level RE. However, in the real world, a large

© Springer Nature Switzerland AG 2021
J. Lladós et al. (Eds.): ICDAR 2021, LNCS 12821, pp. 517–531, 2021.
https://doi.org/10.1007/978-3-030-86549-8_33

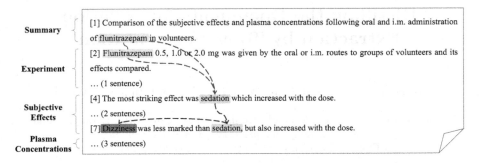

Fig. 1. An example of document-level RE excerpted from the CDR dataset. The relation between *flunitrazepam* and *dizziness* can be inferred through the interaction of multiple sentences. By dividing the document by topic, some contents are not important in the inference process, such as the description of Plasma Concentrations.

number of relations are expressed from multiple sentences, the sentence-level RE is inevitably restricted in practice. For example, 30% relational facts in the Biocreative V CDR dataset can only be extracted across sentences, such as the following excerpt in Fig. 1, entities like *flunitrazepam* and *dizziness* are mentioned in the document and exhibit complex interactions. Identifying the relational fact (*flunitrazepam, chemical-induced-disease, dizziness*) requires reading and reasoning over multiple sentences in a document. Thus, moving research from sentence-level to document-level is an emerging trend of RE [19].

Document-level RE methods usually treat the whole document as a long sequence. After encoding the sequence, the information of target entity mentions is aggregated to predict the relation [5,13,16]. However, the long sequence encoding faces a new problem: Longer documents lead to lower computational efficiency, but the existing methods have not explored this problem. In addition, we found that there is always useless content in the document for relation extraction, which not only wastes the coding time, but also brings potential noise information. As shown in Fig. 1, we can determine the relation fact (*flunitrazepam, chemical-induced-disease, dizziness*) based on sentences 1, 2, 4, and 7. Other sentences are unnecessary for the judgment of this relation, so eliminating other sentences as much as possible can improve calculation efficiency and reduce document noise. The simplest way is to only keep the sentences containing the target entity (i.e., sentence 1, 2, 7), but this will lead to the loss of important information (i.e., sentence 4). Thus, we turn to consider the document from the perspective of topics. The content of each document is dominated by several topics. Correspondingly, the contents are highly cohesive in those topics and less correlation between topics. We can observe that both sentence 4 and sentence 7 belong to the topic Subjective Effects. Keeping the topic Summary, Experiment, and Subjective Effects while discarding the Plasma Concentrations can reduce the number of sentences by 3/10 and purify the information for RE.

In this paper, we introduce Topic Segmentation (TS) for document-level RE and propose a novel document-level RE framework. Topic Segmentation divides a long textual document into meaningful segments based on their topical continuity, which has been commonly applied for text summarization and information retrieval [1,6]. To the best of our knowledge, we are the first to introduce topic segmentation for relation extraction. Our framework consists of two layers: distillation layer and relation extraction layer. In the first layer, we divide the document into topic segments, and then merge those segments containing the target entities into a distilled document. This document becomes the input of the second layer. In the second layer, we can adopt most of the existing relation extraction models. To make full use of the topic information in the distilled document, we propose a novel topic enhancement module. Specifically, we employ the Bi-Attention mechanism [14] to obtain topic-to-relation and relation-to-topic attention, and combine the attention information with the relations inferred by the RE model to enhance the result.

Our contribution can be summarized as follows:

1. We propose a novel document distillation framework and first introduce topic segmentation for document-level relation extraction task to obtain concise and comprehensive information of entity relations.
2. We design a topic enhancement module by Bi-Attention mechanism to further utilize the topic information in the distilled documents.
3. In terms of effectiveness, our framework outperforms all baseline document-level RE models. In terms of efficiency, our framework can increase the speed by 3.06 times and reduce the memory consumption by 2.90 times when the maximum F1 score is reached.

2   Related Work

2.1   Document-Level Relation Extraction

Current document-level relation extraction models can be mainly divided into two categories: Some only encode the context in documents and predict the relations directly; Others infer the relations after encoding the context.

In the first category, representative models are CNN-RE [20], RNN-RE [20], Transformer-RE [17] and BERT-RE [18]. They directly encode the original document, then extract the mentions in the encoded document. The mentions are aggregated into entity pairs to predict the corresponding relations. This type of model has become a common baseline in the field of document-level RE.

Based on the encoded context, some approaches concentrate on reasoning between entities. These methods are normally graph-based. [13] constructs a graph with tokens as nodes and the semantic dependencies as edges, and updates the nodes through GCN. EoG [5] expands the types of nodes. It takes mentions, entities, and sentences as nodes, and updates the representation of each edge through an edge-oriented model. Then, EoG extracts the edges between entity nodes to predict the relations. GCGCN [22] proposes a more complex model structure. It updates information in the graph attention network through contextual attention and multi-head attention. Also on the basis of EoG, EoGANE [16] considers the role of graph nodes in relation extraction, and improves the effect of EoG by explicitly computing the representations for the nodes.

Although the existing models encode the context and infer the relations well, they still have a common problem: All the models take the whole document as input, but we normally do not need to consider the information of the entire document to extract the relations. The long documents also make the computation expensive. Hence, we propose a novel document distillation framework to solve this problem.

2.2   Unsupervised Topic Segmentation

Most topic segmentation methods follow the lexical cohesion theory and identify the topic shifts by the low lexical cohesion. Due to the lack of manual annotation in the initial topic segment datasets, many approaches explore the unsupervised methods. These algorithms are divided into two types: resorting to the lexical similarity and utilizing the probability.

TextTiling [9] was first introduced, which calculates the cosine similarity between different blocks of words. C99 [4] takes another approach. It uses a similarity matrix for ranking and clustering to get segments. MinCut [10] utilizes the normalized minimum-cut graph to obtain the similarity between sentences, while GraphSeg [8] finds the max clique on the document graph, where the nodes are sentences and the edges are connected by highly similar nodes.

Besides the lexical similarity methods, the probabilistic approaches mostly follow the Latent Dirichlet allocation (LDA) [2]. BayesSeg [7] considers that each segment is produced by a distinct lexical distribution and utilizes the dynamic programming to discover the maximum-likelihood segmentation. Then, other works such as TopicTiling [12] combines the probability-based method with the text-similarity-based method. It assigns a topic to each word through LDA and then uses a dense vector to represent each sentence according to the topic of the words. TopicTiling locates the segment boundaries by finding the drop of cosine similarity between adjacent sentences.

Our framework can take any unsupervised topic segmentation algorithms in its Distillation layer. In this paper, we explore which type of these algorithms is more suitable for document-level relation extraction task.

3   Framework

We formulate the document-level relation extraction as a classification task. Given a document D and a set of entities $E = \{e_i\}_{i=1}^{N}$, where $e_i = \{m_j\}_{j=1}^{M}$, m_j corresponds to the j-th mention of the i-th entity, our goal is to identify the relation between all related entity pairs, namely $\{(e_i, r_{ij}, e_j)|e_i, e_j \in E, r_{ij} \in R\}$, where R represents the pre-defined relation set.

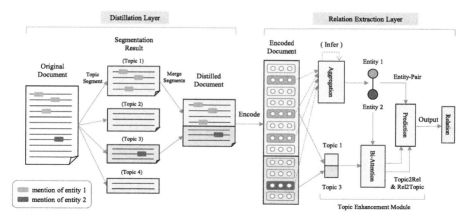

Fig. 2. Overview of our framework. It first splits the document by a topic segmentation algorithm, then merges the segments containing the target entities to form a distilled document. After encoding the document and aggregating entity pairs, the topic enhancement module enhances the entity pairs through the Bi-Attention mechanism.

Figure 2 represents our document distillation framework, which includes two layers: Distillation Layer and Relation Extraction Layer. The distillation layer is responsible for distilling the document based on its latent topics, and retaining those topics that contain the target entities. The relation extraction layer is responsible for classifying relations by making full use of topic information in distilled documents. In the following sections, we introduce the formal definition of these two layers.

3.1   Distillation Layer

Topic Segmentation. As shown in Fig. 2, the first step of the Distillation Layer is to adopt the topic segmentation method to split the original document into several segments. Each segment corresponds to a topic and contains several sentences. The content of these sentences is highly consistent with the topic. Ideally, the segments which contain our target mentions will retain all the useful information (e.g., coreference) of the mentions. Thus, the more accurate the topic segmentation result is, the more conducive to the task of relation extraction. We formalize the document segmentation process as follows:

Consider a document D with K sentences: $D = \{sen_i\}_{i=1}^{K}$, and T latent topics: $\{t_i\}_{t=1}^{T}$. The topic segmentation algorithm separates the document into L segments, and each segment S_i corresponds to one latent topic t_j.

$$\{S_1, S_2, \ldots, S_L\} = Topic\_segmentation(D) \tag{1}$$

where $S_i = \{sen_j\}_{j=1}^{H}(H \leq K)$; $S_i \cap S_j = \oslash(i \neq j)$; $S_1 \cup S_2 \cup, \ldots, \cup S_L = D$

For the document-level RE task, documents are not annotated for topics. Therefore, we utilize unsupervised topic segmentation methods [7–9, 12] to divide the documents. These methods follow the lexical cohesion and divide documents according to the change of latent topic between sentences.

Segment Fusion. After dividing a document, the l mentions of target entities: $M_{target} = [m_1, m_2, \ldots, m_l]$ may scatter in different segments, so we need to merge segments which contain the mentions into a distilled document.

$$D_d : \{S_i \mid \exists m_j \in M_{target}, m_j \in S_i\} \tag{2}$$

3.2   Relation Extraction Layer

Encoder. The distilled document D_d obtained in the previous steps can be regarded as a sequence of words: $D_d = \{x_1, x_2, \ldots, x_w\}$. We can adopt an encoder such as CNN, LSTM, or BERT to encode D_d into hidden states $\{h_1, h_2, \ldots, h_w\}$.

$$D_{encode} = \{h_1, h_2, \ldots, h_w\} = Encoder(D_d) \tag{3}$$

where $h_i \in \mathbb{R}^d$ is the context-sensitive representation for each word. d is the number of hidden dimensions.

Aggregation. After the document is encoded, the words corresponding to the entities and mentions are aggregated for subsequent relation prediction. The most common aggregation method is average pooling. Each mention representation m_i is computed by the average of the words it contains: $m_j = avg_{x_i \in m_j}(h_i)$; Each entity e_k representation is formed by the average of the mentions associated with it: $e_k = avg_{m_j \in e_k}(m_j)$, both m_j and $e_k \in \mathbb{R}^d$.

Inference (Optional). To simulate the reasoning process between relations, several recent models build graphs of entities, mentions, and other information. They update the information in the graph iteratively to obtain higher-order relations. This process can be briefly expressed as follows:

$$G_{reason} = Infer(G) \tag{4}$$

where $G = \{N, Edg\}$, and N and Edg represent multiple types of nodes and edges, respectively. To further explain this module, we take the EoG model as an example, which is one of the important baselines for document-level RE, and many works are based on it. We also implemented it in the experiment.

In the inference layer of EoG, the graph nodes required for inference are composed of entities, mentions, and sentences. We have obtained the representation of entities and mentions in the Aggregation module. The representation of the sentence is similar to that, which is obtained by averaging each word in the sentence: $sen_a = avg_{x_i \in sen_a}(h_i) \in \mathbb{R}^d$

Through 3 types of nodes, the model constructs 5 types of edges, namely mention2mention, mention2entity, mention2sentence, entity2sentence, and sentence2sentence. The representation of the edge mention2mention is aggregated through the attention mechanism, and the other types of edges are directly concatenated by the node representations at both ends of the edge. The dimensions of all edges are $\mathbb{R}^{2d}$.

In the iteration stage, for any two nodes a and b in the graph. The model generates edges or updates existing edges through intermediate nodes p.

$$f(e_{ap}^{(l)}, e_{pb}^{(l)}) = sigmoid(e_{ap}^{(l)} \cdot W \cdot e_{pb}^{(l)}) \tag{5}$$

$$e_{ab}^{(l+1)} = \beta e_{ab}^{(l)} + (1 - \beta) \sum_{p \neq a,b} f(e_{ap}^{(l)}, e_{pb}^{(l)}) \tag{6}$$

where W is a learnable parameter, l is the iteration number, and $\beta \in [0, 1]$ is a scalar used to control the weight of the original edge and updated edge information. After multiple iterations, the model obtains high-level information of the graph through the various types of intermediate nodes.

Topic Enhancement. The above three modules of relation extraction layer constitute the existing relation extraction models. Through these modules, we have been able to predict the corresponding relations. But in order to further utilize the topic information in the distilled document, we propose a topic enhancement module. It is designed by Bi-Attention mechanism [14], which calculates attention from relation to topic (R2T) and topic to relation (T2R).

The reason we use Bi-Attention is that it can describe two sources of information from different aspects. For example, the relation "director" is more likely to appear under the topic of the movie, and less under topics such as politics. The content of movie-related topics is also easier to have the relation "director". These two types of attention can complement each other to further enhance the model's understanding of relations.

To calculate the attention, we need to get the representation of topics and entity pairs firstly. To get the topic representation, we take the average of each word representation in the topic segment: $topic_a = avg_{x_i \in S_a}(h_i) \in \mathbb{R}^d$. And concatenate each topic together. The topic vector $Topics = [topic_a; \ldots topic_z] \in \mathbb{R}^{topic\_num*d}$ represents different topic information of the distilled document.

For the relation representation, we extract every two target entities after aggregation or from the inferred graph, and also concatenate them into entity pairs: $EP_{ij} = [e_i; e_j]/EP_{ij} = Edg_{ij} \in G$. The dimension of all entity pairs $\{EP\}$ is $\mathbb{R}^{pair\_num*2d}$.

The T2R and R2T are calculated as follows, we first calculate the similarity matrix between the topics and the relations:

$$Matrix = tanh(Topics \cdot W \cdot EP^{\mathrm{T}}) \in \mathbb{R}^{topic\_num*pair\_num} \tag{7}$$

where $W \in \mathbb{R}^{d*2d}$ is a learnable parameter. In this two-dimensional matrix, from a horizontal perspective, we can get the current topic's attention to each relation, and from a vertical perspective, we can get the current relation's attention to each topic:

$$T2R = softmax(Matrix_{rows}) \tag{8}$$

$$R2T = softmax(Matrix_{columns}) \tag{9}$$

Prediction. In the prediction module, we need to integrate the two attentions into the representation of the relations. We multiply the two attentions with entity pairs and topics respectively:

$$EP\_T = column\_max(T2R) \cdot EP \tag{10}$$

$$EP\_R = R2T^{\mathrm{T}} \cdot Topics \tag{11}$$

where $column_{max}()$ a function to find the maximum value of each column. According to EP_T and EP_R, and the original EP, a three-layer fusion mechanism is designed to fully mix the entity pair with the two aspects of information.

$$T\_Fusion = MLP([EP; EP\_T; EP - EP\_T; EP * EP\_T]) \tag{12}$$

$$R\_Fusion = MLP([EP; EP\_R; EP - EP\_R; EP * EP\_R]) \tag{13}$$

$$Relation\_Score = MLP_{predict}([EP; T\_Fusion; R\_Fusion]) \tag{14}$$

The first two MLP networks fuse the EP with the information from the topics aspect and the relations aspect and obtain T_Fusion and $R\_Fusion \in \mathbb{R}^{pair\_num*2d}$. Finally, the two fusion representations and the original entity pair representation are concatenated and mapped into the relation classification space $Relation\_Score \in \mathbb{R}^{pair\_num*R}$. Moreover, a sigmoid function is utilized to calculate the probability of each relation.

$$P(r|EP) = sigmoid(Relation\_Score) \tag{15}$$

We consider the binary cross entropy (BCE) as the loss function to train the model and evaluate the model by F1 score.

4   Experiments

In this section, we introduce two document-level relation extraction datasets, as well as the settings of all baseline RE models and our framework. We compare the performance of our document distillation framework with the baseline aimed at answering the following two research questions:

– RQ1: How is the effectiveness of our framework?
– RQ2: How is the efficiency of our framework?

4.1   Datasets

We evaluated our framework on two biochemistry datasets:

Bio-Creative V CDR Task Dataset: This is a document-level Chemical-Disease Relations (CDR) Dataset [19], which is developed for Bio-Creative V challenge including 1500 PubMed abstracts. The abstracts contain the relations between Chemical and Disease concepts. Furthermore, the relations can

Table 1. Implementation of each module of the baseline model.

	CNN-RE	RNN-RE	BERT-RE	EoG
Encoder	CNN	BiLSTM	BERT	BiLSTM
Aggregation	Avg	Avg	Avg	Avg
Inference	None	None	None	Edge-oriented
Batch size for CDR	8	8	8	2
Batch size for I2b2	16	16	8	None
Learning rate	0.005	0.005	0.005	0.002
Word dimension	100	100	None	200
Position dimension	20	20	None	10
Hidden dimension	140	140	768	200
Number of inference	None	None	None	2
Max iterations (TopicTiling)	50	50	50	50
Topic_num (TopicTiling for CDR)	500	500	500	750
Topic_num (TopicTiling for I2b2)	48	48	48	None

be extracted from intra or inter sentences. The CDR dataset is equally split for training, developing, and testing.

I2b2 2012 Challenge Dataset: The dataset of 2012 informatics for integrating biology and the bedside (i2b2) challenge [15] contains 310 documents of over 178,000 tokens. It is annotated by events, temporal expressions, and temporal relations in clinical narratives. It contains 8 categories of relations and each article contains an average of 86.6 events, 12.4 temporal expressions, and 176 temporal relation instances.

4.2 Baseline Setting

To evaluate our framework, we considered four models as the baseline:

Two traditional baseline models CNN-RE [20] and RNN-RE [20], a pretrained-based model BERT-RE [18], and a graph-based model EoG [5]. For all baseline models, we used their official public code. The corresponding implementation of each module and the hyper-parameter settings are in Table 1. We introduced the specific details of Avg and Edge-oriented in Sect. 3.1 and Sect. 3.2 respectively. We have noticed that several other graph-based RE models have appeared recently [16,22]. But they can also be regarded as variants of EoG in different directions. GCGCN [22] is to achieve better reasoning by transforming the graph network structure and EoGANE [16] is to enrich graph information by explicitly compute the representation of nodes. Therefore, to measure the generality of our framework, we chose to experiment on their basic model: EoG.

When processing the datasets, we found there are only event and time expressions in i2b2 2012 dataset, which can be regarded as the mention-level representations. Hence, there is no annotated entity-level concept in this dataset. The graph-based model with inference module is easy to overfit in this case. Thus, we test the other three baselines without inference module in i2b2 2012 dataset.

4.3   Framework Setting

Our framework can adopt any unsupervised topic segmentation algorithms in its distillation layer, and we choose two text-similarity-based methods: TextTiling [9] and GraphSeg [8] and two probability-based methods: BayesSeg [7] and TopicTiling [12] for distillation layer to explore which type of method is more suitable for our document-level relation extraction framework.

To compare with the topic segmentation methods, we also utilized a sentence-level segmentation method, which split the document sentence by sentence. In this method, each segment is a sentence, so it can be utilized to verify the validity of distilling documents from the perspective of topics.

To control variables during the experiment, we first disable the topic enhancement module and only verify which segment method is more suitable for document-level RE on the baseline models. After getting the conclusion, we enable our topic enhancement module on the best performing framework settings to observe whether the module can further improve the performance.

5   Results

5.1   Results on RQ1

Table 2 represents the performance of the five segment methods combined with RNN-RE, CNN-RE, BERT-RE, and EoG[1]. The first row shows the F1 scores

Table 2. The comparison of F1 scores (%) between document distillation framework with various topic segmentation algorithms and baseline RE models on CDR dataset.

	CNN-RE (%)	RNN-RE (%)	BERT-RE (%)	EoG (%)
Origin	55.50	56.93	62.12	67.32*
SentSeg + Origin	54.75	56.16	61.86	68.47
TextTiling + Origin	55.67	57.34	63.06	68.05
GraphSeg + Origin	55.60	57.09	62.54	67.59
BayesSeg + Origin	55.72	_57.96_	62.88	67.87
TopicTiling + Origin	_56.00_	57.71	_63.56_	_68.68_
+ Bi-Att	**57.29**	**58.86**	**64.35**	**69.86**

---

[1] We used the latest bug-free version officially released by the author: https://github.com/fenchri/edge-oriented-graph, so the result marked by * is higher than that reported in the original paper.

of the four baseline models on CDR test set. We observe that all the distilled documents constructed by topic segment methods enhance the results of baseline models. The improvement of F1 score varies from 0.50% to 1.36%, indicating that topic segmentation can discard irrelevant information and make RE models extract relation easier.

The underlined result is the highest score among the 5 segmentation algorithms. The TopicTiling yields the best result with the most times, and BayesSeg also got the best result once when Bi-LSTM is used as the encoder. It is worth noting that both BayesSeg and TopicTiling belong to the probability-based segmentation and the other two (TextTiling, GraphSeg) belong to the text-similarity-based segmentation.

We believe that the algorithms based on text similarity only consider the local information between several adjacent sentences in the segmentation process, thus ignoring part of the global information, which result in a slightly poor result. However, the probability-based algorithms can take the global information into account by calculating the Document-Topic co-occurrence matrix and Topic-Word co-occurrence matrix, so the overall effect is better than the text-similarity-based ones.

In most cases, TopicTiling is better than BayesSeg. This is because BayesSeg only utilizes global information through co-occurrence matrices, while TopicTiling not only calculates the topic of each word by global information but combines a local comparison pattern of topic similarity. As a result, TopicTiling can benefit from both global and local information.

We also observe that the sentence-level segmentation method (SentSeg) did not improve the performances of most baseline models. We believe that sentence-level segmentation loses a lot of synonyms information or coreference information, such as (this, that) required for relation extraction. However, the topic segmentation methods can keep this type of information in the same segment through text similarity or probability. Therefore, it makes sense to distill information from a thematic perspective.

Under the setting of using TopicTiling to segment documents and distill information, we enable the topic enhancement module. We can observe that the result has been further improved from 0.79% to 1.29%. This shows that the

Table 3. The comparison of F1 scores (%) between document distillation framework with various topic segmentation algorithms and baseline RE models on I2B2 dataset.

	CNN-RE (%)	RNN-RE (%)	BERT-RE (%)
Origin	69.65	70.64	73.29
SentSeg + Origin	70.85	71.38	73.31
TextTiling + Origin	69.94	71.20	73.72
GraphSeg + Origin	70.28	70.83	73.87
BayesSeg + Origin	70.86	71.81	74.15
TopicTiling + Origin	<u>71.22</u>	<u>71.98</u>	<u>74.94</u>
+ Bi-Att	**72.52**	**73.04**	**76.47**

information from topic to relation and from relation to topic is meaningful. It can be effectively used by the enhanced module to better classify relations.

On i2b2 2012 dataset, we also compare the performances of the five segment methods combined with CNN-RE, RNN-RE, and BERT-RE in Table 3.

All topic segmentation methods improve the baseline, and the probability-based algorithms overtake the text-similarity-based algorithms again. TopicTiling reaches the best on all baseline models, which enhance the baseline from 1.34% to 1.80% on F1 score. It indicates that TopicTiling can also retain the most clues for relation extraction with the help of global and local information in i2b2 2012 dataset. The performance of SentSeg is better than the baseline. We found that in this dataset, each event or temporal expression usually only appears once or twice in the article. Hence, there are not many coreference relations in documents. The SentSeg can perform relatively well in this situation.

After enabling the topic enhancement module, the best-performing framework setting (TopicTiling + Origin) has also been further improved, ranging from 1.06% to 1.53%.

Considering the above experimental results, we can conclude that the distilling the documents can effectively improve the results in document-level RE. Moreover, probability-based topic segmentation methods are normally better than text-similarity-based ones since they can obtain more global information by co-occurrence matrices. Besides, the performance of SentSeg is more volatile because it may excessively discard information in the dataset. And the further improvement on the two datasets proves the effectiveness of our topic enhancement module, which can make better use of the topic information contained in the distilled document to assist the model in relation extraction.

5.2   Results on RQ2

Our framework reduces the length of the article during the distillation process, hence, it will also bring efficiency improvements. We first analyze the length

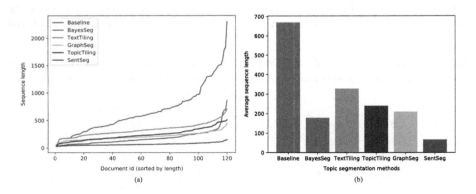

Fig. 3. The comparison of each document's length and the average length of all documents before and after distillation with different topic segmentation methods on i2b2 dataset. (Color figure online)

change of the documents before and after distillation and then analyze the corresponding memory usage and speed improvement.

To explore the lengths of the documents before and after distillation, we calculate the sequence length of both original and distilled documents. The results are shown in Fig. 3.

In Fig. 3(a), the red line represents the original article's length. It can be observed that the other four lines are significantly lower than the red line, which means that our document distillation framework can effectively shorten the length of the article. Moreover, we can observe a trend that as the length of the document increases, our distillation framework can discard more irrelevant information, which is of great significance to the practical application of our framework.

Based on the average length of the documents in Fig. 3(b), our framework can shorten the document by 2 to 10 times. Sentence-level segmentation is the most obvious one. However, the SentSeg is easy to over-abandon the information and reduce the accuracy of relation extraction. A more stable way is to use topic segmentation, the average length of documents distilled from the four topic segmentation algorithms is similar. TextTiling and GraphSeg focus on local information through text similarity, therefore, the distilled document is slightly longer. BayesSeg segments the documents from a global perspective, thus, it distills the shortest document. TopicTiling uses both global and local information, so the length is between BayesSeg and TextTiling.

Table 4. The comparison of running time and memory usage between baseline model (BERT) and various topic segmentation algorithms in document distillation framework on i2b2 dataset.

	Seg Time (s)	RE Time (s)	Total time (s)	VRAM (MiB)
BERT-RE (BR)	0	1465.39	1465.39	28376
GraphSeg + BR	25.35	482.23	507.58 (2.89×)	9365 (3.03×)
TextTiling + BR	4.33	496.49	500.82 (2.93×)	9801 (2.90×)
BayesSeg + BR	1.86	485.36	487.22 (3.01×)	9867 (2.86×)
TopicTiling + BR	3.89	447.72	451.61 (3.24×)	9609 (2.95×)
+ Bi-Att	3.89	475.01	478.90 (3.06×)	9777 (2.90×)
SentSeg + BR	1.05	417.35	418.40 (3.50×)	8855 (3.20×)

Thanks to the shorter distilled documents, our framework can save more memory and run faster. We conducted all the experiments on a machine with an AMD Ryzen CPU and two Titan RTX GPUs. According to Table 4, the total running time of our framework includes segmentation time and relation extraction time. The total running time is 2.89 to 3.50 times faster than the baseline, and the maximum memory consumption is reduced by 2.86 to 3.20 times. When adopting TopicTiling and the topic enhancement module to get the best relation extraction results (Sect. 5.1), the running speed is increased by 3.06 times, and the memory consumption is reduced by 2.90 times.

Therefore, considering both effectiveness and efficiency in document RE task, it is important to incorporate topic segmentation into document distillation framework and make full use of topic information to enhance the results.

6   Conclusion

We presented a novel document-level RE framework to distill the content of documents for relation extraction. By incorporating topic segmentation, the results of our framework on two datasets outperformed all the baseline models. Analysis between different topic segmentation algorithms indicated that the probabilistic ones perform better. Comparing with sentence-level segmentation, topic segmentation could hold more useful information for relation extraction. We also propose a topic enhancement module to make full use of the topic information in distilled documents. Furthermore, our framework can also increase the speed by 3.06 times and reduce the memory consumption by 2.90 times while maximizing the F1 score. In this framework, we concentrated on utilizing the unsupervised topic segmentation algorithms. In future work, we plan to incorporate supervised topic segmentation models in our framework to improve the performance further.

Acknowledgments. This work is partly supported by ICBC Technology.

References

1. Amoualian, H., Lu, W., Gaussier, E., Balikas, G., Amini, M.R., Clausel, M.: Topical coherence in LDA-based models through induced segmentation. In: ACL (2017)
2. Blei, D.M., Ng, A.Y., Jordan, M.I.: Latent dirichlet allocation. JMLR **3**, 993–1022 (2003)
3. Chen, Z.Y., Chang, C.H., Chen, Y.P., Nayak, J., Ku, L.W.: UHop: an unrestricted-hop relation extraction framework for knowledge-based question answering. In: NAACL, pp. 345–356 (2019)
4. Choi, F.Y.: Advances in domain independent linear text segmentation. In: NAACL (2000)
5. Christopoulou, F., Miwa, M., Ananiadou, S.: Connecting the dots: document-level neural relation extraction with edge-oriented graphs. In: EMNLP-IJCNLP (2019)
6. Dias, G., Alves, E., Lopes, J.G.P.: Topic segmentation algorithms for text summarization and passage retrieval: an exhaustive evaluation. In: AAAI (2007)
7. Eisenstein, J., Barzilay, R.: Bayesian unsupervised topic segmentation. In: EMNLP (2008)
8. Glavaš, G., Nanni, F., Ponzetto, S.P.: Unsupervised text segmentation using semantic relatedness graphs. In: STARSEM (2016)
9. Hearst, M.A.: Text tiling: segmenting text into multi-paragraph subtopic passages. Comput. Linguist. **23**, 33–64 (1997)
10. Malioutov, I., Barzilay, R.: Minimum cut model for spoken lecture segmentation. In: ACL (2006)

11. Mesquita, F., Cannaviccio, M., Schmidek, J., Mirza, P., Barbosa, D.: Knowledge-Net: a benchmark dataset for knowledge base population. In: EMNLP, pp. 749–758 (2019)
12. Riedl, M., Biemann, C.: TopicTiling: a text segmentation algorithm based on LDA. In: ACL (2012)
13. Sahu, S.K., Christopoulou, F., Miwa, M., Ananiadou, S.: Inter-sentence relation extraction with document-level graph convolutional neural network. In: ACL (2019)
14. Seo, M.J., Kembhavi, A., Farhadi, A., Hajishirzi, H.: Bidirectional attention flow for machine comprehension. In: ICLR (2017)
15. Sun, W., Rumshisky, A., Uzuner, O.: Evaluating temporal relations in clinical text: 2012 i2b2 challenge. JAMIA **20**, 806–813 (2013)
16. Tran, H.M., Nguyen, T.M., Nguyen, T.H.: The dots have their values: exploiting the node-edge connections in graph-based neural models for document-level relation extraction. In: EMNLP (2020)
17. Verga, P., Strubell, E., McCallum, A.: Simultaneously self-attending to all mentions for full-abstract biological relation extraction. In: NAACL (2018)
18. Wang, H., Focke, C., Sylvester, R., Mishra, N., Wang, W.Y.: Fine-tune Bert for DocRED with two-step process. CoRR (2019)
19. Wei, C.H., et al.: Assessing the state of the art in biomedical relation extraction: overview of the BioCreative v chemical-disease relation (CDR) task. Database (2016)
20. Yao, Y., et al.: DocRED: a large-scale document-level relation extraction dataset. In: ACL (2019)
21. Zhang, Y., Qi, P., Manning, C.D.: Graph convolution over pruned dependency trees improves relation extraction. In: EMNLP (2018)
22. Zhou, H., Xu, Y., Yao, W., Liu, Z., Lang, C., Jiang, H.: Global context-enhanced graph convolutional networks for document-level relation extraction. In: COLING (2020)
23. Zhu, H., Lin, Y., Liu, Z., Fu, J., Chua, T.S., Sun, M.: Graph neural networks with generated parameters for relation extraction. In: ACL (2019)

LAMBERT: Layout-Aware Language Modeling for Information Extraction

Łukasz Garncarek[1]([✉])(ID), Rafał Powalski[1](ID), Tomasz Stanisławek[1,2](ID),
Bartosz Topolski[1](ID), Piotr Halama[1](ID), Michał Turski[1,3](ID),
and Filip Graliński[1,3](ID)

[1] Applica.ai, Zajęcza 15, 00-351 Warsaw, Poland
{lukasz.garncarek,rafal.powalski,tomasz.stanislawek,
bartosz.topolski,piotr.halama,michal.turski,filip.gralinski}@applica.ai
[2] Warsaw University of Technology, Koszykowa 75, 00-662 Warsaw, Poland
[3] Adam Mickiewicz University, 1 Wieniawskiego, 61-712 Poznań, Poland

Abstract. We introduce a simple new approach to the problem of understanding documents where non-trivial layout influences the local semantics. To this end, we modify the Transformer encoder architecture in a way that allows it to use layout features obtained from an OCR system, without the need to re-learn language semantics from scratch. We only augment the input of the model with the coordinates of token bounding boxes, avoiding, in this way, the use of raw images. This leads to a layout-aware language model which can then be fine-tuned on downstream tasks.

The model is evaluated on an end-to-end information extraction task using four publicly available datasets: Kleister NDA, Kleister Charity, SROIE and CORD. We show that our model achieves superior performance on datasets consisting of visually rich documents, while also outperforming the baseline RoBERTa on documents with flat layout (NDA F_1 increase from 78.50 to 80.42). Our solution ranked first on the public leaderboard for the Key Information Extraction from the SROIE dataset, improving the SOTA F_1-score from 97.81 to 98.17.

Keywords: Language model · Layout · Key information extraction · Transformer · Visually rich document · Document understanding

1 Introduction

The sequential structure of text leads to it being treated as a sequence of tokens, characters, or more recently, subword units. In many problems related to Natural Language Processing (NLP), this linear perspective was enough to enable significant breakthroughs, such as the introduction of the neural Transformer architecture [28]. In this setting, the task of computing token embeddings is

---

L. Garncarek, R. Powalski, T. Stanisławek and B. Topolski—Equally contributed to the paper.

© Springer Nature Switzerland AG 2021
J. Lladós et al. (Eds.): ICDAR 2021, LNCS 12821, pp. 532–547, 2021.
https://doi.org/10.1007/978-3-030-86549-8_34

solved by Transformer encoders, such as BERT [6] and its derivatives, achieving top scores on the GLUE benchmark [29].

They all deal with problems arising in texts defined as sequences of words. However, in many cases there is a structure more intricate than just a linear ordering of tokens. Take, for instance, printed or richly-formatted documents, where the relative positions of tokens contained in tables, spacing between paragraphs, or different styles of headers, all carry useful information. After all, the goal of endowing texts with layout and formatting is to improve readability.

In this article we present one of the first attempts to enrich the state-of-the-art methods of NLP with layout understanding mechanisms, contemporaneous with [32], to which we compare our model. Our approach injects the layout information into a pretrained instance of RoBERTa. We fine-tune the augmented model on a dataset consisting of documents with non-trivial layout.

We evaluate our model on the end-to-end information extraction task, where the training set consists of documents and the target values of the properties to be extracted, without any additional annotations specifying the locations where the information on these properties can be found in the documents. We compare the results with a baseline RoBERTa model, which relies on the sequential order of tokens obtained from the OCR alone (and does not use the layout features), and with the solution of [31,32]. LAMBERT achieves superior performance on visually rich documents, without sacrificing results on more linear texts.

1.1 Related Work

There are two main lines of research into understanding documents with non-trivial layout. The first one is Document Layout Analysis (DLA), the goal of which is to identify contiguous blocks of text and other non-textual objects on the page and determine their function and order in the document. The obtained segmentation can be combined with the textual information contained in the detected blocks. This kind of method has recently been employed in [17].

Many services employ DLA functionality for OCR (which requires document segmentation), table detection or form field detection, and their capabilities are still expanding. The most notable examples are Amazon Textract [1], the Google Cloud Document Understanding AI platform [8], and Microsoft Cognitive Services [20]. However, each has limitations, such as the need to create rules for extracting information from the tables recognized by the system, or use training datasets with annotated document segments. More recent works on information extraction using DLA include, among others, [2,3,10,14,19,22,25]. They concentrate on specific types of documents, such as invoices or forms, where the layout plays a relatively greater role: more general documents may contain tables, but they can also have large amounts of unstructured text.

The second idea is to directly combine the methods of Computer Vision and NLP. This could be done, for instance, by representing a text-filled page as a multi-channel image, with channels corresponding to the features encoding the semantics of the underlying text, and, subsequently, using convolutional networks. This method was used, among others, by Chargrid and BERTgrid models [5,15]. On the other hand, LayoutLM [32] and TRIE [34] used the image recognition features of the page image itself. A more complex approach was taken by PICK [33], which separately processes the text and images of blocks identified in the document. In this way it computes the vertex embeddings of the block graph, which is then processed with a graph neural network.

Our idea is also related to the one used in [24], though in a different setting. They considered texts accompanied by audio-visual signal injected into a pretrained BERT instance, by combining it with the input embeddings.

LAMBERT has a different approach. It uses neither the raw document image, nor the block structure that has to be somehow inferred. It relies on the tokens and their bounding boxes alone, both of which are easily obtainable from any reasonable OCR system.

1.2   Contribution

Our main contribution is the introduction of a *Layout-Aware Language Model*, a general-purpose language model that views text not simply as a sequence of words, but as a collection of tokens on a two-dimensional page. As such it is able to process plain text documents, but also tables, headers, forms and various other visual elements. The implementation of the model is available at https://github.com/applicaai/lambert.

A key feature of this solution is that it retains the crucial trait of language models: the ability to learn in an unsupervised setting. This allows the exploitation of abundantly available unannotated public documents, and a transfer of the learned representations to downstream tasks. Another advantage is the simplicity of this approach, which requires only an augmentation of the input with token bounding boxes. In particular, no images are needed. This eliminates an important performance factor in industrial systems, where large volumes of documents have to be sent over a network between distributed processing services.

Another contribution of the paper is an extensive ablation study of the impact of augmenting RoBERTa with various types of additional positional embeddings on model performance on the SROIE [12], CORD [21], Kleister NDA and Kleister Charity datasets [27].

Finally, we created a new dataset for the unsupervised training of layout-aware language models. We will share a 200k document subset, amounting to 2M visually rich pages, accompanied by a dual classification of documents: business/legal documents with complex structure; and others. Due to IIT-CDIP Test Collection dataset [16] accessibility problems[1], this would constitute the largest widely available dataset for training layout-aware language models. It would allow researchers to compare the performance of their solutions not only on the same test sets, but also with the same training set. The dataset is published at https://github.com/applicaai/lambert, together with a more detailed description that is too long for this paper.

---

[1] The link https://ir.nist.gov/cdip/ seems to be dead (access on Feb 17, 2021).

2   Proposed Method

We inject the layout information into the model in two ways. Firstly, we modify the input embeddings of the original RoBERTa model by adding the layout term. We also experiment with completely removing the sequential embedding term. Secondly, we apply relative attention bias, used [11,23,26] in the context of sequential position. The final architecture is depicted in Fig. 1.

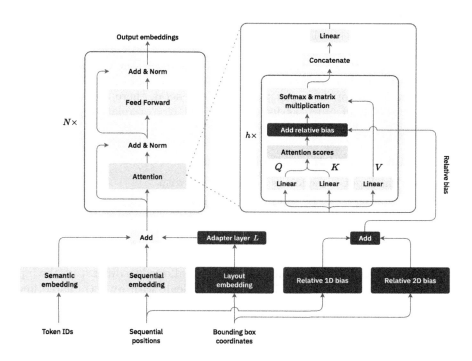

Fig. 1. LAMBERT model architecture. Differences with the plain RoBERTa model are indicated by white text on dark blue background. $N = 12$ is the number of transformer encoder layers, and $h = 12$ is the number of attention heads in each encoder layer. Q, K, and V are, respectively, the queries, keys and values obtained by projecting the self-attention inputs. (Color figure online)

2.1   Background

The basic Transformer encoder, used in, for instance, BERT [6] and RoBERTa [18], is a sequence-to-sequence model transforming a sequence of input embeddings $x_i \in \mathbb{R}^n$ into a sequence of output embeddings $y_i \in \mathbb{R}^m$ of the same length, for the input/output dimensions n and m. One of the main distinctive features of this architecture is that it discards the order of its input vectors. This allows parallelization levels unattainable for recurrent neural networks.

In such a setting, the information about the order of tokens is preserved not by the structure of the input. Instead, it is explicitly passed to the model, by defining the input embeddings as

$$x_i = s_i + p_i, \tag{1}$$

where $s_i \in \mathbb{R}^n$ is the semantic embedding of the token at position i, taken from a trainable embedding layer, while $p_i \in \mathbb{R}^n$ is a *positional embedding*, depending only on i. In order to avoid confusion, we will, henceforth, use the term *sequential embeddings* instead of *positional embeddings*, as the *positional* might be understood as relating to the 2-dimensional position on the page, which we will deal with separately.

Since in RoBERTa, on which we base our approach, the embeddings p_i are trainable, the number of pretrained embeddings (in this case 512) defines a limit on the length of the input sequence. In general, there are many ways to circumvent this limit, such as using predefined [28] or relative [4] sequential embeddings.

2.2 Modification of Input Embeddings

We replace the input embeddings defined in (1) with

$$x_i = s_i + p_i + L(\ell_i). \tag{2}$$

Here, $\ell_i \in \mathbb{R}^k$ stands for *layout embeddings*, which are described in detail in the next subsection. They carry the information about the position of the i-th token on the page.

The dimension k of the layout embeddings is allowed to differ from the input embedding dimension n, and this difference is dealt with by a trainable linear layer $L\colon \mathbb{R}^k \to \mathbb{R}^n$. However, our main motivation to introduce the adapter layer L was to gently increase the strength of the signal of layout embeddings during training. In this way, we initially avoided presenting the model with inputs that it was not prepared to deal with. Moreover, in theory, in the case of non-trainable layout embeddings, the adapter layer may be able to learn to project ℓ_i onto a subspace of the embedding space that reduces interference with the other terms in (2). For instance, it is possible for the image of the adapter layer to learn to be approximately orthogonal to the sum of the remaining terms. This would minimize any information loss caused by adding multiple vectors. While this was our theoretical motivation, and it would be interesting to investigate in detail how much of it actually holds, such detailed considerations of a single model component exceed the scope of this paper. We included the impact of using the adapter layer in the ablation study.

We initialize the weight matrix of L according to a normal distribution $\mathcal{N}(0, \sigma^2)$, with the standard deviation σ being a hyperparameter. We have to choose σ carefully, so that in the initial phase of training, the $L(\ell_i)$ term does not interfere overly with the already learned representations. We experimentally determined the value $\sigma = 0.02$ to be near-optimal[2].

---

[2] We tested the values 0.5, 0.1, 0.02, 0.004, and 0.0008.

2.3   Layout Embeddings

In our setting, a document is represented by a sequence of tokens t_i and their bounding boxes b_i. To each element of this sequence, we assign its layout embedding ℓ_i, carrying the information about the position of the token with respect to the whole document. This could be performed in various ways. What they all have in common is that the embeddings ℓ_i depend only on the bounding boxes b_i and not on the tokens t_i.

We base our layout embeddings on the method originally used in [7], and then in [28] to define the sequential embeddings. We first normalize the bounding boxes by translating them so that the upper left corner is at $(0,0)$, and dividing their dimensions by the page height. This causes the page bounding box to become $(0, 0, w, 1)$, where w is the normalized width.

The layout embedding of a token will be defined as the concatenation of four embeddings of the individual coordinates of its bounding box. For an integer d and a vector of scaling factors $\theta \in \mathbb{R}^d$, we define the corresponding embedding of a single coordinate t as

$$\mathrm{emb}_\theta(t) = (\sin(t\theta); \cos(t\theta)) \in \mathbb{R}^{2d}, \tag{3}$$

where the sin and cos are performed element-wise, yielding two vectors in $\mathbb{R}^d$. The resulting concatenation of single bounding box coordinate embeddings is then a vector in $\mathbb{R}^{8d}$.

In [28, Section 3.5], and subsequently in other Transformer-based models with precomputed sequential embeddings, the sequential embeddings were defined by emb_θ with θ being a geometric progression interpolating between 1 and 10^{-4}. Unlike the sequential position, which is a potentially large integer, bounding box coordinates are normalized to the interval $[0, 1]$. Hence, for our layout embeddings we use larger scaling factors (θ_r), namely a geometric sequence of length $n/8$ interpolating between 1 and 500, where n is the dimension of the input embeddings.

2.4   Relative Bias

Let us recall that in a typical Transformer encoder, a single attention head transforms its input vectors into three sequences: queries $q_i \in \mathbb{R}^d$, keys $k_i \in \mathbb{R}^d$, and values $v_i \in \mathbb{R}^d$. The raw attention scores are then computed as $\alpha_{ij} = d^{-1/2} q_i^T k_j$. Afterwards, they are normalized using softmax, and used as weights in linear combinations of value vectors.

The point of relative bias is to modify the computation of the raw attention scores by introducing a bias term: $\alpha'_{ij} = \alpha_{ij} + \beta_{ij}$. In the sequential setting, $\beta_{ij} = W(i-j)$ is a trainable weight, depending on the relative sequential position of tokens i and j. This form of attention bias was introduced in [23], and we will refer to it as *sequential attention bias*.

We introduce a simple and natural extension of this mechanism to the two-dimensional context. In our case, the bias β_{ij} depends on the relative positions of the tokens. More precisely, let $C \gg 1$ be an integer resolution factor (the number of cells in a grid used to discretize the normalized coordinates). If $b_i = (x_1, y_1, x_2, y_2)$ is the normalized bounding box of the i-th token, we first reduce it to a 2-dimensional position $(\xi_i, \eta_i) = (Cx_1, C(y_1 + y_2)/2)$, and then define

$$\beta_{ij} = H(\lfloor \xi_i - \xi_j \rfloor) + V(\lfloor \eta_i - \eta_j \rfloor), \tag{4}$$

where $H(\ell)$ and $V(\ell)$ are trainable weights defined for every integer $\ell \in [-C, C)$. A good value for C should allow for a distinction between consecutive lines and tokens, without unnecessarily affecting performance. For a typical document $C = 100$ is enough, and we fix this in our experiments.

This form of attention bias will be referred to as *2D attention bias*. We suspect that it should help in analyzing, say, tables by allowing the learning of relationships between cells.

3 Experiments

All experiments were performed on 8 NVIDIA Tesla V100 32 GB GPUs. As our pretrained base model we used RoBERTa in its smaller, base variant (125M parameters, 12 layers, 12 attention heads, hidden dimension 768). This was also employed as the baseline, after additional training on the same dataset we used for LAMBERT. The implementation and pretrained weights from the `transformers` library [30] were used.

In the LAMBERT model, we used the layout embeddings of dimension $k = 128$, and initialized the adapter layer L with standard deviation $\sigma = 0.02$, as noted in Sect. 2. For comparison, in our experiments, we also included the published version of the LayoutLM model [32], which is of a similar size.

The models were trained on a masked language modeling objective extended with layout information (with the same settings as the original RoBERTa [18]); and subsequently, on downstream information extraction tasks. In the remainder of the paper, these two stages will be referred to as, respectively, *training* and *fine-tuning*.

Training was performed on a collection of PDFs extracted from *Common Crawl* made up of a variety of documents (we randomly selected up to 10 documents from any single domain). The documents were processed with an OCR system, `Tesseract 4.1.1-rc1-7-gb36c`, to obtain token bounding boxes. The final model was trained on the subset of the corpus consisting of business documents with non-trivial layout, filtered by an SVM binary classifier, totaling to approximately 315k documents (3.12M pages). The SVM model was trained on 700 manually annotated PDF files to distinguish between business (e.g. invoices, forms) and non-business documents (e.g. poems, scientific texts).

In the training phase, we used the Adam optimizer with the weight decay fix from [30]. We employed a learning rate scheduling method similar to the one used in [6], increasing the learning rate linearly from 0 to 1e−4 for the warm-up period of 10% of the training time and then decreasing it linearly to 0. The final model was trained with batch size of 128 sequences (amounting to 64K tokens) for approximately 1000k steps (corresponding to training on 3M pages for 25 epochs). This took about 5 days to complete a single experiment.

After training our models, we fine-tuned and evaluated them independently on multiple downstream end-to-end information extraction tasks. Each evaluation dataset was split into training, validation and test subsets. The models were extended with a simple classification head on top, consisting of a single linear layer, and fine-tuned on the task of classifying entity types of tokens. We employed early stopping based on the F_1-score achieved on the validation part of the dataset. We used the Adam optimizer again, but this time without the learning rate warm-up, as it turned out to have no impact on the results.

The extended model operates as a tagger on the token level, allowing for the classification of separate tokens, while the datasets contain only the values of properties that we are supposed to extract from the documents. Therefore, the further processing of output is required. To this end, we use the pipeline described in [27].

Every contiguous sequence of tokens tagged as a given entity type is treated as a recognized entity and assigned a score equal to the geometric mean of the scores of its constituent tokens. Then, every recognized entity undergoes a normalization procedure specific to its general data type (e.g. date, monetary amount, address, etc.). This is performed using regular expressions: for instance, the date July, 15th 2013 is converted to 2013-07-15. Afterwards, duplicates are aggregated by summing their scores, leading to a preference for entities detected multiple times. Finally, the highest-scoring normalized entity is selected as the output of the information extraction system. The predictions obtained this way are compared with target values provided in the dataset using F_1-score as the evaluation metric. See [27] for more details.

4   Results

We evaluated our models on four public datasets containing visually rich documents. The Kleister NDA and Kleister Charity datasets are part of a larger Kleister dataset, recently made public [27] (many examples of documents, and detailed descriptions of extraction tasks can be found therein). The NDA set consists of legal agreements, whose layout variety is limited. It should probably be treated as a plain-text dataset. The Charity dataset on the other hand contains reports of UK charity organizations, which include various tables, diagrams and other graphic elements, interspersed with text passages. All Kleister datasets come with predefined train/dev/test splits, with 254/83/203 documents for NDA and 1729/440/609 for Charity.

The SROIE [12] and CORD [21] datasets are composed of scanned and OCRed receipts. Documents in SROIE are annotated with four target entities to be extracted, while in CORD there are 30 different entities. We use the public 1000 samples from the CORD dataset with the train/dev/test split proposed by the authors of the dataset (respectively, 800/100/100). As for SROIE, it consists of a public training part, and test part with unknown annotations. For the purpose of ablation studies, we further subdivided the public part of SROIE into

Table 1. Comparison of F_1-scores for the considered models. Best results in each column are indicated in bold. In parentheses, the length of training of our models, expressed in non-unique pages, is presented for comparison. For RoBERTa, the first row corresponds to the original pretrained model without any further training, while in the second row the model was trained on our dataset. [a]result obtained from relevant publication; [b]result of a single model, obtained from the SROIE leaderboard [13]

Model	Params	Our experiments				External results	
		NDA	Charity	SROIE*	CORD	SROIE	CORD
RoBERTa [18]	125M	77.91	76.36	94.05	91.57	92.39[b]	–
RoBERTa (16M)	125M	78.50	77.88	94.28	91.98	93.03[b]	–
LayoutLM [32]	113M	77.50	77.20	94.00	93.82	94.38[a]	94.72[a]
	343M	79.14	77.13	96.48	93.62	97.09[b]	94.93[a]
LayoutLMv2 [31]	200M	–	–	–	–	96.25[a]	94.95[a]
	426M	–	–	–	–	97.81[b]	**96.01**[a]
LAMBERT (16M)	125M	80.31	79.94	96.24	93.75	–	–
LAMBERT (75M)	125M	**80.42**	**81.34**	**96.93**	**94.41**	**98.17**[b]	–

training and test subsets (546/80 documents; due to the lack of a validation set in this split, we fine-tuned for 15 epochs instead of employing early stopping). We refer to this split as SROIE*, while the name SROIE is reserved for the original SROIE dataset, where the final evaluation on the test set is performed through the leaderboard [13].

In Table 1, we present the evaluation results achieved on downstream tasks by the trained models. With the exception of the Kleister Charity dataset, where only 5 runs were made, each of the remaining experiments were repeated 20 times, and the mean result was reported. We compare LAMBERT with baseline RoBERTa (trained on our dataset) and the original RoBERTa [18] (without additional training); LayoutLM [32]; and LayoutLMv2 [31]. The LayoutLM model published by its authors was plugged into the same pipeline that we used for LAMBERT and RoBERTa. In the first four columns we present averaged results of our experiments, and for CORD and SROIE we additionally provide the results reported by the authors of LayoutLM, and presented on the leaderboard [13].

Since the LayoutLMv2 model was not publicly available at the time of preparing this article, we could not perform experiments ourselves. As a result some of the results are missing. For CORD, we present the scores given in [31], where the authors did not mention, though, whether they averaged over multiple runs, or used just a single model. A similar situation occurs for LayoutLM; we presented the average results of 20 runs (best run of LAMBERT attained the score of 95.12), which are lower than the scores presented in [31]. The difference could be attributed to using a different end-to-end evaluation pipeline, or averaging (if the results in [31,32] come from a single run).

For the full SROIE dataset, most of the results were retrieved from the public leaderboard [13], and therefore they come from a single model. For the base variants of LayoutLM and LayoutLMv2, the results were unavailable, and we present the scores from the corresponding papers.

In our experiments, the base variant of LAMBERT achieved top scores for all datasets. However, in the case of CORD, the result reported in [31] for the large variant of LayoutLMv2 is superior. If we consider the best scores of LAMBERT (95.12) instead of the average, and the scores of LayoutLM reported in [32], LAMBERT slightly outperforms LayoutLM, while still being inferior to LayoutLMv2. Due to the lack of details on the results of LayoutLM, it is unknown which of these comparisons is valid.

For Kleister datasets, the base variant (and in the case of Charity, also the large variant) of LayoutLM did not outperform the baseline RoBERTa. We suspect that this might be the result of LayoutLM being better attuned to the evaluation pipeline used by its authors, and the fact that it was based on an uncased language model. In the Kleister dataset, meanwhile, performance for entities such as names may depend on casing.

5   Hyperparameters and Ablation Studies

In order to investigate the impact of our modifications to RoBERTa, we performed an extensive study of hyperparameters and the various components of the final model. We investigated the dimension of layout embeddings, the impact of the adapter layer L, the size of training dataset, and finally we performed a detailed ablation study of the embeddings and attention biases we had used to augment the baseline model.

In the studies, every model was fine-tuned and evaluated 20 times on each dataset, except for Kleister Charity dataset, on which we fine-tuned every model 5 times: evaluations took much longer on Kleister Charity. For each model and dataset combination, the mean score was reported, together with the two-sided 95% confidence interval, computed using the corresponding t-value. We considered differences to be significant when the corresponding intervals were disjoint. All the results are presented in Table 2, which is divided into sections corresponding to different studies. The F_1-scores are reported as *increases* with respect to the reported mean baseline score, to improve readability.

5.1   Baseline

As a baseline for the studies we use the publicly available pretrained base variant of the RoBERTa model with 12 layers, 12 attention heads, and hidden dimension 768. We additionally trained this model on our training set, and fine-tuned it on the evaluation datasets in a manner analogous to LAMBERT.

Table 2. Improvements of F_1-score over the baseline for various variants of LAMBERT model. The first row (with grayed background) contains the F_1-scores of the baseline RoBERTa model. The other grayed row corresponds to full LAMBERT. Statistically insignificant improvements over the baseline are grayed. In each of three studies, the best result together with all results insignificantly smaller are in bold. [a]filtered datasets; [b]model with a disabled adapter layer

Train epochs and pages	Embeddings dimension	Sequential	Seq. bias	Layout	2D bias	NDA	Charity	SROIE*	CORD
		•				$78.50_{\pm1.16}$	$77.88_{\pm0.48}$	$94.28_{\pm0.42}$	$91.98_{\pm0.62}$
		•	•			$2.42_{\pm0.61}$	$0.52_{\pm0.64}$	$0.79_{\pm0.17}$	$0.03_{\pm0.57}$
				•		$1.25_{\pm0.59}$	$\mathbf{2.62_{\pm0.80}}$	$\mathbf{1.86_{\pm0.15}}$	$0.89_{\pm0.83}$
					•	$-0.49_{\pm0.62}$	$2.02_{\pm0.48}$	$0.53_{\pm0.28}$	$-0.17_{\pm0.62}$
				•	•	$0.88_{\pm0.50}$	$\mathbf{3.00_{\pm0.37}}$	$1.94_{\pm0.16}$	$0.68_{\pm0.62}$
8×2M	128	•		•		$1.74_{\pm0.67}$	$0.06_{\pm0.93}$	$1.94_{\pm0.18}$	$1.42_{\pm0.53}$
		•			•	$1.73_{\pm0.60}$	$2.02_{\pm0.53}$	$\mathbf{2.09_{\pm0.22}}$	$\mathbf{1.93_{\pm0.71}}$
		•		•	•	$0.54_{\pm0.85}$	$1.84_{\pm0.42}$	$\mathbf{2.08_{\pm0.38}}$	$\mathbf{2.15_{\pm0.65}}$
		•	•	•		$1.66_{\pm0.76}$	$0.32_{\pm1.35}$	$\mathbf{1.75_{\pm0.35}}$	$1.06_{\pm0.54}$
		•	•		•	$0.85_{\pm0.91}$	$1.84_{\pm0.27}$	$2.01_{\pm0.24}$	$\mathbf{1.95_{\pm0.46}}$
		•	•	•	•	$\mathbf{1.81_{\pm0.60}}$	$\mathbf{2.06_{\pm0.69}}$	$1.96_{\pm0.16}$	$\mathbf{1.77_{\pm0.46}}$
	128	•		•		$1.74_{\pm0.67}$	$0.06_{\pm0.93}$	$1.94_{\pm0.18}$	$1.42_{\pm0.53}$
8×2M	384	•		•		$0.90_{\pm0.54}$	$0.70_{\pm0.40}$	$1.86_{\pm0.22}$	$1.51_{\pm0.60}$
	768	•		•		$0.71_{\pm1.04}$	$0.50_{\pm0.85}$	$\mathbf{2.18_{\pm0.25}}$	$\mathbf{1.54_{\pm0.51}}$
	768[b]	•		•		$0.77_{\pm0.58}$	$\mathbf{2.30_{\pm0.20}}$	$0.37_{\pm0.15}$	$\mathbf{1.58_{\pm0.52}}$
8×2M		•	•	•	•	$\mathbf{1.81_{\pm0.60}}$	$2.06_{\pm0.26}$	$1.96_{\pm0.18}$	$1.77_{\pm0.46}$
8×2M[a]	128	•	•	•	•	$\mathbf{1.86_{\pm0.66}}$	$1.92_{\pm0.19}$	$2.60_{\pm0.18}$	$1.59_{\pm0.61}$
25×3M[a]		•	•	•	•	$\mathbf{1.92_{\pm0.50}}$	$\mathbf{3.46_{\pm0.21}}$	$\mathbf{2.65_{\pm0.13}}$	$\mathbf{2.43_{\pm0.19}}$

5.2   Embeddings and Biases

In this study we disabled various combinations of input embeddings and attention biases. The models were trained on 2M pages for 8 epochs, with 128-dimensional layout embeddings (if enabled). The resulting models were divided into three groups. The first one contains sequential-only combinations which do not employ the layout information at all, including the baseline. The second group consists of models using only the bounding box coordinates, with no access to sequential token positions. Finally, the models in the third group use both sequential and layout inputs. In this group we did not disable the sequential embeddings. It includes the full LAMBERT model, with all embeddings and attention biases enabled.

Generally, we observe that none of the modifications has led to a significant performance deterioration. Among the models considered, the only one which reported a significant improvement for all four datasets—and at the same time, the best improvement—was the full LAMBERT.

For the Kleister datasets the variance in results was relatively higher than in the case of SROIE* and CORD. This led to wider confidence intervals, and reduced the number of significant outcomes. This is true especially for the Kleister NDA dataset, which is the smallest one. In Kleister NDA, significant improvements were achieved for both sequential-only models, and for full LAMBERT. The differences between these increases were insignificant. It would seem that, for sequential-only models, the sequential attention bias is responsible for the improvement. But after adding the layout inputs, it no longer leads to improvements when unaccompanied by other modifications. Still, achieving better results on sequential-only inputs may be related to the plain text nature of the Kleister NDA dataset.

While other models did not report significant improvement over the baseline, there are still some differences between them to be observed. The model using only 2D attention bias is significantly inferior to most of the others. This seems to agree with the intuition that relative 2D positions are the least suitable way to pass positional information about plain text.

In the case of the Kleister Charity dataset, significant improvements were achieved by all layout-only models, and all models using the 2D attention bias. Best improvement was attained by full LAMBERT, and two layout-only models using the layout embeddings; the 2D attention bias used alone improved the results significantly, but did not reach the top score. The confidence intervals are too wide to offer further conclusions, and many more experiments will be needed to increase the significance of the results.

For the SROIE* dataset, except for two models augmented only with a single attention bias, all improvements proved significant. Moreover, the differences between all the models using layout inputs are insignificant. We may conclude that passing bounding box coordinates in any way, except through 2D attention bias, significantly improves the results. As to the lack of significant improvements for 2D attention bias, we hypothesize that this is due to its relative nature. In all other models the absolute position of tokens is somehow known, either through the layout embeddings, or the sequential position. When a human reads a receipt, the absolute position is one of the main features used to locate the typical positions of entities.

For CORD, which is the more complex of the two receipt datasets, significant improvements were observed only for combined sequential and layout models. In this group, the model using both sequential and layout embeddings, augmented with sequential attention bias, did not yield a significant improvement. There were no significant differences among the remaining models in the group. Contrary to the case of SROIE*, none of the layout-only models achieved significant improvement.

5.3   Layout Embedding Dimension

In this study we evaluated four models, using both sequential and layout embeddings, varying the dimension of the latter. We considered 128-, 384-, and 768-dimensional embeddings. Since this is the same as for the input embeddings of RoBERTa$_{BASE}$, it was possible to remove the adapter layer L, and treat this as another variant, in Table 2 denoted as 768^b.

In Kleister NDA, there were no significant differences between any of the evaluated models, and no improvements over the baseline. On the other hand, in Kleister Charity, disabling the adapter layer and using the 768-dimensional layout embeddings led to significantly better performance. These results remain consistent with earlier observations that in Kleister NDA the best results were achieved by sequential-only models, while in the case of Kleister Charity, by layout-only models. It seems that in the case of NDA the performance is influenced mostly by the sequential features, while in the case of Charity, removing the adapter layer increases the strength of the signal of the layout embeddings, carrying the layout features which are the main factor affecting performance.

In SROIE* and CORD all results were comparable, with one exception, namely in SROIE*, the model with the disabled adapter layer did not, unlike the remaining models, perform significantly better than the baseline.

5.4   Training Dataset Size

In this study, following the observations from [9], we considered models trained on 3 different datasets. The first model was trained for 8 epochs on 2M unfiltered (see Sect. 3 for more details of the filtering procedure) pages. In the second model, we used the same training time and dataset size, but this time only filtered pages were used. Finally, the third model was trained for 25 epochs on 3M filtered pages.

It is not surprising that increasing the training time and dataset size, leads to an improvement in results, at least up to a certain point. In the case of Kleister NDA dataset, there were no significant differences in the results. For Kleister Charity, the best result was achieved for the largest training dataset, with 75M filtered pages. This result was also significantly better than the outcomes for the smaller dataset. In the case of SROIE* the two models trained on datasets with filtered documents achieved a significantly higher score than the one trained on unfiltered documents. There was, in fact, no significant difference between these two models. This supports the hypothesis that, in this case, filtering could be the more important factor. Finally, for CORD the situation is similar to Kleister Charity.

6   Conclusions and Further Research

We introduced LAMBERT, a layout-aware language model, producing contextualized token embeddings for tokens contained in formatted documents. The model can be trained in an unsupervised manner. For the end user, the only difference with classical language models is the use of bounding box coordinates as additional input. No images are needed, which makes this solution particularly simple, and easy to include in pipelines that are already based on OCR-ed documents.

The LAMBERT model outperforms the baseline RoBERTa on information extraction from visually rich documents, without sacrificing performance on documents with a flat layout. This can be clearly seen in the results for the Kleister NDA dataset. Its base variant with around 125M parameters is also able to compete with the large variants of LayoutLM (343M parameters) and LayoutLMv2 (426M parameters), with Kleister and SROIE datasets achieving superior results. In particular, $LAMBERT_{BASE}$ achieved first place on the Key Information Extraction from the SROIE dataset leaderboard [13].

The choice of particular LAMBERT components is supported by an ablation study including confidence intervals, and is shown to be statistically significant. Another conclusion from this study is that for visually rich documents the point where no more improvement is attained by increasing the training set has not yet been reached. Thus, LAMBERT's performance can still be improved upon by simply increasing the unsupervised training set. In the future we plan to experiment with increasing the model size, and training datasets.

Further research is needed to ascertain the impact of the adapter layer L on the model performance, as the results of the ablation study were inconclusive. It would also be interesting to understand whether the mechanism through which it affects the results is consistent with the hypotheses formulated in Sect. 2.

Acknowledgments. The authors would like to acknowledge the support the Applica.ai project has received as being co-financed by the European Regional Development Fund (POIR.01.01.01–00–0605/19–00).

References

1. Amazon: Amazon Textract (2019). https://aws.amazon.com/textract/. Accessed 25 Nov 2019
2. Bart, E., Sarkar, P.: Information extraction by finding repeated structure. In: DAS 2010 (2010)
3. Cesarini, F., Francesconi, E., Gori, M., Soda, G.: Analysis and understanding of multi-class invoices. IJDAR **6**, 102–114 (2003)
4. Dai, Z., et al.: Transformer-XL: Attentive language models beyond a fixed-length context. In: ACL (2019)
5. Denk, T.I., Reisswig, C.: BERTgrid: contextualized Embedding for 2D document representation and understanding. In: Workshop on Document Intelligence at NeurIPS 2019 (2019)
6. Devlin, J., Chang, M.W., Lee, K., Toutanova, K.: BERT: pre-training of deep bidirectional transformers for language understanding. In: NAACL-HLT (2019)
7. Gehring, J., Auli, M., Grangier, D., Yarats, D., Dauphin, Y.N.: Convolutional sequence to sequence learning. In: ICML (2017)

8. Google: Cloud Document Understanding AI (2019). https://cloud.google.com/document-understanding/docs/. Accessed 25 Nov 2019

9. Gururangan, S., Marasović, A., Swayamdipta, S., Lo, K., Beltagy, I., Downey, D., Smith, N.A.: Don't stop pretraining: adapt language models to domains and tasks. In: Proceedings of the 58th Annual Meeting of the Association for Computational Linguistics, pp. 8342–8360. Association for Computational Linguistics (2020). https://doi.org/10.18653/v1/2020.acl-main.740

10. Hamza, H., Belaïd, Y., Belaïd, A., Chaudhuri, B.: An end-to-end administrative document analysis system. In: 2008 The Eighth IAPR International Workshop on Document Analysis Systems, pp. 175–182 (2008)

11. Huang, Y., et al.: Gpipe: Efficient training of giant neural networks using pipeline parallelism. In: NeurIPS (2019)

12. ICDAR: Competition on Scanned Receipts OCR and Information Extraction (2019). https://rrc.cvc.uab.es/?ch=13. Accessed 21 Feb 2021

13. ICDAR: Leaderboard of the Information Extraction Task, Robust Reading Competition (2020). https://rrc.cvc.uab.es/?ch=13&com=evaluation&task=3. Accessed 7 Apr 2020

14. Ishitani, Y.: Model-based information extraction method tolerant of ocr errors for document images. Int. J. Comput. Process. Orient. Lang. **15**, 165–186 (2002)

15. Katti, A.R., et al.: Chargrid: towards understanding 2D documents. In: EMNLP (2018)

16. Lewis, D., Agam, G., Argamon, S., Frieder, O., Grossman, D., Heard, J.: Building a test collection for complex document information processing. In: Proceedings of the 29th Annual International ACM SIGIR Conference on Research and Development in Information Retrieval (2006)

17. Liu, X., Gao, F., Zhang, Q., Zhao, H.: Graph convolution for multimodal information extraction from visually rich documents. In: NAACL-HLT (2019)

18. Liu, Y., et al.: RoBERTa: A Robustly Optimized BERT Pretraining Approach. ArXiv arXiv:1907.11692 (2019)

19. Medvet, E., Bartoli, A., Davanzo, G.: A probabilistic approach to printed document understanding. IJDAR **14**, 335–347 (2011)

20. Microsoft: Cognitive Services (2019). https://azure.microsoft.com/en-us/services/cognitive-services/. Accessed 25 Nov 2019

21. Park, S., et al.: CORD: A Consolidated Receipt Dataset for Post-OCR Parsing. In: Document Intelligence Workshop at Neural Information Processing Systems (2019)

22. Peanho, C., Stagni, H., Silva, F.: Semantic information extraction from images of complex documents. Appl. Intell. **37**, 543–557 (2012)

23. Raffel, C., et al.: Exploring the limits of transfer learning with a unified text-to-text transformer. J. Mach. Learn. Res. **21**(140), 1–67 (2020)

24. Rahman, W., et al.: Integrating multimodal information in large pretrained transformers. In: ACL (2020)

25. Rusinol, M., Benkhelfallah, T., Poulain d'Andecy, V.: Field extraction from administrative documents by incremental structural templates. In: ICDAR (2013)

26. Shaw, P., Uszkoreit, J., Vaswani, A.: Self-attention with relative position representations. In: NAACL-HLT (2018)

27. Stanisławek, T., et al.: Kleister: A novel task for information extraction involving long documents with complex layout (2021) . ArXiv arXiv:2105.05796 Accepted to ICDAR 2021

28. Vaswani, A., et al.: Attention is all you need. In: Advances in Neural Information Processing Systems 30 (2017)

29. Wang, A., Singh, A., Michael, J., Hill, F., Levy, O., Bowman, S.R.: GLUE: a multi-task benchmark and analysis platform for natural language understanding. In: Proceedings of ICLR (2019). https://gluebenchmark.com/. Accessed 26 Nov 2019

30. Wolf, T., et al.: Transformers: state-of-the-art natural language processing. In: Proceedings of the 2020 Conference on Empirical Methods in Natural Language Processing: System Demonstrations, pp. 38–45. Association for Computational Linguistics, Online (October 2020). https://www.aclweb.org/anthology/2020.emnlp-demos.6

31. Xu, Y., et al.: LayoutLMv2: Multi-modal pre-training for visually-rich document understanding. arXiv arXiv:2012.14740 (2020)

32. Xu, Y., Li, M., Cui, L., Huang, S., Wei, F., Zhou, M.: LayoutLM: pre-training of text and layout for document image understanding. In: Proceedings of the 26th ACM SIGKDD International Conference on Knowledge Discovery & Data Mining, pp. 1192–1200 (2020)

33. Yu, W., Lu, N., Qi, X., Gong, P., Xiao, R.: PICK: Processing key information extraction from documents using improved graph learning-convolutional networks. In: 2020 25th International Conference on Pattern Recognition (ICPR), pp. 4363–4370 (2021). https://doi.org/10.1109/ICPR48806.2021.9412927

34. Zhang, P., et al.: TRIE: end-to-end text reading and information extraction for document understanding. In: Proceedings of the 28th ACM International Conference on Multimedia (2020)

ViBERTgrid: A Jointly Trained Multi-modal 2D Document Representation for Key Information Extraction from Documents

Weihong Lin[1]([✉]), Qifang Gao[2], Lei Sun[1], Zhuoyao Zhong[1], Kai Hu[3], Qin Ren[2], and Qiang Huo[1]

[1] Microsoft Research Asia, Beijing, China
{weihlin,lsun,zhuoyao.zhong,qianghuo}@microsoft.com
[2] School of Software and Microelectronics, Peking University, Beijing, China
{v-qgao,v-qinren}@microsoft.com
[3] Department of EEIS., University of Science and Technology of China, Hefei, China
v-kahu1@microsoft.com

Abstract. Recent grid-based document representations like BERTgrid allow the simultaneous encoding of the textual and layout information of a document in a 2D feature map so that state-of-the-art image segmentation and/or object detection models can be straightforwardly leveraged to extract key information from documents. However, such methods have not achieved comparable performance to state-of-the-art sequence- and graph-based methods such as LayoutLM and PICK yet. In this paper, we propose a new multi-modal backbone network by concatenating a BERTgrid to an intermediate layer of a CNN model, where the input of CNN is a document image and the BERTgrid is a grid of word embeddings, to generate a more powerful grid-based document representation, named ViBERTgrid. Unlike BERTgrid, the parameters of BERT and CNN in our multi-modal backbone network are trained jointly. Our experimental results demonstrate that this joint training strategy improves significantly the representation ability of ViBERTgrid. Consequently, our ViBERTgrid-based key information extraction approach has achieved state-of-the-art performance on real-world datasets.

Keywords: Key information extraction · Multimodal document representation learning · Joint training of BERT and CNN · ViBERTgrid

1 Introduction

Key Information Extraction (KIE) is the task of extracting the values of a number of predefined key fields from documents such as invoices, purchase orders, receipts, contracts, financial reports, insurance quotes, and many more, which is illustrated in Fig. 1. KIE is an essential technology for many large-scale document processing scenarios such as fast indexing, efficient archiving, automatic financial auditing and so on.

W. Lin and Q. Gao—Equal contribution.
Q. Gao, K. Hu and Q. Ren—This work was done when Qifang Gao, Kai Hu and Qin Ren were interns in Speech Group, Microsoft Research Asia, Beijing, China.

© Springer Nature Switzerland AG 2021
J. Lladós et al. (Eds.): ICDAR 2021, LNCS 12821, pp. 548–563, 2021.
https://doi.org/10.1007/978-3-030-86549-8_35

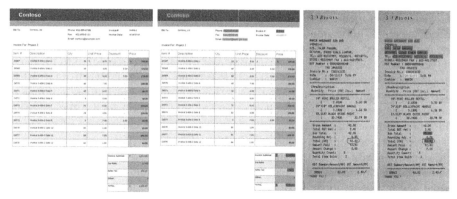

Fig. 1. Two examples of KIE task. Some key fields, including company name, address, phone number, date, email, fax number, invoice number, subtotal, tax and total amount,, are extracted from a fake invoice (https://azure.microsoft.com/en-us/services/cognitive-services/form-recognizer/ - feature) and a receipt from the SROIE [1] dataset.

With the rapid development of deep learning, many deep learning based KIE approaches have emerged and significantly outperformed traditional rule-based and template-based methods in terms of both accuracy and capability. Most of these approaches simply treat KIE as a token classification problem and use different deep learning models to predict the field type of each document token which could be a character, a subword, or a word. According to their used document representation methods, these approaches could be roughly categorized into three types: sequence-based, graph-based, and grid-based. Sequence-based approaches need to serialize a document into a 1D text sequence first, then use existing sequence tagging models in NLP (e.g., [2–6]) to extract field values. To reduce the influence of serialization, earlier methods [7–9] tried to encode the 2D position information of tokens into their token embeddings, but they still relied on the accurate serialization of the text segments so that it was hard to apply them to documents with complex layouts where the serialization step is nontrivial. Two recent works, LayoutLM [10] and LAMBERT [11], proposed to add a 2D position embedding to the input token embedding of the BERT model. After being pre-trained on a large-scale document dataset, its output token embedding could capture the spatial relationship among tokens within a document, which makes LayoutLM and LAMBERT less sensitive to the serialization step. Pramanik et al. [12] and LayoutLMv2 [13] extended the idea of LayoutLM further by integrating the image information in the pre-training stage to learn a stronger multimodal document representation. However, these pre-training based methods rely on large datasets and computation resources to achieve good performance. Graph-based approaches model each document page as a graph, in which text segments (words or text-lines) are represented as nodes. The initial representation of each node could combine visual, textual and positional features of its corresponding text segment [14]. Then graph neural networks or self-attention operations [15] are used to propagate information between neighboring nodes in the graph to get a richer representation for each node. After that, some methods like [16] fed these enhanced node embeddings to a classifier to do field type classification directly, while some other methods [14, 17–22] concatenated each node embedding to all token embeddings in its corresponding text segment, which were then fed into sequence tagging models to extract field values. Recent graph-based methods like PICK [14], TRIE [21] and VIES [22] have achieved superior performance on the public SROIE dataset [1]. Grid-based approaches like Chargrid [23], Dang et al. [24], CUTIE [25], BERTgrid [26] represented each document as a 2D grid of token embeddings and then use standard instance segmentation models to extract field values from the 2D grid. These grid representations have preserved the textual and layout information of documents but missed image texture information. So VisualWordGrid [27] combined these grid representations with 2D feature maps of document images to generate more powerful multimodal 2D document representations, which could simultaneously preserve visual, textual and layout information of documents. However, we find that these grid-based approaches cannot achieve comparable performance against the state-of-the-art methods like LayoutLM, PICK, TRIE and VIES due to two reasons: 1) Methods including Chargrid, Dang et al. [24], CUTIE and VisualWordGrid don't leverage state-of-the-art contextualized word embedding techniques like BERT to extract strong enough token embeddings; 2) Although BERTgrid incorporates BERT into the grid representation, the parameters of pretrained BERT are fixed during model training so that the potential of BERT-based token embedding is not fully exploited.

In this paper, we propose two effective techniques to significantly improve the accuracy of grid-based KIE methods. First, we propose a new multi-modal backbone network by combining the best of BERTgrid and CNN to generate a more powerful grid-based document representation, named ViBERTgrid, to simultaneously encode the textual, layout and visual information of a document in a 2D feature map. Second, we propose a new joint training strategy to finetune the parameters of both BERT and CNN so that the representation ability of ViBERTgrid is significantly improved. Consequently, our ViBERTgrid-based KIE approach has achieved superior performance on real-world datasets.

2   Related Works

Key information extraction from documents has been studied for decades. Before the advent of deep learning based methods, early works [28–33] mainly depended on some rules or human-designed features in known templates, therefore they usually failed on unseen templates and were not scalable in practical applications. With the development of deep learning, significant progress has been made in KIE. As mentioned above, most of modern deep learning based approaches formulate KIE as a token classification problem. In addition to the above-mentioned works, another work [34] proposed a CNN-based method to jointly perform handwritten text detection, transcription and named entity recognition from input document images. Other than this formulation, KIE can also be formulated as other problems. Majumder et al. [35] proposed a representation learning approach to extracting the values of key fields with prior knowledge. For each field, some candidate words were firstly selected. Then, the feature of each word was embedded with its contextual information and the cosine similarity between this embedding and the embedding of target field was calculated as a similarity score. SPADE [36] formulated

KIE as a spatial dependency parsing problem. It constructed a dependency graph with text segments and fields as the graph nodes first, then used a decoder to extract field values from identified connectivity between graph nodes. BROS [37] improved SPADE further by proposing a new position encoding method and an area-masking based pre-training objective. Another category of methods [38–40] adapted sequence-to-sequence models used in other NLP or image captioning tasks to directly predict all the values of key fields without requiring word-level supervision. Our proposed multimodal 2D document representation can be easily integrated into most of these approaches. Exploring the effectiveness of our document representation for other KIE frameworks will be one of our future works.

Other than KIE, grid-based document representations have also been studied in other document understanding tasks. Xiao et al. [41] constructed a 2D text embedding map with sentence embeddings and combined this textual map and visual features with a fully convolutional network for pixel-level segmentation of image regions such as table, section heading, caption, paragraph and so on. Raphaël et al. [42] proposed a multimodal neural model by concatenating a 2D text embedding to an intermediate layer of a CNN model for a more fine-grained segmentation task on historical newspapers. Unlike our approach, these methods didn't leverage state-of-the-art BERT models to generate text embedding maps and the parameters of their text embedding models and CNN models were not jointly trained.

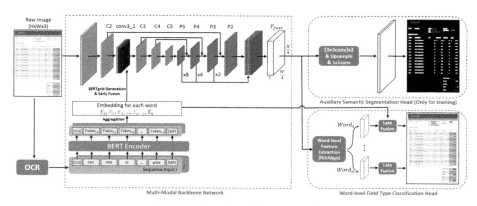

Fig. 2. Overview of our approach.

3 Methodology

As illustrated in Fig. 2, our approach is composed of three key components: 1) A multimodal backbone network used to generate a ViBERTgrid feature map; 2) An auxiliary semantic segmentation head used to perform pixel-wise field type classification; 3) A word-level field type classification head used to predict the field type of each word extracted by an OCR engine. Details of these modules and the proposed joint training strategy will be described in detail in the following subsections.

3.1 ViBERTgrid

3.1.1 BERTgrid Generation

We follow BERTgrid [26] to construct a 2D word grid, where word embeddings are generated by a pretrained BERT-style language model. Given an input document image, words and their bounding boxes are extracted with an OCR engine. These words are then organized as a unidimensional sequence of length N, denoted as $D = (w^{(1)}, w^{(2)}, \cdots, w^{(N)})$, by reading them in a top-left to bottom-right order. The quadrilateral bounding box of $w^{(i)}$ is denoted as $B^{(i)} = [(x_1^i, y_1^i), (x_2^i, y_2^i), (x_3^i, y_3^i), (x_4^i, y_4^i)]$, where $(x_j^i, y_j^i), j \in \{1, 2, 3, 4\}$ represents the coordinates of its corner points in the input document image. Following BERT [5], we tokenize D into a sub-word token sequence of length M, denoted as $T = (t^{(1)}, t^{(2)}, \cdots, t^{(M)})$. As the maximum input sequence length of BERT is restricted to 512, we use a sliding window to slice long sequences (with a length greater than 510) into several fixed-length subsequences whose lengths are set to 510. Neighboring subsequences are not overlapped with each other. For each subsequence, two special tokens [CLS] and [SEP] are inserted as the first and the last tokens, respectively. Some [PAD] tokens are added to the end of the last subsequence if its length is less than 512. Then we encode each subsequence with a BERT encoder to get the embedding of each token, denoted as $e(t^{(i)})$, and obtain the word embedding of $w^{(j)}$, denoted as $E(w^{(j)})$, by averaging the embeddings of its tokens. The spatial resolution of our used BERTgrid is $1/S$ of the original document image, where S denotes the stride of the convolutional feature map which is concatenated to. The BERTgrid representation of the document is defined as

$$G_{x,y,:} = \begin{cases} E(w^{(i)}) & if \ (x * S, \ y * S) \prec B^{(i)} \\ 0_d & others \end{cases} \tag{1}$$

where $\prec$ means "be located in", d is the embedding dimensionality, 0_d denotes an all-zero vector used for background whose dimension is d, G is a rank-3 tensor, i.e., $G \in R^{(H/S) \times (W/S) \times d}$, H and W are the height and width of the original image, respectively.

3.1.2 Visual and Textual Feature Fusion

As shown in Fig. 2, we concatenate the generated BERTgrid G to an intermediate layer of a CNN to construct a new multi-modal backbone network. To save computation, a lightweight ResNet18-FPN network [43], which is built on top of a ResNet18-D network [44], is used as the CNN module. ResNet18-D is an improved version of ResNet18 [45]. The output of ResNet18-FPN is a set of down-sampled feature maps, i.e., P_2, P_3, P_4, P_5, whose resolutions are $1/4, 1/8, 1/16, 1/32$ of the original document image. All of these maps have 256 channels. We set the stride of BERTgrid to 8 by default and conduct an early fusion operation, which first concatenates G to the output of the residual block conv3_1 in ResNet [45] and then performs dimensionality reduction with a 1×1 convolutional layer to make channel number the same as that of CNN feature map. Finally, after resizing P_3, P_4 and P_5 to the size of P_2, these four feature maps are concatenated and fed to a 1×1 convolutional layer to generate a feature map P_{fuse} with 256 channels. Here, P_{fuse} is a powerful multi-modal 2D feature map for document understanding tasks.

3.2 Word-Level Field Type Classification Head

For each word $w^{(i)}$, ROIAlign [46] is used to convert the features inside the word bounding box on the P_{fuse} map into a small $7 \times 7 \times 256$ feature map X_i. Then we use an additional skip connection to pass the embedding of each word $E\left(w^{(i)}\right)$ to the word-level field type classification head so that the error gradients could be propagated backward directly from the output layer to the BERT encoder layers. We find that this skip connection could slightly improve the accuracy of our KIE approach. We use a late fusion module to fuse X_i and $E\left(w^{(i)}\right)$ for each word. Specifically, each feature map X_i is convolved with two 3x3 convolutional layers to generate a new $7 \times 7 \times 256$ feature map, which is fed into a 1024-d fully connected layer to generate a 1024-d feature vector x_i. After that, x_i and $E\left(w^{(i)}\right)$ are concatenated and fed into an additional fully connected layer to generate a 1024-d new feature vector x_i'. As some words could be labeled with more than one field type tags (similar to the nested named entity recognition task), we design two classifiers to input x_i' and perform field type classification for each word. The first classifier is a binary classifier used to predict whether the input word is the value of any pre-defined field. If a word is not pruned by the first classifier, it will be classified again by the second classifier, which is composed of C independent binary classifiers (C is the number of pre-defined field types). The ith binary classifier predicts whether the input word is the value of the ith field, where $i \in [1, 2, \ldots, C]$. Each binary classifier is implemented with a Sigmoid function. Let $o_1(w_i)$, $o_{2,k}(w_i)$ denote the output of the first classifier and the kth binary classifier of the second classifier, respectively. The loss L_1 of the first classifier is a binary cross-entropy loss computed by

$$L_1 = -\frac{1}{N_1} \sum_i y_1(w_i) log(o_1(w_i)) + (1 - y_1(w_i)) log(1 - o_1(w_i)) \qquad (2)$$

where $y_1(w_i)$ means the ground truth of $w^{(i)}$ for the first classifier and N_1 is the number of words sampled from all the words in this document image. Similar to L_1, the loss L_2 of the second classifier is as follows:

$$L_2 = -\frac{1}{N_2} \sum_{i,k} y_{2,k}(w_i) log\left(o_{2,k}(w_i)\right) + \left(1 - y_{2,k}(w_i)\right) log\left(1 - o_{2,k}(w_i)\right) \qquad (3)$$

where $y_{2,k}(w_i)$ is 1 if $w^{(i)}$ has a label of the kth field, and is 0 if this word does not belong to this field, N_2 denotes the mini-batch size of the second classifier. We then calculate the total loss L_C of this head by

$$L_C = L_1 + L_2 \qquad (4)$$

3.3 Auxiliary Semantic Segmentation Head

While training this end-to-end network, we find that adding an additional pixel-level segmentation loss can make the network converge faster and more stably. Similar to the word-level field type classification head, this semantic segmentation head also includes two classifiers designed for two different classification tasks. The first one is used to classify each pixel into one of three categories: (1) Not located in any word box; (2) Inside a word box which belongs to any pre-defined field; (3) Inside other word boxes. If a pixel is classified into the second category by the first classifier, it will be classified again by the second classifier, which contains C independent binary classifiers for C field types, to get its field type. Specifically, followed by two 3×3 convolutional layers (256 channels), an up-sampling operation and two parallel 1×1 convolutional layers on P_{fuse}, two output feature maps $X_1^{out} \in R^{H \times W \times 3}$ and $X_2^{out} \in R^{H \times W \times C}$ are generated. Each pixel on X_1^{out} and X_2^{out} is classified by a Softmax function and C Sigmoid functions to get the results of the first and second classifiers, respectively. Let $o_1'(x, y)$ and $o_{2,k}'(x, y)$ denote the output of the first classifier and the kth binary classifier of the second classifier for a pixel (x, y), respectively. The softmax loss L_{AUX-1} of the first classifier is as follows:

$$L_{AUX-1} = -\frac{1}{N_1'} \sum_{(x,y) \in S_1} CrossEntropyLoss\left(o_1'(x, y), y_1'(x, y)\right) \tag{5}$$

where $y_1'(x, y)$ is the ground truth of pixel (x, y) in the first task, N_1' is the number of sampled pixels, S_1 is the set of sampled pixels. We calculate the loss L_{AUX-2} of the second task like Eq. (3):

$$L_{AUX-2} = -\frac{1}{N_2'} \sum_{(x,y) \in S_{2,k}} y_{2,k}'(x, y)log\left(o_{2,k}'(x, y)\right) + \left(1 - y_{2,k}'(x, y)\right)log\left(1 - o_{2,k}'(x, y)\right) \tag{6}$$

where $y_{2,k}'(x, y) \in \{0, 1\}$ is the ground truth of pixel (x, y) for the kth binary classifier in the second task, $y_{2,k}'(x, y)$ is 1 only if the word, which covers pixel (x, y), has a label of field k, N_2' is the mini-bch size of this task, S_2 is the set of sampled pixels.

The auxiliary segmentation loss L_{AUX} is the sum of L_{AUX-1} and L_{AUX-2}:

$$L_{AUX} = L_{AUX-1} + L_{AUX-2} \tag{7}$$

3.4   Joint Training Strategy

Joint training of a BERT model and a CNN model is not easy as these two types of models are fine-tuned with different optimizers and hyperparameters. Pre-trained BERT models are usually fine-tuned with an Adam optimizer [47] with a very small learning rate, while CNN models are fine-tuned with a standard SGD optimizer with a relatively larger learning rate. So, we train the pre-trained BERT model and the CNN model with an AdamW [48] optimizer and an SGD optimizer, respectively. Hyperparameter settings for these two optimizers are also independent.

Finally, the objective function of our whole network is

$$Loss = L_C + \lambda L_{AUX} \tag{8}$$

where λ is a control parameter, which is set to 1 in our experiments.

4 Experiments

4.1 Dataset

SROIE: The ICDAR SROIE dataset is a public dataset for information extraction which contains 626 receipts for training and 347 receipts for testing. It predefines 4 types of entities, including company, date, address and total. Following [10, 14], we use the ground-truth bounding boxes and transcripts of text segments provided by the organizer in our experiments.

INVOICE: This dataset is our in-house dataset which contains 24,175 real-world invoice pages for training and 643 invoice pages for testing. These invoices are generated from several hundred templates of complex layout structures. 14 types of key fields are annotated by human labelers, including CustomerAddress, TotalAmount, DueDate, PONumber, Subtotal, BillingAddress, CustomerName, InvoiceDate, TotalTax, VendorName, InvoiceNumber, CustomerID, VendorAddress and ShippingAddress. Statistics of these 14 key fields are listed in Table 1. We extract the transcripts and their corresponding word-level bounding boxes from these invoice images with Microsoft Azure Read API[1].

Table 1. Statistics of INVOICE dataset. All the numbers here mean word numbers.

Field name	Training	Testing	Field name	Training	Testing
Customer Address	150042	4097	Customer Name	117567	3115
Billing Address	63108	1680	Vendor Name	135774	3509
Vendor Address	208544	5454	Customer ID	32106	866
Shipping Address	18386	535	PO Number	2878	70
Invoice Date	36407	971	Subtotal	6692	180
Due Date	25601	1442	Total Tax	11767	323
Invoice Number	22550	601	Total Amount	34832	955

4.2 Implementation Details

We implement our approach based on PyTorch and conduct experiments on a workstation with 8 NVIDIA Tesla V100 GPUs (32 GB memory). In training, the parameters of the transformer encoder module and CNN module are initialized with BERT/RoBERTa and ResNet18-D pretrained on the ImageNet classification task, respectively. The parameters of the newly added layers are initialized by using random weights with a Gaussian distribution of mean 0 and standard deviation 0.01. The batch size we use is 16, so that the BatchNorm layers in the CNN module can be trained with Synchronized BatchNorm. As

---

[1] https://docs.microsoft.com/en-us/azure/cognitive-services/computer-vision/concept-recognizing-text.

mentioned in Sec. 3.4, we use an AdamW optimizer and an SGD optimizer to optimize the parameters of the BERT encoder and CNN module, respectively. Following the widely used setting of hyperparameters during finetuning, for the AdamW optimizer, the learning rate, betas, epsilon and weight decay are set as 2e-5, (0.9, 0.999), 1e-8 and 1e-2, while for the SGD optimizer, the learning rate, weight decay and momentum are set as 0.016, 5e-4 and 0.9, respectively. All models are trained for 33 epochs and a warmup strategy is applied in the first epoch. The learning rate of SGD is divided by 10 every 15 epochs. In each training iteration, we sample a mini-batch of 64 positive and 64 negative words, a mini-batch of 32 hard positive and 32 hard negative words, a mini-batch of 256 Category-1, 512 Category-2 and 256 Category-3 pixels, a mini-batch of 512 hard positive and 512 hard negative pixels, to calculate L_1, L_2, L_{AUX-1}, L_{AUX-2}, respectively. The hard positive and negative words or pixels are sampled with the OHEM [49] algorithm. As each receipt in the SROIE dataset only contains a small number of words, the numbers of sampled words or pixels mentioned above are reduced to half. Note that for many document pages in INVOICE dataset, the number of input sequences can be very large, which could lead to the out-of-memory issue due to the limited memory of GPU. To solve this problem, for each document page, we just randomly select at most L ($L = 10$) sequences to propagate gradients so that the memory of intermediate layers belonging to the other sequences can be directly released after acquiring output token embeddings. We adopt a multi-scale training strategy. The shorter side of each image is randomly rescaled to a number from {320, 416, 512, 608, 704} while the longer side should not exceed 800. In inference, we set the shorter side of each testing image as 512.

4.3   Comparisons with Prior Arts

In this section, we compare our proposed approach with several state-of-the-art methods on INVOICE and SROIE. On INVOICE, we adopt a word-level F1 score as evaluation metric, while on SROIE, we use the official online evaluation tool to calculate the field-level F1 score directly to make our result comparable to others.

INVOICE. We first compare our approach with several sequence tagging methods based on BERT, RoBERTa and LayoutLM. These methods directly use the output word embeddings of the transformer encoder to perform token-level classification. For fairer comparisons, we use the same sliding window strategy to slice the long input sequences. As shown in Table 2, BERT and RoBERTa cannot achieve satisfactory results because they only use textual information, while LayoutLM can outperform them by a big margin owing to the encoding of 2D position embeddings and large-scale pre-training. As a grid-based approach, our proposed ViBERTgrid can encode the textual, layout and visual information of a document in a 2D feature map simultaneously. Even without large-scale pre-training, our approach can achieve a comparable result against LayoutLM-Large when only using BERT-Base as the transformer encoder. When replacing BERT-Base with more powerful RoBERTa-Base or LayoutLM-Base, the performance of our model can be further improved to 92.60% and 92.84% in Macro F1 score, respectively. We also compare our approach with BERTgrid, which is another grid-based approach. In our current implementation, we set the stride of BERTgrid as 1 and use it as the input of ResNet18-FPN network. It can be seen that our approach outperforms BERTgrid significantly by improving the Macro F1 score from 88.78% to 92.15% when using BERT-Base as the transformer encoder. The effectiveness of our proposed ViBERTgrid representation and the joint training strategy is clearly demonstrated.

Table 2. Performance comparison on INVOICE dataset.

Entities	BERT-Base	BERT-Large	RoBERTa-Base	RoBERTa-Large	LayoutLM-Base	LayoutLM-Large	BERTgrid	ViBERTgrid (BERT-Base)	ViBERTgrid (RoBERTa-Base)	ViBERTgrid (LayoutLM-Base)
CustomerAddress	87.52	90.16	89.34	91.87	91.60	92.88	91.23	92.44	93.01	92.86
TotalAmount	89.97	92.07	92.56	93.24	92.69	93.52	91.20	92.69	93.74	94.57
DueDate	92.52	95.75	96.38	96.23	96.89	96.64	64.10	96.71	95.64	96.38
PONumber	60.43	65.22	59.68	69.57	78.83	87.22	88.11	83.69	83.45	85.71
Subtotal	86.83	86.13	86.83	86.86	89.32	88.46	87.08	87.57	88.52	89.73
BillingAddress	91.24	92.47	91.76	93.99	96.02	95.68	95.52	95.87	96.91	95.70
CustomerName	88.35	89.24	88.59	88.79	89.61	89.82	89.62	90.38	91.50	91.42
InvoiceDate	93.21	94.70	92.40	94.33	93.04	94.23	92.75	93.70	94.27	95.49
TotalTax	90.40	91.61	90.46	91.16	92.42	90.88	85.80	92.33	92.33	93.19
VendorName	90.76	91.99	91.06	91.91	92.30	92.49	88.62	92.58	92.99	92.45
InvoiceNumber	88.64	90.11	88.63	87.22	90.79	91.08	89.08	91.64	91.95	91.40
CustomerID	89.16	89.12	88.51	89.53	91.51	91.89	88.84	92.03	91.67	91.37
VendorAddress	95.67	96.80	96.74	97.17	97.62	97.61	97.37	97.58	97.67	98.03
ShippingAddress	80.67	89.17	82.18	87.05	92.66	91.64	93.61	90.87	92.71	91.43
Micro F1	90.77	92.46	91.79	92.82	93.51	93.81	90.93	93.79	94.24	94.22
Macro F1	87.53	89.61	88.22	89.92	91.81	92.43	88.78	92.15	92.60	92.84
# of Parameters	110M	340M	125M	355M	113M	343M	142M	142M	157M	145M

Table 3. Performance comparison on SROIE.

Model	Need domain-specific pretraining	# of Parameters	F1 Score
LayoutLM-Large* [10]	✓	343M	96.04
PICK [14]		---	96.12
VIES [22]		---	96.12
TRIE [21]		---	96.18
LayoutLMv2-Base [13]	✓	200M	96.25
LayoutLMv2-Large [13]	✓	426M	96.61
ViBERTgrid (BERT-Base)		142M	96.25
ViBERTgrid (RoBERTa-Base)		157M	96.40

* This result is from https://github.com/microsoft/unilm/tree/master/layoutlm.

SROIE. Since SROIE only provides the ground truth transcriptions of key information, we pre-process the training data to generate labels of each text segment by matching their OCR results with these transcriptions. In inference phase, following [10, 14], we adopt a lexicon built from training data to auto-correct the predicted results. We compare our approach with other most competitive methods, including LayoutLM, LayoutLMv2, PICK, TRIE and VIES. As shown in Table 3, our approach achieves better results than LayoutLM, PICK, TRIE and VIES. Even without using millions of document images to do bi-modal domain-specific pretraining, our ViBERTgrid can still achieve comparable results with LayoutLMv2-Base when using BERT-Base and RoBERTa-Base as text embedding models. Although LayoutLMv2-Large achieves slightly higher accuracy than ViBERTgrid (RoBERTa-Base), it contains much more parameters.

4.4 Ablation Studies

In this section, we conduct a series of ablation experiments to evaluate the effectiveness of each module of our approach and explore the impacts of different experimental settings. Since experimental results tend to be more stable on the much larger INVOICE dataset, all experiments here are conducted on it.

Effectiveness of Joint Training Strategy. We compare our joint training strategy with several other training strategies to demonstrate its effectiveness. The results are shown in Table 4. "Fixed T + V" means the parameters of the BERT encoder are fixed during fine-tuning, which can be considered as a re-implementation version of VisualWordGrid by using BERT. "T + V" means that the parameters of BERT and CNN are trained jointly. "Lr-T" and "Lr-V" are the learning rates of BERT and CNN, respectively. The experimental results demonstrate that our proposed joint training strategy, namely training the BERT model and the CNN model with an AdamW optimizer (learning rate 2e-5) and an

Table 4. Effectiveness of joint training strategy. "XXX" means that it does not converge.

Model	BERT module optimizer	CNN & CLS Head optimizer	Lr-T	Lr-V	Micro F1	Macro F1
Fixed T + V	---	SGD	---	0.016	91.33	89.52
T + V	AdamW	AdamW	0.016	0.016	XXX	XXX
T + V	AdamW	AdamW	2e–5	2e–5	93.37	91.77
T + V	AdamW	AdamW	2e–5	0.016	91.89	88.48
T + V	SGD	SGD	0.016	0.016	89.97	87.72
T + V	SGD	SGD	2e–5	2e–5	76.84	61.82
T + V	SGD	SGD	2e–5	0.016	91.95	90.15
T + V	AdamW	SGD	2e–5	0.016	**93.79**	**92.15**

SGD optimizer (learning rate 0.016) respectively, can significantly improve the representation ability of grid-based document representations and achieve the best performance among all these joint training strategies.

Effect of Multi-Modality Features on the KIE Task. We conduct the following ablation study to further examine the contributions of visual and textual features to the KIE task and present the results in Table 5. To remove visual information in our ViBERTgrid, we replace the input image with a zero tensor whose shape is (H, W, 3). To use visual information only, we remove the BERTgrid module from ViBERTgrid and use the remaining CNN layers to generate a 2D feature map directly. The results show that textual information is more important than visual information for the KIE task and combining both visual and textual information achieves better results than using visual or textual information only no matter whether the joint training strategy is used or not.

Table 5. Effect of multi-modality features on the KIE task.

Use visual information	Use textual information	Joint training	Micro F1	Macro F1
✓			87.51	83.78
	✓		91.06	89.19
✓	✓		91.33	89.52
	✓	✓	93.52	91.68
✓	✓	✓	**93.79**	**92.15**

Table 6. Effectiveness of CNN module. BERT Mini [50] is a pretrained miniature BERT with 4 layers, 4 heads and 256 hidden embedding size.

Model	Micro F1	Macro F1
BERT-Mini	86.98	81.96
BERT-Mini + CNN	**91.36**	**89.00**
BERT-Base	92.47	89.02
BERT-Base + CNN	**93.79**	**92.15**
RoBERTa-Base	92.87	89.84
RoBERTa-Base + CNN	**94.24**	**92.60**
LayoutLM-Base	94.03	92.49
LayoutLM-Base + CNN	**94.22**	**92.84**

Effectiveness of the CNN Module. To evaluate the effectiveness of the CNN module, we exclude it and directly use output word embeddings of the transformer encoder to do word-level classification while the training strategies described in Sec. 3.2 are used. As shown in Table 6, no matter which transformer encoder is used, the CNN module can consistently improve the performance. The improvement becomes more significant when the transformer encoder is relatively weaker, e.g., when BERT-Mini is used as the transformer encoder, the CNN module can improve the Macro F1 score by 7.04% absolutely.

Table 7. Comparisons of early fusion at different feature stages.

Stride of G	Feature stage	Micro F1	Macro F1
4	C2	93.34	91.27
8	C3	**93.79**	**92.15**
16	C4	93.41	91.69

Comparison of Early Fusion at Different Feature Stages. We train three models by fusing BERTgrid G with the output of the residual block conv2_1 (C2), conv3_1 (C3) and conv4_1 (C4) in ResNet-18D, whose strides are 4, 8, 16, respectively. As shown in Table 7, the second model which fuses G with C3 achieves the best performance.

Effectiveness of Early Fusion and Late Fusion. In this section, we train different models by using early fusion, late fusion, and both. Moreover, to evaluate the impact of the representation ability of different text embeddings, we construct BERTgrid G from the 3rd layer, 6th layer and 12th layer of BERT-Base, respectively. As shown in Table 8, the combination of early fusion and late fusion can achieve better performance than early fusion or late fusion only under all experimental settings. Furthermore, compared with late fusion, early fusion plays a more important role when the text embeddings are relatively weaker.

Table 8. Effectiveness of early fusion ("Early") and late fusion ("Late"). BERT-Base is used as the transformer encoder here.

Layer number	Fusion mode	Micro F1	Macro F1
12	Early	93.64	91.97
12	Late	93.54	91.71
12	Early + Late	**93.79**	**92.15**
6	Early	93.00	91.40
6	Late	92.95	91.21
6	Early + Late	**93.40**	**92.02**
3	Early	91.88	90.26
3	Late	90.52	87.55
3	Early + Late	**92.17**	**90.67**

5 Conclusion and Future Work

In this paper, we propose ViBERTgrid, a new grid-based document representation for the KIE task. By combining the best of BERTgrid and CNN, and training the parameters of BERT and CNN jointly, the representation ability of our ViBERTgrid is much better than other grid-based document representations such as BERTgrid and VisualWord-Grid. Consequently, our ViBERTgrid-based KIE approach has achieved state-of-the-art performance on two real-world datasets we experimented.

The proposed ViBERTgrid can be easily integrated into some other KIE frameworks introduced in Sect. 2. Therefore, exploring the effectiveness of our document representation for other KIE frameworks will be one of our future works. Moreover, we will explore the effectiveness of our ViBERTgrid-based document representation for other document understanding tasks where both visual and textual information are useful, such as layout analysis and table structure recognition.

References

1. Huang, Z., et al.: ICDAR 2019 competition on scanned receipt ocr and information extraction. In: ICDAR, pp. 1516–1520 (2019)
2. Lample, G., Ballesteros, M., Subramanian, S., Kawakami, K., Dyer, C.: Neural architectures for named entity recognition. In: NAACL, pp. 260–270 (2016)
3. Chiu, J.P., Nichols, E.: Named entity recognition with bidirectional LSTM-CNNs. TACL **4**, 357–370 (2016)

4. Ma, X., Hovy, E.: End-to-end sequence labeling via Bi-directional LSTM-CNNs-CRF. In: ACL, pp. 1064–1074 (2016)
5. Devlin, J., Chang, M.-W., Lee, K., Toutanova, K.: BERT: pre-training of deep bidirectional transformers for language understanding. In: NAACL, pp. 4171–4186 (2019)
6. Liu, Y., et al.: RoBERTa: A Robustly Optimized BERT Pretraining Approach. arXiv preprint arXiv:1907.11692 (2019)
7. Palm, R.B., Winther, O., Laws, F.: CloudScan - a configuration-free invoice analysis system using recurrent neural networks. In: ICDAR, pp. 406–413 (2017)
8. Sage, C., Aussem, A., Elghazel, H., Eglin, V., Espinas, J.: Recurrent neural network approach for table field extraction in business documents. In: ICDAR, pp. 1308–1313 (2019)
9. Hwang, W., et al.: Post-OCR parsing: building simple and robust parser via BIO tagging. In: Workshop on Document Intelligence at NeurIPS (2019)
10. Xu, Y., Li, M., Cui, L., Huang, S., Wei, F., Zhou, M.: LayoutLM: pre-training of text and layout for document image understanding. In: SIGKDD, pp. 1192–1200 (2020)
11. Garncarek, Ł., et al.: LAMBERT: layout-aware (language) modeling for information extraction. arXiv preprint arXiv:2002.08087 (2020)
12. Pramanik, S., Mujumdar, S., Patel, H.: Towards a multi-modal, multi-task learning based pre-training framework for document representation learning. arXiv preprint arXiv:2009.14457 (2020)
13. Xu, Y., et al.: LayoutLMv2: Multi-modal pre-training for visually-rich document understanding. arXiv preprint arXiv:2012.14740 (2020)
14. Yu, W., Lu, N., Qi, X., Gong, P., Xiao, R.: PICK: processing key information extraction from documents using improved graph learning-convolutional networks. In: ICPR (2020)
15. Vaswani, A., et al.: Attention is all you need. In: NeurIPS, pp. 6000–6010 (2017)
16. Lohani, D., Belaïd, A., Belaïd, Y.: An invoice reading system using a graph convolutional network. In: Carneiro, G., You, S. (eds.) ACCV 2018. LNCS, vol. 11367, pp. 144–158. Springer, Cham (2019). https://doi.org/10.1007/978-3-030-21074-8_12
17. Qian, Y., Santus, E., Jin, Z., Guo, J., Barzilay, R.: GraphIE: a graph-based framework for information extraction. In: NAACL, pp. 751–761 (2019)
18. Liu, X., Gao, F., Zhang, Q., Zhao, H.: Graph convolution for multimodal information extraction from visually rich documents. In: NAACL, pp. 32–39 (2019)
19. Wei, M., He, Y., Zhang, Q.: Robust layout-aware IE for visually rich documents with pre-trained language models. In: ACM SIGIR, pp. 2367–2376 (2020)
20. Luo, C., Wang, Y., Zheng, Q., Li, L., Gao, F., Zhang, S.: Merge and recognize: a geometry and 2D context aware graph model for named entity recognition from visual documents. In: TextGraphs Workshop at COLING, pp. 24–34 (2020)
21. Zhang, P., et al.: TRIE: end-to-end text reading and information extraction for document understanding. In: ACM Multimedia, pp. 1413–1422 (2020)
22. Wang, J., et al.: Towards robust visual information extraction in real world: new dataset and novel solution. In: AAAI (2021)
23. Katti, A.R., et al.: Chargrid: towards understanding 2D documents. In: EMNLP, pp. 4459–4469 (2018)
24. Dang, T.N., Thanh, D.N.: End-to-End information extraction by character-level embedding and multi-stage attentional U-Net. In: BMVC (2019)
25. Zhao, X., Niu, E., Wu, Z., Wang, X.: CUTIE: learning to understand documents with convolutional universal text information extractor. arXiv preprint arXiv:1903.12363 (2019)
26. Denk, T.I., Reisswig, C.: BERTgrid: contextualized embedding for 2D document representation and understanding. In: Document Intelligence Workshop at NeurIPS (2019)
27. Kerroumi, M., Sayem, O., Shabou, A.: VisualWordGrid: information extraction from scanned documents using a multimodal approach. arXiv preprint arXiv:2010.02358 (2020)

28. Dengel, A.R., Klein, B.: SmartFIX: a requirements-driven system for document analysis and understanding. In: Lopresti, D., Hu, J., Kashi, R. (eds.) DAS 2002. LNCS, vol. 2423, pp. 433–444. Springer, Heidelberg (2002). https://doi.org/10.1007/3-540-45869-7_47

29. Cesarini, F., Francesconi, E., Gori, M., Soda, G.: Analysis and understanding of multi-class invoices. IJDAR **6**, 102–114 (2003)

30. Medvet, E., Bartoli, A., Davanzo, G.: A probabilistic approach to printed document understanding. IJDAR **14**, 335–347 (2011)

31. Esser, D., Schuster, D., Muthmann, K., Berger, M., Schill, A.: Automatic indexing of scanned documents - a layout-based approach. In: DRR (2012)

32. Schuster, D., et al.: Intellix -- End-User trained information extraction for document archiving. In: ICDAR, pp. 101–105 (2013)

33. Rusinol, M., Benkhelfallah, T., Poulain d'Andecy, V.: Field extraction from administrative documents by incremental structural templates. In: ICDAR, pp. 1100–1104 (2013)

34. Carbonell, M., Fornés, A., Villegas, M., Lladós, J.: A neural model for text localization, transcription and named entity recognition in full pages. Pattern Recogn. Lett. **136**, 219–227 (2020)

35. Majumder, B.P., Potti, N., Tata, S., Wendt, J.B., Zhao, Q., Najork, M.: Representation learning for information extraction from form-like documents. In: ACL, pp. 6495–6504 (2020)

36. Hwang, W., Yim, J., Park, S., Yang, S., Seo, M.: Spatial dependency parsing for semi-structured document information extraction. arXiv preprint arXiv:2005.00642 (2020)

37. Teakgyu Hong, D.K.M.J., Hwang, W., Nam, D., Park, S.: BROS: a pre-trained language model for understanding texts in document. In: Submitted to ICLR (2021)

38. Palm, R.B., Laws, F., Winther, O.: Attend, copy, parse - End-to-end information extraction from documents. In: ICDAR, pp. 329–336 (2019)

39. Guo, H., Qin, X., Liu, J., Han, J., Liu, J., Ding, E.: EATEN: entity-aware attention for single shot visual text extraction. In: ICDAR, pp. 254–259 (2019)

40. Sage, C., Aussem, A., Eglin, V., Elghazel, H., Espinas, J.: End-to-End extraction of structured information from business documents with pointer-generator networks. In: SPNLP Workshop at EMNLP, pp. 43–52 (2020)

41. Yang, X., Yumer, E., Asente, P., Kraley, M., Kifer, D., Giles, C.L.: Learning to extract semantic structure from documents using multimodal fully convolutional neural network. In: CVPR, pp. 4342–4351 (2017)

42. Barman, R., Ehrmann, M., Clematide, S., Oliveira, S.A., Kaplan, F.: Combining visual and textual features for semantic segmentation of historical newspapers. J. Data Min. Digital Humanit. HistoInformatics, jdmdh:7097 (2021)

43. Lin, T.-Y., Dollár, P., Girshick, R., He, K., Hariharan, B., Belongie, S.: Feature pyramid networks for object detection. In: CVPR, pp. 2117–2125 (2017)

44. He, T., Zhang, Z., Zhang, H., Zhang, Z., Xie, J., Li, M.: Bag of tricks for image classification with convolutional neural networks. In: CVPR, pp. 558–567 (2019)

45. He, K., Zhang, X., Ren, S., Jian, S.: Deep residual learning for image recognition. In: CVPR, pp. 770–778 (2016)

46. He, K., Gkioxari, G., Dollár, P., Girshick, R.: Mask R-CNN. In: ICCV, pp. 2961–2969 (2017)

47. Kingma, D.P., Ba, J.: Adam: a method for stochastic optimization. In: ICLR (2015)

48. Loshchilov, I., Hutter, F.: Decoupled weight decay regularization. In: ICLR (2019)

49. Shrivastava, A., Gupta, A., Girshick, R.: Training region-based object detectors with online hard example mining. In: CVPR, pp. 761–769 (2016)

50. Turc, I., Chang, M.-W., Lee, K., Toutanova, K.: Well-read students learn better: on the importance of pre-training compact models. arXiv preprint arXiv:1908.08962 (2019)

Kleister: Key Information Extraction Datasets Involving Long Documents with Complex Layouts

Tomasz Stanisławek[1,2], Filip Graliński[1,3](✉), Anna Wróblewska[2],
Dawid Lipiński[1], Agnieszka Kaliska[1,3], Paulina Rosalska[1,4],
Bartosz Topolski[1], and Przemysław Biecek[2,5]

[1] Applica.ai, 15 Zajęcza, Warsaw 00351, Poland
{tomasz.stanislawek,dawid.lipinski,paulina.rosalska,
bartosz.topolski}@applica.ai
[2] Warsaw University of Technology, Koszykowa 75, Warsaw, Poland
anna.wroblewska@pw.edu.pl
[3] Adam Mickiewicz University, 1 Wieniawskiego, Poznan 61712, Poland
{filip.gralinski,agnieszka.kaliska}@amu.edu.pl
[4] Nicolaus Copernicus University, 11 Gagarina, Torun 87100, Poland
[5] Samsung R&D Institute Poland, Plac Europejski 1, Warsaw, Poland
przemyslaw.biecek@samsung.com

Abstract. The relevance of the Key Information Extraction (KIE) task is increasingly important in natural language processing problems. But there are still only a few well-defined problems that serve as benchmarks for solutions in this area. To bridge this gap, we introduce two new datasets (*Kleister NDA* and *Kleister Charity*). They involve a mix of scanned and born-digital long formal English-language documents. In these datasets, an NLP system is expected to find or infer various types of entities by employing both textual and structural layout features. The Kleister Charity dataset consists of 2,788 annual financial reports of charity organizations, with 61,643 unique pages and 21,612 entities to extract. The Kleister NDA dataset has 540 Non-disclosure Agreements, with 3,229 unique pages and 2,160 entities to extract. We provide several state-of-the-art baseline systems from the KIE domain (Flair, BERT, RoBERTa, LayoutLM, LAMBERT), which show that our datasets pose a strong challenge to existing models. The best model achieved an 81.77% and an 83.57% F1-score on respectively the Kleister NDA and the Kleister Charity datasets. We share the datasets to encourage progress on more in-depth and complex information extraction tasks.

Keyword: Key information extraction, visually rich documents, named entity recognition

1 Introduction

The task of Key Information Extraction (KIE) from Visually Rich Documents (VRD) has proved increasingly interesting in the business market with the recent

© Springer Nature Switzerland AG 2021
J. Lladós et al. (Eds.): ICDAR 2021, LNCS 12821, pp. 564–579, 2021.
https://doi.org/10.1007/978-3-030-86549-8_36

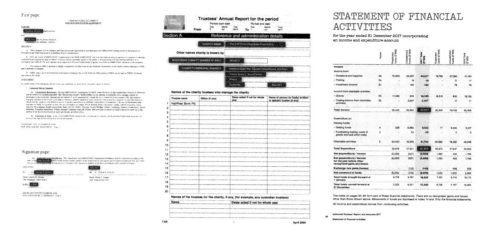

Fig. 1. Examples of a real business applications and data for *Kleister* datasets. (Note: The key entities are in blue.) (Color figure online)

rise of solutions related to Robotic Process Automation (RPA). From a business user's point of view, systems that, fully automatically, gather information about individuals, their roles, significant dates, addresses and amounts, would be beneficial, whether the information is from invoices or receipts, from company reports or contracts [9,12,13,16,18,21,22]. There is a disparity between what can be delivered with the KIE domain systems on publicly available datasets and what is required by real-world business use. This disparity is still large and makes a robust evaluation difficult. Recently, researchers have started to fill the gap by creating datasets in the KIE domain such as scanned receipts: *SROIE*[1] [18], form understanding [11], NIST Structured Forms Reference Set of Binary Images (*SFRS*)[2] or Visual Question Answering dataset *DocVQA* [15].

This paper describes two new English-language datasets for the Key Information Extraction tasks from a diverse set of texts, long scanned and born-digital documents with complex layouts, that address real-life business problems (Fig. 1). The datasets represent various problems arising from the specificity of business documents and associated business conditions, e.g. complex layouts, specific business logic, OCR quality, long documents with multiple pages, noisy training datasets, and normalization. Moreover, we evaluate several systems from the KIE domain on our datasets and analyze KIE tasks' challenges in the business domain. We believe that our datasets will prove a good benchmark for more complex Information Extraction systems.

---

[1] https://rrc.cvc.uab.es/?ch=13&com=evaluation&task=3.
[2] https://www.nist.gov/srd/nist-special-database-2.

The main contributions of this study are:

1. *Kleister* – two novel datasets of long documents with complex layouts: 3,328 documents containing 64,872 pages with 23,772 entities to extract (see Sect. 3);
2. our method of collecting datasets using a semi-supervised methodology, which reduces the amount of manual work in gathering data; this method has the potential to be reused for similar tasks (see Sect. 3.1 and 3.2);
3. evaluation over several state-of-the-art Named Entity Recognition (NER) architectures (Flair, BERT, RoBERTa, LayoutLM, LAMBERT) employing our *Pipeline* method (see Sect. 4.1 and 5);
4. detailed analysis of the data and baseline results related to the Key Information Extraction task carried out by human experts (see Sect. 3.3 and 5).

The data, except for the test-set gold standard, are available at https://github.com/applicaai/kleister-nda.git and https://github.com/applicaai/kleister-charity.git. A shared-task platform where submissions can be evaluated, also for the test set, is available at https://gonito.applica.ai.

2   Related Work

Our main reason for preparing a new dataset was to develop a strategy to deal with challenges faced by businesses, which means overcoming such difficulties as complex layout, specific business logic (the way that content is formulated, e.g. tables, lists, titles), OCR quality, document-level extraction and normalization.

2.1   KIE from Visually Rich Documents (publicly Available)

A list of KIE-oriented challenges is available at the International Conference on Document Analysis and Recognition ICDAR 2019[3] (cf. Table 1). There is a dataset called SROIE[4] with information extraction from a set of scanned receipts. The authors prepared 1,000 whole scanned receipt images with annotated entities: company name, date, address, and total amount paid (a similar dataset was also created [18]). Form Understanding in Noisy Scanned Documents is another interesting dataset from ICDAR 2019 (*FUNSD*) [11]. FUNSD aims at extracting and structuring the textual content of forms. However, the authors focus mainly on understanding tables and a limited range of document layouts, rather than on extracting particular entities from the data. The point is, therefore, to indicate a table but not to extract the information it contains.

2.2   KIE from Visually Rich Documents (publicly Unavailable)

There are also datasets for the Key Information Extraction task based on invoices [9,12,16,17]. Documents of this kind contain entities like 'Invoice date,'

---

[3] http://icdar2019.org/competitions-2/.

[4] https://rrc.cvc.uab.es/?ch=13.

'Invoice number,' 'Net amount' and 'Vendor Name', extracted using a combination of NLP and Computer Vision techniques. The reason for such a complicated multi-domain process is that spatial information is essential for properly understanding these kinds of documents. However, since they are usually short, the same information is relatively rarely repeated, and therefore there is no need for understanding the more extended context of the document. Nevertheless, those kinds of datasets are the most similar to our use case.

2.3   Information Extraction from One-Dimensional Documents

The *WikiReading* dataset [8] (and its variant *WikiReading Recycled* [6]) is a large-scale natural language understanding task. Here, the main goal is to predict textual values from the structured knowledge base, Wikidata, by reading the text of the corresponding Wikipedia articles. Some entities can be extracted from the given text directly, but some have to be inferred. Thus, as in our assumptions, the task contains a rich variety of challenging extraction sub-tasks and it is also well-suited for end-to-end models that must cope with longer documents.

Key Information Extraction is different from the Named Entity Recognition task (the *CoNLL 2003* NER challenge [20] being a well-known example). This is because: (1) retrieving spans is not required in KIE; (2) a system is expected to extract specific, actionable data points rather than general types of entities (such as people, organization, locations and "others" for CoNLL 2003).

Table 1. Summary of the existing English datasets and the Kleister sets. (*) For detailed description see Sect. 3.3.

Dataset name	CoNLL 2003	WikiReading	FUNSD	SROIE	Kleister NDA	Kleister charity
Source	Reuters news	Wikipedia	Forms	Receipts	EDGAR	UK charity Com.
Documents	1,393	4.7M	199	973	540	2,778
Pages	–	–	199	973	3,229	61,643
Entities	35,089	18M	9,743	3,892	2,160	21,612
Train docs	946	16.03M	149	626	254	1,729
Dev docs	216	1.89M	–	–	83	440
Test docs	231	0.95M	50	347	203	609
Input/Output on token level(*)	✓	✗	✓	✓	✗	✗
Long document(*)	✗	✓	✗	✗	✓	✓
Complex layout(*)	✗	✗	✓	✓	✓	✓
OCR(*)	✗	✗	✓	✓	✗	✓

3 Kleister: New Datasets

We collected datasets of long formal born-digital documents, namely US non-disclosure agreements (Kleister NDA) and a mixture of born-digital and (mostly) scanned annual financial reports of charitable foundations from the UK (Kleister Charity). These two datasets have been gathered in different ways due to their repository structures. Also, they were published on the Internet for different reasons. The crucial difference between them is that the NDA dataset was born-digital, but that the Charity dataset needed to be OCRed. Kleister datasets have a multi-modal input (PDF files and text versions of the documents) and a list of entities to be found.

3.1 NDA Dataset

The NDA Dataset contains Non-disclosure Agreements, also known as Confidentiality Agreements. They are legally binding contracts between two or more parties, where the parties agree not to disclose information covered by the agreement. The NDAs can take on various forms (e.g. contract attachments, emails), but they usually have a similar structure.

Data Collection Method. The NDAs were collected from the Electronic Data Gathering, Analysis and Retrieval system (EDGAR[5]) via Google's search engine. The original files were in an HTML format, but they were transformed into PDF files to keep processing simple and similar to that of other public datasets. Transformation was made using the `puppeteer` library.[6] Then, a list of entities was established (see Table 1).

Annotation Procedure. We annotated the whole dataset in two ways. Its first part, made up of 315 documents, was annotated by three annotators, except that only contexts with some similarity, pre-selected using methods based on semantic similarity (cf. [3]), were taken into account; this was to make the annotation faster and less-labor intensive. The second part, with 195 documents, was annotated entirely by hand. When preparing the dataset, we wanted to determine whether semantic similarity methods could be applied to limit the time it would take to perform annotation procedures; this solution was about 50% quicker than fully manual annotation. The annotations on all documents were then checked by a super-annotator, which ensured the annotation's excellent quality Cohen's κ (=0.971)[7]. Next, all entities were normalized according to the standards adopted by us, e.g. the `effective date` was standardized according to ISO 8601 i.e. YYYY-MM-DD[8].

Dataset Split. The Kleister NDA dataset contains a relatively small document count, so we decided to add more examples into the test split (about 38%) so as to be more accurate during the evaluation stage (see Table 1 for exact numbers).

---

[5] https://www.sec.gov/edgar.shtml.
[6] https://github.com/puppeteer/puppeteer.
[7] https://en.wikipedia.org/wiki/Cohen%27s_kappa.
[8] The normalization standards are described in the public repository with datasets.

3.2   Charity Dataset

The Charity dataset consists of annual financial reports that all charities registered in England and Wales must submit to the Charity Commission. The Commission subsequently makes them publicly available on its website.[9] There are no strict rules about the format of these charity reports. Some are richly illustrated with photos and charts and financial information constitutes only a small part of the entire report. In contrast, others are a few pages long and only necessary data on revenues and expenses in a given calendar year are given.

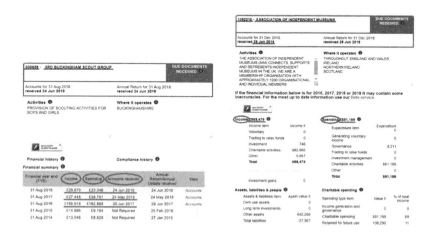

Fig. 2. Organization's page on the Charity Commission's website (left: organization whose annual income is between 25k and 500k GBP, right: over 500k). Note: Entities are underlined in red and names of entities are circled.

Data Collection Method. The Charity Commission website has a database of all the charity organizations registered in England and Wales. Each of these organizations has a separate sub-page on the Commission's website, and it is easy to find the most important information about them there (see Fig. 2). This information only partly overlaps with information in the reports. Some entities such as, say, a list of trustees might not be in the reports. Thus, we decided to extract only those entities which also appear in the form of a brief description on the website.

In the beginning, we downloaded 3,414 reports (as PDF files).[10] During document analysis, it emerged that several reports were written in Welsh. As we are interested only in English, all documents in other languages were identified and

---

[9] https://apps.charitycommission.gov.uk/showcharity/registerofcharities/ RegisterHomePage.aspx.

[10] Organizations with an income below 25,000 GBP a year are required to submit a condoned financial report instead.

removed from the collection. Additionally, documents that contained reports for more than one organization or whose OCR quality was low were deleted. This left us with 2,778 documents.

Annotation Procedure. There was no need to manually annotate all documents because information about the reporting organizations could be obtained directly from the Charity Commission. Initially, only a random sample of 100 documents were manually checked. Some proved low quality: `charity name` (5% of errors and 13% of minor differences), and `charity address` (9% of errors and 63% of minor differences). Minor errors are caused by data presentation differences on the page and in the document. For example, the charity's name on the website and in the document could be written with the term *Limited* (shortened to *LTD*) or without it. These minor differences were corrected manually or automatically. In the next step, 366 documents were analyzed manually. Some parts of the charity's address were also problematic. For instance, counties, districts, towns and cities were specified on the website, but not in the documents, or *vice versa*. We split the address data into three separate entities that we considered the most essential: postal code, postal town name and street or road name. The postal code was the critical element of the address, based on the city name and street name[11]. The whole process allowed us to accurately identify entities (see Table 1) and to obtain a good-quality dataset with annotations corresponding to the gold standard.

Dataset Split. In the Kleister Charity dataset, we have multiple documents from the same charity organization but from different years. Therefore, we decided to split documents based on charity organization into the train/dev/test sets with, respectively, a 65/15/20 dataset ratio (see Table 1 for exact numbers). The documents from the dev/test split were manually annotated (by two annotators) to ensure high-quality evaluation. Additionally, 100 random documents from the test set were annotated twice to calculate the relevant Cohen's κ coefficient (we achieved excellent quality $\kappa = 0.9$).

3.3 Statistics and Analysis

The detailed statistics of the Kleister datasets are presented in Table 1 and Table 2. Our datasets covered a broad range of general types of entities; the `party` entity is special since it could be one of the following types: `ORGANIZATION` or `PERSON`. Additionally, some documents may not contain all entities mentioned in the text, for instance in Kleister NDA the `term` entity appears in 36% of documents. Likewise, some entities may have more than one gold value; for instance in Kleister NDA the `party` entity could have up to 7 gold values for a single document. `Report_date`, `jurisdiction` and `term` have the lowest number of unique values. This suggests that these entities should be simpler than others to extract.

---

[11] Postal codes in the UK were aggregated from www.streetlist.co.uk.

Table 2. Summary of the entities in the NDA and charity datasets. (*) Based on manual annotation of text spans.

Entities	General entity type	Total count	Unique values	(*) Avg. entity count/doc	(*) Avg. token count/entity	Example gold value
NDA dataset (540 documents)						
Party	ORG/PER	1,035	912	19.74	1.62	Ajinomoto Althea Inc.
Jurisdiction	LOCATION	531	37	1.05	1.21	New York
Effective_date	DATE	400	370	1.95	3.10	2005-07-03
Term	DURATION	194	22	1.03	2.77	P12M
Charity dataset (2 788 documents)						
Post_town	ADDRESS	2,692	501	1.12	1.06	BURY
Postcode	ADDRESS	2,717	1,511	1.12	1.99	BL9 ONP
Street_line	ADDRESS	2,414	1,353	1.12	2.52	42–47 MINORIES
Charity_name	ORG	2,778	1,600	13.80	3.67	Mad Theatre Company
Charity_number	NUMBER	2,763	1,514	2.47	1.00	1143209
Report_date	DATE	2,776	129	10.58	2.96	2016-09-30
Income	AMOUNT	2,741	2,726	1.95	1.01	109370.00
Spending	AMOUNT	2,731	2,712	2.03	1.01	90174.00

Manual Annotation of Text Spans. To give more detailed statistics we decided to annotate small numbers of documents on text span level. Four annotators annotated 60/55 documents for, respectively, the Kleister Charity and Kleister NDA. In Table 2, we observe that 5 out of 12 entities appear once in a single document. There are also three entities with more than ten counts on average (**party**, **charity_number** and **report_date**). Annotation on the text-span level

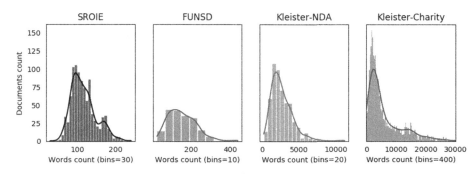

Fig. 3. Distribution of document lengths for kleister datasets compared to other similar datasets (note that the x-axes ranges are different).

could prove critical to checking the quality of the training dataset for methods based on a Named Entity Recognition model, something which an *autotagging* mechanism produces (see Sect. 4.1).

Comparison with Existing Resources. In Table 1, we gathered the most important information about open datasets (which are the most popular ones in the domain) and the Kleister datasets. In particular, we outlined the difference based on the following properties:

- **Input/Output on token level**: it is known which tokens an entity is made up from in the documents. Otherwise, one should: a) create a method for preparing a training dataset for sequence labeling models (subsequently in the publication, we use the term *autotagging* for this sub-task); b) infer or create a canonical form of the final output in order to deal with differences between the target entities provided in the annotations and their variants occurring in the documents (e.g. for `jurisdiction` we must transform a text-level span *NY* into a document-level gold value **New York**).
- **Long Document**: Fig. 3 presents differences in document lengths (calculated as a number of OCRed words) in the Kleister datasets compared to other similar datasets. Since entities could appear in documents multiple times in different contexts, we must be able to understand long documents as a whole. This leads, of course, to different architectural decisions [2,4]. For example, the `term` entity in the Kleister NDA dataset tells us about the contract duration. This information is generally found in the *Term* chapter, in the middle part of a document. However, sometimes we could also find a `term` entity at the end of the document, the task is to find out which of the values is correct.
- **Complex Layout**: this requires proper understanding of the complex layout (e.g. interpreting tables and forms as 2D structures), see Fig. 1.
- **OCR**: processing of scanned documents in such a way as to deal with possible OCR errors caused by handwriting, pages turned upside down or more general poor scan quality.

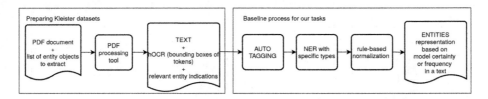

Fig. 4. Our preparation process for Kleister datasets and training baselines. Initially, we gathered PDF documents and expected entities' values. Then, based on textual and layout features, we prepared our pipeline solutions. The pipeline process is illustrated in the second frame and consists of the following stages: autotagging; standard NER; text normalization; and final selection of the values of entities.

4 Experiments

Kleister datasets for the Key Information Extraction task are challenging and hardly any solutions in the current NLP world can solve them. In this experiment, we aim to produce strong baselines with the *Pipeline* approach (Sect. 4.1) to solve extraction problems. This method's core idea is to select specific parts of the text in a document that denote the objects we search for. The whole process is a chain of techniques with, crucially, a named entity recognition model: once indicated in a document (multiple occurrences are possible), entities are normalized, then all results are aggregated into the one value specified for a given entity type.

4.1 Document Processing Pipeline

Figure 4 presents the whole process, and all the stages are described below (a similar methodology was proposed [17]).

Autotagging. Since we have only document-level annotation in the Kleister datasets, we need to generate a training set for an NER model which takes text span annotation as the input. This stage involves extracting all the fragments that refer to the same or to different entities by using sets of regular expressions combined with a gold-standard value for each general entity type, e.g. date, organization and amount. In particular, when we try to detect a `report_date` entity, we must handle different date formats: 'November 29, 2019', '11/29/19' or '11-29-2019'. This step is performed only for the purpose of training (to get data for training a sequence labeler; it is not applied during the prediction). The quality of this step varies across entity types (see details in Table 3).

Named Entity Recognition. We trained a NER model on the autotagged dataset using one of the state-of-the-art (1) architectures working on plain text such as Flair [1], BERT-base [5], RoBERTa-base [14], or (2) models employing layout features (LayoutLM-base [23] and LAMBERT [7]). Then, at the evaluation stage, we use the NER model to detect all entity occurrences in the text.

Normalization. At this stage objects are normalized to the canonical form, which we have defined in the Kleister datasets. We use almost the same regular expression as during autotagging. For instance, all detected `report_date` occurrences are normalized. So 'November 29, 2019', '11/29/19' and '11-29-2019' are rendered in our standard '2019-11-29' form (ISO 8601).

Aggregation. The NER model might return more than one text span for a given entity, sometimes these are multiple occurrences of the same correct information. Sometimes these represent errors of the NER model. In any case, we need to produce a single output from multiple candidates detected by the NER model. We take a simple approach: all candidates are grouped by the extracted entities' normalized forms and for each group we sum up the scores and finally we return the values with the largest sums.

4.2   Experimental Setup

Due to the Kleister document's length, most currently available models limit input size and so are unable to process the documents in a single pass. Therefore, each document was split into 300-word chunks (for Flair) or 510 BPE tokens (for BERT/RoBERTa/LayoutLM/LAMBERT) with overlapping parts. The results from overlapping parts were aggregated by averaging all the scores obtained for each token in the overlap.

For the Flair-based pipeline, we used implementation from the Flair library [1] in version 0.6.1 with the following parameters: *learning rate* = 0.1, *batch size* = 32, *hidden size* = 256, *epoch* = 30/15 (resp. NDA and Charity), *patience* = 3, *anneal factor* = 0.5, and with a CRF layer on top. For pipeline based on BERT/RoBERTa/LayoutLM, we used the implementation from *transformers* [10] library in version 3.1.0 with the following parameters: *learning rate* = 2e−5, *batch size* = 8, *epoch* = 20, *patience* = 2. For pipeline based on LAMBERT model we used implementation shared by authors of the publication [7] and the same parameters as for the BERT/RoBERTa/LayoutLM models. All experiments were performed with the same settings.

Moreover, in our experiments, we tried different PDF processing tools for text extraction from PDF documents to check the importance of text quality for the final pipeline score:

- **Microsoft Azure Computer Vision API (Azure CV)**[12] – commercial OCR engine, version 3.0.0;
- **pdf2djvu/djvu2hocr**[13]– a free tool for object and text extraction from born-digital PDF files (this is not an OCR engine, hence it could be applied only to Kleister NDA), version 0.9.8;
- **Tesseract**[19] – this is the most popular free OCR engine currently available, we used version 4.1.1-rc1-7-gb36c.[14];
- **Amazon Textract**[15] – commercial OCR engine.

5   Results

Table 3 shows the results for the two Kleister datasets obtained with the Pipeline method for all tested models. The weakest model from our baselines is, in general, BERT, with a slight advantage in Kleister NDA over the Flair model and a large performance drop on Kleister Charity in comparison to others. The best model is LAMBERT, which improved the overall F_1-score with 0.77 and 2.04 for,

---

[12] https://docs.microsoft.com/en-us/azure/cognitive-services/computer-vision/concept-recognizing-text.

[13] http://jwilk.net/software/pdf2djvu, https://github.com/jwilk/ocrodjvu.

[14] run with --oem 2 -l eng --dpi 300 flags (meaning both new and old OCR engines were used simultaneously, with language and pixel density set to English and 300dpi respectively).

[15] https://aws.amazon.com/textract/ (API in version from March 1, 2020 was used).

Table 3. The detailed results (average F_1-scores over 3 runs) of our baselines for Kleister challenges (test sets) for the best PDF processing tool. Autotagger F_1-scores were calculated based on results from our regexp mechanism and manual annotation on the text span level (see Sect. 3.3). Human performance is a percentage of annotators agreements for 100 random documents. We used the Base version of the BERT, RoBERTa, LayoutLM and LAMBERT models.

Kleister NDA dataset (pdf2djvu)							
Entity name	Flair	BERT	RoBERTa	LayoutLM	LAMBERT	Autotagger	Human
Effective_date	79.37	80.20	81.50	80.50	**85.27**	79.00	100%
Party	70.13	71.60	**80.83**	76.60	78.70	33.15	98%
Jurisdiction	93.87	95.00	92.87	94.23	**96.50**	54.10	100%
Term	**60.33**	45.73	52.27	47.63	55.03	74.10	95%
ALL	77.83	78.20	81.00	78.47	**81.77**	60.09	97.86%
Kleister Charity dataset (Azure CV)							
Post_town	83.07	77.03	77.70	76.57	**83.70**	66.04	98%
Postcode	89.57	87.10	88.40	88.53	**90.37**	87.60	100%
Street_line	69.10	62.23	72.03	70.92	**74.30**	75.02	96%
Charity_name	72.97	75.93	78.03	**79.63**	77.83	67.00	99%
Charity_number	96.60	**96.67**	95.37	96.13	95.80	98.60	98%
Income	70.67	67.30	69.73	70.40	**74.70**	69.00	97%
Report_date	95.93	96.60	96.77	96.40	**96.80**	89.00	100%
Spending	68.13	64.43	68.60	68.57	**74.20**	73.00	92%
ALL	81.17	78.33	81.50	81.53	**83.57**	78.16	97.45%

respectively, NDA and Charity. It is worth noting that for born-digital documents in Kleister NDA this difference is not substantial. This is due to the fact that only for `effective_date` entity does the LAMBERT model have a clear advantage (about 4 points gain of F_1-score) over other baseline models. For Kleister Charity LAMBERT achieves the biggest improvement over sequential models on `income` (+4.03) and `spending` (+5.60) which appears mostly in table-like structures.

The most challenging problems for all models are entities (`effective_date`, `party`, `term`, `post_town`, `postcode`, `street_line`, `charity_name`, `income`, `spending`) related to the properties described in Sect. 3.3.

Input/Output on Token Level (Autotagging). As we can observe in Table 3, our autotagging mechanism with information about entity achieves, on the text span level, a performance inferior to almost all our models on the document level. It shows that, despite the fact that the autotagging mechanism is prone to errors, we could train a good quality NER model. Our analysis shows that there are some specific issues related to a regular-expression-based mechanism, e.g. `party` in the Kleister NDA dataset has the lowest score because organization names often occur in the text as an acronym or as a shortened form; for instance for `party` entity text *Emerson Electric Co.* means the same as *Emerson*. This is not easy to capture with a general regexp rule.

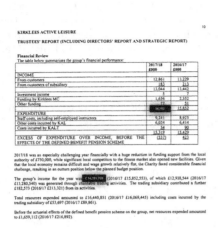

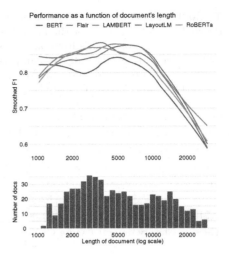

Fig. 5. Normalization issues for an `income` entity (amount in the table should be multiplied by 1000).

Fig. 6. Relationship between F_1-scores and document length in the Kleister Charity test set for the Azure CV OCR.

Input/Output on Token Level (normalization). We found that we could not achieve competitive results by using models based only on sequence labeling. For example, for the entities `income` and `spending` in the Kleister Charity dataset, we manually checked that in about 5% of examples we need to also infer the right scale (thousand, million, etc.) for each monetary value based on the document context (see Fig. 5).

Long Documents. It turns out that, for all models, worse results are observed for longer documents, see Fig. 6.

Complex Layout. The LAMBERT model has proved the best one, which proved the importance of using models employing not only textual (1D) but also layout (2D) features (see Table 3). Additionally, we also observe that the entities appearing in the sequential contexts achieve higher F_1-scores (`charity_number` and `report_date` entities in the Kleister Charity dataset).

OCR. We present the importance of using a PDF processing tool of good quality (see Table 4). With such a tool, we could gain several points in the F_1-score. There are two main conclusions: 1) Commercial OCR engines (Azure CV and Textract) are significantly better than Tesseract for scanned documents (Kleister Charity dataset). This is especially for true for 1D models not trained on Tesseract output (Flair, BERT, RoBERTa); 2) If we have the means to detect born-digital PDF documents, we should process them with a dedicated PDF tool (such as pdf2djvu) instead of using an OCR engine.

Table 4. F_1-scores for different PDF processing tools and models checked on Kleister challenges test sets over 3 runs with standard deviation. (*) pdf2djvu does not work on scans. We used the Base version of the BERT, RoBERTa, LayoutLM and LAMBERT models.

PDF tool	Flair	BERT	RoBERTa	LayoutLM	LAMBERT
Kleister NDA dataset (born-digital PDF files)					
Azure CV	$78.03_{\pm0.12}$	$77.67_{\pm0.18}$	$79.33_{\pm0.68}$	$77.43_{\pm0.29}$	$80.57_{\pm0.25}$
pdf2djvu	$77.83_{\pm0.26}$	$78.20_{\pm0.17}$	$81.00_{\pm0.05}$	$78.47_{\pm0.76}$	$\mathbf{81.77_{\pm0.09}}$
Tesseract	$76.57_{\pm0.49}$	$76.60_{\pm0.30}$	$77.81_{\pm0.97}$	$77.70_{\pm0.48}$	$81.03_{\pm0.23}$
Textract	$77.37_{\pm0.08}$	$74.83_{\pm0.45}$	$79.49_{\pm0.32}$	$77.40_{\pm0.40}$	$77.37_{\pm0.08}$
Kleister Charity dataset (mixture of born-digital and scanned PDF files) (*)					
Azure CV	$81.17_{\pm0.12}$	$78.33_{\pm0.08}$	$81.50_{\pm0.23}$	$81.53_{\pm0.23}$	$\mathbf{83.57_{\pm0.29}}$
Tesseract	$72.87_{\pm0.81}$	$71.37_{\pm1.25}$	$76.23_{\pm0.15}$	$77.53_{\pm0.20}$	$81.50_{\pm0.07}$
Textract	$78.03_{\pm0.12}$	$73.30_{\pm0.43}$	$80.08_{\pm0.15}$	$80.23_{\pm0.41}$	$82.97_{\pm0.21}$

6 Conclusions

In this paper, we introduced two new datasets Kleister NDA and Kleister Charity for Key Information Extraction tasks. We set out in detail the process necessary for the preparation of these datasets. Our intention was to show that Kleister datasets will help the NLP community to investigate the effects of document lengths, complex layouts, and OCR quality problems on KIE performance.

We prepared baseline solutions based on text and layout data generated by different PDF processing tools from the datasets. The best model from our baselines achieves $81.77/83.57$ F_1-score for, respectively, the Kleister NDA and Charity, which is much lower in comparison to datasets in a similar domain (e.g. 98.17 [7] for SROIE). This benchmark shows the weakness of the currently available state-of-the-art models for the Key Information Extraction task.

Acknowledgments.. The Smart Growth Operational Programme supported this research under project no. POIR.01.01.01-00-0605/19 (*Disruptive adoption of Neural Language Modelling for automation of text-intensive work*).

References

1. Akbik, A., Blythe, D., Vollgraf, R.: Contextual string embeddings for sequence labeling. In: Proceedings of the 27th International Conference on Computational Linguistics, pp. 1638–1649. Association for Computational Linguistics, Santa Fe, New Mexico, USA (August 2018), https://www.aclweb.org/anthology/C18-1139
2. Beltagy, I., Peters, M.E., Cohan, A.: Longformer: The long-document transformer. ArXiv arXiv:2004.05150 (2020)

3. Borchmann, L., et al.: Contract discovery: Dataset and a few-shot semantic retrieval challenge with competitive baselines. In: Cohn, T., He, Y., Liu, Y. (eds.) Proceedings of the 2020 Conference on Empirical Methods in Natural Language Processing: Findings, EMNLP 2020, Online Event, 16–20 November 2020, pp. 4254–4268. Association for Computational Linguistics (2020)
4. Dai, Z., Yang, Z., Yang, Y., Carbonell, J., Le, Q., Salakhutdinov, R.: Transformer-xl: Attentive language models beyond a fixed-length context. In: Proceedings of the 57th Annual Meeting of the Association for Computational Linguistics (2019). https://www.aclweb.org/anthology/P19-1285
5. Devlin, J., Chang, M., Lee, K., Toutanova, K.: BERT: pre-training of deep bidirectional transformers for language understanding. ArXiv arXiv:1810.04805 (2018)
6. Dwojak, T., Pietruszka, M., Borchmann, L., Chłędowski, J., Graliński, F.: From dataset recycling to multi-property extraction and beyond. In: Proceedings of the 24th Conference on Computational Natural Language Learning, pp. 641–651. Association for Computational Linguistics, Online (November 2020). https://doi.org/10.18653/v1/2020.conll-1.52, https://www.aclweb.org/anthology/2020.conll-1.52
7. Garncarek, Ł., et al.: LAMBERT: Layout-Aware (Language) Modeling using BERT for information extraction. ArXiv arXiv:2002.08087 (2020)
8. Hewlett, D., et al.: WikiReading: a novel large-scale language understanding task over Wikipedia. In: Proceedings of the 54th Annual Meeting of the Association for Computational Linguistics (Volume 1: Long Papers), pp. 1535–1545. Association for Computational Linguistics, Berlin, Germany (2016)
9. Holt, X., Chisholm, A.: Extracting structured data from invoices. In: Proceedings of the Australasian Language Technology Association Workshop 2018, pp. 53–59. Dunedin, New Zealand (December 2018). https://www.aclweb.org/anthology/U18-1006
10. Hugging Face: Transformers. https://github.com/huggingface/transformers (2020)
11. Jaume, G., Kemal Ekenel, H., Thiran, J.: FUNSD: A dataset for form understanding in noisy scanned documents. In: 2019 International Conference on Document Analysis and Recognition Workshops (ICDARW), vol. 2, pp. 1–6 (2019)
12. Katti, A.R., Reisswig, C., Guder, C., Brarda, S., Bickel, S., Höhne, J., Faddoul, J.B.: Chargrid: Towards Understanding 2D Documents. ArXiv arXiv:1809.08799 (2018)
13. Liu, X., Gao, F., Zhang, Q., Zhao, H.: Graph convolution for multimodal information extraction from visually rich documents. In: Proceedings of the 2019 Conference of the North American Chapter of the Association for Computational Linguistics (2019). http://dx.doi.org/10.18653/v1/N19-2005
14. Liu, Y., et al.: RoBERTa: A Robustly Optimized BERT Pretraining Approach. ArXiv arXiv:1907.11692 (2019)
15. Mathew, M., Karatzas, D., Jawahar, C.V.: DocVQA: A Dataset for VQA on Document Images. ArXiv arXiv:2007.00398 (2021)
16. Palm, R.B., Laws, F., Winther, O.: Attend, copy, parse end-to-end information extraction from documents. In: International Conference on Document Analysis and Recognition (ICDAR) (2019)
17. Palm, R.B., Winther, O., Laws, F.: Cloudscan - a configuration-free invoice analysis system using recurrent neural networks. In: 14th IAPR International Conference on Document Analysis and Recognition (ICDAR) (2017). https://doi.org/10.1109/icdar.2017.74
18. Park, S., et al.: CORD: a consolidated receipt dataset for post-OCR parsing. In: Document Intelligence Workshop at Neural Information Processing Systems (2019)

19. Smith, R.: Tesseract Open Source OCR Engine (2020). https://github.com/tesseract-ocr/tesseract
20. Tjong Kim Sang, E.F., De Meulder, F.: Introduction to the CoNLL-2003 Shared Task: Language-Independent Named Entity Recognition. In: Proceedings of the Seventh Conference of the North American Chapter of the Association for Computational Linguistics (2003)
21. Wellmann, C., Stierle, M., Dunzer, S., Matzner, M.: A framework to evaluate the viability of robotic process automation for business process activities. In: Asatiani, A., et al. (eds.) BPM 2020. LNBIP, vol. 393, pp. 200–214. Springer, Cham (2020). https://doi.org/10.1007/978-3-030-58779-6_14
22. Wróblewska, A., Stanisławek, T., Prus-Zajączkowski, B., Garncarek, Ł.: Robotic process automation of unstructured data with machine learning. In: Position Papers of the 2018 Federated Conference on Computer Science and Information Systems, FedCSIS 2018, Poznań, Poland, 9–12 September 2018, pp. 9–16 (2018). https://doi.org/10.15439/2018F373
23. Xu, Y., Li, M., Cui, L., Huang, S., Wei, F., Zhou, M.: LayoutLM: pre-training of text and layout for document image understanding. In: Proceedings of the 26th ACM SIGKDD International Conference on Knowledge Discovery & Data Mining (2020). https://doi.org/10.1145/3394486.3403172

Graphics, Diagram, and Math Recognition

Towards an Efficient Framework for Data Extraction from Chart Images

Weihong Ma[1], Hesuo Zhang[1], Shuang Yan[2], Guangshun Yao[2], Yichao Huang[2], Hui Li[3], Yaqiang Wu[3], and Lianwen Jin[1,4(✉)]

[1] South China University of Technology, Guangzhou, China
{eeweihong_ma,eehesuo.zhang}@mail.scut.edu.cn, eelwjin@scut.edu.cn
[2] IntSig Information Co. Ltd., Shanghai, China
{shuang_yan,guangshun_yao,charlie_huang}@intsig.net
[3] Lenovo Research, Beijing, China
{lihuid,wuyqe}@lenovo.com
[4] Guangdong Artificial Intelligence and Digital Economy Laboratory (Pazhou Lab), Guangzhou, China

Abstract. In this paper, we fill the research gap by adopting state-of-the-art computer vision techniques for the data extraction stage in a data mining system. As shown in Fig. 1, this stage contains two subtasks, namely, plot element detection and data conversion. For building a robust box detector, we comprehensively compare different deep learning-based methods and find a suitable method to detect box with high precision. For building a robust point detector, a fully convolutional network with feature fusion module is adopted, which can distinguish close points compared to traditional methods. The proposed system can effectively handle various chart data without making heuristic assumptions. For data conversion, we translate the detected element into data with semantic value. A network is proposed to measure feature similarities between legends and detected elements in the legend matching phase. Furthermore, we provide a baseline on the competition of Harvesting raw tables from Infographics. Some key factors have been found to improve the performance of each stage. Experimental results demonstrate the effectiveness of the proposed system.

Keywords: Data extraction · Box detection · Point detection · Data conversion

1 Introduction

Chart data is one of the important information transmitted medium that clarifies and integrates difficult information concisely [26]. In recent years, an increasing number of chart images have emerged in multimedia, scientific papers,

This research is supported in part by NSFC (Grant No.: 61936003, 61771199), GD-NSF (no. 2017A030312006).

J. Lladós et al. (Eds.): ICDAR 2021, LNCS 12821, pp. 583–597, 2021.
https://doi.org/10.1007/978-3-030-86549-8_37

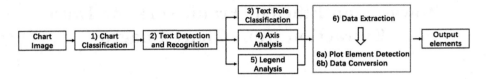

Fig. 1. Generic pipeline for extracting data in a data mining system. We mainly discuss the sixth stage (data extraction) assuming that the previous output has been obtained.

and business reports. Therefore, the issue of automatic data extraction from chart images has gathered significant research attention [3,15,16,19].

As shown in Fig. 1, in general, a chart data mining system [8] includes the following six stages: chart classification, text detection and recognition, text role classification, axis analysis, legend analysis and data extraction. Among all the aforementioned stages, data extraction is conducted as the most crucial and difficult part, whose performance depends on the quality of localization. In this work, we mainly discuss the data extraction stage. The goal in this stage is to detect elements in the plot area and convert them into data marks with semantic value. As shown in Fig. 2, this task has two subtasks: plot element detection and data conversion.

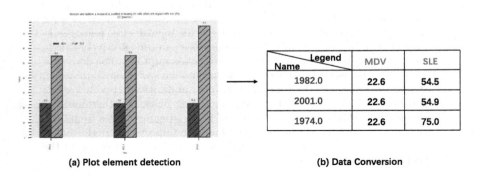

(a) Plot element detection | **(b) Data Conversion**

Fig. 2. Illustration of the data extraction stage.

To build a robust data extraction system, we can learn methods from the field of object detection. However, it should be clear that chart images differ significantly from natural images. As shown in Fig. 3, (a) is an image from COCO dataset [12], and (b) is an image from synthetic chart dataset [8]. First, compared with general object, elements in the chart images have a large range of aspect ratios and sizes. Chart images contain a combination of different elements. These elements can be either very short, such as numerical tick points or long, such as in-plot titles. Second, chart images are highly sensitive to localization accuracy. While the intersection-over-union (IoU) values in the range of 0.5 to 0.7 are acceptable for general object detection, it is unacceptable for chart images. As shown in Fig. 3b, even when the IoU is 0.9, there is still a small numerical deviation on bar images, which shows the sensitivity of chart images to IoU. Therefore, for chart data extraction, highly precise bounding boxes or points, i.e., with high IoU values are required for the detection system.

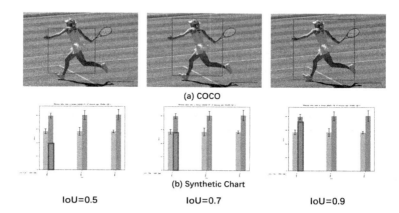

Fig. 3. Visualization when intersection-over-union (IoU) values range from 0.5 to 0.9 on (a) COCO and (b) synthetic chart images. An IoU value of 0.5 is acceptable on natural images; a higher IoU value, such as 0.7, is redundant. However, for chart images, such values are unacceptable. Even if the IoU is 0.9, there is still a small numerical deviation on bar charts.

Currently, state-of-the-art computer vision techniques have not been fully adopted by chart mining approaches. Moreover, there have been very few comparisons using deep learning-based methods for chart mining. It is believed that deep learning-based methods can avoid hard heuristic assumptions and more robust when handling various real chart data. In this study, using published real-world datasets[1], we attempt to fill this research gap in the data extraction stage. In the proposed framework, elements in the main plot area are first detected. Based on axis analysis and legend analysis results from previous stages in a data mining system, we then convert detected elements into data marks with semantic value. The contribution of this work can be summarized as follows. (i) For building a robust box detector, we comprehensively compare different deep learning-based methods. We mainly study whether existing object detection methods are adequate for box-type element detection. In particular, they should be capable of (a) detecting elements with a large aspect ratio range and (b) localizing objects with a high IoU value. (ii) For building a robust point detector, we use a fully convolutional network (FCN) with feature fusion module to output a heatmap mask. It can distinguish close points well while traditional methods and detection-based methods easily fail. (iii) For data conversion, in

---

[1] http://tc11.cvc.uab.es/datasets/ICPR2020-CHART-Info_1.

the legend matching phase, a network is trained to measure feature similarities. It is robust than image-based features when noise exists in feature extracting phase. Finally, we provide a baseline on a public dataset which can facilitate further research. Experimental results demonstrate the effectiveness of the proposed system. The implementation of our pipeline will be available to the public for reproducing these results.

2   Related Work

In this section, we review previous works on data extraction in a chart mining system. We mainly focus on related works of classification, element detection and data conversion.

According to the types of detection data, we can divide the chart data into box-type and point-type data. Box-type data includes bars and boxplots. These charts are commonly used to visualize data series that have a categorical independent variable. For the task of box detection, some methods have been proposed to detect elements through the characteristics of bars [1,2,7,23]. Assuming that the bar mark is solidly shaded using a single color, Savva et al. [23] used connected component (CC) analysis method and heuristic rules to extract the data mark. Balaji et al. [2] and Dai et al. [7] also used image processing methods to detect bars. They first obtained the binary image and used open morphological operation to filter noise. Next, they performed the CC labeling algorithm to find the bars. Rabah et al. [1] used heuristic features based on shape, pixel densities, color uniformity, and relative distances to the axes. However, these methods may fail when detecting small bars. In response to this problem, some methods based on deep neural networks use object detection architectures to locate the bars [5,14,18], which are more robust to extract features.

Point-type data, including charts such as scatter, line, and area, are semantically similar because they present one or more sequences of 2D points on a given Cartesian plane. The scatter chart is the most basic type in these charts. A line chart is created when the points are connected to form curves. An area chart highlights the area under the curve, typically using coloring or texture patterns. Assuming that only detected data and text elements are in the plot area, Khademul et al. [17] proposed a system for extracting data from a line chart. The system first detected axes by projection method and then used CC analysis to filter text elements inside the plot area. Finally, data are extracted using a sequential scanning process. Considering that the chart might have grid lines, Viswanath et al. [21] proposed an image processing-based method and developed a semi-supervised system. However, the method assumes that grid lines should not be more visually distinctive than the data marks. Thus, these approaches make multiple assumptions and often fail when processing images with a complex background. Cliche et al. [6] used object detection model [27] to detect scatter points. This method is more robust than the image processing-based method. However, this method may fail when points are close to each other.

In the data conversion stage, legend analysis and axis analysis should be obtained. If legends exist, the shape and color of elements will be used to identify data marks by data series. Choudhury et al. [20] proposed clustering based on shape and color for line graphs. Each cluster is output as a curve. Using the axes analysis results, these relative coordinates can be projected onto the original data space. To recover data from the bar chart, Savva et al. [23] considered linear mapping and calculated the scaling factor between image space and data space.

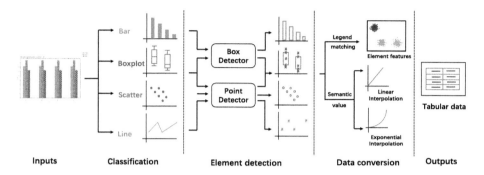

Fig. 4. The overall architecture of the proposed framework. First, the input chart image is classified by the pre-trained classification model. Second, two detectors, named box detector and point detector, are built for different chart types. Third, in legend matching phase, elements are divided into corresponding legends by comparing their features similarities. Their semantic value are calculated by interpolation method. Finally, the value of elements are output into tables.

3   Methodology

The overall architecture of our proposed method is presented in Fig. 4. Functionally, the framework consists of three components: a pre-trained chart classification model, element detection module for detecting box or point, and data conversion for determining element values. In the following sections, we first introduce the details of the box and point detectors. Next, we provide implementation details of the data conversion.

3.1   Box Detector

To extract robust features at different scales, we use a ResNet-50 [10] with a feature pyramid network (FPN) [11]. FPN uses a top-down architecture with lateral connections to fuse features of different resolutions from a single scale input, enabling it to detect elements with a large aspect ratio range. To detect a box with high IoU, we choose the Cascade R-CNN [4] as our box detector. As shown in Fig. 5(a), the box detector has four stages, one region proposal network (RPN) and three for detection with IoU $= 0.5, 0.6, 0.7$. The sampling of the first detection stage follows [22]. In the following stages, resampling is implemented by simply using the regressed outputs from the previous stage.

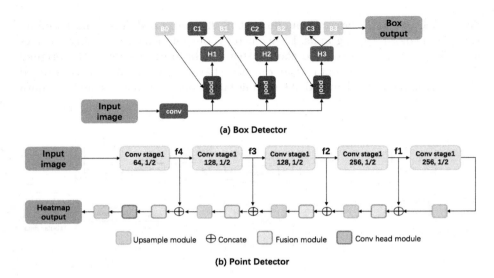

Fig. 5. Network architecture of box detector and point detector.

3.2   Point Detector

Points are another common chart elements in chart data. As mentioned earlier, the corresponding chart types include scatter, line, and area. Generally, points are densely distributed in the plot area, and the data is represented as the format of (x, y). In this work, we use segmentation-based method to detect points, which can help distinguish close points.

Network Architecture. As shown in Fig. 5(b), four levels of feature maps, denoted as f_i, are extracted from the backbone network, whose sizes are $1/16$, $1/8$, $1/4$ and $1/2$ of the input image, respectively. Then, the features from different depths are fused in the up-sampling stage. In each merging stage, the feature map from the last stage is first fed to the up-sampling module to double its size, and then concatenated with the current feature map. Next, the fusion module, which is built using two consecutive $conv_{3\times3}$ layers, produces the final output of this merging stage. Following the last merging stage, the head module, built by two $conv_{3\times3}$ layers, is then used. Finally, the feature map is upsampled to the image size.

Label Generation. To train the FCN network, we generate a heatmap mask. The binary map, which sets all pixels inside the contour to the same value, can not reflect the relationship between each pixel. In contrast to the binary segmentation map, we draw Gaussian heatmap for these points on the mask. The Gaussian value Y_{xy} is calculated using the Gaussian kernel function. If two Gaussians overlap and one point has two values, we use the maximum value.

$$Y_{xy} = e^{-\frac{(x-p_x)^2 + (y-p_y)^2}{2\sigma^2}} \tag{1}$$

where (x, y) is the point coordinate on the mask, (p_x, p_y) is the center of the target point. σ is a Gaussian kernel parameter that determines the size. Here, we set the value of σ as 2.

Post-processing. In the testing phase, the point detector outputs a heatmap mask. We first filter the output noise outside the main plot area. Then, we use a high confidence threshold to output positive region. The final points output are obtained by finding the center of connected components. In the process of connected component analysis, when we detect a larger connected area, it indicates points are close and the model cannot accurately separate these points. Therefore, in order to improve the recall rate, we also randomly sample points inside the connected area, where number of points is linearly related to the size.

3.3   Data Conversion

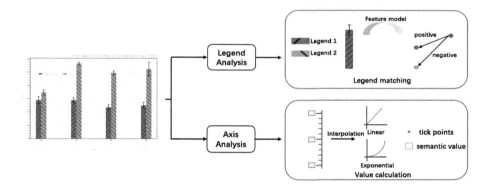

Fig. 6. Pipeline of data conversion.

After detecting the elements, we need to determine the element values. In this stage, the goal is to convert detected elements in the plot area into data marks with semantic value. As shown in Fig. 6, legend matching and value calculation are performed in this stage.

Legend Matching: Based on the legend analysis result, which is obtained from the fifth stage in a data mining system, we can get the position of the legends. If there exists legends, we need to extract features of elements and legends. Then we use L2 distance to measure feature similarities and divide elements into corresponding legends. Image-based features, such as RGB features and HSV features, is not robust when the detection result is not tight enough. Therefore, we propose to train a feature model to measure feature similarity.

The network directly learns a mapping $f(x)$ from patch input image x into embedding vectors $\mathbb{R}^d$. It is comprised of multiple modules, which are built using conv-BN-ReLU layers, and finally outputs a 128-D embedding vector for each patch input. In the training phase, the network is optimized using a triplet-based loss [24]. This loss aims to separate the positive pair from the negative by a distance margin. Embedding vectors of the same cluster should have small distances and different clusters should have large distances. In the testing phase, the cropped legend patch and element patch are fed into the model. For each element, the legend with the smallest distance on feature dimension is the corresponding class.

Value Calculation: Based on the axis analysis result, which is obtained from the forth stage, we can get the position of the detected tick points and their corresponding semantic values. Then, we analyze the numerical relationship between adjacent tick points, including the case of linear or exponential. Finally, we calculate the value of the unit scale and use the interpolation method to determine the element value.

4   Experiments

4.1   Datasets

Table 1. Distribution of the two datasets used for the experiments

Dataset	Chart type	Num. of training set	Num. of validation set	Total num.
Synth2020	Bar	4, 264	536	4, 800
	Boxplot	2, 132	268	2, 400
	Line	1, 066	134	1, 200
	Scatter	1, 066	134	1, 200
UB PMC2020	Bar	781	98	879
	Boxplot	174	22	196
	Line	658	83	741
	Scatter	272	35	307

There are two sets of datasets used in this work, which are named Synth2020 and UB PMC2020, respectively. The earlier version of these two datasets are published in [8]. Thanks to the ICPR2020 Competition on Harvesting Raw Tables from Infographics[2], more challenging images have been added and the size of the dataset has been extended. The first dataset, Synth2020, is the extended version of Synth2019. Multiple charts of different types are created using the

---

[2] https://chartinfo.github.io.

Matplotlib library. The second dataset is curated from real charts in scientific publications from PubMedCentral [8], which has different image resolutions and more uncertainties in the images. We divide the official training dataset into a training set and validation set with a ratio of 8:1 randomly. Details of the split of these two datasets are given in Table 1. The specific training and validation sets will be published.

4.2   Evaluation

We use the competition evaluation script[3] to measure the model performance. Two metrics for measuring the detector performance and data conversion performance are proposed. For different types of chart data, the script has a different evaluation mechanism. Details of the evaluation mechanism are mentioned in the competition. For box detection evaluation, to accurately evaluate the model performance under different IoU, we also refer to the field of object detection and calculate F-measure when IoU is equal to 0.5, 0.7, and 0.9.

4.3   Implementation Details

The synthetic and UB PMC datasets are trained and tested separately. In this section, we introduce the details of our implementation.

In the box detector experiment, we choose bar type data for training. The backbone feature extractor is ResNet-50 pre-trained on ImageNet. In the regression stage, we adopt RoIAlign to sample proposals to a 7×7 fixed size. The batch size is 8 and the initial learning rate is set to 0.01. The model is optimized with stochastic gradient descent(SGD) and the maximum epochs for training is 20. In the inference stage, non-maximum suppression (NMS) is utilized to suppress the redundant outputs.

In the point detector experiment, we choose scatter type data for training. In the training stage, we use MSE loss to optimize the network. Multiple data augmentations are adopted, including random crop, random rotate, random flip, and image distortion, to avoid overfitting. We adopt the OHEM [25] strategy to learn hard samples. The ratio of positive and negative samples is 1:3. The model is optimized with Adam optimizer and the maximum iterations is 30k with a batch size of 4.

In the data conversion experiment, we train the model to extract features for clustering. The input size for training is 24×24, and the embedding dimension is set to 128. The model is optimized with Adam optimizer and the maximum iterations is 50k. The batch size is 8 and the initial learning rate is set to 0.001.

4.4   Result and Analysis

Evaluation of Box Detector. In this section, the performance of the box detector is evaluated in terms of Score_a and F-measure when the value of IoU

---

[3] https://github.com/chartinfo/chartinfo.github.io/tree/master/metrics.

is set to 0.5, 0.7, 0.9, respectively. Score_a uses the evaluation mechanism from ICPR2020 competition. The trained models are tested on the Synth2020 validation set and UB PMC2020 testset, respectively. Since the testset of Synth2020 is currently unavailable, we use validation set to test the model performance on the Synth2020 dataset.

Table 2. Evaluation results of box detector on bar type data

Dataset	Model	IoU = 0.5	IoU = 0.7	IoU = 0.9	Score_a
Synth2020 validation bar	SSD	80.98	69.56	28.79	67.87
	YOLO-v3	85.30	76.54	38.55	90.96
	Faster R-CNN	94.80	92.47	48.92	92.33
	Faster R-CNN+FPN	96.46	94.39	52.30	92.89
	Cascade R-CNN+FPN	**96.86**	**96.25**	**93.97**	**93.36**
UB PMC2020 testset bar	SSD	43.65	26.28	2.67	25.83
	YOLO-v3	58.84	36.14	4.14	60.97
	Faster R-CNN	66.37	60.88	29.13	70.03
	Faster R-CNN+FPN	85.81	78.05	31.30	89.65
	Cascade R-CNN+FPN	**86.92**	**83.53**	**55.32**	**91.76**

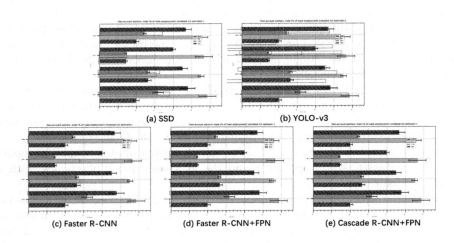

Fig. 7. Detection result of different models on an example bar image from Synth2020.

For comparison, we implement different detection models, including one-stage and two-stage detection models. The one-stage models are SSD [13] and YOLO-v3 [9], whereas the two-stage model is Faster R-CNN [22]. As listed in Table 2, the performance of the one-stage model performs worst, and the multi-stage regress heads help to obtain high accuracy. Furthermore, the additional FPN structure effectively helps to detect elements with a large aspect ratio range. On both Synth2020 and UB PMC2020 dataset, the Cascade R-CNN model with FPN structure performs the best. Therefore, for bar type data detection, models with multiple regression heads and FPN structure achieve impressive performance.

One-stage models output poor results in earlier iterations. At the same time, NMS can not filter these error outputs effectively, which can be best viewed in Fig. 7(b). NMS can not suppress these outputs because the IoU between these long rectangles is smaller than 0.5. Owing to these reasons, the model can not reach the global optimal solution.

Table 3. Evaluation results of point detector on scatter type data

	CC based	Detect based	Pose ResNet	Proposed
Synth2020 validation set	57.84	72.34	78.98	**87.20**
UB PMC2020 validation set	53.36	82.12	82.46	**86.46**
UB PMC2020 testset	51.58	84.54	84.58	**88.58**

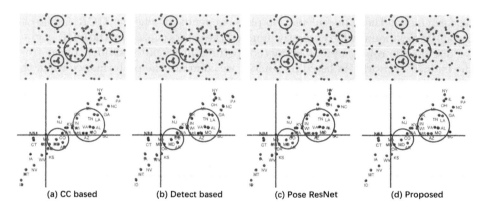

(a) CC based   (b) Detect based   (c) Pose ResNet   (d) Proposed

Fig. 8. Visualization of the detection results on chart data from different models. Patches are shown for visualization. The detected points are drawn as red dots on the input image. The data of first row is from Synth2020 and the data of second row is from UB PMC2020. Circles in red show some key differences between these models. (Color figure online)

Evaluation of Point Detector. In this section, the performance of the point detector is evaluated in terms of the evaluation mechanism published in the competition. The trained models are tested on Synth2020 validation, UB PMC2020 validation and testset.

We compare our method with traditional image processing method, such as connected component analysis and detection-based method. The detection model is based on Faster R-CNN. To train the Faster R-CNN model, we expand the point (x, y) into a rectangle $(x - r, y - r, x + r, y + r)$ whose data format is $(left, top, right, bottom)$. We also implement another segmentation-based method Pose ResNet [28], which is initially proposed for pose keypoint detection. The Pose ResNet model adopts the structure of down-sampling and then up-sampling, without considering the feature fusion of different depths.

As listed in Table 3, the proposed method, which is simple and effective, outperforms other methods on three testsets. As shown in Fig. 8, on the Synth2020 validation set, there are many cases where scatter points are connected and form a larger connected component. On the UB PMC2020 testset, there are many noises in the plot area such as text elements. Traditional image processing method can not distinguish close points which form a large component. The detection-based method fails when the number of points is large or adjacent points are connected. Compared to Pose ResNet, the feature fusion method helps to distinguish adjacent points, as shown in Fig. 8(d). The proposed method can effectively deal with these situations and locate adjacent points accurately.

Robustness of Feature on Data Conversion. We choose line type data to evaluate the performance of data conversion. The performance of data conversion depends on the legend matching phase and value calculation phase. The performance in the value calculation phase depends on whether OCR engine can recognize tick point value correctly. Ignored the errors caused by the OCR engine, we discuss the robustness of extracted features in legend matching phase from the trained network. As listed in Table 4, we compare the performance when the legend matching phase is performed on the groundtruth and prediction result. For short notation, here s1, s2, s3 represent average name score, average data series score, and average score, respectively, which is claimed in the evaluation script.

When using groundtruth as input, the position of the elements is quite accurate. Features extracted from trained network are comparable with features from the concatenation of RGB and HSV features. The performance can be further improved by considering the cascading of features. When prediction detection results are used, the position of elements may not be tight enough, which will bring in noise while extracting features. Experiments show that features from our proposed method are more robust than image-based features.

Evaluation Result of Proposed System. As listed in Table 5, we provide our proposed system performance on ICPR2020 Competition which can serve as a baseline and facilitate further research. For short notation, here s0, s1, s2, and s3 represent visual element detection score, average name score, average data series score and average score, respectively. In this work, no additional data or model ensemble strategy is adopted. It is shown that our system outperforms the Rank1 and Rank2 result of the competition on UB PMC2020 testset, which demonstrate the effectiveness of the proposed system.

Table 4. Evaluation results of data conversion on line type data

Features	Groundtruth			Prediction		
	s1	s2	s3	s1	s2	s3
Baseline (RGB)	83.40	75.32	77.34	83.50	55.61	69.68
RGB+HSV	83.48	77.58	79.06	83.50	52.52	68.16
RGB+HSV+CNN	**83.48**	**78.31**	**79.60**	83.06	53.81	68.83
CNN	82.71	77.78	79.02	**83.53**	**67.19**	**75.36**

Table 5. Evaluation results of proposed system

Chart types	Synth2020 validation set				UB PMC2020 testset			
	s0	s1	s2	s3	s0	s1	s2	s3
Bar	93.36	99.82	99.11	99.29	91.75	96.96	94.20	94.89
Scatter	87.19	100.00	82.97	87.23	88.58	86.61	65.20	70.55
Boxplot	100.00	99.83	98.37	98.73	98.62	92.57	81.62	84.36
Line	99.29	99.09	98.81	98.88	84.03	83.24	67.01	71.06
Average (Rank1*)	–	–	–	–	88.23	**90.42**	76.73	80.15
Average (Rank2*)	–	–	–	–	87.00	78.54	55.40	61.18
Average (proposed)	**94.96**	**99.69**	**94.82**	**96.03**	**90.75**	89.85	**77.00**	**80.22**

* From https://chartinfo.github.io/leaderboards_2020.html

5   Conclusion

In this work, we discuss the data extraction stage in a data mining system. For building a robust box detector, we compare different object detection methods and find a suitable method to solve the special issues that characterize chart data. Models with multiple regression heads and FPN structure achieve impressive performance. For building a robust point detector, compared with image processing-based methods and detection-based methods, the proposed segmentation-based method can avoid hard heuristic assumptions and distinguish close points well. For data conversion, we propose a network to measure feature similarities which is more robust compared with image-based features. In the experiments, we conduct experiments in each stage of data extraction. We find the key factors that improve the performance of each stage. The overall performance on a public dataset demonstrates the effectiveness of the proposed system. Because an increasing number of charts have appeared in recent years, we believe the field of automatic extraction from chart data will develop quickly. We expect this work to provide useful insights and provide a baseline for comparison.

References

1. Al-Zaidy, R.A., Giles, C.L.: A machine learning approach for semantic structuring of scientific charts in scholarly documents. In: Proceedings of the AAAI Conference on Artificial Intelligence, pp. 4644–4649 (2017)

2. Balaji, A., Ramanathan, T., Sonathi, V.: Chart-Text: a fully automated chart image descriptor. arXiv preprint arXiv:1812.10636 (2018)

3. Böschen, F., Scherp, A.: A comparison of approaches for automated text extraction from scholarly figures. In: Amsaleg, L., Guðmundsson, G.Þ, Gurrin, C., Jónsson, B.Þ, Satoh, S. (eds.) MMM 2017. LNCS, vol. 10132, pp. 15–27. Springer, Cham (2017). https://doi.org/10.1007/978-3-319-51811-4_2

4. Cai, Z., Vasconcelos, N.: Cascade R-CNN: delving into high quality object detection. In: Proceedings of the IEEE Conference on Computer Vision and Pattern Recognition, pp. 6154–6162 (2018)

5. Choi, J., Jung, S., Park, D.G., Choo, J., Elmqvist, N.: Visualizing for the non-visual: enabling the visually impaired to use visualization. In: Computer Graphics Forum, vol. 38, pp. 249–260. Wiley Online Library (2019)

6. Cliche, M., Rosenberg, D., Madeka, D., Yee, C.: Scatteract: automated extraction of data from scatter plots. In: Ceci, M., Hollmén, J., Todorovski, L., Vens, C., Džeroski, S. (eds.) ECML PKDD 2017. LNCS (LNAI), vol. 10534, pp. 135–150. Springer, Cham (2017). https://doi.org/10.1007/978-3-319-71249-9_9

7. Dai, W., Wang, M., Niu, Z., Zhang, J.: Chart decoder: generating textual and numeric information from chart images automatically. J. Vis. Lang. Comput. **48**, 101–109 (2018)

8. Davila, K., et al.: ICDAR 2019 competition on harvesting raw tables from info-graphics (chart-infographics). In: Proceedings of the IEEE Conference on Document Analysis and Recognition, pp. 1594–1599. IEEE (2019)

9. Farhadi, A., Redmon, J.: Yolov3: an incremental improvement. In: Proceedings of the IEEE Conference on Computer Vision and Pattern Recognition (2018)

10. He, K., Zhang, X., Ren, S., Sun, J.: Deep residual learning for image recognition. In: Proceedings of the IEEE Conference on Computer Vision and Pattern Recognition, pp. 770–778 (2016)

11. Lin, T.Y., Dollár, P., Girshick, R., He, K., Hariharan, B., Belongie, S.: Feature pyramid networks for object detection. In: Proceedings of the IEEE Conference on Computer Vision and Pattern Recognition, pp. 2117–2125 (2017)

12. Lin, T.Y., et al.: Microsoft COCO: common objects in context. In: Fleet, D., Pajdla, T., Schiele, B., Tuytelaars, T. (eds.) ECCV 2014. LNCS, vol. 8693, pp. 740–755. Springer, Cham (2014). https://doi.org/10.1007/978-3-319-10602-1_48

13. Liu, W., et al.: SSD: single shot multibox detector. In: Leibe, B., Matas, J., Sebe, N., Welling, M. (eds.) ECCV 2016. LNCS, vol. 9905, pp. 21–37. Springer, Cham (2016). https://doi.org/10.1007/978-3-319-46448-0_2

14. Liu, X., Klabjan, D., NBless, P.: Data extraction from charts via single deep neural network. arXiv preprint arXiv:1906.11906 (2019)

15. Liu, Y., Lu, X., Qin, Y., Tang, Z., Xu, J.: Review of chart recognition in document images. In: Visualization and Data Analysis 2013. vol. 8654, p. 865410. International Society for Optics and Photonics (2013)

16. Mei, H., Ma, Y., Wei, Y., Chen, W.: The design space of construction tools for information visualization: a survey. J. Vis. Lang. Comput. **44**, 120–132 (2018)

17. Molla, M.K.I., Talukder, K.H., Hossain, M.A.: Line chart recognition and data extraction technique. In: Liu, J., Cheung, Y., Yin, H. (eds.) IDEAL 2003. LNCS, vol. 2690, pp. 865–870. Springer, Heidelberg (2003). https://doi.org/10.1007/978-3-540-45080-1_120

18. Methani, N., Ganguly, P., Khapra, M.M., Kumar, P.: Data interpretation over plots. In: Proceedings of the IEEE Winter Conference on Applications of Computer Vision (2020)

19. Purchase, H.C.: Twelve years of diagrams research. J. Vis. Lang. Comput. **25**(2), 57–75 (2014)
20. Choudhury, S.R., Wang, S., Giles, C.L.: Curve separation for line graphs in scholarly documents. In: Proceedings of the 16th ACM/IEEE-CS on Joint Conference on Digital Libraries, pp. 277–278 (2016)
21. Reddy, V.K., Kaushik, C.: Image processing based data extraction from graphical representation. In: Proceedings of the IEEE Conference on Computer Graphics, Vision and Information Security, pp. 190–194. IEEE (2015)
22. Ren, S., He, K., Girshick, R., Sun, J.: Faster R-CNN: towards real-time object detection with region proposal networks. IEEE Trans. Pattern Anal. Mach. Intell. **39**(6), 1137–1149 (2016)
23. Savva, M., Kong, N., Chhajta, A., Fei-Fei, L., Agrawala, M., Heer, J.: Revision: automated classification, analysis and redesign of chart images. In: Proceedings of the 24th Annual ACM Symposium on User Interface Software and Technology, pp. 393–402 (2011)
24. Schroff, F., Kalenichenko, D., Philbin, J.: FaceNet: a unified embedding for face recognition and clustering. In: Proceedings of the IEEE Conference on Computer Vision and Pattern Recognition, pp. 815–823 (2015)
25. Shrivastava, A., Gupta, A., Girshick, R.: Training region-based object detectors with online hard example mining. In: Proceedings of the IEEE Conference on Computer Vision and Pattern Recognition, pp. 761–769 (2016)
26. Sricharoen, W.V.: Infographics: the new communication tools in digital age. In: The International Conference on e-technologies and Business on the Web (EBW2013), pp. 169–174 (2013)
27. Stewart, R., Andriluka, M., Ng, A.Y.: End-to-end people detection in crowded scenes. In: Proceedings of the IEEE Conference on Computer Vision and Pattern Recognition, pp. 2325–2333 (2016)
28. Xiao, B., Wu, H., Wei, Y.: Simple baselines for human pose estimation and tracking. In: Proceedings of the European Conference on Computer Vision, pp. 466–481 (2018)

Geometric Object 3D Reconstruction from Single Line Drawing Image Based on a Network for Classification and Sketch Extraction

Zhuoying Wang, Qingkai Fang, and Yongtao Wang$^{(\boxtimes)}$

Peking University, Beijing, China
{wzypku,wyt}@pku.edu.cn, poeroz@bupt.edu.cn

Abstract. Geometric objects in educational materials are usually illustrated as 2D line drawings without depth information, and this is a barrier for readers to fully understand the 3D structure of these geometric objects. To address this issue, we propose a novel method to recover the 3D shape of the geometric object from a single 2D line drawing image. Specifically, our proposed method can be divided into two stages: sketch extraction stage and reconstruction stage. In the sketch extraction stage, we propose a deep neural network to identify the category of the geometric object in the line drawing image and extract its sketch simultaneously. Our network architecture is based on High-Resolution Network (HRNet), which integrates two task-specific decoders: one for classification and the other for vertices detection. With the predicted category and location of vertices, we can easily obtain the sketch of the geometric object in the input line drawing image. Compared with previous methods, our CNN-based method can directly extract the sketch of geometric objects without any hand-crafted features or processes, which gives a more robust performance. In the reconstruction stage, we exploit an example-based method and conduct the reconstruction by optimizing an objective function of reconstruction error. Moreover, we generate a simulated dataset to alleviate the problem caused by unbalanced distribution across different categories in the manually-collected dataset, which greatly improves the performance of our deep neural network model. Extensive experimental results demonstrate that the proposed method performs significantly better than the existing methods in both accuracy and efficiency.

Keywords: Sketch extraction · 3D reconstruction · Keypoint detection

1 Introduction

Educational materials like high school textbooks contain lots of 3D geometric objects, which are illustrated as 2D line drawings. However, a 2D line drawing can only show the projection of the 3D geometric object from a certain perspective and thus lose the depth information. Without depth information, it's difficult for readers (*e.g.* students) to fully understand the structure of 3D geometric objects. Therefore, it is necessary to develop a system to reconstruct the 3D geometric object from a single 2D line drawing

© Springer Nature Switzerland AG 2021
J. Lladós et al. (Eds.): ICDAR 2021, LNCS 12821, pp. 598–613, 2021.
https://doi.org/10.1007/978-3-030-86549-8_38

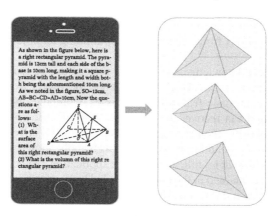

Fig. 1. Illustration of our application on mobile devices. With our application, users can observe the geometric object from any perspective.

image both accurately and efficiently and allow readers to access it with mobile devices such as smartphones or tablets.

Many researchers have focused on reconstructing a 3D geometric objects from a single 2D line drawing over the past three decades. Earlier methods can be divided into three categories: (1) *rule-based methods* [7,9,12,13,15–18,20], (2) *deduction-based methods* [1,4,10,11], (3) *divide-and-conquer-based methods* [2,14,23,24,27]. *Rule-based methods* solve reconstruction problem by optimizing an objective function with some constraints based on geometric rules. *Deduction-based methods* adopt deductive methods with some assumptions. *Divide-and-conquer-based-methods* usually divide the complex geometric object into multiple simple components, then perform reconstruction on them respectively, finally combine the reconstruction results of all components together. However, these methods require a complete and accurate sketch of the line drawing as input. Thus they will fail when only the origin line drawing image is offered. Most recently, some studies [5,6,25,26] worked on recovering the 3D shape from the input line drawing image directly. These methods first extract the sketch from the line drawing image with some image processing techniques, then perform reconstruction like earlier methods. However, all of these sketch extraction algorithms depend on the man-made features of the image, and processes are sensitive to the image contents and qualities. Thus they generate incomplete or inaccurate sketches sometimes, which bring a huge impact on reconstruction accuracy.

In this paper, we propose a robust sketch extraction method based on keypoint detection. To be more specific, we adopt a multi-task learning framework to build an encoder-decoder network architecture based on HRNet [19,22]. First, the encoder generates a group of reliable shared feature maps. Then we exploit two decoders, one for classification and the other for vertices detection. Therefore, our model can predict the category and location of vertices (*i.e.* keypoints) of the geometric object contained in the input line drawing image simultaneously. When the category and location of vertices are obtained, we can easily get the sketch vertices of the geometric object in the input line drawing image. Furthermore, we adopt an example-based method [24] to reconstruct 3D geometric object from 2D sketch by optimizing an objective function of the reconstruction error.

To the best of our knowledge, we are the first to use convolutional neural network (CNN) for extracting the sketch of the geometric object from the line drawing image without any man-made features of processes. Combined with a reconstruction algorithm, our method can recover the 3D shape of geometric objects accurately and efficiently. Extensive experimental results have shown the effectiveness and robustness of our proposed method. In addition, we build a web application to deploy our method, which can be accessed with mobile devices. Our application can accurately reconstruct the geometric object contained in the photo uploaded by users in real-time. As the Fig. 1 shows, users can observe the geometric object from any perspective with our application. A demo video can be found in the supplementary material. In summary, our major contributions are:

- We propose the first CNN-based sketch extraction method of geometric objects, which adopts a multi-task learning framework to build an encoder-decoder model based on HRNet [19,22], which can predict the category and location of vertices of a geometric object simultaneously.
- Based on our sketch extractor, we further exploit an example-based reconstruction algorithm [24] to recover the 3D shape of geometric object from a single 2D line drawing image, which is more accurate and efficient than existing methods.
- We develop a real-time web application that can reconstruct the geometric object in the uploaded line drawing image on mobile devices such as smartphones and tablets thus can help users fully understand the 2D line drawing.

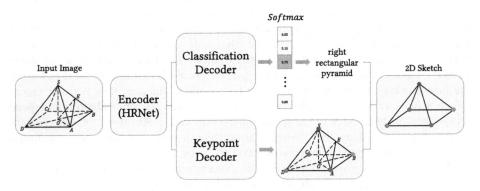

Fig. 2. The framework of the proposed sketch extractor. Our sketch extractor takes a single line drawing image as input and outputs predicted category and location of vertices simultaneously. Combined these information together, we can obtain the 2D sketch of the geometric object based on our prior knowledge about the geometric models.

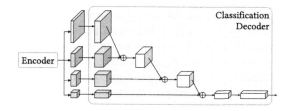

Fig. 3. The architecture of the proposed classification decoder. The inputs are the feature maps of four resolutions produced by the encoder.

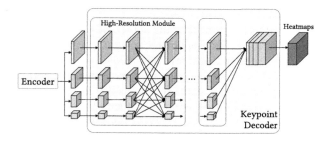

Fig. 4. The architecture of the proposed keypoint decoder. It is composed of several High-Resolution modules, which contain several bottlenecks and a multi-scale fusion layer. Finally we concatenate all (upsampled) representations from all the resolutions and regress the heatmaps.

2   Related Work

2.1   3D Object Reconstruction from Line Drawings

Over the past three decades, numerous studies have attempted to reconstruct the 3D geometric object from a single 2D line drawing. According to the required input, the previous methods can be divided into two categories: earlier methods [1,2,4,7,9–13, 15–18,20,23,24,27] require complete and accurate sketch of the line drawing as input, while recent methods [5,6,25,26] can reconstruct the geometric object from the line drawing image directly.

Earlier methods can be further divided into three categories: (1) *rule-based methods*, (2) *deduction-based methods*, (3) *divide-and-conquer-based methods*. *Rule-based methods* are popular in the earlier research. This type of methods usually define an objective function with some constraints based on geometric rules, then transform the reconstruction problem into an optimization problem. Marill [16] proposed an objective function based on the minimum standard deviation of angles (MSDA). This simple approach can recover 3D shape of a simple 2D line drawing and it was very enlightening at that time. Motivated by MSDA, many researchers did further studies on this basis. Leclerc *et al.* [9] added an planarity constraint to gain a more effective objective function. Lipson *et al.* [12] added more geometric rules such as face planarity, face perpendicularity, line parallelism, line verticality, line collinearity, and obtained a more robust model. Shoji *et al.* [18] proposed a new principle to minimize the entropy of angle distribution (MEAD) between line segments, which is more general than MSDA. Liu *et al.* [7, 15] focused on face identification with optimization methods, which provides important information for the reconstruction of 3D geometric objects. Masry *et al.* [17] proposed a reconstruction algorithm that uses the angular distribution of the strokes and their connectivity to determine an orthogonal 3D axis system. As we can see, many different optimization principles contribute to a better reconstruction result, but too many principles often make the objective function too complex and lead to failure when the input line drawing is complex. Therefore, Liu *et al.* [13] and Tian *et al.* [20] proposed some methods to reduce the complexity of the objective function and accelerate the optimization.

Deduction-based methods usually make some assumptions on the geometric relations of points, lines and faces in line drawing at first, then perform reconstruction with deductive methods to verify the assumptions. Cao *et al.* [1] proposed a novel approach to deduct the invisible edges and vertices, then reconstruct a complete 3D geometric object. Lee *et al.* [10, 11] assumed that there is at least one cubic corner in the geometric object, then identify this corner and perform reconstruction from there. Cordier [4] assumed that the geometric object is symmetric, then identify the symmetry relationship and perform reconstruction. However, *deduction-based methods* also have some limitations that reconstruction may be failed if the assumption is invalid.

Divide-and-conquer-based methods are based on the following ideas: divide the complex geometric object into multiple simple components, then perform reconstruction on each component respectively, and finally combine the reconstruction results of all components together. Chen *et al.* [2], Liu *et al.* [14], Xue *et al.* [23] and Zou *et al.* [27] decomposed the geometric object into multiple simpler components based on the identification result of faces or cuts (a planar cycle on the surface of a manifold). Xue *et al.* [24] proposed an example-based method to reconstruct each simple component of a complex object based on a 3D model from the pre-built database.

However, the above methods have a common limitation: all of them require complete and accurate sketch of the line drawing as input. Therefore these methods will fail when only origin line drawing image is offered. Some recently proposed methods overcome this shortcoming. They can reconstruct the geometric object from the line drawing image directly. SGOR [26] can extract the sketch of the input line drawing image with conventional image processing techniques, then adopt the example-based method to reconstruct the geometric object. CGOR [25] made some improvements on the basis of SGOR, which use not only the geometric information from the line drawing, but also the context information. In addition, CGOR can handle both planar and curved geometric objects while SGOR can not. GO3R [6] adopted a bottom-up and top-down method to classify the geometric object and extract the sketch, which improves both accuracy and efficiency significantly. Later, Guo *et al.* [5] introduced convolutional neural network (CNN) to classify the geometric object contained in the image, which gained further improvement in both accuracy and efficiency.

Above all, many methods of geometric object reconstruction have been proposed during the past three decades. Furthermore, recent methods can accomplish reconstruction from a single line drawing image directly, which means they can extract the sketch from the line drawing image automatically. However, all existing methods extract the sketch depending on the man-made features, which have limited feature types, require expert knowledge to select and may be sensitive to changing image conditions. Considering that the sketch quality has a large impact on final reconstruction accuracy, in this paper, we proposed a more robust sketch extractor based on pose estimation (*a.k.a* keypoint detection) with deep learning, which boosts the performance in both accuracy and efficiency.

2.2 Pose Estimation

Pose estimation is an active research topic which has enjoyed the attention of the computer vision community for the past few decades. It is also known as keypoint detection, which means localizing a given set of interest points. Although most of the current related works focus on the human pose estimation because of its broad application prospects, we can easily transfer this method to estimate the pose of any other objects with a particular set of interest points. Considering that there is only one geometric object contained in our input line drawing image, we should adopt a single-object pose estimator.

Deep learning has achieved great success in single-object pose estimation task. DeepPose [21] firstly introduced CNN to solve pose estimation problem, since then the CNN-based methods have dominated the field. Recently, a novel network architecture named as HRNet [19, 22] has been proposed, which is able to maintain high-resolution representations through the whole process. HRNet achieves state-of-the-art results in pose estimation, and also performs well in many other computer vision tasks, such as image classification, object detection, semantic segmentation, and so on. In this paper, we adopt HRNet as the backbone and perform classification and pose estimation of the input line drawing image simultaneously, which gains improvement in both accuracy and efficiency compared with previous sketch extraction methods.

3   Method

To fully illustrate our proposed method, we make some assumptions about the input line drawing image: (1) there is only one geometric object per line drawing image. (2) the line drawing image is a parallel projection of the 3D geometric object, which means that the parallel lines in the 3D geometric object keep parallel in the 2D line drawing image.

Our method consists of two stages: (1) sketch extraction, (2) 3D reconstruction. Section 3.1 will present how we extract the sketch of the geometric object from the input line drawing image with a neural network. Section 3.2 will state how we reconstruct the geometric object from the 2D sketch by optimizing an objective function of the reconstruction error.

3.1   Sketch Extraction

Sketch means the 2D coordinates of vertices and connection relationship between vertices of the geometric object. For a geometric object of category c, let n_c denote the number of vertices for geometric objects of this category (the ellipse in the curved object can be represented as a triangular following CGOR [25]). The sketch can be represented as $S = (\boldsymbol{V}^{2D}, \mathcal{G})$, where $\boldsymbol{V}^{2D} = (\boldsymbol{v}_1^{2D}, \boldsymbol{v}_2^{2D}, ..., \boldsymbol{v}_{n_c}^{2D}) \in \mathbb{R}^{2 \times n_c}$ are the 2D coordinates of n_c vertices, $\mathcal{G}$ is an undirected connected graph containing n_c vertices, denoting which two vertices are connected.

To obtain the sketch from a single line drawing image, we adopt a multi-task learning framework to predict the category and location of vertices simultaneously. As the Fig. 2 shows, our architecture is a deep convolutional encoder-decoder network. The encoder is based on HRNet [19, 22], which maintains high-resolution representations through the whole process and repeatedly conducts fusions across parallel convolutions. The encoder can produce reliable feature maps on four different scales that are shared by two task-specific convolutional decoders named *classification decoder* and *keypoint decoder*. Specifically, the classification decoder outputs the category of the geometric object in the input line drawing image and the keypoint decoder outputs the 2D location of the vertices of the geometric object.

Classification Decoder. Figure 3 shows the architecture of our classification decoder. At first, the four-resolution feature maps produced by the encoder are fed into a convelutional layer to increase the channels. Then we repeat the downsample and addition operation three times. Next we transform the small resolution feature map into 2048 channels and perform a global average pooling operation. Finally we feed the 2048-dimensional vector into a fully-connected layer to obtain the predicted category.

Keypoint Decoder. Figure 4 shows the architecture of our keypoint decoder. Inspired by HRNet [19, 22], which shows that High-Resolution module (several bottlenecks followed by a multi-scale fusion layer) can be used to effectively generate high quality feature maps, we adopt several High-Resolution modules to build the keypoint decoder on top of the four-resolution feature maps produced by encoder. Then we concatenate the (upsampled) representations from all the resolutions. Finally we regress the heatmaps simply from the concatenated high-resolution representations as the output.

In addition, we find that only one High-Resolution module is enough to transfer the shared features to a group of high quality heatmaps. More High-Resolution modules may even reduce the performance and bring more overhead. More details are shown in Sect. 4.

Ground-Truth Heatmaps. The ground-truth heatmaps are generated by applying 2D Gaussian with constant variance centered at the ground-truth location of each keypoint. We define the keypoints of a geometric object as all of its vertices. For a geometric object of category c, there are exactly n_c heatmaps which encode the location of its n_c vertices. Let N denote the total number of heatmaps outputed by the keypoint decoder, then N should be equal to $\sum_{c=1}^{C} n_c$, where C is the number of categories. In this way, our model can predict keypoints of geometric objects of several different categories in a single network with just a little additional overhead. Figure 5 illustrates this.

Loss Function. During training, we define a multi-task loss function as $L = L_{cls} + \lambda L_{kpt}$, where λ is used to balance the classification loss Lcls and the keypoint prediction

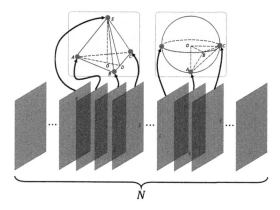

Fig. 5. Illustration of heatmaps for our keypoint decoder. The keypoint decoder outputs N heatmaps in total, where $N = \sum_{c=1}^{C} n_c$. Specifically, for a geometric object of category c, there are exactly n_c heatmaps encoding the location of n_c vertices of it.

loss Lkpt. The classification loss L_{cls} is the cross entropy loss. The keypoint loss L_{kpt} is the mean squared loss, which is applied for comparing the predicted heatmaps and the ground-truth heatmaps. Notice that for a geometric object associated with ground-truth category k, L_{kpt} is only defined on n_k heatmaps in total, which corresponds to n_k vertices of the geometric object (other heatmaps do not contribute to the loss).

After we get the predicted category and location of keypoints (*i.e.* vertices), we can directly obtain the sketch $S = (\boldsymbol{V}^{2D}, \mathcal{G})$ mentioned before. For a geometric object with predicted category c, the predicted location of n_c vertices corresponds to $\boldsymbol{V}^{2D}$, and $\mathcal{G}$ can be constructed based on our prior knowledge about the geometric models. We will empirically demonstrate the effectiveness of our proposed sketch extraction method in Sect. 4.

3.2   3D Reconstruction

We adopt an example-based method [24] to reconstruct the geometric object from the 2D sketch in Sect. 3.1. More specifically, we manually build a database of 3D models, and perform reconstruction by optimizing an objective function to do with 2D sketch and corresponding 3D model in the database.

3D Model Database. Our 3D model database includes 26 categories of geometric objects, 19 planar categories and 7 curved categories, which cover most categories of geometric objects occurred in high school textbooks. As shown in Fig. 6, each 3D model is controlled by a set of parameters, which can be represented as a parameter vector $\boldsymbol{\alpha}$. Obviously, the parameter vector determines the shape of the 3D object. Consequently, we can use a 3D model to match any other 3D geometric objects of the same category just by adjusting the parameters. Formally speaking, a 3D model of category c can be represented as $M = (\boldsymbol{\alpha}, \boldsymbol{V}^{3D}, \mathcal{G})$, where $\boldsymbol{\alpha}$ is the parameter vector mentioned before, $\boldsymbol{V}^{3D} = (\boldsymbol{v}_1^{3D}, \boldsymbol{v}_2^{3D}, ..., \boldsymbol{v}_{n_c}^{3D})$ is a $3 \times n_c$ matrix representing the 3D coordinates of the vertices, whose elements are the linear combination of elements in α, and $\mathcal{G}$ is an undirected graph indicating which two vertices are connected. Here is an example: for right rectangular pyramid, we define a 3D model as follows: $\alpha = (a, b, c)^{\mathrm{T}}$, where a, b, c represent the length, width and height respectively. V^{3D} is defined as follows:

$$V^{\mathrm{3D}} = \begin{pmatrix} a/2 & 0 & a & a & 0 \\ b/2 & 0 & 0 & b & b \\ c & 0 & 0 & 0 & 0 \end{pmatrix},$$

It represents 3D coordinates of the vertices using parameters in α. The edge set of graph $\mathcal{G}$ can be represented as $\{(0, 1), (0, 2), (0, 3), (0, 4), (1, 2), (2, 3), (3, 4), (4, 1)\}$, indicating the connected pair of vertices.

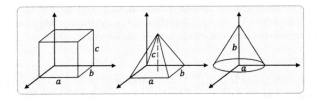

Fig. 6. Some examples of 3D models in the database. Each model is controlled by a set of parameters.

Optimization. Based on the 3D model database, we can construct an objective function as follows:

$$f = \frac{1}{n_c} \sum_{i=1}^{n_c} \| K(R v_i^{\mathrm{3D}}) + t - v_i^{\mathrm{2D}} \|^2, \tag{1}$$

where $K = \begin{pmatrix} 1 & 0 & 0 \\ 0 & 1 & 0 \end{pmatrix}$ is the parallel projection matrix, R is the rotation matrix, $t = (t_0, t_1)^{\mathrm{T}}$ is the translation vector. For a 3D model with difinite parameters α, all possible 2D projections can be obtained by performing three operations: rotation, projection, translation. Thus this objective function f measures the distance between a projection of 3D model and 2D sketch we extracted before. Then the reconstruction problem is transformed into an optimization problem as follows:

$$\tilde{\alpha}, \tilde{R}, \tilde{t} = \arg\min \frac{1}{n_c} \sum_{i=1}^{n_c} \| K(R v_i^{\mathrm{3D}}) + t - v_i^{\mathrm{2D}} \|^2, \tag{2}$$

$$\text{subject to } R^{\mathrm{T}} R = I$$

where $\tilde{\alpha}$ is the optimal parameter vector, $\tilde{R}$ is the optimal rotation matrix, and $\tilde{t}$ is the optimal translation vector. We adopt an alternative minimization algorithm [24] to obtain $\tilde{\alpha}, \tilde{R}$ and $\tilde{t}$. With this values, we can obtain the 3D model instance which matchs our 2D sketch. Thus the reconstruction is completed.

3.3   Method Analysis

Integrating the two components in Sects. 3.1 and 3.2, we finally get a complete algorithm which can reconstruct the geometric object from the line drawing image directly. Compared with previous work [5,6,25,26], our method has the following merits. First, all previous methods depend on the man-made features or processes to extract the sketch, while we adopt a convolutional neural network (CNN) as a sketch extractor, thus our sketch extractor is more accurate and robust. Secondly, our sketch extractor allows us to output the vertices in the same order as in the 3D model in the database, thus we can leave out the process of sub-graph matching algorithm with VF2 [3] in the previous work [5,6,25,26]. Last, due to our method greatly simplifies the whole process, it runs much faster than previous methods. We will give more details in the Sect. 4.

4   Experiments

4.1   Dataset

We conduct experiments on the dataset same as previous work [5], which contains 4,858 line drawing images manually collected from high school geometry studying materials, including 26 categories of geometric objects: 19 planar categories and 7 curved categories. We divide this dataset into 4,018/840 images for training and validation, respectively. In addition, following the method in [5], we generate a simulated dataset to alleviate the problem caused by unbalanced distribution across different categories in the manually-collected dataset. We generate 13000 simulated line drawing images as the extra data, 500 images for each category, to further improve the training performance of our model.

4.2   Evaluation Metrics

Classification. For the classification task, we use classification accuracy to evaluate the performance, which is defined as the number of correctly classified images to the total number of images.

Keypoint Detection. For the keypoint detection task, the evaluation metric is AP measured by Object Keypoint Similarity(OKS): $OKS = \frac{\sum_i \exp(-d_i^2/2s^2k_i^2)\delta(v_i>0)}{\sum_i \delta(v_i>0)}$. Here d_i is the Euclidean distance between the detected keypoint and the corresponding ground truth, v_i is the visibility flag of the ground truth, s is the object scale, and k_i is a per-keypoint constant that controls falloff. We report average precision scores: AP^{50} (AP at OKS $= 0.50$), AP^{75}, AP (the mean of AP scores at 10 positions, OKS $= 0.50, 0.55, ..., 0.90, 0.95$). Note that only images with correct predicted category take part in the calculation of AP.

Reconstruction. For the reconstruction task, which is our ultimate optimization objective. We utilize the same accuracy metric as in previous work [5,6,25,26], defined as $f_a = |\mathbf{F}_{success}|/|\mathbf{F}|$, where $\mathbf{F}_{success}$ is successfully reconstructed (correct classification, correct sketch and small reconstruction residual) geometric object set, and $\mathbf{F}$ is the experimental dataset.

4.3   Implementation Details

For the sketch extraction part, we implement our model in PyTorch. During the training process, the input image is resized to 224×224. Random scale ($[0.7, 1.3]$) is used for data augmentation. we use the Adam optimizer [8] and the base learning rate is set as 10^{-3}, reduced to 10^{-4} and 10^{-5} at the 60th and 80th epochs, respectively. The training process is terminated within 90 epochs. We train our model using 2 NVIDIA TITAN X Pascal GPU cards, and test our model on both GPU and CPU. The batch size is set 64. For the reconstruction part, we implement our algorithm in C++ to obtain faster performance.

Table 1. Results of keypoint detection on validation set with different number of High-Resoluiton modules in the keypoint decoder. The backbone is HRNet-W18-C-Small-v1. No extra data used for training.

Num.	AP	AP^{50}	AP^{75}
0	96.78	**99.88**	98.06
1	97.03	**99.88**	**98.54**
2	**97.20**	99.64	98.42

Table 2. Comparison expriments of extra simulated dataset. The backbone is HRNet-W18-C-Small-v1.

Extra data	Classification	Keypoint			Reconstruction
	Accuracy (%)	AP	AP^{50}	AP^{75}	Accuracy (%)
✗	97.86	97.03	**99.88**	98.54	94.4
✓	**98.93**	**97.51**	99.76	**98.68**	**96.4**

4.4   Results and Analysis

To ensure the efficiency of our method, we use three light-weight versions of HRNet [19,22] with different configurations as the backbone, which are called *HRNet-W18-C-Small-v1*, *HRNet-W18-C-Small-v2* and *HRNet-W18-C*, respectively.

Firstly, we explore the architecture of keypoint decoder. More specifically, we explore the number of High-Resoluiton modules as illustrated in Sect. 3.1. Table 1 shows the results on *HRNet-W18-C-Small-v1*. As we can see, when the number of High-Resoluiton modules increases from 0 to 1, AP and AP^{75} gain a large improvement, which means that directly regress the heatmaps from the concatenation of four-resolution feature maps produced by encoder is not a good choice. However, when we add one more High-Resoluiton module, the effect is limited and even leads to drop of AP^{50} and AP^{75}, meanwhile it brings much more overhead. Therefore, using a single High-Resolution module in the keypoint decoder is the best choice.

Table 3. Comparisons about efficiency with the previous methods.

Method	Backbone	#param. (M)	GFLOPs	Platform	Speed (s/image)
SGOR [26]	–	–	–	CPU	19.881
CGOR [25]	–	–	–	CPU	–
GO3R [6]	–	–	–	CPU	0.586
GOR [5]	ResNet-50	–	–	GPU	0.467
Ours	HRNet-W18-C-Small-v1	12.2	1.9	CPU	0.193
Ours	HRNet-W18-C-Small-v1	12.2	1.9	GPU	**0.023**
Ours	HRNet-W18-C-Small-v2	14.9	2.9	GPU	0.033
Ours	HRNet-W18-C	21.6	4.6	GPU	0.059

Table 4. Comparisons about classification accuracy with the previous methods.

Method	Backbone	Accuracy (%)
GO3R [6]	–	87.62
GOR [5]	ResNet-50	92.91
GOR [5]	ResNet-50	94.71
Ours	HRNet-W18-C-Small-v1	98.93
Ours	HRNet-W18-C-Samll-v2	**99.29**
Ours	HRNet-W18-C	98.69

Then we verify the effectiveness of our simulated dataset based on *HRNet-W18-C-Small-v1*. As is shown in Table 2, using the simulated dataset, the results of both classification and keypoint detection are improved, and the reconstruction accuracy is significantly improved by 2.0%. We find out that the improvement is mainly contributed by those categories containing fewer training samples in the original manually-collected dataset, which demonstrates that our simulated dataset helps to alleviate the problem caused by unbalanced distribution across different categories in the manually-collected dataset. Hence, unless explicitly noted, our model are trained on both manually-collected dataset and simulated dataset in the rest of this section.

Table 5. Results of keypoint detection using three versions of HRNet as backbone.

Backbone	AP	AP^{50}	AP^{75}
HRNet-W18-C-Small-v1	97.51	99.76	98.68
HRNet-W18-C-Small-v2	**97.67**	**99.88**	**98.80**
HRNet-W18-C	**97.67**	**99.88**	98.79

Table 6. Comparisons about reconstruction accuracy with the previous methods.

Method	Backbone	Accuracy (%)
SGOR [26]	–	72.9
CGOR [25]	–	81.2
GO3R [6]	–	85.2
GOR [5]	ResNet-50	93.0
Ours	HRNet-W18-C-Small-v1	**96.4**
Ours	HRNet-W18-C-Small-v2	**96.4**
Ours	HRNet-W18-C	95.8

We conduct experiments on three versions of HRNet as mentioned before. For classification, we compare our method with GO3R [6] and GOR [5], which also perform geometric objects classification before reconstruction. As presented in Table 4, our method booms the accuracy by a large margin on the manually-collected validation set, with about 4.0% improvement. For keypoint detection, to the best of our knowledge, we are the first to use keypoint detection algorithm for extracting the sketch. Table 5 reports the results of keypoint detection. With such a strong high-resolution backbone, we achieves up to 97.67% AP on the validation set.

Finally, we compare the reconstruction accuracy with previous methods [5,6,25,26], which is the ultimate optimization objective. As shown in Table 6, our method achieves up to 96.4% accuracy, which outperforms all previous works. Since the previous sketch extraction algorithms depend on the man-made features of the image or processes, which are sensitive to the image contents and qualities. Hence, they usually produce incomplete or inaccurate sketches, resulting in low reconstruction accuracy. As a contrast, our CNN-based method can directly extract the sketch of geometric objects without any hand-crafted features or processes, which gives a more robust performance. The results in Table 6 demonstrate that our method achieve state-of-the-art reconstruction accuracy. Figure 7 shows some example obtained with our method, from which one can see that our method can successfully reconstruct both planar and curved geometric objects. We further demonstrate the efficiency of our method. As the Table 3 shows, compared with previous methods [5,6,25,26], our model runs much faster than them. Our method based on the smallest model *HRNet-W18-C-Small-v1* accelerates almost 20 times than GOR [5] when running on GPU. And it also runs fast on the platform with pure CPU, which demonstrates its inherent efficiency.

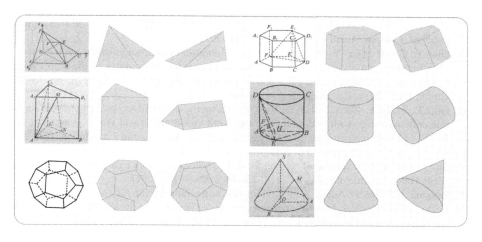

Fig. 7. Some reconstruction example.

5   Conclusion

In this paper, a robust method is proposed to recover the 3D shape of the geometric object directly from a single 2D line drawing image. We first propose a CNN-based method to extract the sketch of the geometric object with keypoint detection algorithm. Then we adopt an example-based reconstruction method, and conduct the reconstruction by optimizing an objective function. Combined these two components together, our method can reconstruct the geometric object from line drawing images accurately. Extensive experimental results demonstrate that our proposed method achieves much better performance than the existing methods in both accuracy and efficiency. In the future, we will explore how to build an end-to-end deep learning model, which can recover the 3D coordinates from the 2D line drawing images directly.

Acknowledgements. This work was supported by National Key R&D Program of China No. 2019YFB1406302. This work was also a research achievement of Key Laboratory of Science, Technology and Standard in Press Industry (Key Laboratory of Intelligent Press Media Technology).

References

1. Cao, L., Liu, J., Tang, X.: What the back of the object looks like: 3d reconstruction from line drawings without hidden lines. IEEE Trans. Pattern Anal. Mach. Intell. **30**(3), 507–517 (2008)
2. Chen, Y., Liu, J., Tang, X.: A divide-and-conquer approach to 3d object reconstruction from line drawings. In: 2007 IEEE 11th International Conference on Computer Vision, pp. 1–8, October 2007
3. Cordella, L.P., Foggia, P., Sansone, C., Vento, M.: A (sub)graph isomorphism algorithm for matching large graphs. IEEE Trans. Pattern Anal. Mach. Intell. **26**(10), 1367–1372 (2004)

4. Cordier, F., Seo, H., Melkemi, M., Sapidis, N.S.: Inferring mirror symmetric 3d shapes from sketches. Comput. Aided Des. **45**(2), 301–311 (2013)
5. Guo, T., Cui, R., Qin, X., Wang, Y., Tang, Z.: Bottom-up/top-down geometric object reconstruction with cnn classification for mobile education. In: Proceedings of the 26th Pacific Conference on Computer Graphics and Applications: Short Papers, PG 2018, pp. 13–16, Goslar, DEU, 2018. Eurographics Association
6. Guo, T., Wang, Y., Zhou, Y., He, Z., Tang, Z.: Geometric object 3d reconstruction from single line drawing image with bottom-up and top-down classification and sketch generation. In: 2017 14th IAPR International Conference on Document Analysis and Recognition (ICDAR), vol. 01, pp. 670–676, November 2017
7. Liu, J., Lee, Y.T.: Graph-based method for face identification from a single 2D line drawing. IEEE Trans. Pattern Anal. Mach. Intell. **23**(10), 1106–1119 (2001)
8. Kingma, D.P., Ba, J.: Adam: A method for stochastic optimization (2014)
9. Leclerc, Y.G., Fischler, M.A.: An optimization-based approach to the interpretation of single line drawings as 3d wire frames. Int. J. Comput. Vision **9**(2), 113–136 (1992)
10. Lee, Y.T., Fang, F.: 3D reconstruction of polyhedral objects from single parallel projections using cubic corner. Comput. Aided Des. **43**(8), 1025–1034 (2011)
11. Lee, Y.T., Fang, F.: A new hybrid method for 3D object recovery from 2D drawings and its validation against the cubic corner method and the optimisation-based method. Comput. Aided Des. **44**(11), 1090–1102 (2012)
12. Lipson, H., Shpitalni, M.: Optimization-based reconstruction of a 3d object from a single freehand line drawing. Comput. Aided Des. **28**(8), 651–663 (1996)
13. Liu, J., Cao, L., Li, Z., Tang, X.: Plane-based optimization for 3d object reconstruction from single line drawings. IEEE Trans. Pattern Anal. Mach. Intell. **30**(2), 315–327 (2008)
14. Liu, J., Chen, Y., Tang, X.: Decomposition of complex line drawings with hidden lines for 3d planar-faced manifold object reconstruction. IEEE Trans. Pattern Anal. Mach. Intell. **33**(1), 3–15 (2011)
15. Liu, J., Tang, X.: Evolutionary search for faces from line drawings. IEEE Trans. Pattern Anal. Mach. Intell. **27**(6), 861–72 (2005)
16. Marill, T.: Emulating the human interpretation of line-drawings as three-dimensional objects. Int. J. Comput. Vision **6**(2), 147–161 (1991)
17. Masry, M., Kang, D., Lipson, H.: A freehand sketching interface for progressive construction of 3d objects. Comput. Graph. **29**(4), 563–575 (2005)
18. Shoji, K., Kato, K., Toyama, F.: 3-d interpretation of single line drawings based on entropy minimization principle. In: Proceedings of the 2001 IEEE Computer Society Conference on Computer Vision and Pattern Recognition, CVPR 2001, vol. 2, p. II, December 2001
19. Sun, K., Xiao, B., Liu, D., Wang, J.: Deep high-resolution representation learning for human pose estimation. In: CVPR (2019)
20. Tian, C., Masry, M.A., Lipson, H.: Physical sketching: Reconstruction and analysis of 3d objects from freehand sketches. Comput. Aided Des. **41**, 147–158 (2009)
21. Toshev, A., Szegedy, C.: Deeppose: human pose estimation via deep neural networks. In: 2014 IEEE Conference on Computer Vision and Pattern Recognition, June 2014
22. Wang, J., et al.: Deep high-resolution representation learning for visual recognition. In: TPAMI (2019)
23. Xue, T., Liu, J., Tang, X.: Object cut: complex 3d object reconstruction through line drawing separation. In: 2010 IEEE Computer Society Conference on Computer Vision and Pattern Recognition, pp. 1149–1156, June 2010
24. Xue, T., Liu, J., Tang, X.: Example-based 3d object reconstruction from line drawings. In: 2012 IEEE Conference on Computer Vision and Pattern Recognition, pp. 302–309 (2012)

25. Zheng, J., Wang, Y., Tang, Z.: Context-aware geometric object reconstruction for mobile education. In: Proceedings of the 24th ACM International Conference on Multimedia, MM 2016, pp. 367–371 (2016). Association for Computing Machinery, New York (2016)
26. Zheng, J., Wang, Y., Tang, Z.: Recovering solid geometric object from single line drawing image. Multimedia Tools and Applications **75**(17), 10153–10174 (2015). https://doi.org/10.1007/s11042-015-2966-x
27. Zou, C., Yang, H., Liu, J.: Separation of line drawings based on split faces for 3d object reconstruction. In: 2014 IEEE Conference on Computer Vision and Pattern Recognition, pp. 692–699, June 2014

DiagramNet: Hand-Drawn Diagram Recognition Using Visual Arrow-Relation Detection

Bernhard Schäfer[1,2]([⊠])(ID) and Heiner Stuckenschmidt[2](ID)

[1] Intelligent Robotic Process Automation, SAP SE, Walldorf, Germany
[2] Data and Web Science Group, University of Mannheim, Mannheim, Germany
{bernhard,heiner}@informatik.uni-mannheim.de

Abstract. Offline hand-drawn diagram recognition is concerned with digitizing diagrams sketched on paper or whiteboard to enable further editing. Our proposed DiagramNet model addresses this recognition problem. We combine shape detection and visual arrow-relation detection to recognize arrows between shape pairs. A shape degree predictor predicts the number of in- and outgoing arrows in each direction. An optimization procedure uses the generated predictions to find the set of globally coherent arrows. Previous offline methods focus on clean images from online datasets with nicely layouted diagrams. We show that our approach is effective in the domain of camera-captured diagrams with chaotic layouts and various recognition challenges such as crossing arrows. To that end, we introduce a new dataset of hand-drawn business process diagrams that originate from textual process modeling tasks. Our evaluation shows that DiagramNet considerably outperforms prior state-of-the-art in this challenging domain.

Keywords: Graphics recognition · Offline recognition · Diagram recognition

1 Introduction

Several online hand-drawn diagram datasets have been published to date, in which each diagram has been drawn using a digital pen or device (e.g., tablet) [1,2,4,10]. These datasets promote research in online diagram recognition methods, where the drawing is provided as a temporal sequence of strokes [4,5,14,15,21,22,27]. Online methods require stroke data to be given. In contrast, so called *offline methods* are more widely applicable in scenarios where such information is not provided (e.g., on scanned documents). The lack of datasets that are offline by nature resulted in comparatively little attention towards offline methods, and existing works evaluated their methods on online datasets [3,13,19,20]. In prior diagram datasets, each writer was asked to copy a set of predefined flowcharts [1,4,10] or finite automata [2] templates, resulting in well-organized sketches. Yet, in practice, diagrams are often sketched on

© Springer Nature Switzerland AG 2021
J. Lladós et al. (Eds.): ICDAR 2021, LNCS 12821, pp. 614–630, 2021.
https://doi.org/10.1007/978-3-030-86549-8_39

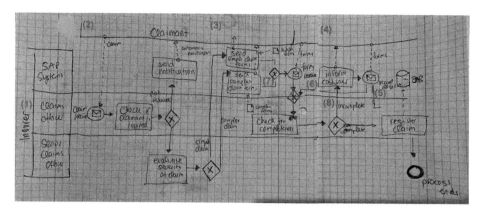

Fig. 1. Exemplary hdBPMN diagram. Recognition challenges include partially drawn rectangular pool and lane shapes (1 & 2), bleed-through ink (3), multiple sheets and paper warping (4), crossed out elements (5), interrupted lines (6), missing arrowheads (7), and crossing lines (8).

paper or whiteboard to graphically model software architectures or business processes [6,9]. Constrained by the size of the physical canvas and the limited ability to reorganize elements, this modeling process results in diagrams with chaotic layouts and arrows [17]. We argue that there is a large gap between the nicely organized flowcharts in commonly used datasets and realistic diagrams sketched on paper.

hdBPMN dataset. To foster research into offline diagram recognition, we introduce the hdBPMN dataset, which contains camera-based images of hand-drawn BPMN diagrams. Business Process Model and Notation (BPMN) is the standard graphical notation for specifying business processes. hdBPMN contains diagrams from over 100 participants and each diagram is the solution to a modeling exercise based on a textual process description. Since the participants were not constrained in how they develop their solution, the dataset has a high degree of diversity, caused by variation in terms of the employed modeling style, paper, pen, and image capturing method. Figure 1 shows a hdBPMN diagram and highlights several recognition challenges associated with the dataset. The dataset is available at https://github.com/dwslab/hdBPMN.

DiagramNet. We propose DiagramNet, a diagram recognition method that consists of five logical stages: 1) shape detection, 2) shape degree prediction, 3) edge candidate generation, 4) edge prediction, and 5) edge optimization. To detect the shapes of the diagram, we use a standard object detection pipeline [18]. We introduce a degree prediction network which predicts the in- and out-degree of each shape along four directions. To avoid considering every directed pair of shapes as a potential edge, we use syntactical rules and shape degrees to prune unlikely shape connections. Since the arrow between two shapes is not always a straight line, we use the predicted shape degrees to create compact arrow-relation bounding boxes that more likely contain the drawn arrow. We propose an edge prediction network that classifies each arrow-relation and predicts the drawn arrow path. Last, we introduce an edge optimization procedure that determines the final diagram using edge probabilities and predicted shape degrees. We validate DiagramNet on the hdBPMN dataset and compare our performance to the state-of-the-art method Arrow R-CNN [19]. We find that our approach achieves an absolute gain of more than 5% on mean F_1 over all arrow classes. We also perform an extensive ablation study to quantify the impact of our modeling choices.

Summary of Contributions. We introduce hdBPMN, the first offline diagram dataset to our knowledge. We propose DiagramNet, a novel method for hand-drawn diagram recognition which combines direction-based shape degree prediction and visual arrow-relation detection. Last, we show that our method outperforms the previous state-of-the-art approach in hand-drawn diagram recognition.

2   Related Work

Hand-Drawn Diagram Recognition. Existing offline diagram recognition methods can be divided into two groups: *stroke-based* [3,23] and *object-based* [13,19]. *Stroke-based* methods assume that the strokes in an image can be reconstructed in a preprocessing step. To that end, prior works either binarize the image using a constant threshold [3], or solve a simplified diagram recognition task by using the ground-truth strokes during inference [23]. After stroke reconstruction, previous works apply techniques inspired by online recognition methods, except that they consider strokes in spatial instead of temporal proximity to be related.

Object-based diagram recognition methods use deep learning object detectors to detect diagram symbols (shapes, arrows, and text phrases) in a given image. Julca-Aguilar and Hirata [13] train the popular Faster R-CNN [18] object detection pipeline to that end. Standard object-based methods are not able to recognize edges, since the arrow bounding box is insufficient to identify the shapes the arrow connects. Schäfer et al. [19] alleviate this limitation with Arrow R-CNN, which extends Faster R-CNN with an arrow keypoint predictor. The keypoint predictor learns to predict the location of arrowhead and tail. An edge is then recognized by identifying the shapes closest to each respective keypoint. On existing diagram datasets, Arrow R-CNN outperforms prior stroke- and object-based methods by a large margin. We therefore use Arrow R-CNN as a baseline in our experiments. As discussed in [19], object-based arrow detection has difficulties in recognizing (1) straight arrows with narrow bounding boxes, and (2) arrows whose bounding box largely overlaps with another arrow. Our work addresses these issues by modeling arrows as a relation between two shapes, and not as standalone objects with bounding boxes. While previous works additionally focus on the detection of text phrases in digitally captured diagrams, our

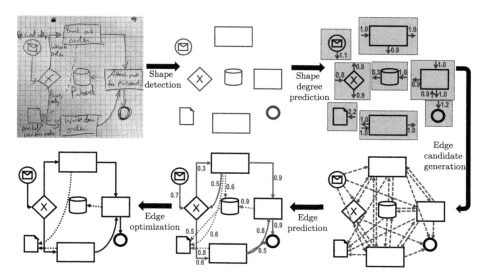

Fig. 2. Overview of our method. Given an image, our approach first detects shapes. The local shape context (green) is then used to predict in- and out-degrees for each direction. Next, a pruned graph of edge candidates is generated. The edge prediction network classifies each candidate and predicts the drawn arrow path as a sequence of keypoints. Finally, the edge optimization procedure uses the predicted shape degrees and edge probabilities to determine the final diagram. (Color figure online)

proposed approach leaves text detection aside. It rather addresses the challenges of recognizing the shapes and arrows in camera-captured, noisy recordings with potentially chaotic layouts.

Visual Relationship Detection. Visual relationship detection is an important topic in computer vision, where the task is to predict the relationship between pairs of objects in natural images. One such application is scene graph generation, where a graph models the objects (nodes) and their relationships (edges) in a scene image. Diagram recognition can also be formulated as graph generation, with shapes as nodes and arrows as edges. Existing scene graph methods [25,26,28] use Faster R-CNN [18] to generate a set of object proposals. The visual relationship features for an object pair are extracted from the union of both object bounding boxes using RoIAlign [11]. The node and edge features are enriched with global context using recurrent or graph neural networks. Scene graph methods are commonly evaluated on datasets with more than 100k images. Our object detection and relationship feature extraction approach is inspired by scene graph methods. However, most hand-drawn diagram datasets have less than 1k images. We therefore opt not to update local shape and edge features based on global context to prevent our method from simply memorizing the graphs in the training set.

Davis et al. [7] is the prior work most similar to ours and uses a visual relationship approach to detect and associate pre-printed and input texts in historical form images. An object detector detects pre-printed and input text lines. Relationship candidates are generated using a line-of-sight approach, which considers spatially close objects whose direct connection path does not cross the bounding box of other objects. To classify object pairs, Davis et al. propose a relationship classifier network with visual and spatial features as input. Further, a neighbor prediction network predicts the number of associated objects for each text object. To determine the final set of relations, a global optimization step combines the relationship probabilities and the predicted number of neighbors. This is similar to our optimization procedure, except that our edge relations are directed. Consequently, our optimization objective consist of two degree terms.

3   Method

This section introduces DiagramNet, our proposed model for recognizing hand-drawn diagrams. As Fig. 2 illustrates, DiagramNet decomposes diagram recognition into a sequence of five stages. Below we introduce each stage in detail.

3.1   Shape Detection

We frame shape detection as an object detection task and use the Faster R-CNN [18] framework to detect the set of shapes V in an image. Each detected shape $v \in V$ is associated with a predicted bounding box $\mathbf{b}_v \in \mathbb{R}^4$, probability p_v, and class $c_v \in \mathcal{C}$, where $\mathcal{C}$ is the set of shape types of a modeling language. Our definition of $\mathcal{C}$ also includes shape types that are not connected to other shapes through arrows, such as the lane element in BPMN. We follow prior work [19] and use Faster R-CNN with the feature pyramid network (FPN) extension [16]. During inference, we keep all shapes with $p_v \geq 0.7$. To eliminate duplicate detections, we also apply non-maximum suppression (NMS) with an intersection over union (IoU) threshold of 0.8 over all shape classes except lane. We exclude lanes since we observe a large bounding box overlap between lanes and their enclosing pools. During training, Faster R-CNN uses a multi-task loss, where the bounding box regression loss L_{loc} is weighted higher than the classification loss L_{cls}. Since we consider accurate classification equally important for shape detection, we decrease the weight for L_{loc} by a factor of 2.5. As a result, the localization and classification losses are in the same range during training.

3.2   Shape Degree Prediction

Given the set of detected shapes V from the previous step, we predict the out-degrees $\mathbf{d}_v^+ \in \mathbb{R}^4$ and in-degrees $\mathbf{d}_v^- \in \mathbb{R}^4$ for each shape v. The vector $\mathbf{d}_v^+$ represents the predicted number of outgoing edges for shape v in directions left, top, right, and bottom. Figure 2 conceptually shows the predicted in- and out-degrees for significant shape directions. We use the predicted shape degrees in all three subsequent edge-centric steps:

1. Edge candidate generation: prune edge candidates with insignificant shape degree
2. Edge prediction: generate more targeted arrow-relation bounding boxes
3. Edge optimization: use shape degrees to create globally coherent solution.

We formulate degree prediction as a regression problem, and propose a *degree prediction network* that predicts the degrees for each shape given a visual shape feature representation. Arrows are not always properly connected to their source and target shapes, i.e., sometimes there is some distance between the drawn arrow and the shapes they connect. We therefore pad each shape bounding box with 50px local context. Given the padded shape bounding box, we use RoIAlign [11] to extract a $28 \times 28 \times 256$ feature map from the image feature pyramid. We concatenate a 28×28 binary mask that encodes the location of the shape bounding box within the padded box. The resulting $28 \times 28 \times 257$ context-enriched shape representation is used as input for the degree prediction network.

Our degree prediction network is inspired by the MobileNet [12] architecture and contains a sequence of 6 depthwise separable convolutions. The spatial resolution of the feature map is downsampled twice by a factor of 2 using strided convolutions in the third and last depthwise convolution. An average pooling layer projects the resulting $7 \times 7 \times 256$ feature map to a 256-d vector. After two fully connected layers with 256 dimensions each, two linear heads predict the direction out-degrees $\mathbf{d}_v^+ \in \mathbb{R}^4$ and in-degrees $\mathbf{d}_v^- \in \mathbb{R}^4$. As in [12], each network layer is followed by batch normalization and ReLU nonlinearity, with the exception of the final fully connected layer. Subsequent steps of our approach also require the total shape out- and in-degree. We define this as the sum over all directions, and denote it as $d_v^+ = \|\mathbf{d}_v^+\|_1$ and $d_v^- = \|\mathbf{d}_v^-\|_1$.

We train the degree prediction network using mean squared error loss. During training, we randomly shift each side of the ground-truth shape bounding box and vary the context padding to improve generalization.

3.3 Edge Candidate Generation

Given the set of detected shapes V, the fully-connected graph $V \times V$ has $O(|V|^2)$ directed edge candidates. To avoid considering all shape pairs, we prune edges by leveraging (1) syntactical rules of the modeling language and (2) predicted shape degrees. The syntactical rules in a modeling language govern how elements of the language can be combined. As we want to recognize unfinished diagrams or diagrams with modeling errors as drawn, we do not apply all syntactical rules of the modeling language. Instead, we empirically consider rules that we don't see violated in the training set, which are typically edges that are very unintuitive to be modeled as such. As an exemplary rule for the BPMN language, we prune all edge candidates between gateway (diamond) and business object (file and database) shapes, as illustrated in Fig. 2. Given the initial fully-connected graphs of the hdBPMN test set, syntactical pruning eliminates 31.8% of all candidate edges, and the resulting graphs have a mean graph density of 0.75.

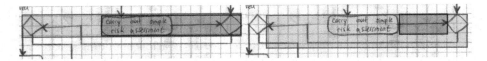

Fig. 3. *Arrow-relation bounding box.* The left image shows arrow-relation bounding boxes generated as the union of both shape boxes. In our direction-based approach (right image), the bottom side of the green box has been padded. Also, the left side of both boxes and the right side of the blue box have been shrinked. The resulting boxes capture the hand-drawn arrows more closely. (Color figure online)

We further prune the candidate graph using the predicted shape degrees from the previous step. We prune each edge (u, v) where $d_u^+ < \alpha$ or $d_v^- < \alpha$, i.e., we prune edges where the degree of at least one shape is below a threshold. We find that $\alpha = 0.05$ sufficiently balances precision and recall. Our degree-based pruning approach is also illustrated in Fig. 2, where the outgoing edges for the data object (file icon) have been pruned. On the hdBPMN test set, degree pruning eliminates an additional 11.3% of all candidate edges, and further reduces the mean graph density to 0.62.

3.4   Edge Prediction

Given the set of candidate edges from the previous step, our edge prediction approach is concerned with three major aspects:

1. *Arrow-relation bounding box*: identify the relevant arrow region for each shape pair.
2. *Classification*: predict a distribution over the arrow classes for each candidate.
3. *Path prediction*: predict the path of each arrow as a sequence of k equidistant points.

Arrow-Relation Bounding Box. As discussed in Sect. 2, visual relationship detection methods commonly define the relation region for an object pair as the union of their bounding boxes. This approach is not well suited for arrow-relation recognition since (1) the respective arrow is not necessarily located within this region and (2) the shape bounding boxes in their entirety are not required to recognize the arrow. We therefore propose an approach that leverages the predicted shape degrees introduced in Sect. 3.2. For each edge (u, v), we use the predicted shape out-degrees $\mathbf{d}_u^+$ and in-degrees $\mathbf{d}_v^-$ to (1) *pad* shape boxes on sides with predicted arrows and (2) *shorten* shape boxes on irrelevant sides. We then generate the arrow-relation bounding box as the union of the transformed shape boxes. Before we go into further detail, Fig. 3 shows an example of this approach.

An in- or outgoing arrow of a shape that does not point towards its opposing shape is a strong indicator that the edge is not drawn as a straight arrow. Therefore, we *pad* the shape bounding box on sides with a predicted arrow.

Since we are only interested in the existence of in- and outgoing arrows in each direction, we binarize the predicted shape degrees using a threshold $T = 0.3$. Given an edge candidate (u, v), we pad u with local context on sides where $\mathbf{d}_u^+ > T$ and pad v on sides where $\mathbf{d}_v^- > T$. We find that a local context of 50px sufficiently covers the distribution of diverse arrow types such as curved and elbow arrows. In Fig. 3, the arrow between the two diamond gateway shapes exits the source shape in bottom direction. After padding the source shape in bottom direction, the arrow-relation box (green) fully contains the drawn arrow.

Further, we *shorten* the shape bounding boxes on sides where there is no predicted arrow on the side itself and its adjacent sides. The rectangular task in Fig. 3 has only one outgoing edge to the right side. Thus, we shorten its left side for arrow-relations with this shape as source. The resulting arrow-relation box in the right image (blue) contains neither its source nor its target shape. We also shorten BPMN pool shapes boxes, which differ from other arrow-connected shapes in that their larger side usually spans most of the diagram. To avoid overly large arrow-relation boxes for shape pairs that involve a pool, we limit the longer side of pools to 100px.

Edge Classification and Path Prediction. We propose an *edge prediction network* that classifies each edge candidate and predicts the drawn arrow path. The network is similar to the shape degree prediction network introduced in Sect. 3.2. We extract a $28 \times 28 \times 256$ feature representation using the previously defined arrow-relation bounding box. Next, we concatenate two 28×28 binary masks, one for the source and one for the target shape. These binary masks indicate the spatial proximity in which to expect the arrowhead and tail. For shapes that are not contained in the arrow-relation box, we set the border pixels closest to the shape to one. We use the same MobileNet inspired architecture to project the resulting $28 \times 28 \times 258$ feature map to a 256-d visual feature vector. The predicted shape classes are then concatenated as one-hot encoded vectors. This semantic information is added so that the network predicts edge classes consistent with the source and target shape classes. After two fully connected layers, a linear layer with a softmax function predicts the edge probability p_e and edge class $c_e \in \mathcal{R}$, where the set of edge classes $\mathcal{R}$ includes a negative "background" class. A second linear layer predicts the encoded coordinates of k arrow keypoints. Following Schäfer et al. [19], we encode arrow keypoints relative to the relation bounding box.

We train the edge classifier using cross entropy, and train the arrow keypoint predictor using smooth L_1 loss as in [20]. Since our model requires a constant number of arrow keypoints, we sample k equidistant keypoint targets from the ground-truth arrow path. We find that $k = 5$ captures the majority of arrow drawing styles, and use this configuration throughout all experiments. We average the loss over all keypoints, and multiply the keypoint loss by a factor of 10 to ensure that the classification and keypoint loss are in the same range. During training, we randomly sample 80% of the ground-truth edges as positive edge candidates, and add twice as many negative candidates from the set of candidate edges from the previous step. As in Sect. 3.2, we randomly shift each side of the ground-truth shape bounding boxes and the generated arrow-relation bounding box to improve generalization.

3.5   Edge Optimization

Given the predicted probability p_e for each edge candidate $e \in E$ from the previous step, a standard postprocessing method is to use a score threshold T and keep each edge e where $p_e \geq T$. However, we can improve upon this baseline by also considering the predicted shape degrees d_v^+ and d_v^- introduced in Sect. 3.2. To that end, we employ a global optimization that minimizes the difference between predicted shape degrees and the number of accepted edges which connect to that shape. We denote $E^+(v)$ as the outgoing and $E^-(v)$ as the incoming edges of shape v. Let $\mathbf{x} \in \{0,1\}^{|E|}$ be a binary vector where $x_e = 1$ indicates that edge e is accepted. We define the degree penalty terms $\deg^+(v)$ and $\deg^-(v)$ as

$$\deg^+(v) = \left(d_v^+ - \sum_{e \in E^+(v)} x_e \right)^2 \qquad \deg^-(v) = \left(d_v^- - \sum_{e \in E^-(v)} x_e \right)^2 \qquad (1)$$

Each term measures the squared difference between the predicted degree of v and the number of accepted edges that connect to v. We formulate the optimization problem as

$$\mathbf{x}^* = \underset{\mathbf{x}}{\operatorname{argmax}} \left[\sum_{e \in E} (p_e - T) x_e - \lambda \sum_{v \in V} \deg^+(v) - \lambda \sum_{v \in V} \deg^-(v) \right] \qquad (2)$$

The score threshold T is now a soft threshold, and λ is a hyperparameter for weighting the degree terms. We set $T = 0.4$ and $\lambda = 0.4$ after experimenting on the hdBPMN validation set and use these in our evaluation. To reduce the number of edge candidates, we already reject all candidates with $p_e < 0.05$ prior to optimization. We obtain three sets of disjunct edges from our proposed edge optimization procedure: the accepted edges $E[\mathbf{x}^*]$, the rejected edges $E[\neg \mathbf{x}^*]$, and the edges rejected before optimization $E[p_e < 0.05]$. The bottom-left diagram in Fig. 2 conceptually shows the final recognition result with the accepted edges after optimization.

4   hdBPMN Dataset

The hdBPMN dataset contains 502 hand-drawn diagrams from 107 participants with more than 20,000 annotated elements (shapes, edges, and text phrases). Each diagram is the solution to a (graded) assignment, where the participants were asked to model a business process given a textual process description. The diagrams in hdBPMN are modeled using BPMN, the established standard for

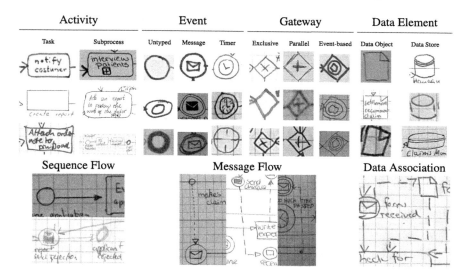

Fig. 4. hdBPMN shape (top row) and edge (bottom row) examples. Pool and lane shape examples are omitted due to their excessive size.

modeling business processes [9]. While BPMN is a complex language with over 100 symbols, a small core set of symbols is sufficient to cover most modeling needs [9]. For this work, we distinguish between 12 shape and 3 edge classes. Figure 4 provides hdBPMN dataset examples for each class, except for pool and lanes due to their excessive size. These two are illustrated in Fig. 5. The obtained diagrams in hdBPMN stem from 10 modeling tasks with varying difficulty. The diagrams for the simplest task have 11.0 shapes and 11.9 edges on average, and the most complex task has 26.7 shapes and 28.2 edges on average. As indicated in Sect. 1, the hdBPMN dataset combines a large number of recognition challenges:

- *Modeling style.* Participants alternate between reading the textual process description and drawing parts of the diagram, which means they only have a vague idea of the final diagram when drawing the first shape [17]. This results in diagrams with chaotic layouts and artifacts such as crossed-out or partially drawn elements.
- *Paper type.* Some diagrams are drawn on squared, dotted or lined paper, which introduces additional lines that complicate stroke recognition.
- *Pen type.* The diagrams are drawn with pen, pencils, fineliners, or a mixture thereof, and thus differ in stroke thickness, clarity, and contrast.
- *Image capturing.* The majority of images are camera-based, which introduces a series of recognition challenges [8].

We split the images by participant into training, validation, and test set. The number of diagrams and elements per dataset split are given in Table 1.

Table 1. hdBPMN dataset statistics

Split	Diagrams	Participants	Shapes	Edges
Train	308	65	5123	5325
Validation	102	21	1686	1734
Test	92	21	1615	1717

5   Experiments

5.1   Experimental Setup

Implementation Details. We implement DiagramNet using the detectron2 [24] framework. We use a ResNet50 with FPN [16] as backbone, and use models pretrained on the COCO dataset. We sum the losses of all DiagramNet components and jointly train the network. We use the same training procedure as in [19], except that we multiply the learning rate with 0.4 after iterations 50k and 70k. Our augmentation pipeline is also similar to [19], we resize images to a longer size of 1333px and randomly shift, scale, rotate and flip the image. To simulate the properties of camera-based images, we randomly add Gaussian noise, change brightness and contrast, and randomly shift hue, saturation, and value.

Arrow R-CNN Baselines. We compare our approach to the state-of-the-art method Arrow R-CNN [19]. To that end, we reimplemented Arrow R-CNN using the detectron2 framework. The image augmentation methods in [19] target online diagram datasets. Thus, we use two Arrow R-CNN baselines, one with the original augmentations and backbone used in [19], and one with our proposed augmentations and backbone for fair comparison. The edges in hdBPMN are annotated as a sequence of waypoints. Since Arrow R-CNN requires edge bounding boxes, we define the edge bounding box as the smallest possible box that includes all waypoints. For straight arrows drawn parallel to an axis, this results in bounding boxes with a width or height of 1. In early experiments, we noticed that this bounding box definition has a large negative effect on Arrow R-CNN's recognition performance. We thus introduce a minimum arrow bounding box width and height of 20px for an image resized to a longer side of 1000px.

Metrics. Prior works use bounding-box centric methods to evaluate shape and edge recognition in hand-drawn diagrams and require at least 80% bounding box overlap between predicted and ground-truth objects for both shapes and edges. In addition, edges need to be matched to the correct source and target shapes, which we refer to as shape-match criterion. The problem with evaluating arrows based on bounding box becomes clear when considering a straight horizontal arrow with a bounding box height of e.g., 4 pixels. Using the aforementioned metric, a predicted arrow with a bounding box off by 1 pixel would be evaluated

Table 2. Overall results on the hdBPMN test set.

Method	Shape				Edge			
	Precision	Recall	F_1	mAP	Precision	Recall	F_1	mAP
Faster R-CNN [18]	93.1	89.5	90.9	88.7	–	–	–	–
Arrow R-CNN [19]	96.6	90.1	92.9	89.8	86.1	83.3	84.6	79.0
Arrow R-CNN[a] [19]	96.4	92.0	94.0	91.7	87.2	83.7	85.4	80.1
DiagramNet (ours)	96.1	93.7	**94.7**	**94.8**	92.2	88.9	**90.5**	**90.8**
Evaluation with ground-truth shapes								
Arrow R-CNN [19]	–	–	–	–	91.0	88.1	89.5	87.7
Arrow R-CNN[a] [19]	–	–	–	–	92.1	88.3	90.1	89.5
DiagramNet (ours)	–	–	–	–	96.0	92.0	**93.9**	**95.8**

[a] Uses our proposed image augmentations and backbone for fair comparison.

Table 3. Ablation study for edge prediction components to quantify the impact of syntax pruning (SP), keypoint prediction (KP), degree prediction (DEG), and direction prediction (DIR).

SP	KP	DEG	DIR	Shape				Edge			
				Precision	Recall	F_1	mAP	Precision	Recall	F_1	mAP
✓	-	-	-	96.0	93.1	94.4	95.5	74.9	69.0	71.6	75.2
✓	✓	-	-	96.2	94.0	95.0	**95.8**	84.6	80.6	82.5	85.1
✓	✓	✓	-	96.0	94.6	**95.2**	95.0	90.9	85.4	88.0	88.3
✓	✓	-	✓	96.1	93.7	94.7	94.8	92.2	88.9	**90.5**	**90.8**
-	✓	-	✓	95.7	93.0	94.2	96.1	92.1	87.7	89.8	89.2

as false positive. We argue that it is more intuitive to evaluate edge recognition only based on the shape-match criterion, and thus do not consider the arrow bounding box. The shape-match criterion has one disadvantage, though: consider a shape with 5 connected edges that has been accurately localized, but misclassified. With the shape-match criterion, all edges connected to this shape are inevitably evaluated as false positives. To separate edge from shape recognition, we also report edge recognition performance when using ground-truth instead of predicted shapes.

For measuring precision, recall, and F_1-score, we keep all shapes with $p_v \leq 0.7$, and use the set of accepted edges from the edge optimization procedure. Since both shape and edge classes are imbalanced, we report macro scores by taking the average over the individual class scores. We also measure mean average precision (mAP) for shape and edges. Since average precision requires a consolidated edge ranking, we concatenate the sets $E[\mathbf{x}^*]$, $E[\neg\mathbf{x}^*]$, and $E[p_e < 0.05]$ introduced in Sect. 3.5 in that order.

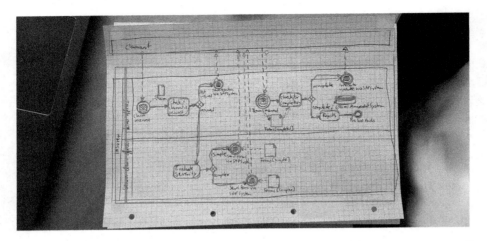

Fig. 5. hdBPMN test set image with predicted diagram overlay (orange). All shapes and edges have been correctly recognized. The BPMN shapes are rendered using the predicted shape classes and bounding boxes. The edges are drawn by connecting the five predicted arrow keypoints. (Color figure online)

5.2   Results

Overall Results. Table 2 shows that DiagramNet has a slightly better shape recognition performance than the Arrow R-CNN baselines, and significantly outperforms Arrow R-CNN in edge recognition. The edge evaluation using groundtruth shapes confirms this finding. For comparison, we also include the shape recognition results of a Faster R-CNN object detector that uses our backbone and standard image augmentations (horizontal flipping and random resizing). For a qualitative assessment of our method, Fig. 5 shows the recognized diagram for a test set image. As demonstrated, DiagramNet is able to recognize a diagram from a camera-based image that includes background objects. The total inference time per image is 319 ms on average, measured on a Tesla V100 GPU with 16 GB memory. This includes the proposed edge optimization procedure, which takes 129 ms on average.

The ablation study in Table 3 shows how each edge component contributes to the end result. Since we only vary edge related components, the shape recognition performance is similar in all settings. In the first experiment, we create the arrow-relation bounding boxes as the union of both shape boxes, which is standard in visual relationship detection. Also, we predict neither arrow keypoints nor shape degrees. We observe that this results in a relatively low edge F_1 score (71.6). Training with arrow keypoint prediction (KP) improves edge F_1 considerably from 71.6 to 82.5, even though the edge metric only takes edge classification into account. This indicates that the edge predictor greatly benefits from the keypoints as additional supervision targets. We also conduct a degree ablation study (DEG) in which the shape degree predictor only predicts the overall shape in- and outdegree, instead of predicting the degree for each direction. Like in our

Table 4. Overall results on the online flowchart datasets FC_A [1] and FC_B [4].

Method	FC_A		FC_B	
	Shape F_1	Edge F_1	Shape F_1	Edge F_1
Arrow R-CNN [19]	**99.8**	96.1	98.8	96.7
DiagramNet (ours)	**99.8**	**96.7**	**99.2**	**97.5**

Table 5. Shape and edge recognition results

Group	Class	Precision	Recall	F_1	AP	Count
Activity	Task	97.7	99.5	98.6	99.6	560
	Subprocess	93.8	71.4	81.1	79.3	21
Event	Untyped Event	96.6	99.4	98.0	99.9	170
	Message Event	96.2	99.2	97.7	98.8	126
	Timer Event	100.0	91.7	95.7	95.8	24
Gateway	Exclusive Gateway	98.7	98.7	98.7	99.9	157
	Parallel Gateway	95.2	97.5	96.4	97.7	122
	Event-based Gateway	90.9	90.9	90.9	82.6	11
Collaboration	Pool	99.2	99.2	99.2	100.0	125
	Lane	91.9	91.0	91.5	94.2	100
Data element	Data Object	96.9	97.5	97.2	98.6	161
	Data Store	95.7	88.0	91.7	91.0	25
Edges	Sequence Flow	92.9	94.6	93.7	94.3	1216
	Message Flow	89.8	84.7	87.2	87.4	177
	Data Association	93.9	87.3	90.5	90.6	315

proposed method, we use these predicted degrees to prune candidate edges and perform global edge optimization. As a result, we observe that edge F_1 improves from 82.5 to 88.0. Next, our proposed shape direction predictor (DIR) allows us to generate more targeted arrow-relation bounding boxes. We observe that this further boosts edge F_1 from 88.0 to 90.5. In the final experiment, we disable syntax pruning. As expected, we observe that this leads to a small decrease in shape and edge F_1.

To demonstrate the general applicability of our method, we also trained and evaluated DiagramNet and Arrow R-CNN on two popular online flowchart datasets. The results in Table 4 show that DiagramNet slightly outperforms the state-of-the-art method Arrow R-CNN on flowcharts. However, the near perfect shape results and the edge F_1 scores of at least 96.7 also motivate the need for more difficult datasets such as hdBPMN.

Shape and Edge Recognition. Table 5 provides detailed results per shape and edge class. For 8 out of 12 shape classes, DiagramNet achieves an F_1 score of at least 95. For three out of the remaining shape classes the number of data points is very low (Subprocess, Event-based Gateway, Data Store). For lanes, we observe that the rectangle shapes are often not fully drawn, as illustrated in Fig. 1. The edge class results show that sequence flows (F_1 of 93.7) are easier to recognize than data associations (90.5) and message flows (87.2). For message flows, besides the smaller number of training data points, the challenge is that the often connect shapes far from each other, as illustrated in Fig. 5. Also, message flows and data associations are both dashed arrows, and we observe a large variety of drawing styles for the dashed lines.

We also measure the accuracy of our shape degree prediction network. To this end, we compare the ground-truth degrees per direction of each correctly detected shape with the rounded predicted degrees. On the test set, we observe a higher accuracy for ingoing edges (97.8%) than for outgoing edges (96.9%), which suggests that arrow heads are easier to recognize than arrow tails.

Discussion. The experimental results demonstrate that DiagramNet considerably outperforms Arrow R-CNN on the challenging hdBPMN dataset. The ablation study shows that this is partly due to our direction-based approach for creating arrow-relation bounding boxes. This approach allows us to also capture arrows that connect their respective shapes with a detour. However, it reaches its limits when the detour is too large, and it also creates overly large relation boxes for arrows that directly connect two shapes. Future work could use a complementary approach, in which the arrows detected by Arrow R-CNN are used to refine the arrow-relation bounding boxes. Another possible research direction is the text modality, which we left aside to focus on the challenges associated with shape and edge recognition. In this regard, future work could detect and recognize text phrases, and relate text phrases with shapes and arrows.

6   Conclusion

In this work, we proposed DiagramNet, a new method for hand-drawn diagram recognition. DiagramNet includes a shape degree and an edge prediction network. The generated predictions are used to 1) prune the candidate graph, 2) generate informed arrow-relation boxes, and 3) formulate edge recognition as a global optimization problem. We also propose a new set of metrics that more intuitively assess shape and edge recognition performance. Further, we introduced hdBPMN, the first offline diagram dataset to our knowledge. Our approach outperforms the prior state-of-the-art diagram recognition method on hdBPMN and two flowchart datasets, and therefore, provides a valuable basis for the automated conversion of camera-captured diagram sketches.

References

1. Awal, A.M., Feng, G., Mouchère, H., Viard-Gaudin, C.: First experiments on a new online handwritten flowchart database. In: Document Recognition and Retrieval XVIII, p. 78740A (January 2011)
2. Bresler, M., Phan, T.V., Průša, D., Nakagawa, M., Hlaváč, V.: Recognition system for on-line sketched diagrams. In: ICFHR, pp. 563–568 (September 2014)
3. Bresler, M., Průša, D., Hlaváč, V.: Recognizing off-line flowcharts by reconstructing strokes and using on-line recognition techniques. In: ICFHR, pp. 48–53 (October 2016)
4. Bresler, M., Průša, D., Hlaváč, V.: Online recognition of sketched arrow-connected diagrams. Int. J. Doc. Anal. Recogn. (IJDAR) **19**(3), 253–267 (2016). https://doi.org/10.1007/s10032-016-0269-z
5. Carton, C., Lemaitre, A., Coüasnon, B.: Fusion of statistical and structural information for flowchart recognition. In: ICDAR, pp. 1210–1214 (August 2013)
6. Cherubini, M., Venolia, G., DeLine, R., Ko, A.J.: Let's go to the whiteboard: how and why software developers use drawings. In: CHI, pp. 557–566 (2007)
7. Davis, B., Morse, B., Cohen, S., Price, B., Tensmeyer, C.: Deep visual template-free form parsing. In: ICDAR, pp. 134–141 (September 2019)
8. Doermann, D., Liang, J., Li, H.: Progress in camera-based document image analysis. In: ICDAR, pp. 606–616, vol. 1 (August 2003)
9. Dumas, M., Rosa, M.L., Mendling, J., Reijers, H.: Fundamentals of Business Process Management, 2nd edn. Springer, Heidelberg (2018). https://doi.org/10.1007/978-3-662-56509-4
10. Gervais, P., Deselaers, T., Aksan, E., Hilliges, O.: The DIDI dataset: digital ink diagram data. arXiv:2002.09303 [cs] (February 2020)
11. He, K., Gkioxari, G., Dollar, P., Girshick, R.: Mask R-CNN. In: ICCV, pp. 2961–2969 (2017)
12. Howard, A.G., et al.: MobileNets: efficient convolutional neural networks for mobile vision applications. arXiv:1704.04861 [cs] (April 2017)
13. Julca-Aguilar, F.D., Hirata, N.S.T.: Symbol detection in online handwritten graphics using faster R-CNN. In: DAS, pp. 151–156 (April 2018)
14. Julca-Aguilar, F., Mouchère, H., Viard-Gaudin, C., Hirata, N.S.T.: A general framework for the recognition of online handwritten graphics. Int. J. Doc. Anal. Recogn. (IJDAR) **23**(2), 143–160 (2020). https://doi.org/10.1007/s10032-019-00349-6
15. Lemaitre, A., Mouchère, H., Camillerapp, J., Coüasnon, B.: Interest of syntactic knowledge for on-line flowchart recognition. In: Graphics Recognition. New Trends and Challenges, pp. 89–98 (2013)
16. Lin, T.Y., Dollar, P., Girshick, R., He, K., Hariharan, B., Belongie, S.: Feature pyramid networks for object detection. In: CVPR, pp. 2117–2125 (2017)
17. Pinggera, J., et al.: Styles in business process modeling: an exploration and a model. Softw. Syst. Model. **14**(3), 1055–1080 (2015)
18. Ren, S., He, K., Girshick, R., Sun, J.: Faster R-CNN: towards real-time object detection with region proposal networks. In: NeurIPS, pp. 91–99 (2015)
19. Schäfer, B., Keuper, M., Stuckenschmidt, H.: Arrow R-CNN for handwritten diagram recognition. Int. J. Doc. Anal. Recogn. (IJDAR) **24**(1), 3–17 (2021). https://doi.org/10.1007/s10032-020-00361-1
20. Schäfer, B., Stuckenschmidt, H.: Arrow R-CNN for flowchart recognition. In: ICDARW, p. 7 (September 2019)

21. Wang, C., Mouchère, H., Viard-Gaudin, C., Jin, L.: Combined segmentation and recognition of online handwritten diagrams with high order Markov random field. In: ICFHR, pp. 252–257 (October 2016)
22. Wang, C., Mouchère, H., Lemaitre, A., Viard-Gaudin, C.: Online flowchart understanding by combining max-margin Markov random field with grammatical analysis. Int. J. Doc. Anal. Recogn. (IJDAR) **20**(2), 123–136 (2017). https://doi.org/10.1007/s10032-017-0284-8
23. Wu, J., Wang, C., Zhang, L., Rui, Y.: Offline sketch parsing via shapeness estimation. In: IJCAI (June 2015)
24. Wu, Y., Kirillov, A., Massa, F., Lo, W.Y., Girshick, R.: Detectron2 (2019). https://github.com/facebookresearch/detectron2
25. Xu, D., Zhu, Y., Choy, C.B., Fei-Fei, L.: Scene graph generation by iterative message passing. In: CVPR, pp. 5410–5419 (2017)
26. Yang, J., Lu, J., Lee, S., Batra, D., Parikh, D.: Graph R-CNN for scene graph generation. In: Ferrari, V., Hebert, M., Sminchisescu, C., Weiss, Y. (eds.) ECCV 2018. LNCS, vol. 11205, pp. 690–706. Springer, Cham (2018). https://doi.org/10.1007/978-3-030-01246-5_41
27. Yun, X.-L., Zhang, Y.-M., Ye, J.-Y., Liu, C.-L.: Online handwritten diagram recognition with graph attention networks. In: Zhao, Y., Barnes, N., Chen, B., Westermann, R., Kong, X., Lin, C. (eds.) ICIG 2019. LNCS, vol. 11901, pp. 232–244. Springer, Cham (2019). https://doi.org/10.1007/978-3-030-34120-6_19
28. Zellers, R., Yatskar, M., Thomson, S., Choi, Y.: Neural Motifs: scene graph parsing with global context. In: CVPR, pp. 5831–5840 (2018)

Formula Citation Graph Based Mathematical Information Retrieval

Ke Yuan[1], Liangcai Gao[1(✉)], Zhuoren Jiang[2], and Zhi Tang[1]

[1] Wangxuan Institute of Computer Technology, Peking University, Beijing, China
{yuanke,glc,tangzhi}@pku.edu.cn
[2] School of Public Affairs, Zhejiang University, Hangzhou, China
jiangzhuoren@zju.edu.cn

Abstract. Nowadays, with the quick availability and growth of formulae on the Web, the question of how to effectively retrieve the relevant documents about formulae, namely formula retrieval, has attracted much attention from the researchers of mathematical information retrieval (MIR). Existing MIR search engines have explored much information of formulae such as characters, layout structure, the formula context. However, little attention has been paid to the link or citation relations of formulae among different documents, while these relations are helpful for searching some related formulae whose appearances are not similar to the query formula. Therefore, in this paper, we design a Formula Citation Graph (FCG) to 'dig out' the link or citation relations between formulae. FCG has two main advantages: 1) The graph could generate the descriptive keywords of formulae to enrich the semantics of formula queries. 2) The graph is employed to balance the ranking results between the text and structure matching. The experimental results demonstrate that the link or citation relations among formulae are helpful for MIR.

Keywords: Mathematical information retrieval (MIR) · Inverted index · Formula citation graph (FCG)

1 Introduction

Hundreds of millions of formulae are available on scholarly documents. This ever-increasing sheer volume has made it difficult for users to find useful complementary documents according to the formula that they encounter during learning and reading [5,26,28]. Therefore, how to effectively and efficiently search for mathematical documents has received increasing research attention in recent years, as witnessed by mathIR tasks at the NTCIR conferences [1,29].

The goal of Mathematical Information Retrieval (MIR) is retrieving a sorted mathematical document (with formula attribution) list according to the math query. Current MIR approaches perform well in identifying formulae that contain the same set of identifiers or have a similar layout [1,29]. Meanwhile, existing MIR systems are not that desirable. For instance, for a formula query, besides the exactly matched formulae, users may also want to find some formulae that share

© Springer Nature Switzerland AG 2021
J. Lladós et al. (Eds.): ICDAR 2021, LNCS 12821, pp. 631–647, 2021.
https://doi.org/10.1007/978-3-030-86549-8_40

the similar semantic meanings with the query or contain related background information about the query. Moreover, the citation relations between formulae are seldom utilized in MIR systems and this kind information can be quite important for formula retrieval, since the citation relations usually denote semantic meanings of formulae [19] and are useful to seek out the related background information of it [16]. For example, when we search '$P(A|B) = \frac{P(B|A)P(A)}{P(B)}$', formula '$p(A,B) = P(A|B)P(B) = P(B|A)P(A)$' usually can not be properly retrieved and ranked as they have diverse layout. But the semantic meaning of the two formulae are all related to 'Bayes' Theorem' concept, since they usually cite the concept in their context. Therefore, without considering the citation relations, only analyzing the structural similarity between formulae can lead to the retrieval of non-relevant results.

Furthermore, several studies [18,23] have proved that by considering the semantic meanings of formula queries, the accuracy of MIR systems could be improved. The keywords of formulae are usually used to represent its semantic meanings [18]. However, it is difficult for non-experts to offer suitable keywords due to the lack of relevant knowledge. In addition, formulae often have multiple meanings, even experienced experts often require much time to recognize all of them [19]. Therefore, how to effectively generate valuable keywords of formulae is a challenge for MIR [23].

In this paper, we design a Formula Citation Graph (FCG) to overcome the aforementioned problems in MIR by analyzing the surrounding citations of the formulae in a large scale dataset. The proposed FCG has two main advantages: 1) FCG can help to 'dig out' the multi-dimensional semantic meanings of a formula query. 2) It can more reasonably balance the ranking results between the text and structure matching. For the proposed system, firstly, we generate multiple descriptive keywords of the formula query according to FCG. Then the keywords and layout of the formula query are used to index and retrieve, the *Term Frequency-Inverse Document Frequency (TF-IDF)* algorithm is used for ranking hits. Finally, a random walk based re-ranking process is utilized for sourcing potentially important formulae, including the formulae which share the similar semantic meanings, or contain the related background information about the query. The experimental results demonstrate that our system has achieved the best performance compared with the state-of-the-art system.

In summary, the main contributions of our work are as follows.

- We design a Formula Citation Graph (FCG) based on quantitative analysis of the formula in a large scale dataset. FCG can 'dig out' the representative keywords (covering multiple fields) for the formula query, and a re-ranking algorithm based on FCG can promote the ranking positions of potentially important formulae.
- We construct a novel FCG-based MIR system. The system is much easier to use and can return better ranking results according to the evaluation criteria *Discounted Cumulative Gain (DCG)* and *Precision (P)*.

2   Related Work

In this section, the related research will be introduced from two aspects: the existing techniques for MIR and the formula description extraction.

2.1   Techniques for MIR

MIR approaches can be categorized into *Text-based*, *Spectral* and *Tree-based* approaches according to the primitives of formulae which are used for indexing [30].

Text-based approaches linearize the math expression's appearance before indexing and retrieval (e.g., '$a - b$' is translated to 'a minus b') [9,13,20]. In these approaches, although they can capture significant amount of structural information about formulae, it is required to evaluate all formulae in the dataset by using a quadratic algorithm.

Spectral approaches extract features of formulae [25,27] or embed formulae in vectors [3,12,15,22] to scoring and ranking the relevant formulae. Yuan et al. [27] re-rank the retrieved formulae according to the appearance features of formulae, then they [25] propose multi-dimensional features of formulae, including appearance features and semantics features. Gao et al. [3] and Thanda et al. [22] propose a formula2vec based MIR approach, they regard each term of formula as a word to train a formula embedding model, and then use the vectors to retrieve and scoring the related information. These approaches also need to evaluate all formulae in the dataset by using a quadratic algorithm.

Tree-based approaches use trees to represent the appearance of formulae firstly, and then use the complete trees to index, along with the subtrees to support the partial matching [2,18,23,30]. Zannibi et al. [30] retrieve formulae according to the *symbol layout tree* matching of formulae. Davila et al. [2] retrieve formulae according to two representation trees (*symbol layout tree* and *operator tree*) of formulae. These approaches perform well in identifying formulae that contain similar layout tree structures. However, they neglect the semantic matching between formulae which is important in MIR. Therefore, several methods [18,23] have been proposed to explore the semantic matching in MIR. Wang et al. [23] use the preceding and following paragraphs to represent the descriptive text of the formula. Ruzicka et al. [18] propose a query expansion algorithm by extending the descriptive keywords of the query, but it's only suitable for the query which already contains keywords. Although, these approaches obtain a improvement by using the semantic matching, the descriptive keywords should be provided by users. It is difficult for users to provide all the descriptive keywords, due to the lack of relevant knowledge. Therefore, we introduce FCG to 'dig out' all descriptive keywords of formula queries without the help of users.

2.2   Formula Description Extraction

Driven by the importance of descriptive keywords of formulae in MIR approaches, the learning of descriptive keywords of a formula has attracted lots

of research attention. Thus, various approaches have been proposed in recent
years [14,17,19]. Quoc et al. [17] extract context of the formula as its descriptions. Pagel et al. [14] use a Mathematical Language Processing framework to
assign formula identifiers to its definientia. Nevertheless, in these approaches, all
descriptions of a specific formula are extracted from one document at a time.
However, a specific formula may contains different concepts in different scientific
fields. Thus, Schubotz et al. [19] extract the descriptions of a specific identifier
by analyzing different documents. In practice, the semantic meanings of a formula are usually more useful than the formula identifiers in MIR. So we explore
FCG to mine the descriptive keywords which represent the semantic meanings of
formulae in a large scale dataset by combining different meanings from different
documents.

3   Our Approach

3.1   Framework of the Proposed System

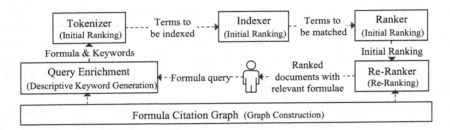

Fig. 1. Framework of the proposed system

Figure 1 shows the framework of the proposed system. The input is a formula
query, the output is a document (with formula attribution) ranking list. The
proposed system includes four processes: *Graph Construction, Descriptive Keywords Generation , Initial Ranking*, and *Re-Ranking*. Firstly, we construct FCG
on a large scale dataset by extracting the citations existing in the the formula
context (see Sect. 3.2). Secondly, we leverage the graph to 'dig out' the potential
descriptive keywords of the query including all concepts which exist in different
fields. Thirdly, the query formula is denoted by the structure tree whose terms
are extracted by the tokenizer with generalization. After the statistical data
(e.g., *tf-idf*, term level) of each term and keywords are calculated and stored in
the inverted index files by the indexer, ranker returns an initial ranking list by
looking up the matched terms from the index files, and calculating the relevant
scores between the semantic enrichment query and documents (see Sect. 3.4).
Finally, the FCG is utilized to seek out the possible documents which contain
the highly related formulae and promote the ranking performance in the process
of *re-ranking*.

3.2   Graph Construction

In this section, we construct a heterogeneous graph, namely Formula Citation Graph (FCG). The citations or references between documents are important to indicate the association between them [6], which motivates us to utilize the citations (hyperlinks) existed in the formula context. The formula context usually is the direct descriptions and contains most related concepts of the formula. Therefore, we build the graph based on two assumptions: *1)* The stronger the association between a formula and a document is, the more likely the document is relevant to the formula. Here, the strength of the association between a formula and a document is indicated by the number of citations existing in the formula context in the dataset; *2)* If there isn't any citation in the formula context in a document, probably this document mainly describes this formula.

The descriptions of a formula which can be extracted from its context (preceding and following sentences) [23] indicate its semantics. While most approaches only use one document at a time to extract the descriptions of a specific formula [8,24], in this paper, we use a large scale dataset to analyse and extract the descriptions of a specific formula, which means we combine the extracted sentences from different documents to construct the descriptions of a specific formula.

However, direct use of the extracted descriptions of a formula can not meet the requirement of MIR, since not all words in the descriptions are meaningful to the formula. In addition, the descriptions of formulae are often very long and contain much redundant information. Thus, instead of directly using the descriptions, we use the concepts of formulae which are extracted from the descriptions. Specifically, we design a FCG-based method to extract the concepts of a formula by analyzing the citations in its descriptions.

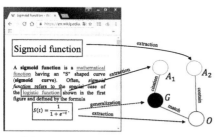

Vertex	Description
A	Title of document
O	Original formula
G	Generalization formula

Edge	Description
$A - O$	The contain relationship between document and original formula
$O - G$	The match relationship between original formula and generalization formula
$G - A$	The citation relationship between generalization formula and document

(a) The process of building FCG

(b) Vertices and edges in FCG

Fig. 2. Formula citation graph

Figure 2(a) illustrates the process of building a formula citation graph. This graph contains three kinds of vertices (A, O, G) which are interconnected by different kinds of edges ($A - O$, $O - G$, $G - A$). A_2 is the title of document.

The citation A_1 is extracted from the descriptions of the formula. The original formula O is directly extracted from the document, and a generalization formula G is generalized from O. More detailed information is depicted in Fig. 2(b).

It is unreasonable if we only use the original formula to extract the descriptive keywords of a formula. Since lots of formulae share the same main layout, the only difference is their symbol identifiers. These formulae are usually supposed to share the same keywords according to the assumptions which have defined in this paper. For instance, formula '$f(x) = \frac{1}{1+e^{-x}}$' and formula '$\varphi(z) = \frac{1}{1+e^{-z}}$' have the similar layout with a slight symbol identifiers difference, but they still contain the same semantic meanings. Based on these considerations, we introduce the generalization formula to denote the formulae which share the same layout.

In order to generate the generalization formula from the layout tree of the original formula, we adopt the formula tree construction algorithm proposed in [23]. In the algorithm, the different formulae with the same main structure are converted into a uniform formula. Variables, constants and the order of operands are all taken into account in the algorithm. Figure 3 illustrates the converting process. The formula '$f(x) = \frac{1}{1+e^{-x}}$' is presented by the layout tree. The constants and variables of the tree are respectively replaced by '$*_c$' and '$*_v$', then we can obtain the generalization formula G as shown in Fig. 3.

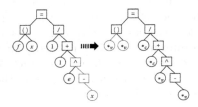

Fig. 3. The process of generating generalization formula

The meanings of edges in the FCG are mainly described in Fig. 2(b). The edge (o_i, g_j) is used to connect the original formula o_i and its generalization formula g_j. The weight of the edge is the probability of converting the generalization formula to an original formula. The probability can be estimated by:

$$w(o_i, g_j) = \frac{count(o_i, g_j)}{\sum_{k=1}^{n} count(o_k, g_j)} \tag{1}$$

where $count(o_i, g_j)$ is the number of times in the dataset that the generalization formula of o_i matches g_j. n is the total number of the original formulae which can be generalized as formula g_j.

In order to obtain more candidate keywords for the formulae which share the same main layout, instead of using the original formula, we use the generalization formula to connect the citing concepts (keywords), and each keyword has a probability to be chosen to denote the semantic meanings of the formula. For instance, when we search a formula with a layout like '$*_v(*_v) = \frac{*_c}{*_c + *_v^{-*_v}}$', the keyword 'Sigmoid function' has larger possibility to represent this formula. The edge (g_i, a_j) is used to denote the probability. The probability $w(g_i, a_j)$ between the generalization formula g_i and keyword a_j is defined the same as Eq. 1.

In a document, the representative formula is usually more important. We use the edge (o_i, a_j) to denote the formula importance, and the method [23] is adopted to calculate the importance. Then, we map the formula importance to the edge (o_i, a_j) and define the weight of the edge as $w(o_i, a_i)$:

$$w(o_i, a_j) = \frac{Importance(o_i, a_j)}{\sum_{k=1}^{n} Importance(o_i, a_k)} \tag{2}$$

where $Importance(o_i, a_j)$ is the value of the original formula o_i's importance to a document a_j, n is the number of documents which contain the formula.

3.3   Descriptive Keywords Generation

As aforementioned, one formula may show different semantic meanings in different scientific fields, causing one keyword usually cannot perfectly cover all the semantic meanings of a formula. Meanwhile, due to the lack of related knowledge, users usually cannot provide all the keywords of a formula when they use MIR systems. In order to better obtain the semantic meanings of the formula query, more related keywords are needed. This section describes the method of how to extract the descriptive keywords of the formula query based on FCG. This method aims to find the most reasonable keywords that can be used to denote the semantic meanings of a formula query. In the process of descriptive keywords generation, we firstly covert the structure of a formula query q into the original formula layout tree o_i and generalization layout tree g_k according to the method mentioned above, respectively. Then we use the edges which directly connect the original and the generalization formula layout trees to look up the potential related descriptive keywords in FCG, the probability of each keyword can be estimated by:

$$P_q(a_j) = \begin{cases} w(o_i, a_j) + w(o_i, g_k) \times w(g_k, a_j) & o_i \in O, g_k \in G \\ w(g_k, a_j) & o_i \notin O, g_k \in G \\ 0 & o_i \notin O, g_k \notin G \end{cases} \tag{3}$$

where $o_i \in O$ means an formula o_i existing in the graph, $g_k \in G$ means g_k existing in the generalization formula G. $w(x)$ is the weight function of each edge describe in Eq. (1)-(3).

The related descriptive keywords are ranked according to their probability. We select the related keywords with probability above a certain threshold and refer them as the semantic meanings of the formula query. In our experiments, the threshold is set to 0.2. After the process of descriptive keywords generation, the query will be represented by a formula and its descriptive keywords. For instance, given a query formula '$f(x) = \frac{1}{1+e^{-x}}$', we will obtain the semantically enriched query '$f(x) = \frac{1}{1+e^{-x}}$, Sigmoid function, Logistic function' after the descriptive keywords generation process.

3.4 Initial Ranking

In this section, we describe the indexing and retrieval according to the semantically enriched query. In the process of initial ranking, it contains three modules: 1) Tokenizer; 2) Indexer; 3) Ranker.

Tokenizer. The goal of tokenizer is to imitate the query understanding process of users and support the substructure matching and fuzzy matching in MIR. Tokenizer consists of two modules: formula tokenizer and keyword tokenizer. Formula tokenizer is used to extract terms from the original and generalization formula, keyword tokenizer is utilized to covert descriptive keywords into terms.

In the process of tokeinzation, the formulae are firstly converted into tree presentation according to the algorithm which is proposed in the literature [23]. Formula tokenizer is used to extract two categories of terms, namely *original terms* and *generalized terms*. The original terms are extracted from the subexpression of the formula. For the sake of fuzzy match, the generalized terms are extracted by changing the variables and constants of the original terms into wildcards, and the wildcard existing in the formula will be preserved in the tokenization. In addition, the wildcard '$*$' can match '$*_v$' and '$*_c$' in the searching process. For instance, formula $f(x) = \frac{1}{1+e^*}$ is firstly converted to the tree presentation similar to Fig. 3, then the original and generalized term are extracted according to the tree presentation. The original and generalization terms are shown in Table 1.

Table 1. Terms of $f(x) = \frac{1}{1+e^*}$

Level	Original terms	Generalized terms
1	$f(x) = \frac{1}{1+e^*}$	$*_v(*_v) = \frac{*_c}{*_c + *_v^*}$
2	$f(x)$	$*_v(*_v)$
2	$\frac{1}{1+e^*}$	$\frac{*_c}{*_c + *_v^*}$
3	$1 + e^*$	$*_c + *_v^*$
4	e^*	$*_v^*$

The keyword terms are extracted by the keyword tokenizer, and these terms are extracted similarly as the traditional text searching engines [11].

Indexer. For the realtime searching, indexer builds the index files by employing the inverted index data structure. In the proposed system, two index files are built to calculate a composite similarity score between a query and a document, one for formula terms and the other for keyword terms.

Table 2 and Table 3 illustrate the storage of formula terms and keyword terms. Table 2 shows the index file of formulae terms. The index of formula terms is denoted as $index_{t-f}$. A list of formulae which contains the terms is recorded. The formula f contains a formula term t which denotes that one of the terms extracted from f is exactly the same as t. In Table 2, $iff(t_i)$ denotes the inverted formula frequency of t_i and the $tf(t_i)$ describes the frequency of t_i occurring in f, $tl(t_i, f_j)$ represents the level of t_i in f_j. $w(f)$ is the importance value of formula in a document. All the statistics information about the formula terms are recorded in $index_{t-f}$.

Table 2. Index of formula terms

Terms	Formula list
$t_1\ iff(t_1)$	$[f_1,\ tf(t_1, f_1),\ tl(t_1, f_1),\ w(f_j)]...[f_j,\ tf(t_1, f_j),\ tl(t_1, f_j),\ w(f_j)]...$
$...$	$...$
$t_i\ iff(t_i)$	$[f_2,\ tf(t_i, f_2),\ tl(t_i, f_2),\ w(f_2)]...[f_k,\ tf(t_i, f_k),\ tl(t_i, f_k),\ w(f_k)]...$

The index file $index_{k-a}$ of keyword terms shown in Table 3 is constructed to calculate the relevant score between the keywords of the query and a document. k_i represents the keyword term, and the $ikf(k_i)$ is the inverted keyword frequency of k_i. $kf(k_i, a_i)$ is the frequency of k_i occurring in document a_i. And all the definitions of keywords are recorded in $index_{k-a}$.

Table 3. Index of keyword terms

Terms	Document list
$k_1,\ ikf(k_1)$	$[a_1,\ kf(k_1, a_1)],[a_2,\ kf(k_1, a_2)]...[a_j,\ kf(k_1, a_j)]...$
$...$	$...$
$k_i,\ ikf(k_i)$	$[a_1,\ kf(k_i, a_1)],[a_2,\ kf(k_i, a_2)]...[a_k,\ kf(k_i, a_k)]...$

Ranker. In the initial ranking process, the query is parsed to terms by tokenizer, and then searched in the index files $index_{t-f}$ and $index_{k-a}$. All the matched terms will be sent to the ranker for the similarity calculation. The composite similarity score $S_q(a_i)$ between the query Q and a document a_i is defined as $S_Q(a_i)$:

$$S_Q(a_i) = (((s_F(a_i) + 1)^2 + s_K(a_i) + 1)/2)^{1/2} - 1 \qquad (4)$$

where F and K are the formula and keywords of query Q, respectively. $s_F(a_i)$ is the similarity score between the formula F and a document a_i, the $s_K(a_i)$ is the similarity score between the keywords K and the document a_i. The $S_Q(a_i)$, which emphasizes formula information, is an approximate derivative average of $s_F(a_i)$ and $s_K(a_i)$. Definitions of $s_F(a_i)$ and $s_K(a_i)$ are given in Eq. 5 and 6.

A document may contain several relevant formulae and each formula f_j existing in the document has a similarity score with the formula F of the query. We choose the highest formula similarity score as its document similarity score, and thus the similarity score between the formula of the query and the document is defined as:

$$s_F(a_i) = \max_j \left\{ W_I(f_j) \times \sum_{t \in F} tf(t, f_j) \times iff^2(t) \times W_L(t, F, f_j) \times (1/t_L) \right\} \qquad (5)$$

where $W_I(f_j)$ is the formula importance value of f_j in the document a_i, proposed in [23]. $tf(t, f_j)$ and $iff(t)$ are directly obtained from the formula index file, $index_{t-f}$. $W_L(t, F, f_j)$ is the minimum level distance between the matched term t in F and f_j referred in [23]. t_L is the level of term t in the query formula F.

The similarity score $s_K(a_i)$ denotes the similarity between keywords K of query and the document a_i based on all the matched word terms in a document.

$$s_K(a_i) = \left(\frac{match(K, a_i)}{count(K)} \right) \times \sum_{k \in K} (kf(k, a_i) \times ikf^2(k)) \qquad (6)$$

where $kf(k, a_i)$ and $ikf(k)$ denote the term frequency and inverse keyword frequency of the keyword term k, respectively. $match(K, a_i)$ is the number of matched terms between keywords K of query and a_i, and $count(K)$ is the number of terms existing in the keywords K of the query.

3.5   Re-Ranking

The re-ranking process aims to balance the text (semantic) matching and the structure matching between query and the hit formulae, in order to generate a more reasonable ranking list. Although, the descriptive keywords of the query have been excavated in the keyword generation process, the ranker of our system still emphasizes the structure matching in the process of initial ranking. Therefore, the hit formulae, which have a similar semantic meaning while performing a different structure with query, cannot be properly ranked.

In order to make the ranking list more reasonable, we introduce the greedy match algorithm to look up the semantically related documents in the formula citation graph, along with calculating the semantically relevant score. The semantically related documents refer to the documents containing the similar semantic meaning or background information of the query.

Firstly, the random walk algorithm [21] is used to find out all the possible paths from original seed formula vertices to candidate related documents. The path which is generated by random walk can be defined as follow:

$$Path = \{(r_1, r_2, \ldots, r_i, \ldots) | RW(r)\} \qquad (7)$$

where $Path$ is the path set from original seed formula vertices to each relevant documents. r_i is ith path generated by random walk $RW(r)$.

Assume $r = (v_{s_1}, v_{s_2}, \ldots, v_{s_n})$, and thus the probability of each path $p(r)$ is:

$$p(r) = \prod w(v_{s_i}, v_{s_j}) \qquad (8)$$

where the $w(v_{s_i}, v_{s_j})$ is the weight of the edge (v_{s_i}, v_{s_j}).

The top-1 formula of the initial ranking is set to the starting vertex. Therefore, the total relevant score from the starting vertices to the candidate document vertex can be defined as:

$$\mathcal{P}(a_j) = P(v_i, v_j) = \sum_{r \in (v_i \rightsquigarrow v_j)} p(r) \qquad (9)$$

Afterwards, the documents, whose relevant scores obtained from top-k formulae are above a certain threshold, are selected. Thus a related candidate document list, along with its relevant score $list\{(a_i, \mathcal{P}(a_i)), (a_j, \mathcal{P}(a_j)), \ldots, (a_k, \mathcal{P}(a_k))\}$ are generated. In our experiments, the threshold is set to 0.1.

The relevant score of the matched document a_i in the initial ranking list is calculated according to Eq. 10.

$$Score_Q(a_i) = (((\mathcal{P}(a_i) + 1)^2 + S_Q(a_i) + 1)/2)^{1/2} - 1 \qquad (10)$$

where $\mathcal{P}(a_i)$ is the semantically relevant score between the query and document a_i, meanwhile $S_Q(a_i)$ is the composite similarity score between them. The $Score_Q(a_i)$ which emphasizes related document is approximate derivative average of $\mathcal{P}(a_i)$ and $S_Q(a_i)$.

Finally, the relevant score of documents in the initial ranking list are upgraded, and we re-rank the documents with the updated scores to generate the final list.

4   Experiments

4.1   Dataset

The dataset used in this paper is the NTCIR-12 Wikipedia dataset which has been released by the NTCIR-12 Math Group[1]. Therefore, the further methods can make a comparative evaluation with our system by using the same dataset. The dataset contains 319,689 webpage documents and includes 592,443 formulae. The heterogeneous graph FCG is built at the backend of our system and the statistic of the graph is shown in Table 4.

Table 4. Vertices and edges in FCG

Vertex	
Type	Count
The title of document	30,836
Original formula	31,012
Generalization formula	228,745
Edge	
Type	Count
The contain relationship between document and original formula	439,569
The match relationship between generalization formula and original formula	31,012
The citation relationship between generalization formula and document	148,889

---

[1] http://ntcir-math.nii.ac.jp/.

4.2 Efficiency

The proposed system is implemented in Scala, with the help of Lucene and Neo4j and runs on a PC with 3.60GHz Intel Core i7, 16GB DDR3 and 1T SATA Disk. In order to make the system to be an acceptable practice system, three index files are built offline. The Fig. 4 shows the time and space consuming of our system. Figure 4(a) and Fig. 4(b) illustrate the time taken to construct the index files whose size are evaluated with the increasing number of webpage documents, respectively. It takes 58 min to generate three index files for the dataset. The total size of index files is less than 650MB. As shown in Fig. 4(a) and Fig. 4(b), both the time consumption for constructing index files and the size of the index files increase steadily with the number of webpage documents.

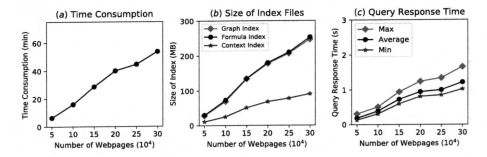

Fig. 4. The time and space efficiency

Figure 4(c) presents the query response time with the increase of the indexed webpage documents. The minimum, the maximum and the average time are tested upon 100 different queries based on increasing number of indexed webpage documents. The 100 different queries are detailedly described in the next section. It can be seen from Fig. 4(c) where the query response time of three categories with increase of the size of indexed webpage documents increase. The average time to response query is 1.36 s on the entire dataset. The maximum response time is costed by the queries containing too much same layout structure, because the more webpage documents containing the same layout formula (e.g., $f(x)$) or the query terms (e.g., a_0), the more time are used to merge the relevant document lists.

4.3 Accuracy

Baseline and Comparison Methods. We compare the proposed method with four baseline systems: one text-based system (KW) and three tree-base systems ($Wikimirs$, $Enrich$, $Rerank$). **KW** is a text-based system based on $TF\text{-}IDF$ algorithms [11]. Note that, the input of **KW** is the keywords or textual descriptions of formulae which are provided by the users, while the input of the other systems are pure formulae. **Wikimirs** is one of the state-of-the-art tree-based indexing mathematics retrieval system [10]. *Wikimirs* system indexes all the substructures of formulae and calculates the similarity of each substructure by considering their levels. In order to verify the effectiveness of the proposed techniques (query enrichment in the process of descriptive keywords generation, the text matching enhancement in the process of re-ranking), two improved system are utilized: **Enrich** is an improved system which adds the query enrichment into the retrieval process; **Rerank** is another improved system which adds the text matching enhancement into the retrieval process.

Query Set. Two query sets are utilized for evaluation, namely a formula query set with its corresponding keywords query set, and each of them contain 100 queries, respectively. The formula query set contains 30 queries which are selected from NTCIR-12 task whose keywords are removed and the other 70 queries are collected from the 'Relevance Page' of Wikipedia, as mentioned in [7]. In the keyword query set, the queries are the keywords which are used to describe the formulae in formula query set.

Evaluation Metrics. For each test query, the top-k documents retrieved by the five different systems are evaluated. The relevance between the retrieved documents and queries is judged by five postgraduates from different majors, such as mathematics, computer science, etc. The average score of evaluators is the final relevance score between the retrieved documents with its query. For each retrieved document, a relevant *score* $\in \{0,1,2,3,4\}$ is given by an evaluator. 0 denotes the document is not relevant to the query, $1 \sim 3$ indicates the document is partially relevant to the query, 4 indicates the document is relevant to the query. In order to evaluate the accuracy of our system, two metrics *Precision* (P) and *Discounted Cumulative Gain* (DCG) [4] are used over all queries. In this paper, we use the TREC evaluation tool to compute the values of *Precision@K* with K in $\{5,10,15\}$. The evaluation tool is used to evaluate the performance of systems participating in the NTCIR-12 competition [29]. $DCG@K$ with $K = \{5,10,15\}$ is calculated based on the relevant score of the hit according to the position in the result list.

Summary of Results. The performance of these five systems with respect to metrics (P and DCG) is presented in Table 5. The text-based system KW performs the worst, since it only utilize the keywords of formula as the query. It neglects the formulae structure information which is the most important information in MIR. Although the input of $Wikimirs$ is formula, the system still performs not good , because it only considers the structure matching between formulae. Two improved tree-based comparison systems, which add the query enrichment strategy ($Enrich$) or semantics matching enhancement strategy ($Rerank$), achieve better performance than $Wikimirs$ system. The results suggest that, the two strategies (query enrichment and semantics matching enhancement) are both useful for MIR task. Because query enrichment strategy can help to find more relevant formulae (the formulae may not structurally similar but share the similar semantic meanings with query), and semantics matching enhancement strategy can help to rank formulae appropriately. Finally, the proposed method ($FCGB$), that integrates all matching strategies, achieves the best performance in terms of all evaluation metrics. This result indicates that the proposed method can integrate all the matching strategies well, so that strategies can cooperate with each other to carry forward its own strength.

Table 5. The P and DCG of the five systems

System	Relevant			Partially relevant			DCG		
	P_5	P_{10}	P_{15}	P_5	P_{10}	P_{15}	DCG_5	DCG_{10}	DCG_{15}
KW	0.0025	0.0012	0.0008	0.0025	0.0025	0.0025	0.0734	0.0777	0.0874
$Wikimirs$	0.3066	0.2400	0.1977	0.5600	0.4700	0.4222	14.2956	17.8668	19.7615
$Enrich$	0.3200	**0.2466**	0.2022	0.5667	0.4733	0.4267	15.1299	18.6904	20.5935
$Rerank$	0.3200	0.2433	0.2022	0.5600	0.4733	0.4267	15.0015	18.5171	20.4682
FCGB	**0.3467**	0.2433	**0.2111**	**0.6000**	**0.4800**	**0.4333**	**15.5284**	**18.6928**	**20.9541**

4.4 Case Study

Table 6. Examples of query keywords extracted by FCG

QUERY	TOP-2 KEYWORDS
$g(x) = \frac{1}{1+e^{-x}}$	Sigmoid function, Logistic function.
$x^2 + y^2 = z^2$	Pythagorean theorem, Formulas for generating Pythagorean triples.
$_2F_1(a, b; c; z)$	K-noid, Generalized hypergeometric function.
$\left(\dfrac{a}{p}\right)$	Legendre symbol, Jacobi symbol.

Table 6 shows four examples of the query's top-2 keywords which are mined by FCG. Taking formula '$g(x) = \frac{1}{1+e^{-x}}$' as an example, the top-2 keywords mined by FCG are 'Sigmoid function' and 'Logistic function'. The two keywords are usually regarded as the descriptive keywords to describe the formula in practice. The keyword 'Sigmoid function' ranking before keyword 'Logistic function' shows that 'Sigmoid function' is more frequently used as the keyword to describe the formulae which share the same main layout with the query '$g(x) = \frac{1}{1+e^{-x}}$'.

Table 7 displays an example of the top-5 results ranked by the proposed system. Based on the query '$F = ma$', the document 'Classical mechanics' and 'Added mass' are ranked within the top-5 only since they contain several formulae whose main structure exactly matches the query formula. However, the fact is that these two documents do not mainly describe the query formula, they just leverage the query formula to assist in explaining some theorems. The documents 'Inertia' and 'History of variational principles in physics' describe more relevantly than the documents 'Classical mechanics' and 'Added mass'.

Therefore, after the process of the re-ranking, documents 'Inertia' and 'History of variational principles in physics' are ranked in top-5, and all the contents of top-5 documents mainly describe the formula query.

Table 7. Top-5 results of the proposed system

QUERY: $F = ma$	
Initial Ranking	Re-Ranked (FCG)
1. Newton (unit)	1. Newton (unit)
2. Classical mechanics	2. Force
3. Added mass	3. Inertia
4. Newton's laws of motion	4. History of variational principles in physics
5. Force	5. Newton's laws of motion

5 Conclusions

In this paper, we have designed a Formula Citation Graph (FCG) to focus on the link or citation relations between formulae. FCG is helpful to promote the ranking accuracy of the semantic-relevant formulae whose appearances are not similar to the query formula. Furthermore, it is useful to dig out the descriptive keywords of the query and balance the ranking results between text and structure matching in MIR system. In addition, we have built a novel MIR system based on FCG and this system achieves the state-of-the-art MIR performance.

Acknowledgements. This work is supported by the projects of National Natural Science Foundation of China (No. 61876003), National Key R&D Program of China (2019YFB1406303), Guangdong Basic and Applied Basic Research Foundation (2019A1515010837) and the Fundamental Research Funds for the Central Universities, which is also a research achievement of Key Laboratory of Science, Technology and Standard in Press Industry (Key Laboratory of Intelligent Press Media Technology).

References

1. Aizawa, A., Kohlhase, M., Ounis, I., Schubotz, M.: Ntcir-11 math-2 task overview. In: Proceedings of the 11th NTCIR Conference, NII (2014)
2. Davila, K., Zanibbi, R.: Layout and semantics: Combining representations for mathematical formula search. In: Proceedings of the 40th International ACM SIGIR Conference on Research and Development in Information Retrieval, pp. 1165–1168 (2017)
3. Gao, L., Jiang, Z., Yin, Y., Yuan, K., Yan, Z., Tang, Z.: Preliminary exploration of formula embedding for mathematical information retrieval: can mathematical formulae be embedded like a natural language? arXiv preprint arXiv:1707.05154 (2017)

4. Järvelin, K., Kekäläinen, J.: Ir evaluation methods for retrieving highly relevant documents. In: ACM SIGIR Forum, vol. 51, pp. 243–250. ACM, New York (2017)
5. Jiang, Z., Gao, L., Yuan, K., Gao, Z., Tang, Z., Liu, X.: Mathematics content understanding for cyberlearning via formula evolution map. In: Proceedings of the 27th ACM International Conference on Information and Knowledge Management, pp. 37–46 (2018)
6. Jiang, Z., Liu, X., Chen, Y.: Recovering uncaptured citations in a scholarly network: A two-step citation analysis to estimate publication importance. JAIST **67**(7), 1722–1735 (2016)
7. Kamali, S., Tompa, F.W.: Retrieving documents with mathematical content. In: Proceedings of the 36th international ACM SIGIR Conference on Research and Development in Information Retrieval, pp. 353–362 (2013)
8. Kristianto, G.Y., Aizawa, A., et al.: Extracting textual descriptions of mathematical expressions in scientific papers. D-Lib Mag. **20**(11), 9 (2014)
9. Libbrecht, P., Melis, E.: Methods to access and retrieve mathematical content in ACTIVEMATH. In: Iglesias, A., Takayama, N. (eds.) ICMS 2006. LNCS, vol. 4151, pp. 331–342. Springer, Heidelberg (2006). https://doi.org/10.1007/11832225_33
10. Lin, X., Gao, L., Hu, X., Tang, Z., Xiao, Y., Liu, X.: A mathematics retrieval system for formulae in layout presentations. In: Proceedings of the 37th International ACM SIGIR Conference on Research & Development in Information Retrieval, pp. 697–706 (2014)
11. Manning, C.D., Raghavan, P., Schütze, H., et al.: Introduction to information retrieval, vol. 1. Cambridge University Press, Cambridge (2008)
12. Mansouri, B., Rohatgi, S., Oard, D.W., Wu, J., Giles, C.L., Zanibbi, R.: Tangent-cft: an embedding model for mathematical formulas. In: Proceedings of the ACM SIGIR International Conference on Theory of Information Retrieval, pp. 11–18 (2019)
13. Miller, B.R., Youssef, A.: Technical aspects of the digital library of mathematical functions. AMAI **38**(1–3), 121–136 (2003)
14. Pagael, R., Schubotz, M.: Mathematical language processing project. arXiv preprint arXiv:1407.0167 (2014)
15. Peng, S., Yuan, K., Gao, L., Tang, Z.: Mathbert: A pre-trained model for mathematical formula understanding. arXiv e-prints pp. arXiv-2105 (2021)
16. Perkiö, J., Buntine, W., Tirri, H.: A temporally adaptive content-based relevance ranking algorithm. In: Proceedings of the 28th Annual International ACM SIGIR Conference on Research and Development in Information Retrieval, pp. 647–648 (2005)
17. Quoc, M.N., Yokoi, K., Matsubayashi, Y., Aizawa, A.: Mining coreference relations between formulas and text using wikipedia. In: ICCL, p. 69 (2010)
18. Rûzicka, M., Sojka, P., Liška, M.: Math indexer and searcher under the hood: Fine-tuning query expansion and unification strategies. In: NTCIR, pp. 7–10 (2016)
19. Schubotz, M., et al.: Semantification of identifiers in mathematics for better math information retrieval. In: SIGIR, pp. 135–144. ACM (2016)
20. Sojka, P., Líška, M.: Indexing and searching mathematics in digital libraries. In: Davenport, J.H., Farmer, W.M., Urban, J., Rabe, F. (eds.) CICM 2011. LNCS (LNAI), vol. 6824, pp. 228–243. Springer, Heidelberg (2011). https://doi.org/10.1007/978-3-642-22673-1_16
21. Spitzer, F.: Principles of Random Walk, vol. 34. Springer, New York (2013). https://doi.org/10.1007/978-1-4757-4229-9
22. Thanda, A., Agarwal, A., Singla, K., Prakash, A., Gupta, A.: A document retrieval system for math queries. In: Proceedings of the 12th NTCIR Conference (2016)

23. Wang, Y., Gao, L., Wang, S., Tang, Z., Liu, X., Yuan, K.: Wikimirs 3.0: a hybrid mir system based on the context, structure and importance of formulae in a document. In: Proceedings of the 15th ACM/IEEE-CS Joint Conference on Digital Libraries, pp. 173–182 (2015)
24. Yokoi, K., Nghiem, M.Q., Matsubayashi, Y., Aizawa, A.: Contextual analysis of mathematical expressions for advanced mathematical search. Polibits **43**, 81–86 (2011)
25. Yuan, K.: Multi-dimensional formula feature modeling for mathematical information retrieval. In: Proceedings of the 40th International ACM SIGIR Conference on Research and Development in Information Retrieval, pp. 1381–1381 (2017)
26. Yuan, K., Gao, L., Jiang, Z., Tang, Z.: Formula ranking within an article. In: Proceedings of the 18th ACM/IEEE on Joint Conference on Digital Libraries, pp. 123–126 (2018)
27. Yuan, K., Gao, L., Wang, Y., Yi, X., Tang, Z.: A mathematical information retrieval system based on rankboost. In: Proceedings of the 16th ACM/IEEE-CS on Joint Conference on Digital Libraries, pp. 259–260 (2016)
28. Yuan, K., He, D., Jiang, Z., Gao, L., Tang, Z., Giles, C.L.: Automatic generation of headlines for online math questions. In: Proceedings of the AAAI Conference on Artificial Intelligence, vol. 34, pp. 9490–9497 (2020)
29. Zanibbi, R., et al.: Ntcir-12 mathir task overview. In: Proceedings of the 12th NTCIR Conference, NII (2016)
30. Zanibbi, R., Davila, K., Kane, A., Tompa, F.W.: Multi-stage math formula search: using appearance-based similarity metrics at scale. In: Proceedings of the 39th International ACM SIGIR Conference on Research and Development in Information Retrieval, pp. 145–154 (2016)

21. Wang, X., Chang, S., Wang, F., Guo, Y., Cao, Y., Wan, X., Li, W. Bat-like Sonar System (...) the carbon dioxide study (...) storage and (...) to improve (...) in a deep learning (...) in Proceedings of the 2016 (...) CTU-SS John (...) in Digital (...) (...).

22. Chang, K., Yang, J., Li, L., Jiang, H., Yu, Y., Chen, (...) to (...) neural network (...) multi (...) approaches to advanced (...) (...) (...) pp. 18–37. (...).

Author Index

Printed in the United States
by Baker & Taylor Publisher Services